Soilborne Microbial Plant Pathogens and Disease Management, Volume Two

Management of Crop Diseases

Soilborne Microbial Plant Pathogens and Disease Management, Volume Two

Management of Crop Diseases

Prof. P. Narayanasamy

CRC Press
Taylor & Francis Group
Boca Raton London New York

CRC Press is an imprint of the
Taylor & Francis Group, an **informa** business

CRC Press
Taylor & Francis Group
6000 Broken Sound Parkway NW, Suite 300
Boca Raton, FL 33487-2742

First issued in paperback 2021

CRC Press is an imprint of Taylor & Francis Group, an Informa business
No claim to original U.S. Government works

Printed on acid-free paper

ISBN-13: 978-0-367-17876-5 (hbk)
ISBN-13: 978-1-03-208732-0 (pbk)

Visit the Taylor & Francis Web site at
http://www.taylorandfrancis.com

and the CRC Press Web site at
http://www.crcpress.com

Dedicated

To

the memory of my parents

for their

Love and affection

Contents

Preface

The earth functions as the habitat, supporting the existence of countless organisms ranging from simple subcellular to complex multicellular forms including human beings. The soil, occupying about 70% of the surface of the earth, has three phases – solid, water and air. Solid phase (minerals and organic matter) generally makes up about 50% of soil volume, whereas soil solution (liquid phase) and air (gaseous phase) constitute the other half of soil volume. Microorganisms present in the soil, water and air are either beneficial or harmful to the development of animals and plants that have been growing wild or domesticated to serve the needs of humans. Soilborne microbial plant pathogens – oomycetes, fungi, bacteria and viruses – are capable of inducing various kinds of diseases in economically important crops, accounting for quantitative and/or qualitative losses that may be enormous when epidemics occur. Microbial plant pathogens exist in the form of several races, varieties or pathotypes differing in their virulence, whereas the crop cultivars and genotypes vary in their levels of susceptibility/resistance to different isolates of the pathogens, necessitating precise identification of the microbial pathogen involved in the disease(s). Techniques based on the morphological, biochemical, immunological and genetic characteristics of soilborne microbial pathogens have been developed to detect, differentiate and quantify them in soil, plants, seeds and water, even when they are present in low populations. Advancements in nanotechnology are likely to improve the sensitivity of pathogen detection and identification in the future. Comprehensive investigations on the biology of microbial plant pathogens to determine the pathogenic potential, host range and the influence of environments on the pathogen population buildup are essentially required to identify effective strategies for reducing incidence and spread of the diseases caused by them.

Soil environments provide both opportunities and challenges to microbial plant pathogens for their survival and perpetuation. They are capable of producing variants or overwintering propagules to overcome the adverse conditions existing in the soils in different agroecosystems. Potato late-blight pathogen *Phytophthora infestans*, with multiple dispersal modes, remains an elusive problem because of its ability to cause epidemics under favorable conditions. This pathogen ruined the livelihood of millions of Irish people, who had to migrate to different countries in the 18th century. Management of diseases caused by soilborne microbial plant pathogens is based on the principles of exclusion, eradication and immunization. Threats to crop cultivation posed by soilborne microbial plant pathogens have to be dealt with firmly by intensification of research on determination of the pathogenic potential of different races/varieties of the pathogens and the susceptibility/resistant levels of crop cultivars/genotypes for formulation of effective strategies for management of the diseases caused by them. Studies on precise identification of microbial plant pathogens and assessment of variability in their virulence have laid the foundation for development of crop disease management strategies. Furthermore, responses of the pathogens to changes in the environment and chemicals, selection of cultivars/genotypes with built-in genetic resistance and deploying biological control agents have received greater attention resulting in reduction in disease incidence/ severity and enhancement of returns for the grower and preservation of the environment. Investigations on the development of integrated disease management systems by combining two or more compatible strategies are on focus for achieving greater levels of pathogen suppression and consequent reduction in the adverse effects of diseases caused by soilborne microbial plant pathogens.

This book is designed to provide information gathered by extensive literature search with a view to providing comprehensive information to the researchers, teachers and graduate level students in the fields of plant pathology, microbiology, plant breeding and genetics, agriculture and horticulture as well as the certification and quarantine personnel, culminating in provision of safe food and environment to all.

P. Narayanasamy

Acknowledgment

The author wishes to place on record his humble salutation with reverence to his alma mater that continues to provide inspiration, knowledge, wisdom and competitive ability to establish his identity and to share his concepts with fellow plant pathologists. The author recognizes the immense benefits derived from the interactions with the staff and students of the Department of Plant Pathology, Tamil Nadu Agricultural University, Coimbatore, India. The author considers himself to be very fortunate to have Mrs N. Rajakumari as his life partner and it would not have been possible to devote his time exclusively for the preparation of the manuscript of this book without her enormous kindness, patience, love and affection.

Furthermore, the author heartily thanks his family members Mr N. Kumar Perumal, Mrs Nirmala Suresh, Mr T. R. Suresh and Mr S. Varun Karthik, who are the fulcrum by which he can smoothly achieve his life goals.

Above all, the author sincerely thanks the various researchers, editors and publishers for the free access to required information included in this publication. Credits have been given for the materials included in the book at appropriate places.

P. Narayanasamy

Author

Prof P. Narayanasamy was awarded a BSc (Ag) in 1958, an MSc (Ag) in 1960, and a PhD in 1963 by the University of Madras. Later, with a Rockefeller Foundation Fellowship, he pursued postdoctoral research on rice virus diseases at the International Rice Research Institute, the Philippines, during 1966–1967. He served as the virus pathologist at the Indian Agricultural Research Institute, New Delhi, during 1969–1970. He returned to his alma mater, which was upgraded to the Tamil Nadu Agricultural University (TNAU), Coimbatore, in 1970. He was appointed associate professor and promoted as professor and Head of the Department of Plant Pathology. He was elected fellow of the Indian Phytopathological Society, New Delhi. He functioned as the editor of the *Madras Agricultural Journal*, published from the Tamil Nadu Agricultural University Campus, and as a member of editorial committees of plant pathology journals published in India. He was invited to participate as the lead speaker and chairman of sessions in national seminars held in India.

As a researcher in plant pathology, he was the leader of the research projects on diseases of rice, legumes and oilseeds. He organized the national seminar for the management of diseases of oilseed crops. His research on antiviral principles yielded practical solutions for the management of virus diseases affecting various crops. He has published over 200 research papers in national and international journals. He was the principal investigator of several research projects funded by the Department of Science and Technology, Government of India, New Delhi and the International Crops Research Institute for Semi-Arid Tropics, Hyderabad. He continues to share his experience and knowledge with the staff and graduate students of the Department of Plant Pathology, TNAU, Coimbatore. As a teacher, he taught courses on plant virology, molecular biology, physiopathology and crop disease management for masters and doctoral programs. Under his guidance, 25 graduate students and 15 research scholars earned masters and doctoral degrees, respectively. With many years of experience and in-depth knowledge on various aspects of microbial plant pathogens and crop disease management, he has authored 18 books published by leading publishers including Marcel Dekker, John Wiley & Sons, Science Publishers, The Haworth Press and Springer Science. These publications cover various aspects of plant pathology and serve as valuable sources of information and have been well received by the intended audience.

He is deeply involved in social welfare activities to help the orphaned, old and infirm, as well as children through Udavum Karangal (Coimbatore), HelpAge (New Delhi), Global Cancer Concern (New Delhi), Children Rights and You (CRY, Bangalore) and The Hindu Mission Hospital (Chennai) to lessen the sufferings of needy persons. As one interested in literature and spirituality, he has composed poems in Tamil and English and published two collections of poems to encourage self-confidence in young people and spiritual exploration for mature people.

1 Management of Soilborne Microbial Plant Pathogens

Exclusion and Prevention Strategies

In an ecosystem, diverse but balanced biological factors arising from long-term coevolution, interact with host plants and soilborne microbial plant pathogens in different environmental conditions. Under wild ecosystem, the plants and the pathogens coexisted in a balanced level. When selection of plant species was made for domestication, resistance to diseases was not given equal importance, as in the case of yield and quality of produce. This attitude resulted in disturbance in natural balance of coexistence of host plant species and microbial pathogens. Ultimately increase in susceptibility of crop plants forced the researchers to search for resistance gene(s) in wild relatives of crop plant species. Thus, the crops became susceptible, not only to the soilborne microbial plant pathogens present in a geographic location, but also to the exotic plant pathogens introduced through seeds and propagules (Narayanasamy 2017). This situation warranted enforcement of regulatory methods to prevent introduction of plant pathogens from other countries or from one region to another region within a country. Such introduction of exotic microbial plant pathogens may lead to epidemics that can be ecologically and economically difficult to contain.

1.1 EXCLUSION OF MICROBIAL PLANT PATHOGENS

Strategies of crop disease management essentially fall into three principles: exclusion of pathogens, eradication of sources of pathogen inoculum and immunization of crop plant species for enhancing the level of resistance to soilborne microbial pathogen(s). Information has to be gathered on pathogen biology, infection process and epidemiological factors that favor disease incidence and spread for development of effective disease management systems. The basic step for development of a disease management system, soon after the incidence of a disease is observed, is the early detection and precise identification of the pathogen(s) by applying rapid, reliable, sensitive and specific diagnostic techniques. Soilborne microbial plant pathogens have to be detected in soils, irrigation water and other environments, in addition to plants and propagative materials. Detection and identification of pathogens in the roots and other below ground organs pose formidable problems, because of the presence of large number of microorganisms along with the pathogens, making their isolation difficult. Crop disease management through exclusion includes three phases: sanitation, quarantine and certification. Sanitation programs are directed toward recovery of healthy plants from

infected plants of selected local cultivar(s). Selection of mother plants/trees of native cultivars, indexing them, recovery of pathogen-free plants by tissue culture techniques, indexing micrografted plants, evaluation of genetic characteristics of healthy plants generated and maintenance of healthy source plants, constitute the different stages of sanitation phase. The responsibility of preventing introduction of new pathogens through seeds, propagules and plants is assumed by domestic and international plant quarantines. The imported plant materials are placed under observation or indexed for the presence of microbial pathogens under controlled greenhouse conditions, using appropriate diagnostic procedures. Certification agencies are responsible for production of disease-free seeds/propagules directly or overseeing the personnel given this responsibility. The certification programs are endowed with the authority to control the quality and sanitary status of nursery plants by enforcing legal regulations at different stages of nursery operations (Navarro 1993).

1.1.1 USE OF DISEASE-FREE SEEDS AND PROPAGULES

Methods of excluding the microbial plant pathogens may be of two kinds: (i) use of disease-free seeds and propagules, and (ii) regulatory methods involving quarantine and inspection for prevention of pathogen introduction. Soilborne pathogens spread in the soil slowly and it may take several crop seasons to infect large number of plants which are generally aggregated into groups in the infested field. In addition, these pathogens can also spread to long distances through infected seeds and propagules that are exported from one country to another country. *Fusarium oxysporum* f.sp. *callistephi* (*Foc*), causing Fusarium wilt disease of China aster (*Callistephus chinensis*) occurred in epidemic proportions in Connecticut and Florida, United States. The pathogen is seedborne, in addition to being soilborne and it could be recovered at a low incidence (<1%) from contaminated seeds. Furthermore, transplant trays, especially made of Styrofoam, when reused, could be a potential source of inoculum for epidemic outbreak of Fusarium wilt disease. Five of 25 commercial packages from three separate distribution companies in Connecticut had seeds contaminated with *Foc*. Farm surveys of two cut-flower farms revealed disease incidences of 32% and 58% in Connecticut, whereas disease incidence ranged from 0.002 to 71.2% in Florida. The results indicated the need to encourage growers to apply sanitation practices (seed treatment with sodium hypochlorite/soaking or spraying Styrofoam) to prevent disease incidence and spread (Elmer and McGovern 2013). Other important microbial pathogens transmitted via seeds/propagules include *Phytophthora infestans* (potato), *Verticillium dahliae* (spinach), *Rhizoctonia solani* (potato), *F. oxysporum* f.sp. *vasinfectum* (cotton), *Tilletia controversa* (wheat), *Plasmodiophora brassicae* (oilseed rape), *Spongospora subterranea* (potato) and *Didymella rabiei* (*Ascochyta rabiei*, chickpea). These pathogens may be transmitted via seeds and seedlings, whereas *Cercospora beticola*, infects roots of sugar beet, as well as leaves from which airborne inoculum is produced. The seeds and seedlings may remain asymptomatic, yet they may function as potential sources of inoculum. It is essential to detect and identify the pathogens in asymptomatic plants and the infected ones should be eliminated. Furthermore, once the pathogen reaches the soil from infected seeds/propagules, it is very difficult to eliminate it from the soil. Hence, seed lots and seedlings free of soilborne diseases have to be planted in the subsequent season. Similarly, the bacterial pathogens, *Ralstonia solanacearum, Streptomyces scabies, Clavibacter michiganensis* subsp. *sepedonicum, Pectobacterium* spp., *Dickeya* spp. and *Agrobacterium tumefaciens*, in addition to being soilborne, are transmitted through infected planting materials such as tubers and cuttings. Among the soilborne viruses, *Tomato mosaic virus* (tomato), *Tobacco ringspot virus* (tobacco) and *Indian peanut clump virus* are transmitted via peanut seeds also. Many of the viruses transmitted by nematode vectors are transmitted also through seeds of crop plant species and weeds. It is superfluous to suggest the importance of using virus-free seeds/propagules to avoid/reduce the incidence of soilborne viruses which are responsible for progressive decline of high value crops like grapevine and strawberry (Narayanasamy 2017).

Grapevine fan leaf virus (GFLV) is transmitted by the nematode, *Xiphinema index* and through vegetative propagation. In order to generate GFLV-free planting materials, indirect embryogenesis was tested as a method for elimination of the virus from three grapevine cultivars. RT-PCR assay was applied to detect GFLV in tissues sampled at various steps of embryogenic process: flower explants, embryogenic and nonembryogenic calli, single somatic embryos and regenerated plants. GFLV was detected in all anthers and ovaries tested, while only one of 63 regenerated plants tested positive for GFLV in the diagnostic assay. GFLV could be eradicated by somatic embryogenesis alone without heat therapy, although the virus was able to invade embryonic calli and embryo-derived plantlets. Successful elimination of GFLV was achieved in almost 100% of the test plants by somatic embryogenesis procedure. GFLV could not be detected in the newly formed leaves of regenerated plants at 2 years after their transfer to greenhouse conditions (Gambino et al. 2009). In another investigation, somatic embryogenesis was established from grapevine anthers. The absence of *Arabis mosaic virus* (ArMV) in 46 regenerated grapevine plant lines was confirmed by ELISA and IC-RT-PCR procedures, repeated after different time intervals in vitro and in vivo, after acclimatization and after one dormancy period under glasshouse conditions. All regenerated grapevine plants appeared true-to-type

and all plants were confirmed to be diploid by flow cytometry technique (Borroto-Fernandez et al. 2009).

There is an increasing demand for potato seed-tubers that are true-to-type, disease-free and high-yielding. As potato seed-tubers were commercially produced in the field, each generation accumulates and further transmits fungal, bacterial or viral disease-causing agents to the next generation(s). Hence, as the number of generations increased, potential of the seed-tubers tended to decline, because of the latent infection (especially viral infection) carried in seed-tubers. In order to overcome this obstacle, use of tissue culture multiplication, also known as 'minituber technology' was adopted by growers. Potato seed-tubers were first multiplied in vitro via nodal cuttings in tissue culture and then planted in the field to obtain true-to-type and disease-free plant materials (Struik 2007). An outbreak of potato powdery scab caused by *S subterranea* subsp. *subterranea (Sss)* occurred in minituber production facility in South Africa. Swab samples were taken from many points in the facility to identify the sources of contamination to apply corrective measures. *Sss*-specific primers (Sps1 and Sps2) were employed in the PCR assay to detect the pathogen. Of the 11 surfaces tested, *Sss* was detected in 6 surfaces and corrective measures were taken to eradicate *Sss* from all contaminated surfaces. Second round PCR assays performed, revealed complete eradication of *S. subterranea* subsp. *subterranea*. Adoption of corrective measures resulted in disease-free potato seed-tubers grown in minituber production channels from 2009 onwards (Wright et al. 2012). Potato crops are affected by several soilborne fungal, bacterial and viral diseases. Most of the pathogens, causing these diseases are carried by potato seed tubers. It is difficult to recognize latent infections in seed tubers, especially by viruses. The mother plants and tubers are indexed to detect the presence of pathogens using immunoassay and/or nucleic acid-based techniques. Likewise, grapevine the host was indexed for several viruses, including soilborne GFLV. Multiplex RT-PCR assay was efficient in detecting multiplex infections by the viruses (Gambino and Gribaudo 2006; Wei et al. 2009). Certification programs are usually applied to maintain genetic purity primarily for breeders' seed through successive generations of multiplication. Detection methods have been useful to assess the extent of infection of seed lots/propagation materials and to reject the ones that contain infection levels above the prescribed standards.

1.1.2 Prevention of Pathogen Entry through Plant Quarantines

Introduction and spread of microbial plant pathogens into a country or state/province within a country may be prevented by applying regulatory methods. Legislative measures have been enacted to regulate the cultivation of crops and distribution of seeds and planting materials between countries or states within the country. Regulatory control is enforced through establishment of plant quarantines, and inspections of crops in the field/greenhouses/warehouses. In certain cases, voluntary or compulsory eradication of host plants of pathogens is a mandatory requirement. Plant quarantines have been established primarily to prevent the introduction and spread of diseases and pests into new areas, where they may have chances for spread under natural conditions. Plant quarantines help protect crop cultivation as well as the environment from avoidable damage that may be caused by pathogens newly introduced by man inadvertently. Following the adoption of the General Agreement on Tariffs and Trade (GATT), a dramatic enhancement of movement of plant products, necessitated the enforcement of phytosanitary measures at the global level. Legislative restrictions were imposed by most countries to prevent or delay the establishment of pathogens in new locations. The International Plant Protection Convention (IPPC) was established on April 4, 1991, following the acceptance of GATT by majority of countries. An Expert Consultation on the Harmonization of Plant Quarantine Principles was convened by the Food and Agriculture Organization (FAO) to identify the basic principles necessary for formulating standards for plant quarantine procedures in relation to international trade (FAO 1991). The Expert Consultation laid down basic principles to be followed by member countries.

The principles of establishment of plant quarantines recognize the sovereignty of the individual country which has the right to implement the phytosanitary measures deemed fit by that country. The necessity of restrictive measures to have minimum impact on the international movement of plant materials has to be ensured by different member countries. The member countries can modify regulations as and when needed and the plant quarantine regulations have to be transparent, so that their phytosanitary measures function harmoniously without any discrimination of exporting countries. The cooperation of member countries forms the corner stone for plant quarantines to function effectively by implementing necessary regulations. Eight regional plant quarantine organizations were set up to cover all countries and to advise and assist member governments on the technical, administrative and legislative measures, concerned with plant quarantine activities. The regional plant protection organizations provide information on changes in the quarantine regulations and requirements and examine the possibility of simplification and unification of phytosanitary regulations. Lists of quarantine pathogens are prepared by the country concerned and the countries from which plants and plant products are imported are informed of the regulations and requirements to be satisfied. International quarantines are established in airports and seaports. The domestic quarantines operating at the boundaries between states are responsible to restrict/control movement of plants and plant products from one state to another, where incidence of defined pathogen(s) may be absent. All precautions have to be taken to restrict the incidence and spread of microbial plant pathogens, particularly, those that can survive in soils and with a wide host range (Narayanasamy 2002, 2017).

1.1.3 IMPACT OF INTRODUCED MICROBIAL PLANT PATHOGENS

Following the adoption of the General Agreement on Tariffs and Trade (GATT) by several countries, the number of plant pathogens carried along with new crops/varieties increased steadily. Consequently, many crops in different countries were affected to alarming levels, as the pathogens had suitable environmental conditions, in addition to the availability of susceptible plants in large areas. This situation emphasized the need for increasing the efficiency of quarantines by providing necessary infrastructure and competent personnel.

1.1.3.1 Potato Late Blight Disease

P. infestans, causal agent of the infamous potato late blight disease, is considered as the primary factor responsible for the Irish potato famine in 1845, resulting in wiping out potato crops in extensive areas. Consequently, one of the greatest migrations of impoverished humans to other European countries and North America occurred. The pathogen would have been introduced with seed potatoes from the origin of potato, Peru to Belgium or France around 1840 (Agarwal and Sinclair 1996). The levels of resistance of potato cultivars to the pathogen(s) were progressively reduced, because the cultivars were selected for high yield and attributes other than resistance to disease(s). Hence, conditions became increasingly favorable for pathogen development, as the area under potato cultivation increased year after year. Further, pathogen genotypes with higher levels of virulence and resistance to fungicides may be introduced through infected tubers, making the control through fungicides ineffective. Clonal propagation and migration through seed tubers contributed to long-distance transport of *P. infestans*. Both A1 and A2 mating types of *P. infestans* were detected in the same seed tuber production areas in Canada. Such a situation is likely to increase the chances for sexual recombination within the pathogen population, paving the way for production of oospores that could survive in the absence potato plants and result in the formation of new strains of *P. infestans* (Peters et al. 2014). *P. infestans* causes late blight disease of tomato also and it is seed-, soil- and air-borne. These characteristics of the pathogen pose great challenges to the growers and researchers to find more effective methods of management (Chowdappa et al. 2015; Fall et al. 2015).

1.1.3.2 Banana Fusarium Wilt Disease

Banana Fusarium wilt disease, also known as Panama Wilt disease, is inflicted by *F. oxysporum* f.sp. *cubense*, and it has worldwide distribution, because of the spread of the pathogen via infected suckers. Sensitive diagnostic assay has to be employed to detect the pathogen in the suckers which remain asymptomatic. In Costa Rica, banana production was drastically reduced from 11 million bunches in 1923 to 1.4 million bunches in 1941 (Schumann 1991). The pathogen can survive for several years in the soil after introduction into new locations, through infected suckers. In addition, pathogen spread may occur through contaminated soil, footwear (or feet) and tools used for handling infected planting materials. The infested fields have to be abandoned, when banana cultivation becomes unprofitable and fallowing for long periods may be required to reduce pathogen populations. Occurrence of Fusarium wilt disease in all banana-growing countries, reflect the ineffective activities of quarantines in preventing the introduction of the pathogen. Strict enforcement of regulatory measures by both domestic and international quarantines in monitoring movement of planting materials might provide rich dividends in containing the banana Fusarium wilt disease.

1.1.3.3 Potato Virus Diseases

Programs for certification of seed potatoes are in operation in Europe and North America, primarily to maintain cultivar purity. Seed certification programs are managed by state-based potato grower associations. Each state in the United States has different regulations and certification agencies are now managed by the state Department of Agriculture, universities or grower groups. Seed potato certification in North America depends on visual inspection of potato crops, because it was considered that large-scale laboratory assays would add cost. Potato varieties that do not express early recognizable symptoms are excluded (de Souza-Dias and Betti 2003). The emergence of soilborne viruses, *Tobacco rattle virus* (TRV) and *Potato mop-top virus* (PMTV) has become a challenge for seed potato certification programs (Johnson 2008). Zero-tolerance standard is followed for all tissue-culture-derived plantlets and plants grown in greenhouses. This standard could be ensured by visual inspection and mandatory laboratory testing for growers of certified seed potatoes. The cost of laboratory assay appears to be prohibitively expensive, although they may provide accurate estimates of virus infection in seed potato seed tuber lots (Halterman et al. 2012).

1.1.4 Removal of Infected Plant Tissues

Removal of infected tissues from live infected plants, followed by protection of the exposed cut ends/surface with fungicides has been effective in elimination of pathogens in certain crop plant species. This practice (tree surgery) is followed in perennial crops. *Armillaria mellea* causes Armillaria root disease of grapevine in California vineyards. The effectiveness of root collar excavation for reducing infection of grapevines by *A. mellea* was assessed. Based on the assumption that root collar excavation, when applied in early stages of root collar infection, might cause mycelia fans of *A. mellea* to recede from the root collar, before development of vascular tissue decay. In one vineyard (N1), excavation significantly increased yield and cluster weight of symptomatic grapevines, reaching the levels as in healthy vines. In another vineyard (K1), excavated root collars frequently refilled with soil, and excavation did not offer any improvement in yield of vines. The root collar excavation seemed to be a promising cultural approach for the management of Armillaria root disease, as long as excavated root collars were kept clear of soil (Baumgartner 2004). Effectiveness of sanitation practices to reduce pathogen population has been shown to be a practical disease management strategy. *Monosporascus cannonballus*, incitant of Monosporascus sudden wilt of melon, produces ascospores, as the primary inoculum, in the infected roots. Hence, the critical step of maintaining soil pathogen population densities as low as possible depends on the destruction

of hyphae of the pathogen in the infected roots very early after harvest, resulting in inhibition of ascospore production. Cultivation practices to lift roots onto the surface of soil for rapid desiccation, significantly arrested ascospore production on infected melon roots. The number of roots bearing perithecia was significantly reduced, compared with controls. Further, ascospore populations in treated plots were significantly reduced ($P <0.05$) than that were present in untreated controls in the 3-year investigation. The results showed that by removing the infected roots, sudden wilt incidence could be significantly reduced (Radewald et al. 2004).

1.2 SOIL HEALTH MANAGEMENT

Microbial pathogens – fungi and bacteria – may exist as free-living organisms in the soil and induce disease symptoms, after entry through roots of susceptible plants. By contrast, viruses, except a few, depend on the soilborne vectors, either fungi or nematodes, for transmission to susceptible plants which develop symptoms, after establishment in the initially infected host cells. The soilborne fungal pathogens may survive for long periods in the soil, as sclerotia, microsclerotia, chlamydospores, oospores or structures that may be resistant to adverse conditions prevailing in the soil. Under optimal environmental conditions, they may exist as saprophytes on organic matter, such as crop residues, organic manures and amendments applied on the soil to improve the growth of crops. Soilborne bacterial pathogens also can proliferate on soil organic matter till they find susceptible crop plants for inducing diseases. In contrast, soilborne viruses require biological vectors to reach the roots of susceptible crop plant species or alternative host plant species. Soils provide the ecological niche for all kinds of organisms, including the microbial pathogens, as well as the antagonists that limit the development of plant pathogens. It is, therefore, essential to assess the soil health by estimating the components of microbial community and their relative biomass. Various techniques are available to precisely determine the dynamics of microbial plant pathogens under different agroecosystems.

Soilborne microbial plant pathogens form an important group that negatively impact crop production by adversely affecting soil health. The capacity of soil to function as a vital living system to sustain biological productivity, maintain environmental quality and promote plant, animal and human health, represents degree of soil health. Cultural practices that promote soil health, include adoption of crop rotation (sequence), cultivation of green manures and cover crops and application of organic amendments, suitable tillage and irrigation practices, which generally have positive effects on the management of soilborne diseases caused by fungal, bacterial and viral diseases. These practices may act through increasing soil microbial biomass, activity and diversity, leading to greater biological suppression of soilborne pathogens and the diseases caused by them. As there are complex interactions among soil components, it is difficult to define ideal healthy soil. Management practices associated with soil health, such as crop rotation, cover crops and organic amendments have been shown to enhance the organic matter contents and the microbial biomass, resulting in reduction of incidence of soilborne diseases (Larkin 2015). The influence of cultural practices on the abundance of *Verticillium dahliae* in the soil and incidence of Verticillium wilt in tomato fields in Ontario, Canada, was investigated. The soil inoculum levels and incidence of wilt disease were generally lower in Kent county fields than in Essex county and *V. dahliae* race 2 was not common in Kent county. Isolates of *V. dahliae* (128) were characterized by RFLP analysis. Isolates with E18 RFLP profiles, highly similar to those of isolates previously collected from potato fields in North America, were detected in Essex county tomato fields but not common in Kent county fields. The lower levels of disease incidence of Verticillium wilt in Kent county were attributed to the adoption of different cultural practices in Kent and possible introduction of virulent isolates of *V. dahliae* into Essex county (Harrington 2000). The type of farming systems has significant effects on the soil microbial community structure and composition in different agroecosystems. Rainfed experiments were conducted continuously for over 80 years in the semiarid and eastern Oregon in the United States. Nature of crop, tillage, crop rotation, soil fertility, crop season and their interactions had large effects on the soilborne pathogens. *Fusarium culmorum* was more dominant than *F. pseudograminearum* in areas where spring crops were grown, whereas opposite trend of dominance was observed in winter wheat crops. *Gaeumannomyces graminis* var. *tritici* was more prevalent in cultivated than in noncultivated soils, while opposite pattern of incidence of *R. solani* AG-8 was observed. Densities of *Pythium* spp. clade F were high, but they were also affected by treatments. The results revealed the variations in different types of root-infecting pathogens in rainfed field crops (Smiley et al. 2016).

1.2.1 CULTURAL PRACTICES

Cultural practices followed in different agroecosystems vary, depending on the soil type, crop requirement and environmental conditions, with the primary purpose of enhancing the yield levels. But these practices have significant effects on incidence and spread of soilborne diseases in the greenhouse and field crops. Quantitative effects of cultural practices may be difficult to assess precisely, since the beneficial effects become perceptible only over a period of several seasons or years. The effectiveness of cultural practices in reducing pathogen inoculum with consequent reduction in disease incidence and subsequent spread and on plant growth and yield has been determined in some pathosystems.

1.2.1.1 Disposal of Infected Plants and Residues

Microbial plant pathogens infect various plant organs from roots to aerial organs, including grains and fruits. The infected plant materials have to be properly disposed of to reduce the inoculum added to the soil and buildup of inoculum capable of infecting subsequent crops. *Peronosclerospora sorghi*, causal agent of sorghum downy mildew, induces extensive shredding of infected leaves. Millions of oospores present

in the plant tissues, are returned to the soil at harvest, if the infected plant debris is not properly disposed of. Infected plants left in the fields, volunteer (self-sown) plants growing from infected seeds and alternative hosts of soilborne pathogens form important sources of inoculum. Removal of all infected plants and plant parts reduces quantum of inoculum added to the soil, leading to appreciable reduction in disease incidence in several crops, especially in crops like banana and sugarcane (Kommedahl and Todd 1991). Potato tubers are infected by many fungal, bacterial and viral diseases. Volunteer plants growing from tubers can be potential sources of inoculum for the subsequent crops. Spread of viral diseases can be restricted by rouging infected plants which are likely to function as foci of infection from which soilborne vectors – fungi and nematodes – may spread the virus(es). Roguing out infected plants, followed by replanting with virus-free plants was adopted as a practical disease management strategy for GFLV transmitted by the nematode vector *Xiphinema index* (Andret-Link et al. 2004).

1.2.1.2 Tillage Practices

Land is prepared for sowing/planting by following suitable tillage practice, in order to preserve the soil status for providing optimum conditions for plant growth and to retain long-term productivity of soils. Various implements may be employed to turn, mix, loosen or compact the soil, resulting in different levels of water retention capacity, aeration and soil temperature. Survival and dispersal of soilborne pathogens depend on the changes brought out by tillage practices, as well as the pathogen-infested crop residues left in the soil after harvest. Most of the fungal and bacterial pathogens complete a part or entire life cycle in the soil or on the organic matter present in the soil. Influence of tillage practices may be variable on the pathogens and antagonists that have to coexist in the soil (Cook and Baker 1983). Tillage practices generally reduce disease risk primarily by reducing pathogen densities by promoting decomposition of infested plant residues, physically separating pathogens from host plants by burying them in deep soil layers or by changing soil conditions (Yang 2002). In several investigations, the combined effects of tillage and other cultural practices have been studied. The beneficial effects of destruction of weeds and volunteer plants that may serve as sources of inoculum of soilborne pathogens is attributed to tillage. Deep plowing may result in burial of pathogens, followed by inactivation of soil microflora or exposure of pathogens to solar radiation and heat, as in the case of *R. solani* and *V. dahliae* (Papavizas and Lewis 1979; Fry 1982). The dispersal of *Phytophthora capsici*, causal agent of Phytophthora blight disease of bell pepper, was significantly reduced in the field, where a fallow-sown, no-till wheat cover crop was raised. The final disease incidence level was significantly low (2.5–43.0%) in no-till plots, whereas pepper planted in bare field showed high incidence (71–72%) where all dispersal mechanisms of *P. capsici* progressed without any obstacle (Ristaino et al. 1997). The effectiveness of tillage practices viz., chisel plowing and disc plowing was compared with no-till, in reducing the incidence of soybean sudden death (SDS) caused by *F. oxysporum* f.sp. *glycines (Fog)*. In the fall of 2000 and 2001, plots were established in a field with a history of severe SDS and with silty clay loam soil type. Tillage practices formed the main plot treatments and 12 soybean cultivars with different levels of resistance/susceptibility formed subplot treatments which were randomized in main plots in the split-plot design. In 2000, chisel tillage plots had less area under disease progress curve (AUDPC) values, compared with no-till or disk-tillage. In 2001, both chisel and disk-tillage treatments showed reduced AUDPC values, compared with no-till. However, in 2000, the rate of infection by *Fog*, and area under root colonization progress curve (AURCPC) were higher in disk- and chisel-tillage than in no-till treatment. In 2001, only chisel-tillage exhibited higher rate of root infection than disk-tillage and no-till treatments. The effects of tillage were variable, based on the foliar and root infections by *F. oxysporum* f.sp. *glycines* (Vick et al. 2006).

The effect of tillage practices on the incidence of wheat Karnal bunt disease caused by *Tilletia indica* in the Indo-Gangetic Plains of India was assessed. Wheat samples from 370 fields adopting no-till (NT) (375), furrow-irrigated rainfed system (FIRBS) (171) and conventional till (CT) (370) were collected. Incidence of Karnal bunt disease, disease incidence index and percent infected samples were determined and analyzed statistically. Lowest mean incidence (9%) of the disease was observed in nil-tillage fields, whereas disease incidence levels were 18.1% and 16.2% in FIRBS and CT fields, respectively. Similar trend was also seen for Karnal bunt disease incidence (Sharma et al. 2007). The impact of tillage on dispersal of *T. indica* teliospores from a concentrated point source was investigated in Arizona, United States. Teliospore suspension (300 ml containing approximately 9.0 x 10^4 teliospores) was sprayed onto a 1 x 3 m soil area. About 400 g of soil was collected before tillage treatments, representing teliospore baseline and after each of five disks passed to a 20 cm depth (approximately) through infestation source. Soil samples were collected along three parallel lines extending from the infested area at increments of 1, 3 or 10 m to a total distance of 10, 30 and 50 m, respectively. Teliospores were recovered from soil samples by a combined size selective sieving sucrose centrifugation technique. An average of 3.6 x 10^3 teliospores per 25 g of soil sample was recovered from the infested area, soon after teliospore infestation. Two different trends in recoverable teliospores were observed at 0- to 10-m sampling distances, following five plow passes: either a decrease in number of teliospores recovered, represented at points 0, 1 and 2 m or an increase in recoverable teliospores found at points 3 to 10 m. Teliospores were recovered at a maximum distance of 24 m. The results indicated that teliospores of *T. indica* were not dispersed in large numbers to long distances by tillage and spatial dispersal of teliospores could not be attributed to the tillage practices adopted in infested wheat fields (Allen et al. 2008).

Tillage and application of manures (organic amendments) precede sowing seeds or planting seedlings. Effects of tilling frequently or surface mulched no-till forming as the main plots and organic amendments, including composted cotton gin trash, composted poultry manure, an incorporated rye-vetch green manure or synthetic fertilizer forming subplots, were studied between 1997 and 2004. Soil samples from different treatments infested with *P. capsici* were analyzed. Both tillage and fertility amendments applied prior to sowing affected disease incidence in pepper and dispersal of the pathogen. Final disease incidence and area under disease progress curve (AUDPC) and extent (distance) of pathogen spread were significantly greater in soils with surface mulch applications than in frequently tilled plots. Final disease incidence, AUDPC and the extent of pathogen spread were also significantly greater in cotton trash amended plots, than in rye-vetch green manure-, poultry manure- or synthetic fertilizer-applied plots. Soil water content, soil porosity, humic matter content and net mineralizable levels of nitrogen were positively correlated and bulk density was negatively correlated with final incidence of Phytophthora blight disease in pepper (Liu et al. 2008). The effects of tillage and crop rotation on production of apothecia by *Sclerotinia sclerotiorum* infecting soybean, were assessed. Minimum tillage, no tillage with residue chopped after harvest or no tillage with no disturbance of residue after harvest in combination with five 2-year rotations of continuous soybean, corn-soybean, soybean-corn and wheat-soybean were the treatments tested under field conditions. Crop rotation had significant effect on the production of apothecia in all years at both sites and in general, lower mean numbers of apothecia and /or clumps of apothecia were recorded in plots planted with corn or winter wheat than with soybean. In all years, and both sites, there were fewer apothecia and clumps of apothecia in the no-tillage treatments than in minimum tillage, although the difference was not statistically different in all years. Interactions between tillage and crop rotations were significant, only in 1996 and 1997 at Mardens, where apothecia production was greatest in minimum-tillage plots planted with soybean, regardless of previous crop. Within the two no-tillage treatments, plantings of continuous soybean produced the highest number of apothecia and clumps of apothecia. Reduction in the numbers of apothecia was 46.4% and 80.6% at Mardens, respectively, in 1996 and 1997 in the no-tillage with chopped residue treatments, when crop rotation was followed, compared with continuous cropping system. The combination of no-tillage and crop rotation was found to be effective in reducing the primary inoculum of *S. sclerotiorum* in infested fields (Garcia-Garza et al. 2002).

The effects of tillage, residue management and crop rotation (sequence) on development of soilborne wheat pathogens in irrigated cropping system were studied for 6 years in east-central Washington, United States. The continuous winter wheat treatment with burning and plowing was compared with a 3-year no-till rotation of winter wheat-spring barley-winter canola (oilseed rape, *Brassica napus*) and three straw management treatments viz., burning, straw removal and leaving the stubble as such after harvest. Take-all disease caused by *G. graminis* var. *tritici* (*Ggt*) incidence in the continuous wheat treatment with burning and plowing increased from 2002 to 2004 and then declined in 2006. In 2003, disease rating was 3.8, but in 2006 it was reduced to 2.0. Pathogen population was quantified, based on DNA contents measured by PCR assays. In the first year, DNA contents did not show variations due to treatments. In the following year, continuous winter wheat had significantly higher DNA contents than in the winter wheat treatment with rotation crops. This difference in DNA contents remained during the subsequent three cropping seasons, although there was a decline in DNA contents of *Ggt* in the continuous winter wheat treatment over time (see Figure 1.1). In the case of *R. solani* AG-8, causal agent of root rot and bare patch, the method of residue management had no effect on the incidence of Rhizoctonia root rot and crown rot on barley and the effect on seminal roots was variable. Inoculum of *R. solani* AG-8 was significantly less in the tilled treatment compared to the no-till treatments. The continuous winter wheat with burning and plowing had lower levels of *Rhizoctonia* in the soil, compared to the no-till treatment. The concentrations of *Rhizoctonia* DNA in no-till soil was approximately 1 log unit higher than in burn and plowing treatment (see Figure 1.2). In 2003 and 2006, the continuous winter wheat had the lowest level of crown and seminal root infection symptoms, compared to other treatments. The continuous wheat treatment also had the lowest Rhizoctonia root rot rating in 2004 and 2005. Another wheat root disease, Fusarium crown rot is caused by *Fusarium pseudograminearum* and *F. culmorum*. Inoculum concentration of *F. pseudograminearum* was greater than that of *F. culmorum* in 1 of the 3 years. *F. pseudograminearum* was at greater population level in treatment with standing stubble and mechanical

straw removal, compared to burn treatments. The investigation revealed the complex interactions that occur, where one disease may be of primary importance, although other pathogens causing root diseases are also present, due to dynamics of host-pathogen interactions (Paulitz et al. 2010).

1.2.1.3 Propagation Practices

The propagation practices, such as grafting and budding have been adopted primarily for improvement of quality of horticultural produce. Grafting with selected resistant rootstocks to combat diseases and pests is an ancient practice widely employed for successful cultivation of a range of woody crop plants. Grafting watermelon (*Citrullus lanatus*) onto squash rootstocks (*Cucurbita moschata*) was done to reduce the loss due to Fusarium wilt disease caused by *Fusarium oxysporum* f.sp. *niveum (Fon)* (Sato and Takamatsu 1930). Grafting has been widely adopted in Korea, Japan, Europe and North America, to overcome the problem of infection by Fusarium wilt pathogen that persists in the soil for several years. Four races of *F. oxysporum* f.sp. *niveum* (0, 1, 2 and 3) were identified by virulence on differential watermelon cultivars. Later several reports indicated the possibility of improving resistance of crop plants to soilborne oomycete, fungal and bacterial diseases by grafting susceptible cultivars to resistant rootstocks. Use of cucurbitaceous rootstocks (nonhosts to most formae speciales of *F. oxysporum*), resulted in reduction in losses due to Fusarium wilt disease of watermelon. *V. dahliae* is another vascular wilt disease affecting Solanaceae and Cucurbitaceae. Both scions and rootstocks appeared to contribute to disease resistance of grafted combinations in watermelon, cucumber and tomato (Paplomatas et al. 2002).

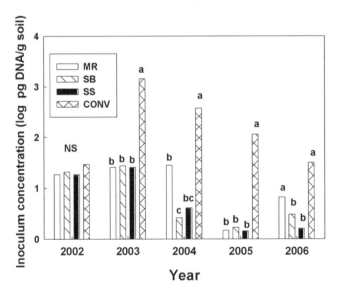

FIGURE 1.1 Effects of crop rotation/residue management treatments on DNA concentrations of *Gauemannomyces graminis* var. *tritici (Ggt)* in winter wheat crop grown soil during 2002–2006 Treatments: winter wheat-spring barley-canola no-till rotation with mechanic stubble removal (MR), stubble burning (SB) or stubble left standing (SS), and continuous winter wheat with stubble burned and moldboard plowed (CONV); differences in the DNA contents in treatments with same letter on bars are not statistically significant (P = 0.05, as per Tuley's mean separation test).

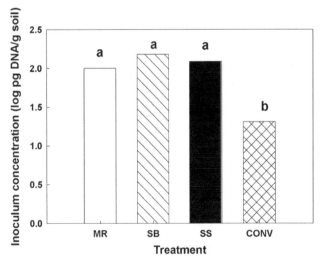

FIGURE 1.2 Effects of crop rotation/residue management treatments on DNA concentrations of *Rhizoctonia solani* AG-8 in winter wheat crop grown-soil averaged across all years (2002–2006) Treatments: winter wheat-spring barley-canola no-till rotation with mechanic stubble removal (MR), stubble burning (SB) or stubble left standing (SS), and continuous winter wheat with stubble burned and moldboard plowed (CONV).

The sudden wilt disease caused by *Monosporascus cannonballus* is soilborne and accounts for heavy losses in melon and watermelon crops grown in hot and semiarid areas. Grafting scions of susceptible melon cultivars onto *Cucurbita maxima* and *C. maxima* x *C. moschata* rootstocks improved resistance of melon (Edelstein et al. 1999). *P. capsici*, causing Phytophthora blight disease seriously reduces the yields of cucumbers. Resistance and yield levels could be improved by grafting cucumber onto bottle gourd (*Lagenaria siceraria*), *C. moschata* and wax gourd (*Benincasa hispida*) as rootstocks (Wang et al. 2004). Watermelons grafted onto selected bottle gourd rootstocks were resistant to *P. capsici* also (Kousik and Thies 2010). Another important disease of tomatoes is the corky root disease caused by *Pyrenochaeta lycopersici*. Grafting tomatoes onto 'Beaufort' rootstocks (*S. lycopersicum* × *S. habrochaites*) resulted in low level of infection, higher yield and larger fruits (Hasna et al. 2009).

F. oxysporum f.sp. *niveum (Fon)*, incitant of Fusarium wilt disease, is host-specific to watermelon and it does not infect other cucurbits. However, *Fon* was found to infect a few other cucurbits, including seedlings and young plants of summer squash (*Cucurbita pepo*) cultivars. Six rootstocks were evaluated for their resistance to races 1 and 2 of *Fon* and the performance of grafted rootstocks was tested for the frequency of infection under field conditions. Grafted and nongrafted watermelon plants and rootstocks were inoculated in the greenhouse with race 1 and 2 or water (control). With both races, the frequency of recovery of *Fon* from scion and rootstock portions of inoculated watermelon plants grafted onto 'Ojakkyo' citron was greater than from watermelon plants grafted onto 'Shintosa Camel' and 'Strong Tosa' interspecific hybrid squash and from plants grated onto 'Emphasis', 'Marcis' and 'WMXP3945' bottlegourd. Recovery of *Fon* from nongrafted plants was at greater levels compared to grafted plants. In spring 2010 and 2011, six rootstocks were grafted onto seedless watermelon 'Tri-X313', which was susceptible to both races of *Fon,* and transplanted in the field infested with races 1 and 2 of *Fon*. Disease incidence for nongrafted and self-grafted Tri-X313 (control) and Tri-X313 grafted onto Ojakkyo citron did not differ significantly. Grafted watermelon plants had greater weights and produced more number of fruits than control plants. Nonpathogenic isolates of *F. oxysporum* and isolates of *Fon* colonized interspecific hybrid squash, bottlegourd and grafted watermelon. The rootstocks restricted movement of *F. oxysporum* f.sp. *niveum* into the watermelon scion, suppressed wilt symptoms and increased fruit yields even in infested field (Keinath and Hassell 2014). In a later investigation, the usefulness of grafting susceptible scions onto resistant rootstocks was assessed to recommend this strategy for the management of Fusarium wilt disease affecting watermelon. The rootstocks, Marathon (*C. maxima* x *C. moschata*) and Macis (*L. siceraria*) were grafted with different scions, including seedless cultivars. Grafted plants had increased yield, compared with nongrafted plants (136% and 159%) in two experiments. Incidence of wilt disease was reduced significantly in grafted plants inoculated with *F. oxysporum* f.sp. *niveum*, resulting in increase in yield levels. Application of this approach led to economic benefits that exceeded the cost involved in grafting (Moreno et al. 2016).

Grafting susceptible scions onto resistant rootstock has been reported to increase resistance also to bacterial and viral diseases affecting vegetable crops. Infection of watermelon by *Melon necrotic spot virus* (MNSV) resulted in appreciable losses in watermelon. Use of resistant stocks of watermelon to tackle melon necrotic spot disease was found to be effective. This approach offered significant advantage over soil fumigation with methyl bromide which has been banned by several countries (Cohen et al. 2007). Tomato bacterial wilt disease caused by *R. solanacearum* is one of the most destructive diseases, impacting the tomato production adversely. Grafting scions of susceptible tomato cultivars onto resistant rootstocks was demonstrated to be effective for reducing the losses due to the bacterial wilt disease. The enhancement of resistance in grafted tomato plants was suggested to be due to reduced colonization of the lower stem by *R. solanacearum*, rather than prevention of the pathogen invasion of xylem tissues (Grimault and Prior 1994; Lin et al. 2008). The efficacy of commercially available rootstocks to reduce bacterial wilt disease incidence in tomato and crop losses was assessed. Tomato plants grafted onto 'Dai Hommei' and 'RST-04-105-T' rootstocks had lower area under disease progress curve (AUDPC) values compared with nongrafted plants (P <0.05).The final bacterial wilt incidence for non- and self-grafted plants was 82 ± 14 to 100%. In contrast, for grafted plants, the incidence was 0 to 65 ± 21%. Final disease incidence in plants grafted onto Dai Hommei rootstock was 0 and 13 ± 3% at two locations in western North Carolina, but 50 ± 3% at a third site in eastern North Carolina. Performance of RST-04-105-T showed similar trend at two of three sites, but it performed poorly in the third site. The total fruit yields were significantly increased by grafting onto resistant rootstocks at all three sites. The yield showed significantly negative relationship with AUDPC values. The results indicated the effectiveness of grafting onto resistant rootstocks for maintaining the economically viable tomato production in soils naturally infested with *R. solanacearaum* (Rivard et al. 2012).

1.2.1.4 Date of Planting/Sowing and Planting Density

Planting at right time in a crop season with optimum density of plants and availability of irrigation sources, supplemented by rainfall, are essential to obtain expected yield levels. For achieving effective reduction in disease incidence and spread, date of sowing/planting has to be determined with the primary objective of reducing the period for which the crop may be at vulnerable stage or viable pathogen propagules are available in large numbers. In the case of virus diseases, the date of sowing/planting has to be so adjusted that the crop is not exposed to the period, when the vector population is high with active mobility. The knowledge on the pattern of incidence of different diseases likely to affect the crop(s) to be grown, would be useful to prepare a realistic plan to avoid the disease by suitable modification of sowing date. However, certain problems imposed by crop requirements have to be addressed effectively (Narayanasamy 2002, 2011).

Reduction in severity of sugar beet yellows and consequent increase in yield could be achieved by early planting which favored rapid crop development. In addition, early sowing permitted seed germination and plant growth, when temperatures were too low for infection by the fungal pathogen *Polymyxa betae*. In contrast, late sowing led to early infection and rapid pathogen development. Further, early sowing provided additional advantage of avoiding rhizomania disease, induced by *Beet necrotic yellow vein virus* (BNYVV) transmitted by *P. betae* (Blunt et al. 1992; Rush and Heidel 1995). The effect of planting date and seed size on the severity of common root rot caused by *Bipolaris sorokiniana* in hard red winter wheat was investigated. Small-, mixed- and large-sized seeds of five wheat cultivars were planted in the first weeks of September and October 1994 and 1995. Disease ratings for incidence and severity of crown and internode infections were assessed in March and at harvest. Overall, seed size had no effect on disease severity or grain yield for both years. However, when sorted by planting date, plants from small seed yielded less than plants from other types of seeds. October plantings showed lower disease indices, compared with September plantings at evaluation done in March. However, reduction in severity was seen in October planting at harvest evaluation only in 1995 (Piccinni et al. 2001).

The impact of planting date, cultivar and stage of plant development on incidence of Fusarium wilt of lettuce caused by *F. oxysporum* f.sp. *lactucae* was studied in 2002–2003 crop seasons. The planting windows adopted were early-season (September), mid-season (October) and late-season (December). Within each planting window, significant differences in disease incidence among lettuce cultivars were observed at crop maturity, with a mean incidence of 92.3, 15.1 and 2.0%, respectively, on crops sown in September, October and December in 2002 and 74.2, 5.1 and 0.7%, respectively, in 2003. The mean soil temperatures at 10-cm depth during September, October and December planting in both years were 26, 14 and 14°C, respectively. Among all cultivars planted in September, only one and two cultivars in 2002 and 2003 respectively showed ≤5% Fusarium wilt disease incidence at maturity, whereas the lowest incidences of disease among Crisphead, Green leaf, Red leaf or Butterhead cultivars were 73.7, 27.0, 20.2 and 65.7% in 2002, respectively, and 62.1, 29.0, 100 and 100% in 2003, respectively. The results indicated the possibility of choosing planting date and lettuce cultivar to reduce disease incidence to the minimum (Matheron et al. 2005). The effect of planting date on the incidence of Ascochyta blight disease in six chickpea cultivars was assessed by planting at three sites, differing in disease pressure. In two sites, disease pressure was increased by spreading infected chickpea plant debris on soil surface at the time of planting and at one of these, sprinkler irrigation was provided. In the third site with dry conditions, artificial inoculum was not applied. Plants from seeds sown in early March, showed maximum disease and in sprinkler-irrigated plots the disease severity ranged from 7.8 on the highly susceptible cv. Camitez to 3.3 on the least susceptible cv. Gokce on the disease rating scale of 1 to 9. An inverse relationship was noted

between disease severity and yield, the resistant cultivar producing twice that of the yield obtained from susceptible cultivar. Delaying planting by 3–5 weeks, reduced disease severity as well as yield in 4 of 6 cultivars. However, in susceptible cv. Camitez and Local, the disease severity was reduced due to delayed planting. Delay of planting by 6–9 weeks, eliminated Ascochyta blight almost, but yields of all cultivars were drastically reduced due to drought stress. Based on the results, early planting of resistant or tolerant cuttings could be recommended in Turkey (Dusunceli et al. 2007).

1.2.1.5 Canopy Management Practices

Crop canopy may exert varying effects on incidence and intensity of diseases due to changes in the microclimate in a field. In carrot crops, foliage side-trimming opens the canopy, permitting greater sunlight penetration and air flow. This situation may result in prevention of moisture buildup and consequent creation of unfavorable conditions for development of *S. sclerotiorum*. The response of carrot to foliage side-trimming and consequent suppression of Sclerotinia rot of carrot (SRC) was assessed. Incidences of SRC at the time of trimming and again at harvest were recorded. In addition, carrots were placed into storage at 10°C for three months to determine the extent of infection in roots. Variations in susceptibility of carrot cultivars to SRC and yield potential differed. Across cultivars, foliage side-trimming significantly reduced total and marketable yield by about 7.5% and 9.1%, respectively. The number of petioles with SRC was significantly reduced at both time of trimming and harvest. Incidence of SRC in stored roots was also reduced significantly by foliage trimming. Selection of carrot cultivars with low levels of susceptibility and foliage side-trimming could be adopted to reduce incidence of Sclerotinia rot disease (Sanderson et al. 2013). Blackleg disease caused by *Leptosphaeria maculans* seriously affects canola, which is grown as an economic break-crop alternative for dual purpose cereals in Australia mixed farming systems. Canola crops were infected by airborne inoculum released throughout the autumn and winter, when crops were grazed. The effect of mechanical defoliation (simulated grazing) on disease severity at plant maturity was investigated both under glasshouse and field conditions. Stem canker severity increased from 4 to 24% in severely defoliated plants, but light defoliation had no effect, compared with undefoliated control plants under glasshouse conditions. Disease severity was increased with defoliation in all field experiments. Defoliation increased crown canker severity from 22.6 to 39.3% at first location and from 3.0 to 7.1% in the second location. Lodging of plants occurred up to 11.9% in a third location in the same set of cultivars assessed at each site. Cultivars with moderate resistance to stem canker showed less disease severity due to defoliation. Canola plants defoliated before stem elongation stage tended to contract less disease, than those defoliated during reproductive phase of plant growth. Growing canola cultivars with high level of resistance to stem canker induced by *L. maculans* and grazing during vegetative stage of plant growth prior to stem elongation might be helpful in reducing disease incidence in canola crop defoliated by grazing animals (Sprague et al. 2010).

1.2.1.6 Irrigation Practices

Irrigation, required to maintain optimum soil moisture, may influence the development of both the host plant and the microbial pathogens to a greater extent, compared with other cultural practices. Absorption of nutrients and maintenance of cell turgidity by host plants are facilitated, when optimum levels of soil moisture are available. Both high and low (water stress) moisture conditions may predispose plants to infection by various microbial plant pathogens. Four types of irrigation systems viz., flooding, furrow-, sprinkler- and drip- irrigation are frequently adopted for cultivation of various crops. When large quantity of water is available, flooding system can be applied, but this system may favor spread of the pathogen propagules to the entire field. However, flooding the soil (flood fallowing) was reported to be effective in eradicating the banana Fusarium wilt disease caused by *F. oxysporum* f.sp. *cubense*. Flooding the field plots drastically reduced the oxygen supply to the soil, making the conditions very unfavorable to the survival of the pathogen. The treatment effectively kept the pathogen at bay for about 4–5 years, after which the pathogen reappeared in the flooding treatment (Kelman and Cook 1977). Open field burning was the primary method of rice residue disposal and of minimizing the carryover inoculum of *Sclerotium oryzae*, causal agent of rice stem rot disease. Alternatives to burning were considered necessary to reduce overwintering sclerotia of *S. oryzae* and consequent reduction in stem rot incidence and severity and increasing the yield levels. In the split-plot design, winter flooding and winter nonflooded treatments formed the main plots, whereas fall incorporation of the straw residue, rolling of the straw to enhance soil contact, balling and removal of residue and fall burning were the subplot treatments. Inoculum of *S. oryzae* and stem rot severity were significantly reduced, while yield registered significant increases in 5 of 6 years in the water-flooded main plots, compared with winter nonflooded plots, over the experimental period. *S. oryzae* inoculum was consistently lower in burn subplots, compared with all other subplots. Consistent differences in disease incidence and severity or yield in the subplots were not seen over the 6-year period. The results suggested that winter flooding might be a suitable alternative to burning for rice stem rot disease management (Cintas and Webster 2001).

Infective propagules of microbial plant pathogens are spread via irrigation water to different parts in an infested field. Fungal pathogens, such as *V. dahliae*, *F. oxysporum* f.sp. *cubense* and *P. infestans,* are known to be dispersed through irrigation water. Incidence of the diseases induced by these pathogens could be reduced by preventing flow of irrigation water from infested field to other fields. The rhizomania disease of sugar beet is transmitted by the vector *P. betae*. Water released into the sewage system by sugar refineries carried *P. betae*. By modifying the irrigation system, it may be possible to reduce the incidence of soilborne diseases to some extent (Sutic and Milovanovic 1980). *P. capsici* zoospores spread from inoculated source plants to healthy potted pepper plants located on separate ebb-and-flow benches, when

the recycled nutrient solution originated from a common reservoir. The surfactant added to the recirculating nutrient solution, resulted in selective killing of pathogen zoospores and the spread of the pathogen could be arrested completely in the ebb-and-flow and a top-irrigated culture system. In contrast, all plants in the ebb-and-flow system were killed within 6 weeks, when the recirculating nutrient solution was not amended with the surfactant. The results showed the possible spread of the pathogen and effectiveness of surfactant in the control of *P. capsici* infecting pepper plants (Stanghellini et al. 2000).

The association between soilborne pathogens and flooded soil conditions was studied, using six soybean genotypes, representing wide range of flood tolerances. Genotypes were planted in single-row plots from 1996 to 1998 with flood treatments of (i) no flood, (ii) flood at emergence (3-day duration) or (iii) flood at fourth leaf node growth stage (7-day duration). The effects of flooding were assessed at 3 days after removing each flood treatment. Roots were examined for the presence of fungi and other filamentous eukaryotic microorganisms. Reduction in plant stands in flood treatment was evident, as compared with nonflooded controls. Adverse effects on plant weight and root discoloration in flooded treatments were observed. *Pythium* was the only genus of filamentous organisms isolated from the experimental plots and their isolation frequency increased with flooding. Of the 60 *Pythium* isolates, 47% were moderately to highly virulent on soybean. Pathogenic species *P. ultimum, P. aphanidermatum, P. irregulare* and *P. vexans* were associated with root infection of soybean. Nonpathogenic *P. oligandrum* (a known biological control agent) was also isolated from soybean. The results suggested that flooding might enhance the chances of infection of soybean by *Pythium* spp. (Kirkpatrick et al. 2006). The effect of flooding on survival of *Leptosphaeria biglobosa* 'brassicae', incitant of blackleg (Phoma stem canker) disease, in the stubble of winter oilseed rape (*B. napus*) was investigated. Basal stems were submerged in water at 16°C continuously or at 20 and 28°C, 28 and 33°C and 33 and 40°C alternatively for 12 and 12 h or kept in dry room temperature (control). In the field experiment, the stem pieces were placed on soil surface in a rice field or in a cotton field and either flooded in water or not flooded respectively. *L. biglobosa* 'brassicae' was isolated on V8-juice agar, from the stem pieces, after 1, 2, 4, 6 and 8 weeks. Selected isolates were identified, using PCR assay. Flooding for 1–2 weeks substantially reduced recovery of *L. biglobosa* 'brassicae', compared with controls, whereas flooding for 4 weeks eliminated the pathogen from the stem pieces almost completely. The identity of the pathogen was established, based on the amplification of a 444-bp DNA fragment specific to *L. biglobosa* 'brassicae'. The results showed that the infected stem pieces were rapidly decomposed by flooding. However, the oilseed rape-rice rotation, a routine cultivation practice followed in central China might be a more effective way to reduce the longevity of *L. biglobosa* 'brassicae' in the stubbles of winter of oilseed rape (Cai et al. 2015).

Development of soybean sudden death syndrome (SDS), induced by *Fusarium virguliforme* (Fv), is generally favored by high soil moisture conditions. The effects of duration of flooding and resultant anaerobic conditions on the soybean-*F. virguliforme* interaction were investigated. Susceptible and resistant soybean cultivars were exposed 3, 5 or 7 days of continuous flooding, repeated short-term flooding for 8 h/ week for 3 weeks and no-flood (control). At 7, 14 and 21 days after flooding (DAF), seedlings in the no-flood, 3-day and repeated short-term flooding showed the highest root rot and foliar symptom severity, whereas seedlings in 7-day flooding showed the lowest severity. *F. virguliforme* inoculum density in soil was the lowest in the 7-day flooding treatment. In the hydroponic system, the steady transcript levels of soybean defense genes were quantified, in response to different oxygen levels, using quantitative PCR assay. Roots of infected plants exposed to 12 h of anaerobic conditions showed down-regulation of the defense-related soybean genes, Laccase, PR3, PR10, phenylalanine-ammonia lyase (PAL), chalcone synthetase (CHS) and the virulence genes of *F. virguliforme*, pectate lyase (PL) and *F. virguliforme* homolog of the pisatin demethylase (PDA). The results suggested that short-term flooding might increase SDS severity, whereas prolonged flooding could negatively impact SDS, due to reduction of pathogen density in soil (Abdelsamad et al. 2017).

The effects of different moisture regimes, in the presence of *V. dahliae*, causal agent of Verticillium wilt of cauliflower on plant growth and disease development were assessed. Root length density at 5 and 25 cm from plant was significantly higher in subsurface drip than in furrow irrigation system. Concomitantly the wilt incidence and severity were higher in excessive and moderate irrigation regimes, regardless of the irrigation system. Stomatal resistance in lower diseased leaves was significantly higher in infested than in fumigated plots, but there was no difference in the upper healthy leaves. Cauliflower yield was not affected by *V. dahlia*, but the deficit irrigation regime resulted in reduction in yield, even though it suppressed wilt in cauliflower. The results showed that higher moisture levels led to increase in root length and density in *V. dahliae*-infested plots, which consequently resulted in higher levels of disease incidence and severity (Xiao and Subbarao 2000). Incidence of Verticillium wilt (VW), caused by *Verticillium albo-atrum* was frequently observed in irrigated alfalfa (*Medicago sativa*) crops, but rarely in dry land crops. In order to assess the effect of irrigation on VW incidence, a line-source irrigation water gradient was adopted on two alfalfa cultivars, Beaver (susceptible) and Barrier (resistant). The irrigation gradient was subdivided into six irrigation level treatments within each cultivar plot. A 1-m wide strip through the center of each subplot was inoculated with a spore suspension of *V. albo-atrum*, before first harvest in 1995. Beaver had more VW-infected plants in 1995 and 1996 than Barrier, and the number of diseased plants of Beaver increased with irrigation level. Ground cover of Beaver declined in 1997 and this response showed negative relationship with VW-diseased plant counts in 1995 and 1996. On the other hand, no significant relationship between disease incidence and ground cover in Barrier, could be seen. Forage yield increased linearly with irrigation level for both cultivars in two harvests during 1995.

Forage yield of Beaver showed decline progressively in 1996 and 1997, as the ground cover was reduced by 40% at higher irrigation level (6), compared with 22% reduction with low irrigation level (1). The results showed that *V. albo-atrum* infection in susceptible alfalfa cultivar might result in proportion to amount of irrigation water applied and resistant cultivars should be grown in irrigated fields (Jefferson and Gossen 2002).

In order to investigate the influence of irrigation frequency on the incidence and development of Verticillium wilt (VW) of olive, caused by *V. dahliae*, a split-plot design in microplots was established in naturally infested soil. Three olive cultivars with varying levels of resistance to VW were included. Final disease incidence (DI) and mortality in cv. Picual plants subjected to daily irrigation treatment (T1) attained values of 100 and 63%, respectively. The area under disease progress curve (AUDPC) values in T1 treatment, were significantly different between December 2012 and July 2013 (14.8–42.8%), compared with the average values of treatments T2 (weekly), T3 (biweekly) and T4 (deficit) (0.4–11.5%). Irrigation treatments did not influence disease incidence and development in cv. Arbequina, although DI progressed consistently to 60% in all treatments. Disease incidence in cv. Frantoiv was very little. The results showed that daily irrigation favored disease incidence and development of symptoms in susceptible cv. Picual in infested soil and reduction in irrigation frequency might result in less infection by *V. dahliae* (Pérez-Rodríguez et al. 2015). In a later investigation, field trials were conducted to determine the effect of irrigation frequency on olive Verticillium wilt disease on cv. Picual. Disease onset (at 61 weeks after planting) and incidence (average 75.6%) did not differ among irrigation treatments viz., daily, biweekly and dry land in both fields. Irrigation consistently increased disease development in dry land treatment, but this effect varied over time. Significant correlation between disease incidence and severity increments during spring and fall with soil water content was observed. The results suggested that scheduling irrigation based on rainfall could be a feasible approach for containing the olive Verticillium wilt disease (Pérez-Rodríguez et al. 2016).

Macrophomina phaseolina produces microsclerotia (MS), which are survival structures and are highly resistant to adverse soil environments. The effects of irrigation and soil stress on MS densities in the soil and roots of soybean were investigated. Soybean cvs. Davis and Lloyd were irrigated until flowering (TAR2), after flowering (IAR2), full season (FSI) or not at all (NI). Microsclerotia were present in the roots of irrigated and also nonirrigated soybean within 6 weeks after planting. By vegetative growth stage (V13), MS densities reached relatively stable levels in the NI and FSI treatment [2.23 to 2.35 and 1.35 to 1.63 log (MS/g of dry root respectively] through reproductive growth stage (R6). MS densities in roots increased in all treatment, after stoppage of irrigation at R6 stage. Initiation (IAR2) or termination (TAR2) of irrigation at R2 stage, resulted in significant changes in root MS densities, which reached levels intermediated between those of FSI and NI treatments. The range of root colonization by *M. phaseolina* was significantly greater in Davis than Lloyd at R5 and R6 stages. The results indicated that irrigation level might limit, but not prevent, colonization of soybean by *M. phaseolina* and soybean cultivars showed differential response to colonization by the pathogen (Kendig et al. 2000). In another investigation, the influence of different irrigation systems – daily drip, 3-day drip and alternate-row-furrow irrigation – on the incidence of *P. capsici* in pepper was assessed. Drip irrigation provided favorable soil moisture conditions for pepper plant growth, resulting in enhanced yields, whereas it was unfavorable for pathogen development (Xie et al. 1999). *P. capsici* produces the asexual spores, zoospores that initiate root infection. Conditions of wet and dry cycles in soil are primarily required for completion of different stages of pathogen life cycle. Rainfall and periodic furrow irrigation provided the wet-dry cycle in the soil, favoring production of sporangia during dry period and release of zoospores during wet period. Plants receiving drip irrigation contracted higher disease level (78.4%) than those irrigated by basin (56.3%) and furrow irrigation (59.9%) systems (Sağir et al. 2005). The effect of drip irrigation on incidence of olive Verticillium wilt disease caused by *V. dahliae* was assessed in three *V. dahliae*-infested olive orchards. Pathogen in soil was quantified in wet zones around the drippers and in dry zones out of them. Inoculum density was higher in wet than in dry areas. Soil population of *V. dahliae* increased appreciably in both wet and dry areas, but inoculum density was higher in wet areas, when assessed after 4 months of irrigation (López-Escudero and Blanco-López 2005).

The influence of furrow irrigation with conventional tillage and subsurface-drip irrigation with minimum tillage on the temporal and spatial dynamics of sclerotia of *Sclerotinia minor* and incidence of lettuce drop disease was investigated in the spring and fall seasons during 1993-1995. Incidence of lettuce drop was significantly higher under furrow irrigation than under subsurface-drip irrigation throughout the experimental period and it was higher on fall crops than on spring crops. Under furrow irrigation, the number of sclerotia at the end of a crop season increased significantly over that at the beginning of the season, but no significant changes were evident over the years. In contrast, the number of sclerotia within a single season did not increase significantly under subsurface-drip irrigation and the year-to-year accumulation of sclerotia was not significant. The conventional tillage after harvest under furrow irrigation decreased the degree of aggregation of sclerotia after each season, but the distribution pattern of sclerotia under subsurface-drip irrigation was not significantly altered by the associated minimum tillage. Analyses of spatial pattern suggested that aggregation of sclerotia of *S. minor* occurred at the rate of less than 1 m and distribution of diseased lettuce plants was random at a rate larger than 1 m. The subsurface-drip irrigation combined with minimum tillage had the potential for application as an effective cultural practice for reducing the incidence of lettuce drop disease caused by *S. minor*, because of the low rate of sclerotial production and their unaltered distribution in the soil (Wu and Subbarao 2003). In a later study, the irrigation practices associated with cool-season vegetable cultivation may have significant influence on the incidence

of lettuce drop disease caused by *S. minor* and *S. sclerotiorum*. The effects of different irrigation frequencies and two bed widths (1 and 2 m) were compared in two separate field experiments over four lettuce crops in 2 years. Treatments were 1- and 2-m bed widths, as main plot treatments and twice-weekly, weekly and biweekly drip-irrigation forming subplot treatments. Incidence of lettuce drop was recorded at weekly intervals, after thinning, and continued till crop maturity. The effects of bed width and irrigation frequency were significant for both pathogen species. Twice-weekly irrigation and 2-m bed width resulted in greater levels of disease incidence than other treatments. For *S. sclerotiorum*, these two treatments had greater number of apothecia per unit area and increased the accumulation of soilborne sclerotia over multiple cropping seasons. The practice of increased bed width (2-m) and increased irrigation frequency followed in Salinas Valley became a serious concern for increased incidence of lettuce drop disease due to *S. sclerotiorum* (Wu et al. 2011).

Provision of supplemental irrigation to potato crops under water stress offers beneficial effect by improving plant growth and enhancing yield of tubers. Supplemental irrigation was provided, based on tensiometer or moisture block readings deployed in field plots. The severity of black scurf (*R. solani*), black dot (*Colletotrichum coccodes*), silver scurf (*Helminthosporium solani*) and common scab (*S. scabies*) diseases was quantified on potato tubers randomly sampled at harvest and stored at 7.2°C before visual assessment. The mean incidence of black scurf, silver scurf and black dot ranged over 3 to 18%, 2 to 33% and 4 to 9%, respectively in best, irrigated and reduced irrigation treatments (Olanya et al. 2010). Development of black dot disease on underground plant parts was monitored in two field trials during 2004–2007 in England and Scotland. Irrigation coupled with fungicide (azoxystrobin) significantly reduced black dot disease on stem and stolons, compared with other treatments – cultivar resistance and crop duration. The disease development progressed on stolons and roots at the same place at English and Scottish sites with irrigation (with or without fungicide), reducing infection of stolons and roots by *C. coccodes*. However, in the English site azoxystrobin had marginal effect on disease incidence, whereas irrigation significantly reduced black dot disease incidence (Brierley et al. 2015).

1.2.1.7 Crop Nutrition

Growth and yield of crops, as well as development of diseases caused by microbial plant pathogens are significantly influenced by the nutrition provided to the crops. As the area of cultivable land is shrinking, because of the industrial development and greater demand for housing of growing human population, crop production management strategies had to be modified. This situation necessitated enhancement of chemical fertilizer use to increase yield and application of chemicals to keep disease incidence at the minimal level. The need for alternative methods to restrict or replace use of fertilizers and chemicals applied against microbial plant pathogens was realized, because of the development of resistance to microbial pathogens. Cultural practices such as application of organic amendments and crop rotation (sequence) have been found to offer beneficial effects by reducing disease incidence, severity and function through altered biological activity by encouraging antagonists present in the soil. In order to maintain a nutritional balance from deficiency to sufficiency, required quantities of organic matter and minerals have to be supplied to crops, facilitating the growers to obtain potential yield levels (Huber and Haneklaus 2007; Narayanasamy 2013).

1.2.1.7.1 Application of Organic Amendments

Organic matter application enhances soil fertility and also encourages development of both antagonists and microbial plant pathogens. Organic matter incorporated into the soil is known as amendment, whereas organic matter spread on the soil surface is designated mulch. Organic amendments may promote biological destruction of pathogen populations through germination-lysis mechanism and intensification of microbial activity, resulting in degradation of propagules of plant pathogens. Some of the organic amendments may either stimulate or inhibit pathogen development and consequent expression of disease symptoms. The organic matter should be composted well before application to the soil, because of possible survival of pathogens on organic amendments and introduction into the soil inadvertently. Application of well-decomposed compost with low NH_4-N concentration and high Ca concentration reduced corky root disease severity caused by *Pyrenochaeta lycopersici*, a serious problem for organic tomato production in Sweden (Hasna et al. 2009).

Various kinds of amendments have been evaluated for their efficacy in reducing the pathogen propagules and disease development. Incorporation of barley straw in the soil effectively reduced populations of *V. albo-atrum*, causing wilt disease in potato. Direct quantitative reduction of pathogen population was considered responsible for reduced disease incidence, since no change in host resistance or pathogenicity of the pathogens could be observed (Harrison 1976). In another study, the effectiveness of application of sudangrass or corn (maize) as green manures in reducing Verticillium wilt disease of potato was demonstrated (Davis et al. 2010a). Incidence and severity of pepper crown and root rot caused by *P. capsici* was markedly reduced, following application of perennial peanut and chitosan (Kim 1989). Suppressive potential of rotation crops was assessed in a field with a history of incidence of severe Verticillium wilt of potato caused by *V. dahliae*. High glucosinolate mustard blend ('Caliente 119') as a mixture of white mustard (*Sinapis alba*) and oriental mustard (*Brassica juncea*) with known biofumigation potential and a sorghum-sudangrass hybrid were grown as a single-season green manure followed by a subsequent potato crop during 2007–2008. All rotation treatments significantly reduced incidence of Verticillium wilt compared to a standard barley rotation. The mustard blend and sudangrass green manure rotations reduced disease incidence by 15 to 19% and 21 to 31% in 2007 and 2008, respectively. However, neither green manure rotation was as effective as the barley rotation in combination with metam sodium fumigation which reduced Verticillium wilt incidence by 28% and 43% in 2007 and

2008, respectively. Averaged over both seasons, sudangrass, mustard and fumigation treatments reduced Verticilliium wilt by 17, 24 and 35%, respectively. In the 2009 potato crop representing the second rotation cycle, Verticillium wilt incidence was much higher in all plots and green manure rotation were less effective, averaging only 8 to 10% reduction, compared to barley control. Fumigation reduced wilt incidence more effectively (35%), but disease incidence was still higher than in previous 2 years. Both green manure rotations significantly reduced Verticillium wilt disease (average reductions of 25% and 18%, respectively). Verticillium wilt incidence in the subsequent potato was at a similar level as in the standard barley (control). But they were not effective as the chemical fumigant (metam sodium). The mustard blend provided additional benefit of reducing black scurf and common scab diseases of potato better than all other rotations and increasing tuber yield as well. However, disease incidence reached high levels in all rotations by the second rotation cycle, whereas chemical fumigation provided marked reduction (35%) in Verticillium wilt incidence and in the populations of *Verticillium* spp. too (see Figure 1.3). Previous cropping history did not affect

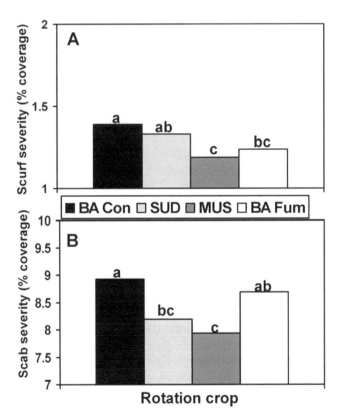

FIGURE 1.3 Disease suppressive effects of rotation treatments on severity (percent surface coverage) induced by potato black scurf (A) and common scab (B) averaged over two potato seasons (2007 and 2008) Treatments: BACon- standard barley rotation (control); SUD- sudangrass green manure; MUS- mustard blend green manure, and BAFum- barley rotation with fumigation by metam sodium; bars with same letter on top are not significantly different as per Fischer's protected least significant difference test (P <0.05).

[Courtesy of Larkin et al. 2011 and with kind permission of the American Phytopathological Society, MN]

significantly the incidence of Verticillium wilt disease. The results indicated the usefulness of disease-suppressive rotations for management of Verticillium wilt and other soilborne diseases causing substantial crop losses (Larkin et al. 2011).

Growing sweet corn cvs. Jubilee sweet corn and Jubilee super-sweet corn, as green manure crops for two or three seasons suppressed potato Verticillium wilt disease and increased potato tuber yields under field conditions. Colonization of potato feeder-roots and potato stem apices by *V. dahliae* was reduced. Feeder-root colonization by *V. dahliae* was positively correlated with Verticillium wilt incidence and negatively correlated with yield. Further, corn green manures increased populations of other soilborne fungi such as *Ulocladium* and *Fusarium equiseti*. Corn cultivars varied in their effectiveness as green manures. The sweet corn increased the biomass (> two folds) and wilt incidence was reduced to a level less than half, compared with super-sweet corn. Suppressive effect of the soil acquired due to green manure application was almost eliminated, following cultivation of potato for 2 consecutive years. However, a single green manure of sweet corn could restore the soil suppressiveness against *V. dahliae*, in spite of enhancement of pathogen population by 4-fold (from 45 to 182 CFU/g of soil). The results demonstrated the effectiveness of green manure application as a management strategy against wilt disease caused by *V. dahliae* (Davis et al. 2010a). In another study, the ecological relationships of suppression of Verticillium wilt of potato by growing green manure crops were investigated. The Verticillium wilt development was suppressed on the first years of Russet Burbank potato cropping, following 2–3 successive years of green manure incorporation. Yields of potato increased with the first year of potato cropping, but were reduced during second consecutive year of cropping with potato. In the second year of continuous cropping with Russet Burbank, total yields were reduced by 10% and these yield reductions were accompanied by increased soil populations of *V. dahliae* by more than five folds. However, with a single season of a green manure (Austrian winter pea, sudangrass, Dwarf Essex rape, Bridger rape, oats, rye or sweet corn), Verticillium wilt development was again suppressed and potato yields registered increases. In addition, green manures stimulated microbial activities which showed negative relationship with disease incidence (Davis et al. 2010b). The effects of incorporation of alfalfa residues into the soil prior to planting potato on the development of Verticillium wilt disease caused by *V. dahliae* were investigated. Microsclerotia of *V. dahliae* were quantified in the field soil amended with alfalfa residues. Bacterial metagenomics approach was applied to characterize soils with and without amendment of alfalfa residues. The number of *V. dahliae* microsclerotia in soil was greater (P = 0.000 3) in fields receiving alfalfa amendment, compared with field without alfalfa amendment, but not in plots receiving fumigation with chloropicrin. Alfalfa residue incorporation did not significantly change the soil bacterial metagenome, compared to field not amended with alfalfa residues in 2 years of experimentation (Frederick et al. 2018).

The potential of hairy vetch (*Vicia villosa*) as a soil amendment for suppressing the development of watermelon Fusarium wilt disease caused by *F. oxysporum* f.sp. *niveum* was assessed. The effect of inoculum density and the host resistance level on hairy vetch-induced suppression was determined on 12 watermelon cultivars with varying levels of resistance and 16 naturally infested soil samples collected from commercial watermelon fields. Development of wilt disease was suppressed to different extent, as the level of resistance of the cultivar increased. Fusarium wilt suppression was 22, 53 and 63% in hairy vetch–amended soil, compared with nonamended soil on cultivars ranked as susceptible, moderately resistant and highly resistant, respectively. Fusarium wilt development was observed in nine soils with pathogen populations below 1,100 CFU/g of soil. The magnitude of disease suppression decreased with increase in inoculum densities in the soils. Induced wilt suppression seemed to depend on the increase in populations of bacteria. The results indicated that hairy vetch-induced Fusarium wilt suppression in watermelon was dependent on the cultivar resistance level and it might be overcome by the increased pathogen inoculum densities in soils (Zhou and Everts 2007). The efficacy of *Brassica* spp. applied as amendment in reducing the populations of *Pythium* spp. and *F. oxysporum* f.sp. *niveum*, causing damping-off and wilt of watermelon respectively, was investigated under field conditions. Incorporating flowering *B. napus* cv. Dwarf Essex canola or *B. juncea* cv. Cutlass, laying black polythene mulch at incorporation or 1 month after incorporation, application of methyl bromide and nontreated control formed the treatments of the field trials conducted for 2 years. The total concentration of glucosinolates incorporated per square meter was significantly higher for *B. juncea* than for *B. napus*. Isothiocyanates were not detected in all amended soil samples and they were absent after 12 days of incorporation of amendment. Amended plots contained higher populations of *F. oxysporum* and *Pythium* spp., than in plots fumigated with methyl bromide and some amended plots had populations greater than in control plots. Presence of fluorescent *Pseudomonas* spp. in greater populations was recorded in some amended plots, compared to other treatments. Incidence of damping-off and severity of Fusarium wilt on susceptible seedless watermelon cv. Tri-X313 were not consistently lower in *Brassica*-amended soils or methyl bromide-treated plots than in nontreated controls. The results indicated that neither biofumigation nor chemical fumigation could be an effective option for the management of Fusarium wilt and damping-off of watermelon in the Coastal South California (Njoroge et al. 2008). Root rot (RR) caused by *Pythium* spp. and *Fusarium* spp. occurred seriously on bean (*Phaseolus vulgaris*), impacting production and the pathogen infected also maize (corn), sorghum (*Sorghum bicolor*), sweet potato and garden pea, which were grown mainly as intercrops/rotation crops. The contribution of the root rot infection in the rotation/intercrop to survival of RR pathogens was investigated in the naturally-infested farmers' fields in Uganda, as well as in soils artificially infested soils under screen house conditions. Incidence and severity of RR were high in bean and sorghum in both farmers' fields and screenhouse conditions. The results indicated that sorghum had the potential of supporting the survival of RR pathogens in the soil, leading to increase in the disease incidence/severity in bean crops (Ocimati et al. 2017).

For the management of potato stem rot caused by *S. sclerotiorum*, the potential of biofumigation with *B. napus*, *B. juncea* and *B. campestris* was assessed in three naturally infested potato fields during three cropping seasons during 2008–2010. Disease incidence and mean percentage of dead plants were significantly reduced by all *Brassica* crops, *B. juncea* providing the highest reduction of disease incidence. Average reduction in disease incidence attained was 55.6% over all fields and years using *B. juncea* as green manure, compared to 31.6% and 45.8% reduction achievable using *B. napus* and *B. campestris* respectively (Ojaghian et al. 2012). The efficacy of *Brassica* green manure soil amendments, as an alternative to chemical management against *R. solani*, infecting impatiens and petunias was assessed. *Brassica* crops, *B. juncea* cvs. Fumis and Bionute and *B. napus* cv. Jetton were grown in microplots at 700, 1400 and 4,200 g/m² fresh weight and infested with *R. solani* to determine the disease incidence on impatiens and petunias. All green manure crops reduced disease symptoms in both ornamentals. Rate of *Brassica* biomass was more important than *Brassica* cultivars. The highest rate of green manure (4,200 kg/m²) applied, reduced crown lesions by 21% and 24%, root discoloration by 9% and 7% and isolation frequency of *R. solani* by 15% and 8% for impatiens and petunias respectively, compared with 700 g/m² application. *Brassica* green manure application following a waiting period of 4 weeks before planting ornamental crops did not induce any phytotoxic effects (Cochran and Rothrock 2015).

R. solanacearum race 3 biovar 2, causal agent of potato brown rot, is capable of occurring in epidemic level in tropical, subtropical and temperate countries in the world. The effects of different amendments on development of *R. solanacearum* in potato tubers grown in different soil types, sand or clay from Egypt desert and Dutch (Netherlands) soils were assessed. Slight reduction in brown rot infection was observed in organically raised potatoes, compared to conventionally grown crops. By contrast, significant increase in disease incidence and survival of *R. solanacearum* were recorded in organic crops on Dutch sandy and clay soils. No correlation could be established between disease incidence or severity and pathogen diversity in the rhizosphere in organic and conventional crops. Cow manure amendment significantly reduced disease in organic Dutch sandy soil, but pathogen population remained unaffected. In Egyptian sandy soil cow manure adversely affected pathogen density. The suppressive effects of cow manure in Dutch and Egyptian sandy soils might be due to release of ammonia, which is toxic to microorganisms or a shift in microbial community structure (Messiha et al. 2007). The efficacy of lopsided oat (*Avena strigosa*) or woolly pod vetch (*V. villosa*) applied as green manures for reducing potato common scab (PCS) caused by *S. scabies*, *S. turgidiscabies* and *S. acidiscabies* was assessed. Lopsided oat or woolly pod vetch were more effective in reducing disease severity than oat and continuous cultivation of potato (P <0.000 1). Furthermore, the

marketable tuber ratio was also more significantly increased (65.3% and 74.0%, respectively), compared with oat and continuous potato cultivation. Population density of *S. turgidiscabies* was positively correlated with disease incidence. Under field conditions, lopsided oat treatment reduced population density of *S. turgidiscabies* more efficiently than other treatments, and also increased marketable tuber ratio to a higher level, compared with sugar beet treatment. The effectiveness of lopsided oat cultivation alone or resistant potato cultivar was observed even under high disease pressure conditions. The results suggested that lopsided oat could be raised as fallow green manure to reduce the severity of potato common scab disease (Sakuma et al. 2011).

1.2.1.7.2 Application of Mulches

Organic and inorganic mulches applied to cover the soil surface have been evaluated for their influence on plant development and incidence of soilborne diseases. The efficacy of wheat straw mulch in reducing the incidence of peanut (groundnut) Cylindocladium black rot (CBR), Sclerotinia blight and Southern blight caused respectively by *Cylindrocladium parasiticum*, *S. minor* and *Sclerotium rolfsii* was assessed. The microplots were infested with the respective pathogens, followed by planting two peanut cultivars NC7 or NC107. Wheat straw was applied to provide coverage for 80 to 90% of the soil surface. Southern blight incidence was not increased by straw mulch, whereas the final inoculum density of *S. rolfsii* reached the highest level in treated plots. Root rot severity, due to *C. parasiticum* and inoculum density, was not affected by wheat straw mulch treatment. Incidence of Sclerotium blight was reduced by wheat straw mulch in 2 of 3 years of experimentation, compared to control plots (Ferguson and Shew 2001). The effects of fall- and spring-killed or spring-sown living crop mulches of pumpkin (*Cucurbita pepo*) were determined on development of Fusarium fruit rot (FFR) disease caused by *Fusarium solani* f.sp. *cucurbitae*. On artificially infested plots, percentages of ground cover at harvest and FRR-infected fruit were 89% and 5% in fall-sown winter rye, 88% and 10% in fall-sown rye, 85% and 5% in fall-sown rye + hairy vetch, 19% and 30% in fall-sown hairy vetch, 23% and 23% in spring-sown oat, 1% and 25 to 39% in spring-sown living annual medics and 0% and 46% in bare soil plots respectively. The results suggested the possibility of reducing FRR in pumpkin by using cover crop mulches, such as fall-sown winter rye, fall-sown rye + hairy vetch or spring-sown, spring-killed oat and left on soil surface (Wyenandt et al. 2011). The effects of mulches (bare soil, wheat straw or plastic), bed height, composted poultry litter and squash cultivars (Cougar or Pay roll) on Phytophthora crown rot caused by *P. capsici* were assessed under field conditions, using split-plot arrangement of a randomized block design. Incidence of plant death (%) was determined up to 35 days postinoculation (*dpi*) with *P. capsici*. Plant death at 35 *dpi* and area under disease progress curve (AUDPC) differed significantly (P <0.000 1) between the cultivars. Mean plant death at 35 *dpi* was 87% for Payroll and 99% for Cougar. Plant death at 35 *dpi* and AUDPC for Payroll was greater in flat beds than in raised beds. Disease incidence was not influenced by bed height, mulch type or application of poultry litter (Meyer and Hausbeck 2012).

1.2.1.7.3 Solarization for Sanitation

Structural solarization of greenhouses for sanitation by closing them involves dry heating to 60°C and higher with a consequent low relative humidity (RH, 15%), thus providing an extended period of thermal inactivation of soilborne pathogens. Various regimes of inoculum moistening were evaluated for their efficacy to enhance pathogen suppression by increasing moisture during hot periods of the day. But, wetting inoculum of *F. oxysporum* f.sp. *melonis* (*Fom*) and *F. oxysporum* f.sp. *radicis-lycopersici* (FORL) resulted in less effective pathogen suppression, compared to that of dry heating. The effective dose (ED$_{50}$, 50%) values of thermal inactivation of wetted and dry inoculum for *Fom* were 18 and 7 days, respectively and 9 and 4 days for FORL, respectively. Wetting resulted in inoculum cooling due to evaporation, which led to drying of inoculum, eventually. Use of a double-tent system reduced the difference between greenhouse and ambient temperatures to 1–2°C from ≈ 10°C, prevailing in wetted greenhouse, resulting in higher level of suppression of development of *F. oxysporum* f.sp. *radicis-lycopersici* (Shelvin et al. 2004). Population of *Phytophthora* spp. increases to high levels in orchard soil planted to citrus. *Phytophthora nicotianae* was the predominant pathogen present in Arizona citrus groves. In three field trials over a 3-year period, *P. nicotianae* could not be detected at a depth of 10 cm, after soil naturally infested with the pathogen was subject to dry summer fallow period of at least 31 days in the desert southwest region of Arizona, United States. The mean soil temperature at 10 cm depth ranged from 37°C to 39°C. In addition, in two of three trials, *P. nicotianae* could not be detected after summer dry fallow periods of 38 and 45 days, at a depth of 15–20 cm. The pathogen was detectable in only one of 19 soil samples at a depth of 25–30 cm. In comparison, *P. nicotianae* was recoverable from a high percentage of soil samples subjected to a dry winter fallow period or maintained in the greenhouse and planted with a seedling of citrus, alfalfa or irrigated without a plant, where the mean soil temperature ranged from 15°C to 30°C. The results suggested that in regions with a hot and dry summer climate, a dry summer fallow treatment of soil, after removal of an existing citrus planting and before establishment of a new citrus grove could provide a rapid and relatively inexpensive method of reducing populations of *P. nicotianae* to virtually nondetectable levels at least a depth of 30 cm (Matheron and Porchas 2009).

The efficacy of soil solarization for killing the teliospores of *T. indica*, causing Karnal bunt in wheat, was investigated, using a replicated field trial for 3 years in Arizona, United States. Teliospores and bunted wheat kernel were buried in Karnal bunt-infested wheat field at depths of 5, 10 and 20 cm. Replicated samples were collected from under a clear plastic solarization cover at 7-day interval and the numbers of viable teliospores were determined. Initial nonsolarized (0 day) mean teliospore germination rates of 43, 71 and 82% were reduced

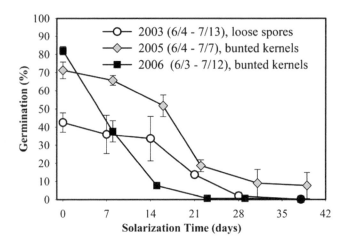

FIGURE 1.4 Effects of soil solarization on germination of teliospores of *T. indica* recovered each year Time data (depth) pooled within year to obtain averages and standard deviation bars.

[Courtesy of Peterson et al. 2008 and with kind permission of the American Phytopathological Society, MN]

to 0.1, 0.7 and 0.2%, after 38 days of solarization across all depths of burial of teliospores/bunted kernels in 2003, 2005 and 2006. There was no appreciable difference in the number of variable teliospores recovered from 5-, 10- and 20-cm depths after 21 days of solarization, in spite of significant difference in soil temperature ranges prevailing at 5 and 20 cm depths (see Figure 1.4). However, temperatures at 20 cm depth remained closer to daytime temperatures at night, resulting in a higher number of cumulative hours at temperatures above 30°C, compared to 5 cm depth. A rapid decline in teliospore viability occurred under clear plastic sheeting at all depths over 38 days of solarization, with efficacy comparable to methyl bromide fumigation. Mean daily maximum soil temperatures at 5 and 20 cm under clear plastic sheeting in 2003, 2005 and 2006 were 67, 53 and 60°C and 43, 38 and 43°C, respectively. The results indicated the possibility of obtaining rapid deregulation of Karnal bunt-affected fields under the USDA disease management strategies (Peterson et al. 2008). Fusarium wilt of lettuce caused by *F. oxysporum* f.sp. *lactucae* (*Fol*) continued to spread and cause considerable loss in Arizona lettuce fields. The potential of summer solarization as a disease management strategy was assessed by solarization for 2 to 8 weeks in soil naturally infested by *Fol* in microplots. Lettuce plant growth was consistently greater in solarized plots than in nonsolarized plots. In four trials within a field containing *Fol*, the incidence of Fusarium wilt on lettuce sown after solarization was reduced (42%), compared to disease incidence in nonsolarized plots (91%). No significant benefit due to solarization for longer periods (2 month) over shorter period (1 month) was observed in soils, where the mean soil temperature at a depth of 5 cm during the one-month period in 2005 and 2006 were 47°C and 49°C, respectively. The results suggested that soil solarization could be an effective tool for the management of lettuce wilt disease, especially in an integrated program (Matheron and Porchas 2010).

Low-cost greenhouse crop production systems are in use for tomato crops in some countries. Late blight disease of tomato caused by *P. infestans*, was found to be a serious problem limiting production significantly. The individual and combined effects of covering the soil with polyethylene mulch before planting and fungicide (Kocide 2000) application, commonly used for organic crops were assessed in three experiments during 2002–2005. Polyethylene mulch was consistently effective by providing high significant level of disease suppression (83.6 ± 5.5%) of the disease, whereas the fungicide application resulted in inconsistent and insufficient disease suppression (34.5 ± 14.3%). The individual effect of mulching with bicolor aluminized, black or clear polyethylene, in suppressing late blight development was significantly greater than that offered by fungicide spraying. The combined effect was additive and reduced disease severity by 96.0%. In a second set of three experiments conducted in the walk-in tunnels, the late blight epidemics almost completely destroyed tomato plants in nonmulched controls and by the end of the season, the disease severity reached 92.5 and 81% in 2003–2004 and 2004–2005, respectively. Fungicide application had no significant effect on disease severity throughout the season or on the relative area under disease progress curve (RAUDPC) values. On the other hand, the bicolor polyethylene mulch reduced disease severity and RAUDPC significantly. The effect was greater in the 2003–2004 experiment. The final disease severity in unsprayed plots in mulched greenhouse was 14.4%, compared to 40.1% in the 2004–2005 experiment. The interaction between mulching and spraying fungicide was not significant (see Figure 1.5) (Shtienberg et al. 2010). The efficacy of cover crops, rapeseed, marigold,

forage, pearl millet, sorghum-sudangrass and corn and solarization in suppressing the development of Verticillium wilt in potatoes caused by *V. dahliae*, was investigated. Cover crops were grown and solarization applied in the first year and potato was planted in the second year. The suppressive effect of cover crops was revealed by treatments with rape and sorghum-sudangrass. Solarization significantly reduced the inoculum levels of *V. dahliae*. In addition, *Pratylenchus penetrans* population was also reduced, due to solarization. The combined effects of cover crops and solarization resulted in a wide range of pathogen population densities. Mean soil inoculum levels showed negative relationship with yield for *V. dahliae* in one of two experiments (MacGuidwin et al. 2012).

Combination of disease management strategies is likely to be more effective in suppressing disease incidence and severity. The combined effects of grafting and use of plastic mulch for solarization on watermelon Verticillium wilt disease caused by *V. dahliae* and plant growth were assessed. Area under disease progress curve (AUDPC) values for grafting treatments generally were higher than nongrafted TriX Palomar (765) than for grafted TriX Palomar onto rootstocks Super Shintosa (132), Tetsukabuto (178) or Just (187). Overall, for mulch, the AUDPC values were higher for plants with black plastic mulch (385) than for plants grown with clear plastic mulch (237). The location effect was significant among three locations tested. Mulches did not have significant influence due *Verticillium* microsclerotia detected by stem assays. After harvest, pathogen density in soil under black plastic increased to different extent in three locations. By contrast, *V. dahliae* soil density under clear plastic mulch was nearly identical to the level at planting at each location (<1.0–27 CFU/g). The results indicated that Verticillium wilt symptoms were expressed in watermelon plants at inoculum level of <3 CFU/g of soil, but the watermelon yield was not affected. However, when *V. dahliae* soil density was >50 CFU/g of soil, yield was greater for grafted plants and for plants grown with clear plastic mulch (Dabirian et al. 2017).

1.2.1.7.4 Application of Inorganic Nutrients

As the organic matter was not available to meet the demands of various crops for enhancing yield levels, application of inorganic fertilizers became indispensable to feed the ever-growing human populations. The inorganic fertilizers have significant influence on plant growth directly and on the development of microbial plant pathogens indirectly by affecting the susceptibility/resistance levels of host plants. Further, the microclimate may be influenced by changes in plant canopy, resulting in differential effects on soil microbial community structure and activities.

Nitrogen may be applied in different forms and it favors vegetative growth the plants in the early stages of growth of crops. However, higher levels of nitrogen supply are known to increase the susceptibility of crops to several diseases. Development of cabbage clubroot disease caused by *P. brassicae* was suppressed effectively by combined application of nitrate fertilizer and lime to increase soil pH which was inhibitory to pathogen development (Smiley 1975). Potato tuber rot

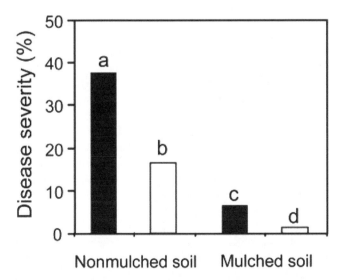

FIGURE 1.5 Effects of bicolor polyethylene and application of Kocide 2000 (0.5%) and Neemguard (2.5%) at weekly intervals on severity of tomato late blight disease caused by *Phytophthora infestans* Treatments: control (filled bars) and fungicide application (open bars); bars with different letters were significantly different, as determined by the highly significant difference test at P = 0.05.

[Courtesy of Shtienberg et al. 2010 and with kind permission of the American Phytopathological Society, MN]

caused by *P. infestans* was significantly increased at higher levels of nitrogen application (Heitefuss 1989). The effects of nitrogen forms and black plastic mulch on Verticillium wilt of eggplant caused by *V. dahliae* were compared. The experimental plots were either mulched with black plastic or grown on bare ground and fertilized with either ammonium sulfate [$(NH_4)_2SO_4$] or calcium nitrate [$Ca(NO_3)_2$] at 224 kg of N/ha in each season during 1996–1998. In 1996 and 1997, no significant interactions could be observed between the mulch and fertilizer treatments on yield or in integrated estimates of the plant canopy growth curve or disease severity. However, treatment effects were additive. Mulching increased plant canopy 3 weeks after planting and reduced the percentage of symptomatic foliage at 8 weeks after planting, compared to bare ground treatment. Ammonium sulfate increased plant canopy after 6 weeks, but did not affect the percentage of symptomatic foliage, compared to calcium nitrate. The rate of nitrogen fertilizer applied at planting or as a side-dress application did not affect growth or disease development. The use of ammonium sulfate and black plastic mulch showed additive effect and might reduce damage from Verticillium wilt on plant (Elmer 2000).

Wheat take-all disease caused by *G. graminis* var. *tritici (Ggt)* accounts for appreciable losses in different countries growing wheat crops. Effect of take-all root lesions on nitrate uptake by wheat was investigated in vitro. Wheat plants were supplied with nutrient solution labeled with ^{15}N during stem elongation and flowering stages to assess the distribution of the isotopic tracer in different plant organs, especially in root segments located on both sides of take-all lesions. The ^{15}N atom percentage excess measured in root segments located below lesions longer than 1 cm was reduced on average by half, compared with that in healthy roots and root segments above lesions, revealing a reduction in nitrogen uptake by these root segments. Severely infected plants exhibited an increase in the uptake rate per unit of efficient root, which appeared to be a compensatory response to reduction of efficient root biomass in order to satisfy demand of shoot for nitrogen. Reductions in nitrogen accumulation in roots and aerial parts at flowering stage were up to 56% and 49%, respectively, for plants with more than 50% of the root system below lesions longer than 1 cm (Schoeny et al. 2003). Severe infection of wheat by take-all pathogen *G. graminis* var. *tritici (Ggt)* resulted in premature ripening and death of plants and this condition could lead to inefficient use of applied nitrogen (N). Fields with different disease severity were selected to determine crop N uptake and the amount of residual mineral N in the soil after harvest. All wheat crops received fertilizer N in spring at recommended rates (190–200 kg N/ha). At every sampling, crops were assessed for severity of take-all infection (TAR) and crop N uptakes and soil nitrate + ammonium (SMN) were determined. Significant negative correlations (P <0.05) were evident between crop N uptake and TAR at anthesis and final harvest. Poor wheat crops had significantly reduced N contents both in plants and grains. Positive relationship between SMN and TAR were observed at anthesis and final harvest. Infection by *Ggt* reduced the efficiency of utilization of applied nitrogen and greater quantities of SMN were present in the 0–50 cm soil layer after harvest. Most of the N (73–93%) was present as nitrate. Localized patches of severe take-all infection decreased the efficiency with which hexaploid wheat plants recovered soil- and fertilizer-derived N and increased the subsequent risk of nitrate leaching (Macdonald and Gutteridge 2012).

Postemergence damping-off of cotton seedlings, considered as a disease complex, was primarily associated with infection by *R. solani* and *Thielaviopsis basicola* and it became a serious threat to cotton cultivation in Spain. Samples of soils and cotton plants were taken from 97 commercial field plots, where postemergence damping-off disease occurred in earlier years. Based on visual examination, infected cotton seedlings were grouped into two, those infected by *R. solani* and *T. basicola* separately. Soil samples were analyzed for availability of N, P, K, Ca, Mg, Cu, Mn and Zn. Seedling infection was recorded in all plant samples. The percentage of seedlings infected by *T. basicola* was 18%, whereas infection by *R. solani* was recorded in 4.1% of plants showing black necrosis symptoms. But in brown necrosis symptomatic plants, *R. solani* was more frequently (12.8%) recovered, than *T. basicola* (10.7%). Incidence of black necrosis was negatively correlated with available N measured as NO_3-N, when corn or sunflower was the preceding crop. Incidence of black necrosis was positively correlated with Fe availability in soil with cotton as the preceding crop, whereas brown necrosis was negatively related with availability of Fe (Delgado et al. 2005). Phytophthora stem rot of soybean caused by *Phytophthora sojae*, is an important factor limiting soybean production. The possibility of reducing disease intensity and pathogen development by applying inorganic nutrients was explored. Various concentrations of KNO_3, $(NH_4)_2SO_4$, $MgSO_4$, $CaCl_2$ and $NaHPO_4$ were used to determine the elements most effective in suppressing the incidence of the disease. A concentration of 24.7–247 mM KNO_3 and 0.1–10.2 mM $CaCl_2$ greatly inhibited infection by *P. sojae*. No significant relationship between inhibition of the growth rate and disease reduction at 2.47 mM KNO_3 and 0.1–5.1 mM $CaCl_2$ application was observed, although mycelia growth of the pathogen isolate was affected by potassium and calcium concentration. Disease suppression under in vitro using pathogen mycelium could be due to the response of plant tissues rather than a direct inhibition of pathogen growth by potassium or calcium. Reduction in disease development was related to an increased potassium and calcium uptake by plants, suggesting that effective elements in reducing soybean Phytophthora stem rot were potassium and calcium. KNO_3 (2.47–247 mM) and $CaCl_2$ (5.1–10.2 mM) decreased the release of zoospores, although lower concentrations of calcium (0.1–2.5 mM) significantly favored zoospore release (Sugimoto et al. 2007). In a later investigation, application of 4–30 mM KNO_3 prior to inoculation was found to markedly reduce incidence of soybean Phytophthora stem rot caused by P. sojae in two soybean cultivars under laboratory conditions. The extent of inhibition of pathogen growth rate by KNO_3 had no relationship with disease reduction, following application of 0.4–10 mM KNO_3. The extent of disease reduction showed positive relationship with increase in concentrations of KNO_3 in both cultivars, with some differences between the cultivars. Observations under scanning electron microscope (SEM), using fresh samples, revealed marked accumulation of potassium at the penetration-stopping sites of *P. sojae* in the cortex layer of soybean plants supplied with 30 mM KNO_3, compared with untreated control plants. Release of zoospores was inhibited by concentrations of 0.4–30 mM KNO_3. The results suggested that application of KNO_3 at 20–30 mM concentration might be effective in decreasing incidence of stem rot disease of soybean crops under field conditions (Sugimoto et al. 2009).

Rates and forms of nitrogen have differing influence on plant growth and disease incidence and intensity. A field trial was conducted to determine the effects of agronomic practices, such as nitrogen fertilizer rates, irrigation intensities (frequencies) and previous crop in rotation on potato powdery scab disease caused by *S subterranea*. Nitrogen application (400 kg N/ha) increased tuber weight/plant by 38% and also the incidence and/or severity of powdery scab disease. Pathogen DNA was quantified to use it as a measure of population of *S. subterranea*. Quantities of pathogen DNA were greater following nitrogen application than in untreated controls, less after providing 'optimum' level of irrigation than 'low' level of irrigation treatment. The DNA contents were greater after a potato/wheat rotation than after potato/pea rotation. Potato crops were raised (without any treatment) for the next two growing seasons to assess the effect of preplanting on *Spongospora* DNA contents. Incidence of powdery scab was generally reduced during next two growing seasons and the effects of all treatments provided earlier were abolished in the second season. However, plots receiving nitrogen had higher concentration of pathogen DNA, compared with untreated control (without N). Weak relationships were observed between amounts of preplanting *Spongospora* DNA in soil and powdery scab incidence in subsequently harvested tubers in the second growing season. In the third season, the relationship was not detectable. Among the agronomic practices, nitrogen application could enhance severity of powdery scab in harvested tubers. Pathogen DNA contents appeared to be an unreliable predictor of incidence and severity of potato powdery scab disease (Shah et al. 2014).

The efficacy of single application of ammonium lignosulfonate (ALS) as soil amendment in reducing potato scab and Verticillium wilt diseases was assessed under field conditions in Ontario during 1998-2000. Potato tubers were planted at 2–4 weeks after ALS application. Treatment with ALS (ca. 6 t solids/ha), reduced potato scab severity by 50 to 80% in the year of application at four sites and reduction in scab severity was significant in the second year at 2 of 4 sites tested. Likewise, incidence of *V. dahliae* was reduced by 40 to 50% by ALS treatment in the first crop season at all sites and only in one site in the second crop. Tuber yield was significantly increased (2.5 times) in one site in 2000. However, marketable tuber yield (with <5% surface scab) was increased (3–7 folds) at all sites in the first year. The soil pH was immediately reduced, following ALS application by 0.4–0.6 units at

three sites and returned to the level in control in the second season in three sites. ALS treatment within weeks enhanced the microbial populations by two to eight folds at all sites, fungal populations showing maximum enhancement. The results indicated the effectiveness of ammonium lignosulfonate in reducing two important diseases affecting potato crops (Soltani et al. 2002).

The major nutrient phosphorus as phosphates is applied to correct deficiency in soil and their effects on disease incidence/severity have not been studied adequately to draw dependable conclusions. Potassium has a key role in the osmoregulation in the cell and maintenance of electrical balance across the plasma membrane. The deficiency of potassium, phosphorus, magnesium and minor elements may result in impairment of processes involved in the maintenance of cell structure and function. Supply of adequate amounts of potassium accelerates formation of cellulose and sclerenchymatous structures, resulting in enhancement of resistance to fungal diseases of cereals, counteracting the adverse effects of excessive application of nitrogen (Heitefuss 1989). The effects of potassium (K) on development of two rice cultivars grown on nutrient solutions containing 0, 50 and 100 mM of K and sheath blight caused by *R. solani* were investigated. Sheath blight development was assessed at 48, 72, 96 and 120 h after inoculation. Concentration of K on rice leaf sheath tissue increased by 61.48 and 116.05% in cvs. BR-IRGA409 and Labelle, respectively, as the K supply increased from 0 to 100 mM. A negative relationship between K rates and area under relative lesion length progress curve (AURLLPC) was observed. The AURLLPC decreased by 29.2 and 21.3% for BR-IRGA409 and Labelle respectively with increase in K rates. The high potassium content of leaf sheath tissues was considered to be responsible for reduced disease severity due to sheath blight of rice (Schurt et al. 2015).

M. phaseolina causes charcoal rot of soybean which is a disease of economic importance in several countries around the world. Two separate field experiments were conducted during 2008–2010 to determine the influence of potassium (K) and phosphorus (P) on severity of charcoal rot disease, using different rates of K (0, 45, 90, 134 and 179 kg K_2O/ha) and P (0, 22, 45, 67 and 90 kg P_2O_5/ha) along with the recommended doses of K or P fertilizer. None of the P or K rates had any observable effect on pathogen development reflected by CFU of *M. phaseolina* in soil in all six locations x year environments. The results showed that increasing K or P rates over the recommended rate and control (0 kg/ha) did not reduce the disease severity significantly (Mengistu et al. 2016). Influence of potassium (K), calcium (Ca) and nitrogen (N) on the susceptibility of four banana cultivars to Banana Xanthomonas wilt (BXW), caused by *Xanthomonas campestris* pv. *musacearum (Xcm)* was investigated. Banana plantlets generated on Murashige-Skoog (MS) medium were exposed to different concentrations of K, Ca and N in nutrient solutions for a period of 8 weeks. Then the plantlets were inoculated with *Xcm*. The plantlets accumulated higher concentrations of K, Ca and N as the concentrations increased. Wilt incidence was reduced and incubation period increased as the nutrient concentrations increased. The results indicated that it might be possible to determine the quantity of fertilizers to be applied to derive benefits of these nutrients in suppressing banana Xanthomonas wilt disease (Atim et al. 2013).

Calcium is applied in different forms to promote plant growth by correcting calcium deficiency in soils. Effects of five formulated compounds and two chemical compounds on the incidence of Phytophthora stem rot of soybean caused by *Phytophthora sojae* were assessed. Application of calcium compounds prior to inoculation significantly suppressed disease incidence. Inhibition of mycelia growth of *P. sojae* determined in vitro did not show any relationship with disease reduction assessed in growth chamber tests. The extent of reduction in disease incidence was related to increased calcium uptake by plants, suggesting its effectiveness in reducing Phytophthora stem rot. Seedling tray experiments using zoospores showed that application of 10 mM of Ca (COOH)$_2$-A was more effective in reducing disease incidence than other concentrations under growth chamber conditions (Sugimoto et al. 2008). In the further study, under infested field conditions, calcium compounds, Ca(HCOO)$_2$-A and Ca(NO$_3$)$_2$, at 4 and 10 mM applied twice (at 0 and 14 days after transplanting soybean), significantly reduced disease incidence and delayed disease onset. Ca(HCOO)$_2$-A was more effective than calcium nitrate in reducing disease incidence and also promoted plant growth and yield. Observations under scanning electron microscope (SEM) on fresh samples, revealed increased accumulation of calcium crystals around the cambium and xylem elements of soybean plants treated with 10 mM Ca(HCOO)$_2$-A and Ca(NO$_3$)$_2$. The results indicated that soybean crops grown in calcium-rich areas might be more resistant to invasion by *P. sojae* and calcium crystals might have a role in calcium ion storage and its availability for those tissues required to maintain long-term field resistance to the pathogen (Sugimoto et al. 2010).

The role of minor nutrients in development of diseases induced by microbial plant pathogens has been investigated less frequently. Incorporation of gypsum as amendment into the soil decreased infection of avocado seedlings by *Phytophthora cinnamomi* infested soils by 71% under greenhouse conditions. In addition, significant reduction in total seedling and root weight caused by *P. cinnamomi* was effectively counteracted by gypsum amendment. However, avocado seedlings grown in gypsum-amended soils followed by root inoculation with zoospore suspensions of the pathogen were not more resistant to infection by *P. cinnamomi*, compared with controls in unamended soils (Messenger et al. 2000). Addition of potassium chloride (KCl) to naturally infested soil at 150 to 400 μg K/g of dry soil increased the incidence of Phytophthora root and stem rot of soybean caused by *Phytophthora sojae* in susceptible cultivar under growth chamber conditions. Application of chloride salts of K, Ca, Mg, Na, Al, Fe and Sr at 250 μg of Cl/g of soil increased disease incidence, compared with distilled water control. Ammonium salts also favored disease incidence. In a leaf-disc baiting bioassay, leaf infection by *P. sojae* zoospores decreased, when 0.01 or 0.02 M KCl was added to soil extracts from flooded soils, but was unaffected by KCl at 150 to 600 μ of K per gram of soil applied to the soil, 5 days before baiting. Disease increased generally with addition of KCl and this might be due to the presence of chloride and association with changes in the micropartioning of root calcium (Canaday and Schmitthenner 2010). Incidence and severity of root disease of oak (*Quercus ilex*) caused by *Phytophthotra cinnamomi* was found to be reduced in soils with medium-high Ca$^+$ content. The ability of calcium fertilizers to induce soil suppressiveness against *P. cinnamomi* was assessed under greenhouse conditions. *P. cinnamomi* cultures exposed to different Ca$^+$ fertilizers showed significant inhibition of sporangial, chlamydospore and zoospore production at millimolar (mM) concentrations, whereas mycelia growth was not affected by the Ca$^+$ fertilizers. Severity of disease on foliage and root of Holm oak was significantly reduced in the seedlings grown on artificially infested soils amended with Ca$^+$ fertilizers. The results suggested that limestone amendments in oak rangelands might enhance soil suppressiveness against *P. cinnamomi* and Ca$^+$ might function by inhibiting sporangial production by the pathogen (Serrano et al. 2012).

Spinach Fusarium wilt disease caused by *F. oxysporum* f.sp. *spinaciae* is the most serious limiting factor affecting spinach seed production in the maritime Pacific Northwest in the United States. The pathogen can survive in the soil for extended periods as chlamydospores or by asymptomatic colonization of roots of nonhost plant species. Pathogen development is favored by acid soils and warm temperatures. Very long period (10–15 years) may be required to minimize the losses due to Fusarium wilt disease, by growing rotation crops. Application of agricultural limestone CaCO$_3$ (97%) may offer partial disease suppression by raising soil pH. A field trial was conducted during 2009-2012, to assess the potential for application of agricultural limestone at 0, 2.24 and 4.48 t/ha for 3 years prior to planting spinach seed crop. Three proprietary female spinach lines, with a range of susceptibility to partially resistant to Fusarium wilt, were planted. Three successive annual applications of limestone at 4.48 t/ha reduced midseason wilt incidence by an average of 20%, increased spinach biomass by 33% and increased marketable spinach seed yield by 45%, compared with plots amended once with the same rate of limestone in the spring planting. The suppressive effect increased with increasing rate of limestone amendment with the greatest difference observed, when limestone was applied up to 2.24 t/ha annually for 3 years. The results indicated the possibility of reducing the rotation interval by annual applications of agricultural limestone on acid soils to increase spinach seed production (Gatch and du Toit 2017).

V. dahliae causes many major yield and quality-limiting vascular wilt diseases across a wide-spectrum of crops. In order to assess the influence of sulfur, which has a dual role of fungicide and nutrition, two near-isogenic tomato genotypes differing in their susceptibility to Verticillium wilt disease were treated with low or supra-optimal sulfur supply. A significant sulfur-induced decrease was observed in the amount of infected vascular cells in both genotypes. However, assimilate transport rates in the phloem sap were enhanced by fungal infection to a greater extent in the resistant geno-type under high sulfur nutrition, suggesting a stronger sink for assimilates in infected plant tissues possibly involved in sugar-induced defense. A SYBR Green-based absolute quantitative real-time PCR assay, using species-specific primer was developed, and it clearly reflected sulfur nutrition-dependent changes in fungal colonization patterns. High sulfur nutrition significantly reduced pathogen spread in the stem in both tomato genotypes. Concentrations of selected sulfur-containing metabolites revealed an increase of the major antioxidative redox buffer glutathione under high sulfur nutrition in response to pathogen colonization. The results revealed the existence of sulfur nutrition-enhanced resistance of tomato against *V. dahliae*, mediated by sulfur-containing defense compounds (Bollig et al. 2013).

1.2.1.8 Effects of Other Crops

Crop rotations (sequences) offer multiple benefits, including maintenance of crop productivity and reduction in buildup of soilborne plant pathogens and consequent incidence of crop diseases induced by them. In contrast, monocropping may result in heavy losses due to most of the soilborne diseases. Several crops may be cultivated in a farm in rotation with primary crop or as intercrops, depending on the water availability, rainfall and other environmental conditions. Rotation crops may help conserve, maintain or replenish soil resources and improve the physical and chemical properties of the soils, whereas cover crops are primarily grown to cover and protect the soil from erosion and nutrient losses between periods of crop production. Incidence of diseases may be influenced by other crops, depending on their levels of susceptibility to pathogen(s) infecting the primary crop, resulting in either positive or negative effect on the incidence and severity of diseases. Crop rotations may reduce soilborne pathogens by (i) interrupting or breaking host-pathogen cycle of inoculum production, (ii) producing root exudates that may either be inhibitory to the pathogen(s) or promote activities of antagonistic organisms and/or (iii) by changing the properties of soils making the environment less conducive for the proliferation of pathogens or survival. Crop rotation or sequence involves the use of appropriate crops in a sequence for a decided duration of time (2–3 years in most cases). By increasing the diversity of rotation crops, additional benefits may be derived. Use of diverse combinations of different types of rotation crops may result in enhancement of microbial biomass, activity and biodiversity. Longer duration to include more number of crops may be more effective in reducing disease incidence/severity. While selecting rotation crops, it is important to choose crops that do not have similar pathogen problems as in the primary crop. Selecting specific crops for their effects on soil microbial communities and development of disease-suppressive soils would be a preferable approach for containing soilborne diseases (Larkin 2015). The effectiveness of crop rotation depends on several factors, such as duration, environment, the nature of the pathogen(s) to be controlled, and requirements of the primary and rotation crops. Nonhosts included in the rotation system, may allow sufficient time for

the decomposition of infected crop residues and/or reduction in pathogen survivability and production of fresh inoculum. Further, in the case of some diseases, crop rotation may be effective, if the sources of inoculum are present only in the field concerned or the movement of the pathogen from adjacent or other fields is limited. Inclusion of legumes in cop rotation may provide additional benefit of having atmospheric nitrogen fixed by rhizobia which can be used by succeeding crop(s). Crop rotation may not be effective for pathogens with multiple mode of transmission such as seed-, air- and water-dispersal (Narayanasamy 2017).

The effectiveness of crop rotation in reducing incidence of diseases caused by *Phytophthora* spp. has been reported to vary, depending on the host-pathogen combinations. The rotation of pepper with peanut (groundnut) was found to be more effective in reducing (by 39%) the Phytophthora blight disease, caused by *P. capsici* than with sesame (by 11%) (Kim 1989). The effect of crop rotation on the survival of *P. capsici* at a naturally infested site in Michigan, United States, planted to cucumber in 1998, corn in 1999 and 2000, tomatoes in 2001 was investigated. Majority of isolates of *P. capsici* (89%) recovered had unique amplified fragment length polymorphism (AFLP) fingerprints and no members of the same clonal lineage were recovered among years. Isolates from this location were more similar to each other than to isolates from other locations in Michigan. Isolates recovered could not be distinguished based on year of isolation. Genetic similarity analyses indicated that isolates from this location were part of a genetically distinct outcrossing population. The results showed that oospores of *P. capsici* persisted for 2 years between cucumbers and tomatoes and crop rotation did not seem to be an effective option for reducing incidence of Phytophthora blight disease in pepper in Michigan State (Lamour and Hausbeck 2003). The selection pressure for a soilborne pathogen may be reduced by including nonhosts in the rotation and/or fallowing to 'starve out' the pathogen, resulting in reduction of pathogen population buildup and consequent reduction in disease incidence/severity. The potato pink rot caused by *Phytophthora erythroseptica* can survive as oospores in the soil for several years in the absence of potato crop and it is likely to become endemic, if monoculture of potato is adopted for long periods. Adoption of crop rotation was observed to reduce disease incidence significantly. The effect of 2- and 3-year crop rotations and conservation tillage (reduced or no-till) practice on the incidence of potato pink rot was assessed in Prince Edward Island, Canada. Spring barley and potato cv. Russet Burbank were included in the 2-year rotation, whereas the 3-year rotation included barley (under sown with red clover), red clover and potato. Development of pink rot was significantly less pronounced in potatoes from the 3-year rotational soils than from 2-year rotational soils. Disease incidence was less in the potato plants growing on 3-year rotational soils than in plants growing on 2-year rotational soils under greenhouse conditions (P = 0.05). Inclusion of red clover in the 3-year rotation might enhance disease suppressiveness in the succeeding potato crop, due to availability of additional nitrogen fixed by red clover, promoting the growth of potato plants (Lambert and Salas 2001; Peters et al. 2005).

Verticillium wilt disease caused by *V. dahliae* is a major limiting factor for strawberry production especially in organic production system. In the post-methyl bromide era, Verticillium wilt is likely to reemerge as a major disease for conventional strawberry fields also. The effect of crop rotation on Verticillium wilt disease incidence and yield of strawberry (*Fragaria* x *ananassa*) was assessed at a site infested with the microsclerotia of *V. dahliae* and at another site with no known history of *V. dahliae* infestation during 1997–2000. The effects of different crop rotations on *V. dahliae* and *Pythium* populations, Verticillium wilt severity, strawberry vigor and fruit yield were compared with the standard methyl bromide + chloropicrin fumigated control treatment at both sites. Populations of the pathogens were not affected by rotations. However, microsclerotia of *V. dahliae* were sufficiently reduced with broccoli and Brussels sprouts rotations, compared with lettuce rotations at the *V. dahliae*-infested site. Reduced propagules resulted in lower Verticillium wilt severity on strawberry plants in the broccoli and Brussels sprouts rotations than in lettuce-rotated plots. Vigor and fruit yield of strawberry were significantly lower in lettuce-rotated plots than in broccoli- and Brussels sprouts-rotated plots. None of the rotation crops was better than fumigated controls for all variables measured. Based on the cost-benefit analysis, broccoli-strawberry rotation system could be an economically preferable, viable option, as the regulation on the use of chemical fumigation could prevent its application (Subbarao et al. 2007). In the further study, the effects of broccoli and lettuce rotations on population densities of *V. dahliae* and *Pythium* spp. in soil and Verticillium wilt disease incidence and severity on strawberry grown in conventional and organic production systems in California for 2 years, were assessed. Strawberry was raised after two successive crops of broccoli or lettuce. Densities of *V. dahliae* and *Pythium* spp. declined by 44% in conventional plots at the end of second broccoli crop, although no differences in the preplant densities of *V. dahliae* and *Pythium* spp. were seen. In contrast, densities in organic fields showed a decrease of 47% in 2000 and 25% in 2001 after second crop of broccoli. Lettuce rotations generally did not alter the pathogen densities both in organic and conventional production systems. Crop rotation treatments had no consistent effects on inoculum densities of *Pythium* spp. Verticillium wilt disease incidence on strawberry was 12 to 24% lower in fields rotated with broccoli, compared with lettuce. Verticillium wilt severity was reduced by 22 to 36% in broccoli rotation, compared with lettuce rotation. The results showed that broccoli rotation along with postharvest incorporation of broccoli residue could be adopted as an effective management strategy against Verticillium wilt disease both in organic and conventional strawberry production systems (Njoroge et al. 2009). The effectiveness of crop rotation with broccoli in suppressing Verticillium wilt disease of eggplant caused by *V. dahliae* was assessed. Six field trials were conducted in five fields during 2009–2013 with broccoli rotation coupled with postharvest incorporation before planting

eggplant. Meta-analysis of the data on disease incidence from all trials revealed that combined relative disease risk was 0.53 (95% confidence), indicating that incidence of Verticillium wilt in eggplant with broccoli rotation was reduced by 53% of that in control treatment (without broccoli rotation). Vascular tissue browning due to *V. dahliae* infection in broccoli was limited to the roots and did not reach the aerial plant organs, where *V. dahliae* formed microsclerotia. *V. dahliae* population assessment using quantitative nested real-time PCR showed that *V. dahliae* DNA copy number in soil samples tended to decrease owing to broccoli as rotation crop. The results suggested that broccoli root system might function as a 'decoy' that traps the pathogen in the soil (Ikeda et al. 2015).

Rice crop rotation was evaluated for its effectiveness in reducing Verticillium wilt disease in strawberry nurseries. Eggplant was used as bait (indicator) plant to detect *V. dahliae* in soil. The pathogen could be detected at as low as 1 microsclerotia/g of dry soil using eggplant. *V. dahliae* was detected in 9 of 10 upland fields, but not in any of the 21 rice upland fields during 2000–2007 in field surveys. In Hokkaido, strawberry mother plants were planted and plantlets were produced in upland and rice-upland fields to assess the extent of *V. dahliae* infestation in the soil. Incidence of Verticillium wilt was not observed in one of the 72 rice upland fields tested, compared with 13.2 to 73.9% of plantlets infected with *V. dahliae* in upland fields. Two flooding treatments or two rice cultivation entirely suppressed development of wilt symptoms in eggplants under greenhouse conditions. Likewise, in field trials, one rice cultivation in Chiba and two in Hokkaido totally prevented Verticillium wilt development in eggplant. *V. dahliae* was completely eliminated by one paddy rice cultivation in infested fields in Chiba, as revealed by absence of disease symptoms in strawberry nursery. The number of microsclerotia of *V. dahliae* was significantly reduced under flooded conditions provided for rice cultivation. Based on the results of field experiments, a crop rotation system with rice for 3 years (three times), green manure for 1 year and strawberry nursery for 1 year for Hokkaido, was developed for the management of Verticllium wilt disease in strawberry (Ebihara et al. 2010).

Wheat Rhizoctonia bare patch disease induced by *R. solani* AG-8 is a major fungal root disease in no-till cropping system. In the 8-year investigation carried out in various dry land no-till cropping systems, Rhizoctonia bare patch disease was first observed in year 3 and continued with greater severity through year 8. Crop rotation did not have any effect on disease incidence during the first 5 years. However, from years 6 to 8, both soft white and hard white classes of spring wheat (*Triticum aestivum*) grown in a 2-year rotation with spring barley (*Hordeum vulgare*) resulted in an average of only 7% total land area with bare patches, compared with 15% in continuous annual soft white wheat or hard white wheat (monoculture). In years 6 to 8, the average grain yield was also increased by barley rotation (P <0.001). Monoculture of hard white wheat was more severely affected by *R. solani* than soft white wheat. Soil water levels were higher in bare patches, indicating that roots of healthy plants did not grow into or underneath bare patch areas. Barley rotation might counteract the effects of no-till cropping system on the incidence of bare patch disease caused by *R. solani* AG-8 (Schillinger and Paulitz 2006).

The effect of crop rotation systems on the recovery (populations) of *Rhizoctonia* spp. infecting canola and lupin in the southern and western production areas of Western Cape Province of South Africa was assessed. The pathogen was isolated from canola planted after barley, medic/clover mixture and wheat and lupin planted after barley and wheat with sampling at the seedling, mid-season and seedpod growth stages. Crop rotation and plant growth stage at sampling, all affected the incidence/recovery of *Rhizoctonia*, but with site-specific effects. The binucleate group was more frequently isolated from lupin, whereas the multinucleate group was recovered from canola. The anastomosis group (AG) 2-1 was isolated only from canola and AG-11 only from lupin. The results indicated that crop rotation consistently influenced the incidence and composition of the *Rhizoctonia* spp. associated with canola and lupin (Lamprecht et al. 2011). Effect of continuous cultivation (monoculture) on incidence of brown stem rot (BSR) caused by *Phialophora gregata* f.sp. *sojae (Pgs)* on susceptible and resistant soybean cultivars was assessed in an area with no history of soybean cultivation. Population size and genotype composition of *Pgs* were determined by dilution plating, isolation, PCR and qPCR assays. In general, the population size of *Pgs* in soil was similar regardless of monoculture. The percentage of B genotype of *Pgs* was greater than A genotype in soil, following monoculture of both BSR-susceptible- and -resistant soybean accessions. Overall, *Pgs* populations in stems of susceptible accession were greater than those in stems of a BSR-resistant accession. *Pgs* genotypes A and B were isolated at similar frequencies from stems of BSR-susceptible accession planted, following a BSR-susceptible monoculture. However, the results of qPCR assay were at variance from that of isolation method (Hughes et al. 2009).

Potato crops are affected by soilborne diseases, Rhizoctonia canker and black scurf (*R. solani*), common scab (*Streptomyces scabiei*), silver scurf (*Helminthosporum soalni*) and Verticillium wilt (*V. dahliae*), which cause recurrent problems in various countries. Different 2-year rotations, consisting of barley/clover, canola, green bean, millet/rapeseed, soybean, sweet corn and potato, all followed by potato were applied for over 10 years (1997–2006) to assess their effects on the development of soilborne diseases, yield and microbial communities. Soil populations of culturable bacteria and overall microbial activity reached the maximum in barley, canola and sweet corn rotations, as against the lowest in potato monoculture system. Incidence and severity of stem and stolon canker and black scurf of potato caused by *R. solani* were reduced by most rotations, relative to the continuous potato (monoculture) control. Potato crops following canola, barley or sweet corn showed low levels of infection and gave high quality tubers, whereas potato crops following clover or soybean rotations contracted higher disease incidence in some years. Several microbial parameters, such as

microbial populations, single carbon source utilization (SU) profiles and whole-soil fatty acid methyl ester (FAME) profile characteristics were correlated with Rhizoctonia disease or yield level (Larkin and Honeycutt 2006). Rhizoctonia canker on potato plants (stems and stolons) in the field and black scurf and common scab on harvested tubers occurred predominantly over the entire experimental duration. Overall, canola, rapeseed, sweet corn and barley/clover as rotation crops reduced the severity of Rhizoctonia canker compared to control (continuous potato) (P = 0.001). Canola and rapeseed rotations were the most effective with a reduction in disease incidence ranging from 15 to 45% in individual years and averaging 20 to 23% over 7 years. Canola and rapeseed rotations were most effective in reducing black scurf on harvested tubers also to the extent of 32 to 38%, relative to the control. In the case of common scab, canola and rapeseed reduced severity by 18 to 28%, compared to control (see Figure 1.6). Effectiveness of other rotation crops was more modest for black scurf. The interaction between year and rotation was not significant for any disease studied. Crop rotations exhibited different levels of suppression of Verticillium wilt disease in potato. Significant differences in wilt incidence were noted among rotation crops, with barley, rapeseed and sweet corn rotations resulting in lower disease incidence (40–58%) than other rotations (68–75%), barley providing lower level of disease suppression (see Figure 1.7). The results suggested that for effective disease management and sustainable potato production, a 3-year or longer rotation and disease suppressive rotation crop, prior to potato crop, such as a *Brassica* (canola, rapeseed, mustard) crop or small grain (barley) might be preferred. In addition, use of a fall cover crop, such as winter rye or ryegrass could also be a desirable option to reduce soil-borne diseases effectively (Larkin et al. 2010).

Chinese leek rotation with banana effectively reduced Fusarium wilt disease caused by *F. oxysporum* f.sp. *cubense (Foc)* by 88 to 97% and disease severity by 91 to 96%. Further, crop value was improved by Chinese leek rotation by 36 to 86% in heavily infested field. Under greenhouse conditions and artificially inoculated conditions, Chinese leek treatment reduced Fusarium wilt disease incidence and disease severity index by 58% and 62%, respectively, in banana cv. Baxi (AAA) and by 79% and 81% respectively in cv. Guangfen No.1 (ABB). The inhibitory effects of Chinese leek on pathogen development were investigated to determine the mode of action of Chinese leek. Crude extracts of Chinese leek inhibited mycelia growth of *Foc* race 4 on petriplates and spore proliferation was inhibited by 91%. Spore mortality due to Chinese leek was estimated to be 87%. The results indicated that Chinese leek could be rotated with banana and the inhibitory products from Chinese leek might adversely affect pathogen development at different stages of its life cycle (Huang et al. 2012). The potential of crop sequence (rotation) to limit the incidence of Fusarium basal rot of onion, caused by *F. oxysporum* f.sp. *cepae (Foc,)*, was investigated. The extent of *Foc* multiplication in the roots of 13 rotation crops was determined in the greenhouse. The tested rotation crops allowed *Foc* multiplication to different degrees. The lowest population

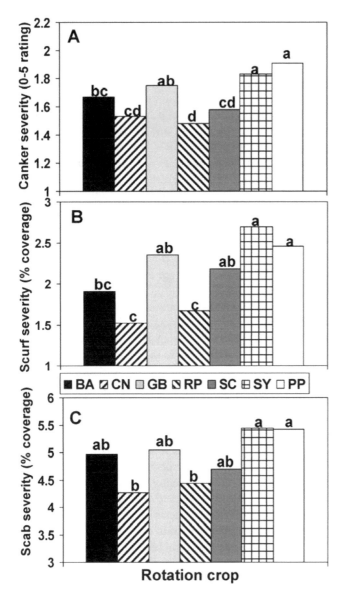

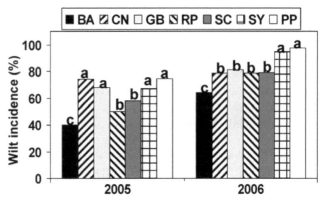

FIGURE 1.7 Effects of rotation crops on incidence of Verticillium wilt (%) in potato during two crop seasons (2005 and 2006) Treatments: barley (BA), canola (CN), millet/rapeseed (RP), green bean (GB), sweet corn (SC), soybean (SY) and potato (PP, control).

[Courtesy of Larkin et al. 2010 and with kind permission of the American Phytopathological Society, MN]

FIGURE 1.6 Effects of rotation crops on incidence of Rhizoctonia canker (A), black scurf (B) and common scab (C) on potato tubers during 2000–2006 Treatments: barley (BA), canola (CN), millet/rapeseed (RP), green bean (GB), sweet corn (SC), soybean (SY) and potato (PP, control) Bars with the same letter at top are not significantly different as per Fischer's protected least significant difference test (<0.05).

[Courtesy of Larkin et al. 2010 and with kind permission of the American Phytopathological Society, MN]

levels of *Foc* were present in the roots of wheat, sunflower, cowpea and millet, the highest pathogen population being in the roots of black bean and sudangrass. Populations of *Foc* in the sequence onion-foxtail millet-wheat-cowpea rotations were 67% lower than in the rotation of onion-sudangrass-black oat-black beans (Leoni et al. 2013).

The effects of broccoli as rotation crop or a fallow period on the incidence of lettuce drop disease caused by *S. minor* and the density of pathogen sclerotia in the soil were assessed under field conditions. Low inoculum density (<7 sclerotia per 100 cm³ of soil) in one site and high inoculum (>7 sclerotia per 100 cm³ of

soil) in the second site were provided. The treatment plots with lettuce (LLL), lettuce rotated with broccoli (LBL) and lettuce followed by fallow (LFL) were established in the first site. In the second site, the treatments included continuous lettuce (LLLL), broccoli-lettuce-broccoli-lettuce (BLBL), broccoli-broccoli-lettuce-lettuce (BBLL) and fallow-lettuce-fallow-lettuce (FLFL). In continuous lettuce at the first site, disease incidence was not increased for at least 2 years, but soilborne sclerotia were increased to a level below threshold at which inoculum density and disease incidence showed positive relationship. Rotation with broccoli decreased disease incidence only in the first year. LFL treatment contained lowest density of sclerotia, whereas the highest density of sclerotia was present in LLL treatment. In the second site, rotation with broccoli reduced both sclerotia density and disease incidence. The number of broccoli crops rather than the sequence of lettuce rotations with broccoli was found to be vital for reducing sclerotial density of *S. minor* in the soil. Sclerotial viability was not affected by different treatments (Hao et al. 2003). The effect of green manure (GM) treatments on *S. rolfsii* sclerotial dynamics was studied under greenhouse and field conditions during 2009–2010. The relative densities of viable sclerotia at 90 days after winter GM (WGM) incorporation were generally lower than that recovered after summer GM (SGM) incorporation, with average recovery values of 60% and 61% for WGM in the field, 66% and 43% for WGM in the greenhouse and 162% to 91% for SGM in the greenhouse, respectively, in 2009 and 2010. Relative decay rates of the sclerotia in SGM amended soil were largest for alfalfa and sudangrass. On the other hand, the largest decay rates of sclerotia were for oat, wheat and alfalfa in WGM amended soil. The effect of cropping sequences – sweet pepper (SWP)-fallow, sweet pepper-black oat (BO) and sweet pepper-onion (SO) – to determine sclerotial dynamics was assessed in microplots. The sequence SWP-O-Fallow-BO resulted in the lowest long-term sclerotia density (7.09 sclerotia/100 g soil) and SWP-Fallow had the highest sclerotial density (17.89 sclerotia/100 g soil) (Leoni et al. 2014).

Aphanomyces euteiches infects leguminous crops, causing root rot diseases accounting for heavy losses when conditions are favorable. The effect of 10 successive monocultural cycles of different legume species/cultivars on the inoculum potential (IP) of soils naturally infested by *A. euteiches* was determined under greenhouse conditions. The inoculum potential of a soil naturally infested by *A. euteiches* could be modified significantly by the nonhost or host status of crop plants, as well as by the levels of resistance of cultivars. Pea, lentil, susceptible cultivars of vetch and faba bean favored the proliferation of *A. euteiches* and continuous cultivation of these crops increased the IP values of a soil with moderate initial IP (from 1.9 to 3.5 after 10 cycles). In contrast, nonhost species and resistant cultivars of vetch and faba bean reduced the IP values of soils, irrespective of the initial IP (from 1.9 to 0.5 and from 4.0 to 2.0, respectively, after 10 cycles). The root rot severity induced by *A. euteiches* on resistant legume species/cultivars were not influenced by the successive cultural cycles. The results indicated that by planting appropriate legumes/cultivars, the inoculum potential of *A. euteiches* in soil could be brought down to have economically acceptable crop yields (Moussart et al. 2013). The effect of *Brassica* crops grown as cover crop with different glucosinolate profiles on the development of *Aphanomyces euteiches* was investigated. *B. juncea, Sinapis alba* and non-*Brassica* species, *Secale cereale* were grown for 11 weeks in pathogen-infested soil at low and high nitrogen (N) fertilizer doses. Peas were grown in the same soil, after removal of all shoots and roots of cover crops, to assess the level of infection in the susceptible pea plants used as bioassay plants. Soil samples were taken before and after harvesting cover crops, and concentrations of volatile compounds in the root-soil environment were estimated. Pea root rot severity was reduced in *Brassica*-grown soil at high nitrogen (N) dose. Growth of *Brassica* spp. did not suppress the abundance of N-cycling-microbial communities [free-living N_2-fixing bacteria and ammonia oxidizing-bacteria (AOB) and arachea (AOA) involved in key soil ecosystem services]. The disease suppressive effect was higher with *S. alba* than with *B. juncea*, and this coincided with a more diverse composition and higher concentration of aliphatic isothiocyanates (ITCs), released from *S. alba* roots. The results showed that *Brassica* cover crops, particularly *S. alba*, could be grown for suppressing development of Aphanomyces root rot in pea crops (Hossain et al. 2015).

The effect of grass species grown either as pure stands or mixed with wheat, after a sequence of wheat crops on the incidence of wheat take-all disease caused by *G. graminis* var. *tritici*. was investigated. Annual brome grasses maintained take-all inoculum in the soil, as well as wheat as a monoculture and much better than cultivated species with a perennial habit. Incidence of take-all increased in wheat grown after *Anisantha sterilis* (barren home) or *Bromus secalinus* (rye brome), with or without wheat, than in continuous, grass-free wheat grown after *Lolium perenne* (rye grass) or *Festuca arundinacea* (tall fescue), although these species supported pathogen development only to a limited extent. Disease severity was high in plots where these two grasses were grown as

mixtures with wheat. The effects of grass species appeared to be transient, since incidence of take-all in a second subsequent winter wheat test crop was similar after all treatments. The results indicated the degree of take-all risk in wheat, when grown after grass weed-infested cereals (Gutteridge et al. 2006). Influence of crop rotation (sequence) in combination with other agricultural practices on the incidence of crop diseases has been investigated. Incidence of wheat take-all disease caused by *G. graminis* var. *tritici (Ggt)* was recorded over three growing seasons in commercial wheat fields in which wheat had been grown for 1–4 years consecutively in New Zealand. The *Ggt* DNA contents (as determined by quantitative PCR assay) and take-all index (TAI) were the least at the commencement of wheat sequence and stabilized in the third and fourth consecutive wheat crops. Median *Ggt* DNA contents in the soil increased by 10 times during the first wheat crop, reaching 78 pg *Ggt* DNA/g in soil after harvest. Rainfall during growing season and position in a continuous wheat crop sequence were most closely associated with TAI and postharvest soil *Ggt* concentrations. Inoculum concentrations, disease incidence and TAI were greater where the frequency of crops susceptible to *Ggt* like wheat, barley or triticale, was greatest in the crop rotation and where crops were irrigated. Irrigation was consistently associated with high postharvest soil *Ggt* populations in the driest of the three growing seasons studied, when the environments were conducive for *Ggt* development. The results indicated that frequency of cereal crop in the rotation, length of breaks between host crops, irrigation and growing season rainfall might interact over the course of a cropping sequence to influence TAI, when wheat may be sown (Bithell et al. 2013).

The effectiveness of employing combination of different components of genes for resistance to crop disease was test-verified by using cultivars with different complements of resistance genes. *Leptosphaeria maculans* causes blackleg (Phoma stem canker) disease in canola (*B. napus*) with worldwide distribution. Under greenhouse conditions, potted canola plants were exposed to blackleg-infested stubbles of canola with different complements of resistance genes and disease incidence was recorded. Plant mortality was reduced, when plants were exposed to stubble from a cultivar with different components of resistance genes, compared to stubble of a cultivar with same resistance gene. The results of 7-year field surveys, showed that changes in selection pressure as a result of extensive sowing cultivars with major gene resistance ('sylvestris resistance') influence significantly the frequency of virulent isolates in the population toward particular resistance genes and hence, disease severity. Sowing canola cultivars with different complements of resistance genes in subsequent years – rotation of resistance genes – might minimize disease pressure through pathogen population manipulation (Marcroft et al. 2012).

R. solanacearum (Rs), incitant of destructive bacterial wilt disease of several crops, has worldwide distribution. The effectiveness of using previous crops as sanitizing crops was evaluated to employ them as environmentally safe method of managing bacterial wilt disease. The ability of *R. solanacearum* to persist in planta and in rhizosphere of Brassicaceae, Asteraceae and Fabaceae grown as previous crops was assessed under controlled conditions and the incidence of bacterial wilt was recorded in the following tomato crop. All tested plant species were latently infected by *R. solanacearum*. The highest pathogen density was detected in planta and in the rhizosphere of *Tagetes erecta*. The density of *R. solanacearum* population in the rhizosphere of radish *(Raphanus sativus)* cv. Karacter was significantly greater than in cv. Melody. In Fabaceae, the pathogen population did not show variations in planta. *Crotalaria juncea* had greater pathogen population than that of *C. spectabilis*. The results showed that C. spcectabilis and *R. sativus* cv. Melody grown as previous crops improved the performance of the following tomato with similar effects on *R. solanacearum* populations in the soil as bare soil. The incidence of tomato bacterial wilt decreased by 86% and 60%, after growing *R. sativus* cv. Melody and *C. spectabilis*, respectively, as rotation crops and the proportion of infected plants also decreased. The results suggested that *C. spectabilis* and *R. sativus* cv. Melody might be grown as previous crops to reduce bacterial wilt incidence in tomato, as an ecologically safe strategy without drastic suppression of *R. solanacearum* populations in stem tissues and in the rhizosphere (Deberdt et al. 2015). Planting nonhosts to break bacterial disease cycle has been suggested as a feasible approach to reduce disease incidence and spread. With the aim of breaking the cycle of banana Xanthomonas wilt (BXW) disease, maize, bean and sweet potato were planted after uprooting banana infected by *Xanthomonas campestris* pv. *musacearum (Xcm)* in Rwanda and DR Congo. A weed fallow (mixed species) served as control. After 1, 2 and/or 4 break-crop or fallow seasons, healthy banana plantlets were replanted and disease incidence was monitored for 12-24 months. The presence of *Xcm* was detected, using PCR assay. Xanthomonas wilt disease occurred in all one-season plots with higher cumulative incidences in maize and bean plots. In two-season plots, disease incidence of 6 to 8% was observed only in bean and sweet potato plots in DR Congo. Lengthening time under break crops to two- and four-seasons reduced disease incidence, respectively, to 3% and 0% in Rwanda and 0 to 8% in two-season plots in DR Congo. Based on the results, two-season with break crop combined with collective Xanthomonas wilt disease management strategies might be a feasible option to reduce the disease incidence and further spread (Rutikanga et al. 2016).

The decoy or catch crops are capable of stimulating the germination of resting spores, resulting in limited expression of disease symptoms. Leafy daikon (radish, *Raphanus sativus* var. *longipinnatus*) was evaluated for its efficacy to reduce incidence of clubroot disease in Chinese cabbage caused by *P. brassicae* under pot culture conditions. Disease indices of Chinese cabbage plants grown in pots, that had leafy daikon, were lower, compared with control plots. Number of resting spores of *P. brassicae* in soil in pots, after growing leafy daikon was reduced by 71%, compared with control pot, when resting spores were recovered and quantified directly. Under field conditions, the numbers of resting spores were reduced by 94%, compared with initial pathogen population, when leafy daikon was raised prior to Chinese cabbage. However,

disease severity was not reduced by leafy daikon treatment. *P. brassicae* infection of root hairs of leafy daikon was observed, but club formation did not occur, as in Chinese cabbage. The results indicated the possibility of using leafy daikon as a decoy crop to manage clubroot disease of cruciferous crops (Murakami et al. 2000). Germination of resting spores of *P. brassicae* may be stimulated by nonhost plant species, while remaining resistant to infection by the pathogen, resulting in reduced persistence of resting spores in soil and consequent reduction in incidence of clubroot disease in crucifers. The effects of nonhosts leek (*Allium porrum*), winter rye (*S. cereale*, and perennial grass (*Lolium perenne*) with ability to stimulate germination of resting spores and red clover (*Trifolium pretense*) were assessed. In the field experiment, none of the nonhost plant species reduced the concentration of *P. brassicae* in soil, as determined by bioassay, using Chinese cabbage (*Brassica rapa* var. *pekinensis*). The results indicated the ineffectiveness of using nonhosts for reducing incidence of clubroot disease in cruciferous crops (Friberg et al. 2006). In a later study, growing canola cultivars resistant to *P. brassicae,* reduced clubroot severity and increased yield significantly. A 2-year break from canola reduced *P. brassicae* resting spore concentration by 90%, relative to growing continuous canola or only a 1-year break in heavily infested field plots. The 2-year break reduced disease severity in susceptible canola cultivar by promoting plant growth. However, the inoculum levels were after 2-year break to obtain commercially acceptable yields in susceptible canola cultivar. In resistant cultivar >2-year breaks increased yields substantially (up to 25%). A 2-year interval with nonhosts between canola crops, together with use of resistant cultivars, might effectively reduce soil inoculum level and increase yield levels (Peng et al. 2014). The effects of fallow periods (F) and bait crops (canola-clubroot susceptible (B), and perennial ryegrass (R) on clubroot severity and populations of *P. brassicae* resting spores were assessed using five sequences of host plant species viz., R-B, B-R, R-F, B-F and F-F. Clubroot severity was reduced by both host and nonhost bait crops in a subsequent crop of susceptible canola cultivar, compared with fallow. Concentrations of resting spores and pathogen DNA content decreased in all treatments, with the lowest levels in R-B and B-R treatments. Under microplot conditions, three crops of susceptible canola cultivars were compared with a 2-year break of oat-pea, barley-pea, wheat-wheat or fallow-fallow. In another assessment, three crops of resistant canola were compared with, two crops of resistant canola with a 1-year break, one crop of resistant canola with a 2-year break and a 3-year break with barley followed by a susceptible canola. Rotations that included nonhost crops of barley, pea or oat reduced clubroot severity and resting spore concentrations, and increased yield, compared with continuous cropping of either resistant, susceptible canola. Cultivation of susceptible canola cultivar enhanced gall mass by 23 to 250-fold, compared with resistant cultivars, indicating the necessity of taking adequate counteracting measures to maintain the yield levels to enable the growers to have reasonable profits (Hwang et al. 2015).

The effects of rotation crops and tillage systems on canola blackleg disease caused by *Leptosphaeria maculans*, were assessed during 1999–2002 in Manitoba, Canada. Rotation crops wheat (W) and flax (F) were raised before canola (C), under conventional tillage (T) or zero-till (Z) systems. Stem infection and severity were significantly lower in 2001 and 2002, when canola was rotated with wheat and flax under both tillage systems. The CCCCT rotation had higher disease incidence than CWCCT, CWFCT and CCWT rotations which were not significantly different from each other. Zero-till plot had significantly greater diseased stem incidence and severity than tilled plots, with or without rotation. Disease incidence was reduced by tillage, when combined with simple crop rotation. However, two-crop rotation was not effective in reducing incidence of canola blackleg disease. The results suggested that appropriate combination of crop rotation and tillage system should be adopted to derive benefits of both strategies (Guo et al. 2005). *Fusarium solani* causes the brown root rot, one of the most economically important diseases of peanuts in Argentina. The effectiveness of crop rotation and tillage in containing the brown root rot disease was assessed under field conditions during 2001-2006. Corn (maize), peanut and soybean and tillage systems – disc harrow, chisel plow, no-till and paratill subsoiler formed the treatments. Corn and soybean were always seeded in no-till system, whereas peanut was planted under all tillage systems. Paratill subsoiler used before planting peanut in the no-till system generally reduced the subsurface layer of compacted soil, which provided a favorable environment for root development, and consequently reduced brown root rot incidence. The extent of reduction in disease incidence was greater in 2-year rotation, including corn-soybean or soybean-corn prior to peanut, than in one-year rotation. Brown root rot incidence in different tillage treatments showed strong negative relationship with pod yield, combined application of paratill subsoiler before peanut seeding and no-till system appeared to be an effective strategy to reduce brown root rot disease in peanut (Oddino et al. 2008). Take-all disease caused by *G. graminis* var. *tritici (Ggt)* seriously damages the root system of infected wheat plants, resulting in poor nutrient uptake and ultimate death of plants. The pathogen is unable to survive in the soil in the absence of susceptible host plant species almost belonging to Gramineae. Consequently, take-all usually does not damage wheat crops that follow nongraminaceous break crops, indicating the possibility of avoiding losses due to take-all by adopting appropriate crop sequence. The effects of different break-crops and set-aside/conservation covers on take-all development were investigated. The break crops per se did not seem to affect take-all development, but the presence of volunteers in crops of oilseed rape, increased the amounts of take-all in the following wheat crop. Severity of take-all was closely related to the numbers of volunteers in the preceding break-crops and covers and was affected by the date of their destruction. Early destruction of set-aside/conservation covers was usually effective in preventing incidence of take-all in the following wheat crop, except sometimes when populations of volunteers were very large (Jenkyn et al. 2014).

Cover crops are grown along with the primary crop to promote plant growth by altering microclimate, adding organic matter biomass and preventing soil erosion. In addition, cover crops may also protect the primary crops against certain microbial plant pathogens, if they are nonhost for the pathogen(s) infecting primary crop. Cover crops also add crop diversity, which may stimulate and diversify soil microbial biomass, resulting in some degree of disease suppression. Cover crops may function in a manner similar to rotation crops, regarding their effects on soil microbial biomass, activity, diversity and potential for disease suppression (Larkin 2015). Velvet bean was grown as a summer cover crop in the southeastern United States. The effect of killed velvet bean as a cover crop mulch left on the soil surface prior to transplantation of collard cabbage in the fall was investigated. Weed-free fallow and killed velvet bean disked into soil formed control treatments. Incidence of wiresterm caused by *R. solani* was 25% in 2000, but reduced to 4% in 2001 in cover crop mulch treatments. But in both years, the infection rate, area under disease progress curve (AUDPC) and final disease incidence were significantly greater in mulch than in the fallow or disked treatments. Wiresterm incidence did not differ between the disked and fallow treatments in either year. Populations of *R. solani* in soil were greater after cover crop mulch than in fallow plots in both years and greater in the disked treatment than in fallow soil in 2000, but not in 2001. The results suggested that velvet bean could be used as a summer cover crop, disked into the soil before transplanting collard (Keinath et al. 2003). Triploid watermelon cultivars (seedless) with little resistance to Fusarium wilt caused by *F. oxysporum* f.sp. *niveum (Fon)* grown in large areas needed effective management system. The effects of four different fall-planted cover crops viz., *V. villosa* (hairy vetch), *Trifolium incarnatum, S. cereale* and *B. juncea* that were tilled in the spring as green manures, and bare ground were evaluated alone and in combination with a biocontrol product Actinovate (containing *Streptomyces lydicus*) on Fusarium wilt severity and watermelon fruit yield and quality in six field trials over 3 years in three locations were recorded. Fusarium wilt development was suppressed significantly by *V. villosa* and *T. incarnatum* to an extent of 21%, compared with nonamended plots. In locations with low disease incidence, disease suppression level was not perceptible. Fruit yield was increased (129%) only by *T. incarnatum* as the cover crop over that of nonamended plots. Actinovate was not effective in reducing disease incidence significantly. Among the cover crops tested, *T. incarnatum* when incorporated as green manure prior to planting watermelon, significantly suppressed Fusarium wilt disease (Himmelstein et al. 2014).

Two or more number of crops may be grown as intercrop, mixed crop or barrier crop in a field to interrupt the continuous availability of the primary crop susceptible to the pathogen. The selected crops may be planted simultaneously or in a staggered manner in such a way that the growth periods of the two crops overlap. Crops planted as intercrop or mixed crop may (i) modify the microclimate of another crop; (ii) intercept or filter out spores of the pathogen; or (iii) restrict movement or reduce the infectivity of the vectors, resulting in reduction in disease incidence or spread. Growing sesame (*Sesamum indicum*) or peanut as intercrops was reported to reduce the incidence of pepper Phytophthora blight caused by *P. capsici* in Korea (Kim 1989). Extracts of soils on which sesame or peanut was raised and the root exudates of these intercrops were inhibitory to mycelial growth, sporangium formation and release of zoospores by *P. capsici*. Further, soils amended with extracts had suppressive activity against *P. capsici*, resulting in reduction of disease incidence significantly (Lee et al. 1990, 1991). The effect of pea-cereal intercropping on Ascochyta blight of field pea, induced by *A. rabiei*, was investigated. Disease pressure was variable during the experimental duration. Development of disease symptom on stipules was affected by intercropping, regardless of variation in disease pressure. But disease severity on pods and stems was appreciably reduced in pea-cereal intercrop, compared with pea monocropping when the epidemic was moderate to severe, resulting in substantial reduction in yield loss. In addition, the quantity of primary inoculum available for subsequent pea crops was reduced by intercropping. The mechanism underlying the beneficial effect of pea-cereal intercropping might be due to modification of microclimate especially reduction in leaf wetness duration, during and after flowering. Further, the investigation under controlled conditions using a rainfall simulator indicated that total dispersal of conidia was reduced by 39 to 78% in pea-wheat canopies, compared to pea canopies. Reduction in primary host plant density and a barrier or relay effect of the nonhost (wheat) might contribute to the reduction in disease severity in pea crop (Schoeny et al. 2010).

The effectiveness of garlic as an intercrop to reduce the bacterial wilt caused by *R. solanacearum* in flue-cured tobacco was investigated in Fujian Province, China, during 2008–2009. Appearance of bacterial wilt in intercropped fields was delayed by about 15 days. The total pathogen population in the rhizosphere soil was significantly lower in intercropped fields than in monocultural fields in 2008. Tobacco leaves from intercropped fields fetched higher monetary returns (14–34%). The results indicated that garlic, as an intercrop could be grown to reduce the ill effects of tobacco bacterial wilt caused by *R. solanacearum* (Lai et al. 2011). The susceptibility levels of the intercrops have to be assessed, prior to growing them in commercial fields. The susceptibility of intercrop species such as *Raphanus sativus, B. juncea, B. rapa, Sinapis alba* and *Phacelia tanacetifolia* to *R. solani*, incitant of sugar beet root rot was assessed in the greenhouse and field with artificial inoculation. All tested intercrop species were susceptible and differences in their susceptibility were consistent. Compared to fallow, growing *B. rapa* and *R. sativus* reduced disease severity in subsequent sugar beet crop, resulting in higher white sugar yield up to 210% and 157% for *B. rapa* and *R. sativus* respectively. Use of efficient and rapid screening tests could be the basis for selecting intercrops resistant to *R. solani* (Kluth et al. 2010). Damping-off and root rot of lentil caused by *R. solani* and *Fusarium solani* was responsible for appreciable loss in Egypt. Intercropping with cumin, anise, onion and garlic significantly reduced damping-off and root rot and increased grain yield. Anise was the most effective, while onion was less effective. Root exudates of intercrops were inhibitory to mycelial growth of the pathogens, indicating the possible mechanism of disease suppression by the intercrops (Abdel-Monaim and Abo-Elyousr 2012).

The mechanism of action of companion crop in limiting incidence of disease in primary crop was studied in melon-Fusarium wilt pathosystem. The effect of root exudates of 10 wheat cultivars on mycelial growth of *F. oxysporum* f.sp. *niveum (Fon)*, incitant of melon wilt disease, was assessed, using petriplate assay. Wheat cv. D123 exudates significantly inhibited mycelial growth of the pathogen. In pot experiments, companion cropping with D123, wheat reduced disease incidence in melon, compared to the monoculture system. Companion cropping with D123 wheat, decreased malondialdehyde content and guaicol peroxidase activities in watermelon roots, following inoculation with *Fon*. Further, polyphenol oxidase activities were enhanced in melon and 15 days after inoculation with *Fon*, whereas slight increases in β-1,3-glucanase and chitinase activities were observed. By contrast, phenylalanine-ammonia-lyase (PAL) activity was reduced in the roots of melon in companion cropping system. The jasmonic acid (JA) and shikimate-phenylpropanoid-lignin biosynthetic genes were induced during early stage of *Fon* infection in the companion cropping, compared to monoculture. The results suggested that companion cropping with wheat might reduce watermelon infection by wilt pathogen, due to inhibition of pathogen growth by root exudates of wheat and changes in the gene expression in watermelon (Xu et al. 2015).

1.3 PHYSICAL TECHNIQUES

Physical and chemical techniques either alone, or in combination with cultural practices have been applied to reduce the pathogen population/virulence and consequent decrease in disease incidence. These methods have been shown to be effective for treatment of whole plants, planting materials and field soils.

1.3.1 Heat Treatments

Different forms of heat, as hot air, hot water and dry heat, have been evaluated for their efficacy in suppressing pathogen development. The effectiveness of heat, as a disinfectant was first demonstrated against *P. infestans*, causing late blight disease of potato. Seed tubers were treated with a current of hot air at 40°C for four hours. Later, hot water was used for control of smuts of barley and oats (Jensen 1888). Drying tomato seeds with discoloration in an air oven at 29.5–37.5°C for 6 h, resulted in elimination of *P. infestans* (Vartanian and Endo 1985). Likewise, the effectiveness of heat treatment for the control of pineapple heart rot disease caused by *Phytophthora cinnamomi* was reported. Pineapple crown buds, lateral buds, suckering buds and rhizomes were exposed to sunlight for 4–8 hours at a temperature of 33°C to suppress the development of *P. cinnamomi* present in plant materials (Yang and Chang 1998). Fusarium root and crown rot of hosta plants grown in containers in nursery, caused by *Fusarium hostae*, affected production seriously, requiring an effective management system for the disease. The disease was most severe on plants grown in 100% aged pine bark, in dry container mix or at a temperature of 18 to 25°C. Disease development was significantly suppressed, when plants were grown in 100% peat, in wet container mix or at 32°C. The results suggested that changes in cultural practices combined with appropriate temperature regime could limit the impact of Fusarium root and crown rot disease in hosta nurseries (Wang and Jeffers 2002).

Phytophthora ramorum causes the destructive sudden oak death disease and also infects *Umbellularia californica* (bay laurel or bay), the leaves of which were found to be the major sources of natural inoculum of *P. ramorum*. The pathogen was highly heat-tolerant and could be reisolated from artificially inoculated bay laurel leaves placed at 55°C for up to 1 week, but not after 2 weeks at 55°C. A heat treatment involving a progressive and gradual heating process combined with application of moderate vacuum eliminated *P. ramorum* in 22 hours. This process did not affect the quality of laurel leaves used as a spice (Harnik et al. 2004). Presence of fungal pathogens, especially *Phytophthora* spp. in recirculated irrigation water has been recognized as a significant health risk to nursery and greenhouse crops. Decontamination of irrigation water could be achieved by heat treatment at 95°C for 30 s. The possibility of reducing the temperature (95°C) to lower level to conserve energy and improve horticultural profitability, while reducing environmental footprint was examined. The effect of water temperature on survival of zoospores and chlamydospores of *P. nicotianae* in vitro and on annual vinca under greenhouse conditions was determined. The zoospores of *P. nicotianae* did not survive and cause any disease on vinca, when exposed to 42°C for 12 h or 48°C for 6 h. The chlamydospores of *P. nicotianae* were killed at 42°C for 24 h or 48°C for 6 h. The oospores of *P. pini* did not survive at 42°C for 12 h or 48°C for 6 h. The results indicated that *Phytophthora* spp. could be eliminated by lowering temperature and extending the duration of exposure, without losing the effectiveness of heat treatment (Hao et al. 2012).

Soilborne fungal pathogens produce different survival structures such as sclerotia and chlamydospores. Agricultural soil samples containing survival structures of *V. dahliae*, *S. sclerotiorum*, *Sclerotium cepivorum* and *Pythium ultimum* were treated with aerated steam in vitro at temperatures ranging from 40 to 80°C in a specially constructed equipment. Steaming at 50 or 60°C for 3 min, followed by an 8-min resting period in the steamed soil and immediate removal of survival structures from soil thereafter, resulted in their complete elimination of diseases caused by these fungal pathogens tested. At 45°C, a significant reduction in the survival of microsclerotia of *V. dahliae* was observed, but not in *S. cepivorum* sclerotia (van Loenen et al. 2003). Effects of temperature and soil moisture on infection and development of root rot in soybean were assessed. All isolates (44) recovered from 35 countries were identified as *R. solani* AG2-2IIB. Five

isolates of *R. solani* could infect and colonize soybean roots and hypocotyls at all temperatures viz., 20, 24, 28 and 32°C tested in growth chamber. The isolates of *R. solani* differed in their ability to induce seed rot, root rot rating and hypocotyl lesions at different temperatures and soil moisture holding capacity (25, 50, 75 and 100%). Varying temperature and soil moisture conditions might not be an effective option for managing the disease in soybean (Dorrance et al. 2003). In a later study, the effect of submerging terminal leafy cuttings of *Rhododendron* cv. Gumpo White azalea on development of web blight disease of azalea was assessed. All 12 azalea cultivars could tolerate 50°C for 20 min. Despite the risk involved in submerging azalea cuttings in water at 50°C, all 12 azalea cultivars were tolerant to submersion duration long enough to eliminate binucleate *Rhizoctonia* species from stem and leaf tissue with only a low likelihood of sustaining damage caused by hot water treatment (Copes and Blythe 2011). Internal discoloration of horseradish roots is induced by *Verticillium* spp. and *Fusarium* spp. The propagative rootstocks (sets) were treated by immersing them into water at 44, 45, 46, 47, 48, 49 and 50°C for 10, 20 and 30 min. Treatments at temperatures <46°C were not effective in eliminating the pathogens. Treatments at 48°C or higher either delayed set germination and reduced plant vigor. Treatment at 47°C for 20 min was the most effective in pathogen elimination with no adverse effect on set germination or plant vigor (Eranthodi et al. 2010).

A nonchemical, physical soil disinfestation technique, flame soil disinfestation (FSD) was developed to disinfest the soil in the greenhouse and field in China. The FSD procedure involves the use of compressed natural gas (NG) as the main fuel and the flame temperature at the nozzle of flame tubes could be increased up to 1,200°C. Treatments in two fields were FSD treated once (FSD1), FSD treated twice (FSD2) and an untreated control arranged in a randomized block design (RBD) with three replicates. The untreated plots were highly infested with *F. oxysporum* and *Phytophthora* sp. and their propagules were significantly greater than in the FSD-treated plots. FSD1 treatment reduced *F. oxysporum* by 44.7% and 73.9% in trial I and II respectively. Similar reduction (62.6% and 78.4%) were also observed in FSD2 treatments. Populations of *Phytophthora* spp. were reduced by FSD1 and FSD2 treatments (56.1% and 47.3% in trial I and 70.2% and 48.9% in trial II). The test plots were heavily infested with *R. solanacearum*. FSD1 and FSD2 treatments reduced the pathogen concentrations by 67.7% and 79.6%, respectively, and no significant differences between treatments could be discerned in the field trials. The FSD treatments were effective also against soilborne nematodes *Meloidogyne incognita*. The results showed that FSD technique could be applied for the control of soilborne fungal pathogens and nematodes (Mao et al. 2016).

Some of the soilborne pathogens are seedborne also, facilitating their long-distance dissemination through infected seeds. The effectiveness of low- and high-temperature dry heat, and hot water treatments in disinfesting cotton seed, of *F. oxysporum* f.sp. *vasinfectum* (*Fov*) was investigated. Naturally infected cotton seeds were air-heated at 30, 35 and

40°C for up to 24 weeks. In addition, freshly harvested seeds from bolls were inoculated with *Fov* race 4 and incubated in dry heat at 60, 70 and 80°C for 2 to 14 days, or were immersed in water at 90°C for 45 s to 3 min. Low- and high-temperature dry heat treatments did not effectively eliminate *Fusarium* spp. from cotton seed, although seed infection declined more rapidly with higher incubation temperatures. High temperature dry treatments significantly reduced seed vigor in both Pima (*Gossypium barbadense*) and Upland (*G. hirsutum*) cultivars, although *Fusarium* spp. was eliminated. Seed from all times of immersion in hot water were less frequently infected with *Fusarium* spp. than nontreated seed. Incidence of seed infection did not vary significantly among immersion times ranging from 75 s to 3 min. Immersion of inoculated seed in water at 90°C did not reduce germination or vigor with exposure times ≤ 20 s and ≤ 50 s for seeds of Pima and Upland cotton respectively. The results suggested that thermotherapy with hot water could be optimized to effectively eliminate *F. oxysporum* f.sp. *vasinfectum*, preventing the spread of the pathogen to uninfested areas through infected seed (Bennett and Colyer 2010).

V. dahliae, incitant of Verticillium wilt of olive, is spread primarily through infected nursery stocks. In order to preventively sanitize the propagation stock, several temperature-exposure time combinations were evaluated, using parameters such as pathogen survival on potato dextrose agar (PDA) medium, pathogen viability on infected shoots and plants, vegetative growth of olive cultivars and rooting ability of cuttings. Colonies of *V. dahliae* were killed at 40 and 47°C after 8 h and 60 min of exposure, respectively. Temperatures >42°C for at least 2 h were lethal for the pathogen infecting shoots. Similarly, moist hot air treatments at 42–44°C for 6–12 h eradicated the pathogen, without affecting viability of plants. Based on the effects of exposure to heat treatments (thermotolerance), the olive cultivars were grouped as sensitive, moderately sensitive and heat-tolerant. The optimized sanitation methods were effective in eliminating *V. dahliae* from all olive cultivars tested, regardless of their thermotolerance. The results showed that hot air treatment could be practiced for production of Verticillium wilt-free olive nursery plants (Morello et al. 2016).

Crown gall induced by *Agrobacterium (tumefaciens* biovar 3) *vitis*, was responsible for progressive loss of vigor and ultimate death of grapevine plants, resulting in heavy production losses. Treatment of dormant cuttings of grapevine cv. Thompson Seedless and rootstock NAZ3 (*Vitis vinifera* x *V. rupestris*) with hot water at 50°C for 30 min, resulted in effective eradication or reduction of pathogen population below detection level. Strains of *A. vitis* showed variation in their sensitivity to heat. They were more sensitive to heat, compared with strains of *A. tumefaciens* biovar 1 and 2. The pathogen population (10^3 CFU/ml) was killed after 30-min heat treatment. Nontumorigenic biovar 1 strains could be recovered from hot water-treated cuttings. Dormant cuttings were treated with hot water and maintained in a field nursery for a period of 9 months. The viability and growth of different cultivars/rootstocks were affected by hot water treatment

to different levels. Hot water treatment increased number and length of canes, root dry weight and diameter of trunks in most cultivars/rootstocks. Treatment time generally affected bud survival and, in most cases, increased the level of callus formation at the base of cuttings. *A. vitis* could be effectively eradicated by hot water treatment which is a simple, effective, economical and environmentally acceptable procedure (Mahmoodzadeh et al. 2003).

Thermotherapy has been shown to be effective in eradicating soilborne virus infections in seeds and plant materials and also in limiting the populations of vectors. Dry heat treatment of tomato seeds at 70°C for 3 days or at 80°C for 1 day effectively eliminated *Tomato mosaic virus* (ToMV) from less deeply seated seeds extracted freshly (Broadbent 1976). Similar dry heat treatment of tomato seeds at 78°C for 2 days eliminated *Tomato mosaic virus*. Heat treatment did not affect seed germination and it was more effective, compared to soaking seeds in trisodium phosphate recommended earlier for seed disinfestation (Green et al. 1987). In a later investigation, the effectiveness of heat treatment of tomato seeds at 85°C for 24 h in eliminating ToMV, without impairing seed germination was demonstrated (Noor-Hassan et al. 1988). The effectiveness of treatment of soil with steam and potassium hydroxide (exothermic conditions) against *Tobacco mosaic virus* (TMV) was assessed. Treatments were applied using a self-propelled soil-steaming machine designed for steam release, after incorporation in soil of a substance that caused exothermic reaction. Quantities of TMV recovered from treated soil were significantly less than that from untreated soil. The infectivity of TMV was also very low, indicating the possible denaturation of virus particles recovered from treated soil (Luvisi et al. 2006). *Beet necrotic yellow vein virus* (BNYVV) is transmitted by *P. betae*. Survival of resting spores of *P. betae* was determined under aerobic (30 min, 4 days and 21 days) and anaerobic (4 days) conditions, under several temperature regimes in a water suspension and in leachate extracted from an anaerobic compost heap. In water under aerobic conditions, the lethal temperature was 60, 55 and 40°C for exposure times of 30 min, 4 days and 21 days, respectively. The effect of compost leachate and/ or anaerobic conditions on survival of *P. betae* depended on temperature. After incubation for 4 days at 20°C under anaerobic conditions, survival of *P. betae* was significantly lower than survival under aerobic conditions in water, as well as in leachate. In leachate taken from an aerobic compost heap, aerobically incubated at 40°C for 4 days, survival of *P. betae* was significantly lower than survival in water at the same temperature (Van Rijn and Termorshuizen 2007).

Acid electrolyzed water (AEW) is a germicidal product of electrolysis of a dilute solution of sodium chloride (0.4% w/v). This solution effectively disinfested wheat seed or soil samples infested with teliospores of *T. indica*, incitant of Karnal bunt of wheat. The effectiveness of a 30-min treatment with AEW was compared with a 2-min treatment with sodium hypochlorite (NaOCl) used earlier to eradicate bacteria and nonsmut fungi. NaOCl and AEW were very effective in eliminating bacteria and fungi from soil extracts. AEW eliminated nearly

100% of bacteria and fungi from soil extract. Free chlorine levels in AEW were very low, suggesting that compounds other than chlorine might have a significant role in sanitation of soil by AEW. A standardized protocol involving a 30-min AEW treatment of wheat seed washes or soil extracts was developed to eliminate contaminating microorganisms. The germicidal properties of AEW revealed the potential of AEW, as an alternative for use as seed disinfectant (Bonde et al. 2003).

1.3.2 Radiation Treatments

Exposure to ultraviolet (UV) radiation may inactivate or kill the microorganisms depending on the dose and duration of exposure. Ultraviolet (UV) disinfection technology has been employed to prevent dispersal of pathogens infecting roots of plants grown in closed hydroponic cultures. The circulating solution was disinfected with different UV-doses (19, 38 and 88 mJ/cm) before circulation. The hydroponic solution was inoculated with *Pythium aphanidermatum* at 6.7 ± 1.5 CFU/ml. After four months, the population density of *P. aphanidermatum* reached 1,030 CFU/ml in the control and 1,028, 970, 610 and 521 CFU/ml in solution treated with 19, 38, 59 and 88 mJ/cm of UV respectively. Significant reduction of both *Pythium* and bacterial populations occurred at all UV doses. However, corresponding decrease in root rot incidence or increase in tomato fruit yield could not be seen. Bacterial populations in the rhizosphere were decreased, following treatment with UV. The results indicated that both pathogen and nontarget bacterial populations were adversely affected by UV treatments, but not to the level required to reduce infection by *P. aphanidermatum* (Zhang and Tu 2000).

The effects of high-power monochromatic (pulsed UV laser) and low-power broad band UV radiation on the survival of *Phytophthora* spp. were investigated. Cysts or zoospores of *Phytophtora capsici*, *P. citrophthora* and *P. nicotianae* were suspended in shallow layers of still water in plastic petridishes with its lid off, and they were exposed to different doses of UV emitted from an excimer laser or a mercury vapor lamp. Survival of the fungal pathogens was assessed by culturing aliquots of the treated or nontreated suspensions onto corn meal agar amended with ampicillin (250 ppm) and rifampicin (10 ppm) and counting the colonies developing on the medium. The UV dose (as energy/unit area) required to kill pathogen propagules was less, when they were suspended in distilled water, than when suspended in recycled nursery water. The reduction in kill-efficiency seemed to be related to UV-absorbing soluble chemicals. Cysts and zoospores were equally sensitive to UV, although the high-peak power pulsed UV-laser source with ultra-short exposure times appeared to have greater kill-efficiency than the conventional Hg-vapor UV-lamp. *P. capsici* and *P. nicotianae* isolates were less sensitive to UV, than isolates of *P. citrophthora* isolated from various hosts and geographical regions. Hyphae of *Phytophthora* spp. were less susceptible to UV than the cysts or zoospores (Banihashemi et al. 2010).

1.4 CHEMICAL TECHNIQUES

1.4.1 Treatment of Irrigation Water

Protective ability of an ozone-generative electrostatic spore precipitator against root-infecting propagules of Fusarium crown and root rot caused by *F. oxysporum* f.sp. *radicis-lycopersici* (FORL) and bacterial wilt caused by *R. solanacearum (Rs)* in tomato was assessed under hydroponic culture conditions. The device consisted of a cylindrical electrostatic spore precipitator (S2 cylinder) in which a positively charged straight conductor wire insulated with a transparent acrylic cylinder originated from a spore-precipitation cylinder (S1 cylinder) designed to physically control airborne conidia in the greenhouse. The S2 cylinder contained two sites for conidial attraction and ozone production. Hydroponic culture troughs to raise tomato seedlings were paralleled with S2 cylinder. The rhizosphere pathogens, FORL and *Rs* could be prevented from entering the hydroponic system during cultivation of tomatoes, whereas the airborne conidia of *Oidium neolycocpersici* were trapped in the spaces between the cylinders. The susceptible tomato plants in culture troughs were not infected by rhizosphere and aerial pathogens throughout the experimental duration (2–3 weeks) (Shimizu et al. 2007). Dissemination of *Phytophthora* spp. through irrigation systems and natural waterways is an important factor affecting plant health requiring attention for mitigation. Pressurized carbon dioxide (CO_2) inactivated zoospores of *P. nicotianae*. In order to determine the effectiveness of CO_2 at low pressure, the influence of injected CO_2 at 63 to 4,000 ppm on survival of zoospores of four *Phytophthora* spp. in irrigation water was investigated. Zoospore survival of *P. nicotianae*, *P. tropicalis* and *P. pini* was reduced by over 90% at 4,000 ppm and was reduced by 40% at 125 to 2,000 ppm, after a 2-h exposure. Survival of *P. megasperma* was less affected by injected CO_2 with a reduction of 37.1% at $\leq$ 4,000 ppm. The CO_2 treatment at 4,000 ppm for 30 or 120 min, of water infested with *P. nicotianae* at zoospore concentrations of 1,000 and 5,000/ml reduced disease incidence on annual vinca (*Carthamus roseus*) respectively by 92% and 75%. Treatment with CO_2 at 2,000 ppm also showed comparable inhibitory effect on zoospores. The CO_2 treatments at <2,000 ppm also significantly reduced disease induced by water-infested with 1,000 zoospores/ml. The results indicated the potential of CO_2 as a safe and effective water disinfectant for *Phytophthora* spp. (Kong 2013).

Disinfestation of irrigation water with different treatments has been attempted to reduce the risk of incidence of crop diseases. The efficacy of disinfestation treatments on a range of microbial plant pathogens at different stages of life cycle and in different water qualities was determined. *F. oxysporum*, *Phytophthora cinnamomi* and *Pythium aphanidermatum* and other five pathogens were exposed to chlorine (sodium hypochlorite), chlorine dioxide and ultra violet (UV) radiation at a range of application rates and dam water. The efficacy of treatments varied depending on exposure time, application rate, water type and pathogen and propagule types (spores, mycelium or cells). *P. cinnamomi* and *P. aphanidermatum* were most sensitive to all treatments, whereas *F. oxysporum* required the highest rates and longest exposure times to chlorine, chlorine dioxide and UV to kill >99% CFUs. Chlorine dioxide applied as a 'shock' treatment at a high rate for a limited time period and UV radiation showed more effective biocidal activity than chlorine levels tested in both water types. The results indicated that optimal rate and exposure time should be determined for each waterborne microbial plant pathogen to effectively disinfest the water used for irrigation (Scarlett et al. 2016).

1.4.2 TREATMENT OF SOIL

Soil alkaline/acidic conditions may be a critical factor limiting the development of some soilborne diseases. Elemental sulfur application reduces the soil pH, making the conditions unfavorable for the development of bacterial wilt of potato caused by *R. solanacearum*. Infection of potatoes by the pathogen could be reduced by applying sulfur to bring down the pH to 3.8 in June and the pH was maintained below 4.0 until November. Excess acidity was neutralized by addition of limestone and the pH was increased to 5.0 or more. Bacterial pathogen population was significantly reduced under such conditions in Florida, United States (Stevens and Stevens 1952). Likewise, development of *S. scabies*, causing potato scab disease could be suppressed by treating the soil with sulfur to reduce the pH, followed by liming the soil (Smiley 1975; Fry 1982). In contrast, enhancing the soil pH was effective in reducing clubroot disease caused by *P. brassicae* in cabbage, cauliflower and broccoli. Addition of lime to raise the pH to 7.2 inhibited the development of clubroot formation in the affected plants. The higher pH reduced spore germination, restricting the ability of the pathogen to induce the disease (Walker 1957). The extent of disease suppression due to lime application was found to be proportional to how thoroughly the lime was mixed with soil (Dobson et al. 1983). Acid soils are generally favorable to development of wilts caused by *F. oxysporum* and they may be suppressed by liming the soils to raise the pH to 7.0 (Cook and Baker 1983).

The mechanism of clubroot suppression by neutral soil pH was investigated by using a highly reproducible germination assay system under soil culture conditions. The germinated spores of *P. brassicae* could be identified by the absence of nucleus, following release of zoospore to infect the root hair of the susceptible plant. *Brassica rapa* var. *perviridis* seedlings were inoculated with a spore suspension of *P. brassicae* (2.0×10^6 spores/g of soil) and grown in the growth chamber for 7 days. The spores were recovered from the rhizosphere and non-rhizosphere soils and stained with both Fluorescent Brightener 28 (cell wall-specific) and SyTo82 orange fluorescent nucleic acid stain (nucleus-specific stain). The percentage of spores without a nucleus (germinated spores) increased significantly in the rhizosphere after 7 days' growth in chamber and the correlation with root-hair infections validated the assay system. Application of calcium-rich compost or calcium carbonate to neutralize the soil significantly reduced the percentage of germinated spores in the rhizosphere, as well as the number of root-hair infections. The results provided a direct evidence to substantiate that inhibition of spore germination was the primary cause of suppression of clubroot disease under neutral soil pH (Niwa et al. 2008). The effect of soil surface pH adjustment using calcium hydroxide on sclerotial germination of *S. minor*, incitant of lettuce drop disease and infection at the collar region of lettuce plants was investigated. Under glasshouse conditions, surface application of 2 to 10 t calcium hydroxide/ha raised the pH of the top 1-2 cm of a duplex sandy loam soil above 8.5 for at least 8 weeks without changing the soil pH within the transplant root zone. A linear relation was observed between the rate of calcium hydroxide applied and extent of disease suppression, with complete suppression at 10 t calcium hydroxide/ha. Under field conditions, two kinds of soil types (duplex sandy loam, pH 6.0 and red ferrosol, pH 6.9), treated with calcium hydroxide at 2.5 t/ha, maintained soil-surface pH above 8.5 for 1–3 weeks and provided up to 58% reduction in lettuce drop disease. Application of polyvinyl alcohol (a soil-conditioning polymer) over the calcium hydroxide layer appeared to reduce loss of calcium hydroxide by wind, but did not enhance the level of suppression of lettuce drop disease. Calcium hydroxide application combined with procyamidone-based fungicide drench was more effective in suppressing disease development than either of them alone (Wilson et al. 2005).

1.4.3 TREATMENT OF TOOLS

Crown gall induced by *A. tumefaciens* develops at grafting wounds in walnut seedlings during the 2-year nursery production. The efficiency of disinfectants was assessed to prevent the spread of *A. tumefaciens* via contaminated tools. The cationic surfactants, benzalkonium chloride (BC), cetyltrimethylammonium bromide (CTAB) and Physan 20 eliminated entirely (100%) of *A. tumefaciens* populations in water, when treated at 7, 5 and 2 ppm concentrations, respectively. Sodium hypochlorite eliminated 100% of *A. tumefaciens* population at 0.5 ppm concentration. However, the efficacy of sodium hypochlorite was reduced by 64% in the presence of total solids (0.7 g/ml), which were present under field conditions. By contrast, the efficiency of cationic surfactants decreased on an average only by 13% at similar concentrations of total solids. The effective elimination of *A. tumefaciens* was achieved after exposure of infested scalpels for 5 s in BC or CTAB at 5,000 ppm (0.5%) solutions. Treatment of scalpels with cationic solutions reduced gall formation on *Datura stramonium* (Yakabe et al. 2012).

REFERENCES

Abdel-Monaim MF, Abo-Elyousr KAM (2012) Effect of preceding and intercropping crops on suppression of lentil damping-off and root rot disease in New Valley-Egypt. *Crop Protect* 32: 41–46.

Abdelsamad NA, Baumbach J, Bhattacharyya MK, Leandro LF (2017) Soybean sudden death syndrome caused by *Fusarium virguliforme* is impaired by prolonged flooding and anaerobic conditions. *Plant Dis* 101: 712–719.

Agarwal VK, Sinclair JB (1996) *The Principles of Seed Pathology*, 2nd edition, CRC-Lewis Publishers, New York.

Allen TW, Workneh F, Steddom KC, Peterson GL, Rush CM (2008) The influence of tillage on dispersal of *Tilletia indica* teliospores from a concentrated point source. *Plant Dis* 92: 351–356.

Andret-Link P, Laporte C, Valat L et al. (2004) *Grapevine fan leaf virus*: Still a major threat to grapevine industry. *J Plant Pathol* 86: 183–195.

Atim M, Beed F, Tusiime G, Tripathi L, van Asten P (2013) High potassium, calcium and nitrogen application reduce susceptibility to banana Xanthomonas wilt caused by *Xanthomonas campestris* pv. *musacearum*. *Plant Dis* 97: 123–130.

Banihashemi Z, MacDonald JD, Lagunas-Solar MC (2010) Effect of high-power broadband UV radiation on *Phytophthora* spp. in irrigation water. *Eur J Plant Pathol* 127: 229–238.

Baumgartner K (2004) Root collar excavation for postinfection control of Armillaria root disease of grapevine. *Plant Dis* 88: 1235–1240.

Bennett RS, Colyer PD (2010) Dry heat and hot water treatments for disinfecting cotton seed of *Fusarium oxysporum* f.sp. *vasinfectum*. *Plan Dis* 94: 1469–1475.

Bithell SL, Butler RCC, McKay AC, Cromey MG (2013) Influences of crop sequence, rainfall and irrigation, on relationships between *Gaeumannomyces graminis* var. *tritici* and take-all in New Zealand wheat fields. *Austr Plant Pathol* 42: 205–217.

Blunt SJ, Asher MJC, Gilligan CA (1992) The effect of sowing date on sugar beet infection by *Polymyxa betae*. *Plant Pathol* 41: 148–153.

Bollig K, Specht A, Myint SS, Zahn M, Horst WJ (2013) Sulphur supply impairs spread of *Verticillium dahliae* in tomato. *Eur J Plant Pathol* 135: 81–96.

Bonde MR, Nester SE, Schaad NW, Frederick RD, Luster DG (2003) Improved detection of *Tilletia indica* teliospores in seed or soil by elimination of contaminating microorganisms with acidic electrolyzed water. *Plant Dis* 87: 712–718.

Borroto-Fernandez EG, Sommerbauer T, Popowich E, Schartl A, Laimer M (2009) Somatic embryogenesis from anthers of the autochthonous *Vitis vinifera* cv. Domina leads to *Arabis mosaic virus*-free plants. *Eur J Plant Pathol* 124: 171–174.

Brierley J, Hilton AJ, Wale SJ et al. (2015) Factors affecting the development and control of black dot on potato tubers. *Plant Pathol* 64: 162–177.

Broadbent L (1976) Epidemiology and control of *Tomato mosaic virus*. *Annu Rev Phytopathol* 14: 75–96.

Cai X, Zhang J, Wu M, Jiang D, Li G, Yang L (2015) Effect of water flooding on survival of *Leptosphaeria biglobosa* 'brassicae' in stubble of oilseed rape (*Brassica napus*) in China. *Plant Dis* 99: 1426–1433.

Canaday CH, Schmitthenner AF (2010) Effects of chloride and ammonium salts on the incidence of Phytophthora root and stem rot of soybean. *Plant Dis* 94: 758–765.

Chowdappa P, Nirmal Kumar BJ, Madhura S et al. (2015) Severe outbreaks of late blight on potato and tomato in South India caused by *Phytophthora infestans* population. *Plant Pathol* 64: 191–199.

Cintas NA, Webster RK (2001) Effects of rice straw management on *Sclerotium oryzae* inoculum, stem rot severity and yield of rice in California. *Plant Dis* 85: 1140–1144.

Cochran KA, Rothrock CS (2015) *Brassicca* green manure amendments for management of *Rhizoctonia solani* in two annual ornamental crops in the field. *HortScience* 50: 555–558.

Cohen R, Burger Y, Horev C, Koren A, Edelstein M (2007) Introducing grafted cucurbits to modern agriculture: The Israeli experience. *Plant Dis* 91: 916–923.

Cook RJ, Baker KF (1983) *The Nature and Practice of Biological Control of Plant Pathogens*, The American Phytopathological Society, St Paul, MN.

Copes WE, Blythe EK (2011) Rooting response of azalea cultivars to hot water treatment used for pathogen control. *HortScience* 46: 52–56.

Dabirian S, Inglis D, Miles CA (2017) Grafting watermelon and using plastic mulch to control Verticillium wilt caused by *Verticillium dahliae* in Washington. *HortScience* 52: 349–356.

Davis JR, Huisman OC, Everson DO, Nolte P, Sorensen LH, Schneider AT (2010a) The suppression of Verticillium wilt of potato using corn as a green manure crop. *Amer J Potato Res* 87: 195–208.

Davis JR, Huisman OC, Everson DO, Nolte P, Sorensen LH, Schneider AT (2010b) Ecological relationships of Verticillium wilt suppression of potato by green manures. *Amer J Potato Res* 87: 315–326.

de Souza-Diaz JAC, Betti JA (2003) A 12-year review of ELISA monitoring of major potato viruses in dormant seed tubers in Brazil. *Acta Hortic* 619: 153–159.

Deberdt P, Goze E, Coranson-Beaudu R et al. (2015) *Crotalaria spectabilis* and *Raphanus sativus* as previous crops show promise for the control of bacterial wilt of tomato without reducing bacterial populations. *J Phytopathol* 163: 377–385.

Delgado A, Franco GM, Páez JI, Vega JM, Carmona E, Avilés M (2005) Incidence of cotton seedling diseases caused by *Rhizoctonia solani* and *Thielaviopsis basicola* in relation to previous crop, residue management and nutrient availability in soils in SW Spain. *J Phytopathol* 153: 710–714.

Dobson RL, Gabrielson RL, Baker AS, Bennett L (1983) Effects of lime particle size and distribution and fertilizer formulation on clubroot disease caused by *Plasmodiophora brassicae*. *Plant Dis* 67: 50–62.

Dorrance AE, Kleinhenz MD, McClure SA, Tuttle NT (2003) Temperature, moisture and seed treatment effects on *Rhizoctonia solani* root rot of soybean. *Plant Dis* 87: 533–538.

Dusunceli F, Meyveci K, Cetin L et al. (2007) Determination of agronomic practices for the management of blight of chickpea caused by *Ascochyta rabiei* in Turkey: 1. Appropriate sowing date. *Eur J Plant Pathol* 119: 449–456.

Ebihara Y, Uematsu S, Nomiya S (2010) Control of *Verticillium dahliae* at a strawberry nursery by paddy-upland rotation. *J Gen Plant Pathol* 76: 7–20.

Edelstein M, Cohen R, Burger S, Shriber S, Pivonia S, Shtienberg D (1999) Integrated management of sudden wilt in melons caused by *Monosporoascus cannonballus*, using grafting and reduced rates of methyl bromide. *Plant Dis* 83: 1142–1145.

Elmer WH (2000) Comparison of plastic mulch and nitrogen form on the incidence of Verticillium wilt of eggplant. *Plant Dis* 84: 1231–1234.

Elmer WH, McGovern (2013) Epidemiology and management of Fusarium wilt of China asters. *Plant Dis* 97: 530–536.

Eranthodi A, Babadoost M, Trierweiler B (2010) Thermotherapy for control of fungal pathogens in propagative root stocks of horseradish. *HortScience* 45: 599–604.

Fall ML, van der Heyden H, Bordeur L, Leclerc Y, Moreau G, Carrise D (2015) Spatiotemporal variation in airborne sporangia of *Phytophthora infestans*: characterization and initiatives towards improving potato late blight risk estimation. *Plant Pathol* 64: 178–190.

Ferguson LM, Shew BB (2001) Wheat straw mulch and its impacts on three soilborne pathogens of peanut in microplots. *Plant Dis* 85: 661–667.

Food and Agriculture Organization (FAO) (1991) *Annex I: Plant Quarantine Principles as Related to International Trade. Report on Expert Consultation on Harmonization of Plant Quarantine Principles*, Food and Agriculture Organization, Rome, Italy, pp. 15–20.

Frederick ZA, Cummings TF, Johnson DA (2018) The effects of alfalfa residue incorporation on soil bacterial communities and the quantity of *Verticillium dahliae* microsclerotia in potato fields in Columbia Basin of Washington State, WA. *Amer J Potato Res* 95: 15–25.

Friberg H, Lagerlöf J, Rämert B (2006) Usefulness of nonhost plants in managing *Plasmodiophora brassicae*. *Plan Pathol* 55: 690–695.

Fry WE (1982) *Principles of Plant Disease Management*, Academic Press, New York.

Gambino G, Di Matteo D, Gribaudo I (2009) Elimination of *Grapevine fan leaf virus* from three *Vitis vinifera* cultivars by somatic embryogenesis. *Eur J Plant Pathol* 123: 57–60.

Gambino G, Gribaudo I (2006) Simultaneous detection of new grapevine viruses by multiplex reverse transcription-polymerase chain reaction with coamplification of a plant RNA as an internal control. *Phytopathology* 96: 1123–1129.

Garcia-Garza JA, Neumann S, Vyn TJ, Boland GJ (2002) Influence of crop rotation and tillage on production of apothecia by *Sclerotinia sclerotiorum*. *Canad J Plant Pathol* 24: 137–143.

Gatch Ew, du Toit LJ (2017) Limestone-mediated suppression of Fusarium wilt in spinach seed crops. *Plant Dis* 101: 81–94.

Green SK, Hwang LL, Kuo YJ (1987) Epidemiology of *Tomato mosaic virus* in Taiwan and identification of strains. *J Plant Dis Protect* 94: 386.

Grimault V, Prior P (1994) Grafting tomato cultivars resistant or susceptible to bacterial wilt: Analysis of resistance mechanisms. *J Phytopathol* 141: 330–334.

Guo XW, Fernando WGD, Entz M (2005) Effects of crop rotation and tillage on blackleg disease of canola. *Canad J Plant Pathol* 27: 53–57.

Gutteridge RJ, Jenkyn JF, Bateman GL (2006) Effects of different cultivated or weed grasses, grown as pure stands or in combination with wheat, on take-all and its suppression in subsequent wheat crops. *Plant Pathol* 55: 696–704.

Halterman D, Charkowski A, Verchot J (2012) Potato viruses and seed certification in the USA to provide healthy propagated tubers. *Pest Technol* 6 (spl issue): 1–14.

Hao J, Subbarao KV, Koike ST (2003) Effects of broccoli rotation on lettuce drop caused by *Sclerotinia minor* and on the population density of sclerotia in soil. *Plant Dis* 87: 159–166.

Hao W, Ahonsi MO, Vinatzer BA, Hong C (2012) Inactivation of *Phytophthora* and bacterial species in water by a potential energy saving heat treatment. *Eur J Plant Pathol* 134: 357–365.

Harnik TY, Mejia-Chang M, Lewis J, Garbelotto M (2004) Efficacy of heat-based treatments in eliminating the recovery of the sudden oak death pathogen (*Phytophthora ramorum*) from infected California bay laurel leaves. *HortScience* 39: 1677–1680.

Harrington MA, Dobinson KF (2000) Influences of cropping practices on *Verticillium dahliae* populations in commercial processing tomato fields in Ontario. *Phytopathology* 90: 1011–1017.

Harrison MD (1976) The effect of barley straw on the survival of *Verticillium albo-atrum* in naturally infected field soil. *Amer Potato J* 53: 385–394.

Hasna MK, Ögren E, Persson P, Mårtensson A, Rämert B (2009) Management of corky root disease of tomato in participation with organic tomato growers. *Crop Protect* 28: 155–161.

Heitefuss R (1989) *Crop and Plant Protection: The Practical Foundations*, Ellis Horwood Limited, Chichester, UK.

Himmelstein JC, Maul JE, Everts KL (2014) Impact of five cover crop green manures and Actinovate on Fusarium wilt of watermelon. *Plant Dis* 98: 965–972.

Hossain S, Bergkvist G, Glinwood R et al. (2015) Brassicaceae cover crops reduce Aphanomyces pea root rot without suppressing genetic potential of microbial nitrogen cycling. *Plant and Soil* 392: 227–238.

Huang YH, Wang RC, Li CH et al. (2012) Control of Fusarium wilt in banana with Chinese leek. *Eur J Plant Pathol* 134: 87–95.

Huber DM, Haneklaus S (2007) Managing nutrition to control diseases. *Landbauforsch Volk* 57: 313–322.

Hughes TJ, Koval NC, Esker PD, Grau CR (2009) Influence of monocropping brown stem rot-resistant and -susceptible soybean accessions on soil and stem populations of *Phialophora gregata* f.sp. *sojae*. *Plant Dis* 93: 1050–1058.

Hwang SF, Ahmed HU, Zhou Q, Turnbull GD, Strelkov SE, Gossen BD, Peng G (2015) Effect of host and nonhost crops on *Plasmodiophora brassicae* resting spore concentrations and clubroot of canola. *Plant Pathol* 64: 1198–1206.

Ikeda K, Banno S, Furusawa A, Shibata S, Nakaho K, Fujimora M (2015) Crop rotation with broccoli suppresses Verticillium wilt of eggplant. *J Gen Plant Pathol* 81: 77–82.

Jefferson PG, Gossen BD (2002) Irrigation increases Verticillium wilt incidence in a susceptible alfalfa cultivar. *Plant Dis* 86: 588–592.

Jenkyn JF, Gutteridge RJ, White RP (2014) Effects of break crops, and of wheat volunteers growing in break crops or in set-aside or conservation covers, all following crops of winter wheat, on the development of take-all (*Gaeumannomyces gaminis* var. *tritici*) in succeeding crops of winter wheat. *Ann Appl Biol* 165: 340–363.

Jensen JL (1888) The propagation and prevention of smut in oat and barley. *J Royal Soc (England)* 24: 397–415.

Johnson DA (2008) *Potato Health Management*, The American Phytopathological Society, St Paul, MN.

Keinath AP, Harrison HF, Marino PC, Jackson DM, Pullaro TC (2003) Increase in populations of *Rhizoctonia solani* and wirestem of collard with velvet bean cover crop mulch. *Plant Dis* 87: 719–725.

Keinath AP, Hassell RL (2014) Control of Fusarium wilt of watermelon by grafting onto bottlegourd or interspecific hybrid squash despite colonization of rootstocks by *Fusarium*. *Plant Dis* 98: 255–266.

Kelman A, Cook RJ (1977) Plant Pathology in the People's Republic of China. *Annu Rev Phytopathol* 15: 409–429.

Kendig SR, Rupe JC, Scott HD (2000) Effect of irrigation and soil water stress on densities of *Macrophomina phaseolina* in soil and roots of two soybean cultivars. *Plant Dis* 84: 895–900.

Kim CH (1989) Phytophthora blight and other diseases of red pepper in Korea: disease and pest problems from continuous cropping. II. Soilborne diseases. *FFTC Extn Bull* 302: 10–17.

Kirkpatrick MT, Rupe JC, Rothrock CS (2006) Soybean response to flooded soil conditions and the association with soilborne plant pathogenic genera. *Plant Dis* 90: 592–596.

Kluth C, Buhre C, Varrelmann M (2010) Susceptibility of intercrops to infection with *Rhizoctonia solani* AG2-2 IIIB and influence on subsequently cultivated sugar beet. *Plant Pathol* 59: 683–692.

Kommedahl T, Todd LA (1991) The environmental control of plant pathogens using eradication. In: Pimentel D (ed.), *Handbook of Pest Management in Agriculture*, CRC Press, Boca Raton, FL.

Kong P (2013) Carbon dioxide as a potential water disinfectant for Phytophthora disease risk mitigation. *Plant Dis* 97: 369–372.

Kousik CS, Thies JA (2010) Response of US bottle gourd (*Lagenaria siceraria*) plant introductions (PI) to crown rot caused by *Phytophthora capsici*. *Phytopathology* 100: S65.

Lai R, You M, Jiang L et al. (2011) Evaluation of garlic intercropping for enhancing the biological control of *Ralstonia solanacearum* in flue-cured tobacco fields. *Biocontr Sci Technol* 21: 755–764.

Lambert DH, Salas B (2001) Pink rot. In: Stevenson WR, Loria R, France GD, Weingartner DP (eds.), Compendium of Potato Diseases, The American Phytopathological Society, St Paul, MN.

Lamour KH, Hausbeck MK (2003) Effect of crop rotation on the survival of *Phytophthora capsici* in Michigan. *Plant Dis* 87: 841–845.

Lamprecht Sc, Tewoldemedhin YT, Hardy M, Calitz FJ, Mazzola M (2011) Effect of cropping system on composition of the *Rhizoctonia* populations recovered from canola and lupin in a winter rainfall region of South Africa. *Eur J Plant Pathol* 131: 305–316.

Larkin RP (2015) Soil health paradigms and implications for disease management. *Annu Rev Phytopathol* 53: 199–221.

Larkin RP, Griffin TS, Honeycutt CW (2010) Rotation and cover crop effects on soilborne potato diseases, tuber yield and soil microbial communities. *Plan Dis* 94: 1491–1502.

Larkin RP, Honeycutt CW (2006) Effects of different 3-year cropping systems on soil microbial communities and Rhizoctonia diseases of potato. *Phytopathology* 96: 68–79.

Larkin RP, Honeycutt CW, Olanya OM (2011) Management of Verticillium wilt of potato with disease-suppressive green manures and as affected by previous cropping history. *Plant Dis* 95: 568–576.

Lee CH, Kim CH, Nam KW (1991) Suppression of Phytophthora blight incidence of red pepper by cropping blight incidence on red pepper by cropping systems. *Korean J Plant Pathol* 7: 140–146.

Lee HU, Kim CH, Lee FJ (1990) Effect of pre- and mixed cropping with nonhost plants on incidence of Phytophthora blight of red pepper. *Korean J Plant Pathol* 6: 440–446.

Leoni C, de Vries M, ter Braak CJF, van Bruggen AHC, Rossing WAH (2013) *Fusarium oxysporum* f.sp. *cepae* dynamics: in-plant multiplication and crop sequence simulations. *Eur J Plant Pathol* 137: 545–561.

Leoni C, ter Braak CJF, Gilsanz JC, Dogliotti S, Rossing WAH, van Bruggen AHC (2014) *Sclerotium rolfsii* dynamics in soil as affected by crop sequences. *Appl Soil Ecol* 75: 95–105.

Lin CH, Hsu ST, Tzeng KC, Wang JF (2008) Application of a preliminary screen to select locally adapted resistant rootstock and soil amendment for integrated management of tomato bacterial wilt. *Plant Dis* 92: 909–916.

Liu B, Gumpertz ML, Hu S, Ristaino JB (2008) Effect of prior tillage and soil fertility amendments on dispersal of *Phytophthora capsici* and infection of pepper. *Eur J Plant Pathol* 120: 273–287.

López-Escudero FJ, Blanco-López MA (2005) Effects of drip irrigation on population of *Verticillium dahliae* on olive orchards. *J Phytopathol* 153: 238–239.

Luvisi A, Materazzi A, Triolo E (2006) Exothermic reactions and steam for the management of *Tobacco mosaic virus* in soil. *Agr Med* 136: 63–69.

Macdonald AJ, Gutteridge RJ (2012) Effects of take-all (*Gaeumannomyces graminis* var. *tritici*) on crop N uptake and residual mineral N in soil at harvest of winter wheat. *Plant and Soil* 350: 253–260.

MacGuidwin AE, Knuteson DL, Connell T, Bland WL, Bartelt KD (2012) Manipulating inoculum densities of *Verticillium dahliae* and *Pratylenchus penetrans* with green manure amendments and solarization influence potato yield. *Phytopathology* 102: 519–527.

Mahmoodzadeh H, Nazemieh A, Majidi I, Kalighi A (2003) Effects of thermotherapy treatments on systemic *Agrobacterium vitis* in dormant grape cuttings. *J Phytopathol* 151: 481–484.

Mao L, Wang Q, Yan D et al. (2016) Flame soil disinfestation: A novel, promising, nonchemical method to control soilborne nematodes, fungal and bacterial pathogens in China. *Crop Protect* 83: 90–94.

Marcroft SJ, Van de Wouw AP, Salisbury PA, Potter TD, Howlett BJ (2012) Effect of rotation of canola (*Brassica napus*) cultivars with different complements of blackleg resistance genes on disease severity. *Plant Pathol* 61: 934–944.

Matheron ME, McCreight JD, Tickes BR, Porchas M (2005) Effect of planting date, cultivar and stage of plant development on incidence of Fusarium wilt of lettuce in desert production fields. *Plant Dis* 89: 565–570.

Matheron ME, Porchas M (2009) Impact of different preplant cultural treatments on survival of *Phytophthora nicotianae* in soil. *Plant Dis* 93: 43–50.

Matheron ME, Porchas M (2010) Evaluation of soil solarization and flooding as management tools for Fusarium wilt of lettuce. *Plant Dis* 94: 1323–1328.

Mengistu A, Yin X, Bellaloui N et al. (2016) Potassium and phosphorus have no effect on severity of charcoal rot of soybean. *Canad J Plant Pathol* 38: 174–182.

Messenger BJ, Menge JA, Pond E (2000) Effects of gypsum soil amendments on avocado growth, soil drainage and resistance to *Phytophthora cinnamomi*. *Plant Dis* 84: 612–614.

Messiha NAS, van Bruggen AHC, van Diepeningen AD, de Vos OJ, Termorshuizen AJ, Tjou-Tam-Sin NNA (2007) Potato brown rot incidence and severity under different management and amendment regimes in different soil types. *Eur J Plant Pathol* 119: 367–381.

Meyer MD, Hausbeck MK (2012) Using cultural practices and cultivar resistance to manage Phytophthora crown rot on summer squash. *HortScience* 47: 1080–1084.

Morello P, Diez CM, Codes M et al. (2016) Sanitation of olive plants infected by *Verticillium dahliae* using heat treatments. *Plant Pathol* 65: 412–421.

Moreno B, Jacob C, Rosales M, Krarup C, Contreras S (2016) Yield and quality of grafted watermelon grown in a field naturally infested with Fusarium wilt. *HortTechnology* 26: 453–459.

Moussart A, Even MN, Lesne A, Tivoli B (2013) Successive legumes tested in a greenhouse crop rotation experiment modify the inoculum potential of soils naturally infested by *Aphanomyces euteiches*. *Plant Pathol* 62: 545–551.

Murakami H, Tsushima S, Akimoto T, Murakami K, Goto I, Shishido Y (2000) Effects of growing leafy daikon (*Raphanus sativus*) on populations of *Plasmodiophora brassicae* (clubroot). *Plant Pathol* 49: 584–589.

Narayanasamy P (2002) *Microbial Plant Pathogens and Crop Disease Management*, Science Publishers, Enfield, NH.

Narayanasamy P (2011) *Microbial Plant Pathogens – Detection and Disease Diagnosis*. Springer Science, Heidelberg, Germany.

Narayanasamy P (2013) *Biological Management of Diseases of Crops*, Volume 2, Springer Science + Business Media, Dordrecht, The Netherlands.

Narayanasamy P (2017) *Microbial Plant Pathogens–Detection and Management in Seeds and Propagules*, Volume 2, Wiley-Blackwell, Chichester, UK.

Navarro L (1993) Status reports: Budwood registration programs. *Proc 12th Conf IOCV*, Riverside, CA, pp. 383–391.

Niwa R, Nomura Y, Osaki M, Ezawa T (2008) Suppression of clubroot disease under neutral pH caused by inhibition of spore germination of *Plasmodiophora brassicae* in the rhizosphere. *Plant Pathol* 57: 445–452.

Njoroge SMC, Kabir Z, Martin FN, Koike ST, Subbarao KV (2009) Comparison of crop rotation for Verticillium wilt management and effect on *Pythium* species in conventional and organic strawberry production. *Plant Dis* 93: 519–527.

Njoroge SMC, Riley MB, Keinath AP (2008) Effect of incorporation of *Brassica* spp. residues on population densities of soilborne microorganisms and on damping-off and Fusarium wilt of watermelon. *Plant Dis* 92: 287–294.

Noor-Hassan R, Krstic B, Tosic M (1988) Contribution to the knowledge of tomato seed infection by mosaic virus and possibility of disinfection. *Zastida-Bilija* 39: 393–399.

Ocimati W, Tusiime G, Opio F, Ugen MA, Buruchara R (2017) Sorghum (*Sorghum biocolor*) as a bean intercrop or rotation crop contributes to the survival of bean root rot pathogens and perpetuation of bean root rots. *Plant Pathol* 66: 1480–1486.

Oddino CM, Marinelli AD, Zuza M, March GJ (2008) Influence of crop rotation and tillage on incidence of brown root rot of peanut caused by *Fusarium solani* in Argentina. *Canad J Plant Pathol* 30: 575–580.

Ojaghian MR, Cui ZQ, Xie G-L, Li B, Zhang J (2012) Brassica green manure rotation crops reduce potato stem rot caused by *Sclerotinia sclerotiorum*. *Austr Plant Pathol* 41: 347–349.

Olanya OM, Porter GA, Lambert DH, Larkin RP, Starr GC (2010) The effects of supplemental irrigation and soil management on potato tuber diseases. *J Plant Pathol* 9: 65–72.

Papavizas GC, Lewis JA (1979) Integrated control of *Rhizoctonia solani*. In: Schippers B, Grams W (eds.), *Soilborne Plant Pathogens*, Academic Press, New York, pp. 425–434.

Paplomatas EJ, Elena K, Tsagkarakou A, Perdikaris A (2002) Control of Verticillium wilt of tomato and cucurbits through grafting of commercial varieties on resistant rootstocks. *Acta Hortic* 529: 445–449.

Paulitz TC, Schroeder KL, Schillinger WF (2010) Soilborne pathogens of cereals in an irrigated cropping system: Effects of tillage, residue management, and crop rotation. *Plant Dis* 94: 61–68.

Peng G, Lahali R, Hwang S-F et al. (2014) Crop rotation, cultivar resistance, and fungicides/biofungicides for managing clubroot (*Plasmodiophora brassicae*) on canola. *Canad J Plant Pathol* 36: (Supl. 1) 99–112.

Pérez-Rodríguez M, Alcántara E, Amaro M et al. (2015) The influence of irrigation frequency on the onset and development of Verticillium wilt of olive. *Plant Dis* 99: 488–495.

Pérez-Rodríguez M, Serrano N, Arquero O, Orgaz F, Moral J, López-Escudero FJ (2016) The effect of short irrigation frequencies on the development of Verticillium wilt in susceptible olive cultivar 'Picual' under field conditions. *Plant Dis* 100: 1880–1888.

Peters RD, Al-Mughrabi KI, Kalischuk ML et al. (2014) Characterization of *Phytophthora infestans* population diversity in Canada reveals increased migration and genotype recombination. *Canad J Plant Pathol* 36: 73–82.

Peters RD, Sturz AV, Carter MR, Sanderson JB (2005) Crop rotation can confer resistance to potatoes from *Phytophthora erythroseptica* attack. *Canad J Plant Sci* 85: 523–528.

Peterson GL, Kosta KL, Glenn DL, Phillips JG (2008) Utilization of soil solarization for eliminating viable *Tilletia indica* teliospores from Arizona wheat fields. *Plant Dis* 92: 1604–1610.

Piccinni G, Shriver JM, Rush CM (2001) Relationship among seed size, planting date, and common root rot in hard red winter wheat. *Plant Dis* 85: 973–976.

Radewald KC, Ferrin DM, Stanghellini ME (2004) Sanitation practices that inhibit reproduction of *Monosporascus cannonballus* in melon root left in the field after crop termination. *Plant Pathol* 53: 660–668.

Ristaino JB, Parra G, Campbell CL (1997) Suppression of Phytophthora blight in bell pepper by a no-till wheat cover crop. *Phytopathology* 87: 242–249.

Rivard CL, O'Connell S, Peet MM, Welker RM, Louws FJ (2012) Grafting tomato to manage bacterial wilt caused by *Ralstonia solanacearum* in the south eastern United State. *Plant Dis* 96: 973–978.

Rush CM, Heidel GB (1995) *Furovirus* diseases of sugar beets in the United States. *Plant Dis* 79: 868–875.

Rutikanga A, Sivirihauma C, Ocimati W et al. (2016) Breaking the cycle of *Xanthomonas campestris* pv. *musacearum* in infected fields through the cultivation of annual crops and disease control in adjacent fields. *J Phytopathol* 164: 659–670.

Sağir A, Eylen M, Pirinc V (2005) Effect of irrigation methods on pepper crown rot disease caused by *Phytophthora capsici*. *Internatl J Agric Biol* 7: 804–806.

Sakuma F, Maeda M, Takahashi M, Hashizune K, Kondo N (2011) Suppression of common scab of potato caused by *Streptomyces turgidiscabies* using lopsided oat green manure. *Plant Dis* 95: 1124–1130.

Sanderson KR, Peters RD, Monaghan FDE, Fillmore SAE (2013) Carrot cultivar response to foliage side-trimming for suppression of Sclerotinia rot. *Canad J Plant Pathol* 35: 279–287.

Sato N, Takamatsu T (1930) Grafting culture of watermelon. *Nogyo sekai* 25: 24–28 (in Japanese).

Scarlett K, Collins D, Tesoriero L, Jewell L, van Ogtrop F, Daniel R (2016) Efficacy of chlorine, chlorine dioxide and ultraviolet radiation as disinfectants against plant pathogens in irrigation water. *Eur J Plant Pathol* 145: 27–38.

Schillinger WF, Paulitz TC (2006) Reduction of Rhizoctonia bare patch in wheat with barley rotations. *Plant Dis* 90: 302–306.

Schoeny A, Devienne-Barret F, Jeuffroy M-H, Lucas P (2003) Effect of take-all root infections on nitrate uptake in winter wheat. *Plant Pathol* 52: 52–59.

Schoeny A, Jumel S, Ronault F, Lemarchand E, Tivoli B (2010) Effect and underlying mechanisms of pea-cereal intercropping on the epidemic development of Ascochyta blight. *Eur J Plant Pathol* 126: 317–331.

Schumann GL (1991) *Plant Diseases: Their Biology and Social Impact*, The American Phytopathological Society, St Paul, MN.

Schurt DA, Lopes UP, Duarte HSS, Rodrigues FA (2015) Rice resistance to sheath blight mediated by potassium. *J Phytopathol* 163: 310–313.

Serrano MS, De Vita P, Fernández-Rebollo P, Hernández MES (2012) Calcium fertilizers induce soil suppressiveness to *Phytophthora cinnamomi* root rot of *Quercus ilex*. *Eur J Plant Pathol* 132: 271–279.

Shah FA, Falloon RE, Butler RC, Lister RA, Thomas SM, Curtin D (2014) Agronomic factors affect powdery scab of potato and amounts of *Spongospora subterranea* DNA in soil. *Austr Plant Pathol* 43: 679–689.

Sharma AK, Babu KS, Sharma RK, Kumar K (2007) Effect of tillage practices on *Tilletia indica* Mitra (Karnal bunt disease of wheat) in rice-wheat rotation of the Indo-Gangetic Plains. *Crop Protect* 26: 818–821.

Shelvin E, Mahrer Y, Katan J (2004) Effect of moisture on thermal inactivation of soilborne pathogens under structural solarization. *Phytopathology* 94: 132–137.

Shimizu K, Matsuda Y, Nonomura T et al. (2007) Dual protection of hydroponic tomatoes from rhizosphere pathogens *Ralstonia solanacearum*, and *Fusarium oxyspoum* f.sp. *radicis-lycopersici* and airborne conidia of *Oidium neolycopersici* with an ozone-generative electrostatic spore precipitator. *Plant Pathol* 56: 987–997.

Shtienberg D, Elad Y Bornstein M, Ziv G, Grava A, Cohen S (2010) Polyethylene mulch modifies greenhouse microclimate and reduces infection of *Phytophthora infestans* in tomato and *Psesudoperonospora cubensis* in cucumber. *Phytopathology* 100: 97–104.

Smiley RW (1975) Forms of nitrogen and pH in the root zone and their importance to root infections. In: Brueh GW (ed.), *Biology and Control of Soilborne Pathogens*, The American Phytopathological Society, St Paul, MN, pp. 55–62.

Smiley RW, Machado S, Rhinhart KEL, Reardon CL, Wuest SB (2016) Rapid quantification of soilborne pathogen communities in wheat-based long-term experiments. *Plant Dis* 100: 1692–1708.

Soltani N, Conn KL, Abbasi PA, Lazarovits G (2002) Reduction of potato scab and Verticillium wilt with ammonium lignosulfonate soil amendment in four Ontario potato fields. *Canad J Plant Pathol* 24: 332–339.

Sprague SJ, Kirkegaard JA, Marcroft SJ, Graham JM (2010) Defoliation of *Brassica napus* increases severity of blackleg caused by *Leptosphaeria maculans*: implications for dual-purpose cropping. *Ann Appl Biol* 157: 71–80.

Stanghellini ME, Nielsen CJ, Kim DH, Ramussen SL, Rorbaugh PA (2000) Influence of sub-versus top irrigation and surfactants in a recirculating system on disease incidence caused by *Phytophthora* spp. in potted pepper plants. *Plant Dis* 84: 1147–1150.

Stevens NE, Stevens RB (1952) *Disease in Plants, Chronica Botanica*, Waltham, MA.

Struik PC (2007) The canon of potato science: 25. Minitubers. *Potato Res* 50: 305–308.

Subbarao KV, Kabir Z, Martin FN, Koike ST (2007) Management of soilborne diseases in strawberry using vegetable rotations. *Plant Dis* 91: 964–972.

Sugimoto T, Watanabe K, Furiki M et al. (2009) The effect of potassium nitrate on the reduction of Phytophthora stem rot disease of soybean, the growth rate and zoospore release of *Phytophthora sojae*. *J Phytopathol* 157: 379–389.

Sugimoto T, Watanabe K, Yoshida S et al. (2008) Select calcium compounds reduce the severity of Phytophthora stem rot of soybean. *Plant Dis* 92: 1559–1565.

Sugimoto T, Watanabe K, Yoshida S et al. (2007) The effects of inorganic elements on the reduction of Phytophthora stem rot disease of soybean, the growth rate and zoospore release of *Phyotphthora sojae*. *J Phytopathol* 155: 97–107.

Sugimoto T, Watanabe K, Yoshida S et al. (2010) Field application of calcium to reduce Phytophthroa stem rot of soybean, and calcium distribution in plants. *Plant Dis* 94: 812–819.

Sutic D, Milovanovic M (1980) Some factors affecting epidemiology of sugar beet rhizomania-like disease. *Proc 5th* Cong Mediterr Phytopathol Union, Patras, Greece, pp. 20–30.

van Loenen MCA, Turbett Y, Mullins CE et al. (2003) Low temperature-short duration steaming of soil kills soilborne pathogens, nematode pests and weeds. *Eur J Plant Pathol* 109: 993–1002.

Van Rijn E, Termorshuizen AJ (2007) Eradication of *Polymyxa betae* by thermal and anaerobic conditions and in the presence of compost leachate. *J Phytopathol* 155: 544–548.

Vartanian VG, Endo RJ (1985) Survival of *Phytophthora infestans* in seeds extracted from tomato fruits. *Phytopathology* 75: 375.

Vick CM, Bond JP, Chong SK, Russin JS (2006) Response of soybean sudden death syndrome to tillage and cultivar. *Canad J Plant Pathol* 28: 77–83.

Walker JC (1957) *Plant Pathology*, McGraw-Hill, New York.

Wang B, Jeffers SN (2002) Effects of cultural practices and temperature on Fusarium root and crown rot of container-grown hostas. *Plant Dis* 86: 225–231.

Wang HR, Ru SJ, Wang LP, Fong ZM (2004) Study on the control of Fusarium wilt and Phytophthora blight in cucumber by grafting. *Acta Agr Zhejiangensis* 16: 336–339 (in Chinese).

Wei T, Lu G, Clover GRG (2009) A multiplex RT-PCR for the detection of *Potato yellow vein virus*, *Tobacco rattle virus* and *Tomato infectious chlorosis virus* in potato with a plant internal amplification control. *Plant Pathol* 58: 203–209.

Wilson CR, DeLittle JA, Wong JAL, Schupp PJ, Gibson LJ (2005) Adjustment of soil-surface pH and comparison with conventional fungicide treatments for control of lettuce drop (*Sclerotinia minor*). *Plant Pathol* 54: 393–400.

Wright J, Lees AK, van der Waals JE (2012) Detection and eradication of *Spongsopora subterrranea* in minituber production in tunnels. *S Afr J Sci* 108 (5/6)

Wu BM, Koike ST, Subbarao KV (2011) Impact of consumer driven changes to crop production practices on lettuce drop caused by *Sclerotinia sclerotiorum* and *S. minor*. *Phytopathology* 101: 340–348.

Wu BM, Subbarao KV (2003) Effects of irrigation and tillage on temporal and spatial dynamics of *Sclerotinia minor* sclerotia and lettuce drop incidence. *Phytopathology* 93: 1572–1580.

Wyenandt CA, Riedel RM, Rhodes LH, Bennett MA, Nameth SGP (2011) Fall- and spring-sown cover crop mulches affect yield, fruit cleanliness, and Fusarium fruit rot development in pumpkin. *HortTechnology* 21: 343–354.

Xiao CL, Subbarao KV (2000) Effects of irrigation and *Verticillium dahliae* on cauliflower root and shoot growth dynamics. *Phytopathology* 90: 995–1004.

Xie J, Cardenas ES, Sammis TW, Wall MM, Lindsey DL, Murray LW (1999) Effects of irrigation method on chile pepper yield and Phytophthora root rot incidence. *Agric Water Manag* 42: 127–142.

Xu W, Liu D, Wu F, Liu S (2015) Root exudates of wheat are involved in suppression of Fusarium wilt in watermelon-wheat companion cropping. *Eur J Plant Pathol* 141: 209–216.

Yakabe LE, Parker SR, Kluepfel DA (2012) Cationic surfactants: Potential surface disinfectants to manage *Agrobacterium tumefaciens* biovar 1. *Plant Dis* 96: 409–415.

Yang RG, Chang Z (1998) Study on the control of pineapple heart disease by sunning the propagative materials. *South China Fruits* 27: 35.

Yang XB (2002) *Timing Tillage for Disease Control*. Available at www.ipm.iastate.edu/ipm/icm/2002/10-21-2002/timetill.html

Zhang W, Tu JC (2000) Effect of ultraviolet disinfection of hydroponic solutions on Pythium root rot and nontarget bacteria. *Eur J Plant Pathol* 106: 415–421.

Zhou XG, Everts KL (2007) Effects of host resistance and inoculum density on suppression of Fusarium wilt of watermelon induced by hairy vetch. *Plant Dis* 91: 92–96.

2 Management of Soilborne Microbial Plant Pathogens

Improvement of Host Plant Resistance

Improvement of host plant resistance to diseases caused by soilborne microbial plant pathogens has immense practical importance. Development of cultivars with acceptable level of resistance to soilborne microbial pathogens is considered as the best approach, because it can reduce or avoid the cost and effects of other chemical, physical, biological, cultural and regulatory control methods, in addition to being eco-friendly. Host resistance indicates the magnitude or capacity of a plant species or cultivar to reduce activity or harmful effects of a microbial pathogen. Plant disease resistance has been studied from different angles to integrate evolutionary, genetic, epidemiologic and economic conceptual framework. Plants are exposed to several microbial pathogens present in the soil environment. However, they are susceptible to only a small proportion of the pathogens, indicating the existence of effective defense mechanisms that prevent infection by large proportions of microbial pathogens. It appears that resistance of plants to microbial pathogens is the rule, while susceptibility is an exception. Plant defense systems operate at different stages of infection by microbial pathogens. The first line of defense is expressed at whole plant level, where cells containing thick cuticle, thickened walls and other physical structures may obstruct pathogen entry/further development. The second line of defense may be exhibited at cellular level, where a well-coordinated, controlled and dynamic regulation of gene activity, results in the synthesis of different kinds of proteins and antimicrobial compounds that arrest the progress of infection. Both passive (preformed) and active nature of defense processes are involved in the prevention of pathogen entry into the plant for deriving nutrition for its growth and reproduction. Two major forms of resistance have been differentiated and they are designated differently, as constitutive and inducible, qualitative and quantitative, race-specific and race-nonspecific and inoculum-reducing and rate-limiting resistance to describe similar phenomena, using overlapping and sometimes confusing terms (Narayanasamy 2002, 2008).

2.1 TYPES OF DISEASE RESISTANCE

A crop plant species contains several varieties or genotypes with variations in different characteristics, such as height, size, duration, yield potential and levels of resistance/susceptibility to microbial pathogens and other biotic and abiotic stresses. Likewise, the microbial plant pathogens exist in the form of *formae speciales*, varieties, races or biotypes, varying in pathogenicity and survivability in soil environments. A

host population, comprising individuals possessing recognizable property of resistance in common is termed pathodeme, and a pathogen population consisting of individuals with distinguishable attribute of virulence in common is designated pathotype (Robinson 1969). Disease resistance is expressed in two forms in plants: host resistance (HR) and nonhost resistance (NHR). Nonhost resistance is defined as the resistance exhibited by a plant species against all pathogens nonadapted to the plant species concerned. This resistance of the plant species is effective against all known isolates (strains) of the nonadapted pathogen. The NHR is the plant resistance that is species-specific and most durable form of resistance of plants (Heath 2001; Mysore and Ryu 2004). Host resistance is the ability of the plant species to exhibit varying responses to different plant pathogen species/varieties/strains/biotypes and varies from high level of resistance to high susceptibility which may be determined by applying standardized disease severity rating scale or by quantifying pathogen DNA contents/populations. Host resistance may be species-, or race-specific and this type of resistance is associated with the gene-for-gene relationship (Flor 1955; Király et al. 2007). A pathogen-specific protein product (effector) produced by the avirulence gene of the pathogen is recognized by the matching gene for resistance in the host plant. Such a recognition leads to induction of signal transduction in the host plant, initiating the defense responses, followed by inhibition of pathogen development in the resistant plant (Tyler 2005).

2.1.1 VERTICAL RESISTANCE

Two broad types of resistance of plants to microbial pathogens, as vertical resistance and horizontal resistance, were distinguished by Vanderplank (1963). These two types of resistance may be differentiated, based on the interactions between the isolates of *Phytophthora infestans*, causal agent of potato late blight disease and potato genotypes (cultivars). When a potato cultivar is inoculated with different races of *P. infestans*, differences in the resistance/susceptibility of the cultivar to the pathogen races are parallel to the vertical axis, representing the resistance levels. The levels of resistance are represented by vertical columns with proportionate heights for the resistance levels. The potato cultivar, having dominant *R1* gene is resistant to one (0) of the races (pathotypes) of *P. infestans* that does not have a matching virulence gene. The unmatched *R1* gene triggers resistance mechanisms in this cultivar. Likewise, the potato cultivar with resistance gene

R2 shows resistance to infection by race that does not have matching virulence gene. *P. infestans* was able to overcome the effects of resistance gene *R1* from *Solanum desmissum* within about 10 years of its incorporation into potato lines (Vanderplank 1963). During the early 1930s, plants with *R1* gene were colonized by *P. infestans*, indicating the formation of a new physiologic race and the new race was named as 'race 1' and the race that could not colonize plants with *R1* gene as 'race 0'. Additional resistance genes were identified in *S. desmissum* and other *Solanum* spp. (Salaman 1949; Black 1960). *P. infestans* produced new races that could match all resistance genes incorporated into potato plants, which could be infected by newly formed races. *P. infestans* seemed to be highly adaptable by producing new races armed with new virulence gene(s) to match the *R* genes of potato plants, making the attempt to employ 'R' genes to protect potato crops to be futile.

P. infestans produced new races at a faster rate than the researchers could produce cultivars with unmatchable resistance gene(s). Vertical resistance may be either stronger or weak, depending on the response of the pathotypes to negative selection pressure. The genes *R1*, *R2* and *R3* provided strong vertical resistance to potato cultivars, whereas cultivars with *R4* gene exhibited weak vertical resistance to *P. infestans* (Vanderplank 1984). Vertical resistance in some crop cultivars has been shown to be durable. But such durable vertical resistance was effective only in certain geographical locations and became ephemeral in other locations. The vertical resistance gene *R1* effective against tomato wilt pathogen *Fusarium oxysporum* f.sp. *lycopersici* protected tomato plants for many decades in the United States, but its effectiveness was lost very rapidly in northern Africa (Robinson 1987). Vertical resistance against potato wart disease caused by *Synchytricum endobioticum* and against cabbage yellows disease caused by *F. oxysporum* f.sp. *conglutinans* was found to be durable in the United Kingdom and Holland and North America, respectively. However, the prevalence of a matching pathotype of *F. oxysporum* f.sp. *conglutinans* in California was reported by Ramirez-Villapudua et al. (1985).

2.1.2 Horizontal Resistance

Two potato cultivars inoculated with different races of *P. infestans* may not show distinct variations in the level of resistance, in contrast to cultivars with vertical resistance. But a moderate level of resistance to two or more races of *P. infestans* may be exhibited by the cultivar with horizontal resistance. One cultivar may be more resistant to all races tested, than the other cultivar. Differences in horizontal resistance will be parallel to the horizontal axis. Resistance is not due to the presence of an unmatched *R* gene in the host, as in the case of vertical resistance. Horizontal resistance is independent of the virulence of the pathogen race. Horizontal resistance can be recognized by exposing cultivars to infection in the field, where virulent races, to which the cultivars are susceptible, are available. The resistance, which remains after elimination of vertical resistance, is horizontal resistance (Vanderplank

1963). Horizontal resistance is essentially due to polygenic action and quantitative inheritance. Grades of resistance are not distinctly defined and vary continuously. Horizontal resistance has been variously named as nonspecific, field, general or polygenic resistance by different researchers. Horizontal resistance may be due to a combination of many mechanisms, resulting in reduction in rates of infection and invasion of host and/or reproduction of the pathogen. As horizontal resistance is governed by polygenes, the level of horizontal resistance is proportional to the concentration (frequency) of polygenes in the cultivar. The complete set of polygenes may be located in different individual plants in a generally mixed population. It may be very difficult to attain high level of horizontal resistance, as the traditional gene transfer techniques cannot be employed as in the case of vertical resistance, which is governed by a single gene. A number of parents with low levels of resistance to the target pathogen may be crossed in all possible combinations. The resistance genes may be gradually accumulated by repeatedly crossing selected progenies over a number of generations. The diallelic selective mating system was proposed for this purpose by Jensen (1970).

Resistance to diseases caused by microbial pathogens may be due to mechanisms controlled by *R* genes at one or a few loci, designated monogenic, oligogenic or polygenic (major gene) resistance. In addition, resistance may be expressed at different stages of plant or organ development. Adult plant resistance develops, when the plant reaches certain growth stage in its development. In general, young plants (seedlings) are highly susceptible to viruses, but the level of resistance to the virus increases, as the plants become older. So, the vegetable seedlings (in the nursery) are susceptible to damping-off caused by *Pythium* spp. and this disease does not occur generally in the main field after transplantation (Narayanasamy 2002). The rapid loss of protection provided by single resistance genes necessitated the use of a combination of effective resistance genes which might protect the plants for longer periods, since the pathogen has to make multiple changes simultaneously to become compatible with the combination of resistance genes. Johnson (1981, 1983) introduced the concept of durable resistance that remains effective in a cultivar, when grown over a large area or for a considerable period of time, in an environment conducive for development of the disease. Systems of disease management aimed at prolonging the useful life of resistance gene(s) require the knowledge of genetics of the host resistance, population genetics, evolutionary biology of pathogen and the interaction of crop management practices with host resistance.

Responses of cultivars and genotypes vary significantly, depending on the race/strains/biotypes of fungal pathogens. It is essential to establish the identity of the isolates/strains for which sources of resistance have to be identified. Isolates of *P. infestans* from Toluca valley of Mexico obtained from five potato varieties with varying levels of disease resistance were compared, based on mating type (MT), isozyme genotype for glucose-6-phosphate isomerase (Gpi) and peptidase (Pep) and sensitivity to metalaxyl fungicide. The host resistance components and specificity of host-pathogen interaction were determined, using detached leaves in the laboratory and attached leaves in the greenhouse for three of five potato varieties. The genetically derived host factors such as efficiency of infection, number and size of lesions, latent period or time required for sporulation after infection and sporulation capacity, were considered as host resistance components. Pathogenic fitness of isolates and levels of host resistance were determined based on the variables (host factors). The isolate from susceptible cv. Alpha did not infect resistant cv. Norteña under laboratory and greenhouse conditions, indicating a gene-for-gene interaction between potato cultivars and *P. infestans* isolates. No relationship between specific mating type and host resistance was observed. The field resistance of cvs. Rosita and Norteña was not expressed under artificial inoculated conditions. The leaf tissues of both cultivars were equally susceptible, as the leaves of susceptible cultivar. However, inhibition of sporulation was the only host resistance component retained by cultivars with field resistance under greenhouse conditions. The cvs. Rosita and Norteña consistently showed field resistance under natural conditions of Toluca Valley. The results cast a shadow of doubt on the validity of laboratory and greenhouse tests for screening potato accessions for selecting genotypes with field resistance (Lozoya-Saldaña et al. 2006).

2.2 HOST PLANT RESISTANCE TO SOILBORNE FUNGAL PATHOGENS

Evaluation of germplasm collections and wild relatives of crop plant species, for their level of resistance to the pathogen(s) concerned, is the basic step for the development of cultivars with built-in resistance and this is the most economical and environmentally safe approach of crop disease management. Cultivars with acceptable level of resistance to disease(s) can be grown without any need for additional cost or expertise. Identification of genotypes possessing resistance genes is the crucial requirement for selecting the dependable sources of resistance to the target disease(s).

2.2.1 GENETIC BASIS OF DISEASE RESISTANCE

Resistance to soilborne pathogen(s) is a heritable characteristic and it may be governed by single dominant gene or one or more loci. The genotypes of a crop plant species may interact with pathogen species/varieties/races, resulting in expression of different levels of disease severity.

2.2.1.1 Screening for Disease Resistance

Assessment of disease severity by visual examination, based on a defined disease-rating scale may be simple and easy to perform, but it lacks accuracy and variations due to assessor's capacity. Further, it is not suitable for determining resistance to individual pathogens involved in a disease complex. Dependence on visual examination alone may result in misleading conclusions on the response of genotypes, when symptoms are nonspecific and likely to be confused with other stress-related symptoms. However, visual assessment

of responses of genotypes to microbial pathogens has been widely practiced under in vitro, greenhouse and field conditions frequently. Molecular techniques have been applied to improve the reliability of assessment of disease severity by quantifying the pathogen biomass and establishing its relationship with disease intensity. Soilborne pathogens may induce disease in susceptible plants at different stages of development from sowing to crop maturity under diverse climatic conditions viz., dry, temperate and humid. Severity of disease may vary with the levels of host resistance and pathogen virulence, geographical location, prevailing environmental conditions and cultural practices adopted. Screening for disease resistance can be effective depending on (i) the availability of large and diverse germplasm collections including wild species, (ii) knowledge on both host plant and pathogen biology, variability, host-pathogen interaction, genetic structure and geographic distribution and (iii) use of precise and accurate screening technique(s). Major limitation in the assessment of diseases caused by soilborne fungal pathogens is due to the possible interactions with other pathogens present in the soil that might influence symptom expression and the destructive method of assessment of resistance of genotypes used for screening. Nondestructive methods using thermography and chlorophyll-inflorescence imaging have been applied to a limited extent to assess the incidence of diseases caused by fungal pathogens (Chaerle et al. 2004).

2.2.1.1.1 Assessment of Disease Resistance Based on Symptom Expression

Assessment of disease resistance by visual examination of disease symptoms is commonly carried out, based on disease incidence, expressed as percentage of infected/dead plants out of total number of plants examined. The evaluation of disease incidence is adopted for assessment of systemic infections leading to death of infected plants, as in the case of wilt diseases. Assessment of disease severity (intensity) is required, when resistance is inherited quantitatively, exhibiting a continuous gradient of symptom severity within a host plant population (Russell 1978). Disease severity scales represented by either 0–5 or 1–9 scale to reflect the increasing intensity of symptoms have been commonly used for different host-pathosystems. The grades of intensity of symptoms are often illustrated with pictures or diagrams to reduce the variations in assessments by different workers. Depending on the disease and accuracy required, scoring may be performed on the above-ground organs, on principal roots or other plant tissues such as crown region, secondary roots and fine lateral rootlets (Infantino et al. 2006). Physiologic races of microbial pathogens, differing in virulence have to be differentiated using differentials providing stable responses. A set of recombinant inbred lines (RILs) were developed and characterized for race differentiation of *Phytophthora capsici*, causing Phytophthora root rot of pepper. The highly resistant *Capsicum annuum* accession Criollo de Morelos-334 was hybridized to a susceptible cv. Early Jalapeno to generate RIL population. The host differentials assigned 17 isolates of *P. capsici* into 13 races. The results indicated that establishment of a set of stable host differential for *P. capsici* and *C. annuum* interaction was essential to investigate the complex inheritance of resistance to Phytophthora root rot and development of pepper cultivars with resistance to the disease (Sy et al. 2008). Sugar beet crops are affected by different soilborne microbial plant pathogens at different stages of growth. They infect seedlings resulting in poor establishment or damage to roots of more mature plants leading to high yield losses. Accessions (580–700) of related cultivated and wild species of *Beta* were assessed for resistance to four soilborne diseases of sugar beet viz., damping-off diseases (*Aphanomyces cochlioides* and *Pythium ultimum*), Rhizoctonia root and crown rot (*Rhizoctonia solani*) and Rhizomania (*Beet necrotic yellow vein virus*, BNYVV) transmitted by *Polymyxa betae*. Disease severity for the diseases was scored, based on the international standardized scale of 1–9. The highest levels of resistance ($\leq$ 2) to *A. cochlioides* and *P. ultimum* were detected in accessions of the more distantly related sections of Corollinae (93% of accessions) and Procumbentes (10%), respectively. Resistance to Rhizoctonia root and crown rot was observed in section Beta (5–7%), depending on field or glasshouses assessments. All sections of Procumbentes and some sections of Corollinae (4%) contained accessions highly resistant to Rhizomania disease complex (Luterbacher et al. 2005).

Screening for resistance of host genotypes will be effective, when high inoculum and favorable environmental conditions for development of disease are available. Accurate simulation of natural environmental conditions, for in vitro evaluation of genotypes for their resistance, has to be done. Appropriate inoculation method and incubation conditions have to be applied for differentiation of different levels of susceptibility/resistance in genotypes. In soybean, two types of resistance – vertical and horizontal – have been differentiated, based on the responses of soybean cultivars and genotypes to a panel of isolates of *Phytophthora sojae*, incitant of Phytophthora root rot (PRR) disease. Vertical or total resistance due to the presence of single resistance gene called 'resistance to *Phytophthora sojae*' (*Rps*) genes was identified first by Bernard et al. (1957). The use of *Rps* genes is a commonly exploited procedure for the management of PRR disease through their introgression into soybean cultivars. As the number of races of *P. sojae* increased, the 'race nomenclature' was replaced by 'pathotype nomenclature'. Isolates of *P. sojae* are characterized by their pathotypes (virulence factors) referring to the compatible interactions obtained after inoculation of differentials (Dorrance et al 2003). In soybean-*P. sojae* pathosystem, several physiologic races of *P. sojae* developed in response to deploying single dominant *Rps* genes for resistance. Accessions PI 273483 to PI427107 of USDA Soybean Germplasm Collections were evaluated for resistance to *P. sojae*, using hypocotyl inoculation technique for *Rps* genes, and the layer test for partial resistance. Of the 1,015 accessions tested, 159 accessions were susceptible to races 7 (*vir1a, 2, 3a, 3c, 4, 5, 6, 7*), 17 (*1b, 1d, 2, 3a, 3b, 3c, 4, 5, 6, 7*) and 25 (*1a, 1b, 1c, 1k, 7*). Among the accessions, 162 were resistant to these three races and 32 were resistant to an additional five races chosen specifically to elicit a susceptible interaction

with two and three *Rps* gene combinations. Further, 55% of the 887 accessions tested, had high levels of partial resistance or tolerance (with scores $\leq$4 in 0–9 scale) to *P. sojae*. Majority of the accessions from Republic of Korea had high levels of partial resistance, indicating that this region might have sources of both specific *Rps* genes and partial resistance to *P. sojae* for further exploitation for improving the resistance of soybean cultivars to this pathogen (Dorrance and Schmitthenner 2000). *P. sojae* causes Phytophthora root rot (PRR), as well as stem rot and damping-off of soybean. The *Rps* genes, *Rps1*, *Rps1k* and *Rps1a* have been incorporated frequently into soybean cultivars for protection against PRR. The Varietal Information Program for Soybean (VIPS) functioning at the University of Illinois evaluated soybean cultivars or advanced lines entered by commercial companies for resistance to PRR. About 600–900 cultivars were evaluated every year from 2004 to 2008 for resistance to either race 17 or 26 of *P. sojae*, employing hypocotyl inoculation method. *P. sojae* single gene resistance genes were reported in 1,808 or 51% of the entries, based on company information. Of these, the most commonly reported resistance genes were *Rps1c* (505), *Rps1k* (40%) and *Rps1a* (10%). On an average, 54% of the entries submitted to VIPS each year were new, reflecting the significant efforts of soybean seed companies by deploying novel genes for protecting the crop against *P. sojae* (Salminko et al. 2010).

Different methods have been employed to identify different pathotypes of *P. sojae*. The hypocotyl inoculation method was used as a standard, because of its ease of application, despite its limitations. The use of the hydroponic bioassay to infect soybean plants by *P. sojae* was developed and this method could overcome the limitations associated with hypocotyl inoculation assay or other assays relying on mycelium slurry (Guérin et al. 2014). Resistance genes (28) such genes have been characterized from *Rps 1* to *Rps 12*, and their different alleles, and others such as *RpsYu25, RPsYD25, Rps* Waseshiroge, *RpsYD29, RpsUN1, RpsB30, RpsZS18, RpsN10, RpsUN2* and *RpsJS* (Sahoo et al. 2017). Horizontal or partial resistance is quantitatively inherited and conditioned by quantitative trait loci (QTLs). The number of effective QTLs may vary (Schneider et al. 2016). The race profile of *P. sojae*, existing in Heilongjiang Province, China and levels of resistance of soybean cultivars for their responses to *P. sojae* races were determined. Single-zoospore *P. sojae* isolates (96) obtained from soil samples collected from 35 soybean fields in 18 counties were included in the investigation conducted during 2005–2007. Eight races of *P. sojae* – 1, 3, 4, 5, 9, 13, 44 and 54 – were identified by inoculating eight differentials, each containing a single gene resistance *Rsp* gene from 80 of 96 isolates. Races 1 and 3 were predominantly detected, representing 60% and 70% of the pathogen population, respectively. Presence of races 4, 5, 44 and 54 was recorded for the first time in Heilongjiang. Of the 62 soybean cultivars evaluated by hypocotyl inoculation procedure, 44 cultivars were resistant (< 30% mortality) to at least one race, indicating their potential for use as sources of resistance to *P. sojae* (Zhang et al. 2010). Soybean lines were screened several times, using isolates of *P. sojae* differing in virulence on cultivars with different *Rps* genes. A set of 14 soybean differential lines, each carrying a specific *Rps* gene, was inoculated on 6 to 10 *Rps* genes with three isolates of *P. sojae* which differed in virulence. Using the mixed inoculum, 1,109 soybean accessions were screened in a blind assay for novel sources of resistance. Accessions of *Glycine max* (17%) and *G. soja* (11%) with resistance (< 30% dead plants) were identified. Combining isolates into a single inoculum source provided advantages like, reduced cost of testing, ability to screen soybean germplasm with an inoculum virulent on all known *Rps* genes and ease of identifying novel resistance sources (Matthiesen et al. 2016). In a later investigation, inoculum of *P. sojae* was mixed into the soil and evaluation of soybean accessions was performed, based on survival rate (%) of soybean seedlings. This rating was significantly correlated (< 0.01) with reduction in root fresh weight and visual root rot severity. The minicore correlations, comprising either 79 accessions originating from Japan (JMC) or 80 accessions collected around the world (WMC) were evaluated. Wide variations in resistance among the individual varieties were observed. Among the accessions tested, 30 accessions of JMC and 41 of WMC showed resistance or moderate resistance to *P. sojae* isolate N1 (with virulence to *Rps1b, RPs3c, Rps4, Rps5* and *Rp6*, with > 50% survival. Of these, 26 JMC and 29 WMC genotypes showed at least moderate resistance to *P. sojae* isolate HR1 (*virRps1a-c, Rps1k, Rps2, Rps3a-c, Rps4-6* and *Rps8*). Two JMC accessions, Daizu and Amagi zairai 90D were resistant to four additional *P. sojae* isolates, indicating their potential for use as sources of resistance to *P. sojae* (Jiang et al. 2017).

Using the hydroponic bioassay, the *P. sojae* infecting soybean, was allowed to express its virulence/avirulence factors and soybean its resistance genes, regardless of the major *Rps* genes or QTLs being involved in partial resistance. The hydroponic bioassay in recirculating systems provided extremely reliable results. However, its requirements of glasshouse facilities, materials for recirculating systems, costs and human resources proved to be limiting factors for small-scale experiments. Hence, maintaining the basic principles, the bioassay was modified by growing plants in small plastic tank containing a hydroponic solution inoculated with zoospores of *P. sojae*. It was possible to reproduce the same results as in recirculating solutions for identification of *P. sojae* pathotypes. The phenotypes could be identified within 7 days and as many as 30 plants could be tested in a simple 10-liter container. This miniature system allowed discrimination of several soybean lines for horizontal resistance. The differences could be assessed by observing the root systems and determining root weight. Inoculation of *P. sojae* isolates on a set of eight differentials resulted in accurate and reproducible identification of pathotypes over a period of 2 years. When applied to test vertical resistance of soybean genotypes with known and unknown *Rps* genes, the bioassay relied on plant dry weight to correctly identify all genes. Both vertical and horizontal resistance in soybean genotypes could be differentiated by overcoming the limitations of hypocotyl inoculation

method, followed earlier. The hydroponic assay reproduced the natural course of infection by *P. sojae*, as zoospores were allowed to infect roots immersed in water. The hydroponic solutions were inoculated with a mixture of three isolates, combining all pathotypes present among 59 lines tested. Five classes ranging from high partial resistance to no partial resistance (class 5) could be discriminated. Among them, seven displayed very high level of horizontal resistance (class 1), whereas 42% of the lines had limited level of horizontal resistance (class 4/class 5). The line PI 144459 and Jack showed high level of resistance, as expected, whereas Conrad showed only intermediate level of partial resistance. The soybean plants in hydroponic solution in containers, rather than in a circulating solution, development of disease symptom could be observed from 5 days postinoculation (*dpi*). Typical symptoms of *P. sojae* infection appeared on roots and aerial parts and many plants were killed between 7- and 14-days dpi. The smaller hydroponic system was as reliable as the recirculating system which was cumbersome and expensive. The results showed that the hydroponic bioassay method could be used for determining the pathotypes of *P. sojae* and levels of vertical and horizontal resistance of soybean lines (Lebreton et al. 2018).

Aphanomyces euteiches causes Aphanomyces root rot of pea which accounts for appreciable losses in many countries. Pea genotypes and cultivars were evaluated for their resistance to *A. euteiches* under controlled conditions. Test seedlings were grown in vermiculite and inoculated with zoospore suspension at 10^3 zoospores/plant. Field assessments of the same set of entries for disease incidence and severity were also made. Positive correlations between assessments were indicated in the analysis (Moussart et al. 2001). The virulence of *A. euteiches* isolates obtained from Iowa and Wisconsin soils was determined by inoculating alfalfa plants. The frequency of isolates virulent on race 1-resistant alfalfa was assessed. *A. euteiches* isolates (14) from different locations in Iowa were inoculated on two race 1-resistant cvs. Paramount and Quantum, a susceptible cv. Agate/Vernal and two resistant breeding populations, WAPH-1 and WAPH-2. Isolates of *A. euteiches* (59) from Wisconsin were inoculated on one susceptible cv. Saranac and WAPH-1 and WAPH-2. Every isolate was virulent on one or more alfalfa cultivars or populations. Alfalfa populations and *A. euteiches* isolate had significant effect on disease severity index (DSI in 1–5 scale). All 14 Iowa isolates were virulent (DSI ≥ 3.0) on Agate, WAPH-1 and two commercial resistant cultivars. The results indicated that race 2 of *A. euteiches* was prevalent in Iowa and Wisconsin soils and it might be a limiting factor adversely affecting alfalfa yield (Munkvold et al. 2001). The levels of resistance/susceptibility of 279 genotypes from major grain legumes grown in temperate climates (faba bean, chickpea, lentil, lupin and common vetch) and three legumes (French bean, clover and alfalfa) grown in France were assessed by inoculating with a pea-infecting isolate of *A. euteiches*. Lentil, alfalfa and French bean were susceptible, whereas some genotypes of common vetch, faba bean and clover showed high levels of resistance. Genotypes of chickpea were highly resistant and lupin genotypes did not show symptoms of infection by pea isolate. The results indicated the need for assessing interactions between pathogen isolates and host genotypes for selection of resistance sources for *A. euteiches* (Moussart et al. 2008).

The levels of resistance of cultivars of soybean (*G. max*) and dry bean (*Phaseolus vulgaris)* to *Sclerotinia sclerotiorum*, causing stem rot disease, were assessed, by employing three inoculation methods viz., mycelial inoculation of cotyledons, cut stem- and detached leaf- inoculations. Six isolates of the pathogen with known relative virulence were inoculated on each of three soybean and dry bean cultivars. All three inoculation methods were effective in identifying virulence levels of pathogen isolates, regardless of cultivar, but identification of susceptible and partially resistant soybean cultivars was influenced by the isolates of the pathogen. The results indicated that cut stem inoculation method was statistically more efficient than cotyledon- and detached leaf-inoculation methods for assessing the levels of resistance of soybean and dry bean cultivars to *S. sclerotiorum* (Kull et al. 2008). Genetic interactions between soybean cultivars (5) and genetically unique isolates of *Sclerotinina sclerotiorum* were investigated by inoculations under controlled environment. Four-week-old soybean plants were inoculated with individual isolates of *S. sclerotiorum*, using straw inoculation method, followed by incubation of inoculated plants for 48 h in continuous leaf wetness. The disease severity in plants was scored at 1 and 2 weeks after inoculation. The cv. NKS08-80 consistently showed the lowest disease severity and no significant differences in disease severity induced by different isolates of the pathogen were discernible. No significant variation in interactions between soybean cultivars and pathogen isolates could be observed (Auclair et al. 2004).

S. sclerotiorum has become endemic, causing serious epidemics across oilseed brassicas, forage brassicas and vegetable brassicas and it can also infect a wide range of other crop plant species. It can survive in soil for several years, as thick-walled resting sclerotia, in the absence of crop hosts. Cultural and chemical methods of managing the diseases caused by *S. sclerotiorum* provide only partial control, in addition to being expensive. Seedling resistance to *S. sclerotiorum* was assessed across 46 diverse cruciferous genotypes from 12 different species by comparing the severity of disease on inoculated cotyledons under controlled conditions. The average size of lesions on cotyledons at 48-h postinoculation (hpi) varied from 0.8 to 7.3 mm. Three genotypes of *Brassica oleracea* (var. *italica* 'Prophet' and var. *capitata* 'Burton' and 'Beverly Hills') were the most resistant to *S. sclerotiorum*. Representatives of *Raphanus raphanistrum, Sinapis arvensis, B. juncea* and *B. carinata* were most susceptible to the pathogen producing largest lesions. The mean lesion size for *B. napus* introgressed with *B. circinata* was 5.6 mm, which was midway between the two parents. In the most resistant genotypes cvs. Prophet and Burton, growth of *S. sclerotiorum* on the cotyledon surface, prior to penetration, was severely impeded; production of appressoria was inhibited and both cytoplasm shrinkage and protoplasm extrusion in the hyphae

were observed. The results indicated the possible mechanism of resistance to *S. sclerotiorum* initiated in the early stages of pathogenesis in resistant genotypes (Uloth et al. 2014).

Plasmodiophora brassicae, an obligate pathogen with worldwide distribution, causes clubroot disease of cruciferous vegetable and oilseed crops, impacting the production considerably. The pathogen survives in the soil for several years, as resting spores which on germination infect roots of susceptible plants, inducing the formation of characteristic clubroots. Genetic resistance to *P. brassicae* may vary from broad spectrum to specific pathotypic levels and both single resistance genes and quantitative trait loci (QTL) have been identified. In order to develop clubroot-resistant cultivars, *Brassica rapa* and *B. juncea* genotypes (71) from China, including cultivars and inbred lines were evaluated for their resistance to three *P. brassicae* pathotypes. Seedlings were inoculated with resting spores (1×10^7 spores/ml) of each pathotype by immersing the roots into the suspension for 20 seconds. Ten cultivars of *B. rapa* were resistant to all three pathotypes. The cv. CR04 was completely resistant to all pathotypes, with all tested plants, showing a disease rating score of 0 in 0–3 rating scale. The genotype CR05 was susceptible to pathotypes 5 and 6 and CR11 was susceptible to pathotype 5. All four of these genotypes were completely resistant to pathotype 3. The results indicated that resistance in some genotypes was pathotype-specific. Seven of eight progenies obtained from selfing of Chinese cabbage cultivars were resistant to pathotype 3, but most were susceptible to pathotypes 5 and 6. Most inbred lines of Chinese cabbage and all inbred lines of pak choi (*B. rapa* subsp. *chinensis*) and mustard were susceptible to all three pathotypes, but their susceptibility was less to pathotype 3 than to pathotypes 5 and 6 (Zhang et al. 2015b). In a later investigation, 15 lines of an F_4 generation and selected six lines of F_5 generations of interspecific hybrids obtained from a cross between a male sterile line of *Brasssica napus* 'MS8', selected from resynthesized oilseed rape (*B. rapa* subsp. *chinensis* x *B. oleracea* var. *gemmifera*) an ecotype of *B. rapa* subsp. *pekinensis* were evaluated for clubroot resistance. A bioassay with P_1–P_5 pathotypes of *P. brassicae* was employed to assess the clubroot resistance levels in the above lines. Resistance to the pathotype P_1 was effectively fixed in the F_5 generation and improved the resistance in some lines to the pathotypes P_2–P_4. Resistance to P_1 pathotype and other tested pathotypes was not linked. Two hybrid lines combined high levels of resistance to clubroot with appropriate plant morphology, good seed quality traits and a stable chromosome number and arrangement (Niemann et al. 2017).

The levels of virulence of isolates of *P. capsici* from 31 bell and hot pepper cultivars and breeding lines were assessed under greenhouse and laboratory conditions. Soilless media were inoculated with *P. capsici* infested millet seed and pepper lines were raised on this media. In a detached fruit assay, fruit rot resistance was evaluated, following inoculation with zoospore suspension (1.75×10^6 zoospores/ml). Four physiologic races were differentiated among four isolates differing in virulence to pepper lines screened for crown and root rot resistance. Pepper lines CM334, NY07-8001, NY07-8006 and

NY07-8006 were resistant to the pathogen isolates tested. *P. capsici* isolates varied in their ability to cause infection on the fruits of different cultivars. In general, pepper fruits were more susceptible to *P. capsici* than roots and crowns (Foster and Hausbeck 2010). In another investigation, tomato cultivars and wild relatives were screened for their resistance to four isolates of *P. capsici* by individually inoculating 6-week-old seedlings with pathogen-infested millet seed (1 g/10 g of soilless medium). *P. capsici* isolates showed differences in their virulence. *Solanum habrochaites* accession LA407 was resistant to all four isolates of *P. capsici*, whereas the genotypes Ha7998, Fla7600, Jolly Elf and Talladega showed moderate resistance to all isolates of *P. capsici*. The pathogen was frequently recovered from root and crown tissues of symptomatic inoculated seedlings, but not from leaf tissue or asymptomatic control plants. Using species-specific PCR assay, the pathogen DNA was detected in resistant and moderately resistant tomato genotypes. Amplified fragment length polymorphism (AFLP) analysis of tomato genotypes showed lack of correlation between genetic clusters and susceptibility to *P. capsici*, indicating the distribution of resistance in several tomato lineages (Quesada-Ocampo and Hausbeck 2010). Among 2,301 accessions of bell pepper screened using a mixture of 6 isolates of *P. capsici*, 77 accessions were identified as resistant to root rot caused by *P. capsici*. Two accessions, PI201237 and PI 640532 showed resistance to six isolates of *P. capsici* consistently (Candole et al. 2010). Resistance of these two accessions to *P. capsici* was later confirmed by McGregor et al. (2011).

Resistance to *P. capsici* of *Cucurbita moschata* (pumpkin) accessions (119) from 30 geographic locations throughout the world and a highly susceptible butternut squash cultivar Butterbush, was assessed by inoculating with a suspension of three highly virulent isolates of the pathogen. Mean disease rating (DR) of *C. moschata* collections ranged from 1.4 to 5 in the disease severity rating scale of 0–5. The accessions of PI176531, PI458740, PI442266, PI442262 and PI634693 showed the lowest rates of crown infection with a mean DR less than 1.0 and/or individuals with no visible symptoms. The results indicated the possibility of utilizing these accessions for breeding for cultivars resistant to crown and root rot caused by *P. capsici* (Chavez et al. 2011). The levels of resistance of *Cucumis melo* (honey dew and muskmelon) plant introductions (308) and two cultivars were determined by inoculating seedlings with 3–4 true leaves with 5-isolate zoospore suspension (1 s 10^4 zoospores/seedling) against *P. capsici* causing crown root rot under greenhouse conditions. All susceptible control plants of cv. Athena were killed within 7 days of postinoculation. Majority of PIs (281 of 308) were highly susceptible to crown rot and died rapidly and only 20% of plants survived. The PIs (87) selected on the basis of first screen were reevaluated twice and PI420180, PI176936 and PI176940 were found to have high levels of resistance to crown rot disease. The results indicated the need for repeated evaluation of host genotypes to shortlist the highly resistant genotypes for use in breeding programs (Donahoo et al. 2013). *P. capsici* causes fruit rot in cucumber, which is

a major constraint in production. Plant introduction (PI) collections (1,076) from 54 different geographic locations around the world along with susceptible cv. Vlaspik were screened for resistance to fruit rot under field conditions by inoculating with zoospore suspensions of *P. capsici* OP97 strain. The PIs (99%) were moderately or highly susceptible, based on disease rating scale of 0–9 scale. Three accessions PI109483, PI178884 and PI214049 showed consistently low mean disease ratings after three screens during 2011-2013 and they could be used as sources of resistance to fruit rot of cucumber caused by *P. capsici* (Colle et al. 2014). *Phytophthora cinnamomi* infects many woody perennial plant species, including highbush blueberry (*Vaccinium corymbosum*). In order to identify genotypes resistant to *P. cinnamomi*, cultivars and advanced selection of highbush blueberry grown in greenhouse were inoculated with pathogen propagules, maintaining uninoculated control plants. Pathogen resistance parameters were, reduction in dry biomass of uninoculated plants and percentage of root infection. Four commercially established cultivars, Aurora, Legacy, Liberty and Reka and two new cultivars, Overtime and Clockwork were resistant to *P. cinnamomi*. Planting susceptible cultivars has to be avoided in pathogen-infested soils to reduce buildup of inoculum over the years (Yeo et al. 2016)

Infection of potato tubers, either wounded or unwounded occurs, when exposed to sporangia from foliar, soil or blighted tuber inoculum. Potato cultivars with foliar blight resistance (*R*-genes) and general resistance were evaluated for tuber blight caused by *P. infestans* (US-1), based on wound-induced and wounded tuber inoculations. Surface lesion diameter, lesion depth and frequency distribution of blighted tubers were determined in in vitro assays and tuber blight incidence was recorded under field conditions. Surface lesion diameter, depth and index ranged from 5 to 40, 2 to 16.3 and 15 to 656 mm, respectively, in wound-inoculated tubers. In non-wounded tuber assays, blight incidence ranged from 0 to 8.7%. Tuber blight infection varied in cultivars between 2 years of field experiments. The results indicated that foliar resistance of cultivars might have limited effect on tuber blight incidence (Nyankanga et al. 2008). Resistance of tubers of 34 potato cultivars to infection by zoospores of *Phytophthora erythroseptica* (causing pink rot) and mycelia of *P. ultimum* (causing tuber rot) was assessed, based on the two parameters viz., incidence of tuber infection (5) and penetration of rot (mm). Tubers of cv. Atlantic appeared to have some resistance, whereas cvs. Russet Norkotah and Snowden were most susceptible to infection by *P. erythroseptica*. Snowden was, however, most resistant to *P. ultimum*. The cvs. FL-1625 and FL-1687 also were less susceptible to P. *ultimum* than other cultivars, while cvs. Superior, Itasca and Dark Red Norland were the most susceptible to *P. ultimum*. The knowledge on the relative susceptibility/resistance of potato cultivars would be necessary to select suitable cultivar(s) for fields infested with soilborne pathogens (Salas et al. 2003). In a later investigation, the relationship between evaluation of resistance to *P. erythroseptica* of potato cultivars under greenhouse and field conditions was studied, by conducting trials for three years with 23 cultivars. Detached tuber evaluations did not suffi-ciently differentiate the potato lines found most resistant. A highly significant correlation of Idaho and Maine field results for nine cultivars suggested that regional variations might not be an important criterion for pink rot screening. The cultivars Atlantic, its progeny Pike and Gem Russell and Snowden were found to be the most resistant in repeated field evaluations. The red-skinned cultivars were, in general, the most suscep-tible to *P. erythroseptica* infection (Fitzpatrick-Peabody and Lambert 2011).

Phytophthora nicotianae isolates (32) collected from tobacco in different regions of South Africa were evaluated for their virulence and race differentiation by employing stem inoculation method. Two new South African races 0 and 1 were identified. Eighty isolates, selected on the basis of geographical origin and virulence, were used to assess the resistance of 11 commercial tobacco cultivars. The cultivars LK3360 and OD1 were highly resistant to race 0 but suscep-tible to race 1. The cultivars Vuma/3/46 and LK3/46 were highly resistant to both races 0 and 1 (Van Jaarsveld et al. 2002). A rapid seedling-based screening technique was com-pared with stem inoculation technique used earlier, for their efficacy for assessing resistance of tobacco cultivars to black shank disease caused by *P. nicotianae*. A strong positive cor-relation was observed between the two assay procedures. Isolates of *P. nicotianae* were characterized as race 0 and 1, using both methods (Van Jaarsveld et al. 2003). The resis-tance levels of 60 accessions/genotypes of diploid *Fragaria* spp. to *Phytophthora cactorum*, causal agent of crown rot and leathery rot of strawberry plants and fruits, were determined, based on the number of weeks of survival, during the first 4 weeks and plants surviving for more than 4 weeks. The results indicated that none of the *Fragaria* spp. could be considered as more resistant or susceptible than others and geographic origin did not have any influence on host resistance (Eikemo et al. 2010). In a later investigation, strawberry cultivars (5) and selections (8) were evaluated for resistance to *P. cactorum*. The cultivars Nobel and Saga and eight selections showed intermediate resistance to highly susceptible reactions for crown rot, whereas cv. Koriona was the most resistant, com-pared to other cultivars and selections. The cultivar Nobel was the most susceptible to leathery rot of fruit and second most tolerant to crown rot. There was no correlation between resis-tance to crown rot and leathery rot of fruits induced by *P. cactorum* (Eikemo and Stensvand 2015).

Melon cultivars (18) were screened for resistance to three soilborne pathogens, *Macrophomina phaseolina* (causing charcoal rot), *Monosporascus cannonballus* (causing root rot) and *R. solani* (causing stem canker) by inoculating each pathogen individually under greenhouse conditions. None of the tested melon cultivar was immune to all three patho-gens. But two cultivars 'Sfidak Khatdar' and 'Sfidak Bekhat' were moderately resistant to all three pathogens. In the sec-ond screening also, these cultivars were moderately resistant to these pathogens. The results indicated the potential of two melon cultivars for use as reliable sources of resistance to the three important soilborne pathogens (Salari et al. 2012).

Sheath blight caused by *R. solani*, is one of the most destruc-tive diseases of rice in Asian countries. The germplasm acces-sions (1,013), including mutants, introgression lines from wild relatives (*Oryza* spp.), A, B, R lines, Tropical *japonica* acces-sions, land races from north eastern India, wild accessions of *O. nivara* and *O. rufipogon* and gall midge biotype differen-tials were evaluated for resistance to *R. solani*, with artificial inoculation under field conditions in 2012. The germplasm accessions showing resistance/moderate resistance were fur-ther evaluated under both glasshouse and field conditions dur-ing 2013-2014. Seven genotypes, SM801 (N22 mutant), 10-3 (introgression line), Ngnololasha, Wazuhophek, Guimdhan and Phougak (land races) and RP2068-18-3-5 (gall midge biotype differential) were resistant to sheath blight patho-gen. Resistance to sheath blight was positively correlated with stem thickness and negatively with tiller number. The results suggested that resistant rice genotypes with medium to semi-dwarf stature might be useful in breeding programs to improve resistance in rice to sheath blight disease (Dey et al. 2016).

Rhizoctonia spp. causes considerable yield losses in wheat crops in Hungary. Responses of 19 wheat cultivars to soilborne *Rhizoctonia* spp. were assessed. Different fac-tors influenced the susceptibility of wheat cultivars to the pathogens. Inhibition of development of survivors of wheat plants in *Rhizoctonia*-infested field correlated with overall susceptibility of the wheat variety. The cvs. Emese, Kikelet and Palotas were less susceptible, but none of the cultivars could be considered as tolerant. The anamorph strains of *Athelia, Ceratobasidium, Ceratorhiza* and *Waitea*, similar to the anamorphs of *Thanatephorus*, selectively infected wheat varieties. However, the symptoms induced by them were not distinguishable by visual examination alone. *R. solani* was more aggressive to emerging wheat seedlings than *R. cerealis* (Oros et al. 2013). *R. solani* (teleomorph, *Thanatephorus cuc-umeris*), causal agent of Rhizoctonia root rot and bare patch disease occurring in dryland cereal crops in several countries, is responsible for significant yield losses. To assess the levels of genetic resistance of wheat and barley cultivars against *R. solani* AG8, three rapid and low-cost assays were developed. Symptom induced by the pathogen was quantified, based on root fresh weight and total root length at 7 and 3 days for tube and coating assays, respectively. In the first assay performed in 50-ml centrifuge tubes, seedlings were grown for 7 days in soil infested with ground oat inoculum. Reductions in differential mean root fresh weight and total root weight were observed. A significant difference was observed, when the mean root surface, the product of root length x root diameter and mean average root diameter were considered (see Figure 2.1). These variables appeared to be better predictors of resistance to *R. solani*. For screening intragenic roots of barley and wheat lines, an assay involving dipping seedling roots in ground oat inoculum of *R. solani* AG8 was developed. Roots of cv. RZ1 coated with pathogen-infested ground oat showed signifi-cant differences from those of cv. Scarlet in mean root fresh weight and total root length at 3 days after inoculation. The roots of noninoculated RZ1 were larger than those of Scarlet.

A Ground oat inoculum   **B** Whole oat kernel inoculum

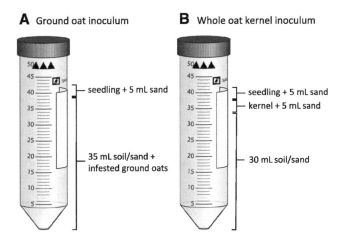

A Wheat seedlings   **B** Barley roots after treatment

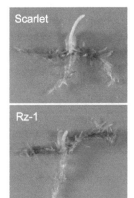

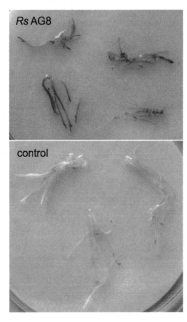

FIGURE 2.1 Comparative efficacy of ground oat soil assay and whole oat kernel soil assay performed in 50-ml tubes for assessing resistance of wheat and barley to *R. solani* AG-8 A: wheat genotypes tested using soil infested with ground oat inoculum of the pathogen B: barley genotypes tested using soil infested with whole oat kernel inoculum of the pathogen Seedlings and the pathogen were placed in 1:1 soil/sand mix and incubated at 15°C.

[Courtesy of Okubara et al. 2016 and with kind permission of the American Phytopathological Society, MN]

The cv. Golden Promise also showed significant reduction in fresh root weight and total root length after inoculation (see Figure 2.2). In the third assay, whole infested kernels in 50-ml tubes were used for evaluating resistance of barley cultivars. Each assay showed the same disease differential between susceptible and partially resistant wheat genotypes. The tube and coating assays required much less space and they may be performed in any incubator maintaining 15°C and equipped with growth lights. In addition, these assays might provide results of screening rapidly by reducing the time and materials needed (Okubara et al. 2016).

A diverse set of wheat germplasm collections from all major wheat producing regions of the United States was screened for resistance to Pythium root rot caused by *P. ultimum* isolate 90038 and *P. debaryanum* isolate 90136. These isolates were used to infest the pasteurized soil, which was then seeded with test genotypes of the wheat and placed in growth chamber at a constant 16°C, with a 12-h photoperiod and ambient humidity. First leaf length and plant height were measured, whereas roots were digitally scanned to create computer files that could be analyzed, using WinRhizo software for length and number of root tips. Differences in plant characteristics among genotypes were significant in the *Pythium*-infested soil. Correlations between plant stunting and root loss were significant (P <0.0001). The genotypes Caledonia, Chinese Spring, MN97695, OH708 and Sunco were the most tolerant to Pythium root rot disease of wheat (Higginbotham et al. 2004). The soybean cultivar Archer (maturity group I) and Hutcheson (maturity group V) were compared for their resistance to *P. ultimum, P. irregulare, P. aphanidermatum* and *P. vexans*. Emergence and establishment assays showed that Archer had greater emergence and fewer disease symptoms at 10 days after

FIGURE 2.2 Efficacy of root coating assay for assessing resistance of wheat and barley to *R. solani* A: root coverage by noninfested ground oat material on roots of wheat cvs. Scarlet (top) and Rz-1 (bottom) B: discoloration and stunting of excised roots of barley cv. Golden Promise at 3 days after dipping in pathogen-infested (top) or noninfested ground oat (bottom).

[Courtesy of Okubara et al. 2016 and with kind permission of the American Phytopathological Society, MN]

inoculation than cv. Hutcheson. The cultivar Archer had greater root weight and fewer disease symptoms than in cv. Hutcheson, after 6 weeks. The results indicated that cv. Archer exhibited higher level of resistance to root rot caused by *Pythium* spp. at all developmental stages tested (Bates et al. 2008).

Several disease resistance genes from wild species of *Solanum* have been incorporated into potato (*S. tuberosum*) through breeding. In order to explore the possibility of identifying resistance to multiple diseases in interspecific hybrids, 32 clones were evaluated for their resistance to Verticillium wilt, black scurf, early blight, soft rot and common scab diseases. To determine the levels of resistance to Verticillium wilt caused by *V. dahliae*, sap from the basal portion of each stem of plants grown in infested field, was squeezed and 100 μl of sap was plated on NPX medium to isolate the pathogen. Average populations of *V. dahliae* in the sap were higher (552 CFU) in 1996 than in 1998 (392 CFU). When the nematode *Pratylenchus penetrans* was also present along with *V. dahliae* in the field, the sap had higher populations of *V. dahliae* (598 CFU) in 1997 than in 1998 (139 CFU). The mean stem colonization by *V. dahliae*, as determined by the pathogen populations in the sap, was significantly higher in the trial with *V. dahliae* + *P. penetrans*. Two clones, C545 and C287 showed consistently higher levels of resistance to *V. dahliae* across years, regardless of the presence of the nematode. The synergistic interaction between *V. dahliae* and *P. penetrans* in the expression of early dying disease was observed. Most of the clones originated from populations that were not initially selected for disease resistance traits. Comparison with the cvs. Atlantic, Russet Norkotah and Russet Burbank indicated that all clones were more resistant than at least one cultivar for at least one disease resistance trait. The clone C545 exhibited improved resistance to early dying disease, soft rot, scab, pitted scab and early blight diseases and appeared to be good source of disease resistance. The use of interspecific hybridization at the diploid level, combined with sexual polyploidization to return to tetraploid level, might provide a procedure to introduce resistance to multiple diseases into advanced clones of potato (Jansky and Rouse 2003).

Soybean plant introductions (PIs) (6,037) were compared for their resistance to sudden death syndrome (SDS) caused by *Fusarium solani* f.sp. *glycines* with a susceptible control, Great Lakes 3302 and two moderately resistant controls, PI520733 and PI567374. At 3 weeks after inoculation, six PIs had significantly less foliar disease ratings, compared with PI567374, while none had lower area under disease progress curve (AUDPC) values. None of the genotypes had smaller lesion lengths on inoculated roots, compared to Great Lakes 3302. Correlation between root lesion lengths and foliar disease severity was not significant. None of the plant morphological characteristics was associated with foliar disease severity. PIs (18) earlier identified as moderately resistant, were inoculated with five isolates of *F. solani* f.sp. *glycines*. Isolate Mont-1 was the most virulent, inducing greatest disease severity ratings. These PIs that showed low foliar disease severity ratings, had potential for use as sources of resistance to SDS caused by *F. solani* f.sp. *glycines* (Mueller et al. 2002a). The ancestral lines (90) of soybean, representing 90% of the genes in modern US cultivars and 55 lines in pedigrees of public cultivars reported to have some resistance, were evaluated for their resistance to *F. solani* f.sp. *glycines*. Nine ancestral lines, Aoda Kim, Jackson, Sioax, Mammoth Yellow, T117, PI171450, PI54615-1 and PI71506 and 12 cultivars or experimental lines had sudden death syndrome (SDS) significantly different (P < 0.05) from PI520733 or PI567374. The putative ancestral line PI54610 was also moderately resistant to SDS disease (Mueller et al. 2003).

Resistance of common bean (*P. vulgaris*) to root rot caused by *F. solani* f. sp *phaseoli* is a quantitative trait that is strongly influenced by environmental factors. Hence, reproducible methods to screen soybean genotypes for plants, the inoculum layer method (ILM) and the traditional liquid inoculum method (LIM) were applied for assessing the resistance levels of 80 inbred lines of soybean. The results of disease rating based on the counts of discolored vascular bundles in the lower stem were correlated with that ILM procedure, compared to the rating of discoloration on root and hypocotyl (Chaudhary et al. 2006). The efficacy of three methods of screening for dry bean (*P. vulgaris*) cultivars/genotypes (11) for resistance to root rot disease was assessed. Sand-corn meal inoculum layer method, spore suspension method and paper towel method were employed under growth chamber and field conditions in 2005. The small red genotype VAX3

showed consistently high level of resistance to Fusarium root rot and it could be a reliable source of resistance for exploitation in breeding programs. Correlation analyses between field and growth-chamber root rot severity ratings indicated that the results of all three growth chamber evaluation methods were well correlated with field assessments, adding required reliability to growth chamber assessments of resistance of dry bean genotypes to Fusarium root rot disease (Bilgi et al. 2008). The germplasm collections of peas were evaluated for their resistance to *F. oxysporum* f.sp. *pisi (Fop)* race 2, using the method combining the parameters, disease incidence, disease rating over time and its related area under disease progress curve (AUDPC). A large variation in the responses of specific pea accessions, ranging from highly resistant to susceptible categories were differentiated within the germplasm collections, indicating the quantitative expression of resistance. The scoring method was found to be reproducible and high level of resistance in 11 accessions was confirmed. Characterization of resistance mechanism in pea-*Fop* interaction was initiated by comparing the external and internal stem symptoms scored, along with the extent of pathogen colonization within the plants in selected pea accessions. The results indicated that in all resistant accessions, the resistance mechanisms efficiently arrested the pathogen invasion at the crown. The identified sources of resistance to pea Fusarium wilt disease may be utilized for improving the levels of resistance in pea cultivars (Bani et al. 2012).

F. oxysporum f.sp. *niveum (Fon)* causal agent of Fusarium wilt disease of watermelon produces chlamydospores that remain infectious in soil for many years, rendering the available disease management strategies ineffective. Watermelon cultivars showed different levels of resistance to *Fon* races 0 and 1, but none of them was found to have resistance to race 2 or 3. Hence, 110 US plant introductions (PI) were screened for resistance, using seed and seedling inoculation methods. The PIs (15) of wild watermelon (*Citrullus lanatus* var. *citrioides*) showed significantly higher resistance to *Fon* race 2 than that of cvs. Sugar Baby or Charleston Grey as well as PI 296341 (belonging to *C. lunatus* var. *citrioides*). These 15 PIs could be used as sources of resistance to *F. oxysporum* f.sp. *niveum* race 2 in breeding programs (Wechter et al. 2012). *F. oxysporum* f.sp. *cubense* tropical race 4 (*Foc*-TR4) occurred in a highly destructive form in banana plantations in the tropics. The genotypes of wild banana were evaluated for resistance to *Foc*-TR4 under greenhouse and field conditions. Among the wild species, *Musa balbisiana* showed the highest levels of disease intensity, followed by *M. acuminata* subsp. *burmannica*. Some individuals of *M. yunnanensis*, *M. nagensium*, *M. ruilensis* and *M. velutina* showed low levels of rhizome discoloration under greenhouse conditions, but they were resistant in the field. *M. basjoo* and *M. itinerans* did not exhibit any visible symptoms of infection, suggesting higher levels of their resistance to *Foc*-TR4. These sources of resistance to Fusarium wilt of banana could be exploited in breeding programs for development of banana cultivars with resistance to the disease (Li et al. 2015).

Resistance to Fusarium wilt diseases has been assessed in 'wilt sick plots' (WSPs), prepared by artificially infesting the soil by incorporating chopped wilt-infected plants into the soil at the end of growing season for at least three seasons, followed by cultivation of a susceptible cultivar in the same plot. In order to increase inoculum density, the pathogen may be grown on sterilized substrates (seeds) and uniformly incorporated into the soil. Evaluation of large number of breeding material is possible under field conditions, using WSPs. The principal advantage of WSPs is the possibility of screening breeding materials easily and reliably. This procedure was followed for screening chickpea against *F. oxysporum* f.sp. *ciceris* (Nene and Haware 1980) and lentil against *F. oxysporum* f.sp. *lentis* (Bayaa et al. 1997). A screening nursery for identification of sources of resistance to *F. oxysporum* f.sp. *pisi (Fop)*, causal agent of pea Fusarium wilt disease was established by depositing several truckloads of soil infested with *Fop* in a defined area. Test entries including breeding lines or germplasm accessions were sown in single rows with a susceptible check line included at regular intervals to ensure uniform distribution of inoculum and the relative infections on test and check lines. Entries showing resistance to *Fop* race 1 and 2 could be identified by using wilt sick plots (Infantino et al. 2006). Pigeon pea germplasm accessions were screened for resistance to wilt disease caused by *F. oxysporum* f.sp. *udum*, using wilt-sick plots and percent of infected plants were calculated for different accessions. The genotype ICEAP 00040 consistently showed high resistance (< 20% infected plants). In addition, ICEAP 00040 gave higher yield with larger grains. Hence, this genotype was released for commercial cultivation by growers in the Southern Africa (Gwata et al. 2006).

The levels of resistance/susceptibility of soybean genotypes were assessed, using a laboratory screening procedure, involving exposure of soybean stem cuttings to cell-free toxic culture filtrate (CF) from *Fusarium virguliforme*, causing soybean sudden death syndrome (SDS). The symptoms induced by the CF were similar to those observed in whole plants inoculated by root inoculation with *F. virguliforme* and those observed in the commercial field under natural conditions. The conditions for production of CF and incubation of stem cuttings with CF for the assay were optimized. The whole plant root inoculation assay was also performed to compare the responses of selected soybean genotypes in both assay methods. Area under disease progress curve (AUDPC) values were the highest, when cuttings were immersed in the CF of the pathogen grown on soybean dextrose broth, in filtrate produced from the fungus grown for 18 or 22 days and when stem cuttings were incubated at 30°C. The AUDPC values and shoot dry weights from the whole plant root inoculations and stem cutting assay showed variations among nine soybean genotypes tested with *F. virguliforme* and *F. tucumaniae* isolates. However, the AUDPC values from both assays were positively correlated (Xiang et al. 2015). A laboratory screening system was developed to identify lily lines showing resistance to *F. oxysporum* f.sp. *lilii (Fol),* incitant of bulb rot disease. Different factors, including culture conditions of *Fol*, inoculum threshold, appropriate plant materials, inoculation method and duration and incubation period for plant materials

after inoculation were optimized. The pathogen was grown in potato dextrose agar (PDA) medium for 6 days at 25°C in darkness. Spore suspensions (1×10^4 spores/ml) were used to inoculate leaf segments (1.5×2.0 cm) by dipping for 22 h at 25°C in darkness. Inoculated leaf tissues were cultured on 0.6% agar in petri plates at 25°C and 50% RH with a photoperiod of 16:8 h (light/darkness). At 7 days after incubation, disease severity was scored, using a 1–6 rating scale. The lines, showing resistance/susceptible reactions to *F. oxysporum* f.sp. *lilii* were evaluated by planting them on pathogen-infested soil. The results of in vitro screening were confirmed by soil inoculation assay (Jang et al. 2015).

Verticillium dahliae, causes Verticillium wilt disease in several crops, resulting in considerable losses in all countries around the world. Strawberry (*Fragaria* x *ananassa*) genotypes (11) identified earlier as resistant to *V. dahliae* were evaluated for investigating the relationship between the extent of pathogen colonization and symptom expression, as the basis of resistance levels. Resistance scores was based on severity of visual symptoms and extent of colonization was quantified as the percentage of petioles free of *V. dahliae*. Both resistance scores and percentages of pathogen-free petioles decreased significantly from May to June (P < 0.05) during two growing seasons. Significant genomic variance (P <0.05) was detected for percentage of pathogen-free petioles, but not for resistance score. The results suggested that overall performance of strawberry genotypes inoculated with *V. dahliae* might be enhanced by both resistance and tolerance, but tolerance might be less stable over the course of a season (Shaw et al. 2010). Among the 11 olive cultivars evaluated for resistance to *V. dahliae* in heavily infested soils, the cultivars Arbequina, Koroneiki, Sevillenca and especially Frantoio, Empeltre and Changlot Real showed high level of resistance to Verticillium wilt disease. The field results confirmed the results obtained under controlled conditions with some discrepancies. The results indicated the usefulness of the identified olive cultivars as sources of resistance to *V. dahliae* (Trapero et al. 2013). The effectiveness of olive cultivars to continuous and natural lighting for assessing the resistance of 12 cultivars to the defoliating pathotype of *V. dahliae* was assessed. The root systems of the test cultivars were dipped into the conidial suspensions of *V. dahliae*. The cv. Frantoio was used as the moderately resistant control cultivar. Six cultivars were moderately resistant. Continuous lighting allowed for the identification of resistant genotypes during a period of three weeks which was shorter than the period normally required for symptom expression in earlier investigations. The reduction in duration was still more distinctly seen in resistant cultivars. The results clearly showed that exposure of inoculated olive genotypes to continuous lighting was an important procedure for adoption in olive breeding programs, as it could reduce the period required for assessment of resistance to Verticillium wilt disease (Garcia-Ruiz et al. 2014).

In order to identify the sources of resistance to *V. dahliae* causing Verticillium wilt disease of cotton, the germplasm accessions were screened in the field and greenhouse conditions. Accessions 1421Bt-4133, O Siv 2, Arcot-1 and GP93 were highly resistant or resistant to Verticillium wilt, whereas Zhong 21371 and Yumian 2067 were moderately resistant. On the other hand, Arcot 4026ne, Arcot 438, Xinluzao-3 and Jimian 11 were highly susceptible to the disease. The lines O Siv2, Arcot1, GP93, Zhong 21371 and Yumian 2067 differed in their response to the pathogen, when artificially inoculated or grown in naturally infested soil. Overall, the line 1421Bt-4133 with required level of resistance to *V. dahliae*, could be used as sources of resistance in breeding programs (Khaskheli et al. 2013). Wild and cultivated *Capsium* accessions (397) were screened for resistance to two isolates (VdCa59 and VdCf45) of *V. dahliae* under greenhouse conditions. Selected accessions (78) were further evaluated in a follow-up experiment. Among the *Capsicum* accessions, 21 (26.9%) and 13 (16.6%) were resistant, when inoculated with isolates VdCa59 and VdCf45, respectively. Eight accessions, Grif 9073, PI 281396, PI 281397, PI 438666, PI 439292, PI 439297, PI 555616 and PI 594125 were resistant to both isolates of *V. dahliae*. Based on the Germplasm Resources Information Network data, two *C. annuum* accessions, Grif 9073 and PI 43927 were identified as resistant to Phytophthora root rot disease, in addition to Verticillium wilt disease (Gurung et al. 2015).

Strawberry genotypes (8) were evaluated for their responses to *V. dahliae* isolates from different hosts viz., strawberry, potato, watermelon, mint, eggplant (aubergine) and cauliflower by inoculating in replicated field trials conducted during 2 successive years. Typical symptoms developed on known susceptible strawberry cultivars and differences in virulence of *V. dahliae* isolates were significantly observed on susceptible cultivars. Two isolates from strawberry were the most virulent, whereas the isolate from cauliflower was the least aggressive. Six strawberry genotypes identified earlier as resistant, were resistant to the isolates of *V. dahliae* tested. Overall, strawberry genotypes represented the largest source of variation with variance components approximately 10-fold greater than those associated with interaction (Gordon et al. 2006). *V. dahliae* causing Verticillium wilt of lettuce exists in the form of race 1 and 2. Cultivation of lettuce cultivars with complete resistance to race 1 may increase the frequency of incidence of race 2 strains. Hence, two independent populations totaling 314 randomly sampled plant introductions (PIs) of lettuce were screened for resistance to race 2 isolates, under greenhouse conditions. None of the accessions showed complete resistance to race 2. Genetic variation for the frequency of *V. dahliae*-colonized plants that remained asymptomatic was detected. Four PIs 169511, 171674, 204707 and 226641 showed significantly lower scores for disease incidence, compared to susceptible control cultivar. The four PIs were morphologically diverse and showed no dependence between rate of bolting and resistance to Verticillium wilt disease (Hayes et al. 2011).

The ability of *Verticillium albo-atrum* and *V. dahliae* strains, infecting *Medicago truncatula* to adopt to new hosts was examined. *M. truncatula* genotypes (25) from a core-collection and six strains of *V. albo-atrum* and *V. dahliae* were used to investigate the potential of nonhost pathogen strains, isolated from different plant species to infect *M. truncatula*

and the levels of susceptibility of the host to pathogen strains. Disease severity was scored, using disease rating scale of 0–4, during 30 days period after inoculation of 10-day-old plantlets. Disease severity was quantified by maximum symptom scores (MSS) and AUDPC, which integrated time course of symptom development. Highly significant differences were observed among plant genotypes and fungal strains and their interaction was also significant. The correlation between MSS and AUDPC was 0.86. The most severe symptoms were induced by the alfalfa strain of *V. albo-atrum* V31-2 and the least disease score by *V. dahliae* JR2, as shown by the mean values obtained on all (25) genotypes inoculated. *M. truncatula* genotype TN8.3 was the most susceptible genotype by mean values obtained with the six pathogen strains. The genotypes F11013-3, F83,005.9 and DZA45.6 were highly resistant to all strains tested. The results indicated the effectiveness of assessing the resistance levels of host genotypes to make appropriate choice for breeding programs (Negahi et al. 2013).

Sclerotinia trifoliorum and *Sclerotium rolfsii*, infecting alfalfa, have similar mechanisms of parasitism. Hence, investigation was taken up to determine, whether selection of quantitative resistance to *S. trifoliorum* in alfalfa might increase resistance to *S. rolfsii* also, as expressed in excised leaf tissues and whole plants. The rate of necrosis in excised leaf tissues and survival of whole plants, following inoculation with *S. rolfsii* were the parameters used as measures of resistance. Three alfalfa populations selected earlier from cv. Delta for quantitative resistance to *S. rolfsii*, as compared to Delta or susceptible populations, in excised leaf tissues. When whole plants of Delta and two of these populations, *S. trifoliorum*-resistant (STR) and Mississippi *Sclerotinia*-resistant (MSR), were inoculated with *S. rolfsii* at 3–8 weeks of age, significant (P = 0.01) differences in survival were observed, attributable to plant age at inoculation and alfalfa populations. Survival of both MSR and STR was significantly (P = 0.05) greater than for Delta. The best differential results were obtained when plants were inoculated at 5–7 weeks of age. Additional populations of alfalfa were selected for resistance to *S. trifoliorum* from four other alfalfa cultivars by leaf inoculation technique and this resistance was confirmed by whole plant inoculations. In excised leaf inoculation, all four of these populations expressed enhanced resistance to *S. rolfsii*, as compared to either parent cultivars or populations of comparable size selected at random. The results indicated that selection of quantitative resistance to *S. trifoliorum* in alfalfa also conferred enhance resistance to *S. rolfsii*, as expressed in excised leaf tissues and whole plant inoculation methods. The genes for quantitative resistance to *S. trifoliorum* and *S. rolfsii* in alfalfa might be synonymous overlapping or closely linked (Pratt and Rowe 2002).

Commercial pepper cultivars and *Capsicum* accessions (110) were evaluated for their resistance to *S. sclerotiorum*, causing stem and fruit rot, using limited-term and ascospore inoculation methods. All commercial varieties were susceptible to stem rot. Following inoculation with ascospores, stem rot incidence in *Capsicum* accessions ranged from 0 to 100%, with 58 accessions, showing significantly less stem rot disease than the cv. Marengo. No relationship between the results of two inoculation methods was evident. The accessions showing higher resistance might be useful as sources of resistance to *S. sclerotiorum* for breeding programs (Yanar and Miller 2003). Resistance of oilseed rape (*Brassica napus*) accessions to *S. sclerotiorum*, causal agent of Sclerotinia stem rot (SSR), was assessed, using petiole inoculation technique and four pathogen isolates, two each from soybean and *B. napus*. The accessions were inoculated in the seedling stage and responses to pathogen isolates were determined based on the number of days required to wilt (DW) and a lesion phenotype (LP) index. No significant difference in virulence of isolates for DW was recorded and only slight differences could be seen for LP. However, significant differences were noted among B. napus accessions for DW and LP, when single or multiple isolates were employed. Higher levels of resistance were recorded in winter-type than in spring-type accessions and among rapeseed quality, compared with canola-quality accessions. The most resistant accessions identified also were the most resistant, when inoculated at flowering stage. The petiole inoculation technique provided the basis for differentiation of oilseed rape *B. napus* germplasm accessions for their levels of resistance to *S. sclerotiorum* and accessions identified as resistant, could be used for improving resistance of *B. napus* cultivars to the SSR pathogen (Zhao et al. 2004). Three methods of evaluation were applied for determining resistance of canola (*B. napus*) cultivars to SSR disease: petiole inoculation technique (PIT), detached leaf assay (DLA) and oxalic acid assay (OAA). The SSR severity varied from low to high among years and locations. Significant differences were observed among cultivars for SSR, using the PIT and OAA methods, but not when DLA method was used. No significant correlations were noted between SSR severities under controlled conditions with SSR severities recorded in field trials. However, significant negative correlations were observed between SSR area under disease progress curve (AUDPC) values for PIT method and yield from one of two locations (Carrington) in 2001 and 2002. The PIT and OAA methods were unable to predict the reactions of cultivars to SSR under field conditions. The results indicated the importance of performance of accessions/cultivars in infested fields for development of canola cultivars resistant to *S. sclerotiorum* (Bradley et al. 2006).

S. sclerotiorum causes SSR disease in soybean significantly reducing production levels in several countries. Two methods of inoculation of soybean cultivars, viz., spray-mycelium method and drop-mycelium method, were developed to screen the soybean cultivars and accessions under greenhouse conditions. Cultured, homogenized suspension of mycelium was sprayed on soybean leaves or a drop of homogenized mycelial suspension was placed on the apex of the main stems and inoculated plants were incubated in a greenhouse chamber with 60 to 80% relative humidity. Plant mortality and area under wilt progress curve (AUWPC) were the parameters determined to reflect disease severity during 3–24 days after inoculation. Significant differences in disease severity ratings of plant mortality and AUWPC values due to SSR were observed among 18 tested soybean genotypes. The results obtained with spray-mycelium and drop-mycelium inoculation methods were significantly correlated (R > 0.73, P < 0.01) with the results of the cut-petiole inoculation method used earlier for both plant mortality and AUWPC. Both inoculation methods required less inoculation time and were low-cost, efficient and reliable than the cut-petiole inoculation method used previously. These methods may be applied for assessing the levels of resistance of soybean genotypes against *S. sclerotiorum* (Chen and Wang 2005). Variability in populations of *S. sclerotiorum* directly is known to affect SSR resistance breeding programs. Isolates of *S. sclerotiorum* (44) were evaluated for variation in different biological characteristics, using a representative subset of nine isolates and soybean genotypes, exhibiting susceptible or resistant responses (determined by using a single isolate), a significant interaction (P = 0.04) was observed between isolates of soybean genotypes, when SSR severity was evaluated. The results suggested that screening of *S. sclerotiorum*-resistant soybean germplasm has to be carried out with multiple pathogen isolates representing overall diversity and existing throughout soybean production areas in a country (Willbur et al. 2017).

A detached stem inoculation method was developed for evaluating resistance levels of *Brassica* crops to *S. sclerotiorum* under controlled conditions. Genotypes (17) from five *Brassica* crops were evaluated in experiment 1, and 71 F2 lines derived from a cross between susceptible and resistant lines of *B. oleracea* were tested in experiment 2. High correlations were observed between stem and branch for lesion length in both experiments and between stem and sections of stem in experiment 1. The lesion length in detached stem inoculation method under controlled conditions was positively correlated with that of toothpick inoculation in experiment 2. However, the variations in lesion length in detached stem inoculation were lower than that in toothpick inoculation in experiment 2. Furthermore, no significant association was detected between lesion length and the diameter of stem or branch. The results suggested that the detached stem inoculation under controlled conditions is a large-scale, flexible and reliable method of screening for resistance of *Brassica* crops against *S. sclerotiorum* (Mei et al. 2012). Chinese genotypes of *B. oleracea* var. *capitata* (52), Indian *Brassica juncea* genotypes (14), carrying wild weedy Brassicaceae introgression(s) and four genotypes carrying B-genome introgression, Australian commercial *B. napus* cultivars (22) and *B. napus* and *B. juncea* genotypes (12) of known resistance were evaluated for their field resistance to SSR induced by *S. sclerotiorum*. All test plants were individually inoculated by placing an agar disc with axenic culture of *S. sclerotiorum* on the stem above the second internode and secured with Parafilm tape. Mean stem lesion length across tested genotypes ranged from < 1 to > 68 mm. *B. oleracea* var. *capitata* genotypes (65%) from China showed highest levels of stem resistance. One Indian *B. juncea* line carrying weedy introgression exhibited significant level of both stem and leaf resistance. Majority of the commercial Australian oilseed *B. napus* cultivars (40%) were the most susceptible among the genotypes tested for stem disease. No correlation appeared to exist between stem and leaf resistance among the genotypes tested, suggesting their independent inheritance. The Chinese *B. oleracea* var. *capitata* genotypes, expressing combined high levels of stem and leaf resistance could be exploited as sources of field resistance against the SSR disease caused by *S. sclerotiorum* in horticultural and oilseed brassicas (You et al. 2016).

Lettuce drop disease is induced by *Sclerotinia minor* and *S. sclerotiorum* which produce two types of inoculum viz., mycelia and ascospores. Hence, cultivars with resistance to infections by both inocula had to be identified. Assessment of resistance of lettuce cultivars was performed in fields infested by *S. sclerotiorum* and/or *S. minor,* with high densities and inoculations with ascospores of both pathogens. Diverse populations of iceberg, romaine, leaf, butterhead, Latin, oilseed lettuce and wild relatives of lettuce were evaluated using mycelial inoculum. The cv. Eruption, a small-statured Latin cultivar had significantly lower level of infection, compared to susceptible cultivars in experiment with high plant densities, indicating that its small size was not the reason for disease escape. Ascospore inoculations confirmed the resistance in cv. Eruption and *L. virosa* SAL012, whereas oilseed lettuce PI 251 246 showed partial resistance of lettuce drop disease (Hayes et al. 2010). Identification of sources of resistance in oilseed rape (OSR, *B. napus*) to *S. sclerotiorum*, causing SSR was accomplished in two stages. Initially, the virulence (aggressiveness) of 18 *Sclerotinia* isolates (17 *S. sclerotiorum* and one *S. subarctica*) was determined, using cultivated representatives of *B. rapa, B. oleracea* and *B. napus*, employing a young plant test. *S. sclerotiorum* isolates from crop hosts were more aggressive than those from wild buttercup (*Ranunculus acris*). The isolates P7 (pea isolate) and DG4 (buttercup isolate), representing aggressive and weakly aggressive strains, were employed to screen 96 *B. napus* lines for SSR resistance in the young plant test. A subset of 20 lines was further evaluated, using stem inoculation test on flowering plants. Five lines showed high level of resistance. One winter OSR (line 3 from Czech Republic) and one rape kale (line 83 from the United Kingdom) were consistently resistant. Further, one line from Sweden 69 (Norway) had outstanding level of high resistance in stem test and they had fewer sclerotia in stems. The results indicated that these lines could be used as sources of resistance to SSR in breeding programs (Taylor et al. 2015).

A two-step procedure was followed to identify sources of resistance in common bean accessions against white mold caused by *S. sclerotiorum*. Common bean accessions (199) of a core collection and 29 cultivars or lines including five lines described earlier as resistant (G112, PC50, A195, Cornell 606 and MO162) were evaluated, using four isolates with differing virulence levels. Some genotypes (19) had a level of resistance equal to or significantly better than G122. Selected genotypes (19) were further evaluated, using a more severe straw test in which 15 genotypes were less susceptible than G122. The genotypes of common bean identified with higher level of resistance to white mold could be used in breeding programs for development of resistant cultivars (Pascual et al.

2010). Variable expression of resistance was a limiting factor for screening common bean accessions for consistent higher level of resistance. *S. sclerotiorum* isolates (10) used in the greenhouse screening and 146 isolates collected from white mold field screening nurseries in the major common bean production areas in the United States were assigned to appropriate mycelial compatibility groups (MCGs) and subjected to an aggressiveness test. Significant differences in the aggressiveness of isolates were detected between some isolates in different MCGs, but isolates within an MCG did not vary in virulence. High isolate variation found within and between field locations could influence disease phenotype of putative white mold germplasm accessions. Further, variability found in and among screening locations did not exhibit variability found in the samples from four growers' fields. The results indicated the need to consider these factors that are likely to influence expression of resistance to *S. sclerotiorum* in common bean accessions (Otto-Hanson et al. 2011).

Responses of lines of *Brassica* spp., including short-term vegetable crops (*B. rapa* subsp. *chinensis* var. *communis*, Shanghai pak choy), Chinese flowering cabbage (*B. rapa* subsp. *pekinensis*), the Rapid Cycling Brassica Collection (RCBC), also known as Wisconsin Fast plants, and spring canola (*B. napus*) to infection by *Plasmodiophora brassicae* pathotype 6, incitant of clubroot disease, were assessed under naturally infested field conditions during 2008–2010. Shanghai pak choi and Chinese flowering cabbage were highly susceptible to the disease. The cultivars of napa cabbage (Deneko, Bilko and Yuki) were highly resistant to pathotype 6. Two RCBC lines *B. napus* and *Raphanus sativus* were resistant to pathotype 6. In addition, two canola cultivars 45H21 and Invigor 5020LL were also highly resistant to *P. brassicae* pathotype 6 (Adhikari et al. 2012). In a later investigation, five RCBC lines (*Brassica carinata, B. juncea, B. napus, B. oleracea* and *B. rapa*), 84 lines of *Arabidopsis thaliana* were screened for their resistance to *P. brassicae* pathotypes 2, 3, 5 and 6. Seedlings were inoculated with resting spores of *P. brassicae*, maintained at 25 and 20°C (day/night), and assessed for disease incidence and severity at 6 weeks after inoculation. Among the RCBC lines, *B. napus* were resistant to all of the pathotypes. *B. oleracea* was resistant to pathotypes 2, 3 and 5, whereas *B. carinata* and B. *rapa* were resistant to pathotypes 2 and 5. *B. juncea* was susceptible to pathotypes 5 and 6 and showed intermediate responses to pathotypes 2 and 3. The line Ct-1 of *A. thaliana* was highly resistant to pathotype 2; Pu2-23 was highly resistant to pathotype 5 and Ws-2 and Sorbo were highly resistant to pathotype 6. Responses of RCBC lines were similar under controlled and field conditions to pathotype 6, but not to pathotypes 3 and 6 of *P. brassicae*. The results indicated the usefulness of RCBC lines, as sources of resistance to *P. brassicae* (Sharma et al. 2013a). In another study, the differential responses of selected *Brassica* vegetable crops to Canadian pathotypes of *P. brassicae* were assessed in fields infested with pathotype 6 (P6), 3 (P3) and 5 (P5). The resistant cvs. Shanghai pack choi and Brussels Sprouts did not show any visible symptoms, whereas all susceptible cultivars had 100% clubroot incidence and very high severity. Under controlled conditions, reaction to P6 was consistent with field assessment, but several cultivars showed differential reactions to P2, P3 and P5. Tillage Radish® (*Raphanus sativus* var. *longe pinnatus*) did not show any visible symptoms in fields infested with the pathotype P5, but developed some clubroot symptoms under controlled conditions (Sharma et al. 2013b). The vegetable crops, Chinese cabbage (*B. rapa*), pak choi (*B. rapa*) and mustard (*B. juncea*) were heavily damaged by clubroot pathogen *P. brassicae* in China. Hence, screening 71 *B. rapa* and *B. juncea* genotypes, including cultivars and inbred lines was taken up for assessing their levels of resistance to three pathotypes of *P. brassicae*. Interactions between pathotypes and cultivars/genotypes were significantly different. Pathotype 3 was less virulent, compared to pathotypes 5 and 6, which were highly pathogenic on most tested genotypes. Chinese cabbage cultivars (10 of 14) were resistant to all three pathotypes, whereas four genotypes showed resistance only to a specific pathotype. The results indicated the possibility of using the genotypes identified as resistant for development of cultivars with required level of resistance to clubroot disease (Zhang et al. 2015a).

Rhizoctonia root and hypocotyl rot (RRHR) disease, caused by *R. solani* (teliomorph, *Thanatephorus cucumeris*), affects soybean severly, reducing yield markedly. As no commercial cultivar showed resistance to RRHR. Ancestral soybean lines, maturing groups (MGs) 000 to X (90) and commercial cultivars, MGs II to IV were evaluated for their responses to *R. solani* in the greenhouse. Most of the ancestral lines and cultivars were susceptible to the disease. However, 21 ancestral lines and 20 commercial cultivars showed partial resistance to the pathogen. Among the ancestral lines, CNS, Mandarin (Ottawa) and Jackson were earlier reported to be partially resistant. No significant relationship between the extent of reduction in dry root weight and disease severity could be seen (Bradley et al. 2001). *R. solani* infects another economically important crop, sugar beet, causing postemergence damping-off of seedlings. The pattern of disease progress in sugar beet seedlings inoculated with a less virulent *R. solani* AG2-2 isolate and highly virulent AG2-2-R-1 isolate were similar during the first 4 days postinoculation. But the progress was relatively constant during the next 7-8 days, followed by decline in the case of less virulent isolate-infected seedlings. On the other hand, in AG-2-2-R-1-infected seedlings, the disease index values continued to increase until seedling death at 13-14 days postinoculation. One genotype EL51, survived seedling inoculation with the highly virulent isolate. In the field nursery, assessment following inoculation was high with EL51, but not with the susceptible hybrid. The results indicated the potential of the genotype EL51 as a reliable source of resistance to *R. solani* (Nagendran et al. 2009). In a later investigation, sugar beet line FC709-2 showed resistance to R. *solani* AG2-2 IIIB strains with differing virulence levels under field conditions. The line FC709-2 was found to be resistant to all strains of AG2-2 IIIB (Strausbaugh et al. 2013). In another investigation aimed to identify sources of resistance in *Brassica* spp. to isolates of *R. solani* AG2-1, AG8, AG10 and binucleate *Rhizoctonia* spp. (*Ceratobasidium* spp.),

none of the genotypes was found to be resistant to AG2-1. But three promising genotypes with moderate resistance to *R. solani* AG10, AG-8 and binucleate *Rhizoctonia* were identified (Babiker et al. 2013).

R. solani, causing stem rot and target spot of tobacco, is responsible for serious problems in nurseries. Genotypes (97) of tobacco and *Nicotiana* spp. were evaluated for seedling resistance to stem rot and target spot. Significant differences in the responses of the genotypes were observed in the initial stages of seedling development for both stem rot and target spot. But resistance to target spot was not recorded, under high disease pressure. Expression of partial resistance to stem rot could be seen in many genotypes in repeated tests, indicating their usefulness in breeding programs for development of tobacco cultivars resistant to *R. solani* (Elliott et al. 2008). The soybean genotypes (14) were evaluated for their levels of resistance to charcoal rot caused by *M. phaseolina*, using cut-stem inoculation technique under greenhouse and growth chamber conditions. Significant differences ($P < 0.05$) between environments and highly significant differences ($P < 0.01$) among soybean genotypes for relative area under disease progress curve (RAUDPC) values were observed. Soybean genotypes DT97-4290, DT98-7553, DT98-17554 and DT99-16864 showed significantly lower RAUDPC values than 7 of 14 genotypes tested. Among the *Phaseolus* spp., the range of RAUDPC values were similar, compared to that of soybean. *Phaseolus lunatus* cv. Bush Baby Lima had significantly lower RAUDPC values than *P. vulgaris* genotypes. Based on the responses assessed by cut-stem inoculation method, differences in the virulence of *M. phaseolina* isolates and levels of resistance of soybean genotypes to *M. phaseolina* could be established in comparison to field tests (Twizeyimana et al. 2012).

Potato accessions (40) from the core collection were evaluated for their resistance to black dot of tubers, caused by *Colletotrichum coccodes* under greenhouse conditions. The accessions PI 189473, PI 230475, PI 161683, PI 243367 and PI 230470 showed less disease on roots and stems ($P < 0.05$) than commercial standard cultivars. The partial resistance of these accessions was retested. Fifteen plants in each genotype from each accession were propagated into multiple clones which were inoculated with *C. coccodes*. These genotypes originated from PI 243367 and PI230475. Clones of the resistant genotypes were added to the collections of the Potato Improvement Laoratory of USDA-ARS, WA, for use in breeding programs for development of black dot disease (Nitzan et al. 2010). Tuber incubation and PCR assays, in addition to visual examination, were employed to assess the levels of infection of potato tubers by *Helminthosporium solani*, causing silver scurf disease and *Colletotrichum coccodes*, causing black dot disease. Tuber incubation and PCR methods were in slight to fair agreement for detection of both pathogens. Most asymptomatic tubers were found to be infected by one or both assays for *H. solani* (75%) or *C. coccodes* (94%). Minituber inoculation assays were performed to screen potato lines for resistance to silver scurf disease. Of the 14 lines tested, a diploid interspecific hybrid, C287 showed consistently low sporulation, suggesting that these lines might have partial resistance to silver scurf disease (Mattupalli et al. 2013).

Levels of susceptibility of 10 potato cultivars to powdery scab disease caused by *Spongospora subterranea* were assessed for 4 years across the European countries, using standardized scoring scales. Soil contamination by *S. subterranea* was determined, using the real-time PCR and enzyme-linked immunosorbent assay (ELISA) procedures. No relationship could be seen between susceptibility of roots and tubers. Assessments of symptom severity of powdery scab symptoms at 1 month prior to harvest were similar to those at 2 months after harvest. Pathogen population levels determined by real-time PCR and ELISA techniques did not reflect the powdery scab symptom severity ratings. Further, differences in the pathogen populations in various locations did not influence the resistance of most of the cultivars with polygenic resistance (Merz et al. 2012). Potato cultivars (30) and advanced clones (83) were evaluated for the levels of resistance to *Spongospora subterranea*, causing root galls and powdery scab in potato plants and tubers in five field trials during 2011–2012 on naturally infested soils in Minnesota and North Dakota States. Higher disease pressure was observed in North Dakota locations across the years. Formation of root galls and development of powdery scab were influenced by environmental conditions, with greatest variability among white- and red-skinned genotypes. Under high disease pressure conditions, the estimates of broad-sense heritability for powdery scab incidence and severity were 0.76 and 0.63, respectively. Russet-skinned genotypes showed less disease on tubers, but root gall incidence was the same as in red-skinned genotypes. The cultivars Dakota Trailblazer, Dakota Russet and Kani were highly resistant, whereas Shepody, Kennebec and Red La Soda were highly susceptible to both root galls and powdery scab phases of the disease induced by *S. subterranea* (Bittara et al. 2016).

Canola *(B. napus)* cultivars with major gene resistance, derived from *B. rapa* subsp. *sylvestris,* were released for cultivation to tackle the blackleg disease caused by *Leptosphaeria maculans* in Australia. But in 2002, plants with stem cankers caused by *L. maculans* were observed indicating the breakdown of major gene resistance in an area of approximately 50,000 ha in South Australia. There was no relationship between disease severity and incidence of blackleg in 2003 and the proximity to the sites where resistance breakdown occurred in 2002. At some locations, frequency of isolates capable of overcoming the monogenic resistance derived from *B. rapa* subsp. *sylvestris* increased between 2002 and 2003. Isolates of *L. maculans* cultured from canola cultivars with either *B. rapa* subsp. *sylvestris*-derived or polygenic resistance showed host specificity, when inoculated onto cultivars with *B. rapa* subsp. *sylvestris* or polygenic resistance, respectively. The results suggested that selection pressure leading to increased frequency might have been mediated by repeated planting cultivars with the major gene resistance in the same locations for a 3-year period, leading to production of large number of sexual and asexual propagules of *L. maculans* (Sprague et al. 2006). New strains of *P. brassicae* emerged during 2013–2014, capable of infecting most clubroot-resistant (CR) canola genotypes grown in Canada. These strains were inoculated onto two susceptible cultivars, one resistant line and six CR cultivars to determine their responses to the pathogen. All cultivars/lines were susceptible to the new strains of *P. brassicae*. But the severity of disease reaction, root hair infection rates and pathogen DNA contents in each genotype differed, depending on the pathogen strain. Furthermore, the effect of inoculum density on disease severity and gall formation was determined for one of the new strains on a universally susceptible Chinese cabbage cultivar and one susceptible and 10 canola genotypes. At inoculum density of 10^3 spores/ml of soil, root galls were induced. But susceptible and resistant reactions could not be clearly differentiated, until the inoculum concentration level reached 10^5 spores/ml of soil. At spore density of 10^6 spores/ml and above all cultivars and lines showed susceptible reaction. The virulence and inoculum density appeared to determine the type of response of canola genotypes to *P. brassicae* (Hwang et al. 2018).

The genotypes of cocoa were evaluated for their resistance to Ceratocystis wilt disease caused by *Certocystis cacaofunesta* by inoculating seedlings or cuttings with propagule suspensions (10^4 to 10^5 spores/ml). Disease incidence, differences in mortality between resistant and susceptible genotypes, disease index and relative lesion heights were the parameters employed to assess genotype response to infection. The genotypes, ICS-1 and TSH-1188 genotypes proved to be standards for susceptibility and resistance to *C. cacaofunesta*, respectively. Other genotypes showed variable levels of resistance. The genotype CEPEC-2008 was resistant; CEPEC-2002 and CEPEC-2007 were moderately resistant and CEPEC-2009, CCN-10, CCN-51, HW-25, PH-16 and SJ-02 were susceptible. The results indicated the possibility of using the identified cocoa genotypes as sources of resistance in breeding programs (Sanchez et al. 2008).

2.2.1.1.2 Assessment of Disease Resistance Using Molecular Techniques

Assessment of pathogen populations in infected plants has been carried out to determine its reliability as a measure of host resistance to crop diseases. Nucleic acid-based techniques are more sensitive, specific, rapid and reliable for detection, differentiation and quantification of microbial pathogens in plants and soils. Real-time polymerase chain reaction (RT-PCR) assays directly estimate fungal DNA from total DNA extracted from infected tissues and this procedure has been applied to determine fungal development even before any visible symptoms appear. A real-time fluorescent PCR assay, employing a set of specific primers and fluorochrome-labeled probe (TaqMan) was developed to quantify fungal pathogen biomass in the alfalfa plants inoculated with *Phytophhora medicaginis*, incitant of root rot disease. The assay provided estimates of fungal DNA in alfalfa plants with different levels of resistance, based on the analysis of DNA extracted from roots of bulked plant samples. The pathogen DNA content was significantly less in highly resistant plants than in susceptible plants, indicating the potential of the real-time fluorescent PCR assay for use in breeding programs (Vandemark and Barker 2003). The real-time PCR technique was applied for the detection and quantification of *Phytophtora capsici*, causal agent of pepper Phytophthora root rot disease. By employing the real-time PCR assay with SYBR Green Supermix (BioRad) and specific primers for *P. capsici*, the pathogen DNA could be detected as early as 8 h postinoculation. The increase in pathogen DNA contents was more rapid in susceptible cultivars and slower in resistant genotypes PI 1201234 and SCM331. The concentration of *P. capsici* DNA quantified in each pepper genotype showed linear relationship with levels of susceptibility (positive) and resistance (negative). The results demonstrated that the real-time PCR with SYBR Green was sensitive and robust enough to determine both pathogen development as well as to assess the levels of resistance of pepper cultivars and genotypes to *P. capsici*, facilitating the identification of sources of resistance for breeding programs (Silvar et al. 2005). Real-time PCR assay was employed to detect and quantify DNA of *A. euteiches* in pea and soil samples. A significant correlation between pathogen DNA contents and disease severity index (DSI) in infected pea for three pathogen isolates was observed. However, such relationship could not be established for two other isolates of *A. euteiches* (Vandemark and Grünwald 2005).

Phytophthora root and stem rot of soybean caused by *P. sojae* is one of the most destructive diseases and it could be effectively managed by growing resistant cultivars. In order to select soybean cultivars resistant to *P. sojae*, a novel real-time quantitative PCR assay was applied for quantification of pathogen DNA in root tissues. QPCR assay was based on the detection of internal transcribed spacer (ITS) gene of *P. sojae* and 18 S ribosomal gene of soybean. Absolute QPCR allowed detection of as low as 10 fg of *P. sojae* DNA in soybean roots. Relative QPCR, employing the threshold cycle (Ct) method, was effective and reliable for quantification of *P. sojae* DNA normalized to plant DNA in infected soybean root tissues. *P. sojae* DNA concentrations detected in both QPCR assays had high correlations with disease severity index (DSI) ratings of soybean varieties. The results showed that the relative resistance of soybean cultivars to *P. sojae* could be assessed by employing the QPCR procedure developed in this investigation (Catal et al. 2013). A low-cost assay was developed for quantification of P. cinnamomi in avocado rootstocks to assess their tolerance to Phytophthora root rot (PRR) disease. A nested real-time PCR assay quantified the pathogen DNA contents in infected avocado roots. Root samples from highly resistant Dusa and less tolerant R0.12 rootstocks were sampled at 0, 3, 7, 14 and 21 days after inoculation with *P. cinnamomi*. Nested primers were specific and detected *P. cinnamomi* in root tissues. The less tolerant R0.12 rootstock had significantly higher pathogen DNA concentration than Dusa at all sampling times. The extent of tolerance of rootstocks to PRR disease could be quantified, based on the DNA contents of *P. cinnamomi* present in the root tissues. The nested real-time PCR assay provided a reliable assessment of levels of resistance of

rootstocks that may be used for reducing Phytophthora root rot disease in avocado (Engelbrecht et al. 2013).

Identification of defense responses in the host plants forms an essential step for understanding disease resistance mechanism operating in plants. Suppression subtractive hybridization (SSH) library was constructed to study the genes involved in response to infection in banana by *F. oxysporum* f.sp. *cubense (Foc)*. The cDNAs from tolerant genotype *Musa acuminata* subsp. *burmannicoides* 'Calcutta-4' infected by *Foc* were used as a tester and cDNAs from uninfected 'Calcutta-4' as driver population. After hybridization and cloning, expressed sequence tag (EST) library of 83 nonredundant clones were obtained. Based on the sequence analysis and homology search, NCBI database, the clones were assigned to different functional categories. The expression patterns of selected eight defense-related genes viz., peroxidase, glutaredoxin, polyphenol oxidase, glutamate synthase, *S*-adenosyl methionine synthetase heat-shock protein and mannose-binding lectin were analyzed through real-time PCR assay for contrasting the genotypes. The results indicated that the expression of these genes during initial disease development (pathogenesis) was higher in tolerant genotype 'Calcutta-4' than in susceptible genotype 'Kadali' (Swarupa et al. 2013). The collections of barrel medic (*Medicago truncatula*), an important pasture legume, were screened for their resistance to *Fusariun oxysporum* f.sp. *medicaginis (Fom)*. The initial screening of collections and accessions (267) under artificial inoculation revealed a wide range of disease response from completely resistant to highly susceptible to one strain of *Fom*, with 26 accessions being resistant, 9 susceptible and others partially resistant. Quantification of *Fom* within plant tissue indicated that the resistance level of the accessions was correlated with the amount of *Fom* within its shoot. Inoculation with a different *Fom* isolate indicated that the resistant phenotype was stable, because accession response to both *Fom* strains followed similar trends. However, grouping accessions according to their geographic origin did not reveal the foci of resistance, suggesting that resistance might have arisen from independent events. The accessions identified as resistant might be useful as sources of resistance to *F. oxysporum* f.sp. *medicaginis* infecting barrel medic (Rispail and Rubiales 2014).

F. oxysporum f.sp. *lycopersici* (FOL) and *F. oxysporum* f.sp. *radicis-lycopersici* (FORL) are the major soilborne pathogens of both greenhouse- and field-grown tomatoes in the warm vegetable growing countries around the world. Development of tomato cultivars resistant to Fusarium wilt disease is the most effective disease management strategy. Use of molecular marker linked to resistance gene would be useful for tomato improvement programs. Random amplified polymorphic DNA (RAPD) and cleaved amplified polymorphic sequence (CAPS) markers were employed to screen tomato lines against resistance genes *Frl 1* and *1-2* respectively. To understand the genotypic structure (homozygous/heterozygous nature), further analysis was performed by digesting the PCR products with *Fok*I and *Rsa*I restriction endonuclease. An *Rsa*I-digested fragment of 500-bp and two restriction fragments of 390- and 500-bp and two restriction fragments of 390- and 410-bp for *Fok*I after digestion of $TAOI_{902}$ were detected in the homozygous resistant plants. The results showed that of the 27 tomato cultivars, 14 were resistant and 13 were susceptible to Fusarium wilt and 20 were resistant and 7 susceptible to Fusarium root rot, respectively. The pathogenicity tests confirmed the results of molecular marker-linked approach for identifying tomato cultivars resistant to diseases caused by *F. oxysporum* (Morid et al. 2012).

Breeding lines of potato were assessed for their levels of tolerance/resistance to *Verticilliium dahliae*. Based on symptom severity, potato clones were ranked as susceptible or moderately and highly resistant. Quantitative PCR assay was applied to quantify *Verticillium* biomass in the bases of stems of potato plants. A clone was considered as tolerant, when the pathogen biomass present in the test plant tissue was equal to, or greater than the collective average amount for all clones in the symptom severity category. Clones were grouped into tolerant and resistant in different trials, but expression of tolerance was usually unstable across runs. The pathogen could be detected in some asymptomatic plants, whereas some symptomatic plants were pathogen-free. The results indicated the need to use sensitive and reliable technique for precise quantification of *V. dahliae* biomass in potato tissues for selecting genotypes to be used as sources of resistance in breeding programs (Dan et al. 2001). The levels of resistance of potato cultivars to Verticillium wilt caused by *V. dahliae*, in relation to the extent of pathogen colonization of host tissues were studied. Eight russet-skinned cultivars were grown in field with low and high populations of *V. dahliae* in soil. They were evaluated for wilt incidence, stem colonization, yield and tuber vascular discoloration. The QPCR assay validated with strong relationships to culture plating assays was employed to determine stem colonization levels and then related to wilt disease and tuber colonization. The results indicated that russet-skinned potato cultivars had different levels of resistance to *V. dahliae* and that QPCR assay could be employed to reliably and rapidly evaluate resistance levels of potato cultivars to Verticillium wilt under field conditions (Pasche et al. 2013b). In the further study, a QPCR assay developed to detect the target trypsin protease gene of *V. dahliae* was compared with the regular methods of *V. dahliae* quantification by isolation and QPCR assay based on amplification of a region of the ß-tubulin gene of *V. dahliae*. The QPCR assay developed in this investigation, was sensitive enough to detect 0.25 pg DNA of *V. dahliae*. Application of the duplex real-time PCR assay, utilizing potato actin gene to normalize quantification, resulted in clear differentiation of levels of resistance among eight russet-skinned potato cultivars inoculated in greenhouse trials, compared with traditional plating assay and QPCR method, resulting in correlation coefficients greater than 0.93. The new procedure detected the pathogen in inoculated stem tissue at higher frequencies than the other two methods tested. The QPCR procedure was rapid, efficient and precise in quantifying *V. dahliae* population in potato cultivars and had the potential for use in breeding programs (Pasche et al. 2013a).

The relationship between the concentration of DNA of *Verticillium albo-atrum* in leaves of infected alfalfa plants and severity of wilt symptoms was examined, using several alfalfa cultivars expressing a range of symptoms. The disease severity index (1 to 5) was scored after visually examining the infected plants. Real-time PCR assay was applied to quantify pathogen DNA in leaves and stems. Differences in pathogen DNA contents or disease severity index were significant for cultivar-pathogen interaction experiment. However, significant differences were recorded between cultivars for quantities of pathogen DNA detected with real-time PCR and also for disease severity index ratings. In both experiments, the highly resistant cv. Oneida VR had significantly less pathogen DNA and showed lower disease severity index ratings than the resistant cv. Samurai, the moderately resistant cv. Vernema and susceptible cv. Saranac. The results suggested that resistance to Verticillium wilt in alfalfa might be due to reduced colonization of resistant genotypes by *V. albo-atrum* (Larsen et al. 2007).

The responses of oilseed rape (*B. napus*) cultivars Darmor (with quantitative resistance) and Eurol (without quantitative resistance) to *Leptosphaeria maculans*, incitant of Phoma stem canker, were assessed under field and controlled conditions. The parameter used were, numbers of Phoma leaf spot lesions in autumn/winter and severity of stem canker in the following summer in three growing seasons. The numbers of leaf lesions did not vary in Darmor and Eurol in autumn/winter. However, stem cankers were less severe on Darmor than Eurol at harvest in the following summer. Under controlled conditions, development of leaf lesions at different temperatures (5–25°C) and leaf wetness durations (12–72 h) was assessed, using ascospore inoculum. Quantitative PCR assay was applied to quantify symptomless growth of *L. maculans* along leaf petioles toward the stem. The pathogen invasion in stem tissues was visualized, using the green fluorescent protein (GFP)-expressing strain of *L. maculans*. The number of lesions determined on Darmor was greater than on Eurol, although no difference in the incubation periods required by the two cultivars for symptom expression was seen. Further, no differences between Darmor and Eurol could be observed in the distances invaded by *L. maculans* along the petioles toward the stem. So also, the DNA contents of *L. maculans* recovered from leaf petioles showed no difference between Darmor and Eurol. But *L. maculans* colonized stem tissues less extensively in Darmor than Eurol. The results suggested that development of *L. maculans* might be inhibited by polygenic resistance during colonization of stem tissues by the pathogen (Huang et al. 2009). The protocol based on quantitative PCR was applied to quantify DNA contents of *P. brassicae*, infecting *Brassica* hosts, in roots of susceptible, intermediately susceptible, intermediately resistant and resistant Brassica hosts and the nonhost wheat at 5, 10, 15, 20 and 42 days postinoculation (*dpi*). The final disease symptom expression in each genotype was scored as index of disease at 42 *dpi*. *P. brassicae* DNA contents showed an increase in susceptible and moderately resistant hosts from 5 to 42 *dpi*, in contrast to a decrease in highly resistant host and nonhost wheat over the same period. Disease index was significantly positively correlated with the pathogen DNA contents in the roots at 5, 15, 20 and 42 *dpi* in one experiment, and at 10, 15, 20 and 42 *dpi* in the experiment repeated later. Further, the DNA contents of *P. brassicae* in the roots at 42 *dpi* were correlated with those at different intervals after inoculation in the two experiments conducted. The results obtained from qPCR assays were validated by visual examination under microscope of the roots inoculated with *P. brassicae*. The qPCR assay could detect DNA of *P. brassicae* in roots as early as 5 days after inoculation, indicating its usefulness for application in investigations on pathogenesis and epidemiology and host plant resistance (Cao et al. 2014).

Alfalfa cultivars showed different levels of disease severity, when inoculated with *A. euteiches,* causing root rot disease. In order to quantify the DNA of the pathogen races 1 and 2 in alfalfa plants, the PCR assay using a set of specific primers and a dual labeled probe (TaqMan) was developed. The assay differentiated between alfalfa populations for resistance, based on the pathogen DNA contents present in bulked samples. Correlation between pathogen DNA contents and disease severity index ratings ($P < 0.005$) were observed. With race 1 isolate, the amount of pathogen DNA present in resistant check WAPH-5 was lower than in Saranac and WAPH-1. Discrimination between commercial cultivars based on quantitative PCR assay results on bulked plant samples was similar to the visual assessments (Vandemark et al. 2002).

2.2.1.1.3 Assessment of Disease Resistance Using In Vitro Techniques

Molecular methods have been applied as a selection tool for assessment of disease resistance in some pathosystems to a limited extent, since it is not suitable for mass scale germplasm assessment. Potato cultivars showing resistance to potato late blight could be identified by measuring β-glucuronidase (GUS) activity (Kamoun et al. 1998). In a later investigation, the effectiveness of GUS activity as a basis for resistance evaluation was verified in tomato lines against wilt pathogen, *F. oxysporum* f. sp. *radicis-lycopersici* (FORL). The β-glucuronidase (*gus*) reporter gene was integrated into FORL in a co-transformation experiment, using hygromycin B resistance *(hph)* gene as a selective marker. Of the 10 stable transformants, F30 showed strong positive correlation between GUS activity and accumulation of biomass in vitro grown FORL. A parallel increase in lesion development and GUS activity was observed in a susceptible and resistant tomato line. However, the levels of GUS activity in resistant line were greater than in susceptible line, indicating that GUS activity could not be considered as a parameter for differentiation of levels of resistance in tomato lines to wilt pathogen (Papadopoulou et al. 2005).

Responses of *Solanum* germplasm to elicitins produced by *P. infestans*, incitant of potato late blight disease, were assessed to facilitate genetic dissection of chitin response in potato plants. Application of *Agrobacterium tumefaciens* (At) binary *Potato virus X* (PVX) expression system (PVX agroinfection) was modified and optimized for potato-*P. infestans* pathosystem. Clones (11) of *Solanum huancabambense* and *S. microdontum* responded to the elicitors INF1, INF2A and INF2B of *P. infestans*. Clones of the wild relatives showed hypersensitivity-like cell death, following infiltration with purified recombinant INF1, INF2A and INF2B, thereby validating the screening technique. The natural variation in response to INF elicitins in the identified *Solanum* accessions was exploited for evaluating the relationship between INF recognition and late blight resistance. Because several INF-responsive *Solanum* plants were susceptible to *P. infestans*, it seems that INF elicitins have to be considered as general elicitors. They did not have a measurable contribution to resistance to late blight pathogen in *Solanum* spp. (Vleeshouwers et al. 2006). *F. oxysporum* f.sp. *melonis* (*Fom*), causal agent of melon Fusarium wilt, induces vascular disease, due to colonization of xylem vessels by any of one of four races (r0, r1, r2 and r1,2, subdivided into r1, 2w and r1,2y). Traditionally, the resistance levels of melon accessions were assessed by macroscopic evaluation of disease symptoms (disease rating, DR) at several days after inoculation with *Fom* spores. Green fluorescent protein (GFP)-tagging of *Fom* isolate to monitor the disease progression from the site of inoculation, was employed to assess the resistance levels in melon accessions to *Fom*. One isolate from each race of *Fom* was transformed by *Agrobacterium tumefaciens* (At) to constitutively express the GFP. The GFP-transformants, as virulent as the wild-type race, were selected to develop an inoculation bioassay, based on the noninvasive evaluation of the fluorescence emitted by *Fom*-GFP strain. Melon root neck (crown) was found to be the appropriate tissue to monitor movement of GFP-tagged strain and a fluorescence signal rating (FSR) was established in parallel to macroscopic disease rating (DR) assessment. Evaluation of GFP signal in the root neck of inoculated melon seedlings at 11–15 days postinoculation (*dpi*) was performed. The GFP signal was scored in 62 melon accessions/breeding lines inoculated with different GFP-tagged strains, followed by evaluation of DR in the aerial part of melon seedlings at 20–28 *dpi*. The results revealed a significant positive relationship between FSR and DR, indicating the potential of GFP-tagging of pathogen strain/race, as an effective and reliable technique for assessing levels of resistance of melon accessions/genotypes to Fusarium wilt disease (Ramos et al. 2015).

Host-pathogen interactions in pepper-*P. capsici*, incitant of root and crown rot disease were investigated, using an isolate of the pathogen constitutively expressing the gene for green fluorescent protein (GFP). Interactions of *gfp*-expressing *P. capsici* with roots, crown and stems of susceptible bell pepper cv. Red Knight, resistant bell pepper cv. Paladin and *Phytophthora*-resistant land race Criollos de Morelos'334 (CM-34) were studied. The numbers of zoospores attached to and germinated on roots of all cultivars were the same at 30 and 120 min postinoculation respectively. However, at 3 days postinoculation (*dpi*), significantly more secondary roots exhibited necrotic lesions on Red Knight than on Paladin and CM-334 plants. At 4 *dpi*, necrotic lesions could be observed on the tap root of Red Knight, but not on Paladin or CM-334 plants. Furthermore, presence of *P. capsici* hyphae only in a few crowns of Paladin, but none was detected on CM-334, whereas hyphae were present in the crowns and stems of all

Red Knight plants at 4 *dpi*. The hyphae were not detected in the stems of any resistant cultivar at 4 *dpi*. The results demonstrated clearly the distinct differences in the responses of susceptible and resistant plants to infection by *P. capsici*, facilitating the reliable selection of resistant sources for development of cultivars resistant to *P. capsici* (Dunn and Smart 2015).

Plants regenerated from organ cultures, calli or protoplasts, exhibit variations in several characteristics, including disease resistance and the variability expressed by regenerated calli is known as somaclonal variations (Larkin and Scowcraft 1981). Somaclones regenerated from potato cv. Russet Burbank protoplasts were resistant to the late blight pathogen *P. infestans* (Doke and Furnichi 1982). *F. oxysporum* f.sp. *cubense (Foc)*, causal agent of Panama (Fusarium) wilt of banana, exists in the form four races, 1, 2, 3 and TR4. The tropical race 4 (TR4), infecting Cavendish cultivar Taiwan had a different host range from other three races. The tissue culture-derived Cavendish plantlets were screened for their resistance to Fusarium wilt disease by planting them in a *Foc*-infested field. Diseased plant tissues were incorporated into the soil to increase the inoculum density (300–1,000 propagules/g of soil). Plantlets (20,000) were grown for 2 months. Suckers from each of the 18 surviving clones were selected after 1 year. Presence of reddish-brown streaks, characteristic of Fusarium wilt was seen after removal of superficial layers of the rhizomes. Asymptomatic suckers were planted in the disease nursery again. All of the first-generation plants from suckers for six surviving clones were healthy after 1 year. The apparently healthy plants were tested in the same way in the second and third generations. The suckers were highly resistant to *Foc* race 4, compared to parent plantlets. Large numbers of plants from these six clones planted in commercial banana farms proved to be highly resistant to the wilt disease. The resistant plants tested under greenhouse conditions providing high inoculum density, were found to be resistant. Among the variants, only one-line GCTCV-2/8 possessed better horticultural parameters, in addition to resistance to Fusarium wilt disease and the fruits had good consumer acceptance characteristics in Taiwan and Japan (Hwang and Ko 2004).

The bioassay method, involving screening *Musa* genotypes in pots, followed by a hydroponic system was time-consuming. Hence, an in vitro inoculation of rooted banana plantlets grown in modified medium was developed to assess reliably and rapidly the levels of resistance of banana genotypes to *F. oxysporum* f.sp. *cubense (Foc)*. Disease severity was scored based on a rating scale (0–6) at 24 days after inoculation with *Foc* tropical race 4 (TR4) at a concentration of 10^6 conidia/ml. The mean final disease severities of cultivars were significantly different (P < 0.05). The plantlets of susceptible cultivars showed higher disease severity than the resistant cultivars. The cv. Brazil Xiangjiao was highly susceptible to TR4 race with a mean final disease severity score of 5.27. The plantlets of moderately resistant cv. Formosana had a mean score of 3.53, lower than that of Guangfen No. 1 (4.33), but higher than that of resistant cv. Nongke No. 1, GCTCV-119, and Dongguan Dajiao (1.87, 1.73 and 1.53 respectively). There

was no acclimatization stage for plants used in the bioassay, enabling improvement of banana breeding efficiency. The results indicated the possibility of acquiring promising resistant clones through nonconventional breeding techniques such as in vitro selection of clones (Wu et al. 2010).

The levels of resistance of *Daphne* genotypes to *Thielaviopsis basicola*, causing sudden death syndrome, were assessed by aseptic propagation on the medium containing WPM mineral salts and vitamins supplemented with NAA, PVP, MES, calcium gluconate and sucrose. Root cultures were initiated from adventitious roots regenerated on micorpropagated shoots. Proliferative root cultures of *D. caucasica*, *D. cneorum*, *D. jasmine* and *D. pontica* were obtained in both solidified and liquid media, supplemented with different concentrations of NAA. The cultured roots were found to be a convenient system for assessing the levels of resistance to *T. basicola*. Susceptibility of cultured organs differed between select resistant/tolerant genotypes to soilborne fungal pathogens infecting roots (Hanus-Fajerska et al. 2014). Somaclonal variation was utilized as an effective approach in strawberry breeding for development of cultivars with resistance to Verticillium wilt disease caused by *V. dahliae*. The clone K40 was selected from in vitro cultures of strawberry cv. Elsanta and it was highly resistant to *V. dahliae*. The stability and transmission of this trait during in vitro shoot proliferation and traditional clonal propagation from runners were evaluated. Elsanta and Senga Sengana plants were propagated in vitro for 45 generations or for four generations of clonal propagation from runners. Under controlled conditions, regardless of the method of propagation, the K40 clonal plants were much more resistant to *V. dahliae*, than the plants of the original cultivar and the plants of Senga Sengana, considered as resistant to Verticillium wilt in field conditions. The results showed that resistance trait elicited in the selected somaclonal variant K40 was stable and transferable both during in vitro micropropagation and clonal plant propagation from runners (Sowik et al. 2015).

Carrot cavity spot disease caused by *Pythium violae* is one of the major diseases inflicting both quantitative and qualitative losses of marketable carrots. As there was no carrot cultivar with significant resistance or less susceptibility, seed progeny from 46 tissue culture-derived carrot somaclones were screened for variability and 19 selected somaclone families were sown under glasshouse conditions, along with four commercial cultivars, Bertan, Nandor, Bolero and Vita Longa, as controls. Mature roots were exposed to *P. violae* in a cavity spot bioassay to determine their response, based on disease incidence and severity. Some somaclones formed fewer lesions than the least susceptible control cultivar, Vita Longa. Seven somaclone families that showed a range of susceptibility were sown under field conditions and the assessment was repeated. No significant relationship could be observed between glasshouse and field assessments of the levels of resistance of the somaclone families. However, one of the somaclones had a mean disease incidence of 1.9 under field conditions, as estimated by transformed data, compared to 37.9 for the most susceptible clones and 3.5 for Bolero, the most resistant among the commercial cultivars tested. The results indicated the extent of genetic variation that could be obtained for susceptibility/resistance to carrot cavity spot disease through somaclones, for further exploitation of this approach for development of carrot cultivars resistant to *P. violae* (Cooper et al. 2006). Potato cultivars are affected by powdery scab disease caused by *Spongospora subterranea*, resulting in considerable yield losses in quantity and quality of tubers. Variations in the resistance of somaclones of potato to *S. subterranea* were assessed by inoculating somaclonal variants. Most of the variants showed higher level of resistance consistently to the pathogen, both under glasshouse and field conditions, compared to the parent cv. Russet Burbank. The highest level of resistance recorded was 51% for disease incidence and 64% for disease severity (based on tuber surface coverage). However, severity of root infection and root galling due to *S. subterranea* remained at the level similar to the parent Russet Burbank, suggesting that different mechanism of defense responses might operate against this phase of disease development. The results suggested operation of a common mechanism against powdery scab and common scab (*Streptomyces scabiei*) affecting potato tubers (Tegg et al. 2013).

Phytotoxins have a vital role in the development of plant diseases (pathogenesis) and attempts have been made to investigate their role in plant disease resistance mechanisms and their possible utilization for screening in vitro cultures to develop cultivars resistant to the pathogens producing the phytotoxins. The toxins are compounds, that are produced by microbial plant pathogens, involved in the induction of part or all of the symptoms induced by the pathogen. They may be of different types such as peptides, glycoproteins, polysaccharides, organic acids and terpenoids (Turner 1984). Resistance of host pants has been shown to be associated with insensitivity to the toxin produced by the pathogen. Tobacco callus from tobacco cultivar susceptible to black shank pathogen *Phytophthora parasitica* var. *nicotianae* survived better, compared to the callus originating from tobacco cultivars resistant to the disease. Correlation between resistance expressed in vitro and in vivo was observed, when regenerated plants were tested under greenhouse conditions. Toxin specificity could be detected in the species, cultivar or genotypes at plant level. Cultivar specificity was expressed to toxic extracellular protein produced by *P. parasitica* var. *nicotianae*, when applied in bioassays using resistant and susceptible tobacco cultivars (Slavov 2005). Rough lemon (*Citrus jambhiri*), although resistant to *Citrus tristeza virus* (CTV) as rootstock, is susceptible to *Phytophthora citrophthora*, incitant of root rot disease of citrus. In order to enhance the level of tolerance of *C. jambhiri*, cotyledon-derived calli were cultured on selective Murashige–Skoog (MS) medium, supplemented with 5 to 100% of culture filtrate (CF) of *P. citrophthora* for determining the critical concentration of the selective agent. The survival of calli under stress were subcultured for mass propagation for 20 days on callus multiplication medium (with 2,4-D 2 mg/l BA 0.75 mg/l) without the CF. After multiplication, these calli were further exposed to other cycles

of selection, which contained the same and the 3-step higher concentration of CF, and this procedure was repeated several times, until the selection regime was completed. The selected tolerant calli were transferred to regeneration medium (MS medium supplemented with BA 2 mg/l and same concentration of CF on which the calli were selected). Regenerated shoots were transferred to rooting medium (1/2 strength of MS medium supplemented with 0.5 mg/l of NAA). Under in vitro conditions, about 81% of the selected regenerants showed resistance to *P. citrophthora*, whereas all control plants were susceptible (Savita et al. 2011).

Guava (*Psidium guajava*), a nutritionally valuable fruit crop, is affected severely by wilt disease caused by *F. solani* and *F. oxysporum* f.sp. *psidi*. *F. solani* is the most virulent among other *Fusarium* spp. associated with guava. The feasibility of employing in vitro selection, as a means of developing wilt-resistant guava cultivars was explored. A recurrent selection system for guava, both at cellular and plant levels was developed. *In vitro* somatic embryogenesis-derived plantlets of guava cv. Allahabad Safeda were screened against pathogen culture filtrate (CF) at various concentrations, 0, 5, 25, 50 and 100% v/v, incorporated into the Murashige–Skoog (MS) basal liquid medium up to four selection cycles. Culture filtrates of five each of the 10 most virulent strains of *F. oxysporum* f.sp. *psidi* (*Fop*) and *F. solani* (*Fs*) were selected for testing their potential for use as selection agent(s). Of the 10 CFs, three of *F. solani* (F2, F15 and F20) were found to be more effective for selecting both explants of callus and regenerated plantlets. The percentage of survival of callus cultured on media containing different concentrations of CF decreased with increasing concentration, reaching 0% survival (at 100%) in undiluted CF. The CFs of *F. solani* F2 and F15 caused maximum mortality of guava plantlets at 50% concentration. The results showed that the CF of *F. solani* could be used effectively as a selection agent for selecting wilt-resistant guava plants (Kamle et al. 2012).

Two tissue culture-derived plants of two strawberry cultivars Filon and Teresa were evaluated for their resistance to *V. dahliae*, incitant of Verticillium wilt disease. Tissue culturing was continued for 16 months in an environmentally-controlled growth room at 18–20°C, 60 to 70% RH and light/darkness for 16/8 h cycle and subcultures were raised at 6-week intervals on modified Murashige–Skoog (MS) medium. Microplants (400 for each cultivar) were inoculated under in vitro conditions at 4-leaf stage with a homogenate for *V. dahliae* in a liquid medium serving as the selecting agent. Intensity of disease symptoms was scored for each plant using a 0–4 rating scale, at 15, 30, 45, 60 and 70 days postinoculation. Percentages of micropropagated plants of cv. Teresa and cv. Filon, showing disease symptoms, were 76.3% and 89.4%, respectively, whereas the disease severity index for them were 0.0962 and 0.1150 respectively. The percentages of surviving plants were respectively 10.54 in cv. Filon and 23.72 in cv. Teresa. The results indicated that micropropagated cv. Teresa showed greater level of genetic resistance than cv. Filon, based on their response to the selecting agent used. Further, the results were consistent with the levels of field resistance of these cultivars to *V. dahliae*. The protocol developed in this investigation was found to be efficient in distinguishing differing levels of genetic resistance of strawberry cultivars and have the potential for effective assessment of resistance of strawberry cultivars to *V. dahliae* in breeding program (Żebrowska 2011).

Allium white rot disease caused by *Sclerotium cepivorum* is a highly destructive disease in Egypt. In order to identify the resistant onion cultivar, in vitro selection procedure was applied, using the phytotoxin oxalic acid (OA) produced by the pathogen. A protocol for in vitro selection for Allium white rot disease tolerance in commercial onion cultivars, Giza 6 and Behari Red, was developed. Seeds of the onion cultivars were germinated on four concentrations of OA (0.0. 0.02, 0.2, 2.0 and 20 mM). The cv. Behari Red had maximum germination frequency (60%) at all concentrations tested, followed by Giza 20 (58.2%) and Giza 6 (53.3%). Cotyledon explants from the Egyptian cultivars were cultured on the MS medium, supplemented with 0, 3, 6 and 12 mM OA. The mean percentage of survival of calli on the OA-supplemented medium (3 mM OA) was 42.1% and at highest OA concentration (12 mM OA), induction of calli was entirely inhibited. Toxicity of OA to onion calli was expressed as an LD80, where a medium supplemented with 3 mM OA reduced 80% of calli growth which formed a suitable in vitro selection protocol. Among 156 Behari Red calli tested, only 23 (14.7%) survived on OA-supplemented medium for 45 days. In the case of Giza 20 and Giza 6, the survival rates were 15.6% and 40.1%, respectively. Plantlets regenerated from surviving calli, were transplanted in the pots in greenhouse conditions and after acclimatization, these plants formed bulbs. The procedure developed in this investigation could be applied for selecting onion lines with resistance to white rot disease (Ahmed et al. 2013).

2.2.1.1.4 Assessment of Resistance of Rootstocks by Grafting Technique

Grafting with rootstocks showing resistance to soilborne pathogens has been adopted as a disease management strategy for diseases caused by soilborne fungal pathogens. Watermelon (*C. lanatus*) was grafted onto squash (*C. moschata*) rootstocks to reduce the incidence of Fusarium wilt disease (Sato and Takamatsu 1930). Since this early report, grafting has been widely employed as an effective tool to tackle soilborne pathogens affecting Solanaceae and Cucurbitaceae vegetable crops. In addition, grafting has been found to be an alternative to soil fumigants, the use of which is either banned or restricted in the United States and other European countries (Kubota et al. 2008). The usefulness of grafting scions of susceptible vegetable crop cultivars to resistant roots, as a cultural practice, has been discussed earlier. The defense mechanisms operating in grafted plants have been studied in certain pathosystems. The levels of resistance to some soilborne oomycete, fungal and bacterial pathogens have been reported to be enhanced in grafted solanaceous and cucurbitaceous crops.

The effectiveness of grafting commercial Dutch type cucumber hybrids onto various cucurbits used as rootstocks was assessed, under growth chamber and greenhouse conditions to reduce the incidence of root and stem rot disease caused by *F. oxysporum* f.sp. *radicis-cucumerinum* (FORC). Of the *Cucurbita* spp. rootstocks evaluated, A27, *Cucurbita ficifolia*, Patron F1 42.91F, TS-1358F1 and TZ-148F were resistant to infection by FORC. These rootstocks were grafted onto the susceptible cv. BrunexF1, Peto42.91F1, TS-1358F1 and TZ-148F1 were found to be superior to other rootstocks, because of their desirable horticultural attributes, under the prevailing environmental conditions during cucumber crop season in Crete, Greece. The results indicated the usefulness of grafting resistant rootstocks onto susceptible cucumber cultivar, as an alternative to the use of methyl bromide (MB) for the management of root and stem rot disease of cucumber (Pavlou et al. 2002). Seedlings of soybean plant introductions (PIs), two each of partially resistant (PR) and susceptible to *F. solani* f.sp. *glycines (Fsg)*, causal agent of sudden death syndrome (SDS) were inoculated with *Fsg* in aeroponic chambers. Root lengths of the four soybean lines differed, but did not correlate with foliar symptom severity. To understand the resistance mechanism by distinguishing between root and plant resistance, three partially resistant PIs, PI520733, PI567374 and PI567650B and one susceptible cv. GL3302 were compared, using different combinations of grafting in aeroponic chambers. The results indicated that resistance to *Fsg* was conditioned by both scion and rootstock. All three PIs showed resistance associated with the scion; resistance in PI567650B was associated with rootstock. The aeroponic system and grafting might be useful to identify new sources of resistance to *F. solani* f.sp. *glycines* with root- or whole plant resistance (Mueller et al. 2002b).

V. dahliae causes Verticillium wilt disease of solanaceous and cucurbitaceous and other tree crops. Both scions and rootstocks contributed to disease resistance of the grafted combinations in watermelon, melon, cucumber and tomatoes (Paplomatas et al. 2002). Interactions between *V. dahliae*, the incitant of Verticillium wilt disease, and two interspecific hybrid pistachio tree rootstocks (UCBI and PGII and standard rootstocks – *Pistacia integerrima* [resistant] and *P. atlantica* [susceptible]) were investigated. These trees were grown in *V. dahliae*-infested soil and the yield, plant growth and disease incidence were determined. The trees were destructively sampled after 10 days for the presence of *V. dahliae* in the xylem at the graft union. The trees on (*P. atlantica* 'KAC' x *P. integerrima*) hybrid UCBI rootstock grew and yielded as well as those on *P. integerrima*. Trees on the hybrid PGII yielded the least. In soils infested with *V. dahliae*, three associations significantly affected pistachio nut yield. Rootstock affected scion vigor and extent of infection. The extent of infection was negatively related to scion vigor. The trees on *P. integerrima* rootstock had the highest vigor based on visual assessment. However, 65% of these trees were infected with *V. dahliae* in the trunk at graft union region, compared with 73% infection in *P. atlantica* and 25% in UCBI. The results suggested operation of different mechanisms of resistance in *P. integerrima* and UCBI rootstocks (Epstein et al. 2004). Phytophthora blight caused by *P. capsici* is one of the most destructive diseases of cucurbits. In *P. capsici*-infested fields, yield of cucumbers grafted onto bottlegourd (*Lagenaria siceraria*), *C. moschata* and wax gourd (*Benincasa hipida*) rootstocks were significantly increased and vegetable growth was more vigorous (Wang et al. 2004). Watermelons grafted onto selected bottle gourd rootstocks were resistant to *P. capsici* (Kousik and Thies 2010). Tomatoes grafted onto 'Beaufort' rootstocks had lower incidence of corky root disease caused by *Pyrenochaeta lycopersici*, and higher yield and larger fruits (Hasna et al. 2009). Tolerance levels of non-grafted, self-grafted and grafted triploid watermelon (*C. lanatus*) cv. 'Crisp'n Sweet' and heirloom tomato (*Solanum lycopersicum*) cv. Cherokee Purple to *V. dahliae* were compared, in addition to assessments of growth, fruit yield and quality. Rootstocks used for watermelon were 'Emphasis' bottlegourd and 'Strong Tosa' interspecific hybrid (*Cucurbit maxima* x *C. moschata*), whereas 'Beaufort' and 'Maxifort' interspecific tomato (*S. lycopersicum* x *S. habrochaites*) were used as rootstocks for tomato. Foliar symptoms of Verticillium wilt disease were not observed on 'Crisp'n Sweet' watermelon at eastern locations (Oregon and Washington) in both 2010 and 2011. However, at eastern location (Mount Vernon), Emphasis and Strong Tosa-grafted plants had significantly lower Verticillium wilt severity, than non-grafted and self-grafted plants in both years. Presence of microsclerotia was observed in all recovered watermelon stems sampled at Eltopia and Mount Vernon. *V. dahliae* could be isolated from nongrafted and Emphasis-grafted 'Crisp'n Sweet' stems at Eltopia and non-grafted stems at Mount Vernon. Foliar symptoms of Verticillium wilt and microsclerotia in stems were observed in Cherokee Purple plants at either location in both years, in spite of disease incidence in earlier years. The results indicated that grafting with Emphasis and Strong Tosa rootstocks might be effective against Verticillium wilt in western Washington. However, grafting Cherokee Purple onto Beaufort and Maxifort was not advantageous for tomato under field conditions (Buller et al. 2013).

The performance of grafted watermelon in different production regions of Washington, United States, was evaluated. Verticillium wilt-susceptible cv. Sugar Baby (diploid) watermelon grafted onto four commercial rootstock cultivars (Marvel, Rampart, Tetsukabuto and Titan) and nongrafted Sugar Baby, as the control were grown in sites with different inoculum densities (varying from <1.0, 5.7 and 18.0 CFU/g of soil at Othello, Eltopia and Mount Vernon respectively). Area under disease progress curve (AUDPC) values differed significantly among treatments at Eltopia and Mount Vernon. Nongrafted Sugar Baby showed the highest AUDPC values at all three sites, whereas Sugar Baby grafted onto Tesukabuto had the lowest AUDPC values at Eltopia and Mount Vernon. Fruit yield was negatively correlated with AUDPC values at both Eltopia and Mount Vernon. Fruit quality was not affected by grafting at Othello and Eltopia in most characteristics. At the end of season, presence of microsclerotia was detected in all samples. The results indicated that watermelon Verticillium wilt disease could be effectively managed by grafting, when the inoculum density of *V. dahliae* exceeded

5.0 CFU/g of soil in Washington, without any adverse effect on fruit quality (Wimer et al. 2015a). In the further study, nongrafted watermelon PI accessions (14) including *Benincasa hipida*, *C. moschata* and *Lagenaria siceraria*, nongrafted commercially available rootstocks (11) and nongrafted Sugar Baby (susceptible control) were evaluated in a naturally-infested field with *V. dahliae* at 17 CFU/g of soil. Sugar Baby had the highest relative area under disease progress curve (RAUDPC) value (26.80), while PI 419060 had the lowest RAUDPC value (1.46). The mean RAUDPC values of PI accessions (5.49) did not vary significantly from that of commercial rootstocks (5.68). Presence of microsclerotia of the pathogen was detected in the stems of all, except PI 181913. A subset of 12 entries was evaluated under greenhouse and field conditions. A strong positive relationship between the assessments under greenhouse and field conditions was observed. The results indicated that PI accessions and commercial rootstocks could be used for grafting onto watermelon susceptible cultivars to manage the Verticillium wilt disease in Washington. After ascertaining assessment of grafting compatibility, it might be possible to predict the performance of rootstock under field conditions (Wimer et al. 2015b).

The trunk drilling inoculation procedure was followed to assess the relative levels of resistance of olive rootstocks to Verticillium wilt disease, caused by *V. dahliae*. The inoculation procedure involved drilling a hole in the trunk of 3-year-old olive tree, followed by injecting a dense conidial suspension of *V. dahliae*. The conidia of the pathogen were translocated and the pathogen colonized the xylem at the same distance above and below the point of injection in both tolerant and susceptible cultivars. However, the pathogen could be subsequently isolated from significantly higher percentage of inoculated susceptible cv. Amphissis, compared to tolerant cv. Kalamon, indicating the development of resistance mechanisms against the vascular infection by the pathogen. Disease severity in the susceptible cultivar was at least sixfold greater than in tolerant cultivar, at 6–11 months after trunk inoculation. The trunk drilling inoculation method was applied to evaluate the levels of resistance of rootstocks. Resistance assessment in Amphissis and Kalamon was verified by root inoculation with various concentrations of microsclerotia. The results showed the effectiveness of trunk drilling inoculation method for determining the levels of resistance of rootstocks to *V. dahliae* (Antoniou et al. 2008).

Chinese wingnut (*Pterocarya stenoptera*) (WN) selections (9) were evaluated as rootstocks for resistance to *Phytophthora cinnamomi* and *P. citricola* and graft compatibility with scions of five Persian walnut (*Juglans regia*) cultivars. Seedlings of Northern California black walnut (NCB) (*J. hindsii*) and Paradox hybrid (PH) (*Juglans hindsii* x *J. regia)* were used as standards. In a commercial walnut orchard infested with *P. cinnamomi*, cv. Hartley survived and grew markedly better on WN selections than on PH. High resistance to *P. cinnamomi* and *P. citricola* was recorded in all WN selections. The results indicated the usefulness of WN selections as rootstocks for cvs. Tulare and Vina in soils infested with *P. cinnamomi* or *P. citricola*. Further, WN selections might contribute to the required level of resistance in walnut rootstock breeding pro-grams to both pathogens (Browne et al. 2011). The Paradox hybrid, used as the most common rootstock for Persian or English walnut was highly susceptible to another root-infect-ing pathogen *Armillaria mellea*, incitant of Armillaria root disease. Hence, the relative levels of resistance to *A. mellea* in six clonally propagated Paradox rootstocks were evaluated and compared with that of *J. hindsii* rootstock selection W17, *J. regia* scion cv. Chandler and Chinese wingnut. The plants were micropropagated in the growth chamber and rooted in vitro, followed by inoculation of culture medium with *A. mellea*. At 2-month postinoculation, the most resistant and susceptible Paradox rootstocks, respectively, were Ax1 and Vx211 of six rootstocks tested, with 9% and 70% mortality. The broad range of resistance within Paradox was consistent with the results of two field trials conducted earlier. In the third trial, some *J. regia* genotypes were found to be more resistant than those of *J. hindsii*. The results indicated the usefulness of growth chamber evaluation to identify putative resistance rootstock (Ax1), the resistance of which could be verified in the field trials later (Baumgartner et al. 2013).

2.2.1.1.5 Identification of Resistance Genes

Genes exhibiting resistance to microbial plant pathogens have been reported to be clustered in the genome of the host plant species. They may be either in groups of genetically separable loci or apparently in multiallelic series (Jørgensen 1992). Different genes within the same cluster may determine resistance to taxonomically diverse pathogens. Advances in molecular genetics have facilitated identification and mapping of resistance genes on plant chromosomes. Cloning disease resistance genes from several plant species has led to the pos-sibility of investigating the components of signal transduction pathways (Bent 1996; Jones and Jones 1997). It has been gen-erally easy to select host genotypes with resistance controlled by genes at one or a few loci–monogenic, oligogenic or major gene resistance encoded by *R* genes. The major drawback of this type of resistance is the ability of the pathogen to over-come this resistance by producing new physiologic race(s)/strains compatible with *R* gene(s) incorporated into a cultivar. Selection for monogenic resistance based on different receptor encoded by different *R* genes, will result in natural selection for mutants without corresponding elicitor in the pathogen. This, in turn, may lead to loss of resistance due to emergence of another new race of the pathogen. The percentage of a set of dominant *R* gene in host with a matching set of dominant *avr* genes in the pathogen has been determined in some patho-systems (Narayanasamy 2002). All known *R* genes can be grouped into five classes, based on their sequence structure and functional domain/motifs. Many *R* genes belong to the nucleotide-binding site, NBS-leucine-rich-repeat (LRR) type (Hammond-Kosack and Jones 1997; Martin 1999).

Disease-resistance genes may contain similar sequence motifs, although they determine resistance to very different pathogens. Because conserved domains are present in resis-tance genes, it has been possible to clone several additional resistance genes from diverse species by polymerase chain

reaction (PCR) assay with degenerate oligonucleotide prim-ers to the conserved motifs. The progenies of crosses may be prescreened by DNA testing to identify those lines/genotypes carrying known resistance gene or chromosomal sequences closely linked to such gene(s). This process designated marker-assisted selection has been efficiently adopted to identify the resistant genotypes/accessions of crop plant species (Henry 1997). The *R* genes 1-1 and 1-2 conferring resistance to races 1 and 2, respectively, of *F. oxysporum* f.sp. *lycopersici* (FOL) were mapped to chromosome 11 in tomato. The *1-2* gene clus-ter has one functional canopy and six nonfunctional homologs of *1-2* gene. Primers designed based on the sequences of the functional copy were employed in a multiplex PCR assay to differentiate tomato genotypes with *1-2* gene from the ones without this gene. Based on the presence or absence of this marker, 39 of 40 tomato genotypes could be differentiated. The cv. Plum Crimson carrying the *1-3* gene was the excep-tion and it exhibited resistance to races 1, 2 and 3 of FOL. The results were validated, and results of the bioassays also cor-roborated the conclusions of the investigation, adding cred-ibility to the effectiveness of marker-assisted selection (MAS) approach in breeding programs (El-Mohtar et al. 2007).

Molecular markers permit dissection of monogenic and quantitative resistance. They are useful for preserving and exploiting germplasm, marker-assisted selection (MAS) of resistance genes and deployment of suitable genes for improv-ing cultivar resistance to target pathogen(s). By adopting MAS approach, effective deployment of resistance genes to provide stable resistance to different economically important diseases affecting major crops, could be accomplished. All regions of the plant genome can be assayed for linkage to resistance, using molecular markers. Resistance to diseases has been found to be linked with undesirable traits. Mapping of the genome, using molecular markers, helps identification of quantitative loci (QTL) for several characteristics, includ-ing resistance to diseases. For rapid mapping of monogenic resistance genes in segregating populations, bulked segregant analysis (BSA) procedure is followed frequently. Locations of many disease resistance genes in clusters in the genome have been detected by employing molecular markers. Development of microarray-based expression profiling methods, in addition to the availability of genomic and/ or expressed sequence tag (EST) data for some plant species, has facilitated rapid prog-ress in characterization of plant pathogenesis-related (PR) responses. It is possible to monitor the expression of even thou-sands of genes simultaneously, by applying DNA microarray technology. New pathogenesis-related genes, coregulated genes and the associated regulatory systems can be identified. Two types of DNA microarrays and oligonucleotide-based arrays are useful to study plant-pathogen interactions. DNA microarray technique allows the examination of the responses of large number of genes simultaneously at a given sampling time. Based on these expression profiles, it is possible to iden-tify the differentially synthesized mRNA species, based on the potential defense-associated functions. Hypothesized *R* gene products are considered to have significant role in initiat-ing many plant defense responses. However, DNA microarray investigations indicated that most of the known *R* genes did not exhibit high levels of transcriptional regulation. It is likely that many *R* genes do not need to be regulated, but instead encode constitutive receptors that may regulate the expression of other genes. Alternatively, *R* gene expression may be required at posttranscriptional levels (Austin et al. 2002; Nishimura and Somerville 2002).

In potato, disease resistance may be governed by one or few loci (monogenic, oligogenic or major gene resistance encoded by *R* genes). New races compatible with host *R* gene(s) are able to overcome the resistance of cultivars with *R* genes. A shift in the objective of breeding became essential and breeding for potato cultivars with polygenic or nonspecific resistance (field resistance) was considered as an effective alternative. Molecular techniques have been employed for characterization of different stages of interaction between the host plant species and microbial pathogens. By using molecular markers, it is possible to assay all regions of the plant genome for linkage of disease resistance with other traits. High resolution of genetic map at the *R3* locus, conferring high level of resistance to avirulent isolates of *P. infestans* was constructed. The *R3* locus consists of two genes R3a and R3b, with distinct specificities which were 0.4 cM (centimorgan) apart. These genes from *Solanum demissum* were incorporated into potato. One accession of *S. demissum* had a natural recombination, between *R3a* and *R3b* (Huang et al. 2009). A new late blight resistance (*R*) locus was identified in S. *bulbocastanum*. A high-resolution genetic map of the new locus was generated, delimiting *Rpi-blb3* to a 0.93 cM interval on chromosome 4. An AFLP marker that cosegregated with *Rpi-blb3* in progeny plants (1,396) of an intraspecific mapping population was identified. In addition, three other late blight resistance genes, *Rpi-abpt2*, *R2* and *R2*-like appear to reside in the same *R* gene cluster on chromosome 4 as the newly identified gene, *Rpi-blb3* (Park et al. 2004).

P. infestans, causing late blight disease in potato and tomato, is a heterothallic species, requiring the presence of two sexually compatible isolates for production of oospores. Pathogen populations in the United States were primarily asexual with limited recombination or soilborne persistent oospores. In order to assess disease development and production of oospores, using US-22 (A2), US-23 (A1) and US-24 (A1) clonal lineages, 15 solanaceous hosts with different resistance genetics, were investigated. Potato and tomato transformed with *RB* gene, and tomato 'Mountain Magic', carrying *Ph-2* and *Ph-3* genes were most tolerant to all pathogen lineages with significantly less disease severity (< 30%) than other cultivars (12) with higher severity (40–100%)). Oospores were formed in greater number at 16°C, following inoculation of susceptible and resistant hosts, with both mating pairs (A1 and A2). The US-24 x US-22 cross produced significantly more oospores than the US-23 x US-22 cross. Oospores were not detected in tomato plants with *Ph-2* and *Ph-3* genes or potato and tomato with *RB* gene, using whole plant assays. Cultivation of varieties with *Ph* and *RB* genes might reduce the risk of oospore formation and sexual recombination, resulting in the production of populations capable

of infecting cultivars showing resistance to *P. infestans* (Sanchez-Perez et al. 2017).

Blackleg disease (Phoma canker) of oilseed rape caused by *Leptosphaeria maculans* accounted for considerable yield losses in Australia. Resistance in oilseed rape cultivars derived from *B. rapa* subsp. *sylvestris*, like cv. Surpass 400, became ineffective within three years after its release for commercial cultivation. The field isolates of *L. maculans* were grouped into three phenotypic classes viz., virulent, intermediate and avirulent. Analysis of crosses between fungal isolates with varying virulence on cv. Surpass 400, showed the presence of two unlinked avirulence genes *AvrLm1* and *AvrLmS*. Complementation of isolates (genotype *avrLm1*) with a functional copy of *AvrLm1*, and genotyping of field isolates, using molecular marker for *AvrLm1*, showed that virulence toward *Rlm1* was necessary, but not sufficient, for expression of a virulent phenotype on cv. Surpass 400. The results suggested that cv. Surpass 400, with 'sylvestris'-derived resistance might possess at least two resistance genes, one of which could be *Rlm1* (Van de Wouw et al. 2009). Canola (*B. napus*) is infected by *Leptosphaeria maculans*, causing blackleg disease in all countries, except in China, where the less aggressive species *L. biglobosa* causes the disease. Chinese *B. napus* cultivars/lines were evaluated for their seedling resistance (major gene resistance/*R* gene resistance) and adult plant resistance (APR). Seedling resistance was determined under controlled conditions, using 11 well-characterized differentials (reference strains). On the other hand, APR was assessed under multiple field environments prevailing in Western Canada. The *R* genes were detected in 40% of the accessions tested. Four specific *R* genes, *Rlm1*, *Rlm2*, *Rlm3* and *Rlm4* were identified. The genes *Rlm3* and *Rlm4* genes were the most frequently detected genes in canola accessions. Significant variations in the responses of canola accessions were recorded in different locations. A large percentage of *B. napus* accessions exhibited high to moderate levels of APR under all environments tested, regardless of their levels of seedling resistance to *L. maculans*. The results indicated the potential of Chinese accessions of *B. napus* for use as sources of resistance against *L. maculans*, when the pathogen might ever become established in China (Zhang et al. 2017).

The performance of 12 soybean cultivars with partial resistance, with or without *Rps* genes to different populations of *P. sojae*, incitant of Phytophthora root and stem rot disease, was analyzed under field conditions with different levels of disease pressure. The soybean cultivars were evaluated in seven field environments with and without metalaxyl application over 4 years. A highly significant genotype-environment interaction was observed partially, because of variable disease pressure. The incidence of *Phytophthora* stem rot in subplots ranged from 0 to 10 plants in the most susceptible cv. Solan, while significantly less stem rot developed in cultivars with high level of partial resistance or partial resistance combined with an *Rps* gene in 3 of 7 environments. When diverse populations of *P. sojae* were present, yields from soybean cultivars with high levels of partial resistance were significantly higher than those with low levels of partial resistance. Soybean cultivars with specific resistance genes *Rps1k*, *Rps1k* + *Rps6*, or *Rps1k* + *Rps3a* had higher yields than those with only partial resistance in environments, where race determination indicated that the populations of *P. sojae* were not capable of causing disease on plants with the *Rps1k* gene. However, in low disease pressure conditions, soybean cultivars with partial resistance yielded at the same level as the cultivars with single *Rps* gene or *Rps* gene combinations. The results showed that genetic traits associated with high levels of partial resistance did not have a negative effect on yield. Soybean cultivars, that had most consistent ranking across environments, were those with moderate levels of partial resistance in combination with either *Rps1k* or *Rps3a* (Dorrance et al. 2003).

In order to characterize the inheritance of resistance to *P. sojae* in soybean plant introductions (PIs), 22 soybean populations from crosses of the 22 PIs and the susceptible cv. Williams were inoculated with isolates of *P. sojae* OH17 (vir1b, 1d, 2, 3a, 3b, 3c, 4, 5, 6 and 7) and OH25 (vir 1a, 1b, 1c, 1k and 7). The isolates were virulent on soybean with all known *Rps* genes and many *Rps* gene combinations. Thirteen of 22 populations had consistent segregation responses following inoculations between two generations. In two PIs, resistance was conferred by two genes to OH17 and three genes to OH25. Resistance to both isolates was conferred by a single gene in PI398440, although the individual families were not resistant to the same isolates. The results indicated that these PIs could be used as sources of novel *Rps* genes for effective management of Phytophthora root and stem rot disease of soybean (Gordon et al. 2007a). The *Rps* genes present in the PIs were examined for their association with eight described *Rps* loci that were mapped to soybean molecular linkage groups F, G, J and N. Nine $F_{2:3}$ soybean populations were genotyped with simple sequence repeat (SSR) markers linked to previously mapped *Rps* loci. The nine PI populations had SSR markers associated with resistance to *P. sojae* isolate OH17 in the *Rps1* region. *Rps1c* might be a candidate in 8 PIs, but novel genes might confer resistance in one PI to *P. sojae* isolate OH17. Two or more *Rps* genes, including some that are potentially novel, may confer resistance to *P. sojae* isolate OH25 in eight populations. However, based on the response to those isolates, virulence could be linked to some of the novel genes identified in this investigation (Gordon et al. 2007b). The components of resistance to Phytophthora root and stem rot disease of soybean caused by *P. sojae*, such as lesion length, number of oospores formed in the plant tissues and infection frequency were measured in 8 soybean genotypes inoculated with two *P. sojae* isolates to differentiate partial resistance from other types of incomplete resistance. The *Rps2* and root rot–resistant genotypes had significantly lower oospore production and infection frequency, compared with the partially resistant genotype and also had significantly smaller lesion lengths. However, the high levels of partial resistance in Jack were distinguishable from *Rps2* in L76-1988, based on the evaluation of these components. Root rot resistance in Ripley and *Rps2* in L76-1988 had similar responses for all components measured. Partial resistance expressed in Conrad, Williams, Jack and General was comprised of various components that

interact for defense against *P. sojae* in the roots, and different levels of each component were found in each of the genotypes. However, forms of incomplete resistance expressed via single genes in Ripley and *Rsp2* in L 76-1988, could not be differentiated from high levels of partial resistance, based on the components of disease resistance, used as basis in this investigation (Mideros et al. 2007).

F. oxysporum f.sp. *ciceris (Foc)* exists in the form of several races with differing virulence levels and they have been identified, using a set of differentials. The effects of temperature maintained during screening chickpea cultivars/genotypes for resistance to races of wilt pathogen *F. oxysporum* f.sp. *ciceris (Foc)* were assessed. The cv. Ayala was moderately resistant to *Foc*, when inoculated plants were maintained at a day/night temperature regime of 24/21°C, but it became highly susceptible at 27/25°C. Field experiment conducted in Israel over three consecutive years indicated that the high level of resistance of Ayala to Fusarium wilt, when sown in mid- to late-January differed from a moderately susceptible reaction under warmer temperatures, when sowing was delayed to February or early March. Growth chamber experiments showed that a temperature increase of 3°C from 24 to 27°C was sufficient to alter resistance levels of cvs. Ayala and PV1 to race 1A of *Foc* from moderately or highly resistant at 24°C to highly susceptible at 27°C. A similar, but less pronounced effect was observed, when Ayala plants were inoculated with *Foc* race 6. On the other hand, the reaction of cv. JG-62 plants inoculated with a variant of race 5 of *Foc* at 27°C, compared well with plants inoculated at 24°C. The results indicated the need for conducting screening tests at appropriate temperature regime to assess the susceptibility/resistance of cultivars/genotypes and also to identify races of *Foc*. In addition, the information on the effect of temperature on chickpea cultivar responses to *F. oxysporum* f.sp. *ciceris*, may be useful to choose the date of sowing of chickpea, as disease management strategy to reduce disease incidence and severity (Landa et al. 2006).

The genetics of resistance to races 1A, 2, 3, 4 and 5 of *Foc* was studied using a population of 100 F_7 recombinant inbred lines (RILs) from a cross of the accessions WR-315 (resistant) and C-104 (susceptible). A population of 26 F_2 plants from a cross between the same two parents was used to study the inheritance of resistance to race 2. Segregations of the RILs for resistance to each of the five races suggested that single genes in WR-315 might govern resistance to each of the five races. The 1:3 resistant to susceptible ratio in the F_2 populations indicated that resistance in WR-315 to race 2 was governed by a single recessive gene. A race-specific slow disease progress reaction was observed in chickpea line FLIP84-92(3) to infection by race 2, a phenomenon designated 'slow wilting', that is different from 'late wilting', with respect to latent period, disease progress rate and final disease rating. The germplasm lines of *Cicer arietinum* (27) and *C. reticulatum* (2), including differentials used earlier, were evaluated for their reactions to infection by the five races. Of the 29 germplasm lines, only eight could differentiate at least one of the five races, based on either susceptible or resistant reactions, whereas the remaining germplasm lines were either susceptible or resistant to all five races or differentiated them by intermediate reactions. A selected set of eight chickpea lines consisted of four genotypes and four F_7 RILs with vertical resistance was developed as differentials for identification of races of *F. oxysporum* f.sp. *ciceris (Foc)*. The selected differentials reacted with early development of wilt symptoms, and clear and consistent disease phenotypes, based on 'no wilt' or 100% wilt incidence, providing unambiguous categorization of responses of test entries, leading to precise identification of races of *F. oxysporum* f.sp. *ciceris* (Sharma et al. 2005). In a later investigation, a second gene conferring resistance to *F. oxysporum* f.sp. *ciceris* race 0 was mapped to linkage group 2 (LG2) of the chickpea genetic map. Resistance to race 0 was governed by two genes which segregated independently. One gene was detected in accession JG62 ($FocO_1/focO_1$) and mapping to LG5 and the second gene was present in accession CA2139($FocO_2/focO_2$), but remaining unmapped. Both genes separately conferred complete resistance to race 0 of *Foc*. Using a recombinant inbred line (RIL) population that segregated for both genes (CA2139 × JG62) and the genotypic information provided by two markers flanking $FocO_1/focO_1$, 10 resistant lines containing the resistant allele $FocO_2/focO_2$ were selected. Genotypic analysis using these 10 resistant lines paired with ten susceptible RILs, selected in the same population, revealed that sequence-tagged microsatellite sites (STMS) markers present on LG2 were strongly associated with $FocO_2/focO_2$. Linkage analysis, using data from two mapping populations (CA2139/JG62 and CA2156/JG62), located $FocO_2/focO_2$ in a region, where genes for resistance to wilt pathogen races 1, 2, 3, 4 and 5 were earlier reported and which were highly saturated with tightly-linked STMS markers that could be selected for marker-assisted selection (MAS) approach (Halila et al. 2009).

The doubled-haploid melon lines (7) parthenogenetically originated, using irradiated pollen were evaluated for resistance to melon wilt pathogen *F. oxysporum* f.sp. *melonis (Fom)*. Two lines Nad-1 and Nad-2 were selected after successive inoculations with *Fom* race 1.2w strain and they were compared with commercial hybrids and parents cv. Isabelle and Giallodi Paceco. The lines Nad-1 and Nad-2 had higher levels of resistance than other genotypes tested. The resistance expressed in the two doubled-haploid lines could be due to their homozygous state that maximized the expression of resistance genes already present in the parent Isabelle. The lines Nad-1 and Nad-2 had the potential for use in breeding programs (Ficcadenti et al. 2002). *F. oxysporum* f.sp. *melonis (Fom)* race 1.2 and its strains could overcome two dominant resistance genes (*Fom1* and *Fom2*) and they are further divided into two types, depending on the symptoms they induce–yellowing or wilting. Recombinant inbred line (RIL) population that was developed by single seed descendent from an F_1 hybrid between Isabelle, a partially resistant line and a susceptible line 'Vedrantais' were evaluated for their partial resistance to *Fom* race 1.2. The plants were inoculated with yellow strain (TST) and wilting strain (D'Oléon 8) and disease severity was scored, using a rating scale of 1–5 and area under disease progress curve (AUDPC) values were also

determined. Phenotypic correlations were highly significant between six different locations and experiments. The heritability of resistance was high from 0.72 to 0.96 and 4 to 14 genetic factors were estimated to confer resistance to *Fom* race 1.2. Thirteen other strains were tested with an RILs subset. Some small strain-specific effects might be involved (Perchepied and Pitrat 2004). Fusarium root and stem rot caused by *F. oxysporum* f.sp. *radicis-cucumerinum* (FORC) affected cucumber yield levels of greenhouse cucumbers seriously. Two melon accessions, HEM (highly resistant) and TAD (partially resistant) were crossed with the susceptible accession DUL. Evaluation of the responses of three accessions and F1 crosses showed that HEM contributed higher resistance to the F1 hybrid than TAD. Roots of susceptible and resistant accessions showed 100% and 79% colonization by FORC respectively, when inoculated artificially. However, colonization of upper plant tissue was observed only in susceptible accession. Microscopic examination of sections of crown region of the susceptible DUL plants revealed the presence of profuse fungal growth in the intercellular spaces of the parenchyma, and none in the xylem or any other vascular tissues (Cohen et al. 2015).

Tomato Fusarium vascular wilt caused by *F. oxysporum* f.sp. *lycopersici (Fol)* has worldwide distribution, impacting production seriously. Development of tomato cultivars resistant to the disease may be the most effective and environmentally safe approach of managing the disease. Resistance genes to *Fol* race 1 (*I-1* gene) and race 2 (*I-2* gene) were mapped to chromosome 11. The *I-2* gene cluster included one functional copy and six nonfunctional homologs of the *I-2* gene. A multiplex PCR assay was developed to differentiate *I-2* genotypes from genotypes without genes, using primers based on the functional gene copy sequence. Of the 40 tomato genotypes tested, 39 with known reactions to race 2 gave positive results. The only exception was the cv. Plum Crimson, carrying the *I-3* gene for resistance to races 1, 2 and 3 of *Fol*, indicating the effectiveness of the multiplex PCR assay in differentiating the tomato genotypes (El Mohtar et al. 2007). Tomato genes *I-3* and *I-7* conferring resistance to *F. oxysporum* f. sp. *lycopersici (Fol)* race 3, were introgressed from *Solanum pennellii*. The gene *I-3* was identified in chromosome 7 and it encoded an S-receptor-like kinase. An RNA-seq and single nucleotide polymorphism (SNP) analysis approach was employed to map *I-7* to a small introgression of *S. pennelli* DNA on chromosome 8 and identified *I-7* gene encoding a leucine-rich-repeat receptor-like protein (LRR-RLP). The *eds1* mutant of tomato was used to show that *I-7* gene, like many other LRR-RLPs conferring resistance to *Fol* in tomato, was EDS1 (enhanced disease susceptibility)-dependent. Using transgenic tomato plants carrying only the *I-7* gene for *Fol* resistance, *I-7* gene also was shown to confer resistance to *Fol* races 1 and 2. As *Fol* race 1 carried *Avr1*, resistance to *Fol* race 1 indicated that *I-7* mediated resistance, unlike *I-2* or *I-3*-mediated resistance, was not suppressed by Avr1. As Avr1 was a bit general suppressor of *Fol* resistance in tomato, Avr1 might be acting against an EDS-1 dependent pathway for resistance activation. Identification of *I-7* gene allowed development of molecular markers for MAS of both genes currently known to confer *Fol* race 3 resistance (*I-3* and *I-7*). Because I-7-mediated resistance was not suppressed by Avr1, the gene *I-7* could be a useful addition for exploitation in breeding programs (Gonzalez-Cendales et al. 2016). *F. oxysporum* causes one of the most important vascular wilt diseases in cape gooseberry (*Physalis peruviana*). As development of cultivars with resistance to wilt disease was considered as the most effective disease management strategy, attempts were made to select promising genotypes of cape gooseberry and to identify single nucleotide polymorphism (SNPs) associated with *Fusarium* resistance as potential markers under greenhouse conditions. Cape gooseberry accessions (21) showed different resistant responses within 100 accessions tested. A total of 60,663 SNPs were identified using Genotyping by Sequencing (GBS) procedure. Model-based population structure and neighbor-joining analyses showed three populations comprising the cape gooseberry panel. The SNP markers associated with resistance response against *F. oxysporum* were identified, based on common tags, using the reference genomes of tomato and potato, as well as the root/stem transcriptome of cape gooseberry. By comparing with their location with tomato genome, 16 SNPs involved in defense/resistance response were identified. Likewise, comparison with potato genome led to the identification of 12 SNP markers. The results revealed the first association mapping in cape gooseberry for identifying promising accessions with resistance phenotypes and a set of SNP markers mapped to defense/resistance genes against *F. oxysporum* (Osorio-Gaurin et al. 2016).

Sweet basil (*Ocimum basilicum* L.) suffers heavily, because of the destructive nature of Fusarium wilt disease caused by *F. oxysporum* f.sp. *basilici* (*Fob*) and the problem became acute, due to intensive monoculture in greenhouse and unavailability of methyl bromide for soil fumigation. To overcome the production loss, the cv. Nufar was used as the resistant parent for breeding for disease resistance. The mode of inheritance of resistance to *Fob* in sweet basil was investigated in progenies derived from crosses of the homogenous resistant cv. Nufar and the susceptible homogenous cv. Chen. Artificial inoculation of seedlings with a high concentration of a microconidial suspension of a virulent isolate of *Fob* showed no variations between reciprocal backcrosses. The nuclear resistance analysis and dominant characteristics of the resistance of backcross progenies of different parental lines and the segregation for resistance in F2 combinations fitted the expected Mendelian ratio for single dominant gene with two alleles that confer resistance to Fusarium wilt in basil. The results indicated the usefulness of the resistant cv. Nufar for breeding programs for combating the Fusarium wilt disease of sweet basil (Chaimovitsh et al. 2006). In Western Australia and Queensland, Fusarium wilt disease outbreaks caused by *F. oxysporum* f.sp. *fragariae* (*Fof*), occurred increasingly in strawberry (*Fragaria* x *ananassa*) adversely affecting production seriously. In order to develop strawberry cultivars with resistance to *Fof*, a partial incomplete diallelic cross involving four parents was performed for glasshouse screening of progenies which were evaluated for their susceptibility to *Fof*. Disease severity ratings were used to identify the best-performing progenies and analyzed using a linear mixed model, incorporating a pedigree to produce best linear unbiased predictions of breeding values. Variations in disease severity, ranging from highly susceptible to resistant, indicated a quantitative effect. The estimate of the narrow-sense heritability was 0.49 ± 0.04 (SE), suggested that the population should be responsive to phenotypic recurrent selection. Several progeny genotypes had predicted breeding values greater than any of the parents (Paynter et al. 2014).

Development of potato cultivars resistant to Verticillium wilt primarily caused by *V. dahliae* and *V. albo-atrum* was considered as a long-term environmentally-safe strategy to contain the disease. High levels of resistance to stem inoculation were identified in two diploid hybrids between the cultivated potato (*Solanum tuberosum*) and wild relative *Solanum* spp. An intercross between two clones produced a 3:1 ratio of resistant to susceptible plants. The results indicated that a two-gene model in which dominant alleles of both genes might be able to confer resistance to *Verticillium* spp. The two-gene model indicated that a simple mode of inheritance might improve the probability of producing resistant progenies, when resistant alleles were used as parents in breeding programs (Jansky et al. 2004). *Verticillium longisporum* causes Verticillium wilt disease of oilseed rape which is widespread in north Europe, as the area cropped to oilseed rape rapidly increased in these countries. As all commercial cultivars were susceptible to the disease, and very little resistance to *V. longisporum* was available, intraspecific genes transfer from related species was considered, as the most desirable approach for broadening levels of resistance in the cultivars. The amphidiploid species *B. napus* could be resynthesized by crossing the two progenitor species *B. oleracea* and *B. rapa*. The resistant accessions of these two diploid species might be used as resistant donors. The potential of *B. rapa*– and *B. oleracea*–resistance donors (43) was tested for their responses to mixture of two virulent isolates of *V. longisporum* and resistance levels of diverse lines were combined by embryo rescue-assisted interspecific hybridization in resynthesized rapeseed lines. Progenies from crosses of two *B. rapa* genebank accessions 13444 and 56515 to *B. oleracea* gene bank accessions BRA1008, CGN14044, 8207, BRA1398 and 7518 showed a broad spectrum of resistance in pathogenicity tests of the 45 tested resynthesized lines. Significantly higher levels of resistance were detected in 41 lines than in the moderately *V. longisporum*-tolerant cultivar Express (Rygulla et al. 2007). Plant pathogenic fungi secrete effector molecules to establish themselves on their hosts, whereas plants use immune receptors to intercept such effectors of the pathogen(s) to prevent pathogen colonization. The tomato cell surface-localized receptor Ve 1 conferred race-specific resistance against race 1 strains of *V. dahliae* which secreted the Ave 1 effector. Cloning and characterization of *Ve 1* homologues from tobacco, potato, wild eggplant (*Solanum torvum*) and hop (*Humulus lupulus*), were performed. The results showed that particular Ve 1 homologues governed resistance against *V.*

dahliae race 1 strains through recognition of the Ave 1 effector. Phylogenetic analysis showed that *Ve 1* homologues are widely distributed in land plants, suggesting an ancient origin of the *Ve 1* immune receptor in the plant kingdom (Song et al. 2017).

Fusarium vascular wilt of cotton, caused by *F. oxysporum* f.sp. *vasinfectum (Fov)* forms a complex, when infected by root-knot nematodes *Meloidogyne incognita*, resulting in a serious situation, requiring concerted efforts to manage the disease. Genetic analysis of two interspecific crosses (*Gossypium barbadense* cv. Pima S-7 x *G. hirsutum* Acala Nem x Pima S-7 x Acala SJ-2) with allele dosage effect conferred resistance to *Fov* race 1 in Pima S-7. Two amplified fragment length polymorphism (AFLP) markers were linked to *Fov 1* in Pima S-7, with genetic distance from the gene of 9.3 and 14.6 centimorgans (cM). Induction of less severe wilt symptoms in Acala Nem X than in Acala SJ-2 indicated that Acala Nem X had one or more minor genes contributing to delayed expression of wilt symptoms. Highly resistant plants in F_2 and F_3 (Pima S-7 x Nema X) families indicated transgressive segregation effects of minor genes in Acala Nem X combined with *Fov 1* from Pima S-7. The effects of wilt and nematode resistance were assessed using two inoculation methods. When both pathogens were present, wilt damage, as reflected by shoot and root weight reductions was the greatest in Acala SJ-2 (susceptible to both pathogens) and least in root-knot nematode-resistant and wilt susceptible Acala Nem X. The disease severity was intermediate in wilt-resistant and nematode-susceptible Pima S-7. The results showed that nematode resistance was more effective than wilt resistance in suppressing wilt symptoms, when either resistance was found alone. Nematode resistance combined with intermediate wilt resistance, as in F_1 (Pima S-7 x Nem X) was highly effective in protecting plants from race 1 of *F. oxysporum* f. sp *vasinfectum* and nematode infection, and consequently prevention of disease complex in cotton (Wang and Roberts 2006).

Most of the vascular pathogens have restricted host ranges. In contrast, *Verticillium longiporum* can infect cruciferous crops, like cabbage, cauliflower and rapeseed, whereas *V. dahliae* and *V. albo-atrum* cause wilt diseases in over 200 dicotyledonous species, including many economically important crops. A locus responsible for resistance against race 1 strains of *V. dahliae* and *V. albo-atrum* was cloned from tomato. This locus *Ve* has two closely linked inversely oriented genes, *Ve 1* and *Ve 2* that encode cell surface proteins of extracellular leucine-rich repeat receptor-like protein class of disease resistance proteins. The results showed that *Ve 1*, but not *Ve 2*, provided resistance in tomato against race 1 strains of *V. dahliae* and *V. albo-atrum* and not against race 2 strains. The requirement of both EDS1 and NDR1 was demonstrated, using virus-induced gene silencing (VIGS) in tomato, the signaling cascade downstream of *Ve 1*. The *Ve 1*-mediated resistance signaling only partially overlaps with signaling mediated by Cf proteins, type members of the receptor-like protein class of resistance (Fradin et al. 2009).

The responses of chickpea (*Cicer arietineum*) to *Didymella rabiei*, causing Ascochyta blight were assessed.

Two reciprocal populations derived from intraspecific crosses between a moderately resistant late flowering Israeli cultivar and a highly susceptible early-flowering Indian accession, were tested at F_3 and F_4 generations in 1998 and 1999, respectively. A quantitative (parametric) assessment (percent disease severity) was used to evaluate response of chickpea under field conditions. The transformed relative area under disease progress curve (tRAUDPC) value was calculated for each experimental unit for further analysis. Heritability estimates of tRAUDPC values were relatively high (0.67 to 0.85) in both generations for both reciprocal populations. The presence of major genes was examined by testing the relationship between F_3 and F_4 family means and within-family variances. Analyses of these relationships suggested that segregation of a single (or a few) quantitative locus with major effect and possibly other minor loci was the predominant mode of inheritance. The correlation of estimates between the resistance and days to flower (r = 0.19 to –0.44) were negative and significantly (P = 0.054 to 0.001) different from zero, representing a breeding constraint in the development of early flowering chickpea cultivars with resistance to Ascochyta blight disease (Lichtenzveig et al. 2002).

Resistance to *P. brassicae* (causal agent of canola clubroot disease) pathotypes in Canada could be identified in the primary and secondary gene pools of spring *B. napus* canola. Some of these sources, such as winter canola 'Mendel', rutabaga and Pak choi (*B. rapa*) 'Flower Nabana', were used in genetic investigations and breeding programs for development of clubroot-resistant canola cultivars. A dominant gene in cv. Mendel and Flower Nabana conferred resistance to *P. brassicae* pathotype 3, whereas a simple or complex genetic control of resistance was detected in rutabaga. The clubroot resistance (CR) gene in Flower Nabana was mapped to chromosome A3, and molecular markers linked to the CR gene were identified for use in marker-assisted selection (MAS). Using the CR genes of Mendel and rutabaga, several clubroot-resistant spring canola lines were generated. Frequently, CR genes of Mendel and rutabaga conferring resistance to pathotype 3 also conferred resistance to other pathotypes of *P. brassicae* prevalent in Canada, including pathotypes 5, 6 and 8. The CR gene of Flower Nabana was introgressed into *B. napus* and *B. rapa* canola through marker-assisted selection. The results suggested that pyramiding different CR genes into *B. napus* canola would facilitate building durable resistance, since single-gene-controlled resistance sooner or later may be overcome by development of new pathogen race(s) (Rahman et al. 2014).

Phytophthora parasitica var. *nicotianae*, incitant of tobacco black shank disease, is responsible for considerable losses. Flue-cured tobacco (*Nicotiana tabacum*) cv. Coke 371-Golf (C371-G) has a dominant gene *Ph*, that confers high level of resistance to the disease induced by race 0. In order to ascertain the origin of *Ph*, breeding lines homozygous for the *Ph* gene were hybridized with NC1071 and L8, flue-cured and Burley genotypes known to have qualitative resistance genes from *N. plumbaginifolia* and *N. longiflora* respectively. The F₁ hybrids were out-crossed to susceptible testers and the progenies were evaluated in the field black shank nurseries and in greenhouse disease tests with *P. parasitica* var. *nicotianae* race 0. Results indicated that the gene *Ph* was allelic to *Php* from *N. plumbiginifolia* in NC1071. Test cross populations of hybrids between Burley lines homozygous for *Ph* and *L8*, possessing *Ph1* from N. *longiflora*, showed that *Ph* and *Ph1* integrated into the same tobacco chromosome during interspecific transfer. Random amplified polymorphic DNA (RAPD) analyses of the test cross progenies confirmed that recombination between two loci occurred. The RAPD markers (48) linked to *Ph* in doubled haploid lines were used in cluster analyses with multiple accessions of N. *longiflora* and N. *plumbaginifolia*, breeding lines L8, NC1071 and DH92-2770, and cultivars K326, Hicks and C371-G. The results of field and greenhouse trials indicated that the gene *Ph* and *Php* were allelic and originated from *N. plumbaginifolia* (Johnson et al. 2002a). Markers linked to the dominant tobacco black shank resistance gene *Ph* from flue-cured tobacco cv. Coker 371-Gold was identified by bulked segregant analysis (BSA) and RAPD technique. The RAPD markers (60), 54 in coupling and 6 in repulsion phase linkage to *Ph*, were identified in a K326-derived BC_1F_1 ($K326-BC_1F_1$) doubled haploid (DH) population. Thirty RAPD markers, 26 in coupling and 4 in repulsion phase linkage to *Ph*, were used to screen 149 K326-BC_2F_1 haploid plants. Complete linkage between the 26 coupling phase markers and Ph was confirmed by screening 149 K326-BC_2F_1 DH lines produced from the haploid plants in the black shank nurseries. RAPD markers $OPZ-5_{770}$ in coupling and $OPZ-7_{370}$ in repulsion phase linkage were used to select plants homozygous for the *Ph* gene for fourth back crossing to the cv. K326. Black shank disease nursery evaluation of 11 lines of 326-BC4S1 and their test cross hybrids to a susceptible tester confirmed linkage between *Ph* and $OPZ-5_{770}$. The results revealed the efficiency of marker-assisted selection (MAS) for *Ph* using a RAPD marker linked in coupling and repulsion. Complete linkage between 26 RAPD markers and the *Ph* gene was confirmed in the K326-BC_5 generation and RAPD phenotypes were stable across generations and ploidy levels. The usefulness of RAPD markers in MAS for *Ph* to develop tobacco cultivars resistant to black shank disease was established in this investigation (Johnson et al. 2002b). Emergence of race 1 of *P. nicotianae*, incitant of tobacco black shank disease, abolished the usefulness of cultivars with complete resistance developed earlier. The search for new sources of resistance led to the identification of cigar tobacco cv. Beinhart 1000 (BH) as highly resistant to all races of *P. nicotianae*. Double haploid (DH) lines from a cross of BH and the susceptible cv. Hicks were evaluated for black shank resistance. Quantitative trait loci (QTL) on linkage groups (LGs) 4 and 8 accounted for > 43% of the phenotypic variation in resistance. DH lines (43) and parents were evaluated and genotypes with one or both QTL from BJ on LGs 4 and 8 had increased incubation periods for symptom expression and decreased root rot, but higher final inoculum levels than genotypes with neither QTL. A low level of stem resistance was recorded in BH and DH lines with the QTL from BH on LG4, but not LG8. Low levels of leaf resistance were

observed for Hicks, BH and DH lines with both QTL from BH on LG 4 and LG 8. The partial resistance in BH could be exploited for developing tobacco cultivars with resistance to *P. nicotianae* (McCorkle et al. 2013).

Resistance to *Phytophthora capsici*, causing Phytophthroa blight of pepper, was reported to be a dominant trait. Inheritance of Phytophthora blight resistance is complex and influenced by several factors, such as site of infection, differences in disease symptoms (root rot, foliar blight, fruit rot and stem blight), each of which is governed by a specific *R* gene. Furthermore, different *R* genes govern resistance phenotype against different physiologic races of *P. capsici* within each type of symptom (Walker and Bosland 1999; Sy et al. 2005; Monroy-Barbosa and Bosland 2008). The mode of inheritance of resistance to *P. capsici* was investigated, by evaluating a unique susceptible accession 'New Mexico Capsicum Accession 10399' (NMCA10399) against several physiologic races of the pathogen. Only susceptible individuals were obtained in the F_1 population from the hybridization of NMCA10399 to 'Coriollode Morelos' (CM-334) which was the most resistant accession known. CM-334 showed resistance to all isolates of *P. capsici* tested, which was considered to be a dominant trait. Susceptibility in all F_1 populations from hybridization of NMCA 10399 to CM-334, suggested that a resistance inhibition mechanism was present in CM-334, involving a gene or genes which were named as 'inhibitor of *P. capsici* resistance' (*lpcr*). This inhibition mechanism was not effective against resistance to other species of *Phytophthora*. In all inoculation trials, the resistant parent CM-334 displayed phenotypic resistance and no lesions were produced. The susceptible parents NMCA10399 and Camelot scored 9 (plant death) for root rot and 5 (complete leaf necrosis) for foliar blight induced by *P. capsici*. The CM-334 × Camelot F_1 population displayed complete resistance, showing the dominant inheritance of *P. capsici* resistance, for root rot and foliar blight symptoms. The segregation ratio of 3:1 (resistant: susceptible) in the CM-334 × Camelot F_2 population showed that Camelot was homozygous recessive for the *R* gene meaning that it lacked any *R* gene for P. *capsici* resistance, whereas CM-334 was homozygous dominant for the *R* gene. The NMCA10399 × Camelot F_1 and F_2 populations were completely susceptible, indicating that neither of the susceptible parents (Camelot/NMCA10399) had any effective *R* gene against P. *capsici*-induced symptoms by the isolate used for inoculation. The CM-334 × NMCA10399 F_1 population showed complete susceptibility to root rot and foliar blight symptoms. The CM-334 × NMCA10399 F_2 populations showed a 3:13 (resistant/susceptible) ratio, indicating that *ipcr* gene had an epistatic dominant effect over the dominant *R* genes for root rot and foliar blight. Then the CM-334 × NMCA10399 F_1 and F_2 populations were separately inoculated with three additional races of *P. capsici* for root rot (race-2, race-12 and race-15) to determine the ability of *ipcr* gene to inhibit resistance against other races of *P. capsici*. In all combinations, the plant populations responded in the same fashion. The CM-334 × NMCA 10399 F_1 populations showed complete susceptibility, whereas the F_2 population showed

3:13 (resistant: susceptible) ratio, following nonhost resistance in NMCA10399 revealed that it displayed a resistance response against all nonhost *Phytophthoa* spp. tested. This indicated that the pathogen- or microbe-associated molecular pattern (PAMP/MAMP) defense was still functional in MMCA10399 against other *Phytophthora* spp., in spite of the presence of the gene *ipcr* which favored the development of *P. capsici* by inhibiting host resistance in pepper. The results showed that the gene *ipcr* could interfere with host specific *R* genes against *P. capsici* but not against nonhost resistance *R* genes (Reeves et al. 2013).

Two isolates of *P. capsici*, causing Phytophthora root rot, were inoculated onto 126 F8 recombinant inbred lines (RIL) derived from a cross between pepper (*C. annuum*) line YCM334 (resistant) and local cv. Tean (susceptible) to identify the quantitative trait loci (QTL) for resistance to the disease. Highly significant effects of pathogen isolate, plant genotype and genotype x isolate were detected. QTL mapping was performed, using a genetic linkage map covering 1486.6 centimorgans (cM) of the pepper genome and consisted of 249 markers including 136 AFLPs, 112 simple sequence repeats (SSRs) and one cleaved amplified polymorphic sequence (CAPS). Fifteen QTLs were detected on chromosomes 5 (P5), 10 (P10), 11 (P11), Pb and Pc using two data processing methods: percentage of wilted plants (PWP) and relative area under disease progress curve (RAUDPC) values. The phenotypic variation due to each QTL ranged from 6.0 to 48.2%. Seven QTLs were common to resistance against the two isolates on chromosome 5 (5) and six QTLs were isolate-specific for the isolate 09-051 on chromosome 10 (P10) and Pc, and two isolates for isolate 07-127 on chromosome 11 (P11) and Pb. The QTLs in common with the major effect on resistance for two isolates accounted for 20.0 to 48.2% phenotypic variation. Phenotypic variation (6.0–17.4%) was due to isolate-specific QTLs. The results revealed the existence of a gene-for-gene relationship between *C. annuum* and *P. capsici* for root rot resistance (Truong et al. 2012). High level of diversity in the pathogenic potential (virulence) of *P. capsici*, was detected. Resistance to the disease is conditioned by many quantitative loci (QTLs). Pyramiding resistance alleles could be accomplished by using molecular markers tightly linked to resistance genes. For this purpose, an F8 recombinant inbred line (RIL) population derived from a cross between YCM 334 and a susceptible cv. Tean was used in combination with bulked segregant analysis (BSA), utilizing RAPD and conversion of AFLP markers linked to Phytophthora root rot resistance into sequence characterized amplified region (SCAR) markers. For conversion, one marker was successfully converted into a codominant SCAR marker SA133_4 linked to the trait. For BSA, three RAPD primers, UBC, 484, 504 and 553, produced polymorphisms between DNA pools among 400 primers screened. Genetic linkage analysis showed that the SA133_4 and UBC553 were linked to Phytophthora root rot resistance. It was possible to precisely identify these markers as resistant or susceptible in nine commercial pepper cultivars and they could be used advantageously in pepper breeding programs for improvement of resistance levels (Truong et al. 2013). The

limited genetic diversity in cultivated tomato necessitated the use of wild relatives of tomato as sources of resistance to diseases. *Solanum habrochaites* accession LA407 showed complete resistance to four isolates of *P. capsici*, causal agent of crown and root rot of tomato, from cucurbitaceous and solanaceous crops. Under greenhouse conditions, 62 lines of a tomato inbred backcross population between LA407 and the cultivated tomato cv. Hunt 100 and Peto 9543 were evaluated for resistance to two highly virulent *P. capsici* isolates 12889 and OP97. Roots of 6-week-old seedlings were inoculated with *P. capsici* isolates individually. Observations on number of plants with wilting and mortality were made thrice in a week for a period of 5 weeks. Disease responses of inbred tomato lines showed significant differences. Most lines were susceptible to *P. capsici* isolate 12889, but resistant to the isolate OP97. Twenty-four tomato lines were resistant to both isolates. Heritability of resistance to Phytophthora root rot was high in this population. Polymorphic molecular markers located in genes related to resistance and defense responses were identified and added to the genetic map generated earlier for the population. The identified resistant lines and molecular markers could be very useful for development of tomato cultivars resistant to *P. capsici* (Quesada-Ocampo et al. 2016).

European oilseed rape cultivars had only low level of resistance to *Verticillium longisporum*, causal agent of Verticillium wilt disease. As a starting point for marker-assisted selection (MAS) breeding of resistant cultivars, the quantitative trait loci (QTL) were identified. Resistance QTLs were localized in a segregating oilseed rape population of 163 doubled haploid (DH) lines derived by microspore culture from the F1 of a cross between two *B. napus* breeding lines, one of which possessed resistance to *V. longisporum* derived by pedigree selection from a resynthesized *B. napus* genotype. A genetic map was constructed, comprising 165 restriction fragment length polymorphism (RFLP) and 94 AFLP and 45 simple sequence repeats (SSR) markers covering a total of 1,739 cM on 19 linkage groups. Seedlings of the DH lines and parents were inoculated with pathogen isolates in four greenhouse experiments performed during autumn 1999. In three of the experiments, the DH lines were inoculated with a mixture of five pathogen isolates, while in the fourth experiment, only one of the isolates was employed. The disease index (DI) by scoring symptom severity on 21 inoculated plants/lines was calculated, in comparison to the same number of uninoculated plants. Using the composite interval mapping procedure, a total of four different chromosome regions could be identified that showed significant QTL for resistance in more than one environment. Two major QTL regions were identified on the C-genome linkage groups N14 and N15, respectively. Each of these QTLs consistently exhibited significant effects on resistance in multiple environments. The presence of flanking markers for the respective QTL was associated with a significant reduction in DI in the inoculated DH lines (Rygulla et al. 2008).

Canola (*B. napus*) crops suffer severely due to the SSR disease caused by *S. sclerotiorum*. An interspecific cross between *B. napus* (susceptible) and *B. carinata* (resistant) was carried out to transfer resistance to SSR and a doubled-haploid (DH) population was generated from BC2S3 plants. DH lines (58), of which two carried B-genome (with high resistance) linkage groups and nine accessions of the diploid and amphidiploid *Brassica* spp. were screened in a growth-chamber for resistance to stem rot. *B. napus* parent (Westar) and all *B. rapa* and *B. oleracea* reference lines were susceptible to SSR disease, whereas the *B. carinata* parent, *B. nigra* and *B. juncea* reference lines were resistant to SSR disease. Variations in the responses of DH lines to SSR were significant. Two DH lines carrying B-genome chromosomes showed significantly higher levels of resistance (P < 0.01), compared to *Brassica* spp. without B-genome chromosomes. However, DH lines without B- chromosomes were also resistant to the disease (Navabi et al. 2010). Physiological resistance and disease resistance conferred by plant architecture-related traits may contribute to common bean (*P. vulgaris*) field resistance to the white mold caused by *S. sclerotiorum*. A comparative map composed of 79 QTLs for white mold resistance (27), disease avoidance traits (36) and root traits (16) were generated. The white mold resistance QTLs (13), six with strong and seven with weak associations with disease avoidance traits were identified. Root length and lodging QTL colocated in three regions. Canopy porosity, height and lodging were highly correlated with disease severity score in the field screening trials, during 2000–2011. Resistance to lodging was important for reduced disease severity, under high disease pressure conditions. Dry bean lines with physiological resistance in combination with disease incidence traits, did not require fungicide treatments. Selection of lines with field resistance to white mold, combined with high yield potential and acceptable maturity traits might prove to be an effective approach for management of common bean white mold disease (Miklas et al. 2013).

Management of Aphanomyces root rot disease of pea caused by *A. euteiches* by development of resistant cultivars was considered to be the most economically feasible approach. In the study of mapping population of 127 recombinant inbred lines (RILs) derived from the cross 'Puget' (susceptible) x 90-2079 (partially resistant), seven genomic regions, including a major QTL, Aph1, associated with partial resistance to the disease was identified. Later, in the same mapping population, the specificity versus consistency of *Aphanomyces* resistance QTL under two screening conditions – greenhouse and field – was evaluated with two isolates of A. *euteiches* originating from the United States and France. The RILs (127) were evaluated in the greenhouse for resistance to pure culture isolates SP7 (USA) and Ae106 (France). Based on the genetic map prepared earlier, 10 QTLs were identified for resistance in the greenhouse to the two isolates. Among these, *Aph1, Aph2* and *Aph3* were detected for partial resistance in the United States. *Aph1* and *Aph3* were detected in both isolates, and *Aph2* was present only in the French isolate. Seven additional QTLs were specifically detected with one of the two isolates and were not identified for partial field resistance in the United States. The results indicated the usefulness of three QTLs consistently detected in pea for development of

cultivars resistant to Aphanomyces root rot (Pilet-Nayel et al. 2005). *Medicago truncatula* is considered as a valuable model genetic system for investigating the genetic basis of resistance to *A. euteiches* in leguminous crops. In order to identify the genetic determinants of resistance to a broad host range-pea infecting strain of *A. euteiches* in *M. truncatula*, 178 F_3 recombinant inbred lines (RILs) and 200 F_3 families from the cross F83,005.5 (susceptible) x DZA045.5 (resistant) were screened for resistance to *A. euteichus*. Distribution of phenotypes suggested a dominant monogenic control of resistance. A major locus *AER1* associated with resistance to *A. euteiches*, was mapped by bulk segregant analysis (BSA) to a terminal end of the chromosome 3 in *M. truncatula* and accounted for 80% of the phenotypic variations. *AER1* was located in a resistance gene-rich region, where resistance gene analogs and genes associated with disease resistance phenotypes were present. Improvement of resistance to *A. euteiches* in cultivated legumes using the gene *AER1* was shown to be an effective approach (Pilet-Nayal et al. 2009). The genetic variability and major resistance quantitative trait loci (QTLs) to *Verticillium albo-atrum*, infecting *Medicago truncatula* were identified. Subsequently the genetic control of resistance to a non-legume isolate of *V. albo-atrum* LPP0323 was investigated, using a population of recombinant inbred line (RIL) F83,005.5 and susceptible line A17 inoculated with the potato isolate. High genetic variability and transgressive segregation for resistance to pathogen isolate LPP0323 were observed among RILs. Heritabilities were found to be 0.63 for AUDPC values and 0.93 for maximum symptoms score (MSS). A set of four QTLs associated with resistance toward LPP0323 was detected for the parameters MSS and AUDPC values. Additive gene effects showed that favorable alleles for resistance were derived from the resistant parent. The four QTLs were distinct from those described earlier for an alfalfa isolate of *V. albo-atrum*, revealing the operation of several mechanisms of resistance in *M. truncatula*. Furthermore, no localization of any of the QTLs with regions involved in resistance against other pathogens of *M. truncatula* could be detected (Negahi et al. 2014).

The possibility of developing chickpea cultivars with resistance to two most serious diseases caused by *Ascochyta rabiei* (Ascochyta blight disease, and *F. oxysporum* f.sp. *ciceris (Foc)* (wilt disease) was explored. Quantitative trait loci (QTLs) or genes for Ascochyta blight resistance and a cluster of resistance genes for several Fusarium wilt pathogen races (*foc1, foc3, foc4* and *foc5*) located on LG2 of the chickpea map were reported independently earlier. The linkage relationship between the loci that confer resistance to blight and wilt, was investigated, using an intraspecific chickpea recombinant inbred lines (RILs) population that segregated for resistance to both diseases. A new LG2 was established, using sequence tagged microsatellite sites (STMS) markers selected from other chickpea maps. Resistance to race 5 of *F. oxysporum* (*foc5*) was inherited as a single gene and mapped to LG2, flanked by the STMS markers TA110 (6.5 cM apart) and TA59 (8.9 cM apart). A QTL for resistance to Ascochyta blight (QTL$_{AR3}$) was also detected on LG2, using the evaluation data gathered separately in two cropping seasons. This genomic region, where QTL$_{AR3}$ is located, was highly saturated with STMS markers. STMS TA194 appeared to be tightly linked to QTL$_{AR3}$ and was flanked by the STMS markers TR58 and TS82 (6.5 cM apart). The genetic distance between *foc5* and QTL$_{AR3}$ peak was around 24 cM, including six markers within this interval. The markers linked to both loci might facilitate the pyramiding of resistance genes for both chickpea disease via marker-assisted selection (Inuela et al. 2007).

The minimal or no-tillage practice has been reported to increase incidence of soilborne diseases. Further, applications of herbicides before direct seeding temporarily increased the levels of inoculum of soilborne pathogens, such as *R. solani* and *Pythium* spp. The dying plants following herbicide application, provided the substrates for pathogen development and this condition was designated 'green bridge' in wheat-based cropping systems in the Pacific Northwest (PNW) region in the United States. In order to overcome the adverse effects of green bridge, efforts were made to improve the genetic diversity of wheat by producing synthetic wheat lines (> 1,000) by the International Maize and Wheat Improvement Center (CIMMYT). The synthetic wheat was developed by crossing tetraploid wheat (*Triticum turgidum* L., 2n = 4 x = 28, A, B genomes) with *Aegilops tauschii* Cos. (2n = 2 x = 14, D genome), resulting in hybrids that contain the A, B and D genomes. Synthetic wheat genotypes were screened in high-inoculum and low-inoculum field environments. Six genotypes had varying levels of resistance/tolerance to Rhizoctonia root rot caused by *R. solani*. One of the lines, SPBC-3104 (Vorobey) exhibited good tolerance in the field was crossed to susceptible PNW-adapted 'Louise' to determine the pattern of inheritance of the trait. A population of 190 BC$_1$-derived recombinant inbred lines were evaluated in two field green bridge environment and in soils artificially infested with *R. solani* AG8. Genotyping by sequencing and composite interval mapping identified three quantitative trait loci (QTL) controlling tolerance. The results showed that beneficial alleles of all three QTLs were contributed by the synthetically derived genotype SPCB-31-4 for further exploitation (Mahoney et al. 2016).

R. solani, causal agent of rice sheath blight (ShB) disease, is known to be responsible for heavy losses in yield and quality of grains. The ability of controlled environment inoculation assays to detect ShB resistance QTLs in a cross derived from cv. Lemont (susceptible) and cv. Jasmine 85 (resistant) was assessed. The disease severity in 250 F5 recombinant inbred lines (RILs) were measured on the seedlings inoculated, using microchamber and mistchamber assays under greenhouse conditions. Ten ShB-QTLs were identified on chromosomes 1, 2, 3, 5, 6 and 9, using these two methods. The microchamber method identified four of five new ShB-QTLs, one on each of chromosomes 1, 3, 5 and 6. Both microchamber and mistchamber methods identified two ShB-QTLs, qShB1 and qShB9-2. Four of the ShB-QTL regions identified on chromosomes 2, 3 and 9 were reported earlier. The major ShB-QTL qShB9-2, which cosegregated with simple

sequence repeat (SSR) marker RM245 on chromosome 9, contributed to 24.3% and 27.2% of the total phenotypic variations in ShB, using microchamber and mistchamber assays respectively. The qShB9-2, a plant-stage-independent QTL, was also verified in nine haplotypes of 10 resistant Lemon/Jasmine 85 RILs, using haplotype analysis. The results suggested that the multiple ShB-QTLs might be involved in ShB resistance and that microchamber and mistchamber methods might be effective for detecting plant-stage-independent QTLs. The two SSR markers, RM215 and RM245 could be used in MAS breeding programs to enhance resistance to sheath blight disease of rice (Liu et al. 2009). In the further study, the 10 rice sheath blight QTLs mapped earlier in a Lemont x Jasmine 85 recombinant inbred line (LJRIL) population had to be confirmed under field conditions for utilization in marker-assisted selection (MAS) procedure to improve ShB resistance in new rice cultivars. The ShB resistance was evaluated, using 216 LJRILs under field conditions in Arkansas, Texas and Louisiana during 2008–2009. The presence of the major ShB-QTL qShB9-2 was confirmed, based on field data. Another new ShB-QTL between markers RM221 and RM112 on chromosome 2 across all three locations was also identified. The results showed that the microchamber and mistchamber assays developed in this investigation were found to be simple, effective and reliable for identification of major ShB-QTLs like qShB9-2 in the greenhouse at early vegetative stages. The markers RM215 and RM245 were closely linked to qShB9-2 in greenhouse and field assays. The usefulness of the markers in breeding programs for improving resistance to sheath blight in rice cultivars using MAS approach was well established (Liu et al. 2013).

P. ultimum causes seed decay and damage to roots in common bean (*P. vulgaris*). The responses of 40 common bean genotypes to *P. ultimum* and inheritance of resistance in the 92 F7 recombinant inbred lines (RILs) developed from a cross between Xana and Cornell 49242 were investigated, using emergence rate and seedling vigor as the parameters. Emergence of the genotypes was significantly associated with white seed coat. The genotypes (11) with colored seeds exhibited high percentages of emergence and seedling vigor was not significantly different (P > 0.05) from noninoculated plants. Responses of the RIL population showed both qualitative and quantitative modes of inheritance. A major gene, *Py-1* governing the emergence rate was mapped in the region of the gene *P*, a basic color gene involved in control of seed coat color, located on LG7. Using the RIL subpopulation with colored seeds, a significant QTL associated with emergence rate (ER3^{xc}) and another with seedling vigor (SV6^{xc}) were identified on the LG3 and LG6 respectively. QTL SV6^{xc} was mapped in the region of the gene *V*, another gene involved in the seed color. QTLs associated with seed position as regions were involved in responses to *P. ultimum* infection (Campa et al. 2010).

P. brassicae causes clubroot diseases in various vegetable and oilseed crops, accounting for heavy losses all over the world. A dominant clubroot resistance gene from *B. rapa* (Chinese cabbage) was fine-mapped and molecular markers were developed for Chinese cabbage. These markers could be used for marker-assisted selection (MAS) in other *Brassica* crops. For transfer of the clubroot resistance gene to canola (*B. napus*), an interspecific hybridization was applied between canola and Chinese cabbage. Then the F_1 was backcrossed to canola recurrent-parent for three generations to produce B_1 to B_3 progenies. Using these populations, simple repeat (SSR) markers flanking clubroot resistance gene were employed to perform MAS in canola, the selected molecular markers were evaluated in 13 different canola and rapeseed quality genotypes in *B. napus* and *B. rapa*. These markers provided highly reliable identification of clubroot resistance in this diverse set of *Brassica* genotypes. The clubroot resistance co-segregated with also SSR markers flanking the clubroot resistance gene in the BC_3 and BC_3S_1. The segregation ratio of resistant and susceptible individuals in the BC_3 was similar to the expected 1:1 ratio for the segregation of a single Mendelian gene in BC_3S_1 families with homozygous clubroot resistance. The resistance sources identified in this investigation could be useful in breeding programs for development of resistant cultivars (Hirani et al. 2016).

2.2.1.1.6   Mutagenesis for Improvement of Disease Resistance

Development of in vitro culturing of plant cells/organs has led to application of in vitro mutagenesis for genetic improvement in many characteristics, including disease resistance/ tolerance. In vitro mutagenesis, consisting of in vitro culture and induction of mutations, offers an alternative for enhancing variability in cultivars of economic importance that can complement conventional breeding methods. In the case of in vitro selection techniques which employ different selective agents, the resulting changes in phenotypic characteristics, such as resistance/tolerance to diseases, could be interpreted as potential mutations in the plant DNA sequence. Selection with phytotoxins and culture filtrate appears to be more effective than the use of pathogen itself. The use of in vitro methods for evaluation of resistance is dependent upon a positive correlation between resistance to culture filtrates and whole plant disease resistance. In vitro techniques have been combined with induction of mutation for generating genetic variability, including novel disease resistant mutants (Kantoglu et al. 2010). Mutagenesis in vitro is relatively simple, less expensive and efficient technique, with the advantages associated with in vitro culture, such as manipulation of different types of explants (axillary buds, organs, tissues and cells), management of a large number of plants in a small space, ability to separate (subculture) mutated and nonmutated sectors, performance of the entire process under highly phytosanitary conditions, and possibility of evaluation of differences in polygenic characteristics easily and precisely (Solís-Ramos et al. 2015).

White rot disease of garlic (*Alium sativum*) caused by *Sclerotium cepivorum* is an important constraint in production and during storage under natural conditions. Mutation breeding for improving the level of resistance to white rot disease was taken up. Cloves of two garlic cultivars Kisswany and Yabroudy, were irradiated with gamma ray. The M3 and M4 generations were tested by artificial inoculation with *S. cepivorum*, followed by planting in infested soil. Twelve lines of cv. Kisswany showed only 3% infection, as against 29% infection in the controls (untreated parent plants). In 12 lines of Yabroudy, disease incidence was reduced from 20% infection to less than 5% in the control plants. In addition, improvement in storability under natural conditions and appreciable decreases in weight loss during storage were also recorded in the mutant lines generated from both garlic cultivars (Al-Safadi et al. 2000). Banana cultivars with resistance to Fusarium wilt resistance have been generated from in vitro induced mutants (somaclonal variants). The cultivars Tai Chiao No.1 and Formosana were selected from somaclonal variants (Hwang and Ko 2004). *Ethyl methanesulfonate* (EMS) was employed as mutagenic agent to induce variability in apical tissues (Yang and Lee 1981; Omar et al. 1989). Micro-cross section-derived from banana leaf bases were used for callus and shoot bud induction with subsequent regeneration of healthy plants (Okole and Schulz 1996). This procedure to regenerate banana plants was applied for inducing mutation in vitro using EMS and for screening for Fusarium wilt-resistant lines of Brazil Banana (*Musa* sp.). The results indicated that the optimum EMS concentration and duration of treatment of micro-cross-sections obtained from the pseudostem of tissue-cultured plantlet were 300 mM and 60 min, respectively. One hundred plantlets generated following treatments, were screened for Fuarium wilt resistance by inoculating them with *F. oxysporum* f.sp. *cubense* race 4 (spores at 5×10^6/ml). The initial disease symptoms of yellowing of lower leaves appeared in susceptible plants at 2 weeks after inoculation. Only six plants survived at two months after inoculation in the preliminary field test. The early screening eliminated 94% of the plant regenerated from micro-cross-sections with induction of mutation by EMS treatment. The plants selected after first screen, showed marked reduction in wilt incidence. The results indicated the usefulness of EMS for inducing mutation in banana through micro-cross-section cultural system, resulting in the improvement in the level of resistance to Fusarium wilt disease in banana (Chen et al. 2013).

The effects of gamma irradiation and in vitro selection with culture filtrate (CF) of *R. solani* (teleomorph, *Thanatephorus cucumeris*), as selection agent on the susceptibility of the common bean (*P. vulgaris*) were assessed. A dose of 20 Gy of gamma rays of ^{60}Co was determined as the optimum for inducing mutations in the embryonic axes (EAs) of common bean cv. Bribri, without any adverse effect on the growth of in vitro plants. The culture filtrate at 20% concentration in BMS media was used to preselect only those possibly mutated individuals showing resistance to *R. solani* from third subculture Mortality of treated explants was high (less than 50% survival). The survived ones had stunted growth, foliar rosette formation with little or no root development. None of the roots was functional, as they were necrotic. The irradiated EAs cultured initially on BMS medium without culture filtrate, but added in the subculture, showed a survival rate greater than 50%. Survival of irradiated EAs in the absence of CF, was greater than that of irradiated EAs immediately exposed to CF. Delaying addition of CF to BMS medium until the second subculture (MV_2) increased the percentage of non-irradiated explants that survived in the presence of CF (> 80%). At 90 days, no significant differences were observed for survival with respect to non-irradiated axes cultured without CF (controls). Survival of nonirradiated EAs in the presence of CF was low, and differences between nonirradiated control and irradiated EAs with CF (treatment 1) and irradiated EAs with CF in the second subculture (treatment 2) were not significant. At 40 days after planting in pots, one nonirradiated plant selected in vitro with CF, had survived and had abundant foliar tissue and good development. Some of the negative control plants (T4-irradiated, grown without CF), as well as positive controls (T6, non-irradiated, grown without CF) were acclimated. The remaining in vitro plants were maintained in MV_3 media for 60 more days, and were acclimated, when they reached desired height and roots were formed. Plants irradiated and selected in vitro did not survive to acclimatization stage, indicating that it might not be necessary to combine the variation generated by irradiation with the in vitro selection using the culture filtrate of *R. solani* (Solís-Ramos et al. 2015).

2.2.2 Mechanisms of Resistance to Fungal Pathogens

Defense mechanisms operating in plants against microbial plant pathogens may be of two types: (i) passive or preexisting defense mechanisms and (ii) active defense mechanisms. The passive mechanism may be due to the presence of structural barriers such as waxy cuticle or reservoirs of antimicrobial compounds located at strategic positions to act on the invading pathogens. The patterns of vascular colonization of tomato cultivars susceptible and resistant to Fusarium wilt disease caused by *F. oxysporum* f.sp. *lycopersici* (*Fol*) races 0 and 1 were studied. Five susceptible and five resistant cultivar-pathogen combinations were included for an experimental duration of 26 days, following inoculation by root immersion method. Propagules of *Fol* spread discontinuously along the stems in all five cultivars at 1-day postinoculation (*dpi*), regardless of levels of cultivar resistance. The pathogen spread was limited to stem bases at 5 *dpi* in all cultivars. Between 5 and 12 *dpi*, stem colonization by the pathogen was arrested in all cultivar-race combinations. In the resistant combinations, further development of the pathogen was contained with discontinuous distribution of inoculum and symptoms were not visible. In contrast, in the susceptible combinations, a gradual upward colonization of the stems was noted with continuous pathogen inoculum distribution, followed by symptom expression. The results suggested that the pathogen might need an incubation period at the base of vascular system prior to further invasion of host tissues in susceptible genotypes. A positive correlation was indicated between the extent of vascular colonization of stem tissue and the time after inoculation of cultivar with pathogen race. The slope of regression line fitted between the height reached by the pathogen up in the stem (*y*) and the time after inoculation (*x*) provided a measure of

the horizontal (polygenic) resistance in tomato cultivars to the tomato Fusarium wilt disease (Rodriguez-Molina et al. 2003).

In order to determine the biochemical basis of tolerance of banana cultivars to Fusarium wilt disease caused by *F. oxysporum* f.sp. *cubense*, the tissue culture-derived plantlets of cv. Goldfinger (tolerant) and cv. Williams (susceptible) were maintained in a hydroponic system. The test plants were inoculated with conidial suspensions to evaluate the degree of tolerance/susceptibility of the two clones. In addition, the defense responses were induced by treating the plants with an elicitor preparation from pathogen mycelial cell walls. Differences in the induction of lignin and callose deposition, phenolics and the defense-related enzymes involved in cell wall strengthening – phenylalanine ammonia lyase, cinnamyl alcohol dehydrogenase, peroxidase and polyphenol oxidase – were assessed. Root tissue of cv. Goldfinger responded to elicitor treatments via a strong deposition of lignin, preceded by induction or activation of enzyme activities involved in the synthesis and polymerization, whereas only slight increases in cv. Williams were observed. Callose content was not altered in both cultivars. Cell wall strengthening due to deposition of lignin was found to be an inducible defense mechanism operating in banana roots to tolerate infection by *F. oxysporum* f.sp. *cubense* race 4 (De Ascensao and Dubery 2000). Development of vascular wilt disease of cabbage caused by *F. oxysporum* f.sp. *conglutinans* (*Foc*) in resistant and susceptible cultivars was studied, using a green fluorescent protein (GFPs)-expressing strain of the pathogen and monitored by laser scanning confocal microscope (LSCM). Pathogen conidia were attached to root hairs and emergence sites of lateral roots at 1–3 days postinoculation (*dpi*) and then penetrated into the epidermal tissues of roots. During this period, clear differences between susceptible and resistant cultivars were evident. But from 4–6 *dpi*, pathogen hyphae invaded progressively from epidermis into cortical tissues and then entered the xylem vessels in the susceptible cultivar. In contrast, colonization of the tissues in resistant cultivar was rarely seen. Massive colonization and sporulation within infected xylem vessels commenced from 7 to 11 *dpi* and hyphae extending into upper hypocotyl of susceptible seedlings were seen. The resistant cultivar plants were symptom-free, whereas chlorosis and wilting symptoms appeared in susceptible cultivar plants. The upper stems and petioles of resistant cultivar did not yield the pathogen, indicating the absence of the pathogen in the aerial plant parts. The results revealed the pattern of pathogen colonization in the interactions with plants of resistant and susceptible cabbage cultivars to *F. oxysporum* f.sp. *conglutinans* (Li et al. 2015).

The role of preformed stem lignin contents in resistance to SSR disease of soybean caused by *S. sclerotiorum* was investigated, since the presence of lignin-degrading enzyme in the pathogen was not known. Six soybean accessions with varying responses to SSR disease were evaluated for their levels of resistance in a series of field experiments. Soybean stems were sampled at reproductive developmental stages that corresponded to specific events in both plant development and SSR disease cycle. Lignin contents of stem composite samples were quantified. Soybean accessions expressed significantly different disease phenotypes in both 2004 and 2006. Lignin contents varied among accessions, growth stages and plant parts. Positively ranked correlations were observed between accessions. SSR severity and lignin contents for all nodes and internodes were determined. For the R3 growth stage, lignin content of the internode between the fourth and fifth trifoliate leaves correlated best with disease severity data from each year (P = 0.005). The results indicated that resistance was related to low stem lignin content which could be considered as a marker to select accessions as sources of resistance for developing soybean cultivars with resistance to SSR disease (Peltier et al. 2009). The effects of compatible and incompatible interactions between *P. sojae* race 1 (causing root and stem rot disease) and hypocotyls of soybean cv. Harsoy (susceptible) and cv. Haro 1272 (resistant) on the arrangement of microtubules were investigated. Both reaction types were similar during the first 3-h period after inoculation with zoospores of the pathogen in terms of number of cells penetrated and depth of penetration into the cortex. Thereafter, distinct differences between the two interaction types could be observed. Incompatible interactions were characterized by a hypersensitive response (HR) confined to single penetrated cells. By contrast, in compatible interactions, the penetrated cells remained without any response. Both types of responses led to autofluorescence of cells or cytoplasm and at 6-h postinoculation, cell cytoplasm was entirely disorganized. Reorientation and loss of microtubules occurred in the early stages of incompatible interaction in association with cellular hypersensitivity, but not in compatible interactions. Further, in cells adjacent to those that reacted hypersensitively, there was little change in microtubule orientation. Treatment of hypocotyls with microtubule depolymerizer oryzalin prior to inoculation did not alter the compatible interaction, but resulted in breakdown of the incompatible response. Changes in microtubule orientation and state were possibly among the first structural changes that were visible within cells during incompatibility in the soybean-P. *sojae* pathosystem (Cahill et al. 2002). In a later investigation, two soybean genotypes, cv. Conrad (with strong partial resistance) and the line OX760-6 (with weak partial resistance) were assayed for the contents of preformed and induced suberin components to determine their role in resistance to *Phyotphthora sojae* in the early stages of infection process. The pathogen hyphae grew through the suberized middle lamellae between epidermal cells. This event required 2 to 3 h longer in Conrad than in OX760-6, providing longer time to cv. Conrad to establish their biochemical defenses. Subsequent growth of hyphae through endodermis was also delayed in cv. Conrad, which contained higher concentrations of aliphatic suberin than the line OX760-6. The resistant plants were induced to form more aliphatic suberin several days earlier, compared to the line OX760-6. However, the induced suberin was formed subsequent to the initial infection process. Eventually, the amount of suberin at 8 days postinoculation, was the same in both genotypes. Preformed root epidermal suberin could be used as a target

compound (marker) associated with resistance, enabling the identification of soybean genotypes/cultivars with higher levels of resistance to *P. sojae* (Ranathunge et al. 2008).

Active defense mechanisms are initiated, soon after the fungal pathogen initiates the process of pathogenesis ie., contact with host plant cell surface, breaching the structural barriers, resulting in prevention of spore germination/colonization by the pathogen. Active defense responses induced by microbial pathogens may be of three types viz., primary responses, secondary responses and systemically acquired responses. Primary responses are restricted to the infected cells or in close proximity with the pathogen and the specific signal molecules produced by the pathogen have to be recognized by the host cells. The outcome of this interaction may rapidly result in the phenomenon known as programmed cell death (PCD). The PCD is defined as a self-destruction process triggered by external or internal factors and it is mediated through an active genetic program which is considered to have a vital role in the development and survival of diverse organisms. Secondary responses are produced by the adjacent cells surrounding the initially infected cell in response to diffusible signal molecules (elicitors) released by the pathogens. Systemically acquired response is due to the action of defense-related compounds that can be translocated throughout the plant, resulting in the development of systemic acquired resistance (SAR) (Hutcheson 1998; Pontier et al 2004). Plant genomes respond to specific pathogen signals and trigger a hypersensitive response (HR), following recognition of pathogen presence by R proteins encoded by cluster of *R* genes. Later, concomitant with the HR at the site of infection, development of systemic acquired resistance (SAR) results in restriction of the spread of pathogens and consequently systemic infection of tissues far away from the site of infection. SAR is a general defense mechanism conferring long-lasting resistance against broad spectrum pathogens. Several molecular changes due to introduction of SAR, including accumulation of PR proteins with antimicrobial properties are observed. In *Arabidopsis* induction of *PR* gene required the signal molecule salicylic acid (SA) and the transcriptional coactivator NPR1 (nonexpressor of PR genes 1, also known as NIM1) (Durrant and Dong 2004). In the *npr1* mutant, SA-induced PR gene expression and SAR were entirely abolished, indicating the involvement of NPR1 in SAR development. Transcription of PR genes in *Arabidopsis* was regulated by the coactivator NPR1 and repressor SNI 1. Pathogen infection often triggered an increase in somatic DNA combination which might result in transmission of changes to the offspring of infected plants. In the *sni1npr1* double mutant, PR gene induction and SAR were restored, suggesting that besides NPR1 and SNI 1, additional SA-dependent regulatory components controlling PR gene expression may be present. The *ssni 1* (suppressor of *sni 1*) mutant was found to have a recessive mutation in the *RAD51D* gene. SNI1 and RAD51D regulated both gene expression and DNA recombination. RAD51 was required for NPR1-independent PR gene expression (Durrant et al. 2007).

2.2.2.1 Gene-for-Gene Interactions in Plant Host–Pathogen Systems

Molecular mechanisms of host plants that differentiate 'self' and 'non-self' are fundamental in innate immunity to prevent potential infections by microbial plant pathogens. Genetic polymorphism and recognition specificity have been brought out by studies on genetics of disease resistance. In addition, these studies suggested that successful disease resistance can be triggered only, if a resistance (*R*) gene product in the plant recognizes a specific avirulence (*Avr*) gene product of the pathogen. Many resistance-associated proteins that function as a surveillance system to detect pathogen AVR proteins (known as effectors) have been identified. Effectors are the molecules of pathogen origin, that alter host cell structure and function to the advantage of the pathogen, facilitating infection (virulence factors or toxins) and/ or triggering defense responses (avirulence factors or elicitors) (Kamoun 2006). Each AVR protein may be detected by a specific R protein in gene-for-gene interaction, which often triggers the hypersensitive response (HR). When an incompatible interaction is initiated, plants can activate a variety of inducible defense responses, consisting of genetically programmed suicide (programmed cell death, PCD) of infected cells (HR) as well as tissue reinforcement and antibiotic production at the site of interaction. Following local responses, a long-lasting systemic response [systemic acquired resistance (SAR)] is triggered, which in turn, primes the plant for resistance against a broad-spectrum pathogens (Dong et al. 2002; Métraux 2001). A substantial commitment of cellular responses, including extensive reprogramming and metabolic reallocation will be required for the multicomponent response of the host plant. This situation indicates that host plant defenses are in tight genetic control and activated as soon as the plant detects the presence of an invader. Because the plant lacks a circulatory system, plant cells have to autonomously maintain constant and effective surveillance system by expressing large number of *R* genes (Dangl and Jones 2001). The number of avirulence genes cloned from fungal pathogens are less, compared to bacterial pathogens. The bacterial pathogens synthesize effector proteins and deliver them into host cells, where they manipulate the host defense system to derive required nutrition (Epinosa and Alfano 2004). On the other hand, fungal pathogens can deliver the effectors to both inside (cytoplasm) and outside (apoplast) of plant cells (Birch et al. 2006). The AVR protein SIX1 from *F. oxysporum* f.sp. *lycopersici* is cysteine-rich (Rep et al. 2004). The predicted AvrLm1 protein encoded by *AvrLm1* of *Leptosphaeria maculans*, infecting oilseed rape (*B. napus*) is composed of 205 amino acids with a single cysteine residue. It contains a peptide signal, suggesting extracellular localization. *AvrLm1* is constitutively expressed, with possible enhanced expression in response to infection by *L. maculans* (Gout et al. 2006).

An array of disease effector proteins is produced by oomycetes, including *Phytophthora* spp. to reprogram the defense circuitry of host plant cells to achieve parasitic colonization. The effectors are delivered into the host apoplast. The apoplastic effectors interact with extracellular targets and surface receptors, whereas cytoplasmic effectors are translocated inside the plant cell, presumably through specialized structures, such as haustoria or infection hyphae (vesicles) that invaginate inside living host cells. The oomycetes effectors, including the avirulence (AVR) proteins are recognized only by their ability to activate defense responses and innate immunity. *P. infestans* secretes inhibitors that target defense proteases in the plant apoplast (Tian et al. 2004). The effectors like glucanase inhibitors and cell death elicitors are also able to function in the host apoplast (Kamoun 2006). Four *Avr* genes have been identified in the oomycetes: *Avr1b* in soybean-infecting *P. sojae*, *ATR13* and *ATR1*^NdWsB in *Hyaloperonospora parasitica*-infecting *A. thaliana* and *Avr3a* in *P. infestans*-infecting potato (Allen et al. 2004; Armstrong et al. 2005; Rehmany et al. 2005). The *Avr* genes were inferred by the presence of cognate R proteins in the host cytoplasm, suggesting that oomycetes can deliver effectors into the host cells, possibly through haustoria. A conserved motif (RXLR) with 32 amino acids of the predicted signal peptides was revealed by sequence alignment of the AVR proteins. The RXLR was hypothesized to function as a signal that may mediate trafficking into host cells (Rehmany et al. 2005). *P. infestans*, *P. sojae* and *P. ramorum* have conserved RXLR motif as indicated by bioinformatic analysis. Another *P. infestans* RXLR-containing protein carries a functional nuclear localization signal and possibly accumulates in host nuclei during infection (Birch et al. 2006). Evidences indicated that R-Avr gene interactions may mediate several types of partial host resistance as well as nonhost resistance (Kamoun 2001; Song et al. 2003). A broad-spectrum of isolates of *P. infestans* was recognized by the *RB* (also known as *Rpi-blb1*) from *Solanum bulbocastanum* and this gene appears to provide durable resistance to *P. infestans* under field conditions (Song et al. 2003).

Phytophthora spp., including *P. infestans*, produce elicitors that form a family of structurally related extracellular proteins that induce HR defense responses in *Nicotiana* spp., but not in potato and tomato. Most species of *Phytophthora* produce the well-characterized 10-kDa canconical elicitors (class I) such as INF1 of *P. infestans*, infecting potato and tomato. Elicitins from *Phytophthora* spp. encode small secreted proteins (< 150 amino acids) with even number of cysteine residues that may induce defense responses, when infiltrated into plant tissues (van't Slot and Knogge 2002). Many effectors secreted by oomycetes belong to this category of cysteine-rich small proteins. PCR amplification with degenerate primers and random sequencing of cDNAs, facilitated isolation of a complex set of elicitin-like genes. Eight elicitin and elicitin-like genes (*inf* genes) were isolated from *P. infestans*. All these genes encode putative extracellular proteins that share the 78 amino acid elicitin domain, corresponding to the mature class I elicitins (INF1) (Bateman et al. 2002). The six genes *inf 2A*, *inf 2B*, *inf 5*, *inf 6*, *inf 7* and *M-25* encode the predicted proteins with a C-terminal domain, in addition to the N-terminal domain common to all elicitins (Kamoun et al. 1997). The class I elicitins have been shown to bind sterols, such as ergosterol and function as sterol carrier proteins (Boissy et al. 1999), suggesting a biological function of vital importance to *Phytophthora* spp. As these fungal pathogens cannot synthesize sterols and they have to assimilate them from external sources, role of elicitins appears to be significant. Furthermore, the elicitin-like proteins of *P. capsici* with significant similarity to INF5 and INF6, exhibited phospholipase activity which may be required for lip binding or a processing role for different members of elicitin family (Osman et al. 2001). The gene *M-25* producing protein with similarity to elicitins was reported to be induced during mating in *P. infestans* (Fabritius et al. 2002).

The roles of certain pathogen genes in pathogenesis have been studied in host-*Phytophthora* pathosystems. The genes *inf 2A* and *inf 2B-* encoding distinct class (class III of chitin-like proteins) were isolated from *P. infestans*. The *inf 2*-like elicitin genes appear to occur as a small genus-specific gene family and are conserved in all tested species of *Phytophthora*. The class III elicitins are cell surface-anchored polypeptides. The expression profiles of *inf 2A* and *inf 2B* genes during infection of tomato by *P. infestans* were determined using semi-quantitative RT-PCR assay. Expression of the genes *inf 1* and *inf 2B* was observed as early as one day after inoculation (*dai*), whereas expression of *inf 2A* could be detected only at 3 *dai*. Both *inf 2* genes were expressed during colonization of tomato tissues by *P. infestans*. Induction of HR by INF1 and INF2 in *Nicotiana benthamiana* was found to be dependent on the ubiquitin ligase-associated protein SGT1. The variation in the resistance of *Nicotiana* spp. to *P. infestans* depended on their response to INF elicitins, as revealed by the positive response of tobacco, but not of *N. benthamiana* to INF2B. Comparative analysis of elicitin activity of INF1, INF2A and INF2B, using *Potato virus X* (PVX) agroinfection and agroinfiltration demonstrated that INF2A and INF2B induced HR-like symptoms on tobacco, like INF1 and other elicitins. However, a significant difference in the specificity of HR induction could be noted for INF2B which could not induce any necrosis on *N. benthamiana*, while INF1 and INF2A were able to do so (Huitema et al. 2005). *P. infestans*-expressed INF1 protein required the ubiquitin ligase-associated protein SGT1 for induction of HR. SGT1 has been shown to be essentially required for *R*-gene mediated HR signaling in various plants like *Nicotiana benthamiana* (Peart et al. 2002). The requirement of SGT1 for induction of HR by INF2 in *N. benthamiana* was found to be similar to that of INF1. Association of the necrotic response elicited by INF2 proteins with the induction of defense response genes was investigated. Leaves of transgenic tobacco line carrying the GUS reporter gene driven by the promoter of pathogenesis-related (PR) gene *Bgl2* (PR2) with *Agrobacterium tumefaciens* strains carrying pGR106-INF1, pGR106-INF2A, pGR106-INF2B were wound inoculated. Both pGR106-INF1 and pGR106-INF2B elicited increased levels of *PR1a* expression, whereas the elicitin INF2A could not induce PR genes in a significant manner. The results suggested that INF2B could induce the expression of PR genes *PR1a* and *Bgl2* in tobacco like INF1 (Huitema et al. 2005).

Resistance of *Nicotiana* spp. to different species of *Phytophthora* may be dependent on the recognition of elicitins produced by the pathogen, which is considered as one of the components of disease resistance. Evidence for this presumption was provided by the study, using *P. infestans* strains engineered to be deficient in the elicitin INF1 by gene silencing. The deficient strains induced disease lesions on *N. benthamiana*, suggesting that INF1 conditions avirulence to *N. benthamiana* (Kamoun et al. 1998). *Phytophthora parasitica* var. *nicotianae*, incitant of tobacco black shank disease, produces the elicitin PARA1. Most of the isolates of the pathogen did not produce PARA1. The isolates producing the elicitin showed downregulation of *para1* gene expression in planta (Colas et al. 2001). The results of these investigation suggested that elicitins were species-specific avirulence factors. However, some INF1-producing isolates of *P. infestans* could colonize *N. benthamiana* plants, even though this host plant species responded to INF1 elicitins. It is possible that elicitors may not always act only as avirulence factors, but may function as general elicitors in some pathosystems (Vleeshouwers et al. 2006). The mitogen-activated protein kinase (MAPK), by which extracellular stimuli are transduced into intracellular responses in eukaryotic cells via phosphorylation. Dephosphorylation of signaling proteins was also identified (Davis 2000). A high-throughput overexpression of *Nicotiana benthamiana* cDNAs identified a gene for a MAPK, as a potent inducer of HR-like cell death. The product of this gene, NbMKK1 protein was found to be localized to the nuclei and the N-terminal putative MAPK docking site of NbMKK1 was required for its functioning as a cell-death inducer. *P. infestans* INF1 elicitin-mediated HR was delayed in the NbMKK1-silenced plants, indicating that NbMKK1 was involved in this HR pathway. The resistance of *N. benthamiana* to a nonhost pathogen *Pseudomonas cichorii* was compromised in NbMKK1-silenced plants. The results indicated that nonhost resistance including HR cell death may be due to MAPK cascades involving NbMKK1 (Takahashi et al. 2007). Changes in the transcript populations produced in habanero pepper (*Capsicum chinenese*) cell suspensions were analyzed by adding whole mycelium homogenates of a pathogenic isolate of *P. capsici* to identify plant cellular processes modified by the effectors of the pathogen. Several defense-like cellular responses such as alkalinization of the medium, a two-step oxidative burst, induction of ß-1,3-glucanases and activation of mitogen-activated protein kinases (MAPKs) were elicited. The elicitation modified the accumulation of transcripts representative of diverse metabolic pathways, including ethylene biosynthetic enzymes, MAPkinases and defense-related products like PR-proteins, but did not affect expression of C. *chinense* NPR1 and WRKY ortholog genes, which are important modulators of plant defense responses. Inoculation of habanero pepper plants at 6-leaf stage with a virulent isolate of the pathogen revealed that apart from some defense-related enzymes, few systemic modifications in the transcript patterns occurred. All plantlets were killed ultimately, although the *in planta* inoculation induced strong accumulation of two MAPK transcripts. The results suggested that either the extent or timing of defense response might be insufficient to establish proper response against initiation of infection in susceptible plants (Nakazawa-Ueji et al. 2010).

A novel class of necrosis-inducing proteins named as Nep1-like proteins (NLPs) was detected in microbial plant pathogens. The canconical 24-kDa necrosis and ethylene-inducing protein (Nep1) was isolated from the purified culture filtrates of *F. oxysporum* f.sp. *erthroxyli*, providing the basis of the novel protein designation (Bailey et al. 1997; Pemberton and Salmond 2004). The NLPs could induce defense responses in both susceptible and resistant plants. The *PsojNIP* gene in *Phytophthora soja* was expressed late during colonization of host soybean tissue during the necrotic phase of infection (Qutob et al. 2002). Three cDNAs from *P. infestans* (PiNPP1.1, PiNPP1.2 and PiNPP1.3) with significant similarity to NLP family proteins were identified. The PiNPP1.1 induced necrosis in tomato and *Nicotiana benthamiana*, as shown by agro-infection with a binary PVX vector procedure. Expression analysis indicated that PiNPP1.1 was upregulated during late stages of infection of tomato by *P. infestans*. The necrosis inducing activity of PiNPP1.1 was compared with INF1 elicitin by employing virus-induced gene silencing (VIGS) technique. The cell-death induced by PiNPP1.1 was dependent on the ubiquitin ligase-associated protein SGT1 and the heat-shock protein HsP90. In addition, PiNPP1.1 required the defense-signaling proteins COI1, MEK2, NPR1 and TGA2.2 for triggering cell death, whereas INF1 was not dependent on these proteins for its activity, suggesting the requirement of signaling pathways for cell death. Enhancement of cell death was observed due to combined expression of PiNPP1.1 and INF1 in *N. benthamiana*, indicating a possible synergistic interplay between the two cell death responses (Kanneganti et al. 2006).

Secretion of elicitors by other *Phytophthora* spp. has been studied. The GP42, a glycoprotein, present abundantly in the cell wall of *P. sojae* triggered defense gene expression and synthesis of antimicrobial phytoalexins in parsley through binding to a plasma membrane receptor (Nürnberger et al. 1994; Sacks et al. 1995). A cellulose-binding elicitor and lectin-like (CBEL) protein is a 34-kDa cell wall protein and it was isolated from *Phytophthora parasitica* var. *nicotianae*. Elicitation of necrosis and defense gene expression in tobacco and assisting the attachment of cellulosic substrates, such as plant surfaces are the dual functions performed by CBEL (Sejalon-Delmas et al. 1997; Villalba-Mateos et al. 1997). The CBEL protein contains two cellulose-binding domains (CBDs), belonging to the Carbohydrate-Binding Module 1 family which is present exclusively in fungi. The necrosis-inducing activity of CBEL depends on ethylene and jasmonic acid (Khatib et al. 2004). The CBEL is involved in organized polysaccharide deposition in cell wall and in adhesion of the mycelium to cellulose substrates (Gaulin et al. 2002). By using modified versions of CBEL protein produced by *Escherichia coli* or synthesized in planta through *Potato virus X* (PVX) expression system, the role of CBDs in its eliciting activity was investigated in tobacco-*Phytophthora parasitica* var. *nicotianae* (*Ppn*) pathosystem. The recombinant CBEL produced in

E. coli elicited necrotic lesions and defense gene expression, when injected into tobacco leaves. Likewise, CBEL production in planta induced necrosis. Localization of CBEL in leaf tissues developing small necrosis was monitored by immuno-gold labeling technique, using the purified CBEL antibody. The protein deposition occurred in the cell wall and junctions of the pGR10:CBEL-treated parenchyma cells. The construct pGR10:CBEL was based on the entire CBEL coding sequence that was amplified by PCR using CBEL cDNA, as a template. The leaf infiltration experiments using synthetic peptides showed that the CBDs of CBEL were essentially required to stimulate defense responses. The pathogen-associated molecular patterns (PAMPs) are recognized by plants through the receptors and discriminate the 'self' and 'nonself'. The CBEL appears to be a PAMP-containing molecule from pathogenic fungi for which a functional investigation was performed by generating CBEL-silenced *Phytophthora* mutants. The results showed that CBEL was involved in exogenous cellulose perception and *Phytophthora* cell wall organization, suggesting that CBDs of *P. parasitica* var. *nicotoianae* might function as PAMPs (Gaulin et al. 2006).

Plant proteases have multiple roles in plant defense against microbial pathogens, as reflected by their involvement in HR. Proteome complexes in the ubiquitin-mediated protein degradation pathway was implicated in programmed cell death (PCD) and disease resistance (Tor et al. 2003). The plant vacuolar processing enzymes (VPEs) were suggested to contribute to resistance to basal defense against pathogens during susceptible interactions (Rojo et al. 2004). Among many proteases that are upregulated during infection by pathogens, the PR-protein P69B of tomato, an apoplastic subtilisin-like Ser protease was found to accumulate upon infection by *Pseudomonas syringae* (Zhao et al. 2003; Tian et al. 2004). Coevolution of diverse defense-counter defense strategies of host and pathogen had become essential in the arms race for survival. Host plants can use PR-proteins, ß-1,3-glucanases for the dissolution of pathogen cell walls, rendering the pathogen more vulnerable to other plant defense responses or oligosaccharides elicitors may be released activating plant defenses (Kamoun et al. 1998). However, *P. sojae*, as a counter defense activity, secretes glucanase inhibitor proteins that effectively suppress the ß-1,3-glucanases of soybean (Rose et al. 2002). The oomycete pathogens *P. infestans*, *P. sojae*, *P. ramorum* and *P. brassicae* are also able to produce Kazal-like extracellular Ser protease inhibitors belonging to a diverse family that includes 35 members (Tian et al. 2004). The Kazal-like inhibitors, EPI 1 and EPI 10 (2 of 14) present in *P. infestans*, bound and inhibited the PR69B subtilisin-like Ser protease of tomato (Tian et al. 2005). Inhibition of P69B by two structurally different protease inhibitors of *P. infestans* suggested that EPI 1 and EPI 10 could function in counter defense types of proteases such as extracellular aspartic protease (AP), and two Cys proteases, CYP and StCathB were upregulated during infection of potato by *P. infestans*. Induction of AP expression occurred faster and reached higher concentration in resistant potato cultivar, compared with the susceptible one (Guevara et al. 2002). Another family of secreted proteins EPI

C1 to EPI C4 with similarity to cystatin-like protease inhibitor domains was identified. Among these, the *epiC1* and *epiC2* genes present in *P. infestans* lacked orthologs in *P. sojae* and *P. ramorum* and were relatively fast evolving within *P. infestans*. They were upregulated during infection of tomato, suggesting a role in *P. infestans*-tomato interactions. Tests to assess biochemical functions indicated that EPI C2B could interact and inhibit a novel papain-like extracellular cysteine protease, designated *Phytophthora* Inhibited Protease 1 (PIP1). The PIP1 was shown to be a PR protein closely similar to Rcr3, a tomato apoplastic cysteine protease that functioned in fungal resistance. PIP1 and Rcr3 might function in different aspects of plant defense, including perception of invading microbial pathogens, mediation of defense signaling and execution of defense responses (Tian et al. 2007).

Isoflavones are considered to function as preformed antibiotics and as precursors of the defense-related coumestan and pterocarpan phytoalexins (Rivera-Vargas et al. 1993). Isoflavone synthase (IFS) is the key enzyme involved in the biosynthesis of isoflavones and it is encoded by two genes *IFS-1* and *IFS-2* in soybean (Jung et al. 2000). The effect of RNAi silencing of genes for this enzyme on infection of both root and cotyledons of soybean by *P. sojae* was assessed. Infection of roots of cv. Williams 82 by race 1 of *P. sojae* normally resulted in a resistant response due to the presence of the *Rps1k* gene for resistance to race 1. Soybean cotyledonary tissues was transformed with *Agrobacterium rhizogenes* carrying an RNAi silencing construct designed to silence expression of both copies of IFS genes. Infection of IFS RNAi-silenced root tissues progressed without any obstruction, because of a near total silencing of isoflavone accumulation, leading to abolition of *R* gene-mediated resistance in transformed tissues. The cv. Harsoy with *Rps7* gene was susceptible to race 1. The IFS silencing increased the susceptibility of cv. Harsoy further and to a higher level, compared to the susceptibility level of silenced cv. Williams 82. Silencing of IFS was demonstrated throughout the entire cotyledon (in tissues distal to the transformation site) by HPLC (analysis of isoflavones) and by real-time PCR assay. A near complete suppression of mRNA accumulation for both *IFS1* and *IFS2* genes was observed, following distal silencing induced by wounding or treatment with cell wall glucan elicitor from *P. sojae*. Silencing of IFS disrupted both *R* gene-mediated resistance in roots and nonrace-specific resistance in cotyledonary leaves (Subramanian et al. 2005).

Plant defense reactions in chickpea-Fusarium wilt [*F. oxysporum* f.sp. *ciceris* (*Foc*)] pathosystem were investigated, using incompatible race 0 of *Foc* and nonhost isolates of *F. oxysporum*. Germinated seeds of Kabuli chickpea ICCV4 were inoculated with a conidial suspension of the incompatible race 0 of *Foc* or nonhost *F. oxysporum* 'resistance inducers'. At 3 days later, they were challenged by root dip with a conidial suspension of highly virulent *Foc* race 5. However, the extent of disease suppression varied with the incompatible *Foc* race 0. Inoculation with inducers resulted in synthesis of maackian and medicarpin phytoalexins in inoculated seedlings. These phytoalexins did not accumulate in plant tissues, but

were released into the inoculum suspension. Treatment with inducers led also to accumulation of chitinase, ß-1,3-glucanase and peroxidase activities in plant roots. The observed defense-related responses were induced more consistently and intensely by nonhost isolates of *F. oxysporum* than by incompatible *Foc* race 0. The phytoalexins and, to a lesser extent, the antifungal hydrolases, were also induced after challenge inoculation with *Foc* race 5. The defense responses in this case were induced in both preinduced and noninduced plants infected by the pathogen. The results showed that suppression of Fusarium wilt, probably involved an inhibitory effect on the pathogen of preinduced plant defenses, rather than an increase in the expression of defense mechanisms of preinduced plants, following a subsequent challenge inoculation (Cachinero et al. 2002). The tropical race 4 (TR4) of *F. oxysporum* f.sp. *cubense*, the incitant of Panama (Fusarium) wilt disease of banana, has a major negative impact on production in all countries. With a view to understanding the molecular mechanisms underlying banana defense responses, a proteomic approach was applied. The responses of a susceptible cultivar Williams and resistant cultivar GCTV-119 to inoculation with TR4 race were investigated, using molecular techniques. Sixteen protein spots were subjected to MALDI-TOF-MS procedure. All of the proteins identified were related to metabolism energy, immunity and defense, and unknown functions. In the resistant cultivar, eight spots were in higher concentrations and six spots had less concentrations, compared to control. Of these, five protein spots showed significant changes in at least two time points. In the susceptible variety, five spots were more abundant, and three spots were at lower concentration, compared to control. Of these, six spots changed significantly after challenge inoculation with TR4 race. Chitinase and ß-1,3-glucanase and superoxide dismutase were estimated quantitatively by employing qRT-PCR assay to assess, whether the observed proteome changes were associated with changes in the mRNA levels or were due to posttranscriptional regulation. ß-1,3-glucanase and chitinase were present only in pathogen-challenged plants of resistant variety, compared with control plants at 6, 12, 24, 48 and 72 h postinoculation, but ß-1,3-glucanase and chitinase were expressed in the roots of both resistant and susceptible banana varieties infected with race TR4 of *F. oxysporum* f.sp. *cubense* (Lu et al. 2013).

V. dahliae, causal agent of potato Verticillium wilt disease, is capable of infecting a large number of plant species, including several crop plant species, indicating its ability to overcome the effects of defense systems operating in the wide range of host plant species. A differential potato-*V. dahliae* pathosystem consisting of two potato cultivars, susceptible (S) and moderately resistant (MR) and two isolates of *V. dahliae* weakly aggressiv (WA) and highly aggressive (HA), were used to evaluate the expression of five defense-related genes PAL-1, PAL-2, PR-1, PR-2 and PR-5, which are generally associated with salicylic acid (SA) defense signaling pathway. Expression levels of these genes were assessed in potato roots and leaves at 0, 4 and 21 h postinoculation (*hpi*) and 3, 7 and 14 days postinoculation (*dai*). In the roots, the expression of PAL-1, PR-1 and PR-2 in the MR cultivar was higher than in S cultivar, in response to inoculation with either one of the isolates of *V. dahliae* tested. PAL-2 gene expression increased early gradually, starting at 21 hpi in the MR cultivar, compared with the susceptible cultivar. Higher expression of PR-1 was detected at 7 *dpi* in roots and both PAL-1 and PAL-2 genes showed higher expression in the MR cultivar in the leaves, relative to the S cultivar. Combined data analyses revealed the differential transcriptional levels in response to infection by *V. dahliae*. The results indicated that SA pathway was involved in potato defense against *V. dahliae* and added to the better understanding of the signaling mechanisms operating in potato-*V. dahliae* pathosystem (Derksen et al. 2013).

The defense responses of pepper (*C. annuum*) infected with compatible and incompatible strains of *P. capsici*, causing Phytophthora blight disease were investigated. The activities of ß-1,3-glucanase (involved in defense mechanisms) increased markedly in the incompatible interaction. The expression patterns of four defense-related genes, *CABPR1*, *CABGLU*, *CAP01* and *CaRGA1* in the leaves and roots of pepper inoculated with different strains of *P. capsici* were determined. All gene expression levels were higher in the leaves than in the roots. Incompatible and compatible host-pathogen interactions showed significant differences in expression patterns. The expression levels of three genes in the incompatible interactions, revealed an increase by factor of 13.2–20.5, but *CaRGA1* gene expression was noted at a lesser degree (by a factor of 6.0). In contrast, the expression levels of the four defense-related genes in compatible interactions increased by the lower factor of 2.0–11.2, compared to controls. The expression levels of the four genes were much lower in the roots than in leaves. The highest levels of mRNA were those of *CABPR1* gene, which increased by a factor of 5.1 at 24 h in the incompatible and by 3.2 at 48 h in the compatible interactions. The results indicated the involvement of defense-related genes in the defense response of pepper to *P. capsici* infection (Wang et al. 2013). Pattern recognition receptors (PRRs) form complexes with proteins, such as receptor-like kinases, to elicit pathogen-associated molecular pattern-triggered immunity (PTI), an evolutionarily conserved plant defense program. In order to understand the role of components of the receptor complex, interaction between tomato and *Phytophthora parasitica* was investigated. In this pathosystem, SISOBIR1 and SISOBIR1-like genes were involved in defense responses to P. *parasitica*. Silencing of SISOBIR1 and SISOBIR1-like genes enhanced susceptibility to *P. parasitica* in tomato. Callose deposition, reactive oxygen species production and PTI marker gene expression were compromised in SISOBIR1- and SISOBIR1-like-silenced plants. *P. parasitica* infection and elicitin (ParA1) treatment induced the relocalization of SISOBIR1 from the plasma membrane to endosomal compartments and silencing of NbSISOSBIR1 kinase domain was indispensible for ParA1 to trigger SISOBIR1 internalization and plant cell death. The results suggested that solanaceous SISOBR1 might have important role in plant basal defense against oomycete pathogens (Peng et al. 2015).

The key biochemical defense mechanisms operating in seedlings of winter oilseed rape cultivars differing in tolerance to infection by *Leptosphaeria maculans*, causing blackleg disease were studied. In the resistant cultivar, a significant increase in the activity of chitinase and ß-1,3-glucanase was observed, indicating a possible systemic defense reaction. In addition, the resistant cultivar had a more efficient antioxidant system, as shown by the higher activity of specific superoxide dismutase (SOD) isoforms and the higher levels of low molecular antioxidants. Only the resistant cultivar had three Fe-SOD sub-isoforms. A correlation between the content of cell wall-bound phenolics and hydrogen peroxide in the resistant cultivar might indicate that an important method of H_2O_2 neutralization was its participation in the process of phenolics incorporation into the cell wall. Higher concentrations of cell wall-bound phenolics significantly enhanced cell wall integrity, which became a more effective firewall against *L. maculans*. Further, high utilization of soluble sugars during synthesis of phenolic compounds in the pathogenesis in the resistant cultivar was revealed. The key factors contributing to a greater tolerance to *L. maculans* were shown to be an effective antioxidant system and higher activity of PR proteins such as chitinase and ß-1,3-glucanase (Hura et al. 2014). Resistance to cabbage blackleg disease caused by *Leptosphaeria maculans*, has been indicated to be a complex and influenced by virulence levels of the pathogen and stage of plant development (seedlings/adult plant). Furthermore, in addition to race-specific resistance offered by *R* genes, secondary metabolites produced by plants also provide more general resistance to plants against microbial pathogens. Glucosinolates are sulfur- and nitrogen-containing secondary metabolites that are the precursors of isothiocyanates and sulfonates, which have some role in the plant resistance. The association between total glucosinolate content and plant resistance to infection by *L. maculans* has been reported earlier. The complex interaction between cabbage (*B. oleracea*) and *L. maculans* infection that led to the selective induction of genes involved in glucosinolate production and subsequent modulation of glucosinolate profiles was studied. Identification of glucosinolate-biosynthesis genes induced by *L. maculans* and any associated changes in glucosinolate profiles were taken up to understand their roles in blackleg resistance in 3-month-old cabbage plants. The defense responses of four cabbage lines, two resistant and two susceptible, were investigated, following inoculation with two pathogen isolates, 03-02s and 00-100s. The different responses of different cabbage lines were determined by visual scores (0–9) of blackleg severity of symptoms (see Figure 2.3). Of the eight combinations, only one combination of cabbage line BN4303 × pathogen isolate 00-100s showed complete resistance with lowest disease severity score of 2 (range 1–3). The combinations of BN4303 × 03-02s (score 3-5) and BN4098 × 00-100s (score 4-6) showed moderate resistance. Overall, line BN4303 was the most resistant, as it was completely resistant to the isolate 00-100s and moderately resistant to 03-02s, whereas line BN4098 showed moderate resistance only to isolate 00-100s and the other two lines were susceptible (Robin et al. 2017).

Changes in the glucosinolate profiles of 3-month old resistant and susceptible cabbage plants inoculated with two isolates

FIGURE 2.3 Responses of four cabbage lines (A) and scoring criteria (B) to *Leptosphaeria maculans* isolates 03-02 and 00-100s Disease rating scale (1–9) based on visual scoring criteria; lines BN4059 and BN4972 were susceptible and lines BN4098 and BN4303 were resistant to the blackleg pathogen.

[Courtesy of Robin et al. 2017 and with kind permission of Frontiers in Plant Science, Open Access Journal]

of *Leptosphaeria maculans* individually were assessed. The line BN4098 was resistant to both isolates 03-025 and 00-100s of *L. maculans* at the seedling stage, but it was susceptible to 03-025 at 3 months of age. A simultaneous increase in the aliphatic glucosinolates glucoiberin (GIV) and glucoerucin (GER) and the indolic glucosinolates glucobrassicin (GBS) and neoglucobrassicin (NGBS) was associated with complete resistance. An increase in either aliphatic (GIV) or indolic (GBS and MGBS) glucosinolates was associated with moderate resistance. Indolic glucobrassicin (GBS) and neoglucobrassicin (NGBS) were increased in both resistant and susceptible interactions. The expression levels of majority of the transcription factors under untreated control conditions were comparatively higher in BN4098, compared to other three lines. Likewise, the expression levels of most of the aliphatic structural biosynthesis genes were comparatively higher in BN4098 than in other cabbage lines under untreated control conditions. None of the genes for aliphatic transcription factors were upregulated in any resistance combination; rather, expression of the genes (5) generally decreased in BN4098 upon infection. Expression of *ST5b* and *GSL-OH*, which were involved in the biosynthesis of aliphatic glucosinolates increased in both resistant lines BN4303 and BN4098, during blackleg infection. In the resistant line BN4303, expression of *ST5b* genes *Bol 026201* and *Bol026202* increased by 4.31- and 2.73-fold respectively after infection by 00-100s and 3.23- and 1.70-fold after infection with 03-025 compared to mock treated plants. Likewise, expression of *GSL-OH* and AOP_2 genes increased due to infection in resistant cabbage plants. The results showed that *L. maculans* infection induced glucosinolate-biosynthesis genes in cabbage, with concomitant changes in individual glucosinolate contents. In resistant lines, both aliphatic GIV and GER and indolic MGBS glucosinolates were particularly

associated with resistance. Elicitation of association between the genes, the corresponding glucosinolates and plant resistance provides an insight into the glucosinolate-mediated defense against *L. maculans* in cabbage (Robin et al. 2017).

2.2.2.2 Nonhost Resistance to Fungal Pathogens

Nonhost resistance (NHR) is expressed by a plant species against the pathogen(s), which are not adapted to the plant species concerned. NHR is the resistance observed at species-specific level and generally, it is the most durable till a new strain that can overcome NHR is produced. Multiple defense components are involved in NHR in a complex manner. Examination using the light microscopy, showed that the non-adapted pathogen ceased to grow during the early stages of infection and several nonhost species exhibited HR cell death at the infection site. Genes involved in NHR were identified in host plant (polygalacturons) and pathogen (INF1) in potato-*P. infestans* pathosystem (Kamoun et al. 1998). A waxy cuticle and plant cell wall generally provide protection to plant epidermis against microbial pathogens. They form a physical barrier preventing the attempts of pathogens to breach them. The pre-invasive barrier is considered as an important obstacle for the pathogen to overcome. Nonadapted pathogens generally fail to penetrate nonhost plant cells, when blocked by preformed physical barriers present on plant surface (Kamoun 2001). The role of metabolic defense in nonhost resistance may be important in some pathosystems. Plants produce secondary metabolites that form chemical barriers to defend themselves against pathogen invasion. These diverse secondary metabolites may be grouped into two classes: (i) phytoanticipins, a group constituting secondary metabolites and (ii) phytoalexins that are synthesized and accumulated rapidly in response to pathogenetic processes (Dixon 2001). Antimicrobial chemicals of host plant origin are inhibitory to a wide range of fungal plant pathogens and detoxification mechanism may be required for successful initiation and progress of infection resulting in symptom expression. In tomato, the phytoanticipin α-tomatine binds to sterols in fungal membranes, resulting in loss of membrane integrity of pathogens and enhancement of resistance to nonadapted pathogens. *Gaeumannomyces graminis* var. *tritici* *(Ggt)*, incitant of wheat-take-all disease, could not infect nonhost oats, because of its sensitivity to the phytoanticipin avenacin produced by oat. However, *G. graminis* var. *avenae*, infecting oats, was capable of producing avenacinase. Hence, it could infect oats by overcoming the avenacin-mediated growth inhibition and induced disease in oats (Osburn et al. 1994; Bowyer et al. 1995). Likewise, the adapted pathogen, *F. oxysporum* f.sp. *lycopersici (Fol)*, was able to overcome the effects of tomatine by producing tomatinase, a glycosyl hydrolase. Tomatinase-deficient mutant of *Fol* was less virulent on tomato than the wild-type strain of *F. oxysporum* f.sp. *lycopersici* (Pareja-Jaime et al. 2008). Accumulation of nicotianamide in the roots of *Amaranthus gangeticus* prevented colonization by nonadapted pathogen *Aphanomyces cochlioides* (Islam et al. 2004).

Phytoalexins synthesis, in contrast to constitutively produced phytoanticipins, is triggered, following recognition of elicitins, pathogen-associated molecular patterns (PAMPs) or damage-associated molecular patterns (DAMPs) or effectors and it is often mediated by activation of the mitogen-activated protein kinase (MAPK) pathway (Graham et al. 1990; Raaymakers and Ackerveken 2016). Furthermore, phytoalexin synthesis is also linked to other defense signaling mechanism. Phytohormones including ethylene (ET), jasmonate (JA), auxin and cytokinin may control synthesis of phytoalexins. JA acts as a positive regulator and induces secondary metabolite accumulation through transcriptional regulation in several plant species (Matsukawa et al. 2013). Pathogen recognition by nonhost may elicit accumulation of phytoalexins. The oligopeptide elicitor Pep-13, highly conserved among species of *Phytophthora* resulted in transcriptional reprogramming and accumulation of phytoalexins furacoumarins in nonhost parsley (Brunner et al. 2002). Furthermore, another necrosis-inducing *Phytophthora* protein 1 (NPP1) was shown to elicit HR-like cell death and it induced responses similar to Pep-13 in parsley, indicating that the multiple elicitors might be involved in induction of phytoalexin production in nonhost parsley by *Phytophthora* spp. (Hahlbrock et al. 2003). The *PENETRATION 2 (PEN2)* encodes a myrosinase involved in hydrolysis of indole glucosinolates. *Arabidopsis pen 2* mutants deficient in accumulation of an indole and cysteine metabolite allowed initiation of invasive growth of several nonadapted pathogens like *P. infestans* (Lipka et al. 2005). Detoxification of antimicrobial compounds produced in plants as a result of evolution of lineage, by adapted pathogens results in successful infection. In contrast, nonadapted *P. infestans* was more sensitive to capsidol induced in nonhost pepper, compared to adapted pathogen *P. capsici*, which was able to detoxify capsidol and establish infection in pepper (Ginnakopoulou et al. 2014).

The ability of plant cells to differentiate self and non-self and consequently to activate defense signaling is an important characteristic to prevent pathogen invasion at different stages. The pathogen-associated molecular patterns (PAMPs) are highly conserved and functionally required for pathogen fitness. Sensing of potential pathogens is mediated by membrane-associated-pattern recognition receptors (PRRs). PAMP recognition and the PAMP-triggered immunity (PTI) constitute critical components of nonhost resistance. In order to suppress PTI and weaken host defense, adapted pathogens secrete a number of effector proteins into host cells. By contrast, plants induce a second line of defense, termed as effector-triggered immunity (ETI) via recognition of pathogen effectors by intercellular immune receptors known as resistance (R) proteins, which frequently belong to the nucleotide-binding domain and leucine-rich-repeat-containing family. ETI is typically accompanied by a hypersensitive response (HR), a rapid programmed cell death (PCD) at the site of infection, restricting further spread of the pathogen in host tissues (Jones and Dangl 2006; Dodds and Rathjen 2010; Maekawa et al. 2011). Nonhost resistance may be partially mediated by effectors and *R* genes, although specific roles and mechanisms of ETI in NHR are yet to be elucidated clearly for fungal pathogens. Some attempts have been made to deploy NHR components to improve disease resistance in cultivars. The receptor-like protein ELR from wild relative *Solanum microdontum* could recognize several elicitins of *Phytopthora* spp. The transgenic *S. tuberosum* carrying ELR was resistant to infection by *P. infestans* (Du et al. 2015). The molecular interactions between *P. capsici* and nonhost plant tobacco (*Nicotiana tabacum*) were investigated. Tobacco plants acted as nonhost for *P. capsici* by responding with hypersensitive response (HR). The *P. capsici Avr3a*-like gene *(PcAvr3a1)* encoding a putative RXLR effector protein produced a HR upon transient expression in tobacco and several other *Nicotiana* species. The HR response correlated with resistance in 19 of 23 *Nicotiana* spp. and accessions tested, and knock-down of *PcAvr3a1* expression by host-induced gene silencing allowed infection of resistant tobacco. The results suggested that many *Nicotiana* species might have the ability to recognize PcAvr3a1 via products of endogenous disease resistance (*R*) genes. The *R* gene-mediated response might be the major component of nonhost resistance against *P. capsici* (Vega-Arreguin et al. 2014).

2.2.2.3 Enhancement of Resistance through Rootstocks

Grafting susceptible scions onto resistant rootstocks has been adopted as an effective management strategy for several soilborne diseases affecting cucurbitaceous and solanaceous vegetable crops. Mechanisms of development of resistance to soilborne diseases in grafted plants have been studied in some pathosystems. Nonhost resistance of rootstocks is one of the mechanisms considered to operate in grafted plants against soilborne pathogens. Solanaceous vegetables and cucurbits are grafted onto rootstocks, although related, that are different species or hybrids of different species. Nonhost resistance was shown to be a viable mechanism for control of watermelon wilt disease caused by *F. oxysporum* f.sp. *niveum* by grafting onto the bottlegourd genotypes which showed resistance to all races (0, 1 and 2) of the pathogen (Yetisir et al. 2007). Breeding programs for development of cultivars with resistance to Fusarium wilt of melon caused by *F. oxysporum* f.sp. *melonis (Fom)* were taken up. Commercial melon cultivars introgressed with two *R* genes (*Fom-1* and *Fom-2*) were resistant to FOM 0, 1 and 2 races, but not to FOM 1,2 race. Later breeding lines and indigenous cultivars of *C. melo* with resistance to FOM 1,2 race were identified. The results showed that use of resistant rootstocks could be adopted for containing the race FOM 1,2 successfully in melon crops, as FOM 1,2 resistance was polygenically inherited (Hirai et al. 2002). The rhizosphere microorganisms may have either beneficial or adverse effects on the plant growth, as well as on the development of soilborne microbial pathogens. Grafting cucumber onto *C. moschata* increased populations of bacteria and actinomycetes, while the fungal population was reduced in the rhizosphere. The actinomycete populations in the rhizosphere of resistant rootstock were higher, compared to population in the rhizosphere of self-rooted scion controls, when plants were inoculated with *F. solani* (Jiang et al. 2010). Grafting eggplant scion onto *Solanum torvum* reduced the incidence of Verticillium wilt disease. The ratios of bacteria and actinomycetes were increased in the rhizosphere of grafted plants (Yin et al. 2008). The rootstocks could influence the structure and composition of microbial community due to variations in the composition of root exudates. Critical investigations may shed light on the role of rhizosphere populations of the microorganisms in the development of soilborne fungal pathogens.

The molecular mechanism of disease suppression by grafting susceptible scion to resistant rootstocks was studied in melon-*F. oxysporum* f.sp. *melonis (Fom)* interaction. The effect of *Fom* race 1 and 1,2 gene expression was assessed using qPCR assay during infection of both resistant and susceptible grafted melon-scion rootstock combinations. Eleven pathogen genes related to pathogenicity were examined. Expression of selected genes varied according to race, susceptible or resistant interaction and time and these genes clustered into six profiles. The infection-related genes, including Zn-Cys transcription factor FOW2, xylanase and its relative transcriptional activator were highly induced particularly in the resistant combination, when infected by race 1,2 at any point time. In turn, ROS degrading catalase/peroxidase enzyme and actin-binding protein were upregulated only at the early stage of infection, in both resistant and susceptible combinations. Genes such as E3 ubiquitin-protein ligase and UTP-glucose-1-phosphate uridylyl transferase showed significant differences between *Fom* races in both grafting combinations. The only gene significantly dependent on fungal race, resistant/susceptible interaction and time, was a histidine kinase. Temporal transcription profiles were consistent with the ability of race 1,2 to develop in resistant host without inducing wilting symptoms. Overall, single *Fom* gene transcription profiles did not show clear differences between rootstock and scion, confirming that at molecular level, the grafted melon plant might react to pathogen, as a single genotype ruled by the rootstock (Haegi et al. 2017).

Apple orchards, in several countries, face the serious problem of apple replant disease (ARD) complex caused by different soilborne fungi and oomycetes and sometimes it may be aggravated by the nematode *Pratylenchus penetrans*. Preplant fumigation with chemicals reduced the severity of the disease, indicating that involvement of biotic factors as the major component of ARD. Establishing the identity and consistency of the complex inciting ARD was difficult, as the pathogens involved were reported to vary in different countries. Based on the frequency of association of pathogens with ARD, species of *Cylindrocarpon, Phytophthora, Pythium* and *Rhizoctonia*, along with *Pratylenchus penetrans* are considered to be the major components of ARD complex. Host tolerance/resistance is an economically attractive approach for managing diseases. Utilization of dual genotype plants in perennial tree crops, where rootstock (resistant) is grafted onto aerial scion (susceptible) allowed a 'divide and conquer' strategy, is an effective approach for tackling root disease problems. Production of a more fibrous root system contributed to the enhanced performance of some ARD tolerant rootstocks like Geneva 210 (Atucha et al. 2014). The innate resistance of the rootstock to ARD may constitute a cost-effective, durable and environment-friendly disease management strategy. Two apple rootstock genotypes G.935 (resistant) and B.9 (susceptible) showed distinct resistant responses, following infection by *P. ultimum*. The genetic regulation of apple root resistance to soilborne pathogens was investigated. Preinoculation variations in transcriptomes in roots of these two rootstock genotypes were suggested to contribute to the observed disease resistance phenotypes. Results of comparative transcriptome analysis showed elevated transcript abundance for many genes that function in a system-wide defense response in root tissues of the resistant genotype G. 935, in comparison with susceptible genotype B.9. These differentially expressed genes encoded proteins that function in several tiers of defense responses, such as pattern recognition receptors for pathogen detection and subsequent signal transduction, defense hormone biosynthesis and signaling, transcription factors with known roles in defense activation, enzymes of secondary metabolism and various classes of resistance proteins. The results suggested that the root tissues of resistant G.935 genotype contained a set of preformed factors to effectively defend infection by *P. ultimum*, compared with the susceptible genotype B.9 apple rootstock (Zhu et al. 2017).

2.2.2.4 Plant Defense Mechanisms

The active defense responses, initiated after recognition of pathogen presence on host plant surface responses, are expressed in host cells that are in contact with the pathogen or infected by the pathogen (as observed in virus infections). These cells recognize the specific signal molecules of pathogen origin and the outcome of this primary response results in programmed cell death (PCD), leading to appearance of necrotic lesions, generation of reactive oxygen species (ROS), the activation of a complex array of defense genes and the production of antimicrobial phytoalexins in the distal infected plant tissues. The adjacent cells surrounding the initially infected cells exhibit secondary responses in response to diffusible signal molecules (elicitors) formed following primary interaction. The elicitors initiate the activation of plant defense response genes in cultivars carrying the matching or complimentary disease resistance gene(s) and many genes have leucine-rich-repeat (LRR) domains which define the specificity or elicitor recognition. The third type of active defense response constitutes the systemic acquired resistance (SAR) observed in organs that are far away from the site of induction of resistance. SAR may be induced throughout the plant by the pathogens or compounds with hormonal activity and it is expressed as increased levels of resistance to a subsequent challenge by microbial pathogens. SAR is mediated either by gene-for-gene interactions between host plant R genes and pathogen avirulence (*Avr*) genes or by binding of nonhost-specific also designated pathogen associated molecular patterns (PAMPs)to their receptors (Dangl and Jones 2001; Ausubel 2005; Boller 2005).

2.2.2.4.1 Basal Host Plant Responses

Various defense mechanisms may be induced in plants exposed to microbial pathogens. The extent of activation in plants may vary depending on the levels of resistance of plants to the pathogen species or race concerned. In a nonhost plant species, the nonself recognition system perceives at the site of attempted penetration, the typical microbe (pathogen)-associated molecular patterns (MAMPs/PAMPs). The MAMPs/PAMPs recognized by nonhost plant cells are chitin fragments/chitooligosaccharides released from fungal cell walls during pathogen attack (Knogge and Scheel 2006). Deposition of mechanical barriers such as carbohydrates and hydroxylproline-rich glycoproteins within cell walls to restrict the development of fungal hyphae, synthesis of small secondary metabolites such as polyphenoloxidases with fungitoxic properties and production of active oxygen species, are the defense mechanisms operating in infected plants. Further, several defense-related enzymes and small peptides with antimicrobial activity such as thionins, defensins and lipid transfer proteins (LTPs), in addition to pathogenesis-related (PR) proteins, have also been produced, in response to infection by fungal pathogens (Narayanasamy 2006, 2008).

Production of reactive oxygen species (ROS) within a few minutes, after an infection by a microbial pathogen is initiated, probably is the first host response that could be detected, whereas ROS can react with other molecules without the energy input, molecular oxygen (O_2) remains relatively unreactive. ROS includes the superoxide anion (O^-_2), H_2O_2 and the hydroxyl radical (OH^-) constituting the oxidative burst and it has been established as a characteristic feature of hypersensitive response (HR) (Bolwell 1999). The H_2O_2 (a stable and less reactive ROS) from the oxidative burst plays a central role in the expression of HR. In order to avoid the adverse effects of oxidation, plants have developed enzymatic systems for scavenging those highly reactive forms of O_2. The superoxide dismutase (SOD) catalyzes the conversion of O^-_2 to O_2 and H_2O_2. Catalase (CAT) and/or ascorbate peroxidase (APX), in turn, convert H_2O_2 into water and O_2. In addition, the extracellular class III peroxidase (POX) catalyses the oxireduction between H_2O_2 and several other reductants and their activity has been correlated with plant defense against pathogens. Avirulent pathogens recognize the action of R gene products in plant immune system, elicit a biphasic ROS accumulation with a low amplitude, transient phase, followed by a sustained phase of much higher magnitude that correlates with disease resistance. On the other hand, virulent pathogens may avoid host recognition and induce only the first phase of response, suggesting a role for ROS in the establishment of host defense. Elicitors of defense responses known as MAMPs, also may trigger an oxidative burst, affecting the pathogen and/or host cells. Alternatively, MAMPs may function as signaling molecules that are not directly involved in the mechanism that actually arrest the pathogen development. The regulatory functions for ROS in defense may occur in conjunction with other plant signaling molecules such as salicylic acid (SA) or nitric oxide (NO) (Torres et al. 2006).

The superoxide anions produced outside the plant cells usually are rapidly converted into H_2O_2, which can cross the plasma membrane and enter plant cells. Then H_2O_2 is eventually removed from cells by conversion to water through action of catalase, ascorbate, peroxidase or glutathione peroxidase. The H_2O_2 may be either directly toxic to the pathogens or able to induce genes for proteins involved in certain cell protection mechanisms, such as glutathione (Hammond-Kosac and Jones 2000). In the tomato-*Colletotrichum coccodes* pathosystem, the involvement of H_2O_2 in defense response was indicated. The initial resistance to *C. coccodes* was not associated with active transcription and translation or hypersensitive cell death and removal of H_2O_2, with catalase allowed immediate penetration and development of symptoms of the disease. H_2O_2 accumulated in the area around the appressorium and it was associated with oxidative cross-linking of the plant cell wall proteins (Mellersh et al. 2002). In compatible host-pathogen interactions involving HR, NO and ROS, H_2O_2 and superoxide (O^-_2) may either directly reduce the ability of the pathogen to colonize its hosts or act as signaling molecules by inducing defense-related genes (Hancock et al. 2002). The role of NO in the latent infection of tomato by *Colletotrichum coccodes* was investigated, using inhibitors of NO. The effects of NO on superoxide (O^-_2), H_2O_2 levels (as measured by oxidatively cross-linked proteins) and callose deposition were assessed. Increased levels of O^-_2 and reduction in callose depositions and oxidative cross-linking at appressorial sites, were noted, when NO was reduced. In contrast, increased NO resulted in a greater percentage of appressorial sites with callose and cross-linked proteins. Catalase also, like superoxide dismutase, reduced the amount of protein cross-linking. The results suggested a role for NO in the initial defense of tomato against *C. coccodes* via an effect on plant cell wall modifications at sites of appressoria formation. There appeared to be a balanced relationship of NO with other reactive oxygen species. The effect is possibly to allow sufficient H_2O_2 to facilitate non-lethal cell wall defenses, which may temporarily stop colonization by *C. coccodes* (Wang and Higgins 2006). The cell wall-associated defenses, that function to restrict initial infection, are regulated by the balance in quantities of NO, O_2^- and H_2O_2 of which the latter appears to be more important in cross-linking of proteins and other structural polymers in the plant cell wall at the point of potential pathogen penetration. The role of NO in coordination with other activated oxygen species in host defense expression is well corroborated (Delledonne 2005; Wang and Higgins 2006).

V. dahliae, incitant of Verticillium wilt disease affecting several crops, produces both low- and high-molecular weight compounds that can function as elicitors capable of enhancing host defense and phytotoxins, facilitating pathogenesis. About 1,000 expressed sequence tags (ESTs) were generated and many were found to encode proteins harboring putative signal peptides for secretion. Heterologous expression resulted in the identification of a protein designated *V. dahliae* necrosis- and ethylene-inducing protein (VdNEP), composed of 233 amino acids. VdNEP acted as a wilt-inducing factor on excised cotton leaves and cotyledons. In addition, VdNEP induced production of phytoalexins and programmed cell death (PCD) of cotton suspension-cultured cells. The bacterially-expressed fusion protein of His-VdNEP also induced necrotic lesions in *Nicotiana benthamiana* leaves and complex defense responses in *A. thaliana* plants. His-VdNEP protein induced *PR-1* and *PDF1.2* gene expression in *A. thaliana* plants in addition to triggering production of ROS. Addition of the fusion protein at low concentration to suspension cultured cotton (*Gossypium arboreum*) cells, elicitation of biosynthesis of gossypol and related sesquiterpene phytoalexins was observed. By contrast, fusion protein induced cell death at higher concentrations. A low level of expression of VdNEP in the mycelium of *V. dahliae* in culture was detected by Northern blot analysis (Wang et al. 2004). Cellular and biochemical responses of cotton cultivars, Sicot 189 (moderately resistant) and Siokra 1-4 (highly susceptible) to infection of wilt pathogen *F. oxysporum* f.sp. *vasinfectum* were investigated. Distinct changes in xylem vessels and adjacent contact parenchyma cells could be visualized in Sicot 189, using light and transmission electron microscopy. Accumulation of amorphous materials in the infected xylem vessels lumen was observed at 3 days after inoculation. Cytoplasmic density of the adjacent parenchyma cells increased and vacuoles became segmented. Staining with antimony trichloride ($SbCl_3$) of fresh tissue and quantification of terpenoids by high performance liquid chromatography (HPLC), revealed an intense accumulation of the terpenoid hemigossypol, deoxyhemigossypol, deoxymethoxyhemigossypol and gossypol ahead of invading hyphae in the xylem vessels of Sicot 189 and pathogen hyphae continued to invade in advance of terpenoid accumulation. The timing, amount and location of terpenoid accumulation, especially deoxyhemigossypol were consistent with their role in the restriction of pathogen invasion in Sicot 189 (Hall et al. 2011).

Plants are exposed to several plant pathogens from seed germination until crop maturity and they have evolved various means to first recognize them and to defend against infection of both aerial and below ground organs. Recognition of the microbial pathogens is related to massive reprogramming of plant cell metabolism. The metabolic responses of tomato and *A. thaliana* to infection by *V. dahliae* were studied, using nontargeted gas chromatography (GC)-mass spectrometry (MS) profiling. The leaf contents of both major cell components, glucuronic acid and xylose were reduced in the presence of the pathogen in tomato, but enhanced in *A. thaliana*. The leaf content of two tricarboxylic acid cycle intermediates, fumaric acid and succinic acid was increased in the leaf of both species, reflecting a likely higher demand for reducing equivalents required for defense responses. A prominent group of affected compounds was amino acids and based on the targeted analysis in the root, the levels of 12 and 4 amino acids were found to be enhanced by *V. dahliae* infection, respectively in tomato and *A. thaliana*, with leucine and histidine being represented in both host plant species. Contents of six free amino acids in the leaf were reduced in *A. thaliana*, whereas contents of two free amino acids were raised in the tomato plants. The results revealed the role of primary plant metabolites in adaptive responses after colonization of plant tissues by *V. dahliae* (Bhutz et al. 2015).

Plants can sense the invading pathogens by employing two kinds of receptors, extracellular receptors and intracellular receptors. The extracellular receptors recognize pathogen molecules on the cellular surface as well as damage-associated host molecules that are released as a consequence of pathogen activity. By contrast, intracellular receptors recognize pathogen molecules that are delivered inside the host cells. The receptor-mediated recognition of pathogen molecules (MAMPs/PAMPs) leads to activation of plant innate immunity that wards off the invading pathogen. Failure to recognize pathogen effectors leads to establishment of infection and subsequent symptom development. The extracellular receptors–MAMP receptors have been characterized, including *Arabidopsis* FLS2 (flagellin-sensitive 2), EFR (elongation factor Tu receptor), CERK1 (chitin elicitor receptor kinase1) and rice CEBiP (chitin elicitor-binding protein). The FLS2 and EFR encode receptor-like kinases (RLKs) that recognize bacterial MAMPs flg22 and EF-Tu, respectively. CERK1 and CEBiP encode LySM domain-containing receptors that recognize chitin, the principal constituent of cell walls. These MAMP receptors are considered to display a low degree of specificity and broadly act in pathogen defense (Yadeta and Thomma 2013).

Extracellular plant receptors may have a role in plant defense against specific vascular wilt pathogens. Tomato Ve 1 is an extracellular plant receptor playing a role in xylem defense. Ve 1 is an extracellular leucine-rich repeat (LRR) receptor-like protein (RLP) that provides resistance to race 1 isolates of *V. dahliae* and *V. albo-atrum*. Interfamily transfer of *Ve 1* gene to *Arabidopsis* conferred resistance against race 1 isolates of *V. dahliae* and *V. albo-atrum* (Fradin et al. 2011). The pathogen ligand that was perceived by *Ve 1* was identified as the Ave 1 effector, a small (134-amino acid) effector protein with four cysteines required for full virulence on tomato plants lacking *Ve 1*. The Ave 1 homologs are present in a few other plant pathogens, including the vascular wilt pathogen *F.*

oxysporum f.sp. *lycopersici*, but a role in virulence is yet to be demonstrated for these homologs. The Ave 1 homolog for *F. oxysporum* f.sp. *lycopersici* was recognized by Ve 1 upon transient coexpression in tobacco and *Ve l* was able to mediate resistance against this pathogen (De Jonge et al. 2012). The presence of a functional *Ve 1* ortholog in *Nicotiana glutinosa*, and *Ave 1* expression in *N. glutinosa*, induced rapid and localized cell death of plant tissue (HR), surrounding the site, where recognition of pathogen effectors by host immune receptors occurred. In addition, *N. glutinosa* exhibited resistance against race 1 of *V. dahliae* that was compromised upon inoculation with an *Ave 1* deletion mutant of race 1 of *V. dahliae* isolate (Zhang et al. 2013).

The amino acid sequences of the proteins encoded by the genes *Ve 1* and *Ve 2* display structural domains reminiscent of cell-surface receptors. The cytosolic domain of *Ve 2* displays a YXXØ motif commonly involved in receptor-mediated endocystosis. Furthermore, Ve 2 concludes with the residues KKF, similar to the KKX motif that signals endoplasmic reticulum (ER) retention in mammalian and plant cells. The fusions of green fluorescent protein (GFP) with proteins of *Ve 2* and a *Ve 2* mutant with an altered YXXØ signal sequence, was generated for expression in tobacco 'Bright Yellow 2' (BY-2) suspension cells and tobacco plants. Both fusion proteins were localized exclusively with ER, as indicated by fluorescence microscopic analysis. Additional constructs with removed terminal KKF residues in the GFP fusion also appeared in the Golgi apparatus, after expression of tobacco cells. The results indicated that the protein encoded by *Ve 2* was predominantly present in ER (Ruthardt et al. 2007). A locus responsible for resistance against race 1 strains of *V. dahliae* and *V. albo-atrum* has been cloned from tomato only. This locus designated *Ve*, comprises two closely linked inversely oriented genes, *Ve 1* and *Ve 2*, that encode cell surface receptor proteins of the extracellular leucine-rich repeat receptor-like protein class of disease resistance proteins. *Ve 1*, but not *Ve 2*, provided resistance in tomato against race 1 strains of *V. dahliae* and *V. albo-atrum* and not against race 2 strains. Using virus-induced gene silencing (VIGS) in tomato, the signaling cascade downstream of Ve 1 was found to require both EDS1 and NDR1. In addition, NRC1, AC1F, MEK2 and DERK3/BAK1 also acted as positive regulators of Ve 1 in tomato. *Ve 1*-mediated resistance signaling was mediated by Cf proteins, type members of receptor-like protein class of resistance proteins (Fradin et al. 2009). The tomato receptor-like protein (RLP) Ve 1 is known to mediate resistance to the vascular pathogen *V. dahliae*. The functional Ve 1-enhanced GFP from *Nicotiana benthamiana* leaves was transiently expressed and immunopurified, followed by mass spectrometry. Peptides originating from endoplasmic reticulum (ER)-resident chaperones HS70 binding proteins (BiPs) and a lectin-type calreticulin (CRT) were identified by this procedure. Knockdown of the different BiPs and CRTs in tomato resulted in compromised Ve 1-mediated resistance to *V. dahliae* in most cases, revealing that these chaperones played an important role in Ve 1 functionality. It was shown that one particular CRT was required for the biogenesis of RLP-type *Cladosporium fulvum* resistance protein

Cf-4 of tomato, as silencing of CRT3a resulted in a reduced pool of complex glcosylated Cf-4 protein. In contrast, knockdown of various CRTs in *Nicotiana benthamiana* or *N. tabacum* did not lead to reduced accumulation of mature complex glycosylated Ve 1 protein. The results showed that the BiP and CRT ER chaperones differentially contribute to Cf-4- and Ve 1-mediated immunity (Liebrand et al. 2014).

The molecular aspects of interaction of *R. solani*, causal agent of foot rot disease of tomato was investigated to determine the role of chitinase and peroxidase gene expression and levels of lignin deposition in basal defense of tomato against the pathogen. Tomato cultivars Sunny 6066 (partially resistant) and Rio Grande (susceptible) were inoculated with *R. solani* M-2.3 belonging to AG-4. The thioglycolic acid (TGA) extractable cell wall complexes considered to be lignin, accumulated in cv. Sunny at 24 h postinoculation (*hpi*) significantly at greater level than in cv. Rio Grande and reached highest level at 72 hpi and then slightly decreased. The level of lignin detected in partially resistant cv. Sunny plants, using TGA assay was significantly higher than that of susceptible cv. Rio Grande at all time points after challenge inoculation with *R. solani*. After contact with tomato plant surface, the mycelia grew and produced infection structures, which directly penetrated into plant tissues. Growth of the pathogen on the epidermis was initiated at 1-day postinoculation (dpi), followed by formation of infection structures. The onset of disease on Rio Grande was faster and stronger, compared with Sunny and it might be linked to lower levels of lignin formation, which may function as structural barrier to pathogen spread in Rio Grande. Time course investigation on defense-related gene expression was performed on inoculated tomato leaves to determine the time point of upregulation of defense-related genes. The expression of peroxidase gene *CEVI-1* increased at 12 hpi and reached its peak at 48 hours postinoculation (*hpi*) in the Rio Grande plants. However, an increase in *PO-C1* expression at 6 hpi and peaking at 18 hpi was recorded in Sunny plants. Both sunny and Rio Grande plants exhibited highest level of chitinase (OC544149) expression at 24 hpi. However, higher level of OC544149 expression was noted in Sunny, compared to Rio Grande at 12 hpi, indicating priming the chitinase gene expression in cv Sunny plants. Chitinase might be a defense gene involved in the basal resistance exhibited by cv. Sunny tomato plants (Taheri and Tarighi 2012).

2.2.2.4.2 Systemic Host Plant Responses

Host plant responses to infection by microbial pathogens, involve induction of expression of a large array of genes, encoding a wide range of proteins that may have some role in development of resistance in infected plants. The process of development of the phenomenon, systemic acquired resistance (SAR), is initiated following recognition of the presence of the pathogen by the host plants. Signals are released from the point of infection/penetration by the pathogen triggering resistance in adjacent and also in distant tissues. Some families of genes collectively termed as 'SAR genes' are activated, following inoculation of leaves/roots of the test plants (Ward et al. 1991). The time required for SAR gene(s) expression may vary, depending on the nature of biotic or abiotic agent inducing resistance. Plant growth-promoting rhizobacteria (PGPRs) applied in the soil, localize at the surface of roots of treated plants, but they are able to induce systemic resistance (ISR) in leaves and stems far away from the root surfaces, where PGPRs colonize. SAR and ISR have two phases of resistance development. All events resulting in the development of resistance are included in the initiation phase which is transient. During the second maintenance phase, quasi-steady-state resistance occurs as result of events of the initial phase (Ryals et al. 1994). Infection by a microbial pathogen can shift both proximal and distal plant tissues to a physiological state of elevated defense against a broad range of pathogens. Although SAR and ISR are related, they are distinct versions of the systemic response of the host plant. They share two components: (i) elevated production of antimicrobial compounds and (ii) potentiation of the defense activation machinery, so that antimicrobial responses are activated more strongly and rapidly in response to subsequent infections (Feys and Parker 2000; Dong et al. 2002). The cellular responses of a plant challenged with a microbial plant pathogen are efficiently orchestrated and full analysis of these responses could be performed, because of advancements made in this aspect of host-pathogen interactions. Two approaches, microarray and proteomic analyses have allowed a global analysis of cellular regulation. Microarray-based expression profiling methods, together with the availability of genomic and/or expressed sequence tag (EST) sequence data for some plant species have enabled characterization of plant pathogenesis-related responses. Large number of genes associated with host plant resistance to diseases has been identified. Expression of thousands of genes simultaneously could be monitored using DNA microarray procedure. In addition, new pathogenesis-related genes, co-regulated genes and associated regulatory systems have been identified and the interactions between different signaling pathways have also been determined (Harmer and Kay 2001; Kazan et al. 2001). Expressed sequence tags (ESTs) are partial sequences of cDNA clones in an expressed cDNA library. They may be employed to identify all unique sequences (genes) to determine their functions (Mekhedov et al. 2000).

The expression patterns of 20 defense-related ESTs were studied. A leucine zipper protein, SNAKIN2, antimicrobial peptide precursor and elicitor induced receptor protein in a highly resistant chickpea (Cicer arietinum) accession ICC3996 and a susceptible cultivar Lasseter, after inoculation with spores of Ascochyta rabiei, the incitant of Ascochyta blight disease. Application of a time-series allowed putative detection of gene induction over the sampling period. Three defense-related ESTs exhibited differential upregulation in the resistant accession, when compared to the susceptible cultivar. The drawback of the accession ICC3996 was its poor agronomic attributes. ICC3996 was used to generate an enriched library of EST sequences. The ESTs (1,201) were clustered and assembled into 516 unigenes, of which 4% were defense-related, encoding lignin and phytoalexin biosynthesis genes, PR proteins, signaling proteins and putative

defense proteins (Coram and Pang 2005). In the further study, the resistance response of four chickpea genotypes, resistant ICC3996 (IC), moderately resistant FLIP94-508C (FL) and ILWC245 (IL, Cicer echinospermum) and susceptible Lasseter (LA) were investigated, using microarray technique and a set of unigenes of chickpea, grasspea (Lathyrus sativus) ESTs and lentil (Lens culianaris) resistance gene analogs (RGAs). Microarray results were validated by qRT-PCR assay. The time course expression patterns of 756 microarray features led to the differential expression of 97genes in at least one genotype at one time point. The transcritptional changes recorded in the early stages of infection (6 -12 hpi) may reflect initial responses, following the recognition of the pathogen contact and the major responses at 24–48 hpi, may be related to pathogen penetration and signaling cascades that led to an oxidative burst, induction of HR and synthesis of antifungal proteins (Coram and Pang 2006). Transcription of several putative PR-proteins was significantly induced in resistant chickpea genotypes at an earlier time point than in susceptible LA genotype. The specific microarrays expression profile of the PR protein, ß-1,3-glucanase, revealed exclusive upregulation in the resistant genotype IC. Quantitative RT-PCR assay confirmed this expression and also indicated upregulation in F2. The SNAKIN 2, an antimicrobial peptide precursor was upregulated in response to infection by Ascochyta rabiei. Since IC was the most resistant genotype, SNAKIN 2 peptides might be integral to the resistance mechanisms. PR-proteins (PRPs) are structural proteins of primary cell wall involved in strengthening of cell wall to restrict pathogen penetration, which was observed to occur at 24 hours postinoculation (hpi). The expression profiles of superoxide dismutase copper chaperone (SDCC) precursor and glutathione-S-transferase (GST) evidenced oxidative burst in this pathosystem. SDCC was significantly downregulated only in IC. On the other hand, GST was downregulated only in IC at 72 hpi. The other proteins whose regulation may be necessary for Ascochyta blight resistance include several PR-proteins, proline-rich proteins, disease resistance response protein DRRG49-C, polymorphic antigen membrane protein an Ca-binding protein. Comparison between resistant and susceptible genotypes revealed potential gene signatures predictive of effective A. rabiei resistance (Coram and Pang 2006).

The apical region (1-2 mm) of roots of pea (Pisum sativum) contains root meristems, required for root growth and cap development. When infection by microbial pathogen is initiated, root development ceases irreversibly within a few hours, even in the absence of severe necrosis. However, root cap meristems are resistant to infection by most pathogens. This resistance seemed to involve formation of a mucilaginous matrix or 'slime' composed of proteins, polysaccharides and detached living cells known as 'border cells'. The extracellular DNA (exDNA), a component of root cap slime was degraded during infection by a fungal pathogen, resulting in loss of root tip resistance to infection. Most root tips (> 95%) escape infection, even when immersed in inoculum of root rot pathogen, Nectria haematococca. In contrast, all (100%) inoculated root tips treated with DNase I developed necrosis.

Treatment with BAL31, an exonuclease capable of digesting DNA more slowly than DNase I also resulted in increased root tip infection, but the onset of infection was delayed. Untreated control root tips or fungal spores treated with nuclease alone showed normal morphology and growth. Pea root tips incubated with [^{32}P]dCTP during one-hour period, when no cell death occurred, produced root cap slime containing ^{32}P-labeled exDNA. The results suggested that exDNA could be an important component of host plant defense against infection by root rot pathogens like *N. haematococca* (Wen et al. 2009).

Microarray and proteomic analyses are the two approaches that are useful to have a clear insight into the entire network of host responses to infection by microbial pathogens. Both approaches allow a global analysis of cellular regulation. Microarray is restricted to the analysis of gene expression whereas the possibility of monitoring the accumulation and modification of proteins is an advantage provided by proteomics over microarray analysis. The level of gene expression does not necessarily correlate with the protein levels in a cell. Further, the genes required for a response are not necessarily the same genes that are differentially regulated as a result of response (Gygi et al. 1999; Birrel et al. 2002). Hence, analysis of protein levels and protein modification profiles may provide a better picture of a cellular response. In plants, protein phosphorylation seems to represent a major control mechanism for protein activity and to be an important posttranslational modification in response to microbial pathogens. Many signaling components such as kinases and transcription factors (TFs) occur only in very low copy numbers, making it difficult to detect them. Proteomics investigations have been employed to estimate over-or-under expression of proteins that are separated by two-dimensional polyacrylamide gel elelctrophoresis (2DE) which can also be coupled with immunodetection of phosphorylated proteins after Western blotting (Thurston et al. 2005). The plasma membrane (PM) proteins are involved in the perception of elicitors in regulating early responses and are often the targets of pathogen signals. The products of many *R* genes are membrane-associated receptors for the interactions of plant cells with pathogen-derived elicitors (Martin et al. 2003). The proteomic studies are effective for the detection of in vivo phosphorylation sites of low abundance membrane proteins that may be missed, when classical methods are employed (Thurston et al. 2005).

Defense responses of partially resistant and susceptible lines of sunflower to *S. sclerotiorum*, causal agent of white mold, were assessed at 12, 24 and 48 h after inoculation (*hai*), using SDS-PAGE technique. Significant differences were observed in the soluble stem protein contents in partially resistant line C71 and susceptible line C146. Soluble protein contents showed increase only in C71 at 12 and 24 *hai*, but not in susceptible line. Changes in polypeptide compositions of the sunflower lines were monitored at 12 and 24 *hai*. Many proteins accumulated in both partially resistant and susceptible plants. The soluble proteins with MW of 27 kDa accumulated in the stems of C71 infected plants with more pronounced accumulation occurring at 12 *hai*. This protein

band was absent in the susceptible C146 plants. Two polypeptides with MW of 55-60 kDa present in control plants disappeared in partially resistant C71 plants, following infection by *S. sclerotiorum*. The differential defense responses of partially resistant and susceptible sunflower lines to *S. sclerotiorum* might be useful in identifying sources resistant to the white mold disease (Davar et al. 2012). Changes in defense-related compounds of common bean cultivars with different levels of resistance to *S. sclerotiorum* were assessed to determine their relationship to pathogen tolerance. Common bean lines were inoculated with *S. sclerotiorum* and enzymatic and non-enzymatic parameters related to plant defense, such as peroxidase (POX), polyphenoloxidases (PPO), catalase (CAT), superoxide dismutase (SOD) and ascorbate peroxidase (APX), total soluble phenol and lignin contents were determined. At the position of 5 cm of stem samples with necrosis caused by the pathogen and at the second position of 5 cm of the stem samples above the first were analyzed at 12, 24, 48, 72, 96 and 120 hours postinoculation (*hpi*). Greater lignin and total soluble phenol contents and greater induction of POX and SOD activities in inoculated plants in the region near the inoculation point (first position) indicated local activation with later signaling for activation of defense mechanism in other regions of the plant. The genotype with greater level of resistance was superior to the susceptible line with regard to lignin production and the activities of POX, APX and SOD defense-related enzymes. The results suggested that a combination of these defense responses of common bean might contribute to greater plant resistance to the pathogen and that these enzymes have the potential for use in selecting disease resistant genotypes of common bean (Leite et al. 2014).

Proteomic analysis was performed to identify proteins involved in the resistance of canola against *S. sclerotiorum*. Comparison of protein expression profiles in the susceptible line with those in a resistant line during interaction of adult plant *B. napus* with *S. sclerotiorum* led to the identification of 20 important proteins related to disease resistance. These proteins were involved in different functions, including pathogen resistance, antioxidation and transcription regulation. These results showed that some proteins involved in resistance–a glycine-rich protein (GRP), a trypsin inhibitor protein (TIP), two heat-shock proteins (HSPs) and a thiol methyltransferase (TMT)–were upregulated or expressed specially in the resistant lines. These proteins contributed to ROS elimination and pathogen defense in the resistant line. As a consequence, the onset of programmed cell death (PCD) was delayed, and the spread of *S. sclerotiorum* was slowed in the resistant line. The results highlighted the role of specific proteins in pathogenesis and may be useful to identify the genes that can be transferred into susceptible *B. napus* cultivars to improve the levels of resistance to *S. sclerotiorum* (Wen et al. 2013). Hypothetical and predicted proteins may be targeted for studies to elucidate fungal infection and virulence mechanisms. A gene, *ssv263*, encoding a hypothetical, novel protein from *S. sclerotiorum* that was orthologous to secreted protein from *Botrytis cinerea* and, which appeared to be unique to these pathogens, was targeted. Mutant strains of *S. sclerotiorum* were generated by disruption of the *ssv263* gene and characterized for virulence on a susceptible canola (*B. napus*) genotype. The virulence of the *ssv263*-disrupted mutants was significantly reduced, compared to the wild-type strain, as reflected by symptom severity induced by the involvement of protein encoded by the hypothetical gene of the pathogen in the virulence of *S. sclerotiorum* and this knowledge may be useful in determining the levels of susceptibility/resistance to variants (strains) of the fungal pathogens (Liang et al. 2013).

Direct roles for the biological roles of specific *WRKY* genes in host resistance was lacking in *B. napus–S. sclerotiorum* interaction. Hence, *B. napus WRKY* gene *BnWRKY33* was isolated and this gene was found to be highly responsive to infection of *S. sclerotiorum*. Transgenic *B. napus* plants overexpressing *BnWRKY33* showed significantly enhanced resistance to *S. sclerotiorum*, due to constitutive activation of the expression of *BnPR1* and *BnPDF1.2* and inhibition of H_2O_2 accumulation in response to pathogen infection. In addition, a mitogen-activated protein (MAP) kinase substrate gene, *BnMKS1* was also isolated. Using the yeast two-hybrid assay BnWRKY33 was shown to interact with BnMKS1, which could also interact with BnMPK4, consistent with their collective nuclear localization. BnWRKT33, BnMKS1 and BnMPK4 were substantially and synergistically expressed in response to infection by *S. sclerotiorum*. In contrast, three genes showed differential expression in response to phytohormone treatments. The results suggested that *BnWRKY33* might have an important role in the defense response of *B. napus* to *S. sclerotiorum* which might be associated with the activation of SA- and JA-mediated defense response and inhibition of H_2O_2 accumulation (Wang et al. 2014). *Brassica* spp. have the glucosinolate-myrosinase system, producing toxic volatile compounds, when infected by pathogen or mechanically injured. *S. sclerotiorum* produces oxalic acid (OA), which is known to be a pathogenic factor. Hydrolysis of glucosinolates in the presence of oxalic acid in the substrate was studied under different conditions. Colonies of *S. sclerotiorum* were exposed to volatiles from hydrated mustard powder used as a myrosinase and glucosinolate source. The glucosinolate-myrosinase (GL-M) system was activated in the presence of OA at a concentration and pH similar to that expected in vivo. Volatile production was inhibited only at the pH level of 3 or below, suggesting that OA might not have significant role in disarming the GL-M system, during infection process of *S. sclerotiorum* on *Brassica* (Rahmanpour et al. 2010). *S. sclerotiorum* secretes oxalic acid (OA), which can be degraded to CO_2 and H_2O_2 by oxalate oxidase (OXO). Barley oxalateoxidase (BOXO, Y14203) gene was introduced into oilseed rape via *Agrobacterium*-mediated transformation to investigate the mechanism by which OXO promotes resistance to *S. sclerotiorum*. At 72 h postinoculation, about 15 to 61% fewer lesions were formed on leaves of transgenic oilseed rape, compared to control, exhibiting possible partial resistance in leaves to the pathogen. Transgenic plants also exhibited decreased oxalate and increased H_2O_2 levels, compared to the control. The expression of defense response genes involved in the H_2O_2 signaling pathway was also induced. The results indicated that the increased resistance in oilseed rape could be attributed to the enhanced OA metabolism, production of H_2O_2 and the H_2O_2-mediated defense levels during infection (Liu et al. 2015).

Salicylic acid (SA) and jasmonic acid (JA), the phytohormones have a role in regulating signal pathways of plant defense responses to microbial plant pathogen infection. JA application induced synthesis of PR-proteins and phytoalexins, following treatment with elicitors of pathogen origin. The response to necrotrophic fungal pathogens like *Pythium* sp. is regulated by JA pathways (Vijayan et al. 1998). Both positive and negative interactions between SA and JA signaling pathways have been suggested. Expression of SA-dependent (acidic) PR-proteins was inhibited by exogenous application of JA on tobacco leaves. In contrast, treatment enhanced the expression of JA-dependent (basic) PR-proteins (Niki et al. 1998). Enhanced SA-mediated defense responses in JA-insensitive mutants of *A. thaliana* were observed (Petersen et al. 2000). Defense responses of pepper (*Capsicum*) cultivar SCM334 resistant to *P. capsici*, causal agent of Phytophthora blight disease, were assessed. Expression patterns of HR-related and JA synthesis genes in resistant and susceptible pepper cultivars were analyzed. Catalase (CAT) and peroxidase (POD) are reductases of H_2O_2 that are generated during HR. Expression of two genes was induced in susceptible cultivar after inoculation and prior to symptom development or penetration of host tissue by the pathogen. On the other hand, expression of these genes was either delayed, or insignificant in SCM334, following inoculation with *P. capsici*. The gene *OPR3* participates directly in the octadecanoid pathway for JA (Strassner et al. 2002). *OPR3* mRNA was detected immediately after inoculation of SCM334 leaves, but the levels of JA in mock-treated SCM334 leaves did not show increase. Expression of *OPR3* mRNA was accompanied by JA accumulation with a time lag that might be due to processing prior to *OPR3* participation in JA synthesis. The early appearance of JA and later accumulation of SA, following inoculation of the resistant cultivar suggested that JA and SA may play different separate roles in host defense response, resulting in HR-mediated cell death (Ueeda et al. 2006).

The pathogenesis-related (PR) proteins are coded by the host plant, but induced only in pathological or related conditions. PR-proteins have a role in the resistance response of host plants especially in the development of acquired resistance, following formation of necrotic lesions. Induction of PR-proteins is associated with the development of systemic acquired resistance (SAR) in tissues/organs far away from the site of inoculation (Ward et al. 1991). The PR-genes and proteins are widely used as marker genes/ proteins to study defense mechanisms of plants. Contact between PR-proteins with antifungal property and pathogen occur in the major intercellular space of leaf tissues (Van Loon and VanStrien 1999). Localization of chitinase (CA Chi[2]) in compatible interactions of pepper stems with *P. capsici* was visualized by employing immunogold labeling technique. In the compatible interaction, the quantity of gold particles in the pathogen cell surface was limited, but even distribution of gold particles over the entire fungal cell wall, could be observed. In contrast, most of the gold particles were seen over the cell wall area of the pathogen interacting with resistant tissues. Degradation of fungal cell wall was evident at the hyphal tips which had dense deposition of gold particles, indicating that the activity of the host chitinase on the pathogen is limited to the cell wall, since the cytoplasm of fungal cells was nearly free of gold particles (Lee et al. 2000). The molecular mechanisms occurring during the establishment of compatible interaction between *Phytophthora parasitica* and *A. thaliana* in the roots were investigated. Cytological and genetic analyses showed that *P. parasitica* penetrated the roots of *A. thaliana* after formation of appressoria. Initial biotrophic growth was accompanied by the formation of haustoria and it was followed by a necrotrophic phase in its life cycle. *A. thaliana* mutants with impaired SA, JA, or ethylene (ET) signaling pathways were more susceptible than the wild-type to infection. The SA- and JA-dependent signaling pathways were precisely activated, when *P. parasitica* penetrated the roots, but were downregulated during invasive growth, when ET-mediated signaling was predominant (Attard et al. 2010).

Root pathogens have not been investigated enough to better understand their interactions with host plant roots and the underlying mechanisms of defense and resistance. Plants are able to recognize microbes-associated molecular patterns (MAMPs) through pattern recognition receptors that specifically bind to their target MAMP, and recognition leads to activation of plant's basal immune response. MAMP detection leads to signal transduction and amplification of kinase cascade that triggers the activation of PR-proteins, production of reactive oxygen species (ROS) and many secondary metabolites, including deposition of callose that acts as a physical and chemical barrier to prevent pathogen invasion. Hormone-controlled defense pathways, such as systemic acquired resistance (SAR), which protects against subsequent infections are also activated. SAR is mediated by salicylic acid (SA) signaling, but has also been shown to require jasmonic acid (JA) in the initial stages. Root-infecting fungal pathogens may be classified into two groups, as biotrophic and necrotrophic, based on their mode of deriving nutrients from the host plants. The SA- and JA-ethylene (ET) signaling pathways are generally considered to be effective against biotrophic and necrotrophic pathogens respectively (Glazebrook 2005; Beckers and Spoel 2006). Hemibiotrophic pathogens like *P. infestans*, commence as a biotrophic pathogen in the early stage, feeding on living cells and establish infection and later shift to necrotrophic phase to complete their life cycle. This kind of lifestyle requires hemibiotrophic pathogen to hijack both host signaling pathways for their colonization of different host tissues. Formae speciales of *F. oxysporum* can infect several crop plant species, including tomato, cotton, banana, and leguminous and cucurbitaceous crops. *F. oxysporum* acts as a hemibiotrophic pathogen in *Arabidiopsis* and application of salicylic acid (SA) on leaves, resulted in a partial increase in resistance. Mutants deficient in SA-mediated defense were more susceptible to *F. oxysporum*. The *sid2* mutant was impaired in SA biosynthesis and was susceptible

to *F. oxysporum* f.sp. *conglutinans*. However, during infection of *F. oxysporum* strongly induced jasmonic acid (JA)-mediated defense in the leaves (Berrocal-Lobo and Molina 2004; Diener and Ausubel 2005). Increased resistance to *F. oxysporum* in plants, showing insensitivity to jasmonic acid (JA), was observed. The *jasmonate insensitive 1* (*jin 1*) mutant (also known as *myc2*) showed increased resistance to *F. oxysporum*. Likewise, the *coi 1* (*coronatine insensitive 1*) and *pft1* (*phyotchrome and flowering time 1*) mutants were also compromised in JA responses and they showed increased resistance to *F. oxysporum*. However, it is possible that different formae speciales of *Fusarium* may adopt different strategies to induce disease in their host plant species (Thatcher et al. 2009; Cole et al. 2014). Seven *R* genes in *Arabidopsis* designated *RESISTANCE TO FUSARIUM* (*RF01–RF07*) were identified, using a cross between resistant wild-type (WT), Columbia-0 (Col-0) and the more susceptible Taynuilt-0 (Ty-0) ecotype. *RF01* encoded a wall-associated kinase-like kinase 22 (WAKL22) and *RF02* encodes a receptor-like protein. The RF0 showed the capacity to protect plants against multiple formae speciales of *Fusarium* (Diener and Ausubel 2005; Shen and Diener 2013).

F. oxysporum infects susceptible host plant species through roots and becomes systemic by invading vasculature tissue causing symptoms in stem and leaves. Microarray analyses as well as several functional genotype analyses have been performed to identify the signaling processes that are required for development of resistance against *F. oxysporum* in *Arabidopsis*. Genome-wide gene expression of *Fusarium*-infected root samples collected at 48 h after infection was profiled to investigate the defense responses in root for better elucidation of host-pathogen interaction. In contrast to gene expression in leaves, the roots of infected plants showed several down-regulated genes as opposed to upregulated gene. Further, the relative absence of defensins or PR-protein gene expression in infected root tissue was in contrast to leaf tissues. In infected roots, only one relatively uncharacterized defensins gene was activated, but the expression was below the 1.5-fold cut-off used for selecting differentially expressed gene. T-DNA insertion mutants for five differentially expressed genes viz., AT3G55970, AT4G22610, AT1G62500, AT3G16770 and AT3G62670 were examined for their role in defense against *F. oxysporum*. One of the mutants, *erf72* containing a T-DNA insertion in the *AP2/ETHYLENE RESPONSE FACTOR 72* gene, showed increased resistance to *F. oxysporum*, suggesting that ERF72 might be a negative regulator of plant defense against *F. oxysporum*. Quantitative RT-PCR assays showed that no significant change was noted in the expression of SA-associated *PATHOGENESIS-RELATED* genes and the PR-1 and PR-5 or JA-associated defense genes *PDF1.2* and *PR-4*. However, the expression of the *BASIC CHITINASE (CH1-B)* gene (*PR-3*), showed increased expression in the *erf72* mutant, compared to wild type under mock conditions. In addition, using *erf72* mutants expressing ß-glucuronidase (GUS) of *F. oxysporum*, no difference in root colonization could be visualized after GUS staining (see Figure 2.4) (Chen et al. 2014).

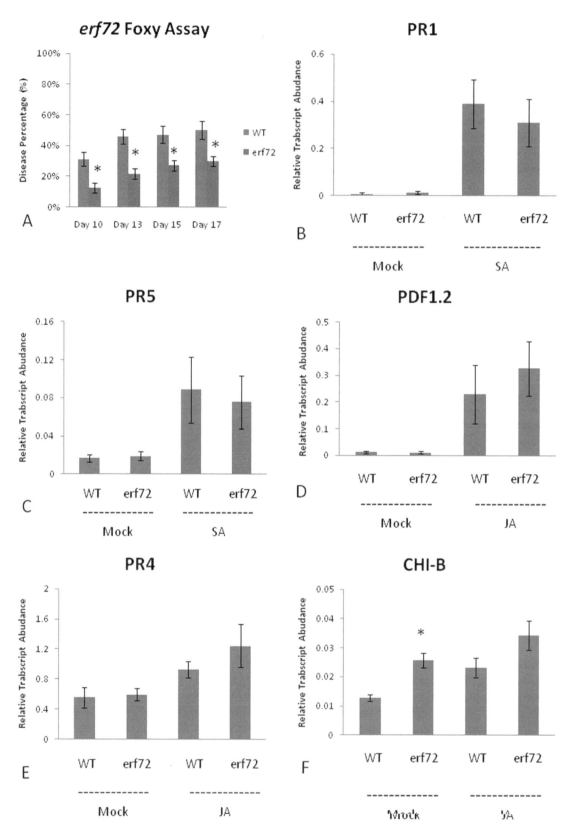

FIGURE 2.4 Effects of SA/JA treatments on disease scores for *erf72* mutants and abundances of different relative gene transcripts Disease scores are indicated by average proportion of symptomatic leaves out of total number of leaves/plant; asterisk (*) indicates significance of *p*-value (< 0.05); bars reflect mean values ± SE independent biological replicates.

[Courtesy of Chen et al. 2014 and with kind permission of Scientific Reports Open Access Journal]

The differentially expressed genes of roots and leaves of *Arabidopsis* showed variations only in three genes, suggesting that the gene expression changes in response to *F. oxysporum* infection were fundamentally similar in the tissues of roots and leaves. *F. oxysporum* required JA signaling components to promote susceptibility and induced JA-associated gene suppression in the roots and shoots. Insensitivity of jasmonate has been implicated in resistance to *F. oxysporum* and hence, the root growth of *erf72* mutants on JeA-containing agar plates was quantified. No difference in root growth of *erf72* mutant and wild-type could be detected. It was, therefore, considered that with the exception of increased CHI-B expression, the *erf72* mutant did not appear to be affected in SA- and JA-associated defense gene expression of MeJA-mediated root inhibition. The enhanced chitinase expression in *erf72* plants prior to infection might contribute toward its increased resistance to *F. oxysporum*. It is likely that *F. oxysporum* might suppress MAMP responses via the JA pathway to allow greater infection level. The pub22/pub23/pub24 triple mutant lacking the PUB22, PUB23 and PUB24 U-box type E3 ubiquitin ligases were inoculated to determine, if an enhanced MAMP response could provide enhanced resistance to *F. oxysporum*. The triple mutant showed increased resistance. The pathogenicity tests indicated the increased level of resistance of the triple mutant to *F. oxysporum*. The results showed that the *pub22/23/24* triple mutant was more resistant to *F. oxysporum* and the root-mediated defenses against soilborne pathogens could be detected at multiple levels (see Figure 2.5) (Chen et al. 2014).

F. oxysporum f.sp. *cepae (Foc)* causes the devastating Fusarium basal plate rot disease of onion and garlic *(Allium sativum)* in several countries. Garlic selection line CBT-As153 was identified as resistant. The *R* genes encode a highly conserved nucleotide-binding site and leucine-rich repeat structure

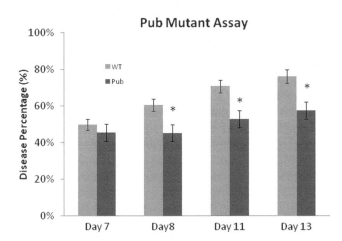

Pub Mutant Assay

FIGURE 2.5 Enhancement of resistance to *F. oxysporum* in *Arabidopsis pub22/23/24* triple mutants, determined based on average proportion of symptomatic leaves /total leaves/ plant Asterisk (*) indicates significance of difference in resistance levels (P < 0.05) as per Student's test; bars represent mean values ± SE of three replicates.

(NBS-LRR), which facilitates isolation of candidate genes linked to FBR resistance in CBT-As153. Degenerate primers based on the NBS conserved motif of NBS-LRR resistance proteins were employed to isolate 28 NBS sequences from CBT-As153 designated *A. sativum* resistance gene analogs (AsRGAs). Sequence analysis grouped AsRGAs into six classes of non-toll interleukin receptor (non-TIR) subfamily. Quantitative real-time polymerase chain reaction (qRT-PCR) showed differential transcript expression of AsRGAs in stem, leaves and roots. AsRGA 29, a putative RGA exhibited 20-fold higher expression of transcript levels in CBT-As153 than that of *Allium fistulosum* and *Allium roylei*, the FBR resistant wild *Allium* species. In addition, AsRGA 29 showed significant induction in the transient levels under *Foc* infection and treatment with four defense signaling molecules SA, MeJ, H_2O_2 and ABA. The results suggested the crucial involvement of AsRGA29 in mediating multiple defense signaling pathways toward protecting garlic against *Foc*. Development of RGA-based markers linked to FBR resistance in garlic and onion might pave way for investigating molecular mechanisms related to *F. oxysporum* f.sp. *cepae* resistance in CBT-As153 (Rout et al. 2014).

The defense responses of root cuttings of two cultivars of carnation *(Dianthus caryophyllus)* with differential responses to vascular wilt caused by *F. oxysporum* f.sp. *dianthi* (FOD) were assessed, using some parameters associated with guaiacol peroxidase activity (GPX, E.C. 1.11.17). The cv. L.P. Candy- (resistant) showed significant increase in GPX activity levels at 12, 24 and 48 h postinoculation (*hpi*), whereas the cv. Tasman (susceptible) showed only a slight increase after a longer interval (96 hpi). The H_2O_2 content in L. P. Candy showed increase at 12 and 24 hpi, with corresponding increase in GPX activity level. However, Tasman exhibited an increase in H_2O_2 content at 12 hpi that had no relationship with GPX activity. The increase in GPX activity in L. P. Candy was due to an increase in levels of constitutive isoforms of peroxidase, as indicated by zymogram analysis. Quantification of transcriptional levels indicated that in the cv. L. P. Candy, the mRNA level of the class III peroxidase DcprxO2 increased by almost eightfold, compared with the control at 6 hpi, whereas the mRNA levels in the cv. Tasman increased approximately by twofold at 12 hpi. The enhanced levels of H_2O_2 showed positive relationship with increase in levels of DcprxO2 transcription. The results suggested that generation of reactive oxygen species (ROS) and subsequent regulation mediated by action of antioxidant enzymes like GPX might be associated with resistance of plants to vascular wilt pathogens (Ardila et al. 2014).

The participation of a class III chitinase in the melon-*F. oxysporum* f.sp. *melonis* (*Fom*) interaction was examined by applying immunoassays in the root and stem base of susceptible cv. Galia and resistant cv. Bredor. Using the specific antibody, the distribution of a class III chitinase was monitored by both Western blotting analysis and by immunolocalization technique. The constitutive expression of chitinase III in stem base tissues of Galia was determined by Western blotting and no chitinase band was detected in the resistant plants. But in Bredor plants, wounded or infected by *Fom*, three bands at

13, 25 and 30 kDa positions were detected. By contrast, following infection in Galia plants, the bands at 30 and 35 kDa positions became faint. Chitinase deposition was more intense in Bredor plants in which presence of pathogen hyphae was not detected. But in susceptible plants, hyphae of *Fom* could be visualized in the xylem vessels and very faint expression of chitinase was seen. It appeared that for melon resistance to *F. oxysporum* f.sp. *melonis*, not only the total chitinase III was required, but rather some of its isoforms might be important for development of resistance, as indicated by Western blotting analysis. The results suggested the association of chitinase III with resistance of melon to Fusarium wilt disease of melon (Baldé et al. 2006). The rice *OsPR3* chitinase cDNA (RC7) was isolated from rice plants inoculated with *R. solani*, causal agent of sheath blight disease and this gene was transgressed into *indica* rice cultivars, IR72, IR64, IR68899B, MH63 and Chinsurah Boro II. These cultivars could be effectively grown to reduce incidence of rice sheath blight disease (Datta et al. 2001). In another investigation, accumulation of a novel *OsPR-4*, a single copy gene in both compatible and incompatible host-pathogen interactions was observed (Agrawal et al. 2003).

Tomato plants contain a tomatinase (phytoanticipin), a steroid glycoalkaloid at a concentration that may inhibit fungal infection of leaves, stems, roots, flowers and green fruit (Arneson and Durbin 1968). Fungal pathogens producing tomatinase that detoxified α-tomatine could colonize the tomato tissues (Ruiz-Rubio et al. 2001). The role of tomatinase in the pathogenicity of *F. oxysporum* f.sp. *lycopersici*, infecting tomato was investigated by performing transformation-mediated disruption of tomatinase gene. Mutants showed increased sensitivity to α-tomatine and decrease in tomatinase activity in vitro. The pathogenicity of the mutants was also reduced as indicated by inoculation assay performed on tomato plants, suggesting an important role of tomatinase. However, Southern blot analysis revealed that the resident tomatinase gene of the mutants was not disrupted, but the gene replacement vector was ectopically integrated into the genome of the mutants. One of the ectopic insertion mutants was examined at both DNA and RNA levels to clarify the mechanisms involved. No methylation of the resident gene was detected. Northern blot analysis revealed a low level of tomatinase gene expression and also the presence of antisense RNA in the ectopic mutant. Thus, the posttranscriptional gene silencing appeared to be involved in the suppression of tomatinase in the ectopic mutant (Ito et al. 2002). *F. oxysporum* f.sp. *lycopersici (Fol)* could degrade α-tomatine to less toxin metabolite, because of the extracellular enzyme tomatinase. The tomatinase gene (*FoTom1*) was also present in some strains of *F. oxysporum* nonpathogenic to tomato plants. Genes encoding Chi9 and GluB, both of which were intracellular and basic PR-proteins, were induced in tomato plants inoculated with nonpathogenic strain, prior to inoculation of tomato plants with nonpathogenic strains. Nonpathogenic strain strongly induced *Chi3* mRNA in tomato plants and suppressed the development of the vascular wilt caused by *F. oxysporum* f.sp. *lycopersici* (Ito et al. 2005).

Resistance in carnation to wilt disease caused by *F. oxysporum* f.sp. *dianthi (Fod)* is polygenic. Anthranilic acid-derivative phytoalexin (AADP) are produced in carnation in response to infection by *Fod*. The possibility of presence of additional resistance factors other than AADP was investigated for their involvement in resistance to Fusarium wilt of carnation. The carnation AADP could not be detected in the roots of resistant genotypes, the natural site of infection by *Fod*. Protective molecules constitutively present in healthy plants are known as phytoanticipins. A constitutive fungitoxic flavonol triglycoside was isolated from the roots and stems of the most resistant carnation cv. Novada, which was found to be an active producer of AADP (Curir et al. 2001). In the further investigation, the biosynthesis of this flavonol phytoanticipins kaempferide 3-0-[2G-b-D-glucopyranosyl]-ß-rutinoside, was stimulated by inoculation with *Fod*, as in the case of phytoalexin. The results suggested that the levels of some preformed antifungal flavonoids might be significantly enhanced following infection and these fungitoxic molecules may actively cooperate with compounds that are synthesized postinfectionally, resulting in higher degree of resistance to the invading fungal pathogen (Curir et al. 2005).

The intracellular plant receptors mediating plant defense against xylem-invading pathogens have been identified. Tomato *I-2* gene is an intracellular receptor that contributes to resistance to race 2 isolates of *F. oxysporum* f.sp. *lycopersici*. It encodes a cytoplasmic coiled-coil (CC)-nucleotide-binding site (NBS)-LRR receptor protein recognizing the effector protein Avr2, which was initially identified in the xylem sap from tomato infected by *F. oxysporum* f.sp. *lycopersici* and it is taken up by tomato cells (Huang and Lindhout 1997; Houterman et al. 2009). The transcriptional changes in tomato induced by *V. dahliae* (vascular pathogen) and *Cladosporium fulvum* (foliar pathogen) were compared. *C. fulvum*-induced transcriptional changes revealed little overlap with those induced by *V. dahliae*. Furthermore, within the subset of genes that were regulated by both pathogens, many genes showed inverse regulation (van Esse et al. 2009). Recognition of vascular wilt pathogens, mediated by either extracellular or intracellular receptors, led to the activation of defense responses in xylem vessels. The physical defense responses (formation of tyloses) halted or contained the pathogen from further spread in the xylem vessels, whereas chemical defense responses killed the pathogen or inhibited its development. Vein-clearing, a typical symptom of *Verticillium* infection was due to *Verticillium*-induced transdifferentiation of chloroplast-containing bundle sheath cells into functional xylem elements (Reusche et al. 2012). Another physical response observed during xylem colonization is vascular coating in the xylem parenchyma cells of tomato and alfalfa infected by *Verticillium albo-atrum*. Infusion of pit membranes, primary walls and parenchyma cells with coating materials could prevent lateral and vertical spread of the pathogen in the xylem vessels (Street et al. 1986; Newcombe and Robb 1988).

The mechanisms of differential regulation of activation of potato defense genes in response to infection by two genotypes, US-1 and US-8 of *P. infestans*, incitant of potato late blight disease, were investigated. The phenylalanine ammonia-lyase (PAL) and 3-hydroxy-3-methylglutaryl CoA reductases (HMGR) are the key enzymes involved in the phenylpropanoid and terpenoid biosynthesis and may have a role in disease resistance mechanisms in many plant species (Kervinen et al. 1998; Kim et al. 2004). Differential expression of *Pal1* and *hmgr2* was investigated by applying northern blot analysis in two potato cultivars, Russet Burbank (RB) and Kennebec (KB) susceptible and moderately resistant respectively, following inoculation with isolates of US-1 and US-2 genotypes of *P. infestans*. Accumulation of *pal 1* transcripts was less in response to US-8, as compared to US-1 and occurred earlier in KB than in RB. This was attributed to stronger defense gene suppression by US-8 isolate. No apparent strong accumulation of *hmgr2* transcripts could be detected in RB, as compared to KB inoculated with either US-1 and US-8 isolates. As the expression of these two genes was stronger in proximal and distal leaflets, when compared to the local site of inoculation of other healthy parts of the potato plants (Wang et al. 2004). Resistance to rice sheath blight (SB) disease caused by *R. solani* is quantitative in nature, and it has been difficult to develop rice cultivars with high level of resistance to this disease. Overexpression of *OsOSM1* gene, encoding osmotin protein that belongs to the PR-5 protein family, was found to enhance the resistance to sheath blight in field assessments. *OsOSM1* was expressed in leaf sheath at the booting stage, coinciding with the critical stage of sheath blight development in the field. Further, *OsOSM1* expression was strongly induced by *R. solani* in SB-resistant rice cv. YSBR1, but not in susceptible varieties. Overexpression of *OsOSM1* (*OsOSM1ox*) in transgenic rice cv. Xudo3 significantly increased resistance to SB in transgenic rice. OsOSM1 RNA levels in different transgenic lines were positively correlated with their levels of SB resistance. As very high levels of *OsOSM1* were harmful to plant development, the levels of expression had to be kept at optimal level. OsOSM1 protein was localized in plasma membrane. OsOSM1 was upregulated by jasmonic acid (JA). In addition, JA-responsive marker genes were induced in *OsOSM1ox* lines. The results showed that *OsOSM1* had an important role in development of resistance to rice sheath blight disease and this gene could be employed as a marker gene in breeding programs (Xue et al. 2016).

The gaseous plant hormone ethylene is involved in regulation of various developmental processes. Further, ethylene production, upon wounding and pathogen attack, is stimulated in the affected tissues, from which it diffuses into surrounding cell layers, before escaping into the atmosphere. In the plant tissues, it functions as a local signal, leading to the activation in neighboring cells by adaptive mechanism that can lessen the effects of stress condition. Enhanced ethylene production is an early, active response of plants to the perception of pathogen attack and appears to be involved in the induction of defense reactions. Strong stimulation of ethylene production is a common characteristic of hypersensitive reaction from the incompatible combination of an avirulent strain of the pathogen and the resistant host in which the pathogen is

rapidly restricted, because of the localized tissue necrosis near the site of pathogen penetration. Hypersensitivity is associated with the accumulation of antimicrobial phytoalexins and pathogenesis-related (PR) proteins, and fortification of cell walls. Depending on the plant species, ethylene can induce or stimulate the enzymes of aromatic biosynthesis necessary for the isoflavonoid phytoalexin production and lignification, as well as promote synthesis of PR proteins (Narayanasamy 2008). A *Tobacco rattle virus* (TRV)-based virus-induced gene silencing (VIGS) system was employed to investigate the role of tomato ethylene receptor ETR4. By comparing wilting symptoms of Verticillium wilt in wild-type ethylene-sensitive, Never ripe (*Nr*) mutant tomato plants and *ETR4*-silenced plants, it was demonstrated that disease severity in the *Nr* and *ETR4*-silenced plants was significantly reduced, compared to wild-type plants, at 33 days postinoculation (*dpi*). In the *ETR4*-silenced plants, disease incidence and severity were reduced by 14% and 15%, respectively, compared to the TRV-only inoculated plants, at 37 *dpi*. Quantification of *V. dahliae* by qPCR assay showed that reduction in symptom severity in the *Nr* plants was associated with significant reduction of growth of the pathogen in the vascular tissues of *Nr* plants, compared to that in the wild-plants, suggesting that impaired perception of ethylene via Never ripe receptor results in increased disease resistance. Pathogen reduction was evident at each sampling day in the *Nr* plants, ranging from 1.5 to 1.75 times less than that in the wild-type plants. Pathogen quantification in *ETR4*-silenced and TRV-only inoculated plants showed similar levels of biomass (Pantelides et al. 2010).

The molecular basis of cotton response to the necrotrophic pathogen *V. dahliae* was studied. *V. dahliae*, which occurs in the form of two pathotypes designated defoliating (D) and nondefoliating (ND), based on the type of symptoms induced by them in infected plants. The RNA isolated from *Gossypium barbadense* was applied to generate highly enriched transcripts by PCR select suppression subtractive hybridization (SSH) and to capture a wide spectrum of differentially expressed genes in the cotton defense response. A total of 211 unique genes was differentially identified and classified into 11 functional categories. The largest groups contained genes involved in metabolism, stress/defense response, cell structure and signal transduction. More than one-third of the genes (38%) were without classification and their functions were not known. A set of disease-related genes involved in the process of the response, including pathogenesis-related genes of different classes, oxidative burst-related genes was identified. The characterization of some transcription factors and kinases enabled to elucidate defense mechanisms clearly. The results suggested that a complex and concerted mechanism involving multiple pathways including salicylic acid (SA), jasmonic acid (JA) and ethylene (ET) might be responsible for the cotton defense response to *V. dahliae*. The expression changes of the ethylene biosynthesis and response genes (ACO1, ACS6, EIN2 and ERF1) in the cultivars to *V. dahliae*, indicated that ethylene played a putative role in the resistant response, as a signal molecule to trigger defense mechanisms and in the development of disease symptoms by interacting with other molecules (Xu et al. 2011). The role of phenolics in olive plant defense against *V. dahliae* was studied, using the susceptible cv. Amfissis and resistant cv. Koroneiki upon infection by D and ND pathotypes. The pathogen DNA levels in vascular tissues were estimated, using quantitative PCR assay. Decrease in symptom severity expressed in Koroneiki was associated with significant reduction in the growth of both pathotypes of *V. dahliae* in the vascular tissues, compared to cv. Amfissis. In Koroneiki trees, the levels of *o*-diphenols and verbascoside were positively correlated with the DNA levels of the D and ND pathotypes. Further, a positive association was observed between the levels of verbascoside and the fungal DNA level in Amfissis trees, whereas a negative relationship between the fungal DNA level and total phenols and oleuropein contents in both cultivars, was observed. The contents of verbascoside were significantly greater in Koroneiki trees, compared with Amfissis trees, indicating the involvement of verbascoside in the resistance mechanism operating in olive trees against *V. dahliae* (Markakis et al. 2010). The role of ß-amylase (BAM) genes in the defense responses of *A. thaliana* to *V. dahliae* was investigated. *A. thaliana* plants impaired in *BAM1, BAM2, BAM3* and *BAM4* genes, along with double, triple and quadruple mutants of these genes were used. Less severe symptoms were induced by *bam* mutants, compared to the wild-type strain. Real-time quantitative PCR (qPCR) assay showed that decrease in symptom severity observed on *bam* plants had positive relationship with corresponding reduction in pathogen growth in plants. Confocal microscopy of the most susceptible *bam* mutants and the wild-type plants showed that there were no differences between them in the number of attached conidia and penetration sites on the roots. *BAM1, BAM2 and BAM3* expression was altered upon *V. dahliae* infection in the aerial tissues of the wild-type. Analysis by qPCR assay of the PR-1 and PDF1.2 expression in the *bam3, bam 1234, bam14* and wild-type plants showed that *PR-1* was upregulated in the roots of *bam* plants upon *V. dahliae* infection (Gkizi et al. 2015).

Infection of xylem vessels by pathogens causes drastic metabolic changes in xylem parenchyma cells which are located adjacent to the infected vessels. Accumulation of different proteins and secondary metabolites in the xylem sap including PR-proteins, peroxidases, proteases, xyloglucan-endotransglycosylase (XET) and xyloglucan-specific endoglucanase inhibitory protein (XEGIP), phenols, phytoalexins and lignin-like compounds occurred. Phenolic compounds such as rutin, oleuropein, luteolin-7 glucoside and tyrosol accumulated at the site of infection in the xylem vessels of olive trees infected by *V. dahliae*. These phenolic compounds were found to be inhibitory to *V. dahliae*. Exogenous application of phenolic compounds on Dutch elm–infected trees induced accumulation of suberin-like compounds in the xylem tissue and consequent increase in resistance to *Ophiolbolus novoulmi*, causal agent of Dutch elm disease (Báidez et al. 2007; Martin et al 2008; Yadeta and Thomma 2013). A differential potato-*Verticilliium dahliae* pathosystem consisting of two cultivars of potato [susceptible (S) and moderately resistant (MR)] and two pathogen isolates [weakly (WA) and highly aggressive

(HA)] was used to evaluate the expression of defense-related genes, *PAL1, PAL2, PR-1, PR-2 and PR-5,* which are associated with salicylic acid (SA) defense signaling pathway. In the roots, expression of *PAL1, PR-1* and *PR-2* in the MR cultivar was higher than in susceptible cultivar, in response to inoculation with either one or the pathogen isolates. *PAL2* gene expression increased gradually starting at 21-hour postinoculation (hpi) in the MR cultivar. Higher expression of *PR-1* was detected at 7 days postinoculation (dpi) in roots. In the leaves, both *PAL1* and *PAL2* genes showed higher expression in the MR cultivar relative to the susceptible cultivar. Expression of *PR-2* was slightly higher in the susceptible cultivar. Analyses of combined data showed that *PAL1, PAL2, PR-1* and *PR-2* genes were regulated at the transcriptional level in response to infection by *V. dahliae*. The results indicated that the SA pathway was also involved in potato defense against *V. dahliae*, further elucidating the signaling mechanism operating in potato-*V. dahliae* pathosystem (Derksen et al. 2013).

The nature of proteins secreted by plants into the apoplast in response to infection by microbial pathogens, may provide vital information to better understand the molecular mechanisms underlying plant's innate immunity. The changes in the root apoplast secretome of island cotton cv. Hai 7124 (*Gossypium barbadense*) resistant to Verticillium wilt, following infection were studied. Differential gel electrophoresis and matrix assisted laser desorption/ionization tandem time-of-flight mass spectrometry analysis identified 68 significantly altered spots, corresponding to 49 different proteins. Gene ontology annotation indicated that most of these proteins functioned in reactive oxygen species (ROS) metabolism and defense response. The ROS-related protein, a thioredoxin, GbNRX1 was characterized and identified among the ROS-related proteins. GbNRX1 protein increased in abundance in response to *V. dahliae* challenge. GbBRX1 functioned in an apoplastic ROS scavenging after ROS burst that occurred upon recognition of the pathogen presence. Silencing of *GbNRX1* resulted in defective dissipation of apoplastic ROS, resulting in accumulation of ROS in protoplasts. Consequently, the *GbNRX1*-silenced plants exhibited reduced wilt incidence, indicating that the initial defense response in the root apoplast required the antioxidant activity of GbNRX1. The results showed that generation and scavenging of ROS might occur in tandem in response to infection by *V. dahliae* and the rapid balancing of redox to maintain homeostasis after ROS burst, involving GBNRX1, was critical for apoplastic immune response (Li et al. 2016).

P. brassicae causes clubroot disease, one of the most destructive diseases affecting *B. juncea* var. *tunida*, the mustard plant, forming the raw material for production of the traditional fermented food in P. R. China. The host-pathogen interaction was investigated, using suppression subtractive hybridization (SSH) to study the complex regulation of resistance mechanism operating in *P. brassicae–B. juncea* var *tunida* pathosystem. A total of 1,842 different genes clones were selected from the forward selected library (using diseased roots as tester and healthy roots as driver) and 224 positive spots were identified following cDNA array dot blotting.

Elimination of polyA tails and sequences shorter than 100-bp generated 196 high quality gene sequences with an average length of 332-bp. Bioinformatic analysis showed that these 196 sequences represented 173 unigenes, comprising 14 contigs and 159 singlets. Of these 146 expressed sequence tags (ESTs) (84.4% of the total) were significantly similar to known sequences in plants, the remaining 23 (13.3%) were of *P. brassicae* origin. Quantitative RT-PCR assay was applied to analyze the six genes most likely to be involved in disease resistance to evaluate the effectiveness of SSH analysis. The results revealed the reliability of the SSH library data and the assay might be useful for breeding programs (Luo et al. 2013). The initial responses of turnip (*B. rapa*) cultivars resistant and susceptible to clubroot caused by *P. brassicae* were investigated, using organ cultures of adventitious roots of seedlings. Primary plasmodia of *P. brassicae* were produced in the root hairs of both susceptible and resistant cultured roots. But secondary plasmodia could proliferate on the susceptible root culture, but not in resistant accession. Clubroots developed in root cultures of susceptible cultivar, following inoculation with 10^4, 10^5, 10^6 spores/ml. By contrast, no clubroot development was observed in resistant cultivar under the same conditions. Cell death was observed in cultured roots from resistant cultivar, but not in root culture of susceptible cultivar, after exposure to *P. brassicae* inoculum. Alkalinization of root culture medium of resistant cultivar was detected at 2 days after treatment with *P. brassicae* spores and this change was not recorded in root culture from susceptible cultivar. The results revealed the usefulness of root culture system as a tool for better understanding of the molecular mechanism of resistance to clubroot disease of turnip (Takahashi et al. 2006). The relationship of lipoxygenase (LOX; EC 1.13.11.12) and patatin, the key storage proteins in potato tubers with resistance to powdery scab disease caused by *Spongospora subterranea* was investigated. Evaluation of potato germplasm in the greenhouse suggested that russet skinned tuber genotypes (Mesa Russet, Centennial Russet and Russet Nugget) with negligible tuber disease severity index (DSI) and 100% marketability were resistant to powdery scab disease. Higher physiological levels of LOX protein (on dry weight basis) were negatively correlated with tuber DSI and positively correlated with tuber resistance to the disease. The physiological levels of LOX protein may be considered as a useful marker for powdery scab resistance in potato tubers for breeding programs (Perla et al. 2014)

The defense responses of wheat lines Neepawa (susceptible) and its sib BW553, that was nearly isogenic for the *bt-10* resistance gene, to differentially virulent races T1 and T27 of common bunt caused by *Tilletia tritici*, was monitored for 21 days after seeding (DAS), using fluorescence and confocal microscopy. Initial perception of the pathogen invasion, after similar germination occurring during 5 to 21 DAS, was similar in both compatible and incompatible interactions during 5-6 DAS, as revealed by autofluorescence in epidermal cells adjacent to appressoria. The total number of sites on a 1-cm segment of coleoptiles adjacent to the seed exhibiting autofluorescence was similar in both compatible and incompatible interactions and rose to a maximum of 35 to 40 per 1-cm length of coleoptiles, following 17 DAS. In the compatible interaction, the autofluorescence became more diffused at 10–12 DAS, emanating in all directions along with pathogen spread. In the incompatible interaction, autofluorescence remained restricted to a small area surrounding the penetration site. Two different reaction zones that extended further in tissues surrounding the penetration point in the incompatible interaction, compared with the compatible interaction, were identified. Although accumulation of callose around invading fungal hyphae was seen in both compatible and incompatible interactions from 8 to 21 DAS, callose accumulation was more extensive and widespread in incompatible interaction. Expression analysis showed that callose synthase transcripts were more abundant in BW553 than in Neepawa and they were upregulated during pathogen infection in both compatible and incompatible interactions (Gaudet et al. 2007). The components of resistance of maize (corn) to *Aspergillus flavus* were differentiated into response to silk, response to developing kernels and response to mature kernels to inoculation with the pathogen. Different inoculation methods were used in in vitro and field experiments on a panel of diverse inbred lines of maize over three years. Significant genotype-environment interactions were found for all components of resistance studied. However, significant variation in maize germplasm for susceptibility to silk and kernel colonization by *A. flavus* was observed under field conditions. Significant correlation of resistance to aflatoxin accumulation with flowering time and kernel composition traits like fiber, ash, carbohydrates and seed weight was observed. Furthermore, correlation analyses indicated that lines that flowered later in the season tended to be more resistant. The components of resistance identified in vitro did not seem to be associated with reduced aflatoxin accumulation in the grains under field conditions (Mideros et al. 2012).

2.3 HOST PLANT RESISTANCE TO SOILBORNE BACTERIAL PATHOGENS

Resistance levels of crop cultivars to soilborne bacterial pathogens, have been assessed under in vitro, greenhouse and field conditions by applying different assessment methods. The responses of the cultivars and genotypes as well as wild relatives of crop plant species differ, based on the resistance levels of the host plant and the levels of virulence of the pathogen strains/isolates, pathogen inoculum density.

2.3.1 GENETIC BASIS OF RESISTANCE TO BACTERIAL PATHOGENS

2.3.1.1 Screening for Resistance to Bacterial Pathogens

Large number of host plant genotypes can be screened under field conditions to shortlist the entries to be tested under greenhouse conditions and also to identify the genotypes showing field resistance, which cannot be recognized under in vitro/greenhouse conditions. Identification of sources of resistance to bacterial wilt disease of tomato caused by *Ralstonia solanacearum* was taken up by screening tomato germplasm, using the temperate race 3, biovar 2 strains. Tomato accessions (82), belonging to different *Solanum* spp. (known earlier as *Lycopersicon* spp.) were evaluated for resistance for race 3 strains in growth chambers. Tomato cv. Roma was used as susceptible control for comparison. None of the accessions showed immunity or complete resistance, indicating race 3 strain was highly virulent to all accessions tested. Partial resistance was exhibited by one accession belonging to *Solanum peruvianum* and one *S. esculentum* var. *cerasiformae* tomato line. Five other genotypes from *S. peruvianum, S. hirsutum* and *S. esculentum* also showed appreciable levels of resistance. The Hawai 7996 line was the best source of resistance to race 3 (2% wilting) identified in this investigation and it could be used as a source of resistance to *R. solanacearum* in breeding programs (Carmeille et al. 2006). In another investigation, *Solanum* accessions (252) and one population of 49 introgression lines of LA716 were screened for resistance to a race 1/biovar 4/phylotype I strain Pss186 of *R. solanacearum*. Most wild tomato accessions were highly susceptible. However, five accessions of *S. pennelli*, LA1943, LA716, LA1656, LA1732 and TL01845 were resistant to strain Pss186. These accessions were then challenged with two other race 1/phylotype I strains Pss4 and Pss190, which were more aggressive. All *S. pennelli* accessions (5) were susceptible to Pss4, but showed high to moderate resistance to Pss190, with 0 to 60% wilted plants, following inoculation with the pathogen. Strain Pss190 was so aggressive, making resistant tomato line Hawaii 7996 susceptible. The results indicated the existence of strain-specific resistance in tomato accessions. LA3501 with an introgression segment on chromosome 6, was resistant to Pss185, among the introgression lines screened, confirming the importance of resistance tract loci on chromosome 6. The resistant sources identified in this investigation, could be exploited in breeding programs (Hai et al. 2008).

Efforts to develop cultivars of tomato, eggplant and pepper (TEP) resistant to bacterial wilt disease caused by *Ralstonia solanacearum* were made and these investigations highlighted the importance of both site environmental conditions and pathogen population variability infecting these three crops. Worldwide collections of accessions of tomato, eggplant and pepper (30) (Core- TEP), that were commonly used as sources of resistance, were included in this investigation. The Core-TEP lines were challenged with a core collection of 12 pathogen strains (Core-Rs 2), representing the phylogenetic diversity of *R. solanacearum*. In six interaction phenotypes, from highly susceptible to highly resistant were recognized. Intermediate phenotypes resulted from the ability of plant to tolerate latent infections (ie., bacterial colonization of vascular elements with limited or no wilting). The Core-Rs 2 strains partitioned into three pathotypes on pepper accessions, five on tomato and six on eggplant. A 'pathoprofile' concept was developed to characterize the strain clusters, which displayed six virulence patterns on the whole set of Core-TEP host accessions. Pathoprofiles with high aggressiveness were mainly observed in strains from phylotypes I, IIB and III. One pathoprofile contained a strain that could infect almost all sources of resistance identified (Lebeau et al. 2011). Tomato accessions (285) from 23 countries were evaluated for their levels of resistance to *Ralstonia solanacearum* by inoculating them with 10^8/ml of inoculum at seedling stage in the greenhouse. Disease index (DI, 0–4 scale) was determined at 14 days after inoculation (see Figure 2.7). Of the 285 accessions, 244 showed disease index greater than 3 (76%). *S. lycopersicum* (3) and *S. peruvianum* (1) accessions were resistant with disease index < 2.0 and one each of these species accessions was moderately resistant. All accessions of S. *corneliomulleri, S. galapagense, S. habrochaites* and *S. pimpenellifolium* were susceptible to bacterial wilt. Four accessions were resistant to *R. solanacearum* (mean DI > 2), following inoculation with high inoculum density (10^8 CFU/ml). IT201664 was the most resistant with a disease index of 1.11. Performance of commercial tomato cultivars Support (resistant) and Hoyong (susceptible) was compared with accessions tested. The four resistant tomato genotypes might be used as sources of resistance against bacterial wilt caused by *R. solanacearum* (Kim et al. 2016).

Genetic analysis of resistance of pepper to bacterial wilt caused by *Ralstonia solanacearum* was carried out on the doubled haploid progeny from a cross between a resistant parental line PM687 and susceptible cv. Yolo Wonder. Following inoculation with a local isolate of *R. solanacearum*, the progeny consisting of 90 lines was transplanted into a naturally infested field. Repeatable results were obtained in the 2-year experiment, with a high heritability of resistance. Two to five genes with additive effects were estimated to control resistance, indicating an oligogenic control, as observed in tomato sources of resistance. Examination of relationships of resistance to bacterial wilt to other soilborne pathogens showed that susceptibility to *Tobacco mosaic virus* (TMV) and to nematodes (*Meloidogyne* spp.) were significantly linked with bacterial wilt. On the other hand, resistance to *P. capsici* or *Leveillula taurica* was linked to bacterial wilt resistance. The similarity to genetics of resistance to bacterial wilt in pepper and tomato and linkage with TMV resistance locus suggested the need for comparative mapping of the resistance quantitative trait loci (QTL) in the genomes of the two crop plant species (Lafortune et al. 2005). In a later investigation, potato cultivars/clones (14) were evaluated for their resistance to bacterial wilt caused by *Ralstonia solancearum* biovar 2 (race 3). The resistance of potato clone MB03 to race 1 and level of latent infection of tubers by *R. solanacearum* were also studied. The cultivars or clones were grown in a naturally infested field with biovar 2 during the spring of 1999 and 2000. The number of potato plants wilted was recorded at weekly interval, and the area under disease progress curve (AUDPC) was plotted. The extent of latent infection of tubers was determined by storing them until budding, followed by testing for the presence of the pathogen. The cultivar Cruza 148 and Clone MB03 were the most resistant among the cultivars tested. However, both of them showed latent infections in the tubers (Silveira et al. 2007). In a later investigation, two pepper cultivars Ujwala and Anugraha showed high level of field resistance to *Ralstonia solanacearum*. The cv. Anugraha was developed via backcross between highly susceptible cv. Pusa Jwala with highly resistant Ujwala, using Pusa Jwala as the recurrent parent. Pusa Jwala and Anugraha were near isogenic lines (NILs), differing only in their level of resistance to *R. solanacearum*. Resistance to bacterial wilt was governed by a homozygous recessive (*rr*) gene action. The F_1s of Anugraha x Pusa Jwala were selfed to generate segregating F_2 population, of which 10 plants were highly susceptible and 10 plants were highly resistant. The DNAs from these plants were bulked separately. Bulked segregant analysis (BSA) was carried out with AFLP primer combination EcoAC + MseCAC, using the DNA from donor parent Ujwala, susceptible parent Pusa Jwala, resistant parent Anugraha, bulked susceptible F_2 plants and bulked resistant F_2 plants. AFLP analysis showed that three fragments, 103-, 117- and 161-bp were linked with the resistant allele, whereas the three fragments, 183-, 296- and 319-bp were linked with the dominant susceptible allele of the bacterial wilt resistance gene (Thakur et al. 2014).

Potato genotypes (36) were evaluated for their resistance to bacterial wilt induced by *Ralstonia solani* under inoculated field conditions. Three genotypes Kenya Karibu, Kenya Sifa and Ingabite were the most resistant among the genotypes tested and they might be used as sources of resistance in Kenya and tropical highlands (Muthoni et al. 2014). In another investigation, utilization of wild relatives of potato as sources of resistance to bacterial wilt was taken up, since they showed high genetic diversity. Lack of screening methods, allowing easy assessment of pathogen colonization of potato tissues and inability to detect latent (asymptomatic) infections were found to be major constraints for evaluation of sources of resistance to bacterial wilt disease. In order to overcome these limitations, *R. solanacearum* UY031 was genetically modified to constitutively generate light (fluorescence) from a synthetic *luxCDABE* operon stably inserted in its chromosome. Life imaging procedure was applied to monitor colonization of this reporter strain on different potato accessions. Detection of the pathogen in planta by this nondisruptive system correlated with the development of wilting symptoms. Furthermore, it was possible to quantify the recombinant strain, using a luminometer with which the latent infection by *R. solanacearum* could be detected. This method showed potential for high throughput evaluation of pathogen colonization in plant populations. The behavior of potato accessions with different levels of resistance could be investigated by this method. The latency in parental lines may be precisely determined, before using them as sources of resistance in breeding programs (Cruz et al. 2014).

Streptomyces spp. are associated with common and netted scab diseases which reduce the quality and quantity of potato tubers. Common scab induced primarily by *S. scabies* and *S. europaeiscabiei* results in deep pustules on tuber surface, whereas *S. reticulosacbiei*, causing netted scab, affects superficial tissues of the tubers, leading to corky alterations of tuber periderm. Some isolates of *S. europaeiscabiei* may induce both common and netted scab symptoms. Thus, three different pathogenicity groups could be differentiated among *Streptomyces* spp., causing scab disease in potato. Repeated experiments with soil artificially infested with isolates of three *Streptomyces* spp. representing the three pathogenicity groups revealed the level and stability of cultivar resistance, as well as the existence of a range of aggressiveness among different isolates. The distribution of scab severity indexes recorded on a collection of 16 potato cultivars and 27 breeding lines grown in soil infested with common scab-inducing isolates was continuous, suggesting isolate nonspecific quantitative resistance. The cultivars Nicola, BF15, Sirtema and Charlotte were least susceptible, whereas cvs. Urgenta, Desiree, Ondine and Bintje were highly susceptible to common scab-causing *Streptomyces* spp. On the other hand, the genotypes Bintje, Desiree or Carmine were highly susceptible or cvs. Charlotte, Sirtema, Monalisa, BF15, Belle de Fontenay were highly resistant to isolates causing netted scab symptoms. The results suggested the existence of isolate-specific qualitative resistance. The ability of some isolates of *S. europaeiscabiei* to induce one or the other types of symptoms was confirmed by this investigation (Pasco et al. 2005). In another investigation, Alta Crown (cv 92028-1), a new russet potato cultivar with resistance to common scab was developed from a cross between breeding clones AC83172-IRU and C086930-IRU at Colorado State University, Canada. This cultivar showed tuber deformities–hollow heart and internal necrosis defects–in addition to common scab disease. Furthermore, cv. Alta Crown possessed moderate resistance to late blight tuber rot caused by *P. infestans* and foliar early blight caused by *Alternaria solani*, but it was susceptible to foliar late blight disease. Moreover, Alta Crown showed moderate resistance to bacterial ring rot caused by *Corynebacterium michiganensis* subsp. *sepedonicus*. Comparison of horticultural characteristics showed that cv. Alta Crown produced similar or lower yields, compared to Russet Burbank and Shepody, but percentage of marketable tuber yield at main crop harvest was greater than the standard cultivars. The average weight of marketable tuber was greater than that Russet Burbank at all sites and to that of Shepody at some sites (Bizimungu et al. 2011). Evaluation of potato breeding lines for resistance to common scab (CS) disease, which is an important disease in the United States, was significantly influenced by high location and season, based on soil and environmental conditions. Hence, efficiency of screening for common scab resistance was assessed within the Wisconsin breeding program across multiple environments from 2006 to 2013. The ability to select (from 60–160 clones/year) for common scab resistance in a set of 18 dedicated CS screening trials (DST) versus 18 similar parallel standard breeding trials (SBT) was determined. Heritability for CS rating across DST was 0.83 vs 0.53 in SBT. Data gathered from DST were effective in separating susceptible cultivars (Atlantic, Snodden) from resistant cultivar (Pike). However, the same data analysis from SBT was unable to separate susceptible from resistant cultivar. Based on the results from six or more DST, five round white clones that outperformed or matched CS tolerant Pike, eleven russet clones that outperformed or matched CS resistance level of Dark Red

Norland could be identified. The approaches employed in this investigation were useful for improving selection efficiency for potato scab resistance (Navarro et al. 2015).

Different species of *Streptomyces* may be responsible for induction of potato common scab disease. Pathogenic strains of *S. scabiei* produce the phytotoxin thaxtomin A, whereas nonpathogenic strains are unable to produce thaxtomin A. Common scab symptoms could be induced by applying thaxtomin on developing tubers in the absence of *S. scabiei*, suggesting that resistance to thaxtomin A in potato seedlings might be correlated with decreased scab disease lesions on tubers (Hiltunen et al. 2006). In a later investigation, the effectiveness of thaxtomin A as a positive selection agent for somatic cell selection of potato with enhanced resistance to common scab disease was assessed. A somatic cell selection procedure was developed to use thaxtomin A as a positive selection agent and 113 potato cell clones of cultivar Iowa were recovered. From these cell clones, 39 plants were regenerated within a single subculture period. Regenerated variants (13) were selected, following inoculation, based on mean tuber coverage with lesions and proportion of disease-free tubers. The best variant had an average reduction of 85% in disease score over that of disease severity in the particular cultivar. Two (65A and 65 B) of the 13 variants produced more yield than the parent cultivar in two of three trials conducted. These two variants had up to an estimated 1.3% lesion tuber surface coverage and showed less tuber surface score (85–86%) than unselected parent cultivar. These disease-resistant variants differed in their response to thaxtomin A in detached leaflet and tuber slice bioassays. The results suggested that thaxtomin A screening had limited reliability in selection for toxin tolerance and that thaxtomin tolerance was not necessarily the critical factor for induction of disease resistance. The majority of disease-resistant variants could yield to the same level as the parent cultivar, indicating the commercial viability of cultivating the resistant variants of potato (Wilson et al. 2009).

Somatic cell selection using thaxtomin A produced by *Streptomyces scabies* as the selection agent was employed to isolate variants of potato cv. Russet Burbank. Exposure of regenerated variants to 4.75 to 6.86 μM thaxtomin A for 1 to 8 days successively inhibited much of the growth of the wild-type cells that would have been expected in the absence of toxin treatment. A total of 253 regenerated plants were obtained from 212 cell colonies. Of the 253 variants, 51 (20%) showed significantly enhanced level of resistance to common scab. The resistant variants were screened in multiple glasshouse experiments and field trials and they maintained consistent resistant responses to the pathogen challenge over a period exceeding 5 years, including multiple tissue culture and tuber generations. Under glasshouse conditions, variants were challenged with a highly virulent pathogen isolate. On the other hand, field testing at two distinct sites, the variants were exposed to a diversity of naturally occurring pathogen strains. Further, selected variants (A155, A168A, A380 and TC-RB8) tested at a third field site provided the opportunity of testing the variants under varied conditions

of pathogen variability and differing disease pressure. The selected disease resistance phenotypes were stable in their response to the pathogen. The resistance of variants obtained from cv. Russet Burbank with moderate resistance was further enhanced to near immunity (up to 97% disease reduction) (Wilson et al. 2010). In another investigation conducted under both pot culture and field conditions, no association between resistance to common scab and tolerance to thaxtomin A could be established. Disease resistant cvs. Russet Burbank and Atlantic were, respectively, sensitive and tolerant to thaxtomin A toxicity. The results indicated that resistance to thaxtomin A was critical to disease expression and reaction to the toxin was one of the components, influencing resistance to common scab disease caused by *Streptomyces* spp. (Tegg and Wilson 2010).

Bacterial soft rot caused by *Pectobacterium* spp. (*Erwinia*) is widely distributed, affecting potato production markedly. In order to identify sources of resistance to the soft rot disease, wild relatives of potato (*Solanum tuberosum*) were examined for their levels of resistance. Association between resistance to soft rot and phenotypic plasticity was indicated. *Solanum paucijugum*, *S. brevicaule* and *S. commersonii* were the most resistant to the soft rot disease among *Solanum* spp. tested (Chung et al. 2011). Blackleg and stem rot are induced by *Dickeya* spp. and *Pectobacterium* spp. (earlier placed under the genus *Erwinia*). In order to identify sources of tolerance, 532 genotypes of potato species were assayed with *P. wasabiae*, *P. carotovorum* and *D. solani* using petiole assay. This assay was optimized using well responding genotypes from the broad screen. The best developmental stage for cell wall degradation test was determined as the fourth to sixth youngest leaf. Only three genotypes were found to be tolerant against all three pathogens under the stringent biotic and climatic screening conditions applied. These genotypes belonged to the series *Yungasena*, which could be readily crossed with cultivated potato (*Solanum tuberosum*) for improving the tolerance level of cultivars (Rietman et al. 2014). As the early tuberization is a critical period for infection of potato tubers by *Streptomyces turgidiscabies*, causing common scab disease, host gene expression responses during this period of tuber development were investigated. The highly susceptible cv. Saturna and the relatively resistant cv. Beate were inoculated with *S. turgidiscabies*. The transcription profiles were obtained by RNA sequencing at two developmental stages viz., the early hook stage and early tuber formation stage. The cv. Beate offered an early and sustained protective response to infection by the pathogen. On the other hand, the defense response of cv. Saturna ceased before the early tuber formation stage. Putative defense-associated genes were expressed in a pronounced manner in cv. Beate. An increase in alternative splicing on pathogen infection at the early hook stage was observed in both cultivars. Significant downregulation of genes involved in the highly energy-demanding process of ribosome biogenesis occurred in infected cv. Beate plants at the early hook stage, which might indicate allocation of resources that favored expression of defense-related genes (Dees et al. 2016).

A reliable and rapid screening method was developed to select banana cultivars resistant to Xanthomonas wilt (BXW) disease, which was a threat to banana production in East Africa. An in vitro screening procedure was applied to evaluate tissue culture-derived plantlets, using sixteen isolates of *Xanthomonas campestris* pv. *musacearum*. Significant differences in the susceptibility of banana cultivars were observed, whereas the levels of pathogenicity (virulence) of the pathogen isolates were similar. The cv. Pisang Awak (Kayinja) was highly susceptible, while *Musa balbisiana* was resistant to the pathogen. The cv. Nikitembe showed moderate resistance and cvs. Mpologoma, Mbwazirume, Sukali Nidiizi, FHIA-17 and FHIA-25 were susceptible to BXW pathogen. Artificial inoculation of the cultivars under pot culture conditions resulted in similar results. The in vitro method of assessing levels of resistance of banana cultivars to BXW disease was found to be a convenient, rapid and cost-effective method to differentiate levels of resistance (Tripathi et al. 2008). Resistance of *Musa balbisiana* to *Xanthomonas vasicola* pv. *musacearum (Xvm)* was examined, based on the expression profiles of *R, NPR1* and *PR* genes. Wilt symptoms developed on *M. balbisiana*, following inoculation with *Xvm*, the incidence and severity attained, were 30% and 20%, respectively, before recovery. The pathogen was limited to the inoculated and the immediately following leaves and the pathogen population declined from 3.06×10^7 CFU at 18 days to 0 CFU by 150 days after inoculation (DAI). In contrast, in susceptible cultivar, the pathogen population increased from 3.41×10^8 to 1.08×10^{11} CFU by 21 DAI with ultimate death of the plant. Expression of *MbNBS, MdNPR1* and *PR3* reduced drastically during first 6 hpi, recovering in later stages. The results showed that mechanism of resistance of *M. balbisiana* to *R. solanacearum* was not due to development of HR, SAR or ISR and other mechanism might be operative, since key pathogen genes were suppressed in *M. balbisiana* (Ssekiwoko et al. 2015). The levels of resistance of 277 cultivars and 70 wild *Dianthus* accessions to bacterial wilt disease caused by *Burkholderia caryophylli* were assessed using conventional inoculation method. *D. capitatus* was identified as highly resistant to the disease. By performing interspecific hybridization, a new bacterial wilt resistant line was developed, the bulked segregant analysis (BSA) procedure was applied to identify RAPD markers linked to the gene governing bacterial wilt resistance. Primers (505) were screened to obtain RAPD markers that could be reliably used for selecting resistant carnation lines. Eight RAPD markers linked to a major resistance gene were identified, using BSA analysis. The marker WG44-1050 was the most effective for selecting lines showing resistance to bacterial wilt. This RAPD marker was converted into sequence-tagged site (STS) marker suitable for marker-assisted selection (MAS) procedure. The WG44-1050 was specifically amplified by combining five primers. One specific pair of STS primers could detect a single, clear DNA fragment tightly linked to a major resistance gene easily and reliably. The STS markers possessed greater degree of reproducibility compared to RAPD markers (Onozaki et al. 2004).

2.3.2 MECHANISMS OF RESISTANCE TO BACTERIAL PATHOGENS

Various mechanisms may be in place in plants to effectively meet the challenges posed by bacterial plant pathogens, which have different levels of virulence, whereas plants are endowed with resistance genes for protecting themselves. The bacterial pathogens have to breach successfully the different layers of protection offered by innate immunity system of the host plants. The exterior surfaces of plants and preformed antimicrobial compounds, in addition to cell walls form barriers to the invading pathogens. In addition, the pathogens have to overcome the formidable plant immunity response. Plant immunity comprises of two components operating on different time scales. The basal defense system operates early in the interaction between host and pathogen, whereas the resistance *R* gene-mediated defense functions on the time scale of hours. As in host-fungal pathogen interactions, the pathogen-associated molecular patterns (PAMPs) that mediate early basal response, include lipopolysaccharides, peptidoglycans, bacterial flagellin and yeast mannans (Nürnberger et al. 2004; Pemberton et al. 2005). The innate immune response may be induced also by the bacterial elongation factor EFTu (Kunze et al. 2004). The receptors located in the plasma membrane recognize PAMPs, activating a phosphorylation cascade upon binding, leading to the induction of early basal resistance. This basal resistance is likely to be responsible for prevention of colonization of plants by nonpathogenic bacteria (Gómez-Gómez and Boller 2000). The PAMP-triggered immunity seems to be sufficient to restrict infection, before the pathogen becomes established. The relationship between inhibition of pathogen growth and recognition of the PAMP flagellin by the receptor FLS2 was demonstrated (Zipfel et al. 2004). Flagellin is the major protein of bacterial flagella and also a well characterized PAMP that was recognized by the leucine-rich repeat receptor kinase FLS2 in *Arabidopsis*. FLS2 was located in the plasma membrane considered to be involved in the early bacterial-plant interaction by recognizing and binding flagellin (Dangl and McDowell 2006). Treatment of plants with flg22, a peptide representing the elicitor-active epitope of flagellin induced the expression of several defense-related genes and resistance to pathogenic bacteria, but not in plants carrying mutations in the flagellin receptor gene FLS2 (Zipfel et al. 2004). Both Flg22 and lipopolysaccharides (LPS) could induce dramatic stomatal closure in the wild-type Col-0 ecotype of *Arabidopsis*. The flg22 peptide did not induce stomatal closure in epidermal peels of fls2 flagellin receptor mutant. However, LPS could induce stomatal closure in the mutant also. The results suggested that guard cell perception of flg22 required the FLS2 receptor, which might be one among several receptors that might enable guard cells to sense multiple MAMPs displayed on the bacterial surface. Defense via stomatal closure was shown to be an integral part of salicylic acid (SA)-regulated innate immune system (Melotto et al. 2006).

The possible relationships between innate immunity, race-specific and nonhost types of resistance responses were investigated, using *Arabidopsis* cell cultures [Landsberg erecta

(Ler)] and seedlings (Col-0) that were treated with flg22. The microarray analysis showed that the majority of the *Flagellin Rapidly Elicited (FLARE)* genes were upregulated and 80% of the encoded proteins were functionally classified into signal transduction-related, signal perception-related, elicitor proteins or other groups. The mRNA for three auxin receptors was found to be negatively regulated by micro-RNA induced by flagellin perception. Many of the signal transduction-related genes-encoded WRKY transcription factors are unique to plants and involved in regulating diverse plant functions including plant defense (Navarro et al. 2004). The lipopolysaccharides (LPS) covering the cell surface of Gram-negative bacteria represent typical PAMP molecules capable of inducing defense-related responses, including suppression HR, the expression of defense genes and systemic resistance in plants. Generation of reactive oxygen species (ROS) was induced by treatment with LPS in addition to induction of expression of PR-proteins (Gerber et al. 2004; Silipo et al. 2005). The LPS treatment of *Arabidopsis* activated nitric oxide (NO) synthase and NO so evolved, activated defense genes, as well as resistance to pathogenic bacteria (Zeidler et al. 2004). The LPS molecules produced by bacterial pathogens, *Erwinia chrysanthemi* and *Ralstonia solanacearum* were able to induce defense responses, such as ROS generation and defense gene expression in host cells. In addition, global analysis of gene expression induced by LPS and chitin oligosaccharides (PAMPs) also showed a correlation between the gene responses induced by them. The LPS of nonpathogens induced programmed cell death (PCD) in rice cells. But PCD induction by LPS treatment did not occur in *A. thaliana* cells (Desaki et al. 2016).

Intracellular recognition of both PAMP and Avr factors is primarily achieved by the nucleotide-binding oligomerization domain (NOD)-protein family. Numerous proteins of animals, plants and microbes origin are included in the NOD family (Inohara and Nunez 2003). Variations in the NOD family members may determine the levels of resistance to fungal, bacterial and viral pathogens, demonstrating the essential role of the NOD-mediated innate immune response in plant and animal biology. The members of the toll-like receptor (TLR) family containing LPR in the extracellular domain and a TLR intracellular domain are involved in the recognition of PAMP in extracellular compartments or at the cell surface (Werling and Jungi 2003). Basal mechanisms are considered to be activated by surface-derived molecules known as pathogen-associated molecular patterns (PAMPs), whereas gene-for-gene based resistance is involved in recognition of the proteins encoded by *avr* genes. The response to lipopolysaccharides (LPS) and other PAMPs may be regarded as an expression of a basal resistance which has to be overcome by the pathogen. The more specific gene-for-gene interactions that control cultivar resistance and HR are superimposed on this PAMP-mediated basal resistance. Resistance induced in tissues far away from the initial infection sites by systemic acquired resistance (SAR), following gene-for-gene recognition between plant resistance proteins and pathogen effector has not been fully understood. Although salicylic acid (SA) participates in the local and systemic response, SAR does not require long distance translocation of SA. However, SAR depends on the accumulation of SA in distal leaves, where it induces change in the cellular redox triggering in the reduction of oligometric disulfide bound NPR1' (nonexpressor PR genes), a central regulation of SAR. The systemically responding leaves rapidly activate a SAR transcriptional signature with strong similarity to local basal defense. A central role for jasmonates in systemic defense has been suggested. Jasmonic acid (JA), but not SA, rapidly accumulated in phloem exudates of leaves challenged with an avirulent strain of *Pseudomonas syringae*. Transcripts associated with jasmonate biosynthesis were upregulated in systemically responding leaves within 4 h and JA increased transiently. The SAR was abolished in mutants defective in jasmonate synthesis or response. The results indicated that jasmonate signaling seemed to mediate the long-distance information transmission (Truman et al. 2007).

The bacterial avirulence gene function is dependent on interactions with HR and pathogenicity (*hrp*) genes. The *avr* gene products are considered to be translocated into the cytoplasm by the *hrp*-encoded protein translocation complex. The *hrp*-encoded protein secretion apparatus also translocates harpins which are glycine-rich proteins produced, when *hrp* genes are expressed. Harpins are produced, in addition to Avr determinants. The virulence determinants harpin (HrpN) and polygalacturonase (PehA) from hrp-positive strain of the soft rot pathogen *Erwinia carotovora* subsp. *carotovora* (Eca) were employed as tools to elucidate plant responses. In the nonhost *Arabidopsis*, establishment of resistance by HrpN was accompanied by the expression of salicylic acid (SA)-dependent, but also jasmonate/ethylene (JA/ET)-dependent marker gene *PR-1* and *PDF1.2*, respectively. Apparently both SA-dependent and JA/ET-dependent pathways were activated. These two elicitors, HrPN and PehA, also cooperated in triggering increased production of superoxide and lesion formation (Kariola et al. 2003). The phytohormones salicylic acid (SA) and jasomonic acid (JA) and ethylene (ET) participate in the regulation of defense responses in plants. SA is predominantly associated with resistance against biotrophic and hemi-biotrophic pathogens and establishment of systemic acquired resistance (SAR) (Grant and Lamb 2006). In contrast, JA- and ET-dependent defense mechanisms are involved in resistance to necrotrophic pathogens, suggesting that signaling network engaged by the host plant is dependent upon the nature of the pathogen and its mode of pathogenicity (Glazebrook 2005). The *NPR1* gene (also known as *NIM1* and *SAI 1*) is the key regulator of SA-mediated SAR in *A. thaliana*. SA-induced PR genes (SAR) and some *R* gene-mediated resistance were not expressed in *npr1* mutants. But overexpression of *NPR1* enhanced resistance to diverse pathogens in a dose-dependent manner in *Arabidopsis* (Friedrich et al. 2001). *NPR1* coordinately induced secretion-related genes required for PR protein secretion during SAR development (Wang et al. 2005). The *NPR1* is functionally conserved in diverse plant species. SA may play a role as defense signal in the model plant, rice, as the transgenic rice expressing the *nahG* gene, encoding a salicylate hydroxylate, failed to accumulate SA and exhibited higher level of susceptibility to rice blast disease (Yang et al.

2004). In addition to SA and JA, the major transcriptional reprogramming associated with the plant defense response required also the action of diverse transcription factors. The WRKY class of transcription regulators appeared to play a major role in the regulations of plant defense response. The conserved elements of the domain are essential for the high binding affinity of WRKY proteins to the sequence designated the W-box (Zhang and Wang 2005). In *Arabidopsis*, 74 WRKY proteins have been classified into three groups (I, II and III). Since the majority of the WRKY genes of *Arabidopsis* were upregulated in defense responses or after treatment with defense-inducing elicitors or hormones, they were considered as crucial regulators of pathogen-induced active defense response (Journot-Catalino et al. 2006).

The mechanism of limiting the movement of *Ralstonia solanacearum* in resistant and susceptible tomato seedlings following inoculation of roots was studied. Distribution and appearance of *R. solanacearum* in the upper hypocotyl tissues of root-inoculated seedlings of resistant cv. LS-89 (a selection of Hawaii 7998) and susceptible cv. Ponderosa were compared. In stems of wilted Ponderosa plants, *R. solanacearum* colonized both the primary and secondary xylem tissues. Pathogen populations were abundant in the vessels, of which the pit membranes were often degenerated. All parenchyma cells adjacent to vessels with bacteria were necrotic and some of them were colonized by the bacteria. By contrast, no recognizable wilting symptoms appeared in the stems of LS-89. Bacteria were present in the primary xylem tissues, but not in the secondary xylem tissues. Necrosis of parenchyma cells adjacent to vessels with bacteria were seen occasionally. The pit membranes were often thicker with high electron density. The inner electron-dense layer of cell wall of parenchyma cells and vessels was thicker and more conspicuous in xylem tissues of infected LS-89 than in xylem of infected Ponderosa or mock-inoculated plants. Electron-dense materials accumulated in or around pit cavities in parenchyma cells in and next to vessels along with bacteria. Many bacterial cells appeared normal in vessels. The results indicated that *R. solanacearum* could move from vessels to vessels in infected tissues through degenerated pit membranes and restricted movement of the bacteria in xylem tissues was characteristic of resistant LS-89 genotype. Restriction of pathogen movement might be related to thickening of the pit membranes and /or the accumulation of electron-dense materials in the vessels and parenchyma cells (Nakaho et al. 2000). In a later investigation, the responses of tomato accessions IT201664 and K177764 resistant and moderately resistant to bacterial wilt pathogen *R. solanacearum* were determined at 14 days after inoculation, maintaining susceptible cv Hoyong and cv. Support as susceptible and resistant controls. Variations in the symptom severity were correlated with the levels of resistance to the pathogen (see Figure 2.6). Histological changes in the hypocotyl tissues of susceptible and resistant tomato plants inoculated with *R. solanacearum* were also determined to bring out the responses of the vascular tissues to infection. In the transverse section of wilting susceptible cv. Hoyong, presence of scattered bacterial population in the vessels of both primary

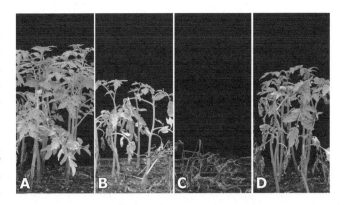

FIGURE 2.6 Responses of pepper accessions of tomato IT201664 (resistant), K177647 (moderately resistant) and cv. Hoyong (susceptible) to *Ralstonia solanacearum* causing bacterial wilt A-resistant IT201664, B- moderately resistant K177647, C- cv. Hoyong and D-resistant control cv. Support.

[Courtesy of Kim et al. 2016 and with kind permission of the Korean Society of Plant Pathology Open access article]

and secondary xylem tissues was visualized. But in resistant tomato genotype IT201664 bacterial colonization was absent in the xylem tissues, when examined under the light microscope (see Figure 2.7). Likewise, electron microscope examination of xylem tissues showed the presence of bacterial cells in the hypocotyl tissues of susceptible plants. Fewer bacterial cells were observed in the resistant plant tissues, compared to susceptible plant tissues. Parenchyma cells adjacent to the vessels colonized by the pathogen were degenerated and contained bacterial cells. The secondary walls and pit membrane were structurally loosened in susceptible plants. The pit membrane of parenchyma cells and vessels were thicker and more conspicuous in the xylem tissue of resistant tomato plants than in susceptible plants. Production of gum in resistant IT201664 plants was noted in the vessel lumen (Kim et al. 2016).

The wild species of *Solanum commersonii* was found to be resistant to *Ralstonia solanacearum*, causing bacterial wilt disease of potato. The expression analysis of the response of a resistant *S. commersonii* genotype against *R. solanacearum*

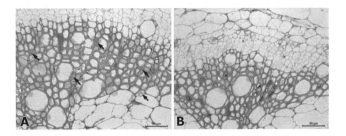

FIGURE 2.7 Histological changes in hypocotyls of susceptible (A) and resistant (B) inoculated with *Ralstonia solanacearum* at 3 days postinoculation Presence of large number of bacteria (arrows) in vessel lumen of susceptible cv. Hoyong (A) and absence of pathogen colonization in resistant accession IT201664 (B) as revealed by light microscopic examination.

[Courtesy of Kim et al. 2016 and with kind permission of the Korean Society of Plant Pathology Open access article]

was performed, using microarrays to elucidate the molecular processes involved in the interaction, to establish the timing of responses and to identify genes related to the resistance. The response to the treatment was initiated at 6 hours post-inoculation (*hpi*) and was established at 24 hpi. During this period, a high number of genes was differentially expressed and several candidate genes for the resistance of *S. commersonii* to *R. solanacearum* were identified. At an early stage, the photosynthetic process was highly repressed and several genes encoding proteins related to reactive oxygen species (ROS) production were differentially expressed. The induction of *ERF* and *ACC-oxidase* genes related to the ethylene pathway and PR1 related to the salicylic acid pathway suggested the hemibiotrophic phase of the pathogen. Five genes related to the plant defense and observed to be differentially expressed at the first two time points were validated by real-time PCR assay (Narancio et al. 2013). The molecular basis of resistance of the wild diploid species *Arachis duranensis* to *Ralstonia solanacearum*, causing the destructive bacterial disease of peanut (*Arachis hypogaea*) was studied. Suppressor subtractive hybridization (SSH) and macroarray hybridization techniques were applied to detect differentially expressed genes (DEGs) in the roots of *A. duranensis*, following inoculation with *R. solanacearum*. Of the 317 unique genes, 265 had homologues and functional annotations. A large proportion of these genes were mainly involved in the biosynthesis of phytoalexins, especially terpenoids, while flavonoids synthesis-associated genes had higher levels of expression in the resistant genotype of *A. duranensis* than in the susceptible genotype, indicating that the terpenoids and flavonoids possibly played a primary role in the development of resistance in roots of *A. duranensis* to *R. solanacearum*. The results revealed an overview of the gene expression profile in the roots of wild relative of peanut, facilitating the identification of candidate resistance genes for exploitation in the breeding programs (Chen et al. 2014).

Large gene families encode WRKY transcription factors across plant kingdom. The biological and molecular functions of several plant species, including pepper (*C. annuum*) and other solanaceous crop plant species are known. A new group of WRKY protein from pepper, CaWRKY58 was functionally characterized. The CaWRKY58 could be localized to the nucleus and activate the transcription of the reporter ß-glucuronidase (GUS) gene driven by the 35S core promoter with two copies of the W-box in its proximal upstream region. In pepper plants infected with *Ralstonia solanacearum*, CaWRKY58 transcript levels showed a biphasic response, manifested in an early/transient downregulation and late upregulation. CaWRKY58 transcription was suppressed by treatment with methyl jasmonate (MeJ) and abscisic acid. Tobacco plants overexpressing CaWRKY58 did not show any obvious morphological phenotypes, but exhibited greater disease severity than the wild-type plants. The increased susceptibility of CaWRKY58-overexpressing tobacco plants correlated with the decreased expression of hypersensitive response marker genes, as well as various defense-related genes. Silencing of CaWRKY58 in pepper plants

by virus-induced gene silencing (VIGS) showed enhanced resistance consistently to the highly virulent strain of *R. solanacearum* FJC100301, and the enhanced resistance levels correlated with enhancement of transcripts of defense-related pepper genes. The results suggested that CaWRKY58 might function as a transcriptional activator of negative regulators in the resistance of pepper to infection by *R. solanacearum* (Wang et al. 2013). Resistance of potato cultivars (12) to soft rot disease was assessed by inoculating potato-tuber slices and small tubers with *Pectobacterium atrosepticum*, *P. carotovorum* subsp. *carotovorum* and *Erwinia chrysanthemi*. The cultivar resistance was expressed as the mean diameter of the rotted tissue area. Grouping of cultivars into classes of relatively high, intermediate, or low resistance was affected by incubation under aerobic or anaerobic conditions and bacterial species used for inoculation, but not by isolates of the same species. The results of experiments with either field-grown seed tubers or glasshouse-grown small tubers for 2 years, were reproducible, except when *E. chrysanthemi* was inoculated and incubated aerobically (Allefs et al. 1995). In a later investigation, partial resistance to *P. atrosepticum*, causing soft rot of tubers and blackleg of potato plants, was observed in some cultivars. The mechanism of resistance was investigated, using the proteomic approach to identify pathways specifically activated during interaction between potato tubers and *P. atrosepticum*. Protein profiles on silver-stained gels in the pH 5–8 range were obtained from healthy and infected tubers from two cultivars with differing resistance levels, analyzed by the two-dimensional electrophoresis (2DE) and nano-, LC-MS/NS procedure. Thirteen proteins were upregulated differentially in the partially resistant cv. Keerpondy. In contrast, no significant difference in protein profiles of inoculated and control tuber could be observed in the susceptible cv. Bintje. Mass spectrometry and database search showed that these proteins were involved in energy metabolism (glyceraldehydes-3-phosphate dehydrogenase, 2-phosphoglycerate dehydratase or enolase, fructose biphosphate, aldolase and ATPase a subunit), cytoskeleton structure (actin), protein catabolism (cysteine protease inhibitor) and patatins or patatin precursors. A proteomic approach appeared to be efficient, providing an insight into the mechanisms and pathways, leading to resistance in potatoes against *P. atrosepticum* (Barzic and Com 2012).

The effects of mineral nutrition of plants on infection by bacterial pathogens and regulation of resistance mechanisms were studied. Tomato plants cultured in nutrient solution with calcium concentrations of 0.5 (low), 5.0 (medium) and 25.0 (high) mM were inoculated with bacterial wilt pathogen *Ralstonia solanacearum* by the root dip method. Disease severity levels in low, medium and high Ca treatments were 100, 77.1 and 56.8%, respectively. Plant growth in high Ca treatment was significantly better than those in low Ca treatment in height, stem diameter and biomass. Tomato plants absorbed significantly greater amounts of Ca in roots and shoots, as the level of Ca increased in the nutrient solution. In addition, H_2O_2 level in high Ca treatment increased more rapidly and reached higher peak with 10.86 UMgFW^{-1} (31.32%)

greater than in medium Ca plants. The activities of peroxidase (PO) and polyphenoloxidase (PPO) also showed greater increase in high Ca treatment with 99.09 UgFW $^{-1}$ and 107.24 UgFW^{-1}, compared to 40.70 UgFW^{-1} and 77.45 UgFW^{-1} in low Ca treatment. A negative correlation was observed between Ca concentration level and H_2O_2, POD and PPO in tomato, and disease severity, indicating that they might have a role in resistance of tomato to bacterial wilt disease. The results suggested involvement of Ca in the regulation of H_2O_2 concentration and activity of POD and PPO in tomato (Jiang et al. 2013). Most of the plants and microbial pathogens require iron for their development. *Dickeya dadantii* has siderophores-mediated iron uptake system required for progress of systemic disease symptoms in several host plant species. The effect of iron status of *Arabidopsis* on severity of disease symptoms caused by *D. dadantii* was assessed. Symptom severity, pathogen fitness and expression of bacterial pectate lyase-encoding genes were reduced in iron-deficient plants. Reduced disease severity correlated with enhanced expression of the salicylic acid (SA) defense marker gene *PR1*. However, levels of the ferritin coding transcript AtFER1, callose deposition and production of reactive oxygen species (ROS) were reduced in iron-deficient infected plants, indicating the noninvolvement of these defenses in limiting the development of the disease and the pathogen. The results showed that the iron status of the plant could influence the outcome of the interaction with the pathogen. Further, iron nutrition strongly affected the disease caused by soft rot plant pathogen with wide host range (Kieu et al. 2012).

2.4 HOST PLANT RESISTANCE TO SOILBORNE VIRAL PATHOGENS

Viruses form a structurally and functionally distinct group of biotrophic pathogens, with their genomes carrying the genes required for their replication (multiplication) and induction of characteristic symptoms in susceptible host plant species. The responses of plants to virus infections have been named in different ways and the terms suggested to represent the levels of resistance to virus diseases have been found to overlap or some being inadequate to reflect the differences in host responses. Some plant species (nonhosts) may be designated immune, because of the absence of any detectable virus replication in inoculated protoplasts or cells of intact plants and in such hosts, no virus progeny is produced. In some apparently immune plants, 'subliminal' infection of inoculated single or a few cells may occur (Fraser 1990; Matthews 1991). The resistance conferred by a gene against virus infection of a cultivar is known as cultivar resistance that may become ineffective, when a virulent (compatible) strain is formed either by mutation or selection from a mixture of strains of the virus concerned. Infection in resistant plants may be limited by a host response to a zone of cells around the initially infected cell, usually by producing visible necrotic local lesions and it is termed as hypersensitive response (HR). Cultivars showing such response are field resistant. Tolerance is the resistance to symptom expression, rather than to virus replication. This type of resistance is not seen in plant interactions with fungal or bacterial pathogens. When plants react with severe disease symptoms, the response is called as sensitive. Plants exhibiting tolerance or sensitive response are considered as susceptible, as the virus becomes systemic and replicates freely in different types of plant tissues. This host response is named as susceptibility (Matthews 1991). The successful deployment of gene(s) into crop plants depends more upon the identification of genetic marker-assisted selection (MAS) breeding and an understanding of how the novel resistance gene will perform in different genetic backgrounds and under varying pathogen pressure in the field. As in the case of fungal and bacterial pathogens, the major virus disease-resistance genes belonging to the NBS-LRR class confer complete resistance (qualitative), but they are not always associated with cell death and /or tissue necrosis (Hayes et al. 2004). Because the viruses depend largely on the vectors for dissemination, genes that confer resistance to vectors and/or block virus transmission may be able to provide additional scope for improvement of genetic resistance (Maule et al. 2007).

Plant viruses, being devoid of enzymes and toxins that facilitate entry into the host plant cells, require activity of biological vectors for initiating infection. The site of virus entry into the host plant differs, according to the feeding behavior of the vectors transmitting the virus. Majority of plant viruses are transmitted into the aerial plant parts by different arthropods (mainly sap-sucking aphids, leafhoppers and whiteflies), whereas some soil-inhabiting zoosporic organisms and root-infecting nematodes are involved in the transmission of soilborne viruses into the roots of susceptible plant species. Hence, the compatibility of the virus with plant tissues or cells, where the virus initially has access, is critical for initiating infection. Plant organs or tissues distinctly differ in their metabolism and physiological characteristics. The characteristics of plant shoots and roots largely are diverged from one another, showing variations in the anatomical structures, cell compositions, gene expression patterns. Furthermore, they are exposed to contrasting environmental conditions between epigeal and hypogeal environments. As a consequence, antiviral defense in roots have to function differently from that of shoots and viruses infecting root tissues may have evolved to adapt to these mechanistic differences existing in roots of susceptible plant species (Hull 2013).

It is generally difficult to manage the diseases caused by soilborne viruses, using conventional cultural practices and chemical application, since the viruliferous vectors may be widespread in the soil at different depths. The viruliferous resting spores of zoosporic vector remain stable and persistent in the infested soil for decades. Hence, use of resistant cultivars may be the more effective strategy for managing the soilborne virus diseases. But emergence of new resistance-breaking strains, becomes a formidable obstacle to sustain the yield at expected levels. However, efforts have been made to build antiviral defense systems, facilitating development of resistant cultivars. This situation is partly due to the fact that only limited number of plant virus-soil-inhabiting vector inoculation systems, established under laboratory conditions, is available. Resistance to virus diseases has a unique aspect. Resistance to the virus has to be differentiated from the resistance to development of visible symptoms. Resistance to the virus ultimately leads to increased resistance to the disease, whereas resistance to disease symptoms does not ensure restriction of virus replication and the infection becomes latent. Later, the virus is able to become systemic in plants and reaches high concentrations in plants showing resistance to symptoms in a manner that is indistinguishable from susceptible plants, even if symptoms are not visible. Such plants are considered as tolerant and this trait is heritable. Tolerance to *Cucumber mosaic virus* (CMV) has been exploited as a management strategy in cucumber production (Hull 2002). However, tolerant plants may be potential sources of infection for susceptible cultivars.

A virus-coded function and the product of host plant resistance gene may be involved in the recognition of the pathogen, leading to susceptibility or resistance. If the product of resistance gene is directly involved in the recognition event, it may be an inhibitor of virus infection and the resistance may be constitutive in nature (constitutive resistance). Alternatively, the resistance gene product produced on interaction with the virus factor, may signal the induction of a host defense mechanism, resulting in the production of compounds inhibitory to virus replication and systemic movement to other cells or plant tissues (active resistance). Plants exhibit different grades of resistance to virus, which may operate at different stages of virus replication, such as uncoating, synthesis of replicase, translation of virus-coded information and production of movement protein. Presence of over 200 virus *R* genes in crops, their wild-relatives and the model plant species *A. thaliana*, has been identified. More than 80% of viral resistance genes identified, are monogenically controlled, the rest being under oligogenic or polygenic control. Although more than half of the reported monogenic resistance traits exhibit dominant inheritance, relatively high proportion of recessive viral *R* genes determine the resistance trait in contrast to fungal or bacterial resistance, that has been shown to be dominant (Kang et al. 2005). The *R* genes have been located and identified, using molecular methods such as random fragment amplified length polymorphism (RFLP), amplified fragment length polymorphism (AFLP), random amplified polymorphic DNA (RAPD) and other polymerase chain reaction (PCR)-based techniques. The genetic markers linked with host resistance *R* genes to viruses have been employed to select cultivars and genotypes showing resistance to the virus concerned via marker-assisted selection (MAS) approach. The quantitative loci (QTLs) that have been tagged for plant viral resistance are less numerous compared to that of fungal and bacterial resistance (Chague et al. 1997; Ben-Chaim et al. 2001; Loannidou et al. 2003).

2.4.1 Genetic Basis of Resistance to Viruses

Presence of a single dominant *N* gene naturally in *Nicotiana glutinosa* results in production of necrotic lesions, when inoculated with *Tobacco mosaic virus* (TMV) and this response

is exhibited by many *Nicotiana* spp. The *N* gene was incorporated into tobacco cultivars Samsun NN, Xanthi nc and Burley NN. Programmed cell death (PCD) occurs at the site of infection, where TMV particles may be detected, but restricted to the region immediately surrounding necrotic lesion (da Graca and Martin 1976). The N gene was transgressed into tomato for conferring resistance to TMV (Whitham et al. 1994). The TMV protein might interact either directly or indirectly with N protein, resulting in the activation of the signal transduction pathway, leading to HR. The *N* gene belongs to the TIR-NBS-LRR class of *R* genes (Dinesh Kumar et al. 2000). The genes conferring HR to TMV in other crops such as *C. annuum* and tomato have been identified. The *L1* gene in *C. annuum*, *L2* in *C. frutescens* and *L3* in *C. chinense* conferred HR resistance (Berzel-Herranz et al. 1995). The HR response to *Tomato mosaic virus* (ToMV) strains is governed by two allelic genes, *Tm-2* and *Tm-2²* in tomato plants. The movement protein (MP), but not the coat protein (CP) of ToMV isolates was found to be the inducer of HR in tomato plants with *Tm-2* or *Tm-2²* (Weber et al. 1993).

Potato mop-top virus (PMTV), and its vector *Spongospora subterranea* the incitant of powdery scab disease, were widespread throughout potato-growing regions of the United States and Canada. The responses of potato cultivars were assessed at three locations in the Peruvian Andes, where the virus and its vector were endemic. All US potato cultivars (21) were susceptible to PMTV infection, especially at La Victoria, where overall incidence of PMTV and powdery scab was high. As the assessment of cultivar responses based on symptom expression alone was unreliable, immunoassay [nitrocellulose membrane (NCM)-ELISA] and nucleic acid-based technique (NASH-RT-PCR) were applied to detect the presence of PMTV in tuber tissue. Field-grown tubers of cvs. Kennebec, Montana and Norland showed infection by PMTV up to 25%. However, no correlation between virus infection and incidence/severity of powdery scab for any cultivar was apparent. Certain percentage of infected tubers exhibited reticulate surface cracking due to PMTV infection, although spraing symptoms were not induced. In hydroponic culture, the proportion of cracked tubers increased dramatically, when the nutrient solution was seeded with viruliferous *S. subterranea*. Tubers of cvs. Monona and Russet Burbank showed less surface cracking, compared to other cultivars, suggesting that these two cultivars might be tolerant to PMTV infection (Tenorio et al. 2006).

The effects of infection of annual small grains by soilborne viruses, *Soilborne wheat mosaic virus* (SBWMV) and *Wheat spindle streak mosaic virus* (WSSMV) were investigated to select less susceptible genotypes. Both SBWMV (genus *Furovirus*) and WSSMV (genus *Bymovirus*) are transmitted by the soilborne protozoan *Polymyxa graminis*. The susceptibility levels were assessed by growing plants of perennial wheat and rye breeding lines from seeds in infested soil. The genotypes tested were AT3425 (putatively *Triticum aestivum* x *Thinopyrum ponticum*, BFPMC (*Thinopyrum intermedium*), Permontra, PI368149, PI368150 (*T. turgidum* x *S. montanum*), Spitzer, TA12252 and Varimontra. The seeds of the genotypes were planted into one-inch balls of moistened soil collected from infested fields or a mixture of two soils for coinoculation with both SBWMV and WSSMV. The soil ball (124/genotype/soil inoculum) were planted into microplots of previously noninfested soil confined by fiberglass barriers in early October 2001. Disease incidence was recorded in May, September and October 2002. All genotypes, except Spitzer, was considered to be at least moderately susceptible to one or both viruses, based on visual symptoms appearing on plants. Symptoms were not observed in Spitzer and randomly sampled leaves from two plants were not positive for either virus by ELISA tests. None of the singly inoculated plants were coinfected by the viruses. But coinoculation resulted in increased infection by both viruses. The results indicated the usefulness of the resistant genotypes identified in this investigation, for management of the soilborne viruses, infecting perennial small grain crops (Cadle-Davidson and Bergstorm 2005).

Mirafiori lettuce big-vein virus (MiLBVV, genus *Ophiovirus*) is the incitant of lettuce big-vein disease and *Lettuce big-vein associated virus* (LBVaV) is also present in infected plants frequently, although its role in disease development is not clear. *Olpidium brassicae*, the soilborne fungus transmits the lettuce big-vein (LBV) disease. Use of cultivars with resistance to LBV disease was considered to be the most acceptable option to contain the disease incidence and spread. Complete resistance to the disease was identified in the accessions of *Lactuca virosa*. Partial resistance in lettuce cultivars expressed as reduced frequency of symptomatic plants or delayed symptom expression until marketing maturity was observed (Bos and Huijberts 1990; Ryder and Robinson 1995). Lettuce cultivars Great Lakes 65, Pavane, Margarita and *L. virosa* accession IVT280 were evaluated for virus incidence and infection in greenhouse trials, using zoospores of *O. brassicae*. Symptom severity was categorized into mild, moderate and severe, based on visual assessment (see Figure 2.8). Different percentages of symptomatic plants were observed among the cvs. Great Lakes 65, Margarita and Pavane in experiment 1. In the experiment 2, Great Lakes 65, Pavane and L. virosa accession IVT280 also had significantly different percentages of symptomatic plants (see Table 2.1). All plants of IVT280 were asymptomatic. Detection of MiLBVV and LBVaV, using RT-PCR assay confirmed infection or coinfection in symptomatic greenhouse-grown lettuce plants. All symptomatic Great Lakes plants (5) were coinfected with both viruses and the number of infected and coinfected plants in other cultivars varied. None of the plants of any cultivar showed infection only by LBVaV. The number of symptomatic plants of Great Lakes and Margarita containing both viruses differed. All plants of *L. virosa* IVT280 remaining asymptomatic did not have any detectable concentration of MiLBVV or LBVaV. The results indicated that both symptomatic and asymptomatic plants of lettuce cultivars could accumulate MiLBVV and LBVaV. The *L. virosa* accession IVT280 could be used as a source of resistance to lettuce big-vein disease, as the inoculated plants remained free of both viruses under greenhouse conditions (Hayes et al. 2006).

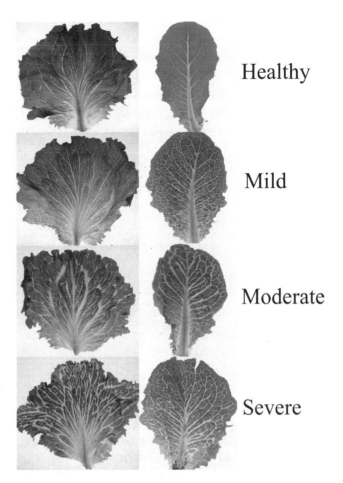

Healthy

Mild

Moderate

Severe

FIGURE 2.8 Degrees of severity of big vein symptoms on cvs. iceberg (left column) and romaine (right column) lettuce leaves differentiated as healthy, mild, moderate, and severe Healthy- no visible symptom; mild-narrow vein clearing and/or symptoms appearing on less than 50% of leaf area; moderate- both wide and narrow vein clearing symptoms observable on more than 51% of the leaf area, slight shortening and stiffening of the leaf, and severe- wide vein clearing seen on all veins of leaf with marked shortening and stiffening of the leaf.

[Courtesy of Hayes et al. 2006 and with kind permission of the American Phytopathological Society, MN]

The levels of resistance of wheat genotypes to *Soilborne wheat mosaic virus* (SBWMV) were assessed under controlled growth chamber conditions. All genotypes that were susceptible in the field, were highly susceptible, when grown on soils infested with viruliferous *Polymyxa graminis* under growth chamber conditions. High titers of the virus were detected by ELISA test in the leaves of inoculated plants. The wheat genotypes could be efficiently tested for resistance to SBWMV by this relatively less-expensive test, providing reliable results rapidly. The main criterion for scoring disease resistance was the absence of SBWMV antigen in the leaf tissues, rather than relying on the appearance of visual symptoms and effect on crop yield. In the growth chamber-based resistance test, the reactions of genotypes to SBWMV could be determined within 6–9 weeks after sowing. In contrast, field screening required at least two growing seasons

TABLE 2.1

Responses of lettuce cultivars and *L. vilosa* accession IVT280 to MiLBVV and LBVaV under inoculated greenhouse conditions (Hayes et al. 2006)

Line	No. of plants tested	No. of symptomatic plants	Percent symptomatic plants
Experiment 1			
Great Lakes 65	132	100	76
Margarita	127	40	31
Pavane	119	35	29
Total	378	175	46
X^2, 2df		24.3[a]	
Experiment 2			
Great Lakes 65	104	91	88
IVT280	36	0	0
Pavane	106	40	38
Total	246	131	53
X^2, 2df		34.9[a]	

[a] = $P < 0.01$.

to complete the assessment of reactions of the entries for resistance to the virus. Further, the growth chamber-based tests provided the advantage of uniform distribution of the populations of the vector *P. graminis*, which was not achievable under field conditions. One *Triticum monococcum* line of the 26 lines tested, did not have detectable virus antigen in the roots. Resting spores of *P. graminis* were also less abundant in the roots of the resistant line. One Bulgarian line was resistant to SBWMV and possibly partially resistant to *P. graminis* also (Kanyuka et al. 2004).

Incidence of *Soilborne wheat mosaic virus* (SBWMV), the type member of the genus *Furovirus*, was first reported as the cause of rosette disease in winter wheat in the United States (McKinney 1925). *Polymyxa graminis*, the obligate biotrophic plasmodiophorid parasite, is the vector involved in the transmission of SBWMV. The disease was considered as one of the most destructive diseases of winter wheat and it was persistent, because the viruliferous resting spores of *P. graminis* could remain viable for decades. Another virus widely distributed in bread wheat, durum wheat and rye crops in France, Italy, Germany, Poland and Denmark, was named as *Soilborne cereal mosaic virus* (SBCMV). Use of resistant cultivars was considered to be the practical, environmental-friendly and sustainable approach for managing these soilborne virus diseases caused by SBWMV and SBCMV. The available disease management strategies were unable to protect the wheat crops effectively and, hence, identification of cultivars/genotypes of wheat with resistance to these viruses was taken up. The wheat genotypes and cultivars (13) were evaluated under controlled conditions to improve the speed and reliability of germplasm screening for resistance to SBWMV, to determine the mode of inheritance of SBCMV resistance

for cv. Cadenza, and to screen diploid *Triticum monococcum* accessions for resistance to SBWMV and its vector *P. graminis*. Different types of genetic control of cultivar resistance to soilborne viruses have been studied. Resistance to *Soilborne wheat mosaic virus* (SBWMV) was reported to be conferred by 1 to 3 genes. Preliminary evidence suggested that resistance prevented systemic movement of SBWMV from roots to foliage, but did not prevent replication and cell-to-cell movement in roots of infected plants. At temperatures above 20°C, resistance to foliar infection was broken down, but symptoms were not expressed. Resistance of genotypes with field resistance became susceptible to mechanical inoculation of foliage. The broad sense heritability estimates of 43 to 55% reflected high heritability of resistance, indicating that about 50% of variation in the number of symptomatic plants was due to genetically inherited host resistance (Cadle-Davidson and Gray 2006). In the later investigation, 115 regionally adapted small grain genotypes were evaluated for SBWMV over four growing seasons. Resistance to SBWMV reduced the percentage of infection of plants that had detectable virus titer and disease symptoms. Logistic regression analysis of disease incidence data facilitated the determination of the effect of small sample size, low disease incidence, as well as irregular disease distribution. By increasing sample size from 20 to 30 stems/replicate, the number of resistance categories was increased through improved resolution of intermediate resistance classes. In environments with low disease incidence, the number of genotypes categorized as susceptible decreased, whereas intermediate genotypes appeared to be resistant in the analysis. None of the genotypes evaluated was immune to the disease in the experiments conducted in multiple years. However, 41 of the regionally adapted genotypes evaluated, repeatedly expressed strong resistance responses to SBWMV, thus providing the growers a wide choice for selecting suitable genotype with desired characteristics (Cadle-Davidson and Gray 2006).

Association studies have been employed to validate and locate quantitative loci (QTLs) or genes for important traits and to map candidate genes (Zhu et al. 2008). Association analysis has the potential to exploit recombination events over multiple breeding cycles and encompasses a diverse array of germplasm, without the need for developing new mapping populations. The association investigation was carried out to identify genes for resistance to *Soilborne wheat mosaic virus* (SBWMV), using a selected panel of 205 elite experimental wheat lines and cultivars from US hard and soft winter breeding programs. Disease severity was assessed twice in SBWMV-infected fields for the panel at Manhattan, KS, in spring 2010 and 2011 and for a sub-panel of 137 hard winter wheat accessions at Stillwater, OK, in spring 2008. ELISA tests were performed to quantify the virus titer based on the absorbance values. Among 205 accessions tested in Manhattan, the frequencies of resistant (R and MR) and susceptible (MS and S) accessions were 70% and 30% in both years. Similar ratios of rating were recorded for hard and soft winter wheat cultivars and the results were repeatable (see Figure 2.9). At Stillwater, more resistant accessions and fewer susceptible accessions were differentiated than from the same set of accessions tested at Manhattan, with frequencies of

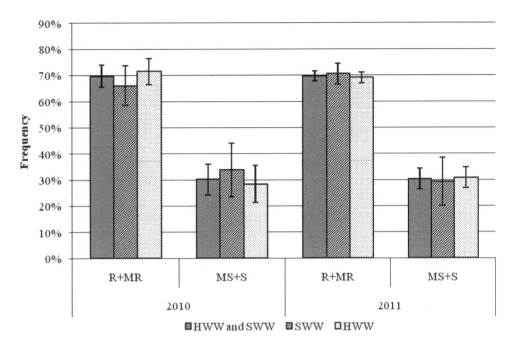

FIGURE 2.9   Assessment of frequency of distribution of reactions of resistant (R and MR) and susceptible (MS and S) wheat accessions to infection by *Soilborne wheat mosaic virus* (SBWMV) in Manhattan, NY. Disease ratings were similar for hard and soft winter wheat cultivars in 2010 and 2011.

[Courtesy of Zhang et al. 2011 and with kind permission of the American Phytopathological Society, MN]

79.5% resistant (R and MR) and 20.5% susceptible (S and MS) (see Figure 2.10). The correlation coefficient was 0.82 (P < 0.0001) for disease scores between the replicates in Stillwater and 0.62 (P < 0.0001) for mean scores between Stillwater and Manhattan. The simple-sequence repeat markers (282) covering all wheat chromosome arms were screened for association in the panel. The marker *Xgwm469* on the long arm of

chromosome 5D (5DL) showed significant association with the disease rating. Association of three alleles viz., *Xgwm469-165* bp, -167bp- and -169bp, with resistance and that of null allele with susceptibility was revealed. Correlations between the marker and the disease rating were highly significant. The alleles *Xgwm469-165bp* and *Xgwm469-169bp* were present mainly in the hard winter wheat group, whereas the allele

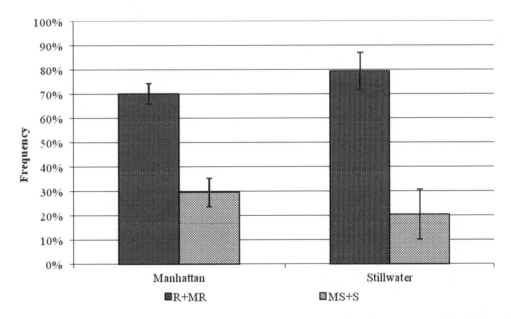

FIGURE 2.10   Determination of frequency of distribution of reactions of resistant (R and MR) and susceptible (MS and S) wheat accessions to infection by *Soilborne wheat mosaic virus* (SBWMV) in Stillwater, OK. Greater number of resistant accessions and fewer susceptible accessions were recognized from the same set of wheat accessions tested in Manhattan, NY.

[Courtesy of Zhang et al. 2011 and with kind permission of the American Phytopathological Society, MN]

TABLE 2.2

Distribution of marker alleles of *Xgwm469* for *Soilborne wheat mosaic virus* (SBWMV) resistance in different phylogenetic groups (Zhang et al. 2011)

Groups/subgroups	Total	165 bp	167 bp	169 bp	R[a]	Other[b]
Hard winter wheat	137	9	20	64	93	44
South[c]	40	3	7	16	26	14
North[d]	36	0	4	16	20	16
Jagger	36	4	5	22	31	5
Other	25	2	4	10	16	9
Soft winter wheat	68	1	28	14	43	25
South	38	1	12	8	21	17
North	30	0	16	6	22	8

[a] Total number of accessions carrying *Xgwm469* alleles associated with SBWMV resistance

[b] Total number of accessions carrying *Xgwm469* alleles associated with SBWMV susceptibility

[c] North subgroup consisting of entries from Illinois, Indiana, and Ohio

[d] South subgroup consisting of entries from Mid-Atlantic and Southern States

Xgwm469-167bp was present predominantly in the soft winter wheat (see Table 2.2). The *169bp* allele could be traced back to the cv. Newton and the *165bp* allele to *Aegillops tauschii*. Further, a novel locus on the short arm of chromosome 4D (4DS) was also found to associate with disease rating. The marker Xgwm469-5DL was closely linked to SBWMV resistance and highly polymorphic across the winter wheat accessions tested and it could be employed in marker-associated selection (MAS) approach in breeding programs to identify resistant wheat genotypes with resistance to *Soilborne wheat mosaic virus* (Zhang et al. 2011). The durum wheat (*Triticum turgidum* var. *durum*) is infected by *Soilborne cereal mosaic virus* (SBCMV) and *Soilborne wheat mosaic virus* (SBWMV), which are closely related viruses belonging to the genus *Furovirus*, vectored by *Polymyxa graminis*. The genetic basis of resistance to SBCMV in the cv. Neodur was analyzed, using a linkage mapping approach. Phenotypic and molecular analyses of 146 recombinant inbred lines (RILs) derived from the cross Crillo (highly susceptible) x Neodur (highly resistant) were performed. A major QTL that explained up to 87% of variability for symptom severity was identified on the short arm of the chromosome 2B, within the 40 cM integral between the markers *Xwmc764* and *Xgwm1128*, with *wPt-2106*, as the peak marker. Three minor QTLs were present on the chromosomes 3B and the minor QTL on chromosome 3B, represented potential candidate genes for the two resistance loci. Microsatellite markers flanking the major QTL were evaluated on a set of 25 durum wheat genotypes that were characterized earlier for SBCMV resistance. The allelic composition of the genotypes at these loci, together with pedigree data, suggested that the old Italian cultivar Cappelli provided the SBCMV-resistance determinants to durum cultivars that were independently bred in different countries during the last century (Russo et al. 2012).

Beet necrotic yellow vein virus (BNYVV), causal agent of sugar beet rhizomania disease, is transmitted by *P. betae*.

Cultivars with monogenic (*Rz1*) partial resistance to the virus were cultivated as a disease management strategy. In order to improve the level and durability of the resistance in these cultivars, search for sources of resistance to *P. betae* was made. Accessions (> 100) of the wild sea beet (*Beta vulgaris* ssp. *maritina*) from European coastal regions were evaluated for resistance to BNYVV under controlled conditions. Recombinant antibodies, raised to a glutathione-S-transferase expressed by *P. betae* in vivo, were employed to quantify the vector biomass in seedling roots. Several putative sources of resistance were identified. Selected plants were hybridized with male-sterile sugar beet breeding line possessing partial resistance to BNYVV (*Rz1*). Evaluation of F_1 hybrid populations identified five, in which *P. betae* resistance had been successfully transgressed from accessions originating from Mediterranean, Adriatic and Baltic coasts. A resistant individual from one of these populations was backcrossed to the sugar beet parent to produce a BC_1 population segregating for *P. betae* resistance. This population was evaluated for resistance to BNYVV. Amplified fragment length polymorphism (AFLP) and single-nucleotide polymorphism markers were used to map resistance QTL to linkage groups, representing specific chromosomes. The QTL for resistance to both *P. betae* and BNYVV were co-localized on chromosome IV in BC_1 population, indicating resistance to rhizomania conditioned by vector resistance. This resistance QTL (Pb2) found on chromosome IX, had a relationship confirmed by general linear model analysis. In BC_1 population, vector-derived resistance from wild sea beet combined additively with the *Rz1* virus resistance gene from sugar beet to reduce BNYVV titers. With partial virus resistance already available in several high yielding sugar beet cultivars, utilization of the simple *Pb1/Pb2* two-gene system may be expected to significantly enhance the resistance of sugar beet cultivars (Asher et al. 2009). Sugar beet crops are seriously affected by viruses transmitted by *Polymyxa graminis* and *P. betae*. Barley is a host for *P. graminis* and nonhost for *P. betae*. The transcriptional reprogramming of barley roots during the earliest nonhost and host interactions with zoospores of these vector species were investigated, using the Barley 1 Gene-Chip® microarray system. At high confidence levels, 20 and 13 genes with increased transcriptional activities were detected in response to *P. betae* and *P. graminis* inoculation, respectively, compared to the unchallenged roots of barley. A majority of the genes from both responses were associated with a classic defense response, as revealed by functional classification of the induced genes. Quantitative RT-PCR analysis showed that all of the genes examined were induced to comparable levels in both nonhost and host responses. The results revealed that the barley defense-associated genes *RAR1*, *ROR1* or *ROR2* were not required for arresting the establishment of *P. betae* infection in barley. Furthermore, the results suggested that in barley roots, the *Polymyxa* spp. might induce similar basal defense responses, whether the interaction is with host or nonhost plant species (McGrann et al. 2009).

2.4.2 Mechanisms of Resistance to Soilborne Viruses

Plant viruses carry a small genome that encodes a limited set of functional proteins (4–10 viral proteins), with which the entire virus replication and proliferation cycle cannot be completed. Hence, viruses recruit host factors to perform infection cycle (Whitham and Wang 2004). The absence of required host factors may interfere with infection process or restrict virus replication/movement, resulting in either mono- or polygenic recessive resistance (Kang et al. 2005). The susceptibility factors conserved in multiple plant-virus systems are the *EUKARYOTIC TRANSLATION INITIATION FACTORS (ETIF), 4E, iso4E, 4G* and *iso4G*. The translation initiation factors may interact directly with viral RNA, where they catalyze the initiation of translation of viral polyproteins (Robagalia and Caranta 2006). In spite of the smaller genome, compared to bacterial and fungal pathogens, viruses undergo a multistep infection process, involving entry into plant cells, uncoating of genomic nucleic acid, translation of viral coat proteins, replication of viral nucleic acid, assembly of progeny virions, cell-to-cell movement, systemic (long-distance) movement and plant-to-plant/plant-to-vector-to-plant transmission. The incompatible interaction between host plant and virus may be predominantly due to dominant resistance and in some cases due to recessive resistance. In certain pathosystems, dominant resistance results from inactive recognition event that may occur between host and viral factors, resulting in induction of host defense responses (Narayanasamy 2008).

RNA silencing is an important mechanism involved in gene regulation in several organisms, including plants. A gene silencing response was first discovered in the nematode, *Caenorhabditis elegans*, by experimentally introducing dsRNA, resulting in the loss of expression of corresponding cellular gene by sequence-specific RNA-degrading mechanism (Fire et al. 1998). Posttranscriptional gene silencing (PTGS) is an RNA silencing-based approach employed to reduce the level of expression of a (viral) gene of interest.

RNA silencing may be further classified into RNA-mediated transcriptional gene silencing (TGS). RNA-mediated TGS occurs, when ds-RNA with sequence homology to a promoter is produced, leading to de novo DNA methylation of promoter region of the structural gene (Mette et al. 2000). By contrast, PTGS leads to reduced steady state levels of targeted host or viral cytoplasmic RNA and to a lesser, but sometimes detectable reduction in nuclear RNA. PTGS may be mediated through a host-encoded RNA-dependent RNA polymerase (RdRp) and RNA helicase (Dalmay et al. 2000, 2001). RNA silencing is conserved in a wide range of eukaryotes and is manifested as 'quelling' in fungi *(Neurospora crassa)*, RNA interference (RNAi) in animals and cosuppressor or posttranscriptional gene silencing (PTGS) in plants. The unifying characteristic of RNA silencing is the presence of 21 to 26-nucleotide (nt) small interfering RNAs (Hamilton and Baulcombe 1999; Hamilton et al. 2002). Posttranscriptional gene silencing (PTGS) in plants inactivates some aberrant or highly expressed RNA in a sequence-specific manner in the host cell cytoplasm and it is an innate antiviral defense in plants. Many RNA and DNA viruses have been shown to stimulate PTGS shortly after infection, because of the formation of double stranded structures (transient replication intermediates). Plants transformed with constructs that produce RNAs capable of duplex formation induce virus immunity with almost 100% efficiency, when targeted against viruses (Smith et al. 2001). RNA silencing is initiated due to the presence of imperfect or true double-stranded RNAs (dsRNAs) derived from cellular sequences or viral genomes and they are processed by a ribonuclease III-like protein in the Dicer family known as 'Dicer-like (DCL) proteins' to generate 21-22 nucleotide (nt) microRNAs (miRNAs) or 21-28 nt interfering RNA (siRNA) duplexes. Each strand of small RNA is then incorporated into the effector complexes designated 'RNA-induced silencing complexes' (RISCs), which contain ARGONAUTE (AGO) proteins to guide the sequence specificity in the down-regulation process (Axtell 2013; Bologna and Voinnet 2014). Plants encode multiple DCL, AGO and RNA-dependent RNA polymerase (RdRp) proteins to cope with diverse endogenous RNA silencing pathways (Zhang et al. 2015).

Mechanism of RNA silencing operating in plant roots was investigated initially less frequently, compared to aerial organs. Later several studies were applied to analyze gene regulation, involving RNA silencing in roots and they revealed certain unique characteristics of RNA silencing in roots. Generally, lower RNA silencing activities were recorded in roots than in leaves, when posttranscriptional gene silencing in transgenic plants was induced by sense transgene. Lower levels of transgene silencing in roots than in leaves of silenced transgenic *Nicotiana benthamiana* leaves carrying the coat protein (CP) read through gene of *Beet necrotic yellow vein virus* (BNYVV) or green fluorescent protein (GFP) gene were observed, as indicated by incomplete degradation of transgene mRNAs and lower levels of transgene siRNAs accumulation (Andika et al. 2005). Accumulation of siRNAs derived from various ssRNA viruses has been detected in the roots of infected plants, including tomato, cucumber, melon and *N. benthamiana*, indicating that viruses could induce antiviral RNA silencing responses in roots. BNYVV siRNA accumulation was lower in roots than in leaves of *N. benthamiana* and inversely related with RNA genome accumulation, suggesting that BNYVV might more efficiently suppress RNA silencing in roots than in leaves (Andika et al. 2015). Sugar beet rhizomania caused by *Beet necrotic yellows virus*, is transmitted by *P. betae*. An RNA silencing mechanism was employed to induce resistance against rhizomania, using intron-hairpin RNA (ihpRNA) constructs. These constructs were based on sequences of the BNYVV 5′-untranslted region of RNA2 or the flanking sequence encoding P21 coat protein, with different lengths and orientations. Both transient and stable transformation methods led to development of effective resistance against rhizomania correlated with the transgene presence. Among the constructs, those generating ihpRNA structures with small intronic loops produced the highest frequencies of resistant events. The inheritance of transgenes and resistance was confirmed over generations in stably transformed plants (Zare et al. 2015).

Soilborne viruses may encode many RNA silencing suppressors. Small cysteine-rich proteins (CRPs) located in a 3′ proximal open reading frame (ORF) of the genome segment of the viruses belonging to the genera *Benyvirus, Furovirus, Pecluvirus* and *Tobravirus* have been identified. *Chinese wheat mosaic virus* (CWMV) and other members of the genus *Furovirus* require cool temperatures (below 20°C) to establish infection in host plants (Ohsato et al. 2003). RNA-dependent RNA polymerase 6 (RDR6) was involved in temperature-dependent antiviral defense against RNA viruses in *Nicotiana benthamiana* (Qu et al. 2005). Knock-down of RDR6 homolog in *N. benthamiana* enabled CWMV accumulation in roots, but not in leaves, after a temperature shift to 24°C and CWMV accumulation was associated with reduced accumulation of viral siRNAs in roots (Andika et al. 2013). Barley yellow mosaic disease caused by *Barley yellow mosaic virus* (BaYMV) and *Barley mild mosaic virus* (BaMMV), both belonging to the genus *Bymovirus* are transmitted by *Polymyxa graminis*. The naturally occurring recessive resistance locus *rym11* conferred complete broad spectrum resistance to all known European isolates of BaMMV and BaYMV. A positional cloning strategy was applied to identify the gene underlying *rym11*-based resistance. This was shown to be a susceptibility factor, belonging to a gene family of *PROTEIN DISULFIDE ISOMERASE* (PDIs) and to be highly conserved across eukaryotic species. Strong correlation existed between natural allelic variation and geographic distribution. The role of PDI in the interaction between BaYMV and barley was investigated. Using the positional cloning strategy, variants of PDI-like 5-1 (*HvPDIL 5-1*) was identified as the cause of the naturally occurring resistance to multiple strains of bymoviruses. Natural variation of *HvPDIL 5-1* was surveyed by resequencing the entire ORF in a geographically referenced collection of 365 wild (*Hordeum vulgare* ssp. *spontaneum*) and 1,452 domesticated barley accessions (*H. vulgare* ssp. *vulgare*). Among these accessions, only a single genotype

HOR1363, carried the *rym11-a*, a resistance allele (haplotype XXVIII). The gene *rym11-b* was detected in 27 accessions (haplotype II). All 28 accessions showed resistance, upon natural and artificial virus infections. The role of wild-type *HvPDIL* 5-1 in conferring susceptibility was confirmed by targeting induced local lesions in genomes for individual mutant alleles, transgene-induced complementation, alleleism tests, using different natural resistance alleles. The geographical distribution of natural genetic variants of *HvPDIL 5-1* revealed the origin of resistance confirming alleles in domesticated barley in Eastern Asia (Yang et al. 2014).

Plants with resistance to target viruses have been generated via RNA silencing using transgenic approach. A portion of genome sequence derived from soilborne viruses has been introduced into either experimental model plants or crop plant species and the transgenic plants were evaluated for their responses to virus infection. Silencing of viral sequences in transgenic plants resulted generally in high degree of protection against the soilborne viruses such as *Wheat yellow mosaic virus* (Dong et al. 2002), *Beet necrotic yellow vein virus* (BNYVV, sugar beet rhizomania disease) (Pavli et al. 2010; Zare et al. 2015) and *Mirafiori lettuce big-vein virus* (Kawazu et al. 2016). However, resistance of silenced plants was less effective in roots than in shoots. Transgenic tobacco carrying 57-kDa read through domain of the replicase gene of *Tobacco rattle virus* (TRV) was highly resistant to manual leaf inoculation, but the virus could be detected in roots following root manual inoculation or nematode inoculation (Vassilakos et al. 2008). Highly durable resistance to *Wheat yellow mosaic virus* infection in the field was developed by transformation with antisense nuclear inclusion b (NIb) replicase of WYMV (Chen et al. 2014).

2.5 TRANSGENIC RESISTANCE TO SOILBORNE MICROBIAL PATHOGENS

Molecular biology of pathogenesis and disease resistance have been studied to have an insight into the functions of virulence and avirulence genes of microbial pathogens and activation of a serries of host responses, resulting in enhancement of resistance in plants to microbial plant pathogens. Genetic engineering techniques provide enormous options for the selection of the desired gene(s) from a wide range of sources not only in the plant kingdom, but also in the animal kingdom and even from pathogens themselves. Presence of disease resistance genes in wild relatives of crop species and other sources has been detected at specific sites, employing restriction endonuleases. The DNA fragment(s) carrying the desired gene(s) are identified and used as molecular markers. Random amplified polymorphic DNA (RAPD) technique in conjunction with polymerase chain reaction (PCR) assay has been applied to identify resistance genes rapidly. These genes are then cloned in test organisms to study the nature of gene products and gene action. The PCR assay is useful to amplify the specific sequences in the DNA of plants and microbial pathogens. Transgenic plants constitutively expressing defense-related genes from selected sources can be generated. The exogenous DNA, introduced artificially via genetic engineering techniques has to pass through the germ line of the plant to be transformed, following incorporation of the foreign DNA into the plant's nuclear material. New crop cultivars with built-in resistance to disease(s) may be developed, by applying different biotechnological methods within a short period, compared to the time required for traditional breeding procedures. Furthermore, classical breeding procedures are hampered by interspecific sterility and linkage of disease resistance to other undesirable traits. In addition, multiplex diverse genes, conferring resistance to diseases may be incorporated simultaneously by transgenic plant technology, which permits a wider genetic diversity to be exploited. Transformation of plants by transferring desired genes from bacteria, fungi, viruses or various species of plants has been achieved mostly by employing the phytopathogenic bacteria *Agrobacterium tumefaciens* (*At*). The *Ti* (tumor-inducing) plasmid of *At* containing T-DNA has been used for gene transfer to plants. The T-DNA can be disarmed, rendering the plasmid nononcognic, but still retaining its ability to integrate with host plant genome and to synthesize opines. The T-DNA can be trimmed down to contain only absolutely essential functions for providing more room for cloning exogenous DNA fragment. Depending on the nature of pathogens and their pathogenic potential (virulence), different techniques have been applied to obtain cultivars of different crops with enhanced resistance to diseases (Narayanasamy 2008, 2017).

2.5.1 Transgenic Resistance to Fungal Pathogens

The fungal pathogens are more complex in structure and are capable of producing many enzymes and toxic metabolites to overcome host responses directed toward inhibition of pathogen development. Different approaches have been applied to develop cultivars with transgenic resistance to fungal pathogens. Development of crop varieties resistant to fungal pathogens may involve (i) activation of expression of some endogenous gene(s) normally induced, following infection; (ii) introduction of gene(s) to direct synthesis of novel low molecular weight antimicrobial compounds; and (iii) introduction of gene(s) to detoxify or deactivate factors of pathogen origin required for pathogenesis.

2.5.1.1 Employing Genes from Wild Relatives for Transformation

The wild relatives of crop plants have been found to be an important source of resistance genes that could be employed for improving the levels of resistance of cultivars to various diseases caused by microbial plant pathogens. A major potato late blight disease resistance gene from *Solanum bulbocastanum* was integrated into potato cultivars (*Solanum tuberosum*) through somatic fusion. The resistance gene *RB* was cloned and found to belong to the largest class of *R* genes that encode proteins with a nucleotide-binding site (NBS) and leucine-rich repeats (NB-LRR). The 5-kb of upstream regulatory sequence containing *RB* gene was stably integrated into *S. tuberosum*, using *Agrobacterium*-mediated transformation

(Song et al. 2003). The somatic fusions and transgenic plants containing *RB* gene were highly resistant to *P. infestans*, causal agent of the potato late blight disease, even under optimal conditions using multiple pathogen isolates. The full-length gene coding sequences, including the open reading frame (ORF) and promoter was integrated into potato cultivar using *Agrobacterium*-mediated transformation. *RB*-containing transgenic plants were challenged with *P. infestans*, under optimal conditions favoring late blight disease in the greenhouse. All transgenic lines containing *RB* gene exhibited strong foliar resistance. Field-grown transgenic tubers did not exhibit increased resistance to late blight pathogen. The yield of transgenic potato plants was not affected by the presence of *RB* gene in several potato cultivars. The broad-spectrum nature of resistance provided by *RB* gene, suggested its ability to recognize an effector that is either essential for the virulence of *P. infestans* or is well conserved among all isolates of the pathogen (Halterman et al. 2008).

Potato late blight resistance genes (resistance to *P. infestans*, *Rpi*) gene confers the capacity of recognizing the effectors specific to *P. infestans* and this recognition is used to trigger activation of defense, which inhibits pathogen development. Mutations in this gene, that encodes this effect, may result in its elimination or alteration. This may enable the pathogen to evade detection by the *Rpi* gene, leading to ineffectiveness of the *Rpi* gene. The *Rpi* genes have been shown to confer resistance to most races of *P. infestans*. The genes *Rpi-blb1*, *Rpi-blb2* and *Rpi-vnt1* were more effective and none of the pathogen race could overcome all three genes (Foster et al. 2009; Vleeshouwers 2011). The prospects for using dominant major *Rpi* genes for disease control have become bright. Multiple *Rpi* genes can be combined on one DNA construct (stacking) and then each gene can reduce the selection pressure against the other genes on the construct. With genetically modified (GM) methods, it is possible to insert the *Rpi* gene stack into the desired potato variety and recover derivatives with all characteristics of the favored variety, along with late blight resistance. The gene *Rpi-vnt1.1* was isolated from *Solanum venturii* and the *Rpi-mcq1* gene from *S. mochiquense*. *Rpi-vnt1.1* conferred ressitance to races 13_A2 and 6_A1 of *P. infestans* in detached leaf assays (DLAs). Transgenic potato cv. Desiree plants, carrying either *Rpi-mcq1* or *Rp-vnt1.1*, were produced. Plant and tuber phenotype traits of *Rpi-vnt1.1* transgenic lines were practically identical to the nontransgenic Desiree plants and total tuber yields were also similar in transgenic and nontransgenic plants. Disease severity was scored, after exposure to weather conditions favorable for late blight disease development. The *Rpi-vnt1.1* transgenic lines showed reduced severity, varying from 50 to 80% of plant tissues affected in different plots as against 100% infection in nontransgenic Desiree plants. The *Rpi-mcq1* gene did not confer resistance under field conditions, as all transgenic plants showed 100% infection (see Figure 2.11 A and B). Even with strong disease pressure due to favorable weather conditions, *Rpi-vnt1.1* transgenic plants remained highly resistant to the disease. No signs of infection were observed either before scoring time or till the end of the experiment, when tubers

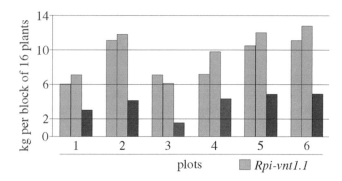

FIGURE 2.12 Effect of late blight on tuber yield of *Rpi-vnt1.1* transgenic and nontransgenic potato plants Yield increase in transgenic plants varied from 50 to 75% in different plots, compared to nontransgenic plants.

[Courtesy of Jones et al. 2014 and with kind permission of the Philosophical Society, England]

FIGURE 2.11 Responses of *Rpi-vnt1.1* transgenic and nontransgenic Desiree potato plants under field conditions A: actively sporulating areas in leaves of potato plants infected by late blight pathogen *P. infestans*; B: development of symptoms of infection on *Rpi-vnt1.1* transgenic and nontransgenic Desiree plants in field trials at 30 days after first symptom of infection appeared; none of the transgenic plants (left) showed infection, while nontransgenic plants (right) were severely affected by late blight disease.

[Courtesy of Jones et al. 2014 and with kind permission of the Philosophical Society, England]

were collected in the first week of October 2012. Disease progressed rapidly on nontransgenic Desiree plants, with 100% infection in August, when almost all plants were already dead and without foliar tissue. Furthermore, the yield of nontransgenic plants was drastically reduced (50–75% reduction) (see Figure 2.12). The dominant isolate of *P. infestans*, occurring in the experimental plots, was the genotype A_A1. The results showed that the *Rpi-vint1.1* gene was functional in the field for three successive seasons in small plots and *Rpi* genes from wild relatives could be stably transformed into cultivar like Desiree without changing its functionality (Jones et al. 2014).

2.5.1.2 Targeting Structural Components of Fungal Pathogens

Plants resistant to fungal pathogens may be developed by successful interference with the assembly of components of cell wall and membrane and/or degrading them, after their assembly by introducing the gene(s) encoding synthesis of antifungal compounds. Introducing genes which encode potential antimicrobial PR protein with chitinase or ß-1,3-glucanase activities has been the frequently applied procedure for transgenic plants resistant to fungal pathogens. Chitinases are considered to play a dual role, firstly by inhibiting fungal growth by cell wall dissolution, and secondly by releasing pathogen-borne elicitors that

may induce further defense-related responses in the host plant. Chitin forms one of the major constituents of the cell walls of many fungal pathogens, such as *R. solani* which has a wide host range and it is hydrolyzed by chitinase. ß-1,3-glucanase is known to degrade glucans present in the fungal cell walls. When both these enzymes are produced, the pathogen growth is more effectively inhibited (Mauch et al. 1988). Tobacco and oilseed rape (*B. napus*) plants were transformed with the chitinase gene from common bean (*P. vulgaris*) for constitutive high level of expression of chitinase. The transformed plants showed higher levels of chitinase activity and significant increase in the level of resistance to *R. solani*, causing seedling rot or damping-off disease. In addition, lysis of hyphal tips of the pathogen was recorded in transgenic plants under the light microscope (Broglie et al. 1991). *B. napus* var. *oleifera* plants transformed with the endochitinase gene from tomato showed greater level of resistance to *S. sclerotiorum* and other fungal pathogens. *Phoma lingam* and *Cylindrosporium concentricum*, compared to nontransformed control plants under field conditions (Grison et al. 1996). The efficacy of chitinase gene transformation, as a strategy for enhancing resistance levels of crops appeared to be influenced by several factors, including plant species, type of chitin protein expressed and sensitivity of the fungal pathogens. Three chitinase genes originating from petunia (acidic), two tobacco (basic) and bean (basic) were used separately to transform three lines of cucumber cv. Endeavor. The transformed lines did not show any improvement in the resistance level to *R. solani*. In contrast, carrot cvs. Nanco and Golden stripe transformed with chitinase gene from bean and tobacco showed resistance to *R. solani*, *S. rolfsii* and also to *Botrytis cinerea*. The transformants expressing chitinase gene from petunia did not exhibit any change in resistance to these pathogens, compared to nontransformed plants. The results suggested that the basic chitinase proteins might be more effective in providing effective protection against fungal pathogens only in some plant species (Punja and Raharjo 1996).

Expression of two plant defense genes may result in more effective protection of plants against microbial pathogens than by single gene. Transgenic tobacco and tomato plants expressing tobacco ß-1,3-glucanase and bean chitinase genes showed greater levels of resistance to fungal pathogens (Ouyang et al. 2003). In the further investigation, the tobacco AP24 osmotin gene and a bean basic chitinase gene were introduced into the tomato A53 cultivar susceptible to the wilt pathogen *F. oxysporum* f.sp. *lycopersici* (FOL) via *Agrobacterium tumefaciens* (At)-mediated transformation to generate genotypes with enhanced resistance to FOL. Molecular characterization by PCR and Southern blot hybridization confirmed the stable integration of the transgenes into the genome at various insertion sites. Of the 44 putative transgenic plants, 41 plants showed the presence of transgenes. The transgenic tomato plants with a single insertion of a dual transgene construct T-DNA showed significantly improved resistance against *F. oxysporum* f.sp. *lycopersici*. The transgenes were stably expressed in two successive generations, indicating the possibility of enhancing the level of resistance to the wilt pathogen by transformation with dual combination of PR genes (Ouyang et al. 2005). Although plants do not have the ability to produce antibodies, they can assemble functional antibody molecules encoded by mammalian antibody genes. *F. virguliforme* causes sudden death syndrome (SDS) disease in soybean. But it has not been possible to isolate the pathogen from diseased foliar tissues, suggesting that the foliar symptoms might be induced by toxin(s) produced by *F. virguliforme*. A possible toxin, FvTox1 was identified in infected soybean plants. The relationship between expression of anti-FvTox1 single-chain variable-fragment (scFv) antibody in transgenic soybean and development of resistance to foliar SDS was investigated. Two scFv antibodies, *Anti-FvTox1* and *Anti-FvTox1.2*, encoding anti-FvTox1 scFv antibodies were produced from RNAs of a hybridoma cell line that expressed mouse monoclonal anti-FvTox1 7E8 antibody. Both anti-FvTox1 scFv antibodies interacted with the antigenic site of FvTox1 that bound to mouse monoclonal anti-FvTox1 7E8 antibody. Binding of TvTox1 by the anti-FvTox1 scFv antibodies expressed in either *Escherichia coli* or transgenic soybean roots, was initially verified on nitrocellulose membranes. Expression of anti-FvTox1 in stable transgenic soybean plants resulted in enhanced foliar SDS resistance, compared with that in nontransgenic control plants. The results suggested that FvTox1 might be an important pathogenicity factor for foliar SDS development and that expression of scFv antibodies against *F. virguliforme* toxins might result in enhancement of resistance to SDS diseases (Brar and Bhattacharyya 2012).

Defensins are widely distributed throughout the plant kingdom and are present in most plant tissues. They are small, basic proteins with typically four disulfide bonds that confer highly stable and conserved structure. The plant defensins NaD1 from *Nicotiana alata* showed high antifungal activity against several fungi, including two important cotton pathogens, *F. oxysporum* f. sp *vasinfectum (Fov)* (causing Fusarium wilt disease) and *V. dahliae* (causing Verticillium wilt disease). Transgenic cotton plants, expressing *NaD1*, were generated and selected plants were evaluated. Homozygous plants were evaluated in greenhouse bioassays for resistance to *Fov*. One line, D1, was selected for field

evaluation over three growing seasons in soils naturally infested with *Fov* and for over two seasons in soils naturally infested with *V. dahliae*. In the field trials with *Fov*-infested soil, the line D1 had 2–3 times the survival rate, a higher tolerance to *Fov* (higher disease rank) and a two- to fourfold increase in lint yield, compared to nontransgenic cv. Coker control. When transgenic line D1 was planted in *V. dahliae*-infested soil, plants had a higher tolerance to Verticillium wilt and up to a twofold increase in lint yield. Line D1 did not have undesirable attributes in comparison with the parent Coker control plants grown in noninfested soil. The results showed that expression of *NaD1* in transgenic cotton plants could be advantageously exploited for protecting cotton crops against two important diseases (Gaspar et al. 2014).

The genomes of mycoparasites, which have evolved specifically to attack other fungi, including plant pathogens, but not plants, form a potential source of powerful antifungal genes. *Trichoderma* spp. produce chitinases and glucanases that could degrade the cell walls of fungal pathogens. The gene *ThEn-42* from *T. harzianum*, encoding a powerful endochitinase with strong antifungal activity against several fungal pathogens was cloned. Constitutive expression of the gene *ThEn-42* was obtained in tobacco and potato. High expression levels of the fungal gene in different tissues of tobacco and potato did not have any detectable adverse effects on the growth and development of transgenic plants. Substantial differences in endochitinase activity were detected among transformants. Selected transgenic lines were highly tolerant or completely resistant to the soilborne pathogen, *R. solani*, as well as to the foliar pathogens *Alternaria alternata, A. solani* and *Botrytis cinerea*. The high level and broad-spectrum of resistance obtained with a single chitinase gene from *T. harzianum* was more effective and overcame the limited efficacy of transgenes from plants and bacteria employed earlier (Lorito et al. 1998). Chitinase gene (*ChiC*) from the actinomycete *Streptomyces griseus* strain HUT6037 was transferred to tobacco lines, adopting the *Agrobacterium*-mediated transformation procedure. One marker-free transgenic tobacco line TC-1 was retransformed with the wasabi defensin gene (WD), isolated from *Wasabia japonica*. Of the retransformed shoots, 37% coexpressed the *ChiC/WD* genes, as confirmed by Western and Northern analyses. Southern blotting showed that no chromosomal rearrangement was introduced between the first and second transformation. Transgenic lines either coexpressing *ChiC* or *WD*, or coexpressing both genes were challenged with *F. oxysporum* f.sp. *nicotianae* (Fon). In vitro plant survival assessments showed transgenic lines coexpressing both genes possessed greater level of resistance against the pathogen with infection indices ranging from 0.0 to 1.2, compared to corresponding isogenic lines, expressing either *ChiC* or *WD* gene with infection indices varying from 2.5 to 9.8. Whole-plant infection indices in transgenic lines were significantly related to the extent of root colonization of tobacco. Extracts of leaves of transgenic tobacco lines inhibited mycelial growth and caused hyphal abnormalities in *F. oxysporum* f.sp. *nicotianae* (Ntui et al. 2011).

Infection of plants by fungal pathogens may be aided by oxalic acid (OA) via a number of routes, including acidification to accelerate cell wall-degrading enzyme activity through pH-mediated tissue damage or via sequestration of Ca^{2+} ions (Dutton and Evans 1996). Oxalate oxidase belongs to germin family of proteins that catalyze degradation of oxalic acid to produce CO_2 and H_2O_2. Oxalate oxidase was considered to have a role in the defense responses of plants to pathogen attack. Transformation of soybean, tobacco and sunflower with wheat oxalate oxidase gene resulted in enhancement of resistance to *S. sclerotiorum* (Donaldson et al. 2001). Induction of plant defense protein was observed in transgenic sunflower plants (Hu et al. 2003). The involvement of oxalic acid in the pathogenesis of *Sclerotinia* spp. has been reported. The ability of transgenic plants expressing an oxalic acid-degrading oxalate oxidase to provide protection against the fungal pathogens was assessed. The expression of a barley oxalate oxidase in transgenic peanut increased resistance to exogenously applied oxalic acid (up to 20 folds to pathophysiological levels). Transgenic peanut plants showed higher level of resistance to Sclerotinia blight disease caused by *S. minor*, based on lesion size formed on detached leaf assays. Peanut transformants did not show any deleterious effect due to the expression of oxalate oxidase and hence, it was used as a sensitive marker for following transformation and regeneration. A sensitive fluorescent enzyme assay was used to quantify expression levels for companion to the colorimetric procedure (Livingstone et al. 2005).

Enzymes capable of degrading oxalic acid (OA) produced by *S. sclerotiorum*, causal agent of white mold disease of soybean, have been utilized to generate transgenic resistant plants. Transgenic soybean lines containing the decarboxylase gene *oxdc* from *Flammulina* sp. were produced by biotic process. Molecular analysis revealed successful incorporation of the gene into the plant genome and showed that the *oxdc* gene was transferred to the progeny lines. Sixteen T2 transgenic events were screened for *S. sclerotiorum* resistance, using the detached leaf assay. The white mold disease progress curve displayed significant delay in symptom development in all *oxdc* events, compared with the nontransgenic genotype. Area under the disease progress curve (AUDPC) values showed reduction in severity of symptoms ranging from 61 to 96%, comparing the *oxdc* events with nontransgenic control plants. One event (OXDC9.21) was present in some plants that did not show any symptoms after 92 h. The RT-PCR analysis for detection of *oxdc* gene transcripts suggested that expression of the *oxdc* gene was associated with resistance of soybean to *S. sclerotiorum* (Cunha et al. 2010). In a later investigation, highly resistant transgenic soybean carrying an *oxo* gene [expressing oxalate oxidase (OxO), which catalyzes the degradation of oxalic acid (OA)] and susceptible line AC Colibri were tested for genome-wide gene expression in response to necrotrophic OA-producing strain of *S. sclerotiorum*, using soybean cDNA microarrays. Many genes and pathways were similarly activated or repressed in both genotypes after inoculation with *S. sclerotiorum*. But OxO genotype displayed a measurably faster induction of basal defense responses, as observed by the different changes in defense-related and secondary metabolite genes, compared

with its susceptible parent AC Colibri. Of the nine genes showing the most extreme opposite directions of expression between genotypes, eight were related to photosynthesis and / or oxidation, highlighting the importance of redox in the control of *S. sclerotiorum* (Calla et al. 2014).

Peanut transgenic lines expressing oxalate oxidase from barley were evaluated for their resistance to *S. minor*, causing Sclerotinia blight disease, under natural field conditions over a 5-year period. The AUDPC values for transgenic lines in single rows planted with seed from single-plant selections averaged 78, 83 and 90% and these values were lower than nontransgenic parents in 2004, 2005 and 2006, respectively. Furthermore, 14 transgenic lines planted with bulked seed in two-row plots showed AUDPC values, 81% and 86% less than that of nontransformed parents in 2005 and 2006, respectively. The pod yields of six transgenic lines were greater (488 to 1,260 kg/ha) than nontransgenic parent in 2005. Similarly, 10 transgenic lines yielded more than that of the nontransgenic parents in 2006. The results showed that transformation of peanut cultivars with the novel gene expressing oxalate oxidase resulted in more effective protection to peanut cultivar, higher yields and also saved costs of chemicals and labor (Patridge-Telenko et al. 2011). The susceptibility of Blight Blocker transgenic peanut lines expressing oxalate oxidase to other peanut diseases caused by fungal pathogens was assessed. The susceptibility of transgenic peanut lines to early leaf spot caused by *Cercospora arachidcola*, Cylindrocladium black rot caused by *Cylindrocladium parasiticum*, Southern stem rot caused by *S. rolfsii*, *Tomato spotted wilt virus* and *Aspergillus flavus* (inducing aflatoxin production in peanut seeds) was similar to that of nontransgenic peanut cultivars, Perry, Wilson and NC7. Transgenic lines consistently offered high levels of protection against *S. minor*, during three years of experimentation. But barley oxalate oxidase had little or no effect on other pathogens causing diseases in peanut (Hu et al. 2016).

2.5.1.3　Employing Genes Expressing Antifungal Proteins

Fungal pathogens produce endopolygalacturonases (endoPGs), which are involved in the early stages of pathogenesis, facilitating the invasion of host tissues in advance of spread of the fungal mycelium. Polygalacturonase-inhibiting proteins (PGIPs) are proteins present in the cell wall of most dicotyledonous plants. PGIP is structurally similar to several resistance gene products, and it belongs to the superfamily of leucine-rich repeat (LRR) proteins (Mattei et al. 2001). The PGIPs accumulating at the site of infection may function as elicitor molecules by facilitating fungal PGs to increase the elicitation of plant defense responses (Cervone et al. 1987). The apple PGIP inhibited PGs secreted by *V. dahliae* grown on potato cell walls and pectin. Hence, the effectiveness of apple PGIP1 to enhance resistance against *V. dalhliae*, infecting potato was assessed. Apple PGIP1 expressed in transgenic tobacco was able to inhibit PGs of *V. dahlia* effectively. Then, transgenic potato cv. BPI plants, containing apple *pgip1* gene under control of the constitutively, expressed CaMV35 promoter were generated. The PCR assay and Southern blot analysis confirmed stable integration of the apple *pgip1* into potato genome. The extracts prepared from all except one transgenic line effectively inhibited PGs of *V. dahliae*. Active PGIP1 was expressed both in leaves and roots of transgenic potato plants. Apoplastic localization of PGIP1 was observed in the *pgip* transgenic potato plants. Six transgenic potato lines showed reduced disease symptoms, compared with controls and other lines grown in the sick-soil. However, the reduction in disease severity could not be attributed to inhibition of PGs of *V. dahliae*, although high level expression of PGIP1 occurred in transgenic potato plants (Gazendam et al. 2004).

Sclerotinia blight disease caused by *S. minor* was a major threat to profitable production of peanut in the southeastern United States. In order to develop peanut cultivars with resistance to Sclerotinia blight disease, transgenic peanut lines expressing antifungal genes of chitinase from rice (*Oryza sativa*) and glucanase from alfalfa (*Medicago sativa*) were generated, using somatic embryos of peanut cultivar Okrun. Transgenic peanut lines were evaluated for their reactions in small field plots (6.1 x 7.6 m) for 3 years. Peanut lines were arranged in a complete randomized block design (RBD) with three replications. Disease pressure was assessed by quantifying the number of sclerotia of *S. minor* present in the soil samples taken from each plot within two weeks after planting. Viable sclerotial density was determined from the top 5 cm of field soil. Sclerotial density was uniform throughout the test plots over 3-year period. Of the 34 transgenic resistant lines tested, line 188 ranked first above the resistant cultivar South Runner, whereas the susceptible cv. Okrun was the last (34th) in the ranking. The line 24 was the most susceptible among the transgenic lines. Overall, transgenic lines consistently showed field resistance in 2000 and 2001 with a correlation efficiency for disease ranking being 0.65 (P = 0.01). Disease incidence ranking was identical over the 3-year period for line 188, Southwest Runner and line 416 with the ranking of 1, 2 and 3, respectively. The line 188 contained a single copy of rice chitinase gene and the transgene expression was 22% above the background level, under nonchallenged conditions. In addition, the upright growth habit of the line 188 might also play a role in the resistance to Sclerotinia blight disease. Lines 540 and 545 contained both rice chitinase and alfalfa glucanase transgenes and both performed equally well with average disease incidence ranking of 4.3 and 4.7 respectively. Both lines had elevated hydrolases activities with chitinase activities measuring 22% above baseline. However, glucanase activity in either line was not significant above background level. The results showed that the transgenic peanut lines showed potential for growing peanuts in regions, where Sclerotinia blight disease casues significant yield losses (Chenault et al. 2005).

2.5.1.4 Use of Genes for Modifying Host Plant Metabolism

The biosynthetic capacity of the host plant may be enhanced by transferring genes from other plant species to effectively arrest invasion of the pathogen into different tissues. Plants produce a wide range of compounds related to disease resistance in response to infection by microbial plant pathogens. Phytoalexins form an important group of compounds considered to be involved in the development of resistance to diseases. Stilbenes, a class of phytoalexins known to be present in grapevines and peanut, are produced by the enzyme stilbene synthase. Tomato plants cv. Vollendung were transformed with two stilbene genes. These transgenes were expressed in transgenic tomato plants, following inoculation with *P. infestans*, wounding, induction with elicitor or UV radiation. Inoculation with *P. infestans* resulted in marked accumulation of stilbene synthase mRNA, which could be detected in leaves after 30 min, reaching the maximum at 72 h after inoculation. Resveratrol accumulation reached the peak at 44 h after inoculation and its presence reduced in transformed tomatoes and it might regulate in a manner similar to that found in transformed tomato plants (Thomzik et al. 1997).

Infection by microbial plant pathogens stimulates production of ethylene in affected plant tissues, from which it diffuses into surrounding cell layers, before escaping into the atmosphere. Dominant genes conferring insensitivity to ethylene have been characterized in *A. thaliana* and tomato and hence, genetic modification of ethylene perception was carried out by using the mutant *etr1-1* gene from *Arabidopsis*. The *ETR1* gene appears to be an ethylene receptor. Because of the dominant character of the mutation, *etr1-1* plants lack several ethylene responses present in wild-type plants. By expressing the mutant *etr1-1* gene under the direction of viral 35S promoters in tomato and petunia, significant delays in different physiological processes were observed. Ethylene production was observed to be associated with both resistance and susceptibility to disease. Transformation of tobacco plants with the mutant *etr1-1* gene from *Arabidopsis*, conferred dominant ethylene insensitivity. Besides lacking known ethylene responses, the transformants (*Tetr*) did not slow down plant growth, compared to neighboring plants and they hardly expressed defense-related basic pathogenesis-related proteins and developed spontaneous stem browning. The *Tetr* plant showed altered responses to microbial pathogens, in that transformed tobacco plants were impaired in expression of basic PR-proteins and became susceptible to normally nonpathogenic soil fungi, such as *Pythium sylvaticum*, *P. splendens*, *Rhizopus* spp. and *Chalara elegans*. In contrast, the hypersensitive response to *Tobacco mosaic virus* (TMV) was not altered. As control of ethylene response is exploited to improve longevity of agricultural products, the possibility of enhancement of susceptibility to disease has to be taken into consideration (Knoester et al. 1998). The gene *GbERF* from *Gossypium barbadense* was cloned and the transcripts accumulated rapidly, reaching high concentrations, following treatment of plants with exogenous ethylene and challenged by *Verticilliium dahliae*, causal agent of Verticillium wilt disease of cotton (Qin et al. 2004).

Broad-spectrum resistance is effective against two or more pathogen species, and /or many different races within a pathogen species. Resistance QTL is considered to confer a variety of traits, including components of basal resistance, developmental or morphological phenotypes that are not conducive to infection, production of antimicrobial compounds by plant (phytoalexins and phytoanticipins), hormonal production and signaling, or other types of defense mechanisms. Due to the specific and sometimes transient nature of *R* gene-mediated resistance, along with the lack of known *R* genes for pathogens such as *R. solani*, infecting large number of plant species, including crops like rice, cotton and legumes, development of cultivars with broad-spectrum and quantitative resistance will be more effective in protecting the crop species. Ethylene (ET) may play a more prominent role than salicylic acid (SA) and jasmonic acid (JA) in mediating rice disease resistance. Endogenous ET levels in rice were manipulated for enhancing resistance to rice sheath blight and blast diseases. Transgenic lines with inducible production of ET were generated by expressing the rice ACS2 (1-aminocyclopropane-1-carboxylic acid synthase, a key enzyme of ET biosynthesis) transgene under control of strong pathogen-inducible promoter. The OsACS2-overexpressing lines showed significantly increased levels of the OsACS2 transcripts, endogenous ET and defense gene expression, compared to the wild-type plant, especially in response to pathogen infection. The transgenic lines exhibited increased resistance to a field isolate of *R. solani* and also to races of *Magnaporthe oryzae*. The plant growth and developmental characteristics were similar in both transgenic lines and wild-type plants. The results suggested that the pathogen-inducible production of ET in transgenic rice could enhance resistance to necrotrophic or hemibiotrophic fungal pathogens without negatively affecting crop production (Helliwell et al. 2013).

Phenylpropanoids involved in plant-pathogen interactions are toxic or inhibitory to microbial pathogens, and they may be either formed constitutively (phytoanticipins) or induced in response to infection by pathogens (phytoalexins). Glyceollin, collective name for soybean phytoalexins is formed following infection by fungal pathogens, *F. virguliforme* and *P. sojae* (Lozovaya et al. 2004; Lygin et al. 2010). Biosynthesis of glyceollin occurs via the isoflavonoid branch of the phenylpropanoid pathway. Involvement of glyceollin in resistance to soilborne pathogens was studied. Rapid glyceollin biosynthesis during infection might be an important component of innate disease resistance in soybean. The response of transgenic lines of soybean with suppressed synthesis of isoflavones and nontransgenic plants to *M. phaseolina* and *P. sojae* was investigated. Transgenic soybean plants were generated via bombardment of embryonic cultures with the phenylalanine ammonia lyase, chalcone synthase and isoflavone synthase (*IFS2*) genes in sense orientation driven by the cotyledon-preferable lectin promoter (to turn on genes in cotyledons), while plants of another line were newly produced, using the *IFS2* gene in sense orientation driven by the *Cassava vein mosaic virus* constitutive promoter (to turn on genes in all plant parts). Isoflavone synthesis was almost entirely inhibited in the cotyledons of young seedlings of transgenic plants transformed with the *IFS2* transgene driven by the cotyledon-preferable lectin promoter, compared with the nontransformed controls, during the 10-day observation period, with the precursors of

isoflavone synthesis being accumulated in the cotyledons of transgenic plants. The lectin promoter could be active not only during seed development, but also during seed germination. Downregulation of isoflavone synthesis on the seed or in the whole soybean plant caused a strong inhibition of the pathogen-inducible glyceollin in cotyledons after inoculation with *P. sojae*, which resulted in increased susceptibility of the cotyledons of both transgenic lines to the pathogen, compared to the inoculated cotyledons of nontransformed plants. When stems were inoculated with M. *phasolina,* suppression of glyceollin synthesis was detected only in stems of transgenic plants expressing the transgene driven by a constitutive promoter, which developed more severe infection. The results revealed that glyceollin accumulation during infection significantly strengthened the innate defense response in soybean (Lygin et al. 2013).

2.5.2 Transgenic Resistance to Bacterial Pathogens

Application of traditional breeding procedures proved to be difficult, leading to the search for alternative methods for development of cultivars resistant to soilborne bacterial diseases affecting different crops. Biotechnological techniques vary in their effectiveness to generate clones with required level of resistance to bacterial pathogens.

2.5.2.1 Transfer of Genes from Nonhost Plants

The ethylene responsive factors (ERF)–transcription factors, have a role in expression of PR genes, which are known to be resistance-related. The possibility of enhancing resistance of *B. rapa* to soft rot disease caused by *Pectobacterium carotovorum* susbp. *carotovorum* (*Pcc*) was explored via transformation with bromelain gene (BL1) of pineapple (*Ananas cosmosus* L.). Three homozygous T2 lines were inoculated with *Pcc*. The BL8-2 line showed the lowest rate of infected leaves (RIL) in both wound inoculation and non-wound inoculation treatments. The nontransformed line showed 100% RIL. The highest expression of *BL1* gene was also detected in BL8-2 homozygous line. The results showed that the overexpression of *BL1* gene conferred enhanced resistance to soft rot disease of *B. rapa* (Koh et al. 2013).

2.5.2.2 Use of Genes Encoding Antibacterial Proteins of Diverse Origin

Genes from different sources, encoding antibacterial proteins have been introduced into susceptible plants to protect them against diseases caused by bacterial pathogens.

2.5.2.2.1 Lysozymes

The lysozymes are proteins with lytic activity against bacterial cell walls. But their lytic activity is at low levels. On the other hand, the mammalian or bacteriophage lysozymes induce more drastic effects on bacterial cell walls. The lysozyme from bacteriophages T4 was highly active against both Gram-negative and Gram-positive bacteria. The lysozymes catalyze degradation of peptidoglycans, an important component of bacterial cell walls. Tobacco plants were transformed with the gene encoding the lysozyme of T4 bacteriophage fused with the signal sequence of barley α-amylase. The target enzyme was produced in the intercellular spaces. Multiplication of *Pectobacterium carotovorum* subsp. *carotovorum* (*Pcc*) was significantly reduced, although only very low concentration of the enzyme could be detected in the transgenic tobacco plants (During et al. 1993). The chicken lysozyme gene *chly* encoding the enzyme was introduced into potato clones of cultivar Desiree via *Agrobacterium tumefaciens* (At)-mediated transformation. The transgenic clones were evaluated in the greenhouse for resistance to blackleg (63 clones) and soft rot (69 clones) diseases caused. Potato clones (21–20%) showed enhanced resistance to infection by *P. carotovorum* ssp. *atroseptica* T7, as shown by reduced severity of blackleg or soft rot symptoms. The number of copies of transgene integrated into the plant genome of these clones was determined by Southern blot analysis. The level of transgene expression detected by Northern blot analysis, correlated with the level of resistance detected in these clones (Serrano et al. 2000).

2.5.2.2.2 Biofilm-Degrading Enzymes

Dispersin B (DspB), an enzyme from the periodontal (oral) pathogen *Aggregatibacter actinomycetemcomitans* degrades the extracellular matrix polysaccharide PGA, and it was evaluated for its ability to inhibit biofilm formation by the soft rot pathogen *Pectobacterium carotovorum* subsp. *carotovorum* through constitutive expression of DspB in tobacco plants. All transformed tobacco plants expressed properly folded and active DspB enzyme, but at different levels of expression. In virulence assays, the transgenic plant line D10 that produced only low level of DspB enzyme, showed significant enhancement of resistance to soft rot disease, compared to nontransgenic lines. The results suggested that DspB expression in plant might be effective to reduce soft rot disease incidence and severity in tobacco and possibly in other crops susceptible to *P. carotovorum* subsp. carotovorum (Raghunath et al. 2012).

2.5.2.2.3 Use of Antibacterial Peptides from Insects

Humoral immunity may be induced in many insect species by injecting viable nonpathogenic or inactivated phytopathogenic bacteria. New proteins are produced in the hemolymph of the treated insects and these proteins have antibacterial properties. Three types of proteins/peptides – lysozymes, cecropins and attacins – were detected in the hemolymph of the giant silk moth *Hyalophora cecropia* (Bonman and Hultmark 1987). The cecropins, belonging to the cytolytic pore-forming peptides group, possesses a broad-spectrum antibacterial activity against both Gram-negative and Gram-positive bacteria. These peptides are small with 35–37 amino acids, strongly basic and consist of three major forms – A, B and C. Transgenic tobacco plants expressing cecropin B anolog gene exhibited enhanced resistance to the bacterial wilt diseases caused by *Ralstonia solanacearum* (Jaynes et al. 1993). Tobacco plants were transformed with a chimeric gene fusion cassette consisting of a secretory sequence from barley α-amylase ligated to a modified cecropin (MB39)

coding sequence and placed under the control of promoter and terminator from potato pectinase inhibitor II (Pi II) gene. Multiplication of the bacterial pathogen was markedly suppressed in the transgenic plants by more than 10-fold, compared to control. Expression of cecropin gene in tobacco resulted in effective protection against another serious disease of tobacco, wildfire, caused by *Pseudomonas syringae* pv. *tabaci* (Huang et al. 1997). Attacins belong to another class of antimicrobial peptides detected in insect hemolymph, in response to bacterial infection. Attacins act on the outer membrane of Gram-negative bacteria, by altering permeability and protein synthesis (Carlsson et al. 1998). Attacin E (*attacin E*) was identified in *H. cecropia*. Potato plants expressing *attacin E* were less susceptible to *Pectobacterium carotovorum* ssp. *atrosepetica* (Arce et al. 1999).

2.5.3 TRANSGENIC RESISTANCE TO VIRAL PATHOGENS

Transgenic plants have been generated to enhance the levels of resistance in plants to viruses transmitted by soilborne vectors. Three different approaches have been followed to develop transgenic plants expressing (i) genes derived from viral sequences, providing pathogen-derived resistance; (ii) genes from other sources that could interfere with target virus at different stages of replication; and (iii) genes from nonhost plant species showing resistance to target virus.

2.5.3.1 Pathogen-Derived Resistance

Transgenic expression of pathogen sequences was hypothesized to provide protection against the respective viral pathogen itself by interfering with its development or replication (Sanford and Johnson 1985). Pathogen-derived resistance (PDR) is the predominant approach by which transgenic plants, expressing viral sequences, have been generated to protect various crops against infection of viruses. The pathogen-derived gene may interfere with the replication process of viruses in their host plant in different ways. The gene product expressed in the transformed plant may inhibit the normal function(s) of the cognate gene in the infecting viral genome or the gene, which may be so modified as to become detrimental to viral replication or maturation. As the functions encoded by different viral genes have been determined, it has been possible to transform plants with different genes encoding structural and nonstructural proteins and the effectiveness of protection offered by virus genes varies with host-virus combinations.

2.5.3.1.1 Use of Coat Protein Gene for Protection

Coat protein (CP) genes of plant viruses have been used to generate transgenic plants to protect them against respective viruses. Transgenic tobacco plants with resistance due to expression of CP gene of *Tobacco mosaic virus* (TMV) were generated. A cloned DNA encoding CP gene of TMV was ligated to the *Cauliflower mosaic virus* (CaMV) 35S promoter and NOS 3′ (nopaline synthase gene 3′ end) polyadenylation signal. This chimeric gene was introduced into tobacco cells on a disarmed Ti-plasmid of *Agrobacterium tumefaciens*

(*At*). The transgenic tobacco expressing CP gene either did not develop symptoms or developed symptoms after a longer incubation period. The titer of TMV was very low or the virus could not be detected in some transgenic tobacco plants. Resistance of the transgenic tobacco was only against the intact virus, but not to the viral nucleic acid (RNA) (Beachy et al. 1990; Nelson et al. 1997).

Grapevine fan leaf virus (GFLV, genus *Nepovirus*), transmitted by the nematode *Xiphinema index* has three particle types viz., the top (T), middle (M) and bottom (B), which occupy the sequential positions in a purified viral suspension, because of particle density. The T component contains empty virus capsid (coat protein), whereas the M and B particles have viral RNA2 and RNA1 + RNA2, respectively. Putatively transgenic individuals (127) of grapevine (*Vitis vinifera*) cv. Russalka were generated by transferring the CP gene, including non-translatable and truncated forms via *A. tumefaciens*-mediated transformation. The transgenic sequences were detected by PCR in all lines. In addition, Southern blot analysis revealed the number of inserted T-DNA copies varying from 1 to 6. Accumulation of GFLV CP could not be detected by ELISA in any one of the 36 transgenic lines analyzed. However, CP mRNA was expressed at variable levels. It was considered that absence of CP in detectable concentration might be an advantage from the safety point of view. However, the level of resistance/tolerance of the transgenic lines of grapevine to GFLV was not determined. Hence, the potential value of the transgenic lines for practical application could not be assessed (Maghuly et al. 2006). In order to develop transgenic lettuce for protection against *Lettuce big-vein associated virus* (LBVaV), the CP gene was used to transform the plant via *A. tumefaciens*-mediated transformation and the presence of the transgene was confirmed by PCR assay. Southern blot analysis showed that one copy of LBVaV CP gene and one copy of the *nptII* gene were present in the transformed lettuce plants. The transgenic A-2 line with CP gene in antisense orientation was resistant to LBVaV. This line showed resistance to *Mirafiori lettuce virus* (MiLV) also primarily associated with lettuce big-vein disease (Kawazu et al. 2006).

2.5.3.1.2 Use of Noncoat Protein Genes for Protection

Cucumber fruit mottle mosaic virus (CFMMV, genus *Tobamovirus*) is soilborne and causes severe wilting, in addition to induction of mosaic and yellow mottling on leaves and fruits in cucumber. The viral gene coding for the putative 54-kDa replicase gene was cloned into *Agrobacterium tumefaciens (At)* binary vector and cotyledons as explants of a parthenocarpic cucumber were transformed. The R1 plants were screened for resistance to CFMMV by symptom expression and back inoculation and enzyme-linked immunosorbent assay (ELISA). Of the total 14 lines containing replicase, eight lines exhibited high level of resistance to CFMMV, and both biological and molecular (RT-PCR) methods did not detect the virus in the transformed plants. The line 144 was immune to infection by mechanical and graft inoculation and to root infection, and it was homozygous for the 54-kDa replicase gene. Further, the line 144 was resistant, when planted on CFMMV-infested

soil. Appearance of symptoms was significantly delayed, when the line 144 plants were infected by three other cucumber-infecting tobamoviruses viz., *Cucumber green mottle virus* (CGMMV-W), *Kyuri green mottle mosaic virus* (KGMMV) and *Zucchini green mottle mosaic virus* (ZGMMV). When used as a rootstock, the line 144 could protect susceptible cucumber scions from soil infection by CFMMV. The accumulation level of the 54-kDa locus transcript in the line 144 was consistently low in spite of being driven by ΔSV, a strong constructive promoter. Inoculation of line 144 plants with CFMMV or ZGMMV did not alter the RNA expression level of the 54-kDa transgene. The results showed that by using transgenic line 144 as rootstock, susceptible cultivars could be protected against CFMMV (Gal-On et al. 2005).

Wheat yellow mosaic virus (WYMV), internalized in the thick-walled resting spores of the vector, *Polymyxa graminis*, persists in the soil for several years. Wheat yellow mosaic disease had spread to several provinces, causing appreciable losses in China. WYMV belongs to the genus *Bymovirus* (family Potyviridae) and its genome is encapsidated in filamentous particles. The genome consists of two positive sense ss-RNA segments RNA1 and RNA2, each of which encodes a polyprotein that undergoes post-translational cleavage. The polyprotein (269-kDa) encoded by RNA1 produces eight proteins, including the CP and nuclear inclusion 'b' protein (NIb) that functions as an RNA-dependent RNA polymerase (RdRp) and it is important for virus replication. RNA2 encodes a polyprotein (101-kDa), that produces two proteins, P1 (28-kDa) and P2 (73-kDa) (Namba et al. 1998). The NIb region (1212-bp) was used as the target sequence, because it was one of the most conserved portions in WYMV genome, providing broad-spectrum resistance. Fourteen independent transgenic events were obtained by co-transformation with the antisense *NIb8* gene (NIb replicase) and a selectable gene bar. The *NIb8* segment amplified from transgenic wheat lines was sequenced, confirming that the *NIb8* inserted in wheat genome was identical to the vector. RT-PCR assay showed that *NIb8* gene was transcribed in some T2 generation transgenic lines, including N12, N13, N14 and N15. These four lines showed high levels of resistance to WYMV. The numbers of virus-infected plants were respectively only 0, 0, 0 and 1 of 19, 48, 42 and 46 plants examined, whereas about half (50%) the susceptible control Yangmai 158 plants were infected. Southern blotting showed that the four transgenic lines had single copy integration of *NIb8* in the T6 generation. The heritability of *NIb8* was studied by crossing lines N12-1 and N14 with susceptible cv. Yangmai 158 as female parent. The F2 segregating population was tested under field conditions. The results suggested that *NIb8* was integrated as a single locus in the N12-1 and N14 transgenic lines and that the resistance could be transferred to other wheat cultivars by hybridization. The PCR assays showed that the *NIb8* gene was present in the plants of each generation, indicating that transformation using antisense *NIb8* might result in transgenic wheat plants with durable resistance. Western blot assays using the antibody specific to WYMV CP were performed in leaves, stems and roots of Yangmai 158 (YM158) and transgenic lines

N12-1 plants grown on WYMV-infested soil. At 4 months after planting, intense staining in tissue printing assay was observed in the upper leaves, lower leaves, stems and roots of YM158 plants, but not in those of N12-1 plants, indicating virus accumulation occurred only in the susceptible nontransgenic control plants. Likewise, strong bands corresponding to WYMV CP were detected by Western blot assays in leaves, stems and roots of YM158 plants, but not in those of N12-1 plants. Northern blot and small RNA deep sequencing analyses showed that there was no accumulation of small interfering RNAs, targeting the *NIb8* gene in transgenic wheat plants, suggesting that transgenic RNA silencing was not involved in the process of development of resistance to WYMV. The durable and broad-spectrum resistance of transgenic plant to WYMV could be advantageously exploited in breeding programs (Chen et al. 2014).

2.5.3.1.3 Use of Viral Nucleic Acid-Based Systems

The complementary RNA molecules may bind to the RNA transcripts of specific genes, resulting in blockage of their translation and these RNAs are designated antisense or micRNAs (messenger RNA-interfering complementary RNAs). *Beet necrotic yellow vein virus* (BNYVV) transmitted by *P. betae* causes rhizomania disease affects sugar beet production drastically. Of the three major groups of BNYVV isolates, the A- and B- types comprise four ss-genomic RNAs, whereas the P-type has an additional fifth RNA species. The P-type was considered to be more virulent than A- and B-types (Heijbroek et al. 1998). Sugar beet plants were transformed with an inverted repeat of the 0.4-kb fragment derived from the BNYVV replicase gene. The transformants displayed high levels of resistance against different strains of BNYVV, following inoculation using *P. betae*. The transgenic plants were highly resistant also under field conditions, and they were equally or more resistant than the conventionally used sources of resistance. The results demonstrated that application of ds-RNA-mediated resistance was more effective against BNYVV, infecting sugar beet crops. Accumulation of siRNA was correlated with high levels of resistance to BNYVV in sugar beet plants. In addition, extremely low (almost undetectable) steady state levels of transgenic mRNA was detected in transgenic plants, indicative of RNA-silencing-mediated resistance (Lennefors et al. 2005). The response of roots of transgenic *Nicotiana benthamiana* plants was studied earlier. The roots of transgenic N. *benthamiana* plants exhibited less resistance to BNYVV, compared to leaves (Andika et al. 2005). In contrast, roots of transgenic sugar beet lines were highly resistant to BNYVV and accumulated only low or nondetectable amounts of BNYVV RNA and no symptoms of rhizomania disease were discernible. This is likely to be due to the use of a part of BNYVV replicase gene duplicated as an inverted repeat, giving rise to a ds-RNA, a conFigureuration known to function as a strong silencing inducer. The transgenic sugar beet plants showed resistance to all the three types of BNYVV isolates. The resistance provided by ds-RNA was more effective than that was provided by the CP gene. The ds-RNA approach appeared to offer an advantage, since RNA silencing does not rely on accumulation of virus proteins. This fact may facilitate the eventual release of transgenic cultivars for large scale use, by eliminating the need for extensive toxicity or allergic tests (Lennefors et al. 2005).

REFERENCES

Adhikari KKC, McDonald MR, Gossen BD (2012) Reaction to *Plasmodiophora brassicae* pathotype 6 in lines of *Brassica* vegetables, Wisconsin Fast Plants, and canola. *HortScience* 47: 374–377.

Agrawal GK, Rakwal R, Jwa NS, Han KS, Agrawal VP (2003) Isolation of a novel rice PR4 type gene whose mRNA expression is modulated by blast pathogen attack and signaling components. *Plant Physiol Biochem* 41: 81–90.

Ahmed ZK, Osman AMS, Abdel-Rahem TA, Mohamed AA (2013) In vitro selection of onion plants resistant to white rot disease in Egypt. *Proc Afr Crop Sci Conf* 11: 543–550.

Allefs JJH, van Dooueweert W, de Jong ER, Prummel W, Hoogendorn J (1995) Factors affecting potato soft rot resistance to pectolytic *Erwinia* species in tuber-slice assay. *J Phytopathol* 143: 705–711.

Allen RL, Bittner-Eddy PD, Briggs LJG et al. (2004) Host-parasite coevolutionary conflict between *Arabidopsis* and downy mildew. *Science* 306: 1957–1960.

Al-Safadi B, Mir Ali N, Arabi MIE (2000) Improvement of garlic (*Allium sativum* L.) resistance to white rot and storability using gamma irradiation-induced mutations. *J Genet Breeding* 54: 175–181.

Andika IB, Kondo H, Tamada T (2005) Evidence that RNA silencing-mediated resistance to *Beet necrotic yellow vein virus* is less effective in roots than in leaves. *Molec Plant-Microbe Interact* 18: 194–204.

Andika IB, Maruyama K, Sun L, Kondo H, Tamada T, Suzuki N (2015) Differential contribution of plant Dicer-like proteins to antiviral defences against *Potato virus X* in leaves and roots. *Plant J* 81: 781–793.

Andika IB, Sun I, Xiang R, Li J, Chen J (2013) Root-specific role for *Nicotiana benthamiana* RdRp in the inhibition of *Chinese wheat mosaic virus* accumulation at higher temperatures. *Molec Plant-Microbe Interact* 26: 1165–1175.

Antoniou PP, Marakis EA, Tjamos SE, Paplomatos EJ, Tjamos EC (2008) Novel methodologies in screening and selecting olive varieties and rootstocks for resistance to *Verticillium dahliae*. *Eur J Plant Pathol* 122: 549–560.

Arce P, Moreno M, Gutierrez M et al. (1999) Enhanced resistance to bacterial infection by *Erwinia carotovora* subsp. *atroseptica* in transgenic plants expressing attacin or cecropin SB-37 genes. *Amer J Potato Res* 76: 169–177.

Ardila HD, Torres AM, Martinez ST, Higuera BL (2014) Biochemical and molecular evidence for the role of class III peroxidases in the resistance of carnation (*Dianthus caryophyllus* L) to *Fusarium oxysporum* f.sp. *dianathi*. *Physiol Molec Plant Pathol* 85: 42–52.

Armstrong MR, Whisson SC, Pritchard L et al. (2005) An ancestral oomycete locus contains late blight avirulence gene *Avr3a*, encoding a protein that is recognized in the host cytoplasm. *Proc Natl Acad Sci USA* 102: 7766–7771.

Arneson PA, Durbin RD (1968) Studies on the mode of action to tomatine as fungitoxic agent. *Plant Physiol* 43: 683–686.

Asher MJC, Grimmer MK, Mutasa-Goettgens ES (2009) Selection and characterization of resistance to *Polymyxa betae*, vector of *Beet necrotic yellow vein virus* derived from wild sea beet. *Plant Pathol* 58: 250–260.

Attard A, Gourgues M, Callemeyn-Torre N, Keller H (2010) The immediate activation of defense responses in *Arabidopsis* roots is not sufficient to prevent *Phytophthora parasitica* infection. *New Phytol* 187: 449–460.

Atucha A, Emmett B, Bauerle TL (2004) Growth rate of the root system influences rootstock tolerance to replant disease. *Plant Soil* 376: 337–347.

Auclair J, Boland GJ, Kohn LM, Rajcan I (2004) Genetic interactions between *Glycine max* and *Sclerotinia sclerotiorum* using a straw inoculation method. *Plant Dis* 88: 891–895.

Austin MJ, Muskett P, Kahn K, Feys BJ, Jones JDG, Parker JE (2002) Regulatory role of SGT1 in early *R* gene-mediated plant defenses. *Science* 295: 2077–2080.

Ausubel FM (2005) Are innate immune signaling pathways in plants and animals conserved? *Nature Immunol* 6: 973–979.

Axtell MJ (2013) Classification and comparison of small RNAs from plants. *Annu Rev Plant Biol* 64: 137–159.

Babiker EB, Hulbert SH, Schroeder KL, Paulitz TC (2013) Evaluation of *Brassica* species for resistance to *Rhizoctonia solani* and binucleate *Rhizoctonia* (*Ceratobasidium* sp.) under controlled conditions. *Eur J Plant Pathol* 136: 763–772.

Báidez A, Gömez P, Del Rio JA, Ortuño A (2007) Dysfunctionality of the xylem in *Olex europaea* L. plants associated with the infection process by *Verticillium dahliae* Kleb.: Role of phenolic compounds in plant defense mechanism. *J Agric Food Chem* 55: 3373–3377.

Bailey BA, Jennings JC, Anderson JD (1997) The 24-kDa protein from *Fusarium oxysporum* f.sp. *erthroxyli*: Occurrence on related fungi and the effect of growth medium on its production. *Canad J Microbiol* 43: 45–55.

Baldé JA, Franscisco R, Queiroz A, Regaldo AP, Ricando CP, Veloso MM (2006) Immunolocalization of a class III chitinase in two musk melon cultivars reacting differentially to *Fusarium oxysporum* f.sp. *melonis*. *J Plant Physiol* 163: 19–25.

Bani M, Rubiales D, Rispail N (2012) A detailed evaluation method to identify sources of quantitative resistance to *Fusarium oxysporum* f.sp. *pisi* race 2 within a *Pisum* spp. germplasm collection. *Plant Pathol* 61: 532–542.

Barzic MR, Com E (2012) Proteins involved in the interaction of potato tubers with *Pectobacterium atrosepticum*: a proteomic approach to understanding partial resistance. *J Phytopathol* 160: 561–575.

Bateman A, Birney E, Cerruti L et al. (2002) The Pfam protein families database. *Nucleic Acids Res* 30: 276–280.

Bates GD, Rothrock CS, Rupe JC (2008) Resistance of the soybean cultivar Archer to Pythium damping-off and root rot caused by several *Pythium* spp. *Plant Dis* 92: 763–766.

Baumgartner K, Fujiyoshi P, Browne GT, Leslie C, Kluepfel DA (2013) Evaluating Paradox walnut rootstocks for resistance to Armillaria root disease. *HortScience* 48: 68–72.

Bayaa B, Erskine W, Singh M (1997) Screening lentil for resistance to Fusarium wilt: Methodology and sources of resistance. *Euphytica* 98: 69–74.

Beachy RN, Loesch-Fries S, Tumer NE (1990) Coat protein-mediated resistance against virus infection. *Annu Rev Phytopathol* 28: 451–471.

Beckers GJM, Spoel SH (2006) Fine-tuning plant defense signaling: salicylate versus jasmonate. *Plant Biol* 8: 1–10.

Ben-Chaim A, Grube RC, Lapidot M, Jahn MM, Paran I (2001) Identification of quantitative trait locus associated with tolerance to *Cucumber mosaic virus* in *Capsicum annuum*. *Theor Appl Genet* 102: 1213–1230.

Bent AF (1996) Plant disease resistance genes-function meets structure. *Plant Cell* 8: 1757–1771.

Bernard RI, Smith PE, Kaufmann MJ, Schmitthenner AF (1957) Inheritance of resistance to Phyotphthora root rot in soybean. *Agron J* 49: 391.

Berrocal-Lobo M, Molina A (2004) Ethylene response factor 1 mediates *Arabidopsis* resistance to the soilborne fungus *Fusarium oxysporum*. *Molec Plant-Microbe Interact* 17: 963–770.

Berzel-Herrenz A, de la Cruz A, Tenllado F (1995) The Capsicum *L3* gene-mediated resistance against tobamoviruses is elicited by the coat protein. *Virology* 209: 498–505.

Bhutz A, Witzel K, Strehmel N, Ziegler J, Abel S, Grosch R (2015) Perturbations in the primary metabolism of tomato and *Arabidopsis thaliana* plants infected with the soilborne fungus *Verticillium dahliae*. *PLoS ONE* 10c9/e013824210.371/journal.pone.o138242

Bilgi VN, Bradley CA, Khot SD, Grafton KF, Rasmussen JB (2008) Response of dry bean genotypes to Fusarium root rot, caused by *Fusarium solani* f.sp. *phaseoli* under field and controlled conditions. *Plant Dis* 92: 1197–1200.

Birch PRJ, Rehmany AP, Pritchard L, Kamoun S, Beynon JL (2006) Trafficking arms: Oomycete effectors enter host plant cells. *Trend Microbiol* 14: 8–11.

Birrel GW, Brown JA, Wu HI, Giaever G, Chu AM, Davis RW (2002) Transcriptional response of *Saccharomyces cerevisiae* to DNA damaging agents does not identify genes that protect against these agents. *Proc Natl Acad Sci USA* 99: 8778–8783.

Bittara FG, Thompson AL, Gudmestad NC, Secor GA (2016) Field evaluation of potato genotypes for resistance to powdery scab on tubers and root gall formation caused by *Spongospora subterranea*. *Amer J Potato Res* 93: 497–508.

Bizimungu B, Holm DG, Kawchuk LM et al. (2011) Alta Crown: A new Russet potato cultivar with resistance to common scab and a low incidence of tuber deformities. *Amer J Potato Res* 88: 72–81.

Black W (1960) Races of *Phytophthora infestans* and resistance problems in potato. *Rep Scotish Soc Res Plant Breeding*, pp. 29–38.

Boissy G, O'Donohue M, Gaudemer O, Peroz V, Pernollet JC, Brunie S (1999) The 2.1A structure of an elicitin ergosterol complex: A recent addition to the Sterol Carrier Protein family. *Protein Sci* 8: 1191–1199.

Boller T (2005) Peptide signaling in plant development and self/nonself perception. *Curr Opin Cell Biol* 17: 116–122.

Bologna NG, Voinnet O (2014) The diversity, biogenesis, and activities of endogenous silencing small RNAs in *Arabidopsis*. *Annu Rev Plant Biol* 65: 473–503.

Bolwell GP (1999) Role of active oxygen species and NO in plant defense responses. *Curr Opin Plant Biol* 2: 287–294.

Bonman HG, Hultmark D (1987) Cell free immunity in insects. *Annu Rev Phytopathol* 41: 103–126.

Bos L, Huijberts N (1990) Screening for resistance to big-vein disease of lettuce. *Crop Prtoect* 9: 446–452.

Bowyer P, Clarke BR, Lunness P, Daniels MJ, Osburn AE (1995) Host range of a plant pathogenic fungus determined by a saponin-detoxifying enzyme. *Science* 267: 371–374.

Bradley CA, Hartman GL, Nelson RL, Mueller DS, Pederson WL (2001) Response of ancestral soybean lines and commercial cultivars to Rhizoctonia root and hypocotyl rot. *Plant Dis* 85: 1091–1095.

Bradley CA, Henson RA, Porter PM, Le Gare DG, del Rio LE, Khot SD (2006) Response of canola cultivars to *Sclerotinia sclerotiorum* in controlled and field environments. *Plant Dis* 90: 215–219.

Brar HK, Bhattacharyya MK (2012) Expression of a single-chain variable fragment antibody against a *Fusarium virguliforme* toxin peptide enhances tolerance to sudden death syndrome in transgenic soybean plants. *Molec Plant-Microbe Interact* 25: 817–824.

Broglie K, Chet H, Holliday M et al. (1991) Transgenic plants with enhanced resistance to the fungal pathogen *Rhizoctonia solani*. *Science* 254: 1194–1197.

Browne GT, Grant JA, Schmidt LS, Leslie CA, McGranahan GH (2011) Resistance to Phytophthora and graft compatibility with Persian walnut among selections of Chinese Wingnut. *HortScience* 46: 371–376.

Brunner F, Rosahl S, Lee J et al. (2002) Pep-13, a plant defence-inducing pathogen-associated pattern from *Phytophthora* transglutaminases. *EMBO J* 21: 6681–6688.

Buller S, Inglis D, Miles C (2013) Plant growth, fruit yield and quality and tolerance to Verticillium wilt of grafted watermelon and tomato in field production in the Pacific Northwest. *HortScience* 48: 1003–1009.

Cachinero JM, Hervás A, Jiménez-Díaz RM, Tena M (2002) Plant defense reactions against Fusarium wilt in chickpea induced by incompatible race 0 of *Fusarium oxysporum* f.sp. *ciceris* and nonhost isolates of *F. oxysporum*. *Plant Pathol* 51: 765–776.

Cadle-Davidson L, Bergstorm G (2005) Susceptibility of perennial small grains to *Soilborne wheat mosaic virus* and *Wheat spindle streak mosaic virus*. *Plant Management Network*, November 2005, 2 pages.

Cadle-Davidson L, Gray SM (2006) *Soilborne wheat mosaic virus*. *The Plant Health Instructor*

Cahill D, Rookes J, Michalczyk A, McDonald K, Drake A (2002) Microtubule dynamics in compatible and incompatible interactions of soybean hypocotyl cells with *Phytophthora sojae*. *Plant Pathol* 51: 629–640.

Calla B, Blahut-Beatty L, Koziol L et al. (2014) Genomic evaluation of oxalate-degrading transgenic soybean in response to *Sclerotinia sclerotiorum* infection. *Molec Plant Pathol* 15: 563–575.

Campa A, Pérez-Vega E, Pascual A, Ferreira JJ (2010) Genetic analysis and molecular mapping of quantitative trait loci in common bean against *Pythium ultimum*. *Phytopathology* 100: 1315–1320.

Candole BL, Conner PJ, Ji P (2010) Screening *Capsicum annuum* accessions for resistance to six isolates of *Phytophthora capsici*. *HortScience* 45: 254–259.

Cao T, Rennie DC, Manolii VP, Hwang SF, Falak I, Strelkov SE (2014) Quantitative resistance to *Plasmodiophora brassicae* in *Brassica* hosts. *Plant Pathol* 63: 715–726.

Carlsson A, Nystrom T, Cock H, Bennich H (1998) Attacin–an insect immune protein–binds LPS and triggers the specific inhibition of bacterial outer membrane protein synthesis. *Microbiology* 144: 2177–2188.

Carmeille A, Prior P, Kodja H, Chiroleu F, Luisetti J, Besse P (2006) Evaluation of resistance to race 3, biovar 2 of *Ralstonia solanacearum* in tomato germplasm. *J Phytopathol* 154: 398–402.

Catal M, Erler F, Fulbright DW, Adams GC (2013) Real-time quantitative PCR assays for evaluation of soybean varieties for resistance to the stem and root rot pathogen *Phytophthora sojae*. *Eur J Plant Pathol* 137: 859–869.

Cervone F, De Lorenzo G, Degra L, Salvi G, Bergami M (1989) Purification and characterization of a polygalacturonase-inhibiting protein from *Phaseolus vulgaris* L. *Plant Physiol* 85: 631–637.

Chaerle L, Hagenbeck D, de Bruyne E, Valcke R, van der Straeten D (2004) Thermal and chlorophyll-fluorescence imaging distinguish plant-pathogen interactions at an early stage. *Plant Cell Physiol* 45: 887–896.

Chague V, Mercier JC, Guenard AD, de Courcel A, Vedel F (1997) Identification of RAPD markers linked to a locus involved in quantitative resistance to TYLCV in tomato by bulked segregant analysis. *Theor Appl Genet* 95: 671–677.

Chaimovitsh D, Dudai N, Putievsky E, Ashri A (2006) Inheritance of resistance to Fusarium wilt in sweet basil. *Plant Dis* 90: 58–60.

Chaudhary S, Anderson TR, Park SJ, Yu K (2006) Comparison of screening methods for resistance to Fusarium root rot in common beans (*Phaseolus vulgaris* L.). *J Phytopathol* 154: 303–308.

Chavez DJ, Kabeka EA, Chaparro JX (2011) Screening of *Cucurbita moschata* Duchesne germplasm for crown rot resistance to Floridian isolates of *Phytophthora capsici*. *HortScience* 46: 536–540.

Chen M, Sun L, Wu H, Chen J, Ma Y, Zhang X, et al. (2014) Durable field resistance to *Wheat yellow mosaic virus* in transgenic wheat containing the antisense virus polymerase gene. *Plant Biotechnol* 12: 447–456.

Chen Y, Wang (2005) Two convenient methods to evaluate soybean for resistance to *Sclerotinia sclerotiorum*. *Plant Dis* 89: 1268–1272.

Chen YC, Wong CL, Muzzi F, Vlaardingerbroek I, Kidd BN, Schenk PM (2014) Root defense analysis against *Fusarium oxysporum* reveals new regulators to confer resistance. *Scientific Reports* 4: 5584

Chen YF, Chen W, Huang X et al. (2013) Fusarium wilt-resistant lines of Brazil banana (*Musa* sp., AAA) obtained by EMS-induced mutation in a micro-cross-section cultural system. *Plant Pathol* 62: 112–119.

Chen YN, Ren XP, Zhou XJ, Huang L, Huang JQ, Yan LY, Lei Y, Qi Y, Wei WH, Jiang HF (2014) Alteration of gene expression profile in the roots of wild diploid *Arachis duranensis* inoculated with *Ralstonia solanacearum*. *Plant Pathol* 63: 803–811.

Chenault KD, Melouk HA, Payton ME (2005) Field reaction to Sclerotinia blight among transgenic peanut lines containing antifungal genes. *Crop Sci* 45: 511–515.

Chung YS, Holmquist K, Spooner DM, Jansky SH (2011) A test of taxonomic and biogeographic predictivity: Resistance to soft rot and wild relatives of cultivated potato. *Physiopathology* 101: 205–212.

Cohen R, Orgil G, Burger Y et al. (2015) Differences in the responses of melon accessions to Fusarium root and stem rot and their colonization by *Fusarium oxysporum* f.sp. *radicis-cucumerinum*. *Plant Pathol* 64: 655–663.

Colas V, Conrod S, Venard P, Keller H, Ricci P, Panabieres F (2001) Elicitin genes expressed in vitro by certain tobacco isolates of *Phytophthora parasitica* are down-regulated during comparable interactions. *Molec Plant-Microbe Interact* 14: 326–335.

Cole SJ, Yoon AJ, Faull KF, Diener AC (2014) Host perception of jasmonates promotes infection by *Fusarium oxysporum* formae speciales that produce isoleucine-, and leucine-conjugated jasomonates. Molec Plant Pathol.

Colle M, Straley EN, Makela SB, Hammar SA, Grumet R (2014) Screening the cucumber plant introduction collection for young fruit resistance to *Phytophthora capsici*. *HortScience* 49: 244–249.

Cooper C, Crowther T, Smith BM, Isaac S, Collin HA (2006) Assessment of the response of carrot somaclones to *Pythium violae*, causal agent of cavity spot. *Plant Pathol* 55: 427–432.

Coram TE, Pang ECK (2005) Isolation and analysis of candidate Ascochyta blight defense genes in chickpea. Part II. Microarray expression analysis of putative defense-related ESTs. *Physiol Molec Plant Pathol* 66: 201–210.

Coram TE, Pang ECK (2006) Expression profiling of chickpea genes differentially regulated during a resistant response to *Ascochyta rabiei*. *Plant Biotechnol J* 4: 647–666.

Cruz APZ, Ferreira V, Pianzzola MJ, Inés Siri M, Coll NS, Vallas N (2014) A novel, sensitive method to evaluate potato germplasm for bacterial wilt resistance using a luminescent *Ralstonia solanacearum* reporter strain. *Molec Plant-Microbe Interact* 27: 277–285.

Cunha WG, Tinoco MLP, Pancoti HL, Ribeiro AE, Aragao FJL (2010) High transgenic resistance to *Sclerotinia sclerotiorum* in transgenic soybean plants transformed to express an oxalate decarboxylase gene. *Plant Pathol* 59: 654–660.

Curir P, Dolci M, Galeotti F (2005) A phytoalexin-like flavonol involved in the carnation (*Dianthus caryophyllus*)-*Fusarium oxysporum* f.sp. *dianthi* pathosystem. *J Phytopathol* 153: 65–67.

Curir P, Dolci M, Lanzotti V, Taglialatela-Scafati O (2001) The new kaempferide 3–0 [2G-ß-D-glucopyranosyl]-ß-rutinoside: a factor of resistance of carnation (*Dianthus caryophyllus*) to *Fusarium oxysporum* f.sp. *dianthi*. *Phytochem* 56: 717–721.

da Graca JV, Martin MM (1976) An electron microscope study of hypersensitive tobacco infected with *Tobacco mosaic virus* at 32°C. *Physiol Plant Pathol* 8: 215–219.

Dalmay T, Hamilton A, Rudd S, Angell S, Baulcombe DC (2000) An RNA-dependent RNA polymerase gene in *Arabidopsis* is required for posttranscriptional gene silencing mediated by a transgene, but not by a virus. *Cell* 101: 543–553.

Dalmay T, Horsefield R, Braunstein TH, Baulcombe DC (2001) *SDE 3* encodes an RNA helicase required for posttranscriptional gene silencing in *Arabidopsis*. *EMBO J* 20: 2069–2077.

Dan H, Alikhan ST, Robb J (2001) Use of quantitative PCR diagnostics to identify tolerance and resistance to *Verticillium dahliae* in potato. *Plant Dis* 85: 700–705.

Dangl JL, Jones JD (2001) Plant pathogens and integrated defense responses to infection. *Nature* 411: 826–833.

Dangl JL, McDowell JM (2006) Two modes of pathogen recognition by plants. *Proc Natl Acad Sci USA* 103: 8575–8576.

Datta K, Tu J, Oliva N et al. (2001) Enhanced resistance to sheath blight by constitutive expression of infection-related rice chitinase in transgenic elite *indica* rice cultivars. *Plant Sci* 160: 405–414.

Davar R, Darvishzadeh R, Majd A (2012) *Sclerotinia*-induced accumulation of protein in the basal stem of resistance and susceptible lines of sunflower. *Not Bor Horti Agrobo* 40: 119–124(online www. notulaebotanicae.ro)

Davis RJ (2000) Signal transduction by the JNK group of MAPKinases. *Cell* 103: 239–252.

De Ascensao ARDCF, Dubery IA (2000) Panama disease: cell wall reinforcement in banana roots in response to elicitors from *Fusarium oxysporum* f.sp. *cubense* race four. *Phytopathology* 90: 1173–1180.

De Jonge R, Van Esse HP, Maruthachalam K et al. (2012) Tomato immune receptor Ve 1 recognizes effector of multiple fungal pathogens uncovered by genome and RNA silencing. *Proc Natl Acad Sci USA* 109: 5110–5115.

Dees MW, Lyspe E, Alsheik M, Davik J, Brurberg MB (2016) Resistance to *Streptomyces turgdiscabies* in potato involves an early and sustained transcriptional reprogramming at initial stages of tuber formation. *Molec Plant Pathol* 17: 703–713.

Delledonne M (2005) No news is good news for plants. *Curr Opin Plant Biol* 8: 390–395.

Desaki Y, Miya A, Venkatesh B et al. (2006) Bacterial lipopolysaccharides induce defense responses-associated programmed cell death in rice cells. *Plant Cell Physiol* 47: 1530–1540.

Dey S, Badri J, Prakasam V et al. (2016) Identification and agro-morphological characterization of rice genotypes resistant to sheath blight. *Austr Plant Pathol* 45: 145–153.

Diener AC, Ausubel FM (2005) *RESISTANCE TO FUSARIUM OXYSPORUM 1*, a dominant *Arabidopsis* disease-resistance gene, is not race-specific. *Genetics* 171: 305–321.

Dinesh Kumar SP, Tham WH, Baker BJ (2000) Structure-function analysis of the *Tobacco mosaic virus* resistance gene *N*. *Proc Natl Acad Sci USA* 97: 14789–14794.

Dixon R (2001) Natural products and plant disease resistance. *Nature* 411: 843–847.

Dodds PN, Rathjen JP (2010) Plant immunity: Towards an integrated view of plant-pathogen interactions. *Nature Rev Genet* 11: 539–548.

Doke N, Furnichi N (1982) Response of protoplasts to hyphal wall components in relationship to resistance of potato to *Phytophthora infestans*. *Plant Pathol* 21: 23–30.

Donahoo RS, Turechek WW, Thies JA, Kousik CS (2013) Potential sources of resistance in U. S. *Cucumis melo* PIs to crown rot caused by *Phytophthora capsici*. *HortScience* 48: 164–170.

Donaldson PA, Anderson T, Lane BG, Davidson AL, Simmonds DH (2001) Soybean plants expressing an active oligometric oxalate oxidase from the wheat *gf-2.8* (germin) gene are resistant to the oxalate-secreting pathogen *Sclerotinia sclerotiorum*. *Physiol Molec Plant Pathol* 59: 297–307.

Dong J, He Z, Han C et al. (2002) Generation of transgenic wheat resistant to *Wheat yellow mosaic virus* and identification of gene silencing induced by virus infection. *Chin Sci Bull* 47: 1446–1450.

Dorrance AE, McClure SA, Martin SKSt (2003) Effect of partial resistance on Phytophthora stem rot incidence and yield of soybean in Ohio. *Plant Dis* 87: 308–312.

Dorrance AE, Schmitthenner AF (2000) New sources of resistance to *Phytophthora sojae* in the soybean plant introductions. *Plant Dis* 84: 1303–1308.

Du J, Verzaux E, Chaparro-Garcia A et al. (2015) Elicitin recognition confers enhanced resistance to *Phytophthora infestans* in potato. *Nat Plants* 1: 15034.

Dunn AR, Smart CD (2015) Interactions of *Phytophthora capsici* with resistant and susceptible pepper roots and stems. *Phytopathology* 105: 1355–1361.

During K, Porsch P, Flading M, Lörz H (1993) Transgenic potato plants resistant to the phytopathogenic bacterium *Erwinia carotovora*. *Plant J* 3: 587–598.

Durrant WE, Dong X (2004) Systemic acquired resistance. *Annu Rev Phytopathol* 42: 185–209.

Durrant WE, Wang S, Dong X (2007) *Arabidopsis* SN1 and RAD51D regulate both gene transcription and DNA recombination during the defense response. *Proc Natl Acad Sci USA* 104: 4223–4227.

Dutton MV, Evans CS (1996) Oxalate production by fungi: its role in pathogenicity and ecology in soil environment. *Canad J Microbiol* 42: 881–895.

Eikemo H, Brurberg MB, Davik J (2010) Resistance to *Phytophthora cactorum* in diploid *Fragaria* species. *HortScience* 45: 193–197.

Eikemo H, Stensvand A (2015) Resistance of strawberry genotypes to leather rot and crown rot caused by *Phytophthora cactorum*. *Eur J Plant Pathol* 143: 407–413.

El Mohtar CA, Atmian HS, Dagher RB, Abou-Jawadh Y (2007) Marker-assisted selection of tomato genotypes with *I–2* gene for resistance to *Fusarium oxysporum* f.sp. *lycopersici* race 2. *Plant Dis* 91: 758–762.

Elliott PE, Lewis RS, Shew HD, Gutierrez WA (2008) Evaluation of tobacco germplasm for seedling resistance to stem rot and target spot caused by *Thanetephorus cucumeris*. *Plant Dis* 92: 425–430.

El-Mohtar CA, Atamian HS, Dagher RB, Abou-Jawdah Y, Salus MS, Maxwell DP (2007) Marker-assisted selection of tomato genotypes with the *I-2* gene for resistance to *Fusarium oxysporum* f.sp. *lycopersici* race 2. *Plant Dis* 91: 758–762.

Engelbrecht J, Duong TA, van den Berg N (2013) Development of a nested quantitative real-time PCR for detecting *Phytophthora cinnamomi* in *Persea Americana* rootstocks. *Plant Dis* 97: 1012–1017.

Epinosa A, Alfanso J (2004) Disabling surveillance: bacterial type III secretion system effectors that suppress innate immunity. *Cell Microbiol* 6: 1027–1040.

Epstein L, Beede R, Kaur S, Ferguson L (2004) Rootstock effects on pistachio trees grown in *Verticilliium dahliae*-infested soil. *Phytopathology* 94: 388–395.

Fabritius AL, Cvitanich C, Judelson HS (2002) Stage-specific gene expression during sexual development in *Phytophthora infestans*. *Molec Microbiol* 45: 1057–1066.

Feys JF, Parker JE (2000) Interplay of signaling pathways in plant disease resistance. *Trend Genet* 16: 449–455.

Ficcadenti N, Sestili S, Annibali S et al. (2002) Resistance to *Fusarium oxysporum* f.sp. *melonis* race 1,2 in muskmelon lines Nad-1 and Nad-2. *Plant Dis* 86: 897–900.

Fire A, Xu S, Montgomery MK, Kostas SA, Driver SE, Mello CC (1998) Potent and specific genetic interference by double-stranded RNA in *Caenorhabditis elegans*. *Nature* 391: 806–811.

Fitzpatrick-Peabody ER, Lambert DH (2011) Methodology and assessment of the susceptibility of potato genotypes to *Phytophthora erythroseptica*, causal organism of pink rot. *Amer J Potato Res* 88: 105–113.

Flor HH (1955) Host-parasite interaction in flax rust–its genetics and other implications. *Phytopathology* 45: 680–685.

Foster JM, Hausbeck MK (2010) Resistance of pepper to Phytophthora crown rot, and fruit rot is affected by isolate virulence. *Plant Dis* 94: 24–30.

Foster SJ, Park T-H, Pel M et al. (2009) *Rpi-vnt 1.1*, a *Tm-2(2)* homolog from *Solanum venturii* confers resistance to potato late blight. *Molec Plant-Microbe Interact* 22: 589–600.

Fradin EF, Abad-El-Haliem A, Masini L, van den Berg GCM, Joosten MHAJ, Thomma BPHJ (2011) Interfamily transfer of tomato *ve1* mediates *Verticillium* resistance in *Arabidopsis*. *Plant Physiol* 156: 2255–2265.

Fradin EF, Zhang Z, Ayala JCJ et al. (2009) Genetic dissection of Verticillium wilt resistance mediated by tomato Ve 1. *Plant Physiol* 150: 320–332.

Fraser RSS (1990) The genetics of resistance to plant viruses. *Annu Rev Phytopathol* 28: 179–200.

Friedrich L, Lawson K, Dietrich R, Willits M, Cade R, Ryals J (2001) *NIM1* overexpression in *Arabidopsis* potentiates plant disease resistance and results in enhanced effectiveness of fungicides. *Molec Plant-Microbe Interact* 14: 1114–1124.

Gal-On A, Wolf D, Antignus Y et al. (2005) Transgenic cucumbers harboring the 54-kDa putative gene of cucumber fruit mottle mosaic tobamovirus are highly resistant to viral infection and protect nontransgenic scions from soil infection. *Transgenic Res* 14: 81–93.

Garcia-Ruiz GM, Trapero C, López-Escudero FJ (2014) Shortening the period for assessing the resistance of olive to *Verticillium dahliae* resistance in tissue culture-derived strawberry somaclones. *HortScience* 49: 1171–1175.

Gaspar YM, McKenna JA, McGinness BS et al. (2014) Field resistance to *Fusarium oxysporum* and *Verticillium dahliae* in transgenic cotton expressing plant defensin NaD1. *J Exptl Bot* 65: 1541–1550.

Gaudet DA, Lu Z-X, Leggett F, Puchalski B, Laroche A (2007) Compatible and incompatible interactions in wheat involving the *Bt-10* gene for resistance to *Tilletia tritici*, the common bunt pathogen. *Phytopathology* 97: 1397–1405.

Gaulin E, Drame N, Laffite C et al. (2006) Cellulose-binding domains of *Phytophthora* cell wall protein are novel pathogen-associated molecular patterns. *Plant Cell* 18: 1766–1777.

Gaulin E, Jauneau A, Villalba F, Rickauer M, Esquerré-Tugaye MT, Bottin A (2002) The CBEL glycoprotein of *Phytophthora parasitica* var. *nicotianae* is involved in cell wall deposition and adhesion to cellulosic substrates. *J Cell Sci* 115: 4565–4575.

Gazendam I, Oelfose D, Berger DK (2004) High-level expression of apple PGIP 1 is not sufficient to protect transgenic potato against *Verticilliium dahliae*. *Physiol Molec Plant Pathol* 65: 145–155.

Gerber IB, Zeidler D, Durner J, Dubeny IA (2004) Early perception responses of *Nicotiana tabacum* cells in response to lipopolysaccharides from *Burkholderia cepacia*. *Planta* 218: 647–657.

Ginnakopoulou A, Schornack S, Bozkurt TO et al. (2014) Variation in capsidol sensitivity between *Phytophthora infestans* and *Phytophthora capsici* is consistent with their host range. *PLoS One* 9: e107462.

Gkizi D, Santos-Rufo A, Rodriguez-Juardo D, Paplomatos EJ, Tjamos SE (2015) The ß-amylase genes: negative regulators of disease resistance for *Verticillium dahliae*. *Plant Pathol* 64: 1484–1490.

Glazebrook J (2005) Contrasting mechanisms of defense against biotrophic and necrotrophic pathogens. *Annu Rev Phytopathol* 43: 205–227.

Gómez-Gómez L, Boller T (2000) FLS2: an LRR receptor-like kinase involved in the perception of the bacterial elicitor flagellin in *Arabidopsis*. *Molec Cells* 5: 1003–1011.

Gonzalez-Candles Y, Catanzariti A-M, Baker B, Mcgrath DJ, Jones DA (2016) Identification of 1–7 expands the repertoire of genes for resistance to Fusarium wilt in tomato to three resistance gene classes. *Molec Plant Pathol* 17: 448–463.

Gordon SG, Berry SA, Martin SK St, Dorrance AE (2007a) Genetic analysis of soybean plant introductions with resistance to *Phytophthora sojae*. *Phytopathology* 97: 106–112.

Gordon SG, Kowitwanich K, Pipatpongpinyo W, Martin SK St, Dorrance AE (2007b) Molecular marker analysis of soybean plant introductions with resistance to *Phytophthora sojae*. *Phytopathology* 97: 113–118.

Gordon TR, Kirkpatick SC, Hansen J, Shaw DV (2006) Response of strawberry genotypes to inoculation with isolates of *Verticillium dahliae* differing in host origin. *Plant Pathol* 55: 766–769.

Gout L, Fudal I, Kuhn ML et al. (2006) Lost in the middle of nowhere: the *AvrLm1* avirulence gene of the Dothideomycete, *Leptosphaeria maculans*. *Molec Microbiol* 60: 67–80.

Graham T, Kim J, Graham M (1990) Role of constitutive isoflavone conjugated in the accumulation of glyceollin in soybean infected with *Phytophthora megasperma*. *Molec Plant-Microbe Interact* 3: 157–166.

Grant M, Lamb C (2006) Systemic immunity. *Curr Opin Plant Biol* 9: 414–420.

Grison R, Grezes-Besset B, Schneider M, Lucante N, Olsen L, Leguary JJ, Toppan A (1996) Field tolerance to fungal pathogens of *Brassica napus* constitutively expressing a chimeric chitinase gene. *Nat Biotechnol* 14: 643–646.

Guérin V, Lberton A, Cogliati EE, Hartley SE, Belzile F, Menzies JG, Belanger RR (2014) A zoospore inoculation method with *Phytophthora sojae* to assess the prophylactic role of silicon on soybean cultivars. *Plant Dis* 98: 1632–1638.

Guevara MG, Oliva CR, Huarte M, Daleo GR (2002) An aspartic protease with antimicrobial activity is induced after infection and wounding in intracellular fluids of potato tubers. *Eur J Plant Pathol* 108: 131–137.

Gurung S, Short DPG, Hu X, Sandoya GV, Hayes RJ, Subbarao KV (2015) Screening wild and cultivated *Capsicum* germplasm reveals new sources of Verticillium wilt resistance. *Plant Dis* 99: 1404–1409.

Gwata ET, Silim SN, Mgonja M (2006) Impact of a new source of resistance to Fusarium wilt in pigeon pea. *J Phytopathol* 154: 62–64.

Gygi SP, Rochon Y, Franza HR, Aebersold R (1999) Correlation between protein and mRNA abundance in yeast. *Molec Cell Biol* 19: 1720–1730.

Haegi A, De Felice S, Scotton M, Luongo L, Belisario A (2017) *Fusarium oxysporum* f.sp. *melonis*-melon interaction: Effect of grafting combination on pathogen gene expression. *Eur J Plant Pathol* 149: 787–796.

Hahlbrock K, Bednarek P, Ciolkowski I et al. (2003) Non-self recognition, transcriptional reprogramming, and secondary metabolite accumulation during plant/pathogen interactions. *Proc Natl Acad Sci USA* 100 (Suppl 2): 14569–14576.

Hai TTH, Esch E, Wang J-F (2008) Resistance to Taiwanese race 1 strains of *Ralstonia solanacearum* in wild tomato germplasm. *Eur J Plant Pathol* 122: 471–479.

Halila I, Cobos MJ, Rubio J, Millán T, Kharrat M, Marrakchi M, Gil J (2009) Tagging and mapping a second resistance gene for Fusarium wilt race 0 in chickpea. *Eur J Plant Pathol* 124: 87–92.

Hall C, Heath R, Guest DJ (2011) Rapid and intense accumulation of terpenoid phytoalexins in infected xylem tissues of cotton (*Gossypium hirsutum*) resistant to *Fusarium oxysporum* f.sp. *vasinfectum*. *Physiol Molec Plant Pathol* 75: 182–188.

Halterman DA, Kramer A, Wielgus S, Jiang J (2008) Performance of transgenic potato containing the late blight resistance gene *RB*. *Plant Dis* 92: 339–343.

Hamilton A, Baulcombe DC (1999) A species of small antisense RNA in posttranscriptional gene silencing in plants. *Science* 286: 940–952.

Hamilton A, Voinnet O, Chappel L, Baulcombe DC (2002) Two classes of short interfering RNA in RNA silencing. *EMBO J* 21: 4671–4679.

Hammond-Kosac KE, Jones JDG (1997) Plant disease resistance genes. *Annu Rev Plant Physiol Plant Molec Biol* 48: 575–607.

Hammond-Kosack K, Jones JDG (2000) Responses to plant pathogens. In: Buchnan B, Gruissen W, Jones R (eds.), *Biochemistry and Molecular Biology of Plants*, Amer Soc Plant Physiol, Rockville, MD, pp. 1102–1156.

Hancock JT, Desikan R, Clarke A, Hurst RD, Neil SJ (2002) Cell signaling following plant/pathogen interaction involves the generation of reactive oxygen and reactive nitrogen species. *Plant Physiol Biochem* 40: 611–617.

Hanus-Fajerska E, Wiszniewska A, Riseman A (2014) Elaboration of in vitro root culture protocols to efficiently limit daphne sudden death syndrome. *Acta Sci Pol Hortorum Cultus* 13: 117–127.

Harmer SL, Kay SA (2001) Microarrays: Determining the balance of cellular transcription. *Plant Cell* 12: 613–615.

Hasna MK, Ögren E, Persson P, Mártensson A, Ramert B (2009) Management of corky root disease of tomato in participation with organic tomato growers. *Crop Protect* 28: 155–161.

Hayes AS, Jeong SC, Gore MA, Yu YG, Buss GR, Tolin SA Saghai Marrof MA (2004) Recombination within a nucleotide binding-site/leucine-rich-repeat gene cluster produces new variants conditioning resistance to *Soybean mosaic virus* in soybean. *Genetics* 166: 493–503.

Hayes RJ, Maruthachalam K, Vallad GE, Klosterman SJ, Subbarao KV (2011) Selection for resistance to Verticillium wilt caused by race 2 isolates of *Verticillium dahliae* in accessions of lettuce (*Lactuca sativa* L.). *HortScience* 46: 201–206.

Hayes RJ, Wintermantel WM, Nicely PA, Ryder EJ (2006) Host resistance to *Mirafiori lettuce big-vein virus* and *Lettuce big-vein associated virus* and virus sequence diversity and frequency in California. *Plant Dis* 90: 233–239.

Hayes RJ, Wu BM, Pryor MB, Chitrampalam P, Subbarao KV (2010) Assessment of resistance in lettuce (*Lactuca sativa* L.) to mycelial and ascospore infection by *Sclerotinia minor* Jagger and *S. sclerotiorum* (Lib.) de Bary. *HortScience* 45: 333–341.

Heath M (2001) Nonhost resistance to plant pathogens: Nonspecific defense or the result of specific recognition events? *Physiol Molec Plant Pathol* 58: 53–54.

Helliwell EE, Wang Q, Yang Y (2013) Transgenic rice with inducible ethylene production exhibits broad-spectrum disease resistance to the fungal pathogens *Magnaporthe oryzae* and *Rhizoctonia solani*. *Plant Biotechnol J* 11: 33–42.

Henry RJ (1997) *Practical Applications of Plant Molecular Biology*, Chapman & Hall, London.

Higginbotham RW, Paulitz TC, Campbell KG, Kidwell KK (2004) Evaluation of adapted wheat cultivars for tolerance to Pythium root rot. *Plant Dis* 88: 1027–1032.

Hiltunen LH, Laakso I, Chobot V, Hakala KS, Weckman A, Valkonen JPT (2006) Influence of thaxtomins, in different combinations and concentrations on growth of micropropagated potato shoot cultures. *J Agric Food Chem* 54: 3372–3379.

Hirai G, Nakazumi H, Yagi R, Nakano M (2002) Fusarium wilt (race 1, 2y) resistant melon (*Cucumis melo*) rootstock cultivars 'Dodai NO. '1 and 'Dodai No.2 '. *Acta Hortic* 588: 155–160.

Hirani AH, Gao F, Liu J (2016) Transferring clubroot resistance from Chinese cabbage (*Brassica rapa*) to canola (*B. napus*). *Canad J Plant Pathol* 38: 89–90.

Houterman PM, Ma L, Van Ooijen G et al. (2009) The effector protein Avr2 of the xylem-colonizing fungus *Fusarium oxysporum* activates the tomato resistance protein I-2 intracellularly. *Plant J* 58: 970–9798.

Hu J, Telenko DEP, Phipps PM, Grabau EA (2016) Comparative susceptibility of peanut genetically engineered for Sclerotinia blight resistance to non-target peanut pathogens. *Eur J Plant Pathol* 145: 177–187.

Hu X, Bidney DL, Yalpani N, Duvick JP, Crasta O, Folkerts O, Lu G (2003) Overexpression of a gene encoding hydrogen peroxide-generating oxalate oxidase evokes defense responses in sunflower. *Plant Physiol* 133: 170–181.

Huang CC, Lindhout P (1997) Screening for resistance in *Lycopersicon* species to *Fusarium oxysporum* f.sp. *lycopersici* race 1 and race 2. *Euphytica* 93: 145–153.

Huang Y, Nordeen RO, Du M, Owens LD, Mc Beath JH (1997) Expression of an engineered cecropin gene cassette in transgenic tobacco plants confers disease resistance to *Pseudomonas syringae* pv. *tabaci*. *Phytopathology* 87: 494–499.

Huang YJ, Pirie EJ, Evans N, Delourme R, King GJ, Fitt BDL (2009) Quantitative resistance to symptomless growth of *Leptosphaeria maculans* (Phoma stem canker) in *Brassica napus* (oilseed rape). *Plant Pathol* 58: 314–323.

Huitema E, Vleeshouwers VGAA, Cakir C, Kamoun S, Govers F (2005) Differences in the intensity and specificity of hypersensitive response induction in *Nicotiana* spp. by INF2A and INF2B of *Phytophthora infestans*. *Molec Plant-Microbe Interact* 18: 183–193.

Hull R (2002) *Matthew's Plant Virology*, 4th edition, Academic Press, New York.

Hull R (2013) *Plant Virology*, 5th edition, Academic Press, Amsterdam.

Hura K, Hura T, Dziurka M (2014) Biochemical defense mechanisms induced in winter oilseed rape seedlings with different susceptibility to infection with *Leptosphaeria maculans*. *Physiol Molec Plant Pathol* 87: 42–50.

Hutcheson SW (1998) Current concepts of active defense in plants. *Annu Rev Phytopathol* 36: 59–90.

Hwang S-C, Ko W-H (2004) Cavendish banana cultivars resistant to Fusarium wilt acquired through somaclonal variation in Taiwan. *Plant Dis* 88: 580–588.

Hwang SF, Strelkov SE, Ahmed HU et al. (2018) Virulence and inoculum density-dependent interactions between clubroot resistant canola (*Brassica napus*) and *Plasmodiophora brassicaePlant Pathol* 66: 1318–1328.

Infantino A, Kharrat M, Riccioni L, Coyne CJ, McPhee KE, Grünwald NJ (2006) Screening techniques and sources of resistance to root diseases in cool season food legumes. *Euphytica* 147: 201–221.

Inohara N, Nunez G (2003) NODs: Intracellular proteins involved in inflammation and apoptosis. *Nature Rev Immunol* 3: 371–382.

Inuela M, Castro P, Rubio J, et al. (2007) Validation of a QTL for resistance to Ascochyta blight linked to resistance to Fusarium wilt race 5 in chickpea (*Cicer arietinum* L.). *Eur J Plant Pathol* 119: 29–37.

Islam M, Hashidoko Y, Ito T, Tahara S (2004) Interruption of the homing events of phytopathogenic *Aphanomyces cochlioides* zoospores by secondary metabolites from nonhost *Amaranthus gangeticus*. *J Pesti Sci* 19: 6–14.

Ito S, Nagata A, Kai T, Takahara H, Tanaka S (2005) Symptomless infection of tomato plants by tomatinase-producing *Fusarium oxysporum* formae speciales nonpathogenic on tomato plants. *Physiol Molec Plant Pathol* 66: 183–191.

Ito S, Takahara H, Kawaguchi T, Tanaka S, Kameya-Iwaki M (2002) Posttranscriptional silencing of the tomatinase gene in *Fusarium oxysporum* f.sp. *lycopersici. J Phytopathol* 150: 474–480.

Jang J-Y, Moon K-B, Ha J-H (2015) Development of an efficient in vitro screening method for selection of resistant lily cultivars against *Fusarium oxysporum* f.sp. *lilii. Korean J Hortic Sci Technol* 33: 883–890.

Jansky S, Rouse DI, Kauth PJ (2004) Inheritance of resistance to *Verticillium dahliae* in a diploid interspecific potato hybrid. *Plant Dis* 88: 1075–1078.

Jansky SH, Rouse DI (2003) Multiple disease resistance in interspecific hybrids of potato. *Plant Dis* 87: 266–272.

Jaynes JM, Nagpala P, Destefano-Beltran L et al. (1993) Expression of cecropin B lytic peptide analogue in transgenic tobacco confers enhanced resistance in bacterial wilt caused by *Pseudomona solanacearum*. *Plant Sci* 85: 43–54.

Jensen NF (1970) A diallele selection mating system in cereal breeding. *Crop Sci* 10: 629–635.

Jiang C-J, Sugano S, Kaga A, Lee SS, Sugimoto T, Takahashi M (2017) Evaluation of resistance to *Phytophthora sojae* in soybean minicore collections using an improved assay system. *Phytopathology* 107: 216–223.

Jiang F, Liu YX, Ai XZ, Zhen N, Wang HT (2010) Study on the relationship among microorganisms, enzymes' activity in the rhizosphere soil and root rot resistance of grafted *Capsicum*. *Scientia Agr Sinica* 43: 3367–3374.

Jiang J-F, Li J-G, Dong Y-H (2013) Effect of calcium nutrition on resistance of tomato against bacterial wilt induced by *Ralstonia solanacearum*. *Eur J Plant Pathl* 136: 547–555.

Johnson ES, Wolff MF, Wernsman EA, Atchley WR, Shew HD (2002a) Origin of the black shank resistance gene, *Ph* in tobacco cultivar Coker 371-Gold. *Plant Dis* 86: 1080–1084.

Johnson ES, Wolf MF, Wernsman EA, Ruffy RC (2002b) Marker-assisted selection for resistance to black shank disease in tobacco. *Plant Dis* 86: 1303–1309.

Johnson R (1981) Durable resistance: Definition of genetic control and attainment in plant breeding. *Phytopathology* 71: 567–568.

Johnson R (1983) Genetic background of durable resistance. In: Lamberti F, Waller JM, Van der Graaf NA (eds.), *Durable Resistance in Crops*, Plenum Press, New York, pp. 5–26.

Jones DA, Jones JDG (1997) The role of leucine-rich proteins in plant defenses. *Adv Bot Res* 24: 89–167.

Jones JDG, Dangl JL (2006) The plant immune system. *Nature* 444: 323–329.

Jones JDG, Witek K, Verweij W et al. (2014) Elevating crop disease resistance with cloned genes. *Philosoph Trans Royal Soc B* 369: 20130087

Jørgensen JH (1992) Discovery, characterization and exploitation of *Mlo* powdery mildew resistance in barley. *Euphytica* 63: 141–152.

Journot-Catalino N, Somssich IE, Roby D, Kroj T (2006) The transcription factors WRKY11 and WRKY17 act as negative regulators of basal resistance in *Arabidopsis thaliana*. *Plant Cell* 18: 3289–3302.

Jung W, Yu O, Law SM et al. (2000) Identification and expression of isoflavone synthetase, the key enzyme for biosynthesis of isoflavones in legumes. *Nature Biotechnol* 18: 208–212.

Kamle M, Kalim S, Bajpai A, Chanra R, Kumar R (2012) In vitro selection for wilt resistance in guava (*Psidium guajava*) cv. Allahabad Safeda. *Biotechnology* 11: 163–171.

Kamoun S (2001) Nonhost resistance to *Phytophthora*: novel prospects for a classical problem. *Curr Opin Plant Biol* 4: 295–300.

Kamoun S (2006) A catalogue of the effector secretome of plant pathogenic oomycetes. *Annu Rev Phytopathol* 43: 41–60.

Kamoun S, Lindqvist H, Govers F (1997) A novel class of elicitin like genes from *Phytophthora infestans*. *Molec Plant-Microbe Interact* 10: 1028–1030.

Kamoun S, van West P, Govers F (1998) Quantification of late blight resistance of potato using transgenic *Phytophthora infestans* expressing ß-glucuronidase. *Eur J Plant Pathol* 104: 521–525.

Kamoun S, van West P, Vleeshouwers VGAA, der Groot KE, Govers F (1998) Resistance of *Nicotiana benthamiana* to *Phytophthora infestans* is mediated by the recognition of the elicitor protein INF1. *Plant Cell* 10: 1413–1426.

Kang BC, Yeam I, Jahn MM (2005) Genetics of plant virus resistance. *Annu Rev Phytopathol* 43: 581–621.

Kanneganti TD, Huitema E, Cakir C, Kamoun S (2006) Synergistic interactions of the plant cell death pathways induced by *Phytophthora infetans* Nep 1-like protein PiNPP1.1 and INF1 elicitin. *Molec Plant-Microbe Interact* 19: 854–863.

Kantoglu Y, Secer E, Erzurum K et al. (2010) Improving tolerance to *Fusarium oxysporum* f.sp. *melonis* in melon using tissue culture and mutation techniques. In: IAEA (ed.), *Mass Screening Techniques for Selecting Crops Resistant to Diseases. Joint FAO/IAEA Programme of Nuclear Techniques in Food and Agriculture*, International Atomic Energy Agency, Vienna, pp. 145–154.

Kanyuka K, Lovell DJ, Mitrofanova OP, Hammond-Kosack K, Adams MJ (2004) A controlled environment test for resistance to *Soilborne cereal mosaic virus* (SBCMV) and its use to determine the mode of inheritance of resistance in wheat cv. Cadenza and for screening *Triticum monococcum* genotypes for SBCMV resistance. *Plant Pathol* 53: 154–160.

Kariola T, Palmäki TA, Brader G, Plava T (2003) *Erwinia carotovora* subsp. *carotovora* and *Erwinia*-derived elicitors Hrp and Peh trigger distinct but interacting defense responses and cell death in *Arabidopsis*. *Molec Plant-Microbe Interact* 16: 179–187.

Kawazu Y, Fujiyama R, Imanishi S, Fukuoka H, Yamaguchi H, Matsumoto S (2016) Development of marker-free transgenic lettuce resistant to *Mirafiori lettuce big-vein virus*. Transgenic Res.

Kawazu Y, Fujiyama R, Sugiyama K, Sasaya T (2006) A transgenic lettuce line with resistance to both *Lettuce big-vein-associated virus* and *Mirafiori lettuce virus*. *J Amer Soc Hortic Sci* 131: 760–763.

Kazan K, Schenk PM, Wilson I, Menners JM (2001) DNA microarray: New tools in the analysis of plant defense responses. *Mol Plant Pathol* 2: 177–185.

Kervinen T, Peltonen S, Teeri TH, Karajailinen R (1998) Differential expression of phenylalanine ammonia-lyase genes in barley induced by fungal infection or elicitors. *New Phytol* 139: 293–300.

Khaskheli MI, Sun JL, He SP, Du XM (2013) Screening of cotton germplasm for resistance to *Verticillium dahliae* Kleb. under greenhouse and field conditions. *Eur J Plant Pathol* 137: 259–272.

Khatib M, Lafitte C, Esquerré-Tugaye MT, Bottin A, Rickauer M (2004) The CBEL elicitor of *Phytophthora parasitica* var. *nicotianae* activities defense in *Arabidopsis thaliana* via three different signaling pathways. *New Phytol* 162: 501–510.

Kieu NP, Aznar A, Segond D et al. (2012) Iron deficiency affects plant defence responses and confers resistance to *Dickeya dadantii* and *Botrytis cinerea*. *Molec Plant Pathol* 13: 816–827.

Kim BR, Nam HY, Kim SU, Pai T, Chang YJ (2004) Reverse transcription quantitative PCR of three genes with homology encoding 3-hydroxy 3-methylglutaryl-CoA reductases in rice. *Biotechnol Lett* 26: 985–988.

Kim SG, Hur O-S, Ro N-Y et al. (2016) Evaluation of resistance to *Ralstonia solanacearum* in tomato genetic resources at seedling stage. *Plant Pathol J* 32: 58–64.

Király L, Barna B, Király Z (2007) Plant resistance to pathogen infection: Forms and mechanisms of innate and acquired resistance. *J Phytopathol* 155: 385–396.

Knoester M, van Loon LC, van den Heuvel J, Hennig J, Bol JF, Linthorst HJM (1998) Ethylene-insensitive tobacco lacks nonhost resistance against soilborne fungi. *Proc Natl Acad Sci USA* 95: 1933–1937.

Knogge W, Scheel D (2006) LysM receptors recognize friend and foe. *Proc Natl Acad Sci USA* 103: 10829–10830.

Koh Y-J, Park J-I, Ahmed NU et al. (2013) Enhancement of resistance to soft rot (*Pectobacterium carotovorum* subsp. *carotovorum*) in transgenic *Brassica rapa*. *Eur J Plant Pathol* 136: 317–322.

Kousik CS, Thies JA (2010) Response of US bottle gourd (*Lagenaria siceraria*) plant introduction (PI) to crown rot caused by *Phytophthora capsici*. *Phytopathology* 100: S65.

Kubota C, McClure MA, Kokalis-Burelle N, Bausher MG, Rosskopf EN (2008) Vegetable grafting: History, use, and current technology status in North America. *HortScience* 43: 1664–1669.

Kull LS, Vuong TD, Powers KS, Eskridge KM, Steadman JR, Hartman GL (2003) Evaluation of resistance screening methods for Sclerotinia stem rot of soybean and dry bean. *Plant Dis* 87: 1471–1476.

Kunze G, Zipfel C, Robatzek S, Nienhaus K, Boller T, Felix G (2004) The N terminus of bacterial elongation factor Tu elicits innate immunity in *Arabidopsis* plants. *Plant Cell* 16: 3496–3507.

Lafortune D, Béramis M, Daubéze A-M, Boissot N, Palloix A (2005) Partial resistance of pepper to bacterial wilt is oligogenic and stable under tropical conditions. *Plant Dis* 89: 501–506.

Landa BB, Navas-Cortés JA, Jiménez-Gasco MDM, Katan J, Retig B, Jiménez-Díaz RM (2006) Temperature response of chickpea cultivars to races of *Fusarium oxysporum* f.sp. *ciceris*, causal agent of Fusarium wilt. *Plant Dis* 90: 365–374.

Larkin PJ, Scowcraft WR (1981) Somaclonal variation: A novel source of variability from cell cultures for plant improvement. *Theoret Appl Genet* 60: 197–214.

Larsen RC, Vandemark GJ, Hughes TJ, Grau CR (2007) Development of a real-time polymerase chain reaction assay for quantifying *Verticillium albo-atrum* DNA in resistant and susceptible alfalfa. *Phytopathology* 97: 1519–1525.

Lebeau A, Daunay MC, Frary A et al. (2011) Bacterial wilt resistance in tomato, eggplant, and pepper: Genetic resources respond to diverse strains in the *Ralstonia solanacearum* species complex. *Phytopathology* 101: 154–165.

Lebreton A, Labbe C, De Ronne M, Xue AG, Marchand G, Belanger RR (2018) Development of a simple hydroponic assay to study vertical and horizontal resistance of soybean and pathotypes of *Phytophthora sojae*. *Plant Dis* 102: 114–123.

Lee YK, Hippe-Sanwald S, Jung HW, Hong JK, Hause B, Hwang BK (2000) In situ localization of chitinase mRNA and protein in compatible and incompatible interactions of pepper stems with *Phytophthora capsici*. *Physiol Molec Plant Pathol* 57: 111–121.

Leite ME, dos Santos JB, Ribeir PM Jr et al. (2014) Biochemical responses associated with common bean defence against *Sclerotinia sclerotiorum*. *Eur J Plant Pathol* 138: 391–404.

Lennefors B, Savenkov EU, Bensefelt J, Wremerth-Weich E, van Roggen P, Tuvesson S, Valkonen JPT, Gielen J (2006) dsRNA-mediated resistance to *Beet necrotic yellow vein virus* infections in sugar beet (*Beta vulgaris* L. ssp. *vulgaris*). *Molec Breeding* 18: 315–325.

Li E, Wang G, Yang Y, Xiao J, Mao Z, Xie B (2015) Microscopic analysis of the compatible and incompatible interactions between *Fusarium oxysporum* f.sp. *inconglutinans* and cabbage. *Eur J Plant Pathol* 141: 597–609.

Li WM, Dita M, Wu W, Hu GB, Xie JH, Ge XJ (2015) Resistance sources to *Fusarium oxysporum* f.sp. *cubense* tropical race 4 in banana wild relatives. *Plant Pathol* 64: 1061–1067.

Li Y-B, Han L-B, Wang H-Y et al. (2016) The thioredoxin GbNRX1 plays a critical role in homeostasis of apoplastic reactive oxygen species in response to *Verticillium dahliae* infection in cotton. *Plant Physiol* 170: 2392–2406.

Liang Y, Yajima W, Davis MR, Kav NVV, Strelkov SE (2013) Disruption of a gene encoding a hypothetical secreted protein from *Sclerotinia sclerotiorum* reduces its virulence on canola (*Brassica napus*). *Canad J Plant Pathol* 35: 46–55.

Lichtenzveig J, Shtienberg D, Zhang HB, Bonfil DJ, Abbo S (2002) Biometric analyses of the inheritance of resistance to *Didymella rabiei* in chickpea. *Phytopathology* 92: 417–423.

Liebrand TWH, Kombrink A, Zhang Z et al. (2014) Chaperones of the endoplasmic reticulum are required for Ve 1-mediated resistance to *Verticillium*. *Molec Plant Pathol* 15: 109–117.

Lipka V, Dittgen J, Bednarek P et al. (2005) Pre- and –post invasion defenses both contribute to nonhost resistance in *Arabidopsis*. *Science* 310: 1180–1183.

Liu F, Wang M, Wen J, Yi B, Shen J, Ma C, Tu J, Fu T (2015) Overexpression of barley oxalate oxidase gene induces partial leaf resistance to *Sclerotinia sclerotiorum*. *Plant Pathol* 64: 1407–1416.

Liu G, Jia Y, Correa-Victoria FJ et al. (2009) Mapping quantitative trait loci responsible for resistance to sheath blight in rice. *Phytopathology* 99: 1078–1084.

Liu G, Jia Y, Mc Clung A, Oard JH, Lee FN, Correll JC (2013) Confirming QTLs and finding additional loci responsible for resistance to rice sheath blight disease. *Plant Dis* 97: 113–117.

Livingstone DM, Hampton JL, Phipps PM, Grabu EA (2005) Enhancing resistance to *Sclerotinia minor* in peanut expressing a barley oxalate oxidase gene. *Plant Physiol* 137: 1354–1362.

Loannidou D, Pinel A, Brucidou C, Albar L, Ahmadi N (2003) Characterization of the effects of a major QTL for the partial resistance to *Rice yellow mottle virus* using a near-isogenic line approach. *Physiol Molec Plant Pathol* 63: 213–221.

Lorito M, Woo SL, Garcia I et al. (1998) Genes from mycoparasitic fungi as a source for improving plant resistance to fungal pathogens. *Proc Natl Acad Sci USA* 95: 7860–7865.

Lozovaya VV, Lygin AV, Li S, Hartman GL, Widholm JM (2004) Biochemical response of soybean roots to *Fusarium solani* f.sp. *glycines* infection. *Crop Sci* 44: 819–826.

Lozoya-Saldana H, Guzmán-Galindo L, Fernandez-Pavia S, Grünwald NJ, McElhinny E (2006) *Phytophthora infestans* (Mont.) de Bary. 1. Host-pathogen specificity and resistance components. *Agrociencia* 40: 205–217.

Lu Y, Liao D, Qi Y, Xie Y (2013) Proteome analysis of resistant and susceptible Cavendish banana roots following inoculation with *Fusarium oxysporum* f.sp. *cubense*. *Physiol Molec Plant Pathol* 84: 163–171.

Luo Y, Yin Y, Liu Y et al. (2013) Identification of differentially expressed genes in *Brassica juncea* var. *tumida* Tsen following infection by *Plasmodiophora brassicae*. *Eur J Plant Pathol* 137: 43–53.

Luterbacher MC, Asher MJC, Beyer W, Mandolino G, Scholten OE, Frese L, Biancardi E, Stevanato P, Mechelke W, Sylvchenko O (2005) Sources of resistance to diseases of sugar beet in related *Beta* germplasm. II. Soilborne diseases. *Euphytica* 141: 49–63.

Lygin AV, Hill CB et al. (2010) Response of soybean pathogens to glyceollin. *Phytopathology* 100: 897–903.

Lygin AV, Zernova OV, Hill CB et al. (2013) Glyocellin is an important component of soybean plant defense against *Phytophthora sojae* and *Macrophomian phaseolina*. *Phytopathology* 103: 984–994.

Maekawa T, Kufer TA, Schulze-Lefert P (2011) NLR functions in plant and animal immune system: So far and yet so close. *Nature Immunol* 12: 817–826.

Maghuly F, Leopold S, Machado ADC et al. (2006) Molecular characterization of grapevine plants transformed with GFLV resistance genes. *Plant Cell Rep* 25: 546–553.

Mahoney AK, Babiker EM, Paulitz TC, See D, Okubara PA, Hulbert SH (2016) Characterizing and mapping resistance in synthetic-derived wheat to Rhizoctonia root rot in a green bridge environment. *Phytopathology* 106: 1170–1176.

Markakis EA, Tjamos SE, Antoniou PP, Paplomatos EJ, Tjamos EC (2010) Phenolic responses of resistant and susceptible cultivars induced by defoliating and nondefoliating *Verticillium dahliae* pathotypes. *Plant Dis* 94: 1156–1162.

Martin GB (1999) Functional analysis of plant disease resistance genes and their downstream effectors. *Curr Opin Plant Biol* 2: 273–279.

Martin GB, Bogdnanove AJ, Sessa G (2003) Understanding the functions of plant disease resistance proteins. *Annu Rev Plant Biol* 54: 23–61.

Martin JA, Solla A, Domingues MF, Coimbra MA, Gil L (2008) Exogenous phenol increase in resistance of *Ulmus minor* to Dutch elm disease through formation of suberin-like compounds on xylem tissues. *Environ Exp Bot* 64: 97–104.

Matsukawa M, Shibata Y, Ohtsu M et al. (2013) *Nicotiana benthamiana calreticulin 3a* is required for the ethylene-mediated production of phytoalexins and disease resistance against oomycete pathogen *Phytophthora infestans*. *Molec Plant-Microbe Interact* 26: 880–892.

Mattei B, Bernalda MS, Federici L, Roepstorff P Cervone F, Boffi A (2001) Secondary structure and posttranscriptional modification of the leucine-rice repeat protein PGIP (polygalacturonase-inhibiting protein) from *Phaseolus vulgaris*. *Biochemistry* 39: 129–136.

Matthews REF (1991) *Plant Virology*, 3rd edition, Academic Press, London.

Matthiesen RL, Abeysekara NS, Ruiz-Rojas JJ, Biyashev RM, Saghai Maroof MA, Robertson AE (2016) A method of combining isolates of *Phytophthora sojae* to screen for novel sources of resistance to Phytophthora stem and root rot in soybean. *Plant Dis* 100: 1424–1528.

Mattupalli C, Genger RK, Charkowski AO (2013) Evaluating incidence of *Helminthosporium solani* and *Colletotrichum coccodes* on asymptomatic organic potatoes and screening potato lines for resistance to silver scurf. *Amer J Potato Res* 90: 369–377.

Mauch F, Mauch-Mani B, Boller T (1988) Antifungal hydrolysis in pea tissue. II. Inhibition of fungal growth by combinations of chitinase and ß-1,3-glucanase. *Plant Physiol* 88: 936–942.

Maule AJ, Caranta C, Boulton MI (2007) Sources of natural resistance to plant viruses: Status and prospects. *Molec Plant Pathol* 8: 223–231.

McCorkle K, Lewis R, Shew D (2013) Resistance to *Phytophthora nicotianae* in tobacco breeding lines derived from variety Beinhart 1000. *Plant Dis* 97: 252–258.

McGrann GRD, Townsend BJ, Antoniw JF, Asher MJC, Mutasa-Götegens ES (2009) Barley elicits a similar early basal defence response during host and nonhost interactions with *Polymyxa* root parasites. *Eur J Plant Pathol* 123: 5–15.

McGregor C, Waters V, Nambeesan S, MacLean D, Candole BL, Conner P (2011) Genotypic and phenotypic variation among pepper accessions resistant to *Phytophthora capsici*. *HortScience* 46: 1235–1240.

McKinney HH (1925) *A Mosaic Disease of Winter Wheat and Winter Rye*. Bull number 1361, USDA, Washington, DC.

Mei J, Wei D, Disi JO, Ding Y, Liu Y, Qian W (2012) Screening resistance against *Sclerotinia sclerotiorum* in *Brassica* crops with use of detached stem assay under controlled environment. *Eur J Plant Pathol* 134: 599–604.

Mekhedov S, de Ilardya MO, Ohlrogge J (2000) Toward a functional catalog of the plant genome: A survey of genes for lipid biosynthesis. *Plant Physiol* 122: 389–401.

Mellersh DG, Foulds IV, Higgins VJ, Heath MC (2002) H2O2 plays different roles in determining penetration failure in three diverse plant-fungal interactions. *Plant J* 29: 257–268.

Melotto M, Underwood W, Koczan J, Nomura K, He SY (2006) Plant stomata function in innate immunity against bacterial invasion. *Cell* 126: 969–980.

Merz U, Lees AK, Sullivan L et al. (2012) Powdery scab resistance in *Solanum tuberosum*: an assessment of cultivars x environment effect. *Plant Pathol* 61: 29–36.

Métraux JP (2001) Systemic acquired resistance and salicylic acid: Current state of knowledge. *Eur J Plant Pathol* 107: 13–18.

Mette MF, Aufsatz W, van der Winden J, Matzke MA, Matzke AJM (2000) Transcriptional silencing and promoter methylation triggered by double-stranded RNA. *EMBO J* 19: 5194–5201.

Mideros S, Nita M, Dorrance AE (2007) Characterization of components of partial resistance, *Rps2*, and root resistance to *Phytophthora sojae* in soybean. *Phytopathology* 97: 655–662.

Mideros SX, Windham GL, William WP, Nelson RJ (2012) Tissue-specific components of resistance to Aspergillus ear rot of maize. *Phytopathology* 102: 789–793.

Miklas PN, Porter LD, Kelly JD, Myers JR (2013) Characterization of white mold disease avoidance in common bean. *Eur J Plant Pathol* 135: 525–543.

Monroy-Barbosa A, Bosland PW (2008) Genetic analysis of Phytophthora root rot race-specific resistance in Chile pepper*J Amer Soc Hortic Sci* 133: 825–829.

Morid B, Mansoor S, Kakvan N (2012) Screening for resistance genes to Fusarium root rot and Fusarium wilt diseases in tomato (*Lycopersicon esculentum*) cultivars using RAPD and CAPS markers*Eur J Exptl Biol* 2: 931–939.

Moussart A, Even MN, Tivoli B (2008) Reaction of genotypes from several species of grain and forage legumes to infection with a French pea isolate of the oomycete *Aphanomyces euteiches*. *Eur J Plant Pathol* 122: 321–333.

Moussart A, Wicker E, Duparque M, Rouxel F (2001) Development of an efficient screening test for pea resistance to *Aphanomyces euteiches*. *Proc 4th European Grain Legume Conf*, Cracow, Poland, pp. 272–273.

Mueller DS, Hartman GL, Nelson RL, Pedersen WL (2002a) Evaluation of *Glycine max* germplasm for resistance to *Fusarium solani* f.sp. *glycines*. *Plant Dis* 86: 741–746.

Mueller DS, Li S, Hartman GL, Pedersen WL (2002b) Use of aeroponic chambers and grafting to study partial resistance to *Fusarium solani* f.sp. *glycines* in soybean. *Plant Dis* 86: 1223–1126.

Mueller DS, Nelson RL, Harman GL, Pedersen WL (2003) Response of commercially developed soybean cultivars and the ancestral soybean lines to *Fusarium solani* f.sp. *glycines*. *Plant Dis* 87: 827–831.

Munkvold GP, Carlton WM, Brummer EC, Meyer JR, Undersander DJ, Grau CR (2001) Virulence of *Aphanomyces euteiches* isolates from Iowa and Wisconsin and benefits of resistance to *A. euteiches* in alfalfa cultivars. *Plant Dis* 85: 328–333.

Muthoni J, Shimelis H, Melis R, Kinyua ZM (2014) Response of potato genotypes to bacterial wilt caused by *Ralstonia solanacearum* (Smith) (Yabuuchi et al.) in the tropical highlands. *Amer J Potato Res* 91: 215–232.

Mysore KS, Ryu C (2004) Nonhost resistance: How much do we know? *Trends Plant Sci* 9: 97–104.

Nagendran S, Hammerschmidt R, McGrath JM (2009) Identification of sugar beet germplasm EL51 as a source of resistance to postemergence Rhizoctonia damping-off. *Eur J Plant Pathol* 123: 461–471.

Nakaho K, Hibino H, Miyagawa H (2000) Possible mechanisms limiting movement of *Ralstonia solanacearum* in resistant tomato tissues. *J Phytopathol* 148: 181–190.

Nakazawa-Ueji YE, Núñez-Pastrana R, Souza-Perera RA, Santana-Buzzy N, Zúñig-Aguilar JJ (2010) Mycelium homogenates from a virulent strain of *Phytophthora capsici* promote a defence-related response in cell suspensions from *Capsicum chinense*. *Eur J Plant Pathol* 126: 403–415.

Namba S, Kishiwazaki S, Lu X, Tamura M, Tsuchizaki T (1998) Complete nucleotide sequence of wheat yellow mosaic bymovirus genominc RNAs. *Arch Virol* 143: 631–643.

Narancio R, Zorrilla P, Robello C, Gonzalez M, Vilaro F, Pritsch C, Rizza MD (2013) Insights on gene expression response of a characterized resistant genotype of *Solanum commersonii*. *Eur J Plant Pathol* 130: 823–835.

Narayanasamy P (2002) *Microbial Plant Pathogens and Crop Disease Management*, Science Publishers, Enfield, NH.

Narayanasamy P (2006) *Postharvest Pathogens and Disease Management*, John-Wiley & Sons, Hoboken, NJ.

Narayanasamy P (2008) *Molecular Biology in Plant Pathogenesis and Disease Management*, Volume 3–Disease Management, Springer Science + Business Media B. V., Heidelberg, Germany.

Narayanasamy P (2017) *Microbial Plant Pathogens–Detection and Management in Seed and Propagules*, Volume 2, John Wiley-Blackwell, London.

Navabi ZK, Strelkov SE, Good AG, Thiagarajah MR, Rahman MH (2010) *Brassica* B genome resistance to stem rot (*Sclerotinia sclerotiorum*) in a doubled haploid population of *Brassica napus* x *B. carinata*. *Canad J Plant Pathol* 32: 237–246.

Navarro FM, Rak KT, Banks E, Bowen BD, Higgins C, Palta JP (2015) Strategies for selecting stable common scab resistant clones in a potato breeding program. *Amer J Potato Res* 92: 326–338.

Navarro L, Zipfel C, Rowland O et al. (2004) The transcriptional innate immune response to *flg22* interplay and overlap with *Avr* gene-dependent defense responses and bacterial pathogenesis. *Plant Pathol* 135: 1113–1128.

Negahi A, Ben C, Gentzbittel L et al. (2014) Quantitative trait loci associated with resistance to a potato isolate of *Verticillium albo-atrum* in *Medicago truncatula*. *Plant Pathol* 63: 308–315.

Negahi A, Sarrafi A, Ebrahimi A et al. (2013) Genetic variability of tolerance to *Verticillium albo-atrum* and *Verticillium dahliae* in *Medicago truncatula*. *Eur J Plant Pathol* 136: 135–143.

Nelson RS, Powell-Abel P, Beachey RN (1997) Lesions and virus accumulation in inoculated transgenic tobacco plants expressing coat protein gene of *Tobacco mosaic virus*. *Virology* 158: 126–138.

Nene YL, Haware MP (1980) Screening chickpea for resistance to wilt. *Plant Dis* 64: 379–380.

Newcombe G, Robb I (1988) The function and relative importance of the vascular coating response in highly resistant, moderately resistant and susceptible alfalfa infected by *Verticillium albo-atrum*. *Physiol Molec Plant Pathol* 33: 47–58.

Niemann J, Kaczmarek J, Książczżyk T, Wojciechowski A, Jedryczka M (2017) Cabbage (*Brassica rapa* ssp. *pekinensis*)–a valuable source of resistance to clubroot (*Plasmodiophora brassicae*). *Eur J Plant Pathol* 147: 181–198.

Niki T, Mitsuhara I, Seo S, Ohtsubo N, Ohashi Y (1998) Antagonistic effect of salicylic acid and jasmonic acid on the expression of pathogenesis-related (PR) protein genes in wounded mature tobacco leaves. *Plant Cell Physiol* 39: 115–123.

Nishimura M, Somerville S (2002) Resisting attack. *Science* 295: 2032–2033.

Nitzan N, Quick RA, Hutson WD, Bamberg J, Brown CR (2010) Partial resistance to potato black dot caused by *Colletotrichum coccodes* in *Solanum tuberosum* Group Andigena. *Amer J Potato Res* 87: 502–508.

Ntui VO, Azadi P, Thirukkumaran G et al. (2011) Increased resistance to Fusarium wilt in transgenic tobacco lines co-expressing chitinase and wasabi defensin genes. *Plant Pathol* 60: 221–231.

Nürnberger T, Brunner F, Kemmerling B, Piater L (2004) Innate immunity in plants and animals: striking similarities and obvious differences. *Immunol Rev* 98: 249–266.

Nürnberger T, Nennstiel D, Jabs T, Sacks WR, Hahlbrock K, Scheel D (1994) High affinity binding of a fungal oligopeptide elicitor to parsley plasma membrane triggers multiple defense responses. *Cell* 78: 449–460.

Nyankanga RO, Olanya OM, Wien HC, El-Bedewy R, Karinga J, Ojiambo PS (2008) Development of tuber blight (*Phytophthora infestans*) on potato cultivars based on in vitro assays and field evaluations. *HortScience* 43: 1501–1508.

Ohsato S, Miyanishi M, Shirako Y (2003) The optimal temperature for RNA replication in cells infected by *Soilborne wheat mosaic virus* is 17 degrees C. *J Gen Virol* 84: 995–1000.

Okole BN, Schulz FA (1996) Micro-cross-sections of banana and plantains (*Musa* spp.) morphogenesis and regeneration of callus and short buds. *Plant Sci* 116: 185–195.

Okubara PA, Leston N, Micknass U, Kogel K-H, Imani J (2016) Rapid quantitative assessment of Rhizoctonia resistance in roots of selected wheat and barley genotypes. *Plant Dis* 100: 640–644.

Omar MS, Novak FJ, Brunner H (1989) In vitro action of ethylmethane sulphonate on banana shoot tips. *Scientia Hortic* 40: 283–295.

Onozaki T, Tanikawa N, Taneya M (2004) A RAPD-derived STS marker is linked to bacterial wilt (*Burkholderia caryophylli*) resistance gene in carnation. *Euphytica* 138: 255–262.

Oros G, Naár Z, Magyar D (2013) Susceptibility of wheat varieties to soilborne Rhizoctonia infection. *Amer J Plant Sci* 4: 2240–2258.

Osburn A, Clarke B, Lunness P, Scott P, Daniels MJ (1994) An oat species lacking avenacin is susceptible to infection by *Gaeumannomyces graminis* var. *tritici*. *Physiol Molec Plant Pathol* 45: 457–467.

Osman H, Mikes V, Milat ML et al. (2001) Fatty acids bind to fungal elicitor cryptogein and compete with sterols. *FEBS Lett* 489: 55–58.

Osorio-Gaurin JA, Enciso-Rodriguez FE, Gonazalez C, Fernández-Pozo N, Mueller LA, Barrero LS (2016) Association analysis for disease resistance to *Fusarium oxysporum* in cape gooseberry (*Physalis peruviana* L.). *BMC Genomics* 17: 248

Otto-Hanson L, Steadman J, Higgins R, Eskridge KM (2011) Variation in *Sclerotinia sclerotiorum* bean isolates from multisite resistance screening locations. *Plant Dis* 95: 1370–1377.

Ouyang B, Chen YH, Li HX, Qian CJ, Huang SL, Ye ZB (2005) Transformation of tomatoes with osmotin and chitinase genes and their resistance to Fusarium wilt. *J Hortic Sci Biotechnol* 80: 517–522.

Ouyang B, Li HX, Ye ZB (2003) Increased resistance to Fusarium wilt expressing bivalent hydrolytic enzymes. *J Plant Physiol Molec Biol* 29: 179–184.

Pantelides IS, Tjamos SE, Paplomatus EJ (2010) Insights into the role of ethylene perception in tomato resistance to vascular infection by *Verticillium dahliae*. *Plant Pathol* 59: 130–138.

Papadopoulou KK, Kavroulakis N, Tourn M, Aggelou I (2005) Use of ß-glucuronidase activity to quantify the growth of *Fusarium oxysporum* f.sp. *radicis-lycopersici* during infection of tomato. *J Phytopathol* 153: 325–332.

Paplomatas EJ, Elena K, Tsagkarkou A, Pedikaris A (2002) Control of Verticillium wilt of tomato and cucurbits through grafting of commercial varieties on resistant rootstocks. *Acta Hortic* 579: 445–449.

Pareja-Jaime Y, Roncero MI, Ruiz-Roldan MC (2008) Tomatinase from *Fusarium oxysporum* f.sp. *lycopersici* is required for full virulence on tomato plants. *Molec Plant-Microbe Interact* 21: 728–736.

Park T-H, Gros J, Sikkema A et al. (2004) The late blight resistance locus *Rpi-blb3* from *Solanum bulbocatanum* belongs to a major late blight *R* gene cluster on chromosome 4 of potato. *Molec Plant-Microbe Interact* 18: 722–729.

Pasche JS, Mallik I, Anderson NR, Gudmestad NC (2013a) Development and validation of a real-time PCR assay for quantification of *Verticillium dahliae* in potato. *Plant Dis* 97: 608–618.

Pasche JS, Thompson AL, Gudmestad NC (2013b) Quantification of field resistance to *Verticillium dahliae* in eight russet-skinned potato cultivars using real-time PCR. *Amer J Potato Res* 90: 158–170.

Pasco C, Jouan B, Andrivon D (2005) Resistance of potato genotypes to common and netted scab-causing species of *Streptomyces*. *Plant Pathol* 54: 383–392.

Pascual A, Campa A, Pérez-Vega E, Giraldez R, Miklas PN, Ferreira JJ (2010) Screening common bean for resistance to four *Sclerotinia sclerotiorum* isolates collected in Northern Spain. *Plant Dis* 94: 885–889.

Patridge-Telenko DE, Hu J, Livingstone-DM, Shew BB, Phipps PM, Grabau EA (2011) Sclerotinia blight resistance in Virginia-type peanut transformed with a barley oxalate oxidase gene. *Phytopathology* 101: 786–793.

Pavli OI, Panopoulos NJ, Goldbach R, Skaracis GN (2010) BNYVV-derived dsRNA confers resistance to rhizomania disease of sugar beet as evidenced by a novel transgenic hairy root approach. *Transgenic Res* 19: 915–922.

Pavlou GC, Vakalounakis DJ, Ligoxigakis EK (2002) Control of root and stem rot of cucumber caused by *Fusarium oxysporum* f.sp. *radicis-cucumerinum* by grafting onto resistant rootstocks. *Plant Dis* 86: 379–382.

Paynter ML, De Faveri J, Herrington ME (2014) Resistance to *Fusarium oxysporum* f.sp. *fragariae* and predicted breeding values in strawberry. *J Amer Soc Hortic Sci* 13: 178–184.

Peart JR, Lu R, Sodanandom A et al. (2002) Ubiquitin ligase-associated protein SGT1 is required for host and nonhost disease resistance in plants. *Proc Natl Acad Sci USA* 99: 10865–10869.

Peltier AJ, Hatfield RD, Grau CR (2009) Soybean lignin concentration relates to resistance to *Sclerotinia sclerotiorum*. *Plant Dis* 93: 149–154.

Pemberton CL, Salmond GPC (2004) The Nep 1-like proteins–A growing family of microbial elicitors of plant necrosis. *Molec Plant Pathol* 5: 353–359.

Pemberton CL, Whitehead NA, Sebaihia M et al. (2005) Novel quorum-sensing-controlled genes in *Erwinia carotovora* subsp. *carotovora*: Identification of a fungal elicitor homologue in a soft rotting bacterium. *Molec Plant Pathol* 5: 353–359.

Peng K-C, Wang C-W, Wu C-H, Huang C-T, Liou R-F (2015) Tomato SOSBIR1/EVR homologs are involved in elicitin perception and plant defense against the oomycete pathogen *Phytophthora parasitica*. *Molec Plant-Microbe Interact* 28: 913–926.

Perchepied L, Pitrat M (2004) Polygenic inheritance of partial resistance to *Fusarium oxysporum* f.sp. *melonis* race 1.2 in melon. *Phytopathology* 94: 1331–1336.

Perla V, Jayanty SS, Holm DG, Davidson RD (2014) Relationship between tuber storage proteins and tuber powdery scab resistance in potato. *Amer J Potato Res* 91: 233–245.

Petersen M, Bordersen P, Naested H et al. (2000) *Arabidopsis* MAPKinase 4 negatively regulates systemic acquired resistance. *Cell* 103: 1111–1120.

Pilet-Nayel ML, Muehlbauer FJ, McGee RJ, Kraft JM, Baranger A, Coyne CJ (2005) Consistent quantitative trait loci in pea for partial resistance to *Aphanomyces euteiches* isolates from the United States and France. *Phytopathology* 95: 1287–1293.

Pilet-Nayel M-L, Prospéri J-M et al. (2009) *AER1*, a major gene conferring resistance to *Aphanomyces euteiches* in *Medicago truncatula*. *Phytopathology* 99: 203–208.

Pontier D, del Pozo D, Lam E (2004) Cell death in plant disease: Mechanisms and molecular markers. In: Noodén LD (ed.), *Plant Cell Death*, Elsevier Academic Press, Amsterdam, pp. 37–50.

Pratt RG, Rowe DE (2002) Enhanced resistance to *Sclerotium rolfsii* in populations of alfalfa selected for quantitative resistance to *Sclerotinia trifoliorum*. *Phytopathology* 92: 204–209.

Punja ZK, Raharjo SHT (1996) Response of transgenic cucumber and carrot plants expressing different chitinase enzymes to inoculation with fungal pathogens. *Plant Dis* 80: 999–1005.

Qin J, Zhao JYKJ, Cao YF, Ling H, Sun XF, Tang KX (2004) Isolation and characterization of an ERF-like gene from *Gossypium barbadense*. *Plant Sci* 167: 1383–1389.

Qu F, Ye X, Hou G, Sato S, Clement TE, Morris TJ (2005) RDR6 has a broad-spectrum but temperature-dependent antiviral defense role in *Nicotiana benthamiana*. *J Virol* 79: 15209–15217.

Quesada-Ocampo LM, Hausbeck MK (2010) Resistance of tomato and wild relatives to crown rot and root rot caused by *Phytophthora capsici*. *Phytopathology* 100: 619–627.

Quesada-Ocampo LM, Vargas AM, Naegele RP, Francis DM, Hausbeck MK (2016) Resistance to crown and root rot caused by *Phytophthora capsici* in tomato advanced backcross of *Solanum habrochaites* and *Solanum lycopersicum*. *Plant Dis* 100: 829–835.

Qutob D, Kamoun S, Gijzen M (2002) Expression of a *Phytophtora sojae* necrosis-inducing protein occurs during transition from biotrophy to necrotrophy. *Plant J* 32: 361–373.

Raaymakers TM, van den Ackerveken G (2016) Extracellular recognition of oomycetes during biotrophic infection of plants. *Front Plant Sci* 7 (published online).

Raghunath C, Shanmugam M, Bendaoud M, Kaplan JB, Ramasubbu N (2012) Effect of a biofilm-degrading enzyme from an oral pathogen in transgenic tobacco on the pathogenicity of *Pectobacterium carotovorum* subsp. *carotovorum*. *Plant Pathol* 61: 346–354.

Rahman H, Peng G, Yu F, Falk KC, Kulkarni M, Selvaraj G (2014) Genetics and breeding for clubroot resistance in Canadian spring canola (*Brassica napus* L.). *Canad J Plant Pathol* 36: 122–134.

Rahmanpour S, Backhouse D, Nonhebel HM (2010) Reaction glucosinolate-myrosinase defence system in *Brassica* plants to pathogenicity factor of *Sclerotinia sclerotiorum*. *Eur J Plant Pathol* 128: 429–433.

Ramirez-Villapudua J, Endo RM, Bosland P, Williams PH (1985) A new race of *Fusarium oxysporum* f.sp. *conglutinans* that attacks cabbage with type A resistance. *Plant Dis* 69: 612–613.

Ramos B, López G, Molina A (2015) Development of a *Fusarium oxysporum* f.sp. *melonis* functional GFP fluorescence tool to assist melon resistance breeding programs. *Plant Pathol* 64: 1349–1357.

Ranathunge K, Thomas RH, Fang X, Peterson CA, Gijzen M, Bernards MA (2008) Soybean root suberin and partial resistance to root rot caused by *Phytophthora sojae*. *Phytopathology* 98: 1179–1189.

Reeves G, Monroy-Barbosa A, Bosland PW (2013) A novel *Capsicum* gene inhibits host-specific disease resistance to *Phytophthora capsici*. *Phytopathology* 103: 472–478.

Rehmany AP, Gordon A, Rose IE et al. (2005) Differential recognition of highly divergent downy mildew avirulence gene alleles *RPP1* resistance genes from two *Arabidopsis* lines. *Plant Cell* 17: 1839–1850.

Rep M, van der Does HC, Meijer M et al. (2004) A small cysteine-rich protein secreted by *Fusarium oxysporum* during colonization of xylem vessels is required for 1–3-mediated resistance in tomato. *Molec Microbiol* 1373–1383.

Reusche M, Thole K, Janz D et al. (2012) *Verticillium* infection triggers VASCULAR-RELATED NAC DOMAIN 7-dependent de novo xylem formation and enhances drought tolerance in *Arabidopsis*. *Plant Cell* 24: 3823–3837.

Rietman H, Finkers R, Evers L, van der Zouwen PS, van der Wolf JM, Visser RGF (2014) A stringent and broad screen of *Solanum* spp. tolerance against *Erwinia* bacteria using a petiole test. *Amer J Potato Res* 91: 204–214.

Rispail N, Rubiales D (2014) Identification of sources of quantitative resistance to *Fusarium oxyporum* f.sp. *medicaginis* in *Medicago truncatula*. *Plant Dis* 98: 667–673.

Rivera-Vargas LI, Schmitthenner AF, Graham TI (1993) Soybean flavonoids effects on and metabolism by *Phytophthora sojae*. *Phytochemistry* 32: 851–857.

Robagalia C, Caranta C (2006) Translation initiation factors: A weak link in plant RNA virus infection. *Trends Plant Sci* 11: 40–45.

Robin AHK, Yi G-E, Laila R, Hossain MR, Park J-I, Kim HR, Nou I-S (2017) *Leptosphaeria maculans* alters glucosinolate profiles in blackleg disease-resistant and -susceptible cabbage lines. *Front Plant Sci* 8: 1–15

Robinson RA (1969) Disease resistance terminology. *Rev Appl Mycol* 48: 593–606.

Robinson RA (1987) *Host Management in Crop Pathosystems*, McMillan Publishing Co, New York.

Rodriguez-Molina MC, Medina I, Torres-Vila LM, Cuartero J (2003) Vascular colonization patterns in susceptible and resistant tomato cultivars inoculated with *Fusarium oxysporum* f.sp. *lycopersici* races 0 and 1. *Plant Pathol* 52: 199–203.

Rojo E, Martin R, Carter C, Zouhar J, Pan S, Plotnikova J, Jin H, Paneque M, Sanchez-Serrano JJ, Baker B, Asubel FM, Raikhel NV (2004) VPE gamma exhibits a capase-like activity that contributes to defense against pathogens. *Curr Biol* 14: 1897–1906.

Rose JK, Ham KS, Darvill AG, Albersheim P (2002) Molecular cloning and characterization of glucanase inhibitor proteins: coevolution of a counter defense mechanism by plant pathogens. *Plant Cell* 14: 1329–1345.

Rout E, Nanda S, Nayak S, Joshi RK (2014) Molecular characterization of NBS encoding resistance genes and induction analysis of a putative candidate gene linked to Fusarium basal rot resistance in *Allium sativum*. *Physiol Molec Plant Pathol* 85: 15–24.

Ruiz-Rubio M, Perez-Espinosa A, Lairini K, Roldan-Arjona T, Di Pietro A, Anaya N (2001) Metabolism of the tomato saponin α-tomatine by phytopathogenic fungi. In: *Studies in National Products Chemistry*, Volume 25, Elsevier Science, Amsterdam, pp. 137–143.

Russell GE (1978) *Plant Breeding for Pest and Disease Resistance*, Butterworths, London, UK.

Russo MA, Fioco BM, Marone D et al. (2012) A major QTL for resistance to *Soilborne cereal mosaic virus* derived from an old Italian durum. *J Plant Interact* 7: 290–300.

Ruthardt N, Fischer R, Emans N, Kawachuk LM (2007) Tomato protein of the resistance gene *Ve 2* to Verticillium wilt (*Verticillium* spp.) is located in the endoplasmic reticulum. *Canad J Plant Pathol* 29: 3–8.

Ryals JA, Ukness SJ, Ward RC (1994) Systemic acquired resistance. *Plant Pathol* 104: 1109–1112.

Ryder EJ, Robinson BJ (1995) Big-vein resistance in lettuce: Identifying, selecting, and testing resistance cultivars and breeding lines. *J Amer Soc Hortic Sci* 120: 741–746.

Rygulla W, Snowdon RJ, Eynck C et al. (2007) Broadening the genetic basis of *Verticillium longisporum* resistance in *Brasssica napus* by interspecific hybridization. *Phytopathology* 97: 1391–1396.

Rygulla W, Snowdon RJ, Friedlt W, Happstadius I, Cheung WY, Chen D (2008) Identification of quantitative trait loci for resistance against *Verticillium longisporum* in oilseed rape (*Brassica napus*). *Phytopathology* 98: 215–221.

Sacks W, Nurenberger T, Hahlborck K, Scheel D (1995) Molecular characterization of nucleotide sequences encoding the extracellular glycoprotein elicitor from *Phytophthora megasperma*. *Molec Gen Genet* 246: 55–65.

Sahoo DK, Abeyasekara NL, Cianzio SR, Robertson AE, Bhattacharya MK (2017) A novel *Phytophthora sojae* resistance *Rps12* gene mapped to a genomic region that contains several *Rps* genes. *PLoS ONE* 12: e0169950.

Salaman RN (1949) *The History and Social Influence of Potato*, Cambridge University Press, London.

Salari M, Panjehkeh N, Nasirpoor Z, Abkhoo J (2012) Screening of *Cucumis melo* L. cultivars from Iran for resistance against soilborne fungal pathogens. *Plant Pathol Microbiol* 3: 6

Salas B, Secor GA, Taylor RJ, Gudmestad NC (2003) Assessment of resistance of tubers of potato cultivars to *Phytophthora infestans* and *Pythium ultimum*. *Plant Dis* 87: 91–97.

Salminko TL, Bowen CR, Hartman GL (2010) Multi-year evaluation of commercial soybean cultivars for resistance to *Phytophthora sojae*. *Plant Dis* 94: 368–371.

Sanchez CLG, Pinto LRM, Pomella AWV, Silva SDVM, Loguerico LL (2008) Assessment of resistance to *Ceratocystis cacaofunesta* in cacao genotypes. *Eur J Plant Pathol* 122: 517–528.

Sanchez-Perez A, Halterman D, Jordan S, Chen Y, Gevens AJ (2017) *RB* and *Ph* resistance genes in potato and tomato minimize risk of oospore production in the presence of mating pairs of *Phytophthora infestans*. *Eur J Plant Pathol* 149: 853–864.

Sanford JC, Jones SA (1985) The concept of parasite derived resistance: Deriving resistance genes from the parasite's own genome. *J Theoret Biol* 115: 395–405.

Sato N, Takamatsu T (1930) Grafting culture of watermelon. *Nogyo sekai* 25: 24–28.

Savita A, Virk GS, Nagpal A (2011) In vitro selection of calli of *Citrus jambhiri* Lush. for tolerance to culture filtrate of *Phytophthora parasitica* and their regeneration. *Physiol Molec Biol Plants* 17: 41–47.

Schneider R, Rolling W, Song Q, Cregan P, Dorrance AE, McHale LK (2016) Genome wide association mapping to plant resistance to *Phytophthora sojae* in plant introduction from the Republic of Korea. *BMC Genomics* 17: 607.

Sejalon-Delmas N, Villalba-Mateos F, Bottin A, Rickauer M, Dargent R, Esquerré-Tugaye MT (1997) Purification, elicitin activity and cell wall localization of a glycoprotein from *Phytophthora parasitica* var. *nicotianae*, a fungal pathogen of tobacco. *Phytopathology* 87: 899–909.

Serrano C, Arce-Johnson P, Torres H (2000) Expression of the chicken lysozyme gene in potato enhances resistance to infection by *Erwinia carotovora* subsp. *atroseptica*. *Amer J Potato Res* 77: 191–199.

Sharma K, Gossen BD, Greenshields D, Selvaraj G, Strelkov SE, McDonald MR (2013a) Reaction of lines of the Rapid Cycling Brassica Collection and *Arabidopsis thaliana* to four pathotypes of *Plasmodiophora brassicae*. *Plant Dis* 97: 720–727.

Sharma K, Gossen BD, Howard RJ, Glaudovacz T, McDonald MR (2013b) Reaction of selected *Brassica* vegetable crops to Canadian pathotypes of *Plasmodiophora brassicae*. *Canad J Plant Pathol* 35: 371–383.

Sharma KD, Chen W, Muehlbauer FJ (2005) Genetics of chickpea resistance to five races of Fusarium wilt and a concise set of race differentials for *Fusarium oxysporum* f.sp. *ciceris*. *Plant Dis* 89: 385–390.

Shaw DV, Gordon TR, Hansen J, Kirkpatrick SC (2010) Relationship between extent of colonization by *Verticillium dahliae* and symptom expression in strawberry (*Fragaria* x *ananassa*) genotypes resistant to Verticillium wilt. *Plant Pathol* 59: 376–381.

Shen Y, Diener AC (2013) *Arabidopsis thaliana RESISTANCE TO FUSARIUM OXYSPORUM 2* implicates tyrosine-sulfated peptide signaling in susceptibility and resistance to root infection. *PLoS Genet* 9, e1003525

Silipo A, Molinaro A, Sturiale L et al. (2005) The elicitation of plant innate immunity by lipopolysaccharide of *Xanthomonas campestris*. *J Biol Chem* 280: 33660–33668.

Silvar C, Díaz J, Merino F (2005) Real-time polymerase chain reaction quantification of *Phytophthora capsici* in different pepper genotypes. *Phytopathology* 95: 1423–1429.

Silveira JRP, Duarte V, Moraes MG et al. (2007) Epidemilogical analyses of clones and cultivars of potato in soil naturally infested with *Ralstonia solanacearum* biovar 2. *Fitopatol Brassiliera* 32: 181–188.

Slavov S (2005) Phytotoxins and in vitro screening for improved disease resistant plantsBiotechnol Biotechnol Eq. 19: 48–55.

Smith N, Sing S, Wang MB, Stoutjesdijk P, Green A, Waterhouse PM (2001) Total silencing by intron-spliced hairpin RNAs. *Nature* 407: 319–320.

Solís-Ramos LY, Valdez-Melara M, Alvarado-Barrantes R et al. (2015) Effect of gamma irradiation and selection with fungus filtrate (*Rhizoctonia solani* Kuhn) on the in vitro culture of common bean (*Phaseolus vulgaris*). *Amer J Plant Sci* 6: 2672–2785.

Song JQ, Bradeen JM, Naess SK et al. (2003) Gene *RB* cloned from *Solanum bulbocastanum* confers broad spectrum resistance to potato late blight. *Proc Natl Acad Sci USA* 100: 9128–9133.

Song Y, Zhang Z, Seidl MF et al. (2017) Broad taxonomic characterization of Verticillium wilt resistance genes reveals an ancient origin of the tomato Ve 1 tomato receptor. *Molec Plant Pathol* 18: 195–209.

Sowik I, Markiewicz M, Michalczuk L (2015) Stability of *Verticillium dahliae* resistance in tissue culture-derived strawberry somaclones. *Hortic Sci (Prague)* 42: 141–148.

Sprague SJ, Marcroft SJ, Hayden HL, Howlett BJ (2006) Major gene resistance to blackleg in *Brassica napus* overcome within three years of commercial production in southern Australia. *Plant Dis* 90: 190–198.

Ssekiwoko F, Kiggundu A, Tushemereirwe WK, Karamura E, Kunert K (2015) *Xanthomonas vasicola* pv. *musacearum* down-regulates selected defense genes during its interaction with both resistant and susceptible banana. *Physiol Molec Plant Pathol* 90: 21–26.

Strassner J, Schaller F, Frick UB et al. (2002) Characterization and cDNA expression analysis of 12-oxophytodienoate reductases reveals differential roles for octadecanoid biosynthesis in the local versus the systemic wound response. *Plant J* 32: 585–601.

Strausbaugh CA, Eujayl IA, Panella LW (2013) Interaction of sugar beet host resistance and *Rhizoctonia solani* AG 2-2-IIIB strains. *Plant Dis* 97: 1175–1180.

Street PFS, Robb J, Ellis BE (1986) Secretion of vascular coating components by xylem parenchyma cells of tomatoes infected with *Verticillium albo-atrum*. *Protoplasma* 132: 1–11.

Subramanian S, Graham Y, Yu O, Graham TL (2005) RNA interference of soybean isoflavone synthase genes leads to silencing in tissues distal to the transformation site and to enhanced susceptibility to *Phytophthora sojae*. *Plant Pathol* 137: 1345–1353.

Swarupa V, Ravishankar KV, Rekha A (2013) Characterization of tolerance to *Fusarium oxysporum* f.sp. *cubense* infection in banana using suppression subtractive hybridization and gene expression analysis. *Physiol Molec Plant Pathol* 83: 1–7.

Sy O, Steiner R, Bosland PW (2005) Inheritance of Phytophthora stem blight resistance in *Capsicum annuum* L. *J Amer Soc Hortic Sci* 130: 75–78.

Sy O, Steiner R, Bosland PW (2008) Recombinant inbred line differential identifies race-specific resistance to Phytophthora root rot in *Capsicum annuum*. *Phytopathology* 98: 867–870.

Taheri P, Tarighi S (2012) The role of pathogenesis-related proteins in the tomato-Rhizoctonia solani interaction. *J Botany*

Takahashi H, Ishikawa T, Kaido M et al. (2006) *Plasmodiophora brassicae*-induced cell death and medium alkalization in clubroot resistant cultured roots of *Brassica rapa*. *J Phytopathol* 154: 156–162.

Takahashi Y, Nasir KHB, Ito A et al. (2007) A high throughput screen of cell-death-inducing factors in *Nicotiana benthamiana* identifies a novel MAPKK that mediates INF1-induced cell death signaling and nonhost resistance to *Pseudomonas cichorii*. *Plant J* 49: 1030–1040.

Taylor A, Coventry E, Jones JE, Clarkson JP (2015) Resistance to a highly aggressive isolate of *Sclerotinia sclerotiorum* in a *Brassica napus* diversity set. *Plant Pathol* 64: 932–940.

Tegg RS, Thangavel T, Aminian H, Wilson CR (2013) Somaclonal selection in potato for resistance to common scab provides concurrent resistance to powdery scab. *Plant Pathol* 62: 922–931.

Tegg RS, Wilson CR (2010) Relationship of resistance to common scab disease and tolerance to thaxtomin A toxicity within potato cultivars. *Eur J Plant Pathol* 128: 143–148.

Tenorio J, Franco Y, Chuquillanqui C, Owens RA, Salazar LF (2006) Reaction of potato varieties to *Potato mop-top virus* infection in the Andes. *Amer J Potato Res* 83: 423–431.

Thakur PP, Mathew D, Nazeem PA, Abida PS, Indira P, Girija D, Shylaja MR, Valsala PR (2014) Identification of allele specific AFLP markers linked with bacterial wilt *Ralstonia solanacearum* [(Smith) Yabuuchi et al.] resistance in hot peppers (*Capsicum annuum* L.). *Physiol Molec Plant Pathol* 87: 19–24.

Thatcher LF, Manners JM, Kazan K (2009) *Fusarium oxysporum* hijacks COI1-mediated jasmonate signaling to promote disease development in *Arabidopsis*. *Plant J* 58: 927–939.

Thomzik JE, Stenzel K, Stöcker R, Schreir PH, Hain R, Stahl DJ (1997) Synthesis of a grapevine phytoalexin in transgenic tomatoes (*Lycopersicon esculentum* Mill.) conditions resistance against *Phytophthora infestans*. *Physiol Molec Plant Pathol* 51: 265–278.

Thurston G, Regen S, Rampitsch C, Xing T (2005) Proteomic and phosphoproteomic approaches to understand plant-pathogen interactions. *Physiol Molec Plant Pathol* 66: 3–11.

Tian M, Benedetti B, Kamoun S (2005) A second Kazal-like protease inhibitor from *Phytophthora infestans* inhibits and interacts with the apoplastic pathogenesis-related protease P69B of tomato. *Plant Physiol* 138: 1785–1793.

Tian M, Huitema E, da Cunha L, Torto-Alalibo T, Kamoun S (2004) A Kazal-like extracellular serine protease inhibitor from *Phytophthora infestans* targets the tomato pathogenesis-related protease P69B. *J Biol Chem* 279: 26370–26377.

Tian M, Win J, van der Hoorn R, van der Knaap E, Kamoun S (2007) A *Phytophthora infestans* cystatin-like targets a novel tomato papain-like apoplastic protease. *Plant Physiol* 143: 364–377.

Tor M, Yemm A, Holub E (2003) The role of proteolysis in *R* gene-mediated defense in plants. *Molec Plant Pathol* 4: 287–296.

Torres MA, Jones JDG, Dangl Jl (2006) Reactive oxygen species signaling in response to pathogen*Plant Physiol* 141: 373–378.

Trapero C, Serrano N, Arquero O, Del Rio C, Trapero A, López-Escudero FJ (2013) Field resistance to Verticillium wilt in selected olive cultivars grown in two naturally infested soils. *Plant Dis* 97: 668–674.

Tripathi L, Odipio J, Tripathi JN, Tusiime G (2008) A rapid technique for screening banana cultivars for resistance to Xanthomonas wilt. *Eur J Plant Pathol* 121: 9–19.

Truman W, Bennett MH, Kubigsteltig I, Turnbull C, Grant M (2007) *Arabidopsis* systemic immunity uses conserved defense signaling pathways and is mediated by jasmonates. *Proc Natl Acad Sci USA* 104: 1075–1080.

Truong HTH, Kim JH, Cho CC, Chae Y, Lee HE (2013) Identification and development of molecular markers linked to Phytophthora root rot resistance in pepper (*Capsicum annuum* L.). *Eur J Plant Pathol* 135: 289–297.

Truong HTH, Kim KT, Kim DW et al. (2012) Identification of isolate-specific resistance QTLs to Phytophthora root rot using an intraspecific recombinant inbred line population of pepper (*Capsicum annuum*). *Plant Pathol* 61: 48–56.

Turner JG (1984) Role of toxins in plant disease. In: Woos RKS, Jellis GJ (eds.), *Plant Diseases: Infection, Damage and Loss*, Blackwell Publishers Oxford, UK, pp 3–12.

Twizeyimana M, Hill CB, Pawlowski M, Paul C, Hartman GL (2012) A cut-stem inoculation technique to evaluate soybean for resistance to *Macrophomina phaseolina*. *Plant Dis* 96: 1210–1215.

Tyler BM (2002) Molecular basis of recognition between *Phytophthora* pathogens and their host. *Annu Rev Phytopathol* 40: 137–167.

Ueeda M, Jubota M, Nishi K (2006) Contribution of jasmonic acid to resistance against Phytophthora blight in *Capsicum annuum* cv. SCM 334. *Physiol Molec Plant Pathol* 67: 149–154.

Uloth M, You MP, Finnegan PM, Banga SS, Yi H, Barbetti MJ (2014) Seedling resistance to *Sclerotinia sclerotiorum* as expressed across diverse cruciferous species. *Plant Dis* 98: 184–190.

Van de Wouw AP, Marcroft SJ, Barbetti MJ et al. (2009) Dual control of avirulence in *Leptosphaeria maculans* towards a *Brassica napus* cultivar with 'sylvestris-derived' resistance suggests involvement of two resistance genes. *Plant Pathol* 58: 305–313.

Van Esse HP, Fradin EF, de Groot P-J, de Wit PJGM, Thomma BPHJ (2009) Tomato transcriptional responses to a foliar and vascular fungal pathogens are distinct. *Molec Plant-Microbe Interact* 22: 245–258.

Van Jaarsveld E, Wingfield MJ, Drenth A (2002) Evaluation of tobacco cultivars for resistance to races of *Phytophthora nicotianae* in South Africa. *J Phytopathol* 150: 456–462.

Van Jaarsveld E, Wingfield MJ, Drenth A (2003) A rapid seedling-based screening technique to assay tobacco for resistance to *Phytophthora nicotianae*. *J Phytopathol* 151: 389–394.

Van Loon LC, Van Strien EA (1999) The families of pathogenesis-related proteins, their activities and comparative analysis of PR-1 type proteins. *Physiol Molec Plant Pathol* 55: 85–97.

van't Slot, Knogge W (2002) A dual role for microbial pathogen-derived effector proteins in plant disease and resistance. *Critic Rev Plant Sci* 21: 229–271.

Vandemark GJ, Barker BM (2003) Quantifying *Phytophthora medicaginis* in susceptible and resistant alfalfa with a real-time fluorescent PCR assay. *J Phytopathol* 151: 577–583.

Vandemark GJ, Barker BM, Gritsenko MA (2002) Quantifying *Aphanomyces euteiches* in alfalfa with fluorescent polymerase chain reaction assay. *Phytopathology* 92: 265–272.

Vandemark GJ, Grünwald NJ (2005) Use of real-time PCR to examine the relationship between disease severity in pea and *Aphanomyces euteiches* DNA contents in roots. *Eur J Plant Pathol* 111: 309–316.

Vanderplank JE (1963) *Plant Diseases: Epidemics and Control*, Academic Press, New York.

Vanderplank JE (1984) *Disease Resistance in Plants*, Academic Press, New York.

Vassilakos N, Bem F, Tzima A et al. (2008) Resistance of transgenic tobacco plants incorporating the putative 57-kDa polymerase readthrough gene of *Tobacco rattle virus* against rub-inoculated and nematode-transmitted virus. *Transgenic Res* 17: 929–941.

Vega-Arreguin JC, Jalloh A, Bos JI, Moffett P (2014) Recognition of an Avr3a homologue plays a major role in mediating non-host resistance to *Phytophthora capsici* in *Nicotiana* species. *Molec Plant-Microbe Interact* 27: 770–780.

Vijayan P, Shockey J, Levesque CA, Cook RJ, Browse JA (1998) A role for jasmonate in pathogen defense of *Arabidopsis. Proc Acad Natl Sci USA* 95: 7209–7214.

Villalba-Mateo F, Rickauer M, Esquerré-Tugaye MT (1997) Cloning and characterization of a cDNA encoding an elicitor of *Phytophthora parasitica* var. *nicotianae* that shows cellulose-binding and lectin-like activities. *Molec Plant-Microbe Interact* 10: 1045–1053.

Vleeshouwers VGAA (2011) Understanding and exploring late blight resistance in the age of effectors*Annu Rev Phytopathol* 49: 507–531.

Vleeshouwers VGAA, Driesprong JD, Kamphuis LG et al. (2006) Agroinfection-based high throughput screening reveals specific recognition of INF elicitins in *Solanum. Molec Plant Pathol* 7: 499–510.

Walker SJ, Bosland PW (1999) Inheritance of Phytophthora root rot and foliar blight resistance in pepper. *J Amer Soc Hortic Sci* 124: 14–18.

Wang D, Weaver ND, Kesarwani M, Dong X (2005) Induction of protein secretory pathway is required for systemic acquired resistance. *Science* 308: 1036–1040.

Wang HR, Ru SJ, Wang LP, Fong ZM (2004) Study on the control of Fusarium wilt and Phytophthora blight in cucumber by grafting*Acta Agr Zhejiangensis* 16: 336–339.

Wang J, Higgins VT (2006) Nitric oxide modulates H2O2-mediated defenses in the *Colletotrichum coccodes*- tomato interaction. *Physiol Molec Plant Pathol* 67: 131–137.

Wang J-E, Li D-W, Zhang Y-L, Zhao Q, He Y-M, Gong Z-H (2013) Defense responses of pepper (*Capsicum annuum* L.) infected with incompatible and compatible strains of *Phytophthora capsici. Eur J Plant Pathol* 136: 625–638.

Wang JY, Cai Y, Gou JY, Mao YB, Xu YH, Jiang WH, Chen XY (2004) VdNEP, an elicitor from *Verticillium dahliae*, induces cotton plant wilting. *Appl Environ Micro* 70: 4989–4995.

Wang X, El Hadrami A, Adam L, Daayf F (2004) US-1 and US-8 genotypes of *Phytophthora infestans* differentially affect local, peroximal and distal gene expression of phenylalanine ammonia-lyase and 3-hydroxy-3-methylglutaryl CoA reductases in potato leaves. *Physiol Molec Plant Pathol* 65: 157–167.

Wang Y, Dang F, Liu Z et al. (2013) CaWRKY58, encoding a group I WRKY transcription factor of *Capsicum annuum*, negatively regulated resistance to *Ralstonia solanacearum* infection. *Molec Plant Pathol* 14: 131–144.

Wang Z, Fang H, Chen Y, Chen K, Li G, Gu S, Tan X (2014) Overexpression of *BnWRKY33* in oilseed rape enhances resistance to *Sclerotinia sclerotiorum. Molec Plant Pathol* 15: 677–689.

Ward ER, Ukness SJ, Williams SC et al. (1991) Coordinate gene activity in response to agents that induce systemic acquired resistance. *Plant Cell* 3: 1085–1094.

Weber H, Schultze S, Pfitzner AJP (1993) Amino acid substitutions in the *Tomato mosaic virus* 30-kilodalton movement protein confer the ability to overcome the *Tm-22* resistance gene in tomato. *J Virol* 67: 6432–6438.

Wechter WP, Kousik C, McMillan M, Levi A (2012) Identification of resistance to *Fusarium oxysporum* f.sp. *niveum* race 2 in *Citrullus lanatus* var. *citroides* plant introductions. *HortScience* 47: 334–338.

Wen F, White GJ, Van Etten HD, Xiong Z, Hawes MC (2009) Extracellular DNA is required for root tip resistance to fungal infection. *Plant Physiol* 151: 820–829.

Wen L, Tan T-L, Shu J-B et al. (2013) Using proteomic analysis to find the proteins involved in resistance against *Sclerotinia sclerotiorum* in adult *Brassica napus. Eur J Plant Pathol* 137: 505–523.

Werling D, Jungi TW (2003) TOLL-like receptors linking innate and adaptive immune response. *Veter Immunol Immunopathol* 91: 1–12.

Whitham S, Dinesh Kumar SP, Choi D, Hehl R, Corr C, Baker B (1994) The product of *Tobacco mosaic virus* resistance gene *N*: similarity to Toll and the interleukin-1 receptor*Cell* 78: 1101–1115.

Whitham SA, Wang Y (2004) Roles for host factors in plant viral pathogenicity. *Curr Opin Plant Biol* 7: 365–371.

Willbur JF, Ding S, Marks ME et al. (2017) Comprehensive Sclerotinia stem rot screening of soybean germplasm requires multiple isolates of *Sclerotinia sclerotiorum. Plant Dis* 101: 344–353.

Wilson CR, Luckman GA, Tegg RS et al. (2009) Enhanced resistance to common scab in potato through somatic cell selection in cv. Iowa with the phytotoxin thaxtomin A. *Plant Pathol* 58: 137–144.

Wilson CR, Tegg RS, Wilson AJ et al. (2010) Stable and extreme resistance to common scab of potato obtained through somatic cell selection. *Phytopathology* 100: 460–467.

Wimer J, Inglis D, Miles C (2015a) Evaluating grafted watermelon for Verticillium wilt severity, yield, and fruit quality in Washington State. *HortScience* 50: 1332–1337.

Wimer J, Inglis D, Miles C (2015b) Field and greenhouse evaluations of cucurbit rootstocks to improve Verticillium resistance for grafted watermelon. *HortScience* 50: 1625–1630.

Wu YL, Yi GJ, Peng XX (2010) Rapid screening of *Musa* species for resistance to Fusarium wilt in an in vitro bioassay. *Eur J Plant Pathol* 128: 409–415.

Xiang Y, Scandiani MM, Herman TK, Hartman GL (2015) Optimizing conditions of a cell-free toxic filtrate stem cutting assay to evaluate soybean genotype responses to *Fusarium* species that cause sudden death syndrome. *Plant Dis* 99: 502–507.

Xu L, Zhu L, Tu L et al. (2011) Differential gene expression in cotton defense response to *Verticillium dahliae* by SSH. *J Phytopathol* 159: 606–615.

Xue X, Cao ZX, Zhang XT et al. (2016) Overexpression of *OsOSM1* enhances resistance to rice sheath blight. *Plant Dis* 100: 1634–1642.

Yadeta KA, Thomma BPHJ (2013) The xylem as battle ground for hosts and vascular wilt pathogens. Front Plant Sci.

Yanar Y, Miller SA (2003) Resistance of pepper cultivars and accessions of *Capsicum* spp. to *Sclerotinia sclerotiorum. Plant Dis* 87: 303–307.

Yang P, Lupken T, Habekuss A, Hensel G, Steuernagel B, Kilian B, Ariyadasa R, Himmelbach A, Kumlehn J, Scholz U, Ordon F, Stein N (2014) *PROEIN DISULFIDE ISOMERASE-LIKE 5–1* is a susceptibility factor to plant viruses. *Proc Natl Acad Sci USA* 111: 2104–2109.

Yang S, Lee S (1981) Multigenic effects of chemical mutagens in bananas. *J Agric Assoc China* 116: 36–47.

Yang Y, Qi M, Mei C (2004) Endogenous salicylic acid protects rice plants from oxidative damage caused by aging as well as biotic and abiotic stress. *Plant J* 40: 909–919.

Yeo JR, Weiland JE, Sullivan DM, Bryla DR (2016) Susceptibility of high bush blue berry cultivars to Phytophthora root rot. *HortScience* 51: 74–78.

Yetisir H, Kurt S, Sari N, Tok FM (2007) Rootstock potential of Turkish *Lagenaria siceraria* germplasm for watermelon: Plant growth, graft compatibility and resistance to *Fusarium. Turk Agric For* 31: 381–388.

Yin YI, Zhou BL, Li YP, Fu YW (2008) Allelopathic effects of grafting on rhizosphere microorganisms populations of eggplants. *Act Hortic Sinica* 35: 1136.

You MP, Uloth MB, Li XX, Banga SS, Banga SK, Barbetti MJ (2016) Valuable new resistances ensure improved management of Sclerotinia stem rot (*Sclerotinia sclerotiorum*) in horticultural and oilseed *Brassica* species. *J Phytopathol* 164: 291–299.

Zare B, Niazi A, Sattari R et al. (2015) Resistance against rhizomania disease via RNA silencing in sugar beet. *Plant Pathol* 64: 35–42.

Żebrowska JI (2011) Efficacy of resistance selection to Verticillium wilt in strawberry (*Fragaria* x *ananassa* Duch.) tissue culture. *Acta Agrobot* 64: 3–12.

Zeidler D, Zahringer U, Gerber I et al. (2004) Innate immunity in *Arabidopsis thaliana*: lipopolysaccharides activate nitric oxide synthase (NOS) and induce defense genes. *Proc Natl Acad Sci USA* 101: 15811–15816.

Zhang C, Wu Z, Li Y, Wu J (2015a) Biogenesis, function, and applications of virus-derived and RNAs in plants. *Front Microbiol* 6: 1237. 2015.01237

Zhang D, Bai G, Hunger RM et al. (2011) Association study of resistance to *Soilborne wheat mosaic virus* in U. S. winter wheat. *Phytopathology* 101: 1322–1329.

Zhang H, Feng J, Zhang S (2015b) Resistance to *Plasmodiophora brassicae* in *Brassica rapa* and *Brassica juncea* genotypes from China. *Plant Dis* 99: 776–779.

Zhang S, Xu P, Wu J et al. (2010) Races of *Phytophthora sojae* and their virulences on soybean cultivars in Heilongjiang, China. *Plant Dis* 94: 87–91.

Zhang X, Peng G, Parks P et al. (2017) Identifying seedling and adult plant resistance in Chinese *Brassica* germplasm to *Leptosphaeria maculans*. *Plant Pathol* 66: 752–762.

Zhang Y, Wang L (2005) The WRKY transcription superfamily: its origin in eukaryotes and expansin in plants. *BMC Evol Biol* 5: 1–12.

Zhang Z, Fradin E, de Jonge R et al. (2013) Optimized agroinfiltration and virus-induced gene silencing to study Ve 1-mediated Verticillium resistance in tobacco. *Molec Plant-Microbe Interact* 26: 182–190.

Zhao J, Meng J, Osborn TC, Grau CR (2004) Evaluation of Sclerotinia stem rot resistance in oilseed *Brassica napus* using a petiole inoculation technique under greenhouse conditions. *Plant Dis* 88: 1033–1039.

Zhao Y, Thilmony R, Bender CL, Schaller A, He SY, Howe GA (2003) Virulence system of *Pseudomonas syringae* pv. *tomato* promotes bacterial speck disease in tomato by targeting the jasmonate signaling pathway. *Plant J* 36: 485–499.

Zhu C, Gore M, Buckler ES, Yu J (2008) Status and prospects of association mapping on plants. *Plant Genome* 1: 5–20.

Zhu Y, Shao J, Zhou Z, Davis RE (2017) Comparative transcriptome analysis reveals a preformed defense system in apple root of a resistant genotype of G.935 in the absence of pathogen. *Internatl J Plant Genomics*, Article ID 8950746, 14 pages

Zipfel C, Robatzek S, Navarro L (2004) Bacterial disease resistance in *Arabidopsis* through flagellin perception. *Nature* 428: 764–767.

3 Management of Soilborne Microbial Plan Pathogens
Biological Management of Crop Diseases

Biological control or biocontrol of crop diseases has been considered both in a narrow sense to indicate the control of one organism by another organisms and also in a broad sense to indicate the use of natural or modified organisms, genes or gene product to reduce the effects of pathogens and to favor development of crop plant species (Beirner 1967; Cook 1987). Biological disease management involves utilization of biotic and abiotic agents that act through one or more mechanisms to reduce the potential of the pathogen directly or indirectly by activating host defense systems to reduce the disease incidence and/intensity. Several investigations have established that the combination of biotic and abiotic agents results in synergism, improving the effectiveness of pathogen suppression. Biological control agents (BCAs) that effectively reduce incidence and spread of crop diseases caused by microbial pathogens may be divided into two major groups: (i) biotic agents and (ii) abiotic agents. They display varying degrees of biocontrol potential, depending on the host-pathogen combination and the environmental conditions in different geographical locations. Biotic BCAs are nonpathogenic, living organisms that exhibit antagonistic potential against microbial pathogens – oomycetes, fungi, bacteria and viruses that are present in a free-state or in vectors living in the soil environments. They are able to suppress pathogen development through one or more mechanisms. Abiotic BCAs are derived from inorganic or organic sources and they may also act directly on the pathogens or indirectly by activating host plant defense systems. They may be classified based on their chemical constitution or on the mode of action on the target pathogens. The biotic BCAs vary in their biocontrol efficiency due to their sensitivity to environmental conditions, capacity to adapt to new locations, rhizosphere competence, quantum of production of metabolites toxic to target microbial pathogen(s). Hence, it is necessary to establish the precise identity of the strains/isolates, based on the cultural, biochemical and genetic characteristics (Narayanasamy 2013).

3.1 ASSESSMENT OF BIOLOGICAL CONTROL POTENTIAL OF BIOTIC AGENTS

The biotic biological control agents (BCAs) have been isolated from soils, plants and water bodies, using appropriate growth media. Polymerase chain reaction (PCR)-based molecular techniques have been more frequently used to identify, differentiate and quantify the BCAs in different substrates. These techniques may be modified, if necessary, and the biological control agents can be identified up to strains. Precise identification of the BCAs is required, since the strains/isolates of one species of the fungus or bacteria may differ widely in their biological control potential. In addition, biological control activity may also vary, depending on the crop plant species/cultivar, type of soil or environmental conditions existing in different ecosystems.

3.1.1 FUNGAL BIOLOGICAL CONTROL AGENTS

The importance of precise identification of the strain of the fungus to be used as a biocontrol agent is underscored by the example of *Trichoderma harzianum (Th)*, which has been registered as commercial biocontrol agent for the management of several crop diseases. But *T. harzianum* is also known to be the incitant of green mold disease of mushroom (Ospina-Giraldo et al. 1999). Furthermore, some isolates of *T. harzianum* were able to produce the mycotoxin belonging to the tricothecene class, which can induce serious ailments in humans and animals, when the contaminated grains are consumed (Sivasithamparam and Ghisalberti 1996). By using the specific tri5 primers in a PCR assay, it was possible to select suitable *T. harzianum* isolate that did not produce the mycotoxin (Gallo et al. 2004). The relationship between functional group within *Trichoderma* spp. and their biocontrol activity based on a combination of physiological, biochemical (enzyme production) and molecular (ITS sequences) criteria, was inferred. The efficacy of particular strains of *T. harzianum* depended on the intended target and the required functions of biocontrol. The results highlighted the importance of selection of the most efficient strains for the target pathogen to be controlled (Grondona et al. 1997). Maintenance of fungal strains to be used as BCAs is essential to have a reliable source of authenticated fungal strains and to ensure consistency of performance of the BCA both in vitro and in vivo. Predominantly, DNA markers that allow authentication of strains and permit monitoring of contamination have been employed (Markovic and Markovic 1998).

The quality control test for *Trichoderma* is based on production of PCR fingerprints, by using semi-random primers designed to primarily target intergenic, more variable areas in the genome (Bulat et al. 1998). Distinct and reproducible fingerprints of strains of *Trichoderma* spp. were generated by applying appropriate universally-primed (UP) primers in the PCR assay. The fingerprints permitted also differentiation of a collection of other strains of *Trichoderma* spp. The UP-PCR analysis was combined with dilution plating method and a semi-selective medium was used to recover *Trichoderma* spp. strains, after application in commercial greenhouses. The isolates from the commercial biocontrol products applied in different greenhouse were also identified by UP-PCR analysis. In addition, the presence of a *Trichoderma* strain in the untreated bench was also detected, indicating the possible spread of the BCA to untreated plants (Lübeck and Jensen 2002). Investigation was taken up to determine diversity of *Fusarium* strains from soils suppressive to banana wilt caused by *Fusarium oxysporum* f.sp. *cubense (Foc)*. Fusarium strains (> 100) isolated from rhizosphere of banana plants were identified up to species level. The PCR-based RFLP analysis of intergenic spacer region of the ribosomal RNA operon was used to characterize nonpathogenic strains. The species-specific primers FOF1 and FOR1, in addition to morphological characteristics were used to establish the identity of the isolates. Twelve different genotypes could be distinguished and identified using a six-letter code allotted to each isolate following digestion with restriction enzymes *Hae*III, *Hha*I, *Hinf*I, *Msp*I and *Scrf*I. Eleven of these included nonpathogenic *F. oxysporum* isolates from South Africa (Nel et al. 2006a).

3.1.1.1 Assessment of Biocontrol Potential of Fungal Isolates

The interactions between microbial plant pathogens and the biological control agents (BCAs) may vary and they may be differentiated into mutualism, antagonism, parasitism, competition and predation. Mutualism refers to the association between two or more organisms deriving benefits from such association. Many of the BCAs can be considered as facultative mutualists, since survival rarely depends on any specific host., Degree of disease suppression may vary depending on the environmental conditions. An association in which one organism is benefited and the other is neither benefited nor harmed, is known as commensalism. If the interaction results in adverse effects exerted by one organism on the other, the interaction is designated antagonism. Organisms exhibit competition to obtain nutrients or occupy available niche for their survival. Such competition may lead to decreased growth, activity or sporulation of the poor competitor. Many BCAs are able to outgrow the slow-growing plant pathogens, which are 'starved out' or prevented from gaining access to the specific host tissues that favor their development. Parasitism indicates the ability of one organism to obtain required nutrition from another organism. Predation is generally seen in the interaction between animals resulting in killing of one organism by another for consumption. Protection to plants may be provided by exploiting suitable forms of interactions between biocontrol agents and the microbial plant pathogens. The biological control potential of the fungus-like and fungal species is determined by screening their isolates/strains under in vitro and in vivo conditions by performing various tests for determining the adverse effects of the BCAs directly or indirectly on the pathogen development.

3.1.1.1.1   Laboratory Tests

The dual culture method, a preliminary test, is followed to select fungi with antagonistic activity against the target microbial pathogen(s). Various isolates of the pathogen and the test fungal species/isolates are grown and maintained in a suitable medium. The mycelial plugs or disks of the test isolates and fungal pathogens are placed at 4 cm apart from each other in the same petriplate with mycelia in direct contact with agar, so that they grow toward each other. Observations on the adverse effects of the test isolates on pathogen development are recorded and the isolates most effectively inhibiting the growth and sporulation and abnormalities induced are tentatively selected for further testing, as in the case of *F. oxysporum* against *Pythium ultimum* infecting cucumber (Benhamou et al. 2002) and *F. oxysporum* f.sp. *cubense* infecting banana (Nel et al. 2006a). Different species of *Trichoderma* possess the ability to parasitize fungal pathogens (mycoparasitism) and to produce antibiotics or other metabolites inhibitory to the pathogens (antagonism). *Trichoderma virens* (known earlier as *Gliocladium virens*) produced glitoxin, inhibitory to *Rhizoctonia solani* (see Figure 3.1). Another antibiotic isolated from *T. virens* was inhibitory to *Pythium ultimum*, but not to *R. solani* (see Figure 3.2). *T. virens* was able to parasitize *R. solani* by forming coils around the pathogen hyphae. The mutant strains of *T. virens* produced by ultraviolet light irradiation (G6-2, G6-15 and G6-57), were unable to parasitize *R. solani* like the wild-type strain, although they were able to produce the antibiotics (see Figure 3.3) (Howell 2003). The basis of antifungal activity of *T. harzianum* SQR-T037, effective against *F. oxysporum* f.sp. *cucumeris (Foc)*, causal agent of cucumber Fusarium wilt disease, was investigated. Presence of 6-pentyl-α-pyrone (6PAP) in the culture filtrate of *T. harzianum* was identified by mass spectrometry and nuclear magnetic resonance spectroscopy. The antifungal activity increased with increasing concentrations of 6PAP, as reflected in inhibition of mycelial growth, spore germination, sporulation and fusaric acid production by the pathogen. Furthermore, addition of 6PAP at 350 mg/kg to

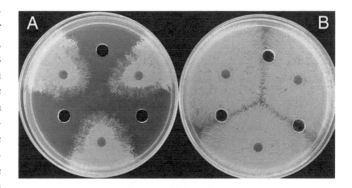

FIGURE 3.1 Inhibition of mycelial growth of *Rhizoctonia solani* by gliotoxin produced by *Trichoderma virens* Restriction of pathogen growth in gliotoxin-amended medium (A) and unrestricted pathogen development in nonamended medium (B).

[Courtesy of Howell 2003 and with kind permission of the American Phytopathological Society, MN]

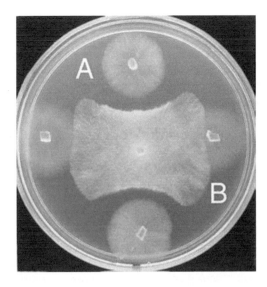

FIGURE 3.2 Differential biocontrol activities of *Trichoderma virens* of wild-strain (A) and mutant strain (B) on *Pythium ultimum* Wild-strain (A) was more efficient than the mutant strain (B) in inhibiting mycelial growth of the pathogen.

[Courtesy of Howell 2003 and with kind permission of the American Phytopathological Society, MN]

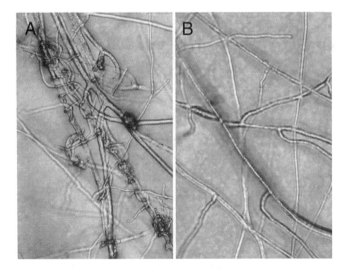

FIGURE 3.3 Relative mycoparasitic activity of wild-strain and mutant strain of *Trichoderma virens* on *Rhizoctonia solani*. The wild-strain forms tight coils around hyphae of the pathogen (A), whereas mycoparasitic activity-deficient mutant (B) could neither coil around or penetrate the pathogen hyphae.

[Courtesy of Howell 2003 and with kind permission of the American Phytopathological Society, MN]

cucumber-continuously cropped soil decreased population of indigenous *F. oxysporum* by 41.2% and reduced disease incidence by 78.1 to 89.6% in pot experiments. The dry weight of cucumber seedlings showed an increase by 60.0 to 92.6%. The results indicated the potential of 6PAP for the management of Fusarium wilt of cucumber (Chen et al. 2012).

Biocontrol efficiency may be assessed by examining the interactions between the BCA and the pathogen under light microscope and/or electron microscope. The interaction between *Rhizoctonia solani* and the mycoparasitic endophyte *Chaetomium spirale* ND35 was investigated under light and electron microscopes. Coiling of *C. spirale* around R. *solani* hyphae and its intracellular growth in *R. solani* was observed. Immunochemistry procedure and transmission electron microscopic observations revealed that contact between the antagonist and the pathogen was mediated by an amorphous ß-1,3-glucan-enriched matrix originated from cell wall of *C. spirale* and stick to *R. solani* cell surface. At the same time, R. *solani* cells reacted by forming hemispherical wall appositions which were intensely labeled by antibodies specific to ß-1,3-glucan. The appositions were formed at sites of potential antagonist entry. However, the antagonist could overcome this barrier effectively, indicating that production of ß-1,3-glucanase might facilitate breaching of the host barrier (Gao et al. 2005). The fungal endophytes, isolated from leaves of Norway maples, were screened for their efficacy in inhibiting the development of *Rhizoctonia solani*, infecting potatoes and several other crops. The mycelial growth inhibition by different fungi was determined. *Trichoderma viride* inhibited the growth of *R. solani* at a faster rate than that of *Phomopsis* sp., *Alternaria longipes* and *Epicocum nigrum*. The culture filtrates also inhibited the pathogen growth to different extent. Culture filtrate of *T. viride* was the most effective in inhibiting the growth of *R. solani*. Confocal microscopic examination of the interaction between *R. solani*, causing black scurf of potato tubers and *Trichoderma viride* revealed that the BCA could establish close contact with the pathogen hyphae by coiling. The coils were very dense and appeared to tightly encircle the hyphae of *R. solani*. Penetration of the pathogen hyphae by *T. viride* could be seen at 7 days after contact, followed by loss of turgor of hyphal cells. *Phomopsis* sp. and *Epicoccum nigrum* did not penetrate pathogen cells, but induced abnormal cell morphology and lysis of cells, possibly due to production of extracellular chitinolytic enzymes produced by these fungal biocontrol agents (Lahlali and Hijri 2010).

Agar drop test was performed as a preliminary screen to evaluate the potential of biological control agents from large number of samples. Five drops of potato dextrose agar (PDA) (0.7 ml/petridish) were dispensed in the petridishes. Mycelial macerate (10 µl) of the 1-week-old culture of the pathogen, *Sclerotium cepivorum* was used to inoculate the agar drops in petridishes. An agar plug (2 mm diameter) from the edges of the culture of each test fungus was added to each of the precolonized agar drops. The plates were then incubated for 5 weeks at 20°C. Sclerotia (100) from each agar drop were examined under the microscope for degradation (soft or collapsed) by squeezing them using a pair of forceps (Clarkson et al. 2001). In another method, petridish-grown test plants were used to assess the biocontrol potential of biological control agents. Fungal endophytes (349) were isolated from root segments of eggplant, melon, barley and Chinese cabbage grown as bait plants in a mixed soil containing forest soils. These isolates were inoculated onto axenically raised Chinese cabbage seedlings grown in petridishes to assess their efficacy in suppressing the development of Verticillium yellows disease of Chinese cabbage caused by *Verticillium longisporum*. Three isolates almost completely suppressed the development of virulent strain of *V. longisporum*. Two of these isolates were identified as *Phialophora fortinii* obtained from the roots of eggplant and Chinese cabbage plants. Third isolate was a dark septate endophyte (DSE) fungus obtained from barley roots. Seedlings grown for one week in the presence of the endophytes were challenged with *V. longipes*. In DSE-treated roots, some cell walls in the epidermal and cortical layers showed cell wall appositions and thickenings, which appeared to limit the ingress of the pathogen in the adjacent cells. Such marked host responses were seen in the root cells of cabbage seedlings preinoculated with the endophytes. Development of Verticillium yellows was more effectively hampered by DSE. External and internal disease symptoms were reduced by 84% and 88%, respectively, by DSE. Chinese cabbage plants treated with DSE yielded desirable level of marketable quality produce (Narisawa et al. 2004).

The antagonistic effects of *Trichoderma asperellum* and *Bacillus* spp. on *Macrophomina phaseolina* and *Fusarium solani*, causing charcoal rot and root rot diseases respectively in strawberry, were assessed by using dual culture (confrontation) assay. Radial growth of *M. phaseolina* and *F. solani* was inhibited by more than 36%. Preventive application of *T. asperellum* by root-dipping reduced the incidence of charcoal rot up to 44% under growth chamber conditions and up to 65% under field conditions. Likewise, crown rot and root caused by *F. solani* was reduced up to 100% under greenhouse conditions and up to 81% under field conditions. Application of *Bacillus* sp. (in formulation) was also effective in reducing charcoal rot incidence, but its effectiveness was variable against *F. solani*, depending on the growth conditions (Pastrana et al. 2016). *Trichoderma virens* strain 110 was transformed with the *gfp* gene from the jelly fish *Aequorea victoria*, coding for green fluorescent protein (GFP), as a reporter gene. The effect of transformation with *gfp* gene on the mycoparasitic activity of *T. virens* was determined by inoculating sclerotia of *Sclerotium rolfsii*, *Sclerotinia sclerotiorum* and transformed *S. minor* and wild-type strains of *T. virens*. Transformation with *gfp* gene did not alter the mycoparasitic activity on the sclerotia of the pathogens tested. Colonization of sclerotia was followed by fluorescent microscopy which revealed intracellular growth of the antagonist in the cortex (in *S. rolfsii*) and intracellular growth in the medulla (in *S. rolfsii* and *S. sclerotiorum*). Uniform distribution of mycelium of *T. virens* just beneath the rind of sclerotia of both *S. rolfsii* and *S. sclerotiorum* suggested that sclerotia became infected at several randomly distributed sites without any preferred point of penetration into the sclerotia (Sarrocco et al. 2006).

The biocontrol potential of *Microsphaeropsis* sp. strain P130A for suppressing the development of tuberborne inoculum of *Rhizoctonia solani*, causing potato black scurf disease was investigated, based on the viability of sclerotia produced in vitro and on both viability and production of sclerotia on potato tubers. Transmission electron microscopy (TEM) technique was employed to study the interactions between *R. solani* and *Microsphaeropsis* sp. The BCA significantly reduced germination, the percent reduction increasing with increasing period of incubation. Similar reduction in sclerotial germination was seen in tuberborne sclerotia, following treatment with the BCA, in addition to reduction in the number of sclerotia present on the tubers during 8 months of storage at 4°C, whereas the number of sclerotia increased in untreated control tubers. TEM observations showed that the BCA-induced rupture of the pathogen plasma membrane and that a chitin-enriched matrix was deposited at the sites of potential antagonist penetration. Host penetration was not associated with pathogen cell wall alterations, which occurred at the time of antagonist development in the pathogen cytoplasm. The cells of *R. solani* treated with crude extract of *Microsphaeropsis* sp. showed disorganization of cytoplasm and breakdown of plasma membranes. The results indicated that *Microsphaeropsis* sp. suppressed the pathogen development through antibiosis and mycoparasitism (Carisse et al. 2001). *Coniothyrium minitans* and *Microsphaeropsis ochracea* did not show any mutual antagonism in dual culture assays, indicating the feasibility of using them together for their suppression of *Sclerotinia sclerotiorum*. Combined application of both mycoparasites increased the sclerotia mortality in a temperature range of 16 to 26°C, compared to application of single BCA. With increasing temperature, the efficacy of *M. ochracea* decreased, but not that of *C. minitans*. Sclerotia degeneration by *C. minitans* was slightly faster than by *M. ochracea*. The BCAs were transformed via *Agrobacterium tumefaciens*-mediated transformation (ATMT), employing reporter genes, encoding GFP and DsRed and the degradation process was visualized using fluorescent microscopy. Both antagonists were able to penetrate effectively the sclerotia rind, and growth forming pycnidia in the cortex and medulla, ultimately degrading the sclerotia entirely in a period of 25 days after single inoculation. The ability to colonize the pathogen sclerotia by both BCAs was revealed by the independent production of pycnidia in the sclerotial medulla and on the sclerotial rind. Combined inoculation of *C. minitans* and *M. ochracea*, although had slightly delayed growth, the degradative effects on sclerotia were additive. The results indicated the advantage of enhancing the biocontrol efficiency by combining both BCAs which show no mutual antagonism to each other (Bitsadze et al. 2015). In confrontation assays, *Trichoderma* isolates with antagonistic activity against *Rhizoctonia solani* and *Sclerotium rolfsii* were identified with degree of antagonism being greater against *R. solani* than against *S. rolfsii*. Production of metabolites in all isolates of *Trichoderma* did not correlate with biocontrol efficacy. However, one *T. viride* isolate T14 produced highest amounts of inorganic phosphate, indole acetic acid (IAA) and siderophores and showed high antagonistic activity and growth-promoting activity (Kotasthane et al. 2015).

3.1.1.1.2 Greenhouse/Growth Chamber Assessments

Laboratory tests may not be useful to select biological control agents (BCAs) that have mechanisms other than mycoparasitism. Hence, various tests are performed to determine the biocontrol potential of BCAs using live plants under greenhouse/growth chamber conditions. The BCAs are applied for treatment of seeds and/or soil to suppress the soilborne pathogens present in the seeds/propagules or soils. Seed treatment with BCAs is preferred, because it is easy and economical. The biocontrol potential of *T. harzianum* KUEN 1585 (a commercial product) was assessed under pot culture conditions. The soil was infested with a pathogenic isolate of *F. oxysporum* f.sp. *cepae* (Foc) and onion seeds were coated with *T. harzianum* at 10 g/kg seed. The basal rot disease incidence was reduced by *T. harzianum* to a level comparable with the fungicides imidazole and prochloraz. Similar results were obtained in the field experiment also. Seed treatment with the BCA enhanced the bulb diameter of sets. Extracts from onion sets grown from treated seeds contained compounds with high antifungal activity against *F. oxysporum* f.sp. *cepae* (Coşkuntuna and Özer 2008). Cotton seeds were treated with *Trichoderma virens* strain G-6, *T. koningii* strain TK-7 and *T. harzianum* strain TH-23 and planted in sterile vermiculite seedling trays. Biocontrol efficacy was determined, based on the percentage of seedlings affected by damping-off diseases caused by *Rhizoctonia solani* (Howell et al. 2000). The efficiency of *T. harzianum* T22 was assessed along with the fungicide by treating spinach seeds against infection by damping-off pathogens, *F. oxysporum* f.sp. *spinaciae*, *Pythium ultimum* and *Rhizoctonia solani*. The BCA was applied as a commercial product, was as effective as the fungicide mefenoxam against *P. ultimum*. None of the treatments was able to reduce the incidence of diseases caused by all three pathogens (Cummings et al. 2009).

Soilborne fungal pathogens have been managed by fumigating soils with chemicals like methyl bromide, the use of which has been either banned or restricted in many countries. As an alternative to the chemical application, the effectiveness of natural volatiles from fungi like *Muscodor* spp. was assessed. Populations of soilborne pathogens, *Rhizoctonia solani*, *Pythium ultimum*, *Aphanomyces cochlioides* were reduced by mycofumigation with *Muscodor albus* and *M. roseus*. Consequently, sugar beet stand establishment was increased, while disease severity was reduced. The mycofumigants were applied as colonized agar strips, ground pasta and alginate formulations. The Stabileze formulation containing a mixture of water-absorbent starch, corn oil, sucrose and fumed silica was applied, and the Verticillium wilt disease incidence and severity showed marked reduction in eggplant (brinjal). Both *M. roseus* and *M. albus* reduced disease severity. *M. albus* mycofungicide reduced the populations of the pathogens significantly (Stinson et al. 2003). A biorational synthetic mixture of organic components mimicking key antimicrobial gases produced by *Muscodor albus* was employed in place of live BCA for the suppression of *Pythium ultimum*, *Rhizoctonia solani* AG2-2 and *Aphanomyces cochlioides*, causing sugar beet seedling diseases. The biorational mixture was as effective as the live M. albus starch-based formulation in suppressing the development of all three pathogens, causing damping-off and also reduced the number of galls induced in the roots of tomato by the nematode, *Meloidogyne incognita* (Grimme et al. 2007).

Phytophthora capsici, incitant of Phytophthora blight disease, is one of the important diseases, accounting for heavy losses in pepper. The effectiveness of *Muscodor albus* as a potential biofumigant was assessed under greenhouse conditions. *P. capsici* infested potting mix was treated with *M. albus* at three concentrations, mefenoxam (Ridomil) and with no treatment (control). Seedlings of five sweet pepper cultivars and one butternut squash cultivar were transplanted into the potting mix. A significant interaction between pepper cultivars and soil treatment was observed. *M. albus* at the highest concentration reduced disease severity slightly, but significantly on cvs. Alliance, Aristotle, Paladin and Revolution, compared with control plants. The extent of reduction in disease severity by *M. albus* on cv. Paladin, the most tolerant cultivar, was up to commercially acceptable levels (Camp et al. 200). The role of antimicrobial volatiles from *M. albus* in suppressing *Rhizoctonia solani* in soil and potting mix was studied. The volatiles reduced damping-off of broccoli seedlings, when pots containing soil or soilless potting mix infested with *R. solani* were placed in the presence of active *M. albus* culture without physical contact in closed containers. In contrast, agar plugs of *R. solani* on PDA were inhibited, when they were placed in the presence of *M. albus* incorporated in garden soil or soilless potting mix. Gas chromatographic analysis with solid-phase micro-extraction showed that isobutyric acid and 2-methyl-1-butanol were released from treated substrates. A significant relationship between production of isobutyric acid and extent of disease reduction was observed. Isobutyric acid was produced only for short time, peaking at 24 h in potting mix and 48 h in soil. Amounts of isobutyric acid released from soil were several times more in soil than in potting mix. The results suggested that soil environment was better for the biological activity or viability of M. albus than that of soilless potting mix (Mercier and Jiménez 2009).

Soil sterilization using chemical fumigants such as dichloropropene, organic sulfur and dazomet (3,5-dimethyl-1,3,5-thiadiazine-2-thione) and biofumigants, has been shown to reduce soilborne pathogens, particularly nematodes. The fungus *Clonostachys rosea* 67-1, a promising mycoparasite was evaluated either alone or in combination with dazomet for their effectiveness in suppressing the development of cucumber Fusarium wilt disease caused by *F. oxysporum* f.sp. *cucumerinum (Foc)* KW2-1. When *C. rosea* 67-1 was applied, after dazomet fumigation, incidence of disease was fully controlled (100%), compared with 88.1% and 69.8% for dazomet and the BCA applied alone treatments, indicating a synergistic effect between the chemical and BCA. The effects of chemical fumigation on colonization and activity of *Foc* and the interaction between the BCA and *Foc* were investigated. The growth of *Foc* decreased with increasing concentration of dazomet. When exposed to dazomet (100 ppm), the pathogen sporulation decreased by 94.4%. Observations under scanning electron microscope revealed severe damage due to fumigation. In the greenhouse, disease incidence was significantly decreased, following fumigation. By contrast, germination of spores of *C. rosea* increased by > sixfold in fumigated soil and its ability to parasitize fumigated *F. oxysporum* f.sp. *cucumerinum* was also significantly enhanced (Tian et al. 2014).

Various mushroom species produce many antimicrobial compounds that can inhibit fungi and bacteria. The spent mushroom substrate (SMS) of *Lentinula edodes* derived from sawdust bag cultivation was evaluated for its control potential against *Phytophthora capsici*, incitant of pepper Phytophthora blight disease. Water extract from SMS (WESMS) of *L. edodes* inhibited mycelial growth of *P. capsici* and also suppressed disease development in pepper seedlings by 65% and promoted plant growth by over 30%. In high performance liquid chromatography (HPLC) analysis, oxalic acid was detected as the principal organic compound in WESMS and it inhibited mycelia growth at a minimum concentration of 200 mg/l. In quantitative real-time PCR assay, the transcriptional expression of CaBPR-1 (PR protein-1), CaBGLU (ß-1,2-glucanase), CaPR-4 (PR protein-4) and CaPR-10 (PR protein-10) were significantly enhanced on WESMS- and DL-ß-amino butyric acid (BABA)-treated pepper leaves, compared to water-treated leaf sample. The results suggested that WESMS of *L. edodes* might suppress pepper Phytophthora blight disease through multiple effects including antifungal activity, plant growth promotion and induction of defense gene(s) (Kang et al. 2017). Avocado white root rot disease is primarily caused by *Rosellinia necatrix*. Both the pathogen isolates (19) and the biocontrol agent *Entoleuca* sp. isolates (21) were obtained by burying bait twigs around avocado escape trees. *Entoleuca* sp. was able to colonize roots of avocado and remained persistent up to 2 years, when it could be recovered from the inoculated avocado roots. Most isolates (86%) of *Entoleuca* sp. reduced disease incidence and hence, they were considered as the effective biocontrol agents with potential for the suppression of avocado white root disease (Arjona-Girona and López-Herrera 2018).

Fungal pathogens belonging to the genera *Rhizoctonia* and *Sclerotium* produce sclerotia, as survival structures resistant to adverse environmental conditions, and they are produced on crop residues and organic matter present in soil. Sclerotia of pathogens in soil samples can be retrieved by floatation and sieving and the biocontrol activity of BCAs may be assessed by sclerotial degradation assay. The degree of degradation of sclerotia was determined by squeezing them with forceps (as reflected by soft or collapsed sclerotia) under the light microscope. The BCAs causing greater level of degradation were further tested in pot trial, as in the onion white rot disease caused by *Sclerotium cepivorum*. Soil amended with medium-grade vermiculite and sclerotia of the pathogen were mixed and the BCA as wheat bran cultures (1 g/100 g) or spore suspension (1×10^7 spores/100 g) was incorporated into the soil in pots. Onion seeds were planted at one seed/pot, maintaining appropriate control. Seedling emergence percentage and appearance of disease symptoms were recorded at weekly intervals up to 14 weeks after planting (Clarkson et al. 2001). *Sclerotinia sclerotiorum* infects oilseed rape (*Brassica napus*) causing stem rot disease. *Penicillium oxalicum* PY-1 was evaluated for its biocontrol potential, using the hyphae-mediated infection procedure. The mycelial homogenate, culture filtrate and spores of PY-1 were sprayed on oilseed rape plants. The inoculated plants were incubated at 20 ± 2°C and 100% RH for 1 week. The number and size of lesions induced by *S. sclerotiorum* were recorded for each treatment. Spore suspension and culture filtrate (10-fold dilution) suppressed infection by *S. sclerotiorum* (Yang et al. 2008).

The biocontrol efficacy of *F. oxysporum* Fo47, nonpathogenic isolates obtained from banana rhizosphere and two field strains of *Trichoderma* T22 and T5 was assessed by treating the banana plantlets with spore suspension of the BCA isolates applied as drench. Pathogenic isolates of *F. oxysporum* f.sp. *cubense (Foc)* was multiplied in PDA (half strength). Mycelial plugs taken from cultures were multiplied in sterilized millet seeds. Pathogen-colonized millet seeds were transferred to steam-sterilized soil at the rate of 15 ml seeds/500 ml of soil. The infested soil was transferred to pots in which banana plants treated with BCAs were planted. Prior to planting, the banana roots were slightly bruised by manually squeezing the root system to facilitate initiation of infection by *Foc*. Disease development was verified by cutting the rhizome open using a scalpel, based on the disease severity rating according to Inihab's Technical Guidelines (Nel et al. 2006a). The combined effects of *T. harzianum* and *T. asperellum* on the development of bean root rot disease caused by *Fusarium solani* were assessed under greenhouse conditions. The roots of bean plants were dipped in fungal BCA biomass suspension (1.8×10^7 CFU) and planted in pots containing wheat bran-corn meal. Development of disease symptoms was recorded up to 6–8 weeks after planting. Stem sections of bean plants were examined for the presence of *F. solani*. Isolation of the pathogen from stem sections was made after sterilization. The BCAs either alone or in combination provided significant protection against the root rot disease (Akrami et al. 2009). The nonpathogenic isolate *F. oxysporum* F221-B inhibited effectively in vitro the mycelial growth of *Rhizoctonia solani*, which causes the serious root rot and wilt disease on lettuce grown under hydroponics system. The strain F221-B reduced the disease incidence and severity by about 60 to 89%, compared to control. The BCA strain also promoted the growth of three lettuce cultivars, increasing the fresh weight by 2 folds over healthy control. The results indicated the effectiveness of nonpathogenic *F. oxysporum* F221-B in reducing the disease and also promoting plant growth (Thongkamngam and Jaenaksorn 2017).

The influence of inoculum density of *Sclerotium cepivorum*, causal agent of onion white rot disease on the efficiency of *Trichoderma koningii* strain Tr5, was investigated. Soil amendment with the BCA grown on autoclaved white millet grain provided 63 to 79% control of white rot disease in soil containing 10, 25, 50 or 100 sclerotia of *S. cepivorum*/ g of soil added at the time of sowing. Pathogen sclerotial density did not have any significant change in the antagonistic activity of *T. koningii*. Rhizosphere colonization by *T. koningii* was determined by incubating onion roots sampled from plants growing in soil with appropriate sclerotial density on a *Trichoderma* selective medium. Isolates of *T. koningii* were identified based on morphologic characteristics and polygalacturonase (PG, EC 3.2.1.15) and pectinesterase (PE, EC 3.1.1.11) isozyme profiles. The extent of disease suppression increased with increase in concentrations of sclerotia of *S. cepivorum* and *T. koningii*. Effect of colonized millet grain was proportional and no further increase in the disease suppression could be seen, when *T. koningii* Tr5-colonized millet was added at > 1,590 kg/ha, at any concentrations of sclerotia of *S. cepivorum* (Metcalf et al. 2004). The effect of *T. harzianum* on the dynamics of *Rhizoctonia solani*, incitant of potato stem canker and black scurf was investigated. *T. harzianum* reduced the severity of symptoms, expressed as 'Rhizoctonia stem lesion index' (RSI), during the first 7 days postinoculation (*dpi*), when the inoculum was placed at certain distances (varying from 30 to 60 mm) from the host. With inoculum at 40 mm from the host, RSI were 6 and 40 with and without *T. harzianum*, respectively. The antagonistic effect was overcome at longer period after inoculation. *T. harzianum* reduced black scurf severity on progeny tubers. The mean number of progeny tubers infected per potato plant was reduced by BCA application, as well as the proportion of small tubers. Furthermore, the number of malformed and green-colored tuber was reduced in pots treated with *T. harzianum* than in the pots without the BCA (Wilson et al. 2008). *Trichoderma* strains have been shown to suppress the development of diseases caused by soilborne pathogens through multiple mechanisms of action, and also possess the ability to promote plant growth significantly. Strains of *Trichoderma atroviride* and *T. harzianum* grown on organic materials were applied as biopreparations in the soil in open-field lettuce cultivation. Populations of the BCAs were monitored over time. The multiplex PCR-*Trichoderma*-identification technique showed that populations of *Trichoderma* spp. increased in the soil and they were not detected in the untreated soil plots. The results of multiplex PCR assays were confirmed by the standard plating method. Indigenous species were identified in the field soil. However, the abundance of the BCAs was estimated to be relatively low (10^3 CFU/g dry soil), after application of biopreparations. *Trichoderma* spp. persisted at this population level even afer two years (Oskiera et al. 2017).

Environmental conditions have marked influence on the development of both the pathogen and the biological control agents (BCAs) in addition to host plants. The influence of different environmental and cropping conditions on nonpathogenic strain of *F. oxysporum* (CS-20 and CS-24) and *F. solani* (CS-1) was assessed under greenhouse and growth chamber conditions. Liquid spore suspensions (10^6/ml) of BCA isolates were applied to soilless potting mix at the time of tomato seeding and seedlings were transplanted in the pathogen-infested field soil 2 weeks later. Temperature regimes ranging from 22 to 32°C significantly affected disease development and plant physiology. Strain CS-20 significantly reduced Fusarium wilt disease incidence at all temperature regimes tested with disease reduction of 59 to 100%, compared to control treatment. Strains CS-24 and CS-1 reduced disease incidence in the greenhouse at high temperatures but were less effective at optimum temperature (27°C) for disease development. Both CS-1 and CS-20 strains were equally effective in suppressing the development of all three races of the wilt pathogen *F. oxysporum* f.sp. *lycopersici*, as well as multiple isolates of each race, reduction in disease incidence being 66 to 80%. Among the three strains, CS-20 could effectively reduce Fusarium

wilt disease of tomato under different environmental conditions (Larkin and Fravel 2002). In a later investigation, the effectiveness of *F. oxysporum* Fo-B2 in suppressing the development of *F. oxysporum* f.sp. *lycopersici* was assessed in three different environments viz., growth chamber with sterile soilless medium, greenhouse with fumigated and nonfumigated soil and nonfumigated field plots. The Fo-B2 strain reduced disease severity significantly, but the efficiency of disease suppression decreased, as the experimental environment became less controlled. The BCA was most effective, when it colonized vascular tissues intensively. The degree of Fo-B2 colonization was markedly reduced in plants grown in nonfumigated soil. The results suggested that indigenous soil microorganisms were a primary factor negatively impacting the efficiency of Fo-B2 (Shishido et al. 2005). Isolates of binucleate *Rhizoctonia* (BNR) obtained from soybean were screened for their biocontrol potential against *Rhizoctonia solani* AG-4 and AG2-2. Eight BNR isolates, when combined with AG-4 or AG2-2, significantly increased seedling emergence and survival of cv. Ozzie and reduced severity of disease, compared with pathogen alone (AG-4 or AG2-2). No interaction between BNR isolates (BNR-4, BNR 8-2 and BNR3) and seven soybean cultivars was evident. With AG-4, BNRs significantly increased emergence and survival of cultivars and reduced disease severity. On the other hand, with AG2-2, BNRs reduced disease severity. BNRs effectively suppressed *R. solani* in both potting soil mix and natural soil. BNRs alone significantly increased plant growth, compared with noninoculated controls. It was possible to isolate BNRs consistently from hypocotyls and roots, indicating colonization of tissues was associated with control. The results indicated the potential of BNR isolates for use against the soilborne *R. solani* which has a wide host range, including many crop plant species (Khan et al. 2005).

Fusarium crown and root rot (FCRR) disease of tomato caused by *F. oxysporum* f.sp. *radicis-lycopersici* (FORL) seriously impacts production in hydroponic rock wool systems. Six isolates of plant growth promoting fungi (PGPF), nonpathogenic *F. oxysporum* and five isolates of bacteria were evaluated for their efficacy in suppressing FCRR development. *F. equisetti* was the most effective and disease reduction rate by this BCA was consistently high and significant. Stem extracts from *F. equisetti*-treated and pathogen-challenged tomato plants inhibited the germination and germ tube length of FORL microconidia. Similar inhibitory activity of stem extracts of BCA-treated and uninoculated control plants was also observed (Horinouchi et al. 2007). Nonpathogenic *F. oxysporum* endophytes isolated from healthy banana roots were evaluated for their biocontrol potential against Fusarium (Panama) wilt of banana caused by *F. oxysporum* f.sp. *cubense*. The identity of the isolates was established using PCR-RFLP analysis. Under greenhouse conditions, 10 nonpathogenic *F. oxysporum* isolates significantly reduced disease incidence in banana plantlets generated via tissue culture technique. *Pseudomonas fluorescens* WCS417 and the most effective nonpathogenic *F. oxysporum* isolate were tested under field conditions. None of the fungal and bacterial BCAs could reduce the disease development, indicating the need for investigating factors contributing to protection of banana plants under field conditions (Belgrove et al. 2011). Two soilborne fungal endophyte strains, almost entirely suppressed development of virulent strain of *F. oxysporum* f.sp. *melonis*, causing melon Fusarium wilt disease, when inoculated into axenically raised melon seedlings in petridishes. The endophytes were identified as *Cadophora* sp., based on morphological characteristics and ITS1-5.8S rDNA-ITS2 sequences. The hyphae of *Cadophora* sp. grew along the surface of the root and colonized root cells of the cortex and reduced the ingress of the pathogen into adjacent cells. Endophyte-treated seedlings grown in soil amended with valine showed decrease in Fusarium wilt incidence in field plots (Khastini et al. 2014).

Coniothyrium minitans isolate Conio, grown on maize meal perlite and ground maize meal-perlite medium produced high sporulation (1.6×10^7 conidia/g inoculum). Preplanting soil incorporation of Conio at 10^{11} CFU/m^2 significantly reduced Sclerotinia rot in a sequence of three lettuce crops raised in the glasshouse. Reduced dosages (10^8/m^2) were not effective in suppressing disease development. After harvest of the second and third crops, application at full rate (10^{11} CFU) of the BCA reduced the number and viability of sclerotia recovered on soil surface and increased infection of the pathogen by *C. minitans*, compared with spore suspension or reduced rate of maize meal-perlite inocula. In the pot culture, *C. minitans* decreased carpogenic germination, recovery and viability of sclerotia and increased infection of sclerotia by the BCA, compared to the spore suspension, as in the glasshouse experiments. In addition, reduced maize meal-perlite treatment also decreased apothecial production, recovery and viability of sclerotia. The inoculum density of *C. minitans* appeared to be the key factor influencing the effectiveness of the biological management of Sclerotinia rot of lettuce (Jones and Whipps 2002). *Sclerotinia sclerotiorum*, causing white mold disease of common bean, produces sclerotia as survival structures in the soil, which contribute significantly to development of epidemics. The mycoparasitic activity of *Coniothyrium minitans* on pathogen sclerotia was investigated. Even a single conidium could infect pathogen sclerotia. Under optimal conditions, use of 2 conidia/sclerotium produced maximum infection of sclerotia (c. 90%) and produced up to 1,000 conidia. Similar results were obtained with stem pieces infected by *S. sclerotiorum*. Under field conditions, application of conidial suspensions of *C. minitans* to the bean crop, soon after white mold outbreak, resulted in higher percentage of sclerotial infection than later applications. Infection of sclerotia was high (90%), when conidial suspensions were applied within 3 weeks. Inoculum concentration or the BCA isolate did not have significant influence on sclerotial infection by *C. minitans*. The results showed that a suspension of 10^6 conidia/ml in 1,000 l/ha sprayed immediately after first symptom appearance led to > 90% infection of sclerotia of *S. sclerotiorum*. Infection of sclerotia, capable of preventing pathogen carryover, occurred within a broad range of inoculum quality (Gerlagh et al. 2003). Effects of different inocula of *Coniothyrium minitans* on carpogenic germination of

sclerotia of *Sclerotinia sclerotiorum* applied at different times of the year were investigated. Five spore-suspension inocula of *C. minitans*, including three different isolates (Conio, IVT1 and Contans) with a standard maize meal-perlite inoculum were evaluated. Maize meal-perlite inoculum at 10^7 CFU/cm^3 soil reduced sclerotial germination and apothecial production in a series of three glasshouse bioassays, with decrease in sclerotial recovery and viability in the second bioassay and increasing *C. minitans* infection of sclerotia in the first bioassay. Spore suspensions were less effective and inconsistent. Sclerotial germination was delayed or inhibited, when bioassays were made in the summer. High temperatures inhibited infection of sclerotia by *C. minitans* also, although the BCA could survive the high temperatures. Apothecial production is critically influenced by inoculum level of *C. minitans* (Jones et al. 2004). Lettuce drop disease caused by *Sclerotinia minor* accounts for appreciable losses. *Coniothyrium minitans* effectively suppressed the development of lettuce drop, but a commercial product of *Trichoderma* or an isolate of *T. virens* was ineffective. *C. minitans* significantly increased the percentage of sclerotia infected and reduced the percentage of viable sclerotia, when applied to lettuce plants exhibiting initial symptoms of the disease. *C. minitans* could be detected always in untreated control plants in the experiments conducted in greenhouse, indicating the likelihood of *S. minor*, being a natural resident in the soils and its natural suppressing effect on the pathogen (Isnaini and Keane 2007).

Polymyxa betae is the vector of one of the destructive diseases of sugar beet rhizomania caused by *Beet necrotic yellow vein virus* (BNYVV). In addition, *P. betae* infects the sugar beet roots causing enlarged roots (sori) containing the resting spores internalized with BNYVV. As the resting spores can remain viable for several years, BNYVV is persistent in infested soil. In order to eliminate resting spores of *P. betae*, the antagonistic fungi, including *T. harzianum* and *Talaromyces flavus* were isolated from soil samples from infested fields. The antagonistic potential of the BCAs was assessed in split plot trial under greenhouse conditions. The cystosorus population of *P. betae* was determined by staining seedling roots with lactic acid and fuchsine (lactofuchsine) at 60 days after sowing, under light microscope. Different methods of application of antagonists viz., soil treatment, seed treatment and combination of both were tested. The number of cystosori in one gram of the roots varied with treatments. Soil application of the BCAs was more effective in reducing cystosori population than seed treatment and root weight consequently showed increase over the control (Naraghi et al. 2014).

3.1.1.1.3 Field Trials

Many factors in different combinations influence the competence and capacity of biocontrol agents (BCAs) to inhibit the development of microbial plant pathogens and disease incidence/severity. The availability of suitable ecological stages in the life cycle of the pathogens, favorable environmental conditions for rapid buildup of BCA populations and persistence for long periods in the absence of crop hosts by producing resistant spores (resting) are important factors affecting the performance of BCAs under field conditions and they are considered for development of commercial products for large scale application. Field trials are conducted to determine the efficacy of the selected species/strains/isolates of the BCA that are found to be effective in the in vitro and/or greenhouse tests. Field experiments are carried out for two or more seasons/years at as many locations as possible. Treatments are arranged in suitable statistical designs, such as randomized block design (RBD) or split-plot design, depending on the number of treatments to be tested and compared. In the case of soilborne pathogens, sites that have past history of incidence of target disease at high levels are selected for testing. Alternatively, high populations of target pathogen are incorporated into the field soil, using pathogen biomass raised on artificial media or large quantities of infected plant materials are used for amending the soils. The efficacy of the biocontrol agents may be affected by the form of BCA preparations – spore suspensions, mycelial cultures with media, pellet or powder/granular – formulations (Narayanasamy 2013).

A high percentage of microorganisms isolated from different substrates, exhibited antagonistic activity under laboratory conditions, proved to be ineffective in the greenhouse and/or field assessments. Such a situation may probably be due to poor competitiveness and unavailability of environmental conditions required for the growth and proliferation rapidly. The putative BCAs should be fast-growing and aggressive against target pathogen(s) and possess a wide spectrum of activity. The putative BCA with two or more mechanisms may be preferred, because of the possibility of obtaining more effective suppression of pathogen development, compared with the BCA functioning via single mode of action on the pathogen. The fungal BCA *Talaromyces flavus* (anamorph–*Penicillium dangeardii*) was able to suppress *Verticillium dahliae*, incitant of Verticillium wilt disease of tomato, potato and eggplant (brinjal) and could also parasitize *Sclerotinia sclerotiorum*, *Rhizoctonia solani* and *Sclerotium rolfsii*. Chitinase produced by *T. flavus* effectively arrested the development of *S. rolfsii* and *V. dahliae*. Spore germination, hyphal growth and melanization of newly formed sclerotia of *V. dahliae* were significantly retarded by the antifungal compounds produced by *T. flavus*. Microsclerotia were killed, when treated with culture filtrate (CF) of *T. flavus* and toxicity of the CF was considered to be due to the glucose oxidase activity of *T. flavus* (Madi et al. 1997). Rapid colonization of plant tissues by the biocontrol agent (BCA), making them unavailable for infection by the fungal pathogen is one of the effective mechanisms of action by the BCAs. The biocontrol potential of *Talaromyces flavus* against *Verticillium albo-atrum*, causing Verticillium wilt disease of potato was assessed. *T. flavus* was applied as tuber treatment and soil application and the treatments were randomized in complete block design with four replications. The isolate Tf-Po-V52 was the most effective in suppressing disease development, when applied as tuber treatment with minimum infection index. Under field conditions, the BCA-treated plots had an infection index of 0.15 as against 3.5 in the untreated control plots. The results indicated the effectiveness of *T. flavus* for reducing the incidence and severity of Verticillium wilt disease in potato (Naraghi et al. 2010).

Sclerotinia sclerotiorum with wide host range, produces sclerotia externally on affected plant parts and also internally in the stem pith cavities. Sclerotia from infected plant tissues reach the soil, where they overwinter and produce apotheica or mycelium during the next season. Pathogens of this nature have to be managed by biological destruction at sites, where the inoculum for new infections is likely to be produced by establishing the biological control agents at the source of inoculum production. Sclerotia of *S. sclerotiorum* could be killed on root surface, inside roots and stems of sunflower plants by applying *Coniothyrium minitans* (Huang 1977). Application of *C. minitans* reduced apothecial production from sclerotia buried in soil, as well as increased parasitism on sclerotia produced on bean plants (Mc Laren et al. 1996). Exogenous nutrients from senescent petals or plant tissues and pollen grains are required for ascospore germination and elongation of germ tubes or mycelia of *S. sclerotiorum* to infect other plant tissues of alfalfa (Li et al. 2005). Senescent petals of alfalfa usually remain attached to pods during pod development, facilitating infection of alfalfa pods and seeds by *S. sclerotiorum*. Field experiments with *C. minitans* and *Trichoderma atroviride* showed that *C. minitans* effectively suppressed Sclerotinia seed rot in all three years of experimentation, whereas *T. atroviride* was ineffective in reducing seed rot in all field trials. Application of *C. minitans* was as effective as as the fungicide benomyl. The results revealed the potential of *C. minitans* to be used as an alternative to the commonly used fungicide (Li et al. 2005).

The interactions between *Sclerotinia minor*, causing lettuce drop disease and *Coniothyrium minitans* were investigated to determine the most susceptible stage in culture to attack by *C. minitans* and to determine the consistency of responses of *S. minor* isolates belonging to four major mycelial compatibility groups (MCGs) to the biocontrol agent. Four isolates of *S. minor* MCG1 and 5 each from MCGs 2 and 3 and one from MCG4 were treated in culture at purely mycelial, a few immature sclerotial and fully mature sclerotial phases with a conidial suspension of *C. minitans*. *S. minor* formed the fewest sclerotia in plates that received *C. minitans* at the mycelial stage; *C. minitans* was recovered from nearly all sclerotia from this treatment and sclerotial mortality was total. However, the responses of MCGs were inconsistent and variable. Field tests to determine the efficacy of *C. minitans* relative to the registered fungicide, Endura, on lettuce drop incidence and soil inoculum dynamics were conducted from 2006 to 2009. Contans (commercial product of *C. minitans*) treatments had significantly lower numbers of sclerotia than Endura and unsprayed controls, and drop incidence was as low as in Endura-treated plots ($P > 0.005$). The lower levels of lettuce drop in Contans treatment were correlated with significantly lower levels of sclerotia. But the lower levels of lettuce drop, despite the presence of higher inoculum level in the Endura treatment, could be attributed to prevention of infection by *S. minor*. It was suggested that for effective management of lettuce drop disease, application of Contans to reduce sclerotial inoculum in soil and fungicide Endura to prevent new infection by *S. minor* might be a practical strategy to be adopted (Chitrampalam et al. 2011).

The effects of a multicomponent treatment, consisting of oilseed rape seedcake fertilizer (amendment) and rice straw component augmented with *Trichoderma* sp. Tri-1 (Tri-1) on the development of Sclerotinia stem rot in oilseed rape and yield were assessed, in comparison with fungicide, carbendazim. Oilseed rape seed yield was significantly increased by the multicomponent treatment, compared with the fungicide in the fields infested with *Sclerotinia sclerotiorum*. Disease incidence in the field trials with residual rice straw + formulated Tri-1 (in oilseed rape seedcake) was significantly lower than that of untreated plots, suggesting that the increase in oilseed rape seed yield was at least in part was associated with biological control of the pathogen (see Table 3.1). Disease suppression was more effective than that provided by control + fungicide spray treatment. Field experiments with sclerotia applied to soil in mesh bags, allowing both recovery of sclerotia and the colonization of sclerotia by Tri-1 and the indigenous microflora showed that inhibition of sclerotial germination by straw + formulated Tri-1 treatment was greater than the untreated control at 90, 120 and 150 days, but the difference between treatments was not statistically significant. The persistence of Tri-1 in bare soil, on oilseed rape and in the presence of rice straw was determined in the greenhouse pot experiment. Populations of hygromycin-resistant Tri-1-like fungi were high in the control straw treatment, but significantly lower than that was found in the straw + Tri-1 treatment. Oilseed rape appeared to support populations of Tri-1-like fungi, although not to the same extent as the rice straw. Populations of hygromycin-resistant Tri-1-like fungi declined very rapidly in the soil. Only straw + Tri-1 treatment, had population below 10^2 CFU/g of soil, after 30 days. The

results indicated that the multicomponent treatment had the potential to reduce the incidence of Sclerotinia stem rot and also increase seed yield of oilseed rape crops (Hu et al. 2015). Among the *Trichoderma* spp. isolated from rhizosphere of various crops, *T. asperellum* (NVTA2) applied as talc formulation for root dipping and soil treatment reduced carnation stem rot disease caused by *Sclerotinia sclerotiorum* incidence by 11.8%, compared to control. In addition, treatment with NVAT2 strain, stimulated growth of plants and increased the flower yield (Vinodkumar et al. 2017).

The efficacy of three strains of *Trichoderma viride* (L4, S17A, 99-27) in suppressing the development of onion white rot disease caused by *Sclerotium cepivorum* was assessed in a three-stage screening system to degrade sclerotia of the pathogen on agar medium and in soil and to reduce white rot disease incidence on onion seedlings. The strains L4 and S17A were tested under field conditions, because of the greater efficacy shown in the laboratory and greenhouse assessments. These strains of *T. viride* were consistent in their antagonistic activity by reducing white rot symptoms, when they were used for seed treatment in 2000 and 2001. The BCA application was equally effective as the fungicide tebuconazole which induced phytotoxic symptoms. Application of *T. viride* strains L4 and S17A could be preferred because of the precise placement of BCA using a special drilling equipment and this requirement might limit the BCA application (Clarkson et al. 2001). In a later investigation, the effects of combination of onion waste compost (OWC) and spent mushroom compost (SMC) and *Trichoderma viride* S17A on viability of sclerotia of *Sclerotium cepivorum* were assessed under glasshouse and field conditions. Incorporation of OWC into the soil reduced the viability of sclerotia and also the incidence of the disease. In two field trials, OWC reduced sclerotial viability and also reduced white rot disease incidence, as effectively as the fungicide tebuconazole (Folicur). Addition of *T. viride* S17A to SMC facilitated proliferation of *T. viride* S17A in the soil and increased the healthy onion bulb yield. The results indicated that infection of onion plants from sclerotia could be prevented by amendment of soil with OWC, SMC or *T. viride* S17A (Coventry et al. 2006).

Phytophthora ramorum and *P. pini* cause the destructive sudden oak death (SOD) disease, which killed over a million oak trees in the California and Oregon States in the United States. *Trichoderma asperellum* was evaluated for its potential of suppressing the pathogen development in the container nursery beds. The effectiveness of soil solarization for eradicating *P. ramorum* in the surface soil of nursery beds and its effect on the establishment of *T. asperellum* in the solarized soil were assessed. In the field trials, the leaf inoculum of *P. ramorum* was buried at 5, 15 and 30 cm below the soil surface. Solarization for 2- or 4-weeks during summer of 2012, eliminated the inoculum buried at all depths in one trial, at 5 and 15 cm in another trial conducted in California, but only at 5 cm in Oregon trial. *T. asperellum* was applied in the solarized soil. BCA application did not result in reduction in pathogen recovery. Population of densities of the introduced *T. asperellum* at 5cm depth, often increased by two- to fourfold in solarized soil, compared to nonsolarized plots and the effect

TABLE 3.1

Extent of Suppression of Take-All and Rhizoctonia Root Rot Diseases of Wheat by Different Mutants and Wild-Type Strain of *Pseudomonas fluorescens* HC1-07 (Yang et al. 2014)

| | Disease severity rating[y] | | | |
| | Take-all | | Rhizoctonia root rot | |
Treatments[x]	Pasteurized	Raw	Pasteurized	Raw
HC1-07 nf	2.8d	3.2c	2.5c	2.4c
HC1-07 viscB	3.5c	3.6b	3.3b	2.9b
HC1-07 viscB-	3.1d	3.4c	2.6c	2.6c
HC1-07prtR2	3.9b	3.7b	2.8c	2.8bc
HC1-07prtR2-1	3.0d	3.3c	2.6c	2.5bc
Control + pathogen	4.4a	3.9a	4.3a	4.3a
Control + MC +pathogen	4.1ab	4.1a	4.1a	4.2a

x- nontreated; control + MC (seed treated with methyl cellulose).

y- disease rating scale: 0 to 8; means followed by the same letter are not significantly different at P = 0.05 according to the Kruskal-Wallis all-pairwise comparison test.

of enhanced BCA population was not significantly reflected on pathogen recovery from the soil. The results suggested that soil solarization might be employed to disinfest the upper layers of soil in container nurseries in some locations, where suitable environmental conditions prevail (Funhashi and Parke 2016).

Verticillium dahliae, causal agent of strawberry Verticillium wilt disease, has a wide host range that includes several crops. As the use of fumigants was restricted and reliable sources of resistance to Verticilliium wilt pathogen were not available, the need for finding out alternate strategies for restricting wilt disease was realized. The biocontrol potential of nonpathogenic strains of *Verticillium dahliae* was assessed by infesting the soil with test strains for two years. Application of nonpathogenic strains had positive effect on 20% of plants, whereas 30% of plants remained unaffected. But 50 to 60% of plants were impacted negatively, as they showed severe wilting symptoms, resulting in total loss. The results indicated that other factors such as use of certified plant materials, and presence of other pathogens infecting strawberry have to be considered for enhancing effectiveness of nonpathogenic strains of *V. dahliae* as biological control agent to protect the strawberry crops (Diehl et al. 2013).

An atoxigenic strain of *Aspergillus flavus,* that did not produce aflatoxin, was evaluated for its biocontrol potential against toxigenic strain of *A. flavus,* contaminating corn grains and cotton seeds. Atoxigenic strains could reduce aflatoxin contamination by displacing or excluding aflatoxin-producing strains of *A. flavus*. The strain AF36 has been applied as a BCA in commercial corn and cotton production in Texas and Arizona to reduce aflatoxin contamination. The efficiency of the strain AF36 in reducing adverse effects of *A. flavus* in pistachio orchards were assessed during 2008–2011. AF36 strain was applied as hyphae-colonized steam-sterilized wheat seed. In all pistachio orchards, application of the wheat-AF36 product substantially increased the proportion of vegetative compatibility group (VCG) YV36, the VCG group to which AF36 belonged, within *A. flavus* soil communities. Application of AF36 product in the subsequent years, further increased YV36 population in the soil, until it reached 93% of the *A. flavus* isolates in treated commercial orchards. Application of AF36 product resulted in reduction in the percentage of samples of nuts (20–45%) contaminated with aflatoxin in treated orchards. The extent of reduction in aflatoxin contamination of nuts was considered to be significant and valuable, because of the high value of pistachio nuts and the costs associated with risk of rejection of shipments (Doster et al. 2014). Powdery scab disease of potato is a serious problem in hydroponic system. Soil fungi (508) were isolated from potato roots grown in suspensions of soils from four potato fields in Hokkaido, Japan. They were screened for their biocontrol potential against *Spongospora subterranea* f.sp. *subterranea (Sss)* in hydroponic culture system under greenhouse and field conditions. The isolate Im 6-50, identified as *Aspergillus versicolor* was the most efficient in suppressing root infection by *Sss* and development of powdery scab on potato tubers. In a 3-year field trial, *A. versicolor* Im 6-50 suppressed powdery scab with a protection value of 54–70 (100

= complete protection), when applied directly on seed tubers, compared with the fungicide fluazinam that provided 77–93 protection value. The BCA could be detected from the surface of daughter tubers and from the soil in which *Sss*-inoculated tubers were planted, employing species-specific primers in PCR assay. The results indicated the ability of *A. versicolor* to establish on the stolon of inoculated potato plants and in the rhizosphere, contributing to the biocontrol potential against *S. subterranea* subsp. *subterranea* (Nakayama 2017).

The biological control potential of the fungal BCA *Muscodor albus* with biofumigation activity was investigated, for suppressing the development of common bunt (CB) disease of wheat caused by *Tilletia caries* by applying as a seed treatment or an in-furrow soil treatment. Dry rye grain culture of *M. albus* was ground into powder and applied as seed treatment at 125 mg/g seed to wheat seed infested with *T. caries* teliospores. For soil treatment, the culture was broken into small particles and applied in the furrow at 48 mg/m of row, along with teliospores-infested seed during planting. Treatments were evaluated during two growing seasons and two planting dates beginning in early spring, when soil temperatures were optimal for disease development (5–10°C) and then about 3 weeks later. In the first year, treatments in the first seeding date reduced common bunt (CB) from 44% diseased spikes in untreated controls to 12% and 9% in seed and in-furrow treatments, respectively, and from 6% in controls to 0% in both treatments in the second seeding date. In the second year, CB was reduced from 8% in controls to 0.5% and 0.25% for seed and in-furrow treatments in the second seeding date. The usefulness of *M. albus* application might be greater for organic wheat production, where it is obligatory to apply nonchemical methods (Goates and Mercier 2011). The higher fungi, Ascomycetes and Basidiomycetes have been underexplored for the presence of antimicrobial compounds effective against plant pathogens, particularly bacterial pathogens. Several proteins have been isolated from mushrooms and used in medicine. Protein extracts (150) from 94 different basidiomycete- and ascomycete-wild mushroom species were evaluated for antibacterial activity against *Ralstonia solanacearum*, which infects a wide range of plants, including several crops using microtiter plate assays. The mushroom extracts varied in their effectiveness in suppressing the development of *R. solanacearum* from partial inhibition to complete inhibition. Three extracts reduced disease progress and severity in artificially inoculated tomato and potato plants. The in vitro inhibitory activities of the extracts did not always correlate with in vivo disease suppression. The extracts from *Amanita phalloides* and *Clitocybe geotropa* showed that the active substances were proteins (180 kDa size). The results indicated the possibility of identifying effective biocontrol agents in higher fungi, which have not received attention of researchers (Erjavec et al. 2016).

3.1.2 Mycorrhizal Biological Control Agents

Mycorrhizal fungi have symbiotic associations with roots of several plant species and such associations are common benefitting both partners. Several plant species have coevolved with these symbionts, indicating their dependence on the association with mycorrhizae. Plants provide the fungi with photosynthates and they, in turn, obtain mineral nutrients from mycorrhizal fungi. In contrast, pathogenic fungi proliferate at the expense of host plants which ultimately succumb to the damages inflicted by the fungal pathogens. However, in some cases, the mycorrhizal association might be less mutualistic or even parasitic to the plants (Klironomos 2003). Two major types of mycorrhizal associations have been differentiated: (i) arbuscular mycorrhizal (AM) symbiosis and (ii) ectomycorrhizal (ECM) symbiosis. The AM symbiosis is formed between the roots of the higher plants and zygomycete fungi belonging to the order Glomales. The AM fungi are obligate biotrophs that depend on their ability to colonize suitable host plant species for completion of their life cycle. The mycelia of AM fungi have an extraradical phase that grows out into the soil and an intraradical phase that proliferates inside the root. These two phases grow in very different environments and distinctly differ in their morphology and physiology (Dodd et al. 2000). AM fungi transfer organic nutrients and water to plants and obtain carbohydrates in exchange. The role of AM fungi in providing protection to the plants against microbial plant pathogens is an important contribution of these fungi for the survival of the host plants. Ectomycorrhizal (ECM) symbiosis is between fungi belonging to Basidiomycetes and Ascomycetes and plants belonging to the taxons of Gymnosperms and Angiosperms. Large trees, shrubs and sometimes herbs develop ECM symbiosis. ECM fungi often form an extensive, fanlike network of hyphal structures that grow from ECM root. The ECM fungi can be grown in pure cultures. These fungi, while colonizing plant roots, form a fungal sheath known as mantle, which covers the rootlets. The hyphae can penetrate between root cells and form a network of intercellular hyphae, which do not penetrate into plant cells. Extraradical mycelium produced from the mantle spreads on the soils and also absorbs nutrients from the soil. The extraradical hyphae get aggregated, forming a rootlike structure called rhizomorphs. Hyphae of the mantle form the Hartig net behind the root cap cells. The Hartig net forms the interface between the plant and ECM fungus and is involved in the bidirectional nutrient transfer. The ECM mycelium forms a physical barrier, preventing the entry of root pathogens (Campbell 1989).

3.1.2.1 Assessment of Biological Control Potential

The biocontrol potential of various species and strains of the mycorrhizal biocontrol agents (BCAs) has been assessed under in vitro, greenhouse/growth chamber and/or field conditions. In vitro experiments have been used to select the most effective BCAs from a large number of isolates of species/strains in suppressing the development of the target pathogen(s) and to study the mechanisms of biocontrol actions of the selected BCAs. Greenhouse/growth chamber investigations have been useful to investigate the biocontrol potential under different environmental and growth conditions and to forward the ones for testing for their efficacy under natural field conditions.

3.1.2.1.1 Laboratory Tests

Biochemical changes, following colonization of roots by arbuscular mycorrhizal fungi (AMFs) and fungal pathogen were determined. Polyacrylamide gel electrophoresis (PAGE) technique was employed to assess the protective effects of *Glomus mosseae (Gm)* and *G. intradices (Gi)* against *Phytophthora parasitica*, causing tomato root rot disease. *P. parasitica* was inoculated on non-mycorrhizal and mycorrhizal tomato plants precolonized for four weeks with either AMF species. In non-mycorrhizal plants two acidic ß-1,3-glucanase isoforms were constitutively expressed and their activity was higher in mycorrhizal roots. Two additional acidic forms were detected in extracts from *Gm*-colonized tomato roots, but not in *Gi*-colonized roots. Roots infected by *P. parasitica* exhibited greater enzyme activities. However, *P. parasitica* did not induce the isoforms related to *Gm* colonization. In tomato plants colonized by *Gm*, infection by *P. parasitica* induced two additional basic isoforms that could be easily recognized. The results indicated that differences in the expression of isoforms of ß-1,3-glucanases in AMF-colonized and non-mycorrhizal plants following infection by *P. parasitica* might have a role in development of root rot disease in tomato (Pozo et al. 1999).

The molecular basis of the bioprotective effect of *Glomus mosseae (Gm)* on tomato root infection by *Phytophthora nicotianae* var. *parasitica (Pnp)* was investigated, using immuno-enzyme labeling technique on whole root segments. Infection intensity by the pathogen was lower in mycorrhizal roots. Immunogold labeling of *Pnp* in cross-section of infected tomato roots showed that inter- or intracellular hyphae developed mainly in the cortex and their presence induced necrosis of host cells. The cell walls and the contents showed strong autofluorescence in reaction to the pathogen. The hyphae of *Gm* and *Pnp* were present in most cases in different root regions and sometimes in the same root tissues. The number of *Pnp* hyphae, growing in the root cortex, was greatly reduced in mycorrhizal root systems. In mycorrhizal tissues infected by the pathogen, arbuscule-containing cells surrounded by intercellular *Pnp* hyphae did not show any necrosis. These host cells showed only weak autofluorescence (Cordier et al. 1998). The role of hydrogen peroxide (H_2O_2) in disease development was studied. It is difficult to study the mechanism of H_2O_2 generation and relieving its stress in intact plant. The role of mycorrhizal inoculation in chilli plants challenged with *Phytophthora capsici* was investigated to study the effect on hypersensitive response. In the control (without mycorrhizal BCA) T3 and with mycorrhizal BCA (T4), visible disorders were detected at two days after inoculation with *P. capsici*, but in the following days, T3 plants developed 25% more necrotic lesions on the leaves than plants in T4 treatment. Leaf necrosis correlated with H_2O_2 accumulation and the greater damage observed in T3 plants coincided with larger accumulation of H_2O_2, at 12 h after inoculation, accompanied by an increase in peroxidase (POX) and superoxide dismutase (SOD) activity. The T4-infected and mycorrhizal plants exhibited an earlier accumulation of H_2O_2 starting at 6 h after inoculation with lower levels, compared to T3 plants. The results suggested a smaller accumulation of reactive oxygen species (ROS), leading to decrease in the wounds observed and slightly reducing the advance of the pathogen. Mycorrhizal colonization appeared to contribute significantly to the maintenance of redox balance, during oxidative stress (Alejo-Iturvide et al. 2008).

Arbuscular mycorrhizal fungi (AMF) have the ability to suppress the development of soilborne pathogens like *F. oxysporum* f.sp. *lycopersici (Fol)*, incitant of tomato Fusarium wilt disease. The dynamics of root exudation of tomato in any intercropping system due to the AM fungus, *Glomus mosseae (Gm)* and/ or *Fol*, its effect on *Fol* development in vitro and effects of compounds identified in the root exudates were investigated. Gas chromatography-mass spectrometry (GC-MS) analyses showed an AMF-dependent increase in sugars and decrease in organic acids, mainly glucose and malate. In high-performance liquid chromatography (HPLC) analyses, an increase in chlorogenic acid was observed in the combined treatment of AMF and *Fol*, indicating the ability of AMF and *Fol* to induce changes in the composition of root exudates of tomato. Root exudates of AMF-colonized tomato roots stimulated spore germination rate of *Fol*, whereas coinoculation with AMF and *Fol* resulted in reduction in spore germination rate. In vitro assays revealed that citrate and chlorogenic acid in the root exudates were likely to be responsible for reduced spore germination rate in the combined AMF + *Fol* treatment, since these compounds occurring at the concentration naturally in the root exudates were inhibitory to *F. oxysporum* f.sp. *lycopersici*, infecting tomato plants (Hage-Ahmed et al. 2013).

Following infection by *F. oxysporum,* plants respond innately via a complex and integrated set of defenses, encompassing both constitutive and induced responses. They include production of signaling compounds such as salicylic acid (SA), jasmonic acid (JA), ethylene (ET) and abscisic acid (ABA) or reactive oxygen species (ROS). Changes in the concentration of the plant hormones in melon shoots, as a consequence of interaction between the plant, *F. oxysporum*, the antagonistic *T. harzianum* and the AMF *Glomus intraradices* were studied. Infection by *F. oxysporum* activated a defense response in melon plants, mediated by SA, JA, ET and ABA, similar to the one induced by *T. harzianum*.

When inoculated with the pathogen, both *T. harzianum* and *G. intraradices* attenuated the plant response mediated by the hormones ABA and ET, elicited by pathogen infection. *T. harzianum* was also able to attenuate the SA mediated response. In the three-way interaction (pathogen-antagonist-AMF), a synergistic effect on the modulation of the hormone disruption induced by the pathogen, was observed. The results suggested that induction of plant basal defense response and attenuation of hormonal disruption caused by *F. oxysporum* could be both mechanisms by which *T. harzianum* might control Fusarium wilt in melon plants, whereas the mode of action of *G. intraradices* appeared to be independent of SA and JA signaling (Martínez-Medina et al. 2010).

Systems for investigating the interactions among plants, arbuscular mycorrhizal fungus (AMF) and fungal pathogens under in vitro conditions offer advantages over pot culture experiments. Avoidance of unwanted contaminants and the possibility of highly controlled and nondestructive dynamic observations on the interactions are the main advantages. An in vitro system was developed for relating infection of premycorrhized soybean plantlets by *Fusarium virguliforme*, causing sudden death syndrome (SDS) in soybean and for monitoring the early step of *F. virguliforme* infection process in the presence of AMF *Rhizophagus irregularis*. Plantlets premycorrhized with *R. irregularis* were inoculated with *F. virguliforme* either locally using a plug of gel supporting medium (method 1) or using macroconidial suspension of the pathogen applied on the medium surface (method 2). In method 1, the hyphae of *F. virguliforme* reached the surface of the roots at 2 days after inoculation (dai), independently of the presence/absence of AMF associated with soybean plantlets. After contact with roots, the pathogen developed profusely and formed a dense network of hyphae during the following 48 h. *F. virguliforme* grew both inter- and intracellularly within the roots. Many hyphae were observed in the root tip zone also. Fungal structures with a swollen and pigmented cell could be visualized under the light microscope (see Figure 3.4). Faster and denser hyphal development occurred, when conidial suspension was used as inoculum. The root system was infected by the pathogen, regardless of the method of inoculation. Small necrotic

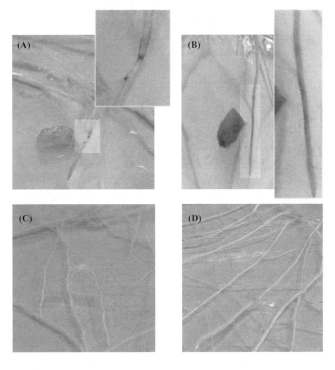

FIGURE 3.4 Biocontrol activity *of Rhizophagus irregularis* against *Fusarium virguliforme* determined using premycorrhized soybean plantlets A: premycorrhized soybean plantlets inoculated with the pathogen (inset showing root necrotic symptoms); B: nonmycorrhized soybean plantlets inoculated with the pathogen (inset showing root necrotic symptoms); C: uninoculated premycorrhized soybean plants and D: uninoculated nonmycorrhizal soybean plants.

[Courtesy of Giachero et al. 2017 and with kind permission of Frontiers in Plant Science Open Access Journal]

areas of 1–3 mm were produced near the point of penetration by the pathogen. The results showed that in the presence of AMF, the intensity of symptoms was reduced and root infection level was also less, compared with the control. The in vitro cultivation system showed the potential for investigating interactions involving AMF, pathogen and plant root system (Giachero et al. 2017).

3.1.2.1.2 Greenhouse Tests

The effects of AM fungi *Glomus fasciculatum* and *G. eutunicatum* on the growth of three strawberry cultivars and root infection by *Phytophthora fragariae* were investigated. Root necrosis induced by the pathogen was reduced in cv. Cambridge Favourite and Elsanta in mycorrhizal plants by about 60% and 30%, respectively, compared with nonmycorrhizal plants. No significant reduction in infection by *P. fragariae* in mycorrhizal plants of the least susceptible cv. Rhapsody was observed (Norman et al. 1996). In another investigation, non-mycorrhizal tomato plants showed widespread root necrosis induced by *Phytophthora nicotianae* var. *parasitica*, as well as significant reduction in growth parameters. In contrast, *Glomus mosseae*-colonized tomato plants showed less infection and reduction in growth parameters. Reduction in necrosis of adventitious roots, ranging from 63 to 89% was recorded in AMF-colonized tomato plants, compared with non-mycorrhizal plants (Trotta et al. 1996). In a later investigation, tomato plants colonized by *Glomus mosseae* were inoculated with zoospores of *P. nicotianae* var. *parasitica* with nonmycorrhizal inoculated control plants. At harvest, the control plants showed extensive root necrosis, reaching 61% infection, whereas infection was limited to 31% in mycorrhizal plants with increase in root system by 50%, indicating the additional advantage through AMF association (Vigo ete al. 2000). The effects of association of asparagus with three AM fungal species were assessed under glasshouse conditions. Incidence of Fusarium wilt disease caused by *F. oxysporum* f.sp. *asparagi* was reduced significantly in AMF colonized asparagus plants, indicating enhancement of tolerance to Fusarium wilt disease. No significant differences could be detected in phosphorus contents of plants due to association with any of the AM fungi, *Gigaspora margarita*, *Glomus fasciculatum* and *Glomus* sp. R10 tested (Matsubra et al. 2001).

Biological control potential of *Glomus intraradices* in suppressing the development of Fusarium wilt disease caused by *Fusarium solani* f.sp. *phaseoli (Fsp)* in bean plants and pathogen population in rhizosphere soil was assessed, using a compartmentalized experimental system. Specific primers in real-time PCR assay, culture-dependent methods and microscopic determination procedures were applied to quantify the pathogen and the AMF in the plant tissues, soil regions of mycorrhizosphere (rhizosphere + mycosphere) and the bulk soil. Nonmycorrhizal bean plants infected by *Fsp* exhibited characteristic disease symptoms, whereas infected plants and colonized earlier by *G. intraradices* appeared normal. The content of genomic DNA of Fsp was significantly reduced in mycorrhizal bean plants and in mycorrhizosphere of soil

compartment. Presence of *G. intraradices* in the mycorrhizosphere was not significantly modified, although the mycorrhizal colonization of roots was slightly increased in the presence of the pathogen. Reduction in pathogen population, as reflected by DNA contents and the root symptom might be due to biotic and/or abiotic modification of the mycorrhizosphere, as a result of colonization with *G. intraradices* (Filion et al. 2003). The effects of AMF *Glomus intraradices* and *G. claroideum* on the pea root rot disease, caused by *Aphanomyces euteiches* were investigated under greenhouse pot culture conditions, over the duration of three harvests, using oospores as the inoculum. Incidence of root rot disease was reduced in AMF-colonized plants, the beneficial effect being provided more effectively by *G. intraradices* than by *G. clarioideum*. At the final harvest, percentage of root length with oospores did not show variation between treatments. The mycorrhizal plants suffered less in terms of shoot growth and disease severity, compared to nonmycorrhizal plants. The degree of induction of tolerance varied, depending on the AMF species (Thygesen et al. 2004).

The AM fungi *Glomus mosseae*, *G. eutunicatum*, *G. fasciculatum* and *Gigaspora margarita* were evaluated for their ability to suppress the development of Phytophthora blight in pepper seedlings, caused by *Phytophthora capsici* and to promote the growth of pepper plants. Among the AM fungi tested, *G. mosseae* was the most effective in reducing the disease severity due to *P. capsici* by 91.7, 43.0 and 57.2% under pot, greenhouse and field conditions, respectively. The phytoalexin, capsidol concentration was increased in pepper plants preinoculated with P. capsici. The results showed that the AM fungus could improve the plant growth, in addition to suppression of Phytophthora blight disease in pepper plants (Ozgonen and Erkilic 2007). Plant growth depression in tomato + AM fungi is known in certain plant-AMF combinations. The AM fungi *Glomus intraradices*, *G. mosseae* and *G. claroideum* were evaluated for their efficiency in enhancing tolerance of tomato plants to Pythium root rot caused by *Pythium aphanidermatum* by employing fully factorial experiment. Two weeks after inoculation of tomato seedlings with AM fungi, roots were challenged with *P. aphanidermatum*. All AMF species caused significant growth suppression, but did not affect PR-1 gene expression or the phosphorus concentration in tomato plants. *G. intraradices* only reduced root infection by *P. capsici*, measured by ELISA and pathogen could be isolated by selective media at harvest time. Likewise, root infection by *P. aphanidermatum* reduced levels of colonization by *G. intraradices*, but not by the other two species of AM fungi. Although adverse effect on plant growth was known, the beneficial effect of AMF may be considered as an advantage of employing AM fungi for protecting tomato plants (Larsen et al. 2012).

The AM fungi *Glomus mosseae*, *G. monosporum*, *G. deserticola*, *G. intraradices* and two other unidentified species were evaluated for their efficacy in reducing the adverse effects of tomato wilt pathogen *F. oxysporum* f.sp. *lycopersici (Fol)* under greenhouse conditions. *G. mosseae* and *G. monosporum* improved the plant growth parameters such as plant height and fruit yield. In mycorrhizal plants challenged with *Fol*, root infection by the pathogen was significantly reduced, compared with nonmycorrhizal plants inoculated with *Fol* (Utkhade 2006). The combined effects of *Glomus mosseae* BEG12 and the bacterial biological control agent *Pseudomonas fluorescens* A6R1 on the development of root rot disease of tomato caused by *Rhizoctonia solani* 1556 were assessed under greenhouse conditions. The root and shoot growth of tomato plants infected by *R. solani* were compared, when protected or not by the BCAs. Microscopical examination of epiphytic and parasitic growth of the pathogen was studied in the presence and absence of the BCAs, allowing the quantification of roots with hyphae appressed to epidermal cells (epiphytic growth) and of roots with intraradical infection (parasitic growth). *G. mosseae* BEG12 and *P. fluorescens* A6r1 could fully overcome the growth depression of tomato caused by *R. solani* 1556. The suppression of disease development was associated with a significant decrease of the epiphytic and parasitic growth of the pathogen, along with increase in root length and number of root tips of inoculated tomato plants (Berta et al. 2005).

The comparative biocontrol potential of *Glomus mosseae, T. harzianum* and *Pseudomonas fluorescens* in suppressing the development of banana Panama (Fusarium) wilt disease caused by *F. oxysporum* f.sp. *cubense (Foc)* was assessed by inoculating in single, dual and tripartite combinations. The BCAs were allowed to colonize banana plants at 0, 45 and 90 days prior to challenge inoculation with the pathogen (1.5 × 10⁶ CFU/g). Appropriate control treatments were maintained. ELISA tests were performed to determine the populations of *Foc* every month. Banana plants preinoculated with *G. mosseae* + *T. harzianum* and challenged with *Foc* could sustain 61% and 70% improvement in plant height and girth respectively and 75% in bunch weight over plants with pathogen inoculation alone. These controls finally succumbed to the disease. Pathogen populations as determined by ELISA tests, were reduced to 0.58 OD in seven months in *G. mosseae* + *T. harzianum* treatments, compared to the level of 1.90 OD in control plants. The results suggested that the protection to banana by the BCAs might be due to physical modification in the cell wall, growth promotion and through induction of resistance to the Panama wilt disease (Mohandas et al. 2010). Isolates of *T. harzianum* and fluorescent *Pseudomonas* spp. were evaluated for their biocontrol potential for suppressing the development of tomato wilt disease caused by *F. oxysporum* f.sp. *lycopersici*. Application of *T. harzianum* and *Pseudomonas* sp. increased seed germination by 22 to 48%. All BCAs reduced the incidence of wilt disease in pot and field trials and combinations of BCAs were more effective than application of single BCA. Combination of *Pseudomonas, T. harzianum* and *Glomus intraradices* protected the plants more effectively, compared to unprotected control plants. Disease incidence and severity were reduced by 74% and 67% respectively in pots and field experiments. The combination treatments increased the yield also by 20%. Further, combinations of all three BCAs with cowdung compost (CDC) reduced the disease incidence by 81% and 74% in

pot and field experiments respectively and increased the yield by 33% (Srivastava et al. 2010).

3.1.2.1.3 Field Tests
Commercial formulations containing vesicular arbuscular mycorrhiza (VAM) *Glomus intraradices* were assessed for its effectiveness in suppressing Allium (onion) white rot (WR) in organic soils in comparison with the fungicide Folicur 3.6 F (430 g a.i./l, tebuconazole) under field conditions during 2000-2001. The AM product MIKRO-VAM used in transplanted onions reduced the incidence of white rot by about 50%, compared with untreated control and the extent of disease reduction was comparable to the fungicide Folicur 3.6F. Onion cultivar Hoopla was more susceptible than cv. Fortress to white rot disease in 10 of 13 field trials. A significant negative correlation between disease incidence and AM root colonization was recorded, indicating the effectiveness of AM root colonization in suppressing development of onion white rot disease (Jaime et al. 2005). The effects of treatment of soybean seeds with mefenoxam, fludioxonil, mefenoxam + fludioxonil, were assessed, maintaining controlwithout seed treatment. Soil fumigation with a mixture of 1,3-dichoropropene and chloropicrin was used as a base to determine the direct effect of the fungicide on plant growth and yield parameters. A significant fumigation x seed treatment interaction was indicated in 2005. Seed treatment with fludioxonil supported AM colonization on nonfumigated soil, where fludioxonil-treated plants had double the root colonization, as in the control and five times root colonization than in plants treated with mefenoxam. In the fumigated soil, plants treated with mefenoxam alone, or in combination with fludioxonil had lower AM colonization than the control and fludioxonil-treated plants. Fumigation did not significantly reduce mycorrhizal colonization across locations. No differences in grain yield, final stand or grain composition were recognized among seed-treated fungicides or between nonfumigated and fumigated soil. With the exception of mefenoxam in fumigated soil in 2005, no evidence of reduction in mycorrhizal colonization of soybean roots due to seed treatment with fungicide could be obtained under field conditions (Murillo-Williams and Pedersen 2008).

3.1.3 Bacterial Biological Control Agents
Bacterial species existing in various substrates such as seeds, propagules, plants, soil and water have been shown to possess antagonistic properties against soilborne microbial plant pathogens, resulting in suppression of disease development. The bacterial species are screened in vitro to eliminate less effective ones and the more efficient isolates are advanced to further testing in the greenhouse/growth chamber, followed by field testing under natural/inoculated conditions, where the putative biocontrol agents have to adapt to variations in environmental conditions and to compete with other microorganisms present on the plant surface and soil. The in vitro assessments are useful to demonstrate the ability of the bacterial species/strain to secrete enzymes, toxic metabolites or

siderophores, hormones involved in pathogen suppression and plant growth promotion. Caution has to be exercised in not eliminating the species/isolates that are not inhibitory directly to the target pathogen(s). Such species/isolates may act indirectly by stimulating innate defense systems of the host plant.

3.1.3.1 Assessment of Biological Control Potential of Bacterial Isolates

Various kinds of assessment methods have been applied to determine the biocontrol activities of the bacterial species/isolates against microbial pathogens.

3.1.3.1.1 Laboratory Methods

The inhibitory (antagonistic) activities of the bacterial biocontrol agents (BCAs) against fungal/bacterial pathogens are demonstrated by measuring the extent of inhibition of spore germination and mycelial growth of fungal pathogens, whereas the colony development/morphology of bacterial pathogens as affected by the bacterial BCAs are determined. The conventional dual culture or confrontation assay is performed in petriplates containing potato dextrose agar (PDA) or other media, favoring the differential development of the pathogen and the test bacterial isolates. The bacterial isolates are streaked on the medium at 2 cm from the periphery of the plate. Agar plates containing the mycelium of the fungal pathogens are placed at the rate of one plug/plate at the center of the plate and incubated at room temperature for 5–7 days, depending on the rate of growth of the fungal pathogen. An inhibition zone is formed between the bacteria and the fungus, if the bacterial isolate has antagonistic properties. The percentage of inhibition is calculated. The effectiveness of the myxobacteria *Myxococcus coralloides*, *M. flavescens*, *M. flavus* and four other species entirely inhibited the mycelial growth of *Sclerotinia sclerotiorum*, *Pythium ultimum*, *Rhizoctonia* spp. and *Phytophthora capsici*, whereas *Verticillium dahliae* and *V. albo-atrum*, *Cylindrocarpon* spp. and *F. oxysporum* f.sp. *apii* were less sensitive to inhibition by the myxobacteria (Bull et al. 2002).

Different variants of dual culture method have been developed to assess the antagonistic properties of bacterial biocontrol agents. The endorhizosphere bacteria obtained from root tips of tomato were screened for their antagonistic activity against *Verticillium dahliae*, causal agent of Verticilliuim wilt disease. Selected bacterial isolates (53 of 435) were antagonistic to *V. dahliae* and several other soilborne pathogens in dual cultures. Three efficient antagonistic isolates were identified as *Bacillus* sp. Two of the most effective bacterial isolates designated K-165 and 5-127 were rhizosphere colonizers and very efficient in inhibiting mycelial growth of *V. dahliae* and successfully suppressed the development of Verticillium wilt of solanaceous hosts (Tjamos et al. 2004). A variant of dual culture method was applied to select *Rhizobioum* isolates effective against *F. oxysporum* f.sp. *ciceris* race O, infecting chickpea. The bacterial isolates were streaked across the petriplates at the center and the second streak was made at right angles to the first streak. Discs (5 mm diameter) of the mycelium of a 7-day old culture were placed at

each side of the bacterial streak maintaining a distance of 2.5 cm between the reactants. Percent inhibition was calculated after an incubation period of 7 days (Arfaoui et al. 2006). The biocontrol potential of *Bacillus* strains (400) isolated from roots of cucumber plants grown in the greenhouses and fields was assessed against *F. oxysporum* f.sp. *cucumerinum*, causing cucumber wilt disease. The strain BO68150 was the most effective in suppressing the pathogen development under laboratory and greenhouse conditions. At seedling stage, the biocontrol efficiency was 50.68%. The results indicated that the mechanism of action of the strain BO68150 was likely to be other than antagonism against *F. oxysporum* f.sp. *cucumerinum* (Li et al. 2012).

Paenibacillus polymyxa strains (25) isolated from rotten ginseng roots were screened for their antifungal activity against *Phytophthora capsici* in vitro. The strain GBR-462 was the most efficient in suppressing the development of the pathogen. Antimicrobial activity was influenced by the initial inoculum density. No inhibitory activity was observed on mycelial growth and zoospore germination of the pathogen, when tested at lower inoculum density of 10^6 CFU/ml of *P. polymyxa* GBR-462. However, sporangium formation and zoospore release were significantly inhibited at the lower inoculum density. Furthermore, light and electron microscopic observations showed aberrant sporangia with no or few nuclei, indicating that the sporangium and zoospore formation were inhibited even at lower inoculum density. Application of *P. polymyxa* GBR-462 into potted soil suppressed the disease development and severity by 30%, compared with untreated control (Kim and Knudsen 2009). Isolates of *Bacillus amyloliquefaciens*, *B. licheniformis*, *Paenibacillus pabuli*, *B. atrophaeus*, *B. subtilis*, *B. pumilus* and *B. endophyticus* were evaluated for their antagonistic activities against *Phytophthora parasitica* var. *nicotianae*, using microbioassay. The assay involved the use of tobacco seedlings grown in petriplates for quantification of initial zoospore inocula and high throughput screening of antagonistic bacterial isolates. Zoospore inocula (10^2–10^4 spores/petri dish) were applied on 14 days old tobacco seedlings for susceptibility test. The optimum inoculum for infection of tobacco seedlings was 10,000 spores. Fifteen isolates of the above species were tested and four of them exhibited 100% protective activity in planta, as in petridishes. The microbioassay was rapid, reproducible and efficient for screening the bacterial isolates for their potential to protect tobacco seedlings under hydroponic conditions in petridishes (Wang et al. 2012).

The antagonistic activity of *Paenibacillus polymyxa* strains against *Phytophthora palmivora* and *Pythium aphanidermatum* was observed, using agar plates, liquid media and soil. *P. polymyxa* reduced colonization of pathogens in liquid assays. Most plants treated with *P. polymyxa* survived and infection by *P. aphanidermatum* was well correlated with mycocidal substance production by the BCA (Timmusk et al. 2009). Isolates of lactic acid bacteria (LAB) were evaluated for their efficacy against *Pythium ultimum* infecting tomato. The LAB isolates (294) obtained from soils and rhizospheres of maize, rye, carrot, garden soil and compost were tested by confrontation assay. About 75% of the isolates showed inhibitory effect on mycelial growth of *P. ultimum*. The most promising isolates protected tomato plants in the pot culture also, indicating their potential for application against *P. ultimum* (Lutz et al. 2012). *Paenibacillus elgii* HOA73 was isolated from soil samples and its identity was established, based on the 16S rRNA gene sequence. *P. elgii* was antagonistic to *F. oxysporum* f.sp. *lycopersici* and other plant pathogens. The bacterial culture filtrate (CF) was highly effective in inhibiting the mycelial growth (by 86.1%) at 50% concentration. The bacterial crude extract was able to inhibit pathogen growth by 72.5%. An antifungal compound was purified from bacterial crude extract, using different chromatographic techniques and it was identified as butyl-2,3-dihydroxybenzoate (B2,3DB). This compound displayed potent antifungal properties and inhibited pathogen growth by 83.2%, when applied at 0.6 mg. The minimum concentration of B2,3DB that inhibited any visible growth of *F. oxysporum* f.sp. *lycopersici* was 32 μg/ml, and the conidial germination of the pathogen was also inhibited at the same concentration of B2,3DB produced by *P. elgii*. The results showed the potential of *P. elgii*, as a biological control agent that could be applied against tomato wilt disease (Nguyen et al. 2015).

The comparative efficacy of the bacterial BCAs *Bacillus subtilis* QST713 and *Streptomyces lydicus* WYEC108 and fungal BCAs *Coniothyrium minitans* CON/M/91-08 and *T. harzianum* T-22 in reducing the survival of sclerotia and production of apothecia of *Sclerotinia sclerotiorum* was assessed under the growth chamber conditions. In general, the efficacy was positively correlated with the rate of application of BCAs. *B. subtilis* reduced production of apothecia and sclerotia by 81.2% and 50%, respectively. *S. lydicus* could inhibit production of apothecia completely (100%), but sclerotial production by 29.6% only. *T. harzianum* was less effective than other BCAs, in reducing apothecia production, but equally effective in reducing sclerotial production as the other BCAs. The results showed the variations in the biocontrol potential of the BCAs, indicating the need for selecting suitable BCA for application to effectively suppress the development of the pathogen and consequently incidence of the disease(s) (Zeng et al. 2012). *Bacillus amyloliquefaciens* Y1 was evaluated for its efficacy in suppressing the Fusarium wilt disease of tomato caused by *F. oxysporum* f.sp. *lycopersici (Fol)*, under in vitro and greenhouse conditions. The strain Y1was effective in inhibiting the growth of *Fol* and it also produced indole acetic acid (IAA), both in the presence and absence of tryptophan. The culture developed in Black White (BW) medium (Y1), BW medium amended with a commercial fungicide (BW + F) and BW medium alone were tested in the greenhouse conditions. Application of Y1 culture and BW + F resulted in significant reduction in disease incidence than in the BW. The shoot length and fresh and dry weight of both roots and shoots were greater in Y1 than in either BW or BW + F treatments. A similar trend was observed for chitinase and ß-1,3-glucanase activities in roots and leaves of tomato plants in Y1 culture treatment during most of the experimental duration. Presence of Y1 strain in the rhizospheric soils of Y1-treated plants

resulted in significant reduction in the populations of other bacteria. The results indicated the usefulness of *B. amyloliquefaciens* Y1 strains for disease suppression and growth promotion in tomato (Maung et al. 2017).

The biocontrol potential of eight isolates of *Bacillus* obtained from rhizosphere soil samples against *Sclerotinia sclerotiorum*, incitant of white mold disease of mustard was assessed. All isolates were identified as *Bacillus amyloliquefaciens* subsp. *plantarum (Bap)*, based on cultural, biochemical and molecular analyses of 16S rDNA and gyrase subunit A (gyrA). These isolates inhibited mycelial growth and suppressed formation of sclerotia in vitro. Deformities and cell wall lysis of mycelia, abnormalities of apothecia and inhibition of germination of ascospores, following interaction with *Bacillus* isolates were observed under light and scanning electron microscopes, indicating the antagonistic potential of *Bap*. Seed bacterization with *Bap* isolates provided protection to the mustard seedlings up to 98% against *S. sclerotiorum* under in vitro conditions (Rahman et al. 2016). In another investigation, strains of *Bacillus* spp. (105) isolated from soils were screened for their antagonistic activities against soilborne pathogens. The strain B14 most effectively inhibited the pathogens, *Sclerotium rolfsii, Sclerotinia sclerotiorum, Rhizoctonia solani, Fusarium solani* and *Macrophomina phaseolina*, with percentages of inhibition varying from 60 to 80. In addition, the strain B14 was able to produce siderophores and also auxins, which promoted plant growth. Based on 16S rRNA gene sequence, the strain B14 was identified as *B. amyloliquefaciens* (Sabate et al. 2017). Pink root disease caused by *Setophoma terrestris* is a major disease of onion, accounting for considerable yield losses in Argentina. *Bacillus subtilis* subsp. *subtilis* (BSS) isolated from the rhizosphere soil from onion plants was evaluated for its antagonistic activity against *S. terrestris*. The BSS strain showed strong inhibitory capacity specifically affecting pathogen growth, but it was not inhibitory to other onion pathogens *F. oxysporum* f.sp. *cepae* and *F. proliferatum*. The cell-free culture filtrate (CF) of BSS showed high growth inhibition of *Foc* in petriplates. Electron microscopy of cocultures of *S. terrestris* and BSS revealed thickened, tortuous, coiled fungal hyphae with granules and globule-like terminations. The results indicated the potential of *B. subtilis* ssp. *subtilis* as a biocontrol agent against *S. terrestris* (Orio et al. 2016).

3.1.3.1.2 Greenhouse Tests

The isolates of putative biological control agents that are highly antagonistic to soilborne microbial pathogen(s) under in vitro conditions are advanced to the next stage of determining their efficacy in suppressing disease development in whole plants under greenhouse/growth chamber conditions. These bioassays are useful in identifying the BCAs which may function directly by inhibiting pathogen proliferation or indirectly on the pathogen by activating host plant defense systems, resulting in enhancement of resistance to target disease(s). Take-all decline (TAD) of wheat is the resultant of natural suppressiveness of soils occurring in different geographical regions. The key role of fluorescent *Pseudomonas* spp., producing the antibiotic 2,4-diacetylphloroglucinol (2,4-DAPG), contributing to two Dutch TAD soils was demonstrated. The 2,4-DAPG-producing fluorescent *Pseudomonas* spp. were present on roots of wheat plants in both the TAD soils at densities at or above the threshold density required to suppress take-all of wheat caused by *Gaeumannomyces graminis* var. *tritici* (Ggt) and in a complementary take-all conducive soil, population density of 2,4-DAPG-producing *Pseudomonas* spp. were below the threshold level. Furthermore, introduction of 2,4-DAPG-producing strain SSB-17 into the take-all conducive soil suppressed development of take-all to the level observed in the TAD soil. Furthermore, a mutant of strain SSB-17, deficient in 2,4-DAPG production was not able to suppress the take-all development, indicating the key role of 2,4-DAPG-producing *Pseudomonas* spp. in TAD soils. The results indicated that the genotypic composition of 2,4-DAPG-producing *Pseudomonas* spp. varied between the Dutch TAD soils and the TAD soils from Washington State, although quantitatively both were similar (de Souza et al. 2003a).

Two isolates of endorhizospheric bacteria K-165 and 5-127, which most effectively suppressed the development of *Verticillium dahliae*, were tested in the glasshouse experiments. Root dipping or soil drenching of eggplants with bacterial suspensions (10^7 CFU/ml) resulted in reduced severity of wilt symptoms induced by *V. dahliae*, compared with untreated controls, under high pathogen inoculum level (40 microsclerotia/g of soil). In potato fields heavily infested with *V. dahliae*, seed tuber treatment with a bacterial talc formulation (10^8 CFU/g formulation) showed significant reduction in symptom development, expressed as percentage of diseased potato plants and the yield was increased by 25% in comparison with untreated control plots. The antagonistic bacterial isolates preferentially colonized the endorhizosphere of tomatoes and eggplants. The fatty acid analysis indicated that the isolate K-165 could be *Paenibacillus alvei*, whereas the isolate 5-127 might be *Bacillus amyloliquefaciens* (Tjamos et al. 2004). Antagonistic efficiency of *Brevibacillus brevis* in suppressing development of tomato wilt disease caused by *F. oxysporum* f.sp. *lycopersici (Fol)* was assessed, using plants raised in petridish microcosms and in pots in the greenhouse. Coinoculation of *Fol* and *B. brevis* resulted in marked decrease in symptom development, compared with inoculation with *Fol* alone. In addition, beneficial effect on tomato growth was also observed in plants coinoculated with the pathogen and BCA, indicating the usefulness of *B. brevis* as a potential biocontrol agent that could be considered for further investigation (Chandel et al. 2010). Antagonistic *Paenibacillus xylanilyticus* YUPP-1, *P. polymyxa* YUPP-8 and *Bacillus subtilis* YUPP-2 were isolated from cotton respectively at seedling, squaring and boll-setting stages. The effects of combined application of these bacterial species on Verticillium wilt disease development were assessed in pot experiments. Cotton plants in the combined BCA application were not infected before squaring stage, whereas infection rates of cotton treated with a single strain at seedling stage were 6.7, 6.7 and 13.3%, respectively for strains YUPP-8, YUPP-1 and YUPP-2. The control plants showed infection rate of 80%.

Field evaluations also showed that combined application of three BCAs was more effective than the individual strains. Verticillium wilt mortality rate and disease index of cotton at the boll-setting period were 9.4% and 6.5%, respectively, for combined application, whereas the control group had 47% and 32.8%. The results showed the effectiveness of combined application of three BCAs to keep the disease incidence and severity under check (Yang et al. 2013).

The biocontrol potential of isolates of *Mitsuaria* and *Burkholderia* obtained from rhizospheric soils of soybean and tomato against soilborne fungal pathogens was assessed, using seedling bioassays. Formation of lesions on developing seedlings was suppressed by isolates of *Mitsuaria*. Severity of lesion production in soybean challenged with *Pythium aphanidermatum* and tomatoes challenged with *P. aphanidermatum* and *Rhizoctonia solani* was significantly reduced by all isolates of *Mitsuaria*. Infection by *Phytophthora sojae* was also reduced by *Mitsuaria*. By contrast, *Burkholderia* isolates suppressed lesion production by *R. solani* in soybean (15%) and tomato (20%) (Benitez and Mc Spadden Gardener 2009). The knowledge of the genetic mechanisms of biocontrol activity provides the basic information for targeting bacterial population of interest, using gene-specific primers. Sequence analyses of those genes are employed to identify the strains of bacterial strains of interest. PCR-based approach to select novel biocontrol bacterial strains was applied by characterizing 2,4-diacetylphloroglucinol (2,4-DAPG)-producing *Pseudomonas* populations, based on amplification of *phlD* gene sequences. This procedure was employed to quantify the abundance and directly characterize the genotype of most abundant *phlD*⁺ populations inhabiting the rhizosphere of various crops. The use of functional markers revealed information on the ecology of bacterial biocontrol agents. On corn and soybean, native populations of bacterial strains exceeded levels required for in situ pathogen suppression (McSpadden Gardener et al. 2005).

The biological control potential may be assessed by applying the test isolates as soil application, seed treatment or root treatment to determine the extent of disease suppression. Strains of *Pseudomonas* spp. isolated from agricultural soils, river silt and rhizosphere soils, were evaluated for their antagonistic activities against *Rhizoctonia* spp. and *Pythium* spp., causal agents of root rot of wheat. Strains 14B2r, 15G2R, 29G9R, 39G2R, 48G9R and Wood 3R reduced the severity of symptoms induced by *R. solani* AG-8 and *P. ultimum*, under greenhouse conditions. The latter two strains suppressed the development of *R. oryzae* and *P. irregulare* also. In addition, four strains promoted growth of wheat plants, which correlated with disease suppression. Based on the 16S rDNA typing, the strains were identified as *Pseudomonas borealis*, *P. chlororaphis*, *P. fluorescens*, *P. marginalis*, *P. poae*, *P. putida*, *P. syringae* and *P. vranovensis* (Mavrodi et al. 2012). In a later investigation, of the 420 bacterial strains tested, *Bacillus subtilis* strain JN2, *Myroides odorantimimus* 3YW8, *Bacillus amyloliquefaciens* 5YN8 and *Stenotrophomonas maltophila* 2JW6 were more efficient (>50%) in suppressing the development of ginger wilt disease caused by *Ralstonia solanacearum* under greenhouse conditions (Yang et al. 2012). The biocontrol potential of 12 bacterial isolates against *Pythium* sp. and *Rhizoctonia* spp. infecting plants in soilless system was assessed. One strain of *Pseudomonas* sp. Pf4 closely related to *P. protegens* (earlier known as *Pseudomonas fluorescens*) was highly antagonistic against two strains of AG1-IB. Strains Pf4 was tested for its effectiveness in protecting lamb's lettuce against *Rhizoctonia solani*, causing root rot in small-scale hydroponics with significant reduction in root rot infection. The assessment of survival and population density of Pf4 on root showed that a density of BCA above the threshold value of 10^5 CFU/g of root tissues was required for effective disease suppression. PCR assays detected the presence of known loci in the Pf4 strain of *Pseudomonas* sp. The gene clusters *plt, phl, ofa* and *fit-rzx* required for biosynthesis of secondary metabolites very similar to those of *P. protegens* Pf5 were identified in Pf4 strain of *Pseudomonas* sp., revealing the potential of the strain for the management of root pathogens *P. aphanidermatum* and *R. solani* (Moruzzi et al. 2017).

Soilborne actinomycetes are known to produce antibiotics that have been used in plant disease management. The biocontrol potential of four isolates of soilborne *Streptomyces* spp. viz., J-2, B-11, B-5 and B-40 was assessed against *Sclerotium rolfsii*. Culture inoculum of isolates was more effective in inhibiting sclerotial germination than the culture filtrates. The isolate J-2 was the most effective in inhibiting sclerotial germination to the extent of 88 to 93%. The disease severity was significantly reduced in seedlings grown from BCA-treated seeds, the isolate J-2 being the most effective than the other isolates of *Streptomyces* spp. (Errakhi et al. 2007). Chickpea is seriously attacked by black root rot disease caused by *Fusarium solani* f.sp. *pisi*. Isolates of actinomycetes (100) obtained from soil samples were evaluated for their antagonistic activity against *Fusarium solani* f.sp. *pisi*. Based on dual culture test assessments in vitro, three most effective isolates S3, S12 and S40 were selected for greenhouse tests. The identity of these isolates was established based on 16S rDNA sequences. The isolates S3 and S12 were most similar to *Streptomyces antibioticus*, whereas the isolate S40 was very similar to *S. peruviensis*. Under greenhouse conditions, these isolates effectively suppressed the development of black root rot symptoms. The isolate S12 provided highest reduction on severity of the disease caused by *F. solani* f.sp. *pisi* (Soltanzadeh et al. 2016).

The biocontrol potential of *Burkholderia cepacia* strain 5-5B and Pesta formulations of binulcleate *Rhizoctonia* B (BNR) isolates (BNR621 and P9023) was assessed against *Rhizoctonia solani*, causing Rhizoctonia stem and root rot of poinsettia was assessed. Application of *B. cepacia* suppressed stem infection during propagation, whereas application of either isolate of BNR did not affect the disease development. In contrast, after transplanting rooted poinsettia, one application of either BNR isolate was more effective in suppressing stem and root rot than the application of *B. cepacia*. Sequential application of *B. cepacia* at propagation followed by a BNR isolate at transplanting was more effective over the crop production cycle than the multiple applications of both BCAs individually or combined application of both BCAs. Root colonization by both BCAs after transplanting rooted poinsettias was affected by application method. The least root colonization by both biocontrol agents occurred in the combined application. The highest root colonization by the BNR isolates was observed in the sequential application that provided the most effective disease control. The results showed that application of different biocontrol agents during the different production phases of poinsettia was effective for disease control (Hwang and Benson 2002). *Pseudomonas* CMR12a strain was evaluated for its potential to suppress the development of Rhizoctonia root rot of bean. The involvement of phenazines and cyclic lipopeptides (CLPs) in the antagonistic activity of the BCA was investigated under growth chamber conditions. The wild-type strain CMR12a significantly reduced disease severity caused by the intermediately aggressive AG2-2 and the highly aggressive AG4HG1 strains of *R. solani*. A CLP-deficient and phenazine-deficient mutant of CMR12a could also protect bean plants, but less efficiently, compared to the wild-type strain. Two mutants deficient in both phenazine and CLP production lost the entire biocontrol activity. Disease-suppressive capacity of CMR12a decreased, after washing the bacteria, before application to soil and thereby removing the metabolites produced during growth on the plate. In addition, microscopic observations, revealed pronounced branching of hyphal tips of both strains of *R. solani* in the presence of CMR12a. More branched and denser mycelium was also observed for phenazine-deficient mutant treatment. However, neither the CLP-deficient mutant nor the mutants deficient in both CLPs and phenazines influenced hyphal growth. The results indicated the involvement of phenazines and CLPs during *Pseudomonas* CMR12a-mediated biocontrol of Rhizoctonia root rot of bean (D'aes et al. 2011).

Bacterial isolates (1,487) obtained from rhizosphere, phyllosphere, endorhiza and endosphere of field-grown pepper were evaluated for their antagonistic activity against *Phytophthora capsici*, causal agent of pepper Phytophthora blight disease, by dual culture assay. The ability of the isolates to produce chitinase, siderophores, cellulase and protease was also tested. Based on the in vitro tests, 40 isolates were advanced for greenhouse test. The pathogen suppressiveness of the isolates ranged from 0.7 to 92.3%, with three isolates, B1301, R98 and PX35, showing maximum biocontrol potential, leading to reduction in disease severity up to 83.5, 92.3 and 83.5%, respectively. Based on 16S rDNA sequences, these isolates were identified as *Bacillus cereus* (B1301), *Chrysobacterium* sp. (R98) and *B. cereus* (PX35). The results indicated that the bacterial BCAs could be considered for large scale application after field tests (Yang et al. 2012). *Bacillus subtilis* SB24 was tested for its biocontrol potential under greenhouse conditions against *Sclerotinia sclerotiorum*, causing soybean stem rot disease. Cell suspension (5 x 10^8 CFU/ml), culture filtrate and broth culture effectively suppressed the development of the pathogen. All three preparations reduced the severity of disease by 60 to 90% (Zhang and Xue 2010). *Bacillus cereus* (SC-1 and P-1) and *B. subtilis* (W-67) were able to significantly reduce mycelial growth and sclerotial germination of *Sclerotinia sclerotiorum*, causing canola stem rot disease. These strains significantly reduced the viability of sclerotia in pot experiments in the greenhouse. Spray application of the BCA strains reduced disease incidence. The strains SC-1, when applied in the field at 10% flowering stage, reduced the disease incidence more efficiently than the strain W-67. However, the BCA strains were less effective than the fungicide Prosaro® 420SC (Kamal et al. 2015).

Streptomyces isolates (717) obtained from the rhizosphere of cucumber were evaluated for their antagonistic activity against *Phytophthora drechsleri*, incitant of damping-off disease of cucumber. Two isolates, C201 and C801 of *Streptomyces*, showing high inhibitory activity against the pathogen (> 70%) and cellulase activity in the presence and absence of NaCl, were tested under glasshouse conditions. The strains C201 and C801 reduced seedling damping-off disease incidence by 77% and 80%, respectively, in artificially infested soils. Strain C201 increased the dry weight of seedlings up to 21%. Analyses of 16S rRNA sequences revealed the close relationship of C201 and C801 to *S. rimosus* and *S. monomycini* respectively. Induction of systemic resistance (ISR) in *Streptomyces*-treated cucumber plants was inferred by increased activity of polyphenoloxidase (PPO) and peroxidase (POX) enzymes (Sadeghi et al. 2017). Two isolates J-2 and B-11 of *Streptomyces* sp. were tested under greenhouse conditions for their ability to protect sugar beet seedlings. Soil application of biomass and culture filtrate (CF) mixture of the isolates significantly reduced the root rot incidence. The isolate J-2 was more effective and increased fresh weight of sugar beet, indicating its potential for large scale use for the management of sugar beet root rot caused by *S. rolfsii* (Errakhi et al. 2009). In a later investigation, the antagonistic activities of species of *Streptomyces*, viz., *S. mycarofaciens* SS-2-243, *S. philanthi* RL-1-178 and *S. philanthi* RM-1-138 against *Sclerotium rolfsii* (stem rot) and *Ralstonia solanacearum* (bacterial wilt), infecting pepper (chilli) were investigated under greenhouse conditions. *S. philanthi* RL-178 suppressed Sclerotium root rot and stem rot as efficiently as the fungal BCA *T. harzianum* NR-1-52 and the fungicide carboxin. *S. mycarofaciens* SS-2-243 and *S. philanthi* RL-1-178 showed antagonistic activity against the pepper bacterial wilt pathogen *R. solanacearum* and the degree of disease suppression was similar to that of the antibiotic streptomycin sulfate. Under field conditions, soil inoculation with both pathogens was carried out. *S. philanthi* RL-1-178 protected the chilli plants agasint *S. rolfsii* and *R. solanacearum* more effectively than *S. mycarofaciens* SS-2-243 or *T. harzianum*, resulting in the survival of 58.75% of the chilli plants growing in the infested soil. The efficacy of the BCA was in equivalence to carboxin or streptomycin treatment (Boukaew et al. 2011). Actinomycetes isolated from groundnut (peanut) rhizospheric soils were screened by dual culture technique for their antagonistic activity against *Sclerotium rolfsii*, incitant of groundnut (peanut) stem rot disease. The effective isolates were evaluated under greenhouse conditions, using their culture filtrates and crude extracts to determine the extracellular antifungal

activity and characterize their biocontrol and plant growth-promoting traits. The most effective isolate RP1A-12 exhibited high level of antagonism against *S. rolfsii*. The isolate RP1A-12 produced hydrogen cyanide (HCN), lipase, siderophores and indole-acetic acid (IAA). Production of oxalic acid, a pathogenicity factor of *S. rolfsii*, was inhibited by crude extracts of RP1A-12, resulting in reduction of stem rot severity. RP-1A-12 was identified as closely related to *Streptomyces flocculus*, based on 16S rRNA gene sequencing (Jacob et al. 2016).

The biocontrol potential of *Rhizobium leguminosarum* bv. *viciae* (*Rlv*), applied as seed treatment, against damping-off disease caused by *Pythium* spp. affecting pea and lentil was evaluated. Treatment of pea seeds with *Rlv* strains R12, R20 and R21 increased root nodule mass. Seed treatment with the strain R21 was the most effective among the strains of *Rlv* in providing disease suppression to a level in equivalence to that of the fungicide Thiram™. By contrast, the strain 12 was the most effective in protecting lentil plants against damping-off and the treatment was as effective as the fungicide thiram. The strain 12 also enhanced plant growth, root nodule mass and shoot biomass. The disease suppression was strain-specific, as the strain R21 was more effective against damping-off affecting pea, whereas strain 12 was more effective in protecting lentil against damping-off disease (Huang and Erickson 2007). The efficacy of *Lysobacter capsici* strain PG4 in suppressing tomato Fusarium wilt disease caused by *F. oxysporum* f.sp. *radicis-lycopersici* (FORL) was assessed under greenhouse conditions. The strain PG4 effectively suppressed the development of the disease and it was found to be a good root colonizer. High population of PG4 (ca. 10^5 CFU/g of roots) could be recovered from the roots of plants growing from PG4-coated seeds. Disease incidence in treated plants was 24%, as against 86% in untreated plants inoculated with FORL. Further, PG4-treatment enhanced plant fresh weight, indicating the growth-promoting effect of *L. capsici* PG4 (Puopolo et al. 2010). Four isolates of *Streptomyces* sp. exhibited strong antagonistic activities against potato soft rot pathogens, *Pectobacterium carotovorum* and *P. atroseptica* in vitro and they were evaluated for their efficacy in suppressing soft rot symptoms in potato slices of cvs. Bintje, Yokon Gold, Russet and Norland. The biomass inoculum and culture filtrates of the BCA isolates were applied on tuber slices. The isolate identified as *Streptomyces* sp. OET reduced the disease symptom severity by 65 to 94% caused by both bacterial pathogens inducing soft rot in potato tubers (Baz et al. 2012).

The nonpathogenic strain of *Agrobacterium vitis* VAR03-1 was evaluated for its efficacy in suppressing the crown gall, by soaking the roots of grapevine, rose and tomato to protect against *A. vitis, A. rhizogenes* and *A. tumefaciens* respectively infecting these crops. The soil was infested with these pathogens, before planting the root-treated plants. Treatment with the strain VAR03-1 significantly reduced the number of plants with tumors and decreased disease severity in grapevine, rose and tomato. The degree of protection provided by VAR03-1 was greater than that by strain K84 used earlier. Furthermore, the strain VAR03-1 provided protection to a wide range of host plants against three tumorigenic *Agrobacterium* spp. (Kawaguchi et al. 2008a, 2008b). The nonpathogenic strains ARK-1, ARK-2 and ARK-3 isolated from graft unions were identified as *Agrobacterium vitis*. Stems of grapevine seedlings were inoculated with cell suspensions of strains of *A. vitis* (Ti), strain VAR03-1 and one of the three nonpathogenic strains, as competitors to assay the suppression of tumor formation by *A. vitis*. The ratio 1:1 of cells of pathogen/strains ARK-1, ARK-2 and ARK-3 reduced tumor incidence, the strain ARK-1 being the most effective in inhibiting tumor formation by *A. vitis*. The strain ARK-1 established its populations well on roots of grapevine tree rootstock and persisted on roots for one year. The strains ARK-1, -2 and -3 did not inhibit pathogen in YMA medium and the culture filtrate of ARK-1 also did not suppress tumor development. The results suggested that the nonpathogenic strains might suppress tumor development in grapevines through a mechanism different from that of VAR03-1 strain. The strain ARK-1 appeared to have potential for suppressing crown gall disease in grapevine effectively (Kawaguchi and Inoue 2012). In a later investigation, the nonpathogenic *Agrobacterium vitis* strain ARK-1 an endophyte in grapevine-controlled grapevine crown gall more effectively than the nonpathogenic strain VAR03-1 used earlier. The biocontrol potential of ARK-1 strain to reduce the crown gall disease affecting apple, Japanese pear, peach, rose and tomato was assessed. The roots of these plants were soaked in a cell suspension of ARK-1 before planting into the soil infested with tumorigenic *Agrobacterium* spp. The number of plants developing crown gall tumor was reduced due to treatment with the BCA. Analysis of results of greenhouse experiments showed that the integrated risk ratio (IRR) after treatment with ARK-1 was 0.29 for rose crown gall and 0.16 for tomato crown gall, indicating the effectiveness of ARK-1 treatment against *Agrobacterium* spp. The results of six field trials with peach conducted during 2010-2013, showed that IRR, was 0.38 for apple crown gall, 0.16 for Japanese pear crown gall and 0.20 for peach crown gall after treatment with ARK-1, indicating that the crow gall disease incidence could be significantly reduced by treatment with nonpathogenic ARK-1 strain of *Agrobacterium vitis* (Kawaguchi et al. 2013).

The biocontrol potential of *Bacillus subtilis* HS93, *B. licheniformis* LS674 and *T. harzianum* was assessed for suppressing Phytophthora blight (*Phytophthora capsici*) and root rot (*Rhizoctonia solani*) affecting pepper. Seed treatment and root drenching with suspension of HS93 with 0.5% chitin was more effective against *P. capsici* and *R. solani* infection than the bacteria alone. Likewise, enhancement of biocontrol activity of LS674 and *T. harzianum* by combining with 0.5% chitin was significant against *R. solani*, but this combination was ineffective against *P. capsici*. In the presence of chitin, the antagonistic activity appeared to increase against some fungal pathogens (Ahmed et al. 2003). The biocontrol potential of bacterial strains (231) isolated from soils and roots of cucumber, pepper and tomato plants from different locations were screened by seedling assay. Two-week old pepper seedlings were inoculated with zoospore suspension of *Phytophthora capsici*, causing Phytophthora blight disease. Four strains,

KJ1R5, KJ2C12, KJ9C8 and 11S16, were consistently effective in protecting pepper plants against *P. capsici*. These BCA strains could protect pepper plants under field conditions also (Kim et al. 2008). Strains of *Pseudomonas putida* FC-6B, *Pseudomonas* sp. FC-7B, *Pseudomonas* sp. FC-24B, *P. putida* FC-8B, isolated from used rockwool soilless substrates, were evaluated for their efficacy in suppressing Fusarium wilt disease of tomato caused by *F. oxysporum* f.sp. *lycopersici* (*Fol*). The pathogen was mixed with soil (at 5 x 10^4 chlamydospores/ml) and distributed in pots. Roots of tomato seedlings were dipped in 100 ml of bacterial suspension (10^8 and 10^9 CFU/ml) for 10 min and planted in pots. *Pseudomonas chlororaphis* MA342 strain applied at 7.5 x 10^6 CFU/ml, as root dipping treatment reduced wilt disease incidence by 20 to 55%. *Pseudomonas* sp. FC-9B strain (at 10^9 CFU/ml) was the most effective in reducing disease incidence and this strain promoted plant growth also (Srinivasan et al. 2009). Another bacterial BCA, *Brevibacillus brevis* suppressed tomato wilt disease incidence significantly compared with untreated and inoculated control treatment (Chandel et al. 2010).

The biocontrol potential of soilborne *Streptomyces* spp. against *Sclerotium rolfsii*, incitant of damping-off disease of sugar beet, was assessed. Four isolates of *Streptomyces* spp. J-2, B-11, B-5 and B-40 inhibited sclerotial germination in infested soil, compared to control. The isolate J-2 was the most effective, inhibiting sclerotial germination by 93% and 88%, respectively, when mycelial inoculum and culture filtrate (CF) were applied. Disease severity was reduced in seedlings growing from the BCA-treated seeds (Errakhi et al. 2007). The disease suppressive effect of putative BCA isolates in reducing the severity of potato scab disease caused by *Streptomyces turgidiscabies* was assessed. Of the 26 isolates, five actinomycetes isolated from either rhizosphere of soil of wild oats grown earlier in potato fields or soil adhering to potato stolons and tubers were effective antagonists. These isolates were identified as *Streptomyces* spp. The isolate WsRs-501 exhibited stronger inhibition of growth of *S. turgidiscabies*, compared to other isolates. Furthermore, WoRs-501 was the most effective in suppressing potato scab disease in pot experiments. A 10% (v/v) mix of WoRs-501 (6.2 × 10^8CFU/g dry mass) reduced the disease severity by 78 to 94%, in comparison to untreated control at a pathogen concentration of 5 × 10^4 to 5 × 10^6 CFU/g dry soil. The isolate WsoRs-501 could tolerate a wide range of pH levels and temperatures of the soil, indicating that this isolate with ability to adapt to different soil environmental conditions, might be an effective candidate suitable for a large-scale field application (Kobayashi et al. 2012). *Bacillus velezensis* BAC03 effectively suppressed the development of scab disease of radish caused by *Streptomyces scabies*. The strain BAC03 applied at 5 days before planting significantly reduced *S. scabies* population and completely suppressed scab disease development. But delaying the BCA application adversely affected its efficiency in proportion to the time interval between BCA application and planting. The strain BAC03 at 10^5 CFU/cm^3/potting mix or higher concentration was effective in reducing radish scab. Increasing the frequency of BCA application had no advantage in terms of disease suppression. The strain BAC03 increased the plant growth, whether the pathogen was present or not in the soil (Meng and Hao 2017).

Plant growth-promoting rhizobacteria (PGPRs) *Bacillus pumilus* SE34 and *Pseudomonas putida* SE89B61 were evaluated for their efficacy in suppressing the bacterial wilt disease caused by *Ralstonia solanacearum*, by treating the tomato seeds cv. Solar Set. After sowing the BCA-treated seeds on flats, the soil was drenched with suspensions of *R. solanacearum* (5 ml at 6 x 10^7 CFU/ml). Tomato plants were transplanted at 3 days after challenge inoculation into pots containing moist soil. Treatment of tomato plants with *B. pumilus* and *P. putida* reduced wilt disease incidence significantly, compared to controls. Two applications of the BCAs were more effective than single application (Amith et al. 2004). *Pseudomonas fluorescens* strain 1100-6 effective against *Agrobacterium vitis*, causing crown gall disease in grapevine was transformed with *gfp* gene encoding green fluorescent protein (GFP). The transformed mutant *P. fluorescens* 1100-6-*gfp* was injected into 3-year-old potted grapevine cuttings, rooted in perlite and then transferred to pasteurized potting mix. The disposition of genetically tagged strain 1100-6-*gfp* was evaluated, after 6 months by PCR amplification of the *gfp* gene in extracts from grapevine inoculated with tagged strain at the time of planting. Roots of plants inoculated with tagged strain either did not have or showed detectable populations near the soil line. But the *gfp*-gene sequence was detectable in stem sections cut from vines up to 6 cm above the soil line. Extracts of soil revealed detectable concentrations of the *gfp*-gene sequence in (3/5) containers, as well as the stem extracts from plants that cohabited with an inoculated vine. Observations on inoculated tissues, using epifluorescence microscope, revealed the presence of the BCA predominantly in the xylem tissues. Both the pith and xylem vessels had the BCA at the point of inoculation. Occasionally, persistence of the BCA in the external surface of grapevines was seen (Eastwell et al. 2006).

3.1.3.1.3 Field Tests

The putative bacterial biological control agents selected based on their effectiveness under laboratory and greenhouse/growth chamber conditions, are advanced to field evaluation for assessing their biocontrol potential under prevailing natural environmental conditions. The selected BCAs have to adapt to the various environmental conditions existing in the soils and also compete with numerous microorganisms for nutrients and appropriate niche, where they can proliferate to the required population levels to suppress microbial pathogens.

The effects of treatment of seed with strains of rhizosphere bacterial strains on development of damping-off disease of sugar beet, canola, safflower and pea caused by *Pythium* spp. were assessed. Based on in vitro assessments, 12 strains belonging to *Pseudomonas fluorescens* (708, 1-2, 1105, 1809, 2106 and 2201), *Bacillus cereus* PS1, *Bacillus megaterium* SB6, *Arthrobacter* sp. 2101, *Pantoea agglomerans* (909, 2-2) and *Erwinia rhapontici* A123 were found to be effective in reducing the incidence of sugar beet damping-off. Strains of *P. fluorescens* 708, 1-2, 2202, *B. cereus* PS1, *E. rhapontici* A123 and *Pantoea agglomerans* 2-2 were efficient in suppressing the development of damping-off of canola, dry pea and sugar beet in fields naturally infested with *Pythium* spp. when applied as seed treatment (Bardin et al. 2003). Isolates (14) of *Rhizobium* were evaluated for their biocontrol efficacy against chickpea Fusarium wilt disease caused by *F. oxysporum* f.sp. *ciceris*. The isolates Pch43 and Rh4 of *Rhizobium* were the most effective in reducing wilt incidence to less than 8% in chickpea, as against 48.7% infection in control plots. In addition, growth-promoting activity of the isolates was also clearly revealed under field conditions (Arfaoui et al. 2006).

The biological control agents (BCAs) have been applied as liquid cultures or formulated products on the soil at required concentrations. In general, soil application of BCAs is both expensive and difficult to achieve uniform coverage of pathogen-infested patches present in a field. *Pseudomonas putida* 06909-rif/nal was applied in irrigation water for suppressing the development of *Phytophthora parasitica*, infecting citrus. Application with irrigation water increased the BCA population, as well as its biocontrol efficacy against *P. parasitica* over that of single yearly applications at the commencement of irrigation season (Steddom and Merge 1999). The efficacy of *Pseudomonas fluorescens* SS101 and its surfactant (cyclic lipopeptides surfactant massetolide A)-deficient mutant *massA 10.24* to suppress the population and root infection of apple and wheat seedlings by *Pythium* sp. was assessed. Both parent (wild-type) strain and the mutant effectively suppressed resident *Pythium* populations to an equivalent level in the presence or absence of plant roots and ultimately suppressed *Pythium* infection to the same degree on all host plants investigated. The split-root plant assays were conducted, using strain SS101 or mutant 10.24 in the orchard soil to study the role of induced resistance in suppressing disease development. Strain SS101 or the mutant 10.24 significantly reduced the infection by *Pythium* spp. on the component of wheat or root system of plants grown on the soil treated with respective bacterial strain. Infection of wheat roots and Gala apple roots and Gala apple seedlings roots was reduced to 11% and 15%, respectively, as against 34 and 60–70% in untreated soil, following application of wild-type and mutant strains of *Pseudomonas fluorescens*. The results of split-root assays indicated that strain SS 101 did not limit root infection by *Pythium* spp. via induced systemic resistance (Mazzola et al. 2007).

The actinomycete *Streptomyces misionensis* (Sm) strain PMS101, *Bacillus thermoglucosidasis* (Bt) strain PMB207 and *S. sioyaensis* (Ss) strain PMS502 were evaluated for their biocontrol potential against *F. oxysporum* f.sp. *lilli,* causing Fusarium seedling blight and basal rot of lily, in comparison with the fungicide Sporgon (50% procloraz-Mn complex). A large-scale trial in an automated and environment-controlled commercial greenhouse showed that treatment of scale bulblets of lily with *B. thermoglucosidasis* (Bt) or Sporgon (100 µg/ml) and Sm resulted in a significant reduction in the incidence of seedling blight. The difference between these two treatments was not significant (P > 0.05). The results of greenhouse and field trials showed that the treatment of scale bulblets or 1-year-old bulbs of lily with Bt strain PMS101 at $1–1.2 \times 10^8$ CFU/ml or Sm strain PMS101 at $1–1.4 \times 10^8$ CFU/ml) without Sporgon was also effective in suppressing the development of basal rot disease of lily. The results showed the effectiveness of *B. thermoglucosidasis* and *S. misionensis* as BCAs for the management of Fusarium seedling blight and basal rot of lily (Chung et al. 2011). The protective effects of compatible endophytic bacterial strains *Bacillus subtilis* EPCO16 and EPC5, and rhizobacterial strains *Pseudomonas fluorescens* Pf1 for the protection of pepper (chilli) against wilt pathogen *Fusarium solani* were assessed. Application of the bacterial strains singly or in combination under greenhouse and field conditions reduced the disease incidence by enhancing host plant resistance. Higher levels of activities of peroxidase (PO), polyphenoloxidase (PPO), phenylalanine ammonia lyase (PAL), ß-1,3-glucanase, chitinase and phenolics indicated the induction of systemic resistance (ISR) in treated pepper plants. Combinations of the three strains were more effective in reducing disease incidence than the individual strains (Sundaramoorthy et al. 2012).

The biocontrol potential of four bacterial strains *Pseudomonas chlororaphis* PA-23, *Bacillus amyloliquefaciens* BS5, *B. amyloliquefaciens* E16 and *Pseudomonas* sp. DF41 was assessed for suppressing the development of stem rot disease of canola caused by *Sclerotinia sclerotiorum* under field conditions. The strains PA-23 and BS6 were as effective in suppressing disease development as the fungicide Rovral Flo® (iprodione). Double-spray application of PA-23 and BS6 did not provide enhancement of protection level to canola plants, compared with plants receiving single sprays of BCA strains. The results suggested that *P. chlororaphis* and *B. liquefaciens* could be used for the management of canola stem rot disease under field conditions (Fernando et al. 2007). *Pseudomonas fluorescens* biotype F isolate DF37 and the Canada milk vetch extract (MVE) were selected, based on their effectiveness in suppressing the development of potato Verticillium wilt disease under in vitro and growth room conditions. *Bacillus pumulis* M1 was able to reduce the disease incidence on cv. Kennebec (highly susceptible), whereas isolate DF37 effectively reduced Verticillium wilt parameters (disease incidence, severity and vascular discoloration), in both cvs. Russet Burbank (moderately susceptible) and Kennebec. The bacterial isolate and plant extract were evaluated under field conditions, using potato cultivars Russet Burbank and Kennebec. In the first year of field testing, isolate DF37 and plant extract MVE effectively reduced the disease on Russet Burbank and Kennebec, respectively. The extent of reduction in per cent infection and vascular discoloration, due to DF 37 application were respectively 26% and 67%, and 45% and 55% for MVE application, compared to control. In the second-year trial, DF37 and M1 and MVE treatments reduced all wilt parameters (ranging from 19 to 31%) and increased the yield (18%) in cv. Kennebec. The isolate DF37 could reduce wilt incidence by 29 to 43% and increase the yield by 24% on cv. Russet Burbank (Uppal et al. 2008).

Bacillus subtilis QST713 strain produced lipopeptides and surfactant antibiotics, which did not have inhibitory effect on resting spores of *Plasmodiophora brassicae*, incitant of clubroot disease of canola. The biofungicide Serenade (formulation containing *B. subtilis* QST713) suppressed clubroot on canola under controlled conditions, but its performance in the field was inconsistent. The effect of timing of application of Serenade or its components (product filtrate and bacterial suspension) on infection by *P. brassicae* was assessed under controlled conditions, by applying as a soil drench at 5% concentration (v/v). The BCA and its components were applied to the planting mix infested with *P. brassicae* at seeding or at transplanting 7 or 14 days after seeding (DAS) to target primary and secondary zoospores of *P. brassicae*. QPCR assay was used to assess root colonization by *B. subtilis*, as well as *P. brassicae*. Serenade was more effective in reducing infection by *P. brassicae* than the individual components. Two applications of Serenade were more effective than one application, providing complete suppression of disease development. By contrast, the individual components could reduce the disease severity only to the extent of 62 to 86%. The pathogen DNA content was reduced in canola roots by 26 to 99% by Serenade at 7 and 14 DAS. The results of QPCR assay were strongly correlated with root hair infection (%) assessed at the same time (r = 0.84 to 0.95). Genes encoding jasmonic acid (BnOPR2), ethylene (BnACO) and phenylpropanoid (BnOPCL and BnCCR) pathways were upregulated by 2.2- to 2.3-fold in plants treated with the biofungicide, relative to the control plants. The results suggested the possibility of involvement of mechanisms, antibiosis and induced systemic resistance, in suppression of clubroot disease development by Serenade (Lahlali et al. 2013).

The ethanol extract of the bacterial biocontrol agent (BCA) *Serratia marcescens* N4-5 strain, applied as seed treatment effectively suppressed the development of damping-off of cucumber caused by *Pythium ultimum* in potting mix and in sandy clay loam soil. The ethanol extract of N4-5 strain was found to be compatible with isolates of *T. harzianum*. The Th23::*hph-egfp* isolates of *T. harzianum* and GL-21showed no inhibitory effect of ethanol extract of N4-5 in vitro. Isolate Th23::*hph-egfp* consistently colonized the cucumber seed coat (100%) and emerging root system (> 94%), when combined with ethanol extract of N4-5 strain. Microscopic observations showed that N4-5 ethanol extract-treated seed coat was more extensively colonized by Th23::*hph-egfp* strain than on thriam-treated seed (see Figure 3.5). The efficiency of colonization of cucumber rhizosphere by Th23::*hph-egfp* strain was determined by applying this strain as a drench to the soil area in combination with N4-5 ethanol extract and Thiram. There was no evidence of inhibition of colonization of Th23::*hph-egfp* by seed treatment with N4-5 ethanol extract. Similar populations of Th23::*hph-egfp* were detected in the cucumber rhizosphere over the period of experimentation (3 weeks) (see Figure 3.6). The inoculum sandwich technique was applied to determine biocontrol potential of suppressing damping-off in natural soil (sandy loam). The N4-5 ethanol extract applied as seed treatment was very effective, as indicated by greater

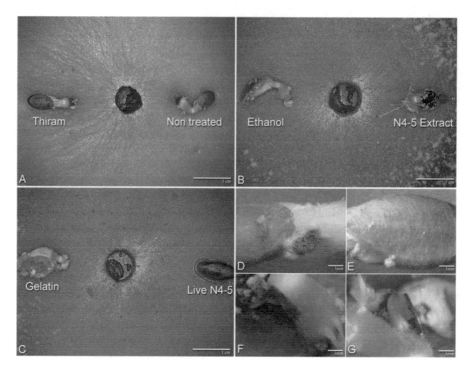

FIGURE 3.5   Effects of different treatments on colonization of cucumber seeds on by *T. harzianum* Th23::*hph-egfp* strain A: seed treated with thiram; B: seed treated with ethanol and *Serratia marcescens* N4-5 ethanol extract; C: seed treatment with gelatin and live N4-5 strain; D: magnified view of seed treated with thiram and colonization by Th23::*hph-egfp* strain; E:magnified view of untreated seed; F: magnified view of seed treated with N4-5 ethanol and G: live N4-5 strain.

[Courtesy of Roberts et al. 2016 and with kind permission of the American Phytopathological Society, MN]

plant stand, compared with untreated control. The ethanol extract of N4-5 strain, as seed treatment was generally more effective than the in-furrow application of Th23::*hph-egfp* and GL-21 strains and the commercial product Mycostop, containing *Streptomyces griseoviridis* strain K61. Combining N4-5 extract with isolates of *T. harzianum* isolates did not result in reduction in effectiveness of disease suppression, but improved in another field soil. The results indicated the potential of ethanol extract of N4-5 strain of *S. marcescens* for suppression of damping-off disease of cucumber either alone or in combination with isolates of *T. harzianum* (Roberts et al. 2016).

Four strains of *Pseudomonas* sp. FP22, FP23, FP30 and FP35 and strain of *Serratia plymuthica* HRO-C48 were evaluated for the impact on cotton Verticillium wilt development, parameters of cotton growth and yield in a naturally infested field. The cotton seed bacterization with BCA isolates reduced the AUDPC values, ranging from 22.1 to 50.9% in field trials in 2005 and 2006. The growth parameters were also beneficially impacted by the BCA treatments. The increase in seed cotton yield by treatment with BCA isolates ranged from 13.1 to 23.0% in cv. Sayar314 and from 4.2 to 12.8% in cv. Acala Maxxa in 2005, whereas there was no significance in seed cotton yield among the treatments in 2006 (Erdogan and Benlioglu 2010). *Pseudomonas fluorescens* strains HC1-07 and HC9-07 produce, respectively, cyclic lipopeptide (CLP) and phenazine-1-carboxylic acid (PCA), which determine the biocontrol potential of the strains against wheat take-all

pathogen *Gaeumannomyces graminis* var. tritici (*Ggt*) The seven-gene operon for the synthesis of PCA from *P. synxantha* 2-79 was introduced into the *P. fluorescens* HC1-07 rif strain to enhance the biocontrol activity of the recombinant strain HC1-07PH2 against *Ggt*. The strain HC1-07PHZ consistently inhibited the hyphal growth of three isolates of *Ggt*. The strain HC1-07PH2 applied at a dose of 10^2 CFU/kg seed suppressed development of take-all disease more effectively than the strain HC1-07rif and H09-07rif applied either individually or in combination. At 10^4 CFU/seed concentration, the strain combination HC1-07rif + HC9-07rif provided significantly greater level of protection to wheat plants than the individual strains including HC1-07PHZ (Yang et al. 2017).

Bacillus amyloliquefaciens IUMC7 and its culture filtrate (CF) showed potential for suppressing the development of *Ralstonia solanacearum*, causal agent of bacterial wilt disease affecting several crops. The mushroom compost infested with the strain IUMC7 was incorporated into the soil and it reduced the bacterial wilt disease severity in tomato plants under greenhouse conditions, compared to the control. Population of *R. solanacearum* decreased in soil inoculated with IUMC7 strain. Thin layer chromatography-bioautoradiography assay showed that one of the antimicrobial compounds produced by the strain IUMC7 might be an iturin-like lipopeptides. The mushroom compost amended with *B. amyloliquefaciens* IUMC7 might be useful for reducing the bacterial wilt disease severity in tomato and possibly in other crops also (Sotoyama et al. 2017).

A

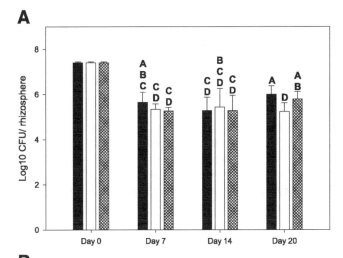

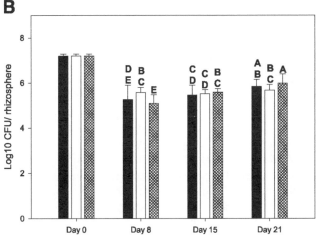

FIGURE 3.6 Effects of application of *T. harzianum* isolate Th23::*hph-egfp* in-furrow drench to the seed area in combination with seed treatment with thiram and N4-5 ethanol extract, on colonization of cucumber by the biocontrol agent A: experiment 1 and B: experiment 2 Bars with the same letter in the same subfigure (A/B) are not significantly different (P ≤ 0.05); LSD for experiment 1 = 0.3963 and LSD for experiment 2 = 0.2951 based on the mean log$_{10}$ CFU/rhizosphere.

[Courtesy of Roberts et al. 2016 and with kind permission of the American Phytopathological Society, MN]

3.1.4 VIRAL BIOLOGICAL CONTROL AGENTS

3.1.4.1 Viruses Infecting Fungal Pathogens

Reduction in virulence of fungal pathogens, following infection by viruses, is termed as hypovirulence. The phenomenon of hypovirulence is considered to have a role in counterbalancing plant diseases in nature. Mycoviruses are able to infect all major taxa belonging to the Kingdom, Fungi. The mycoviruses may have double-stranded (ds)-RNA or single-stranded (ss)-RNA as their genome. The ds-RNA mycoviruses are classified into three families, based on the number of genome segments, capsid structure and nucleotide sequences and some ds-RNA viruses remain unclassified. Isometric ds-RNA mycoviruses are classified into two families, Totiviridae and Partitiviridae, which include viruses with nonenveloped isometric particles of 25–50 nm in diameter, inducing latent

infections frequently in their host fungi. Members of the genus *Totivirus* have nonsegmented genome, while members of the genus *Partitivirus* genus have segmented genomes (Ghabrial 1998). Viruses included in the family Hypoviridae lack conventional virions and their ds-RNAs are enclosed in the host-encoded vesicles (Dawe and Nuss 2001).

Many fungal species have been shown to intraspecifically transmit hypovirulence-associated ds-RNA by anastomosis, following transfection. The ascospore progeny derived from the debilitated strain Ep-1PN of *Sclerotinia sclerotiorum* exhibited normal growth rate and typical colony morphology, indicating the failure of *S. sclerotiorum* debilitation-associated RNA virus (SsDRV) passing through sexual cycle. The homothallic nature of sexual reproduction of *S. sclerotiorum* would also impede the transmission of SsDRV through hyphal anastomosis. The ascospore progeny had similar virulence levels as the wild-type strain Sunf-M, when allowed to colonize detached leaves of oilseed rape. The debilitated strain Ep-1PN could survive on leaves of oilseed rape for more than 1 week and it could protect leaves from attack by Ep-1PNA1, a virulent ascospore progeny of EP-1PN with normal colony morphology. The debilitation phenotype of EP-1PN also could be transmitted to EP-1PNA1 in the soil and subsequently protected seedlings against invasion by normal virulent strains (Xie et al. 2006). A geminivirus-related DNA mycovirus conferring hypovirulence to *Sclerotinia sclerotiorum* was isolated and characterized. Intraspecific transmission of a hypovirulence-associated ds-RNA and hypovirulent phenotype from hypovirulent isolate Ss275 of *S. sclerotiorum* (with high mycelial incompatibility) to five virulent isolates of *Sclerotinia minor* (with fewer mycelial compatibility groups) was observed. The hypovirulent phenotype was successfully transmitted to one isolate of *S. minor* Sm10. Three putatively converted isolates of Sm10 had all cultural characteristics of the hypovirulent phenotypes and pathogenicity as well. In northern hybridizations, ds-RNA isolated from one of the converted isolates Sm1OT hybridized with a DIG-labeled cDNA probe prepared from ds-RNA isolated from Ss275. The RAPD analysis confirmed that the isolate Sm10T was derived from Sm10 and not from Ss275. The results revealed that intraspecific transmission of ds-RNA was possible (Melzer et al. 2002). The later investigation showed that *S. sclerotiorum* hypovirulence-associated DNA virus (SsHADV-1) could be transmitted from strain DT-8 to vegetatively incompatible *S. sclerotiorum* strain with relatively high frequency (Yu et al. 2010).

Phytophthora root rot disease of citrus is induced by *Phytophthora nicotianae* and *P. palmivora* in Florida. *P. nicotianae* isolate Pn117 was characterized as hypovirulent on citrus roots. The biocontrol efficacy of this hypovirulent isolate was indicated by nonrequirement for additional applications to sustain rhizosphere activity for 7 months, after citrus trees were planted. The isolate Pn117 induced less severe symptoms, compared to virulent isolate of *P. nicotianae* Pn 198 and *P. palmivora* Pp99. Inoculation of all rootstocks with hypovirulent Pn117 strain at 3 days prior to inoculation with virulent isolates, resulted in significantly less disease and production of greater amounts of roots, compared to seedlings inoculated with virulent isolates alone. Recovery of virulent strain Pp99 from coinoculated citrus was reduced, if Pn117 was preinoculated, compared with single inoculation with *P. palmivora* alone, indicating the possible competition between the hypovirulent and virulent isolates for space and nutrients (resources) available in host roots. Preestablishment of the hypovirulent isolate in root cortex might lead to loss of non-structural carbohydrates, reducing subsequent colonization by *P. palmivora*. The results indicated that the isolate Pn117 had the potential for use as an effective BCA against virulent isolates of *P. nicotianae* and *P. palmivora*, infecting citrus, since the hypovirulent isolate effectively colonized the host roots and persisted with soil amendments or repeated application. By contrast, the BCAs *Pseudomonas putida* and *Trichoderma viride* required weekly augmentations or addition of organic amendments that favor differential prolification of the biological control agents (Colburn and Graham 2007).

Rosellinia necatrix, incitant of white root rot disease, infects fruit trees and other woody plants. *R. necatrix* isolate W370 contains ds-RNA with12 segments, considered to represent a possible member of the family Reoviridae. The strain W370 was weakly virulent and its hyphal-tips became ds-RNA-free and strongly virulent. The 12 segments of W370 ds-RNA could be transmitted to hygromycin B-resistant strain RT37-1, derived from a ds-RNA-free strain of W370 in all or none fashion through hyphal contact with W370. The W370 ds-RNA-transmitted strains were less virulent than their parent strain RT37-1 on apple seedlings, with mortality ranging between 0% and 16.7% in apple seedlings that were inoculated with the W370 ds-RNA-containing strains and 50 to 100% for seedlings inoculated with the ds-RNA-free strains. Some W370 ds-RNA-containing strains killed more than 16.7% seedlings, but these became free of the ds-RNA in planta. The results indicated that W370 ds-RNA was the hypovirulence factor in R. *necatrix*. Furthermore, one strain lost one segment (S8) of W370 ds-RNA during subculture and the S8-deficient mutant strain also exhibited hypovirulence in *R. necatrix* (Kanematsu et al. 2004). In the further study, a member of the genus *Mycoreovirus* within the family Reoviridae isolated from *Rosellinia necatrix* was named as *Rosellinia necatrix mycoreovirus* 3 (W370) (RnMYRV-3) and established as the hypovirulence factor of the white rot pathogen. Two virus-free fungal isolates (W37 and W97) that were somatically incompatible with virus-harboring field isolate (W370) were transferred with purified RnMTRV-3 particles. Electrophoresis and northern hybridization analysis confirmed infection of *R. necatrix* by RnMYRV-3, which could be back-inoculated to respective virus-free isolates via hyphal anastomosis. Virus-infected strains produced smaller lesions on apple fruits than the virus-free isolates. Virus-cured strains were indistinguishable from wild-type strains in culture morphology and displayed similar level of virulence on apples. The level of virus accumulation varied among virus-transfected subcultures and within its single colonies (Sasaki et al. 2007). The host range of two viruses infecting *S. sclerotiorum* was investigated. Two mycoviruses, the partitivirus

RnPV1-W8 (RnPV1) and the mycovirus RnMyRV3/W370 (MYRV3) from *Rosellinia necatrix* were inoculated onto the protoplasts of fungal plant pathogens with purified virus particles. The presence of ds-RNA of viral genomes in regenerated mycelia of *Diaporthe* sp., *Cryphonectria parasitica* and *Valsa ceratosperma* was confirmed. Horizontal transmission of both viruses from newly infected strains to virus-free, wild-type strains through hyphal anastomosis was readily demonstrated by dual culture method. However, vertical transmission through conidia was rarely observed. The results showed that protoplast inoculation method was useful in extending the host range of mycoviruses (Kanematsu et al. 2010).

3.1.4.2 Viruses Infecting Bacterial Pathogens

Bacteriophages or phages are one of the most abundant biological entities present in the biosphere. Phages are viruses specifically infecting bacteria that subvert the metabolism of their bacterial hosts for their replication. The phages are classified into three different groups, Podoviridae, Myoviridae and Siphoviridae with different head sizes and tail lengths. The biocontrol potential of the selected phages has to be confirmed by further tests under field conditions. Bacteriophage therapy was applied for the control of dysentery in human beings in the preantibiotic era. Use of phages, as a management approach, was indicated by the investigation on cabbage black rot caused by *Xanthomonas campestris* pv. *campestris* (Mallmann and Hemstreet 1924). Application of bacteriophages for the management of bacterial plant diseases commenced in the early 19th century. Kotila and Coons (1925) demonstrated that bacteriophages isolated from the soil suppressed the growth of *Pectobacterium carotovorum* subsp. *carotovorum*, causing potato blackleg disease. Coinoculation of the phage and pathogen prevented rotting of potato tubers. In addition, they isolated phages effective against *P. carotovorum* subsp. *carotovorum* and *Agrobacterium tumefaciens* were isolated from a number of environmental sources such as river water and soil (Coons and Kotila 1925). The potential of phages for suppressing the development of bacterial diseases of crops has been exploited, because of the relative ease of preparing the phage isolates and low cost of production of phages. The effectiveness of phages in suppressing soft rot caused by *Pectobacterium atrosepticum* and *P. carotovorum* subsp. *carotovorum* was demonstrated, using slices of carrot and potato tuber, respectively (Coons and Kotila 1925; Kotila and Coons 1925). Use of bacteriophage biocontrol for suppressing some soilborne bacterial pathogens has been reported to be effective in some investigations. The principal factor for determining the applicability of bacteriophage biocontrol is the exclusive lytic cycle (virulent) nature of the phage. Virulent phages are those that cause infections resulting in lysis (death) of the host bacterium ultimately, leading to the release of progeny phage particles. Ideally, a phage for biocontrol applications should be exclusively lytic and possess a host range, which allows productive infection on all strains of the pathogen species being targeted. Unlike chemical biocides, phages occur naturally in the environment without any harm. After application, the number of phages increases, if the susceptible bacterial species are available to them. They tend to persist in high numbers in any environment as long as the host bacterial species is present, since they are obligate pathogens. Phages generally have a narrow host range, typically being limited to strains within a species of bacteria. Many soil factors such as pH, moisture, organic matter content and soil type may cause phage inactivation either individually or in combination. However, under favorable conditions, phages have been shown to persist in the soil at relatively stable concentrations for several weeks (Buttimer et al. 2017).

Soil samples for phage isolation were collected from tomato, pepper and tobacco fields. After sieving, the soil samples were placed in flasks and the moisture level was adjusted to 40% waterholding capacity with distilled water. Each flask was seeded with overnight cultures of strains of *Ralstonia solanacearum*, causal agent of bacterial wilt disease and incubated for 48 h. The soil samples were suspended in phosphate buffered saline, centrifuged and the supernatant containing the phages was assayed by soft agar overlay procedure. The phages produced plaques (the clear areas formed due to lysis of the susceptible bacterial culture) on the lawns formed by host bacterial pathogen colonies developed on CPG soft agar. The plaques are proportional to the concentration of the phage tested. Plaques were formed on the lawn of *R. solanacearum* culture at 30°C, after incubation for 24 h (Murugaiyan et al. 2010). Nine strains of *R. solanacearum*, including race 1 (biovars 3 and 4) and race 3 isolated from pepper, tomato and tobacco were used to test the host specificity of the phages obtained from the rhizospheres of the three crop plants infected by *R. solanacearum*. A filamentous phage PE26 had a relatively wide host range of bacterial hosts. PE226 and TM227 generated clear plaques on all nine strains of *R. solanacearum* tested. The morphological properties and the genomic characteristics of the two phages showed that they were probably related, although they were isolated from pepper and tomato, respectively. Both PE226 and TM227 had long, filamentous shape, which was different from other phages, infecting *Ralstonia* spp. A filamentous phage ØRSS1, infecting *R. solanacearum* strains was reported earlier by Kawasaki et al. (2007). But the genome organization of PE226 was different from that of phage ØRSS1. As the phage PE226 consistently formed clear plaques on nine strains of *R. solanacearum*, it might be a typical temperate phage carrying the properties of lysis and lysogeny of life cycles (Murugaiyan et al. 2010).

Ralstonia solanacearum can survive in the soil for long time. Furthermore, *R. solanacearum* is easily disseminated via soil, contaminated irrigation water, surface water, farm equipment and infected propagules, requiring an effective management system for containing the bacterial wilt disease induced by this pathogen. Phage therapy in agricultural settings, faced two major problems: (i) extracellular polysaccharides produced by bacterial pathogens, prevents phage adsorption and (ii) levels of susceptibility of bacterial strains to phages differed significantly. Different kinds of phages that specifically infected strains of *R. solanacearum*, belonging

to different races and/or biovars, were isolated and characterized. Phage ØRSA1, P2-like head-tail virus (Myoviridae) with a very wide host range infected all 15 species, strains of race 1, 3 or 4 and biovar 1, N2 or 4. Phage ØRSL1, another mycovirus, lysed 10 of 15 tested strains. Another phage ØRSB1 with T7-like morphology (Podoviridae) was able to lyse 14 of 15 strains from race 1, 3 or 4 and formed very large plaques (10–15 mm diameter) on assay plates (Yamada et al. 2007; Fujiwara et al. 2008; Kawasaki et al. 2009).

Application of bacteriophages for the management of bacterial diseases of crops appeared to be a practical proposition. The efficacy of three lytic phages ØRSA1, ØRSB1 and ØRSL1 in suppressing the development of tomato bacterial wilt disease was assessed. Infection with ØRSA1 and ØRSB1 either alone or in combination with other phages resulted in a rapid increase in the host cell density. The cells resistant to infection by these phages could be recognized at 30 h after addition of phage to the bacterial cell culture. In contrast, infection by ØRSL1, resulting in lysis of bacterial cells and lower host cell density (1/3 of control) were maintained over a long period, indicating the susceptibility of *Ralstonia solanacearum* to ØRSL1. Pretreatment of tomato seedlings with ØRSL1 phage drastically limited penetration, growth and movement of root-inoculated bacterial cells. Tomato plants treated with ØRSL1 phage did not exhibit wilting symptoms during experimental duration of 18 days, whereas the control plants without phage treatment wilted. Phage ØRSL1 was relatively stable in soil, especially at high temperatures (37–50°C). Phage could be recovered from roots of tomato plants and soil at 4 months postinoculation, indicating its persistence in plants and soil. As all *R. solanacearum* cells were not killed by ØRSL1, the coexistence of bacterial cells and phage might provide effective prevention of wilt disease incidence (Fujiwara et al. 2011). The filamentous phage ØRSM infected *R. solanacearum* and its virulence was also reduced. Inoculation of ØRSM3-infected bacterial pathogen cells into tomato plants did not induce bacterial wilt disease. But these cells enhanced expression of pathogenesis-related (PR) genes, including PR-1, PR-2b and PR-7 in tomato plants. Furthermore, pretreatment with ØRSM3-infected pathogen cells protected tomato plants against infection by virulent *R. solanacearum* strains. The effective dose of ØRSM3-infected cells for disease prevention was approximately 10^5 CFU/ml. As the bacterial cells infected by ØRSM3 could grow and produce infections, under appropriate conditions phage particles were produced continuously. It might be possible to develop the approach of employing phages as a management strategy for tomato bacterial wilt disease (Addy et al. 2012).

Persistence of phages at high populations in the phyllosphere and rhizosphere in close proximity to the target bacterial pathogens is of critical importance, as the efficacy is significantly influenced by inoculum densities of both phages and target pathogens. In order to determine the survival of phages in the rhizosphere and translocation into stems, the phages were applied to soil surrounding tomato plants. Phages were detected in foliar plant tissues at levels as high as 10^6–10^7 plaque forming units (PFU)/g plant tissue in the upper leaves and stems at two days after application. Phage populations decreased drastically and plummeted below the limit of detection by the seventh day in plants with damaged roots and by the 15th day in plants with undamaged roots. Suspension of phage specific to *Ralstonia solanacearum*, causing tomato bacterial wilt disease was applied to soil surrounding tomato plants as preinoculation and postinoculation pathogen treatments. Minimum disease reduction was observed, when the phage suspension was applied at 3 days before inoculation and at time of inoculation. Phage application was ineffective, if applied at 3 days after pathogen inoculation, indicating the lack of therapeutic activity of phage against tomato bacterial disease pathogen (Iriarte et al. 2012). *Pectobacterium* spp. cause soft rot diseases of potato and carrot. The effectiveness of bacteriophages in suppressing the development of soft rot disease was investigated. Bacteriophages capable of lysing diverse *Pectobacterium* species and isolates from plant and soil were identified. Repeated isolations from plaques resulted in the isolation of 189 single phages showing wide spectrum of host specificity. Core phages (24) were selected and a dendrogram was generated. Phylogenetic analysis based on DNA fingerprints of phage isolates indicated the extent of genetic diversity of the selected phages for investigation. The phages were stable at 16°C–40°C and pH 6.7. The viability (stability) of the phages differed significantly, depending on the phage isolate (Lee et al. 2017).

3.1.4.3 Cross-Protection with Mild Strains of Plant Viruses

Nonpathogenic isolates or strains of fungal and bacterial pathogens are known to protect plants against infection of virulent strains of the pathogens. The phenomenon of cross-protection involves application of mild strains of a virus to protect the plants against infection by severe strain or related virus. The mild strain-inoculated plants do not develop symptoms of infection by the severe strain of the same virus or related virus. The extent of protection depends on the relatedness of the mild strain with the severe strain. Cross-protection as a disease management strategy has been effectively applied for protection of perennial crops like citrus and annuals like tomato. Mild strains may be selected from naturally occurring strains or they may be artificially produced by exposing viruses or virus-infected plants to chemicals or unusual temperature (high or low) regimes (Narayanasamy 2002, 2013, 2017).

Grapevine fan leaf virus (GFLV) causes grapevine degeneration responsible for substantial losses in production. GFLV is transmitted by the soilborne ectoparasitic nematode *Xiphinema index* from vine to vine. Mild strains of GFLV and the closely related virus *Arabis mosaic virus* (ArMV) were identified by comparative performance analysis of grapevines infected with different strains and comparative severity of symptoms on *Chenopodium quinoa*, a systemic herbaceous host of GFLV (Huss et al. 1989; Legin et al. 1993). Protection with mild GFLV strains against GFLV infection on *Gomphrena globosa* was earlier reported (Bianco et al. 1988). In a later investigation, healthy scions were grafted onto

rootstocks that were healthy or infected with mild protective strains of GFLV-G Hu or ArMV-Ta. Challenge inoculation with GFLV was performed using *X. index*. Development of disease symptoms on test plants was monitored over 9 consecutive years in control plants. Presence of GFLV in vines cross-protected by ArMV-Ta was verified, by employing GFLV-specific antibodies in DAS-ELISA tests. In the case of GFLV-G Hu-protected vines, GFLV infection was monitored by characterizing coat protein gene of superinfecting isolates by immunocapture (IC)-RT-PCR-RFLP analysis. Cross-protected vines had significantly reduced challenge infection rate consistently. However, fruit yield was reduced by 9% and 17%, respectively, by ArMV-Ta and GFLV-G Hu strains, but the quality of the fruit was not affected. The results indicated that cross-protection of grapevines with mild strains might not be an economically viable approach for managing the GFLV disease (Komar et al. 2008).

3.1.5 MECHANISMS OF ACTION OF BIOTIC BIOCONTROL AGENTS

Soilborne microbial plant pathogens may have additional modes of dispersal, in addition to the dissemination via soil. Some of them are transmitted through seeds and propagules, irrigation water and farm equipments. The biological control agents applied for suppression of development of the pathogens and diseases induced by them have to be stable, persistent in soil environment and versatile in action against the target microbial pathogen(s). It is necessary to understand various mechanisms of action of the biocontrol agents on the pathogens and their requirements and limitations for exploiting their potential in the most effective manner for reducing disease incidence and spread within and outside the field.

3.1.5.1 Fungal Biological Control Agents

Suppression of development of fungal pathogens by the biocontrol agents may be the resultant of different mechanisms of action such as parasitism, antibiosis, competition for nutrients and/ or space, prevention of colonization of specific host (root) tissues by the pathogen and induction of local and/or systemic resistance to the target pathogens. Further, host plant growth may be promoted, resulting in enhancement of levels of resistance to microbial pathogens (Narayanasamy 2002, 2013). Microbial plant pathogen development may be suppressed by fungal biocontrol agents (BCAs) via three types of antagonism: (i) direct antagonism, (ii) indirect antagonism and (iii) mixed-path antagonism (Pal and Gardener 2006). The ability of the BCA to parasitize and kill the pathogen represents direct antagonism, whereas in the case of indirect antagonism, no physical contact is made by the BCA with the fungal pathogen, but the level of host resistance is enhanced by activating host defense system. Several fungal BCAs induce resistance against microbial plant pathogens. Competition between the BCA and the pathogen for space and/or nutrients may also inhibit the pathogen development indirectly by starving-out the pathogen or making required plants inaccessible. Mixed-path antagonism includes antagonistic activities based on the ability of the BCA to produce various different kinds of enzymes, antibiotics or metabolites toxic to pathogens.

The genus *Trichoderma* includes many species found in different ecosystem. Some strains can reduce the severity of crop diseases by inhibiting the development of fungal pathogens, primarily in the soil or on plant roots through their antagonistic and mycoparasitic potential. The competitive genome sequence analysis of two important biocontrol agents (BCAs) was performed. The mechanism of mycoparasitism was studied by using *T. atroviride* and *T. virens* to understand how the BCAs suppressed the development of fungal pathogens (Kubicek et al. 2011). The presence of fungal pathogens as the prey and availability of root-derived nutrients are the factors that facilitate the establishment of *Trichoderma* spp. in the rhizosphere. The ability of *Trichoderma* spp. to suppress a broad range of plant pathogens, including oomycetes, fungi, bacteria and viruses, through elicitation of induced systemic resistance (ISR) or localized resistance has been revealed by various investigations. Furthermore, some *Trichoderma* rhizosphere-competent strains are able to promote the growth and yield of crop plants through enhancement, nutrient use efficiency, seed germination and stimulation of plant defenses against biotic and abiotic damage (Shoresh et al. 2010). *Trichoderma* strains are present in the rhizospheres of many plants and plant-derived sucrose is an important resource available to *Trichoderma* sp., facilitating root colonization, coordination of defense mechanisms and increased rate of leaf photosynthesis. Adherence of *Trichoderma* to the root surface may be mediated by hydrophobins, which are small hydrophobic proteins of the outer-most cell wall layer that coat the fungal cell surface and expansin-like proteins related to cell wall development. *Trichoderma asperellum* produces class I hydrophobin Tas Hyd1, which supports colonization of plant roots, possibly by enhancing its attachment to the root surface and protecting the hyphal tips from plant defense compounds (Viterbo and Chet 2006). Plant cell wall-degrading enzymes are also involved in active root colonization, as it occurs with endopolygalacturonase, the PG1 from *T. harzianum* (Morán-Diez et al. 2009). The proteome analysis showed that a small secreted cysteine-rich protein (SSCP) was present in *T. harzianum* and *T. atroviride* and it was a homologue of the avirulence protein Avr4 from *Cladosporium fulvum*. It is possible that binding of Avr4 to chitin could protect *Trichoderma* against plant chitinases (Stergiopoulos and de Wit 2009).

3.1.5.1.1 Mycoparasitism

Parasitization of fungal pathogen by fungal biocontrol agents (BCAs) is required for obtaining nutrition from the host pathogen. Among fungi that have potential for use as biocontrol agents, *Trichoderma* spp. appears to be the leader, as reflected by the availability of more than 50 different *Trichoderma*-based agricultural products registered in different countries for use to protect and improve yield of vegetables, ornamentals and fruit trees (Woo et al. 2006). Strains of *Trichoderma* secrete various enzymes and antimicrobial compounds. *T. harzianum* (Th) produces trichodermin and a small peptide that inhibit *Rhizoctonia solani*, which in

turn, secretes a coumarin-derivative, capable of inhibiting the mycelial growth of *T. harzianum*. However, the antifungal compound produced by *T. harzianum* is a more powerful inhibitory compound, effective even at low concentration than those produced by *R. solani* (Bertagnolli et al. 1998). Several fungal BCAs like *Trichoderma virens* function as aggressive mycoparasites penetrating the hyphae, as well as resting bodies (sclerotia), reducing the survivability/viability of the pathogen in the soil. Destruction of sclerotia by *T. virens* is likely to result in reduction in the inoculum potential of *R. solani* (Howell 1987). Penetration of hyphae of *Rhizoctonia solani* by haustoria of *T. virens* could be visualized under light microscope (Howell 2003). *Coniothyrium minitans* (Cm) is a mycoparasite of *Sclerotinia sclerotiorum* (Ss). The effect of oxalic acid (OA) degradation on the ß-1,3-glucanase activity of *C. minitans* involved in the mycoparasitism was assessed. OA was degraded by *C. minitans* to an extent of 86 to 92% at 20°C on potato dextrose broth (PDB) medium and the pH of cultures was increased from 3.4–4.8 to 8.3–8.6. The spread of *C. minitans* toward the colonies of *S. sclerotiorum* was correlated with increase in the ambient pH from 2.9 to 6.6. OA degradation was correlated with enhanced production of ß-1,3-glucanase by *C. minitans* and the stimulated activity of this enzyme. Degradation of OA by *C. minitans* might also be a mechanism by which the BCA might protect host plants, in addition to parasitism of *S. sclerotiorum* (Ren et al. 2007). In a later investigation, amendment of synthetic oxalate in PDA (0.25–2.0/g) suppressed the aggressiveness of parasitism by *C. minitans* on colonies of *S. sclerotiorum* strain PB. The results suggested that infection of hyphae of *S. sclerotiorum* was negatively affected by the presence of oxalate. The role of oxalate degradation by *C. minitans* in its mycoparasitism on *S. sclerotiorum* provided a key clue for improvement of the biocontrol potential of *C. minitans* (Huang et al. 2011).

T. harzianum antagonistic to *Phytophthora capsici*, either alone or in combination with a compatible bacterial BCA, *Streptomyces rochei* was evaluated for the efficacy in suppressing the development of the Phytophthora blight in pepper. *T. harzianum* not only arrested the spread of mycelial growth of *P. capsici* in the petridish, but also invaded the whole surface of the pathogen colony and sporulate over it. The hyphae of *P. capsici* were surrounded by those of *T. harzianum*, resulting in their subsequent disintegration and eventual suppression of the growth of *P. capsici*, as revealed by observation under scanning electron microscope (SEM). By contrast, *S. rochei*, secreted an antifungal compound (1-propanone-4-chlorophenyl), primarily responsible for its biocontrol activity against *P. capsici* (Ezziyyani et al. 2007). In a later study, mycoparasitism of *Sclerotinia sclerotiorum* by *T. harzianum* was investigated by nucleic acid-based techniques to detect and quantify the genomic DNAs of both the BCA and pathogen. Sclerotia of *S. sclerotiorum* were incubated on *T. harzianum* culture. Germination of sclerotia by producing mycelium was reduced by 50% within one day, and the decrease in germination continued with increase in interval after incubation. Quantification of *Sclerotinia* DNA in older sclerotia by qPCR assay revealed a decrease in the genomic

DNA, indicating decrease in pathogen population. In contrast, the *Trichoderma* DNA increased, and the increase persisted in the older sclerotia, reflecting the higher BCA population. Fresh sclerotia did not appear to be affected by *T. harzianum* (Kim and Knudsen 2009). The mechanism of antagonism of *T. harzianum* (Ths97) against *Fusarium solani* (Fso14), causing Fusarium root rot of olive trees was investigated. Optical microscopic examination of the confrontation zone in petriplates used for dual culture procedure showed the strain Ths97 grew alongside Fso14 with numerous contact points, suggesting parasitic activity on the pathogen. The Ths97 strain exhibited a strong protective role against root infection by *F. solani* Fso14 whether inoculated before or after inoculation with the pathogen (Amira et al. 2017).

The mechanism of antagonism of *Trichoderma atroviride* against *Rhizoctonia solani* AG3, causal agent of potato black scurf disease was studied, employing confocal microscopy. The mycelium of *T. atroviride* established close contact with those of *R. solani* by coiling. The coils were very dense encircling the pathogen hyphae very tightly. At 7 days after establishing contact, the hyphae of the BCA penetrated *R. solani* hyphae, resulting in loss of turgor (Lahlali and Hijri 2010). Another fungal BCA, *Trichoderma asperellum* showed antagonistic activity against *Rhizoctonia solani* in confrontation assay. The effects of mitogen-activated protein kinase-encoding gene *task1* on morphological development, mycoparasitic interaction, production of cell wall-degrading enzymes and secondary metabolites were investigated in *T. asperellum*. The Δ*task1* mutant of *T. asperellum* showed altered growth morphology and lost its ability to parasitize R. solani. The mutant also showed increased expression of several cell wall-degrading enzymes during confrontation with *R. solani*. *T. asperellum task1* expression was negatively correlated with cell wall-degrading enzyme activities during inducing assays. In antibiosis assays, *task1* gene deletion enhanced output of 6-pentyl-α-pyrone and inhibition of pathogen growth (Yang 2017).

Acremonium strictum is a mycoparasite of *Helminthosporium solani*, causing silver scurf disease of potato tubers. Both *A. strictum* and *H. solani* invariably occurred together in the tubers. Axenic culture of *H. solani* was obtained by repeated hyphal tip isolation. *A. strictum* was tightly linked to and partially dependent on *H. solani* in culture. It appeared that *A. strictum* was dependent on *H. solani* for its survival and for its growth in culture. However, growth and sporulation and germination of *H. solani* was reduced in the presence of *A. strictum*. Observations under scanning electron microscope (SEM) showed shriveled and shrunken conidia of *H. solani*, when present along with *A. strictum*. The adverse effects of *A. strictum* on *H. solani* might be due to either direct parasitism or inhibition due to antifungal compounds secreted by *A. strictum* (Rivera-Varas et al. 2007). The mechanism of biocontrol activity of *Pythium oligandrum* against *Rhizoctonia solani* AG-3, causing potato black scurf was investigated. Seed tubers infected by R. *solani* were dipped for a few seconds in a suspension of *P. oligandrum* oospores, followed by air-drying. Confocal laser scanning microscopic examination with an immmuno-enzymatic staining technique revealed that the hyphae of *P. oligandrum* colonized the sclerotia of *R. solani* and established close contact by coiling around hyphae of the pathogen present on the surface of seed tubers, in a manner similar to that was observed in dual culture assay. Quantification of *R. solani* DNA on seed tubers by PCR showed that *R. solani* population was reduced on seed tubers treated with *P. oligandrum*, compared with untreated control tubers (Ikeda et al. 2012).

Molecular biology of mycoparasitic interaction between the fungal pathogen and biocontrol agents has been studied in some pathosystems. The role of cell wall-degrading enzymes (CWDEs) and antibiotics in mycoparasitism of BCAs was investigated. The strain P1 of *T. harzianum* was genetically modified by target disruption of the single copy of *ech-42* gene encoding for the secreted 42-kDa endochitinase (CHIT 42). The stable mutants lacked the *ech-42* transcript, the protein and endochitinase activity in culture filtrates. Other chitinolytic and glucanolytic enzymes expressed during mycoparasitism were not affected by disruption of *ech-42*. The mutant was as effective as P1 strain against *Pythium ultimum*, whereas its effectiveness against the foliar pathogen *Botrytis cinerea* on bean leaves, was reduced by 33%. However, the endochitinase deficient mutant was more effective against another soilborne pathogen *Rhizoctonia solani* than the wild-type P1 strain. The results indicated that the biocontrol activity of *T. harzianum* might be significantly influenced by the nature of fungal pathogen involved in the interaction (Woo et al. 1999). The activities of several cell wall-degrading enzymes including proteases, chitinases and glucanases of the BCAs may be required for mycoparasitism of the fungal pathogens. Purified host cell walls, compounds secreted by hosts and also live host may stimulate the expression of the genes encoding these enzymes. Such enhanced gene expression may improve the biocontrol potential of the BCAs. Expression of novel genes in *Trichoderma hamatum* effective against *Sclerotinia sclerotiorum, S. minor, Rhizoctonia solani* and *Pythium* spp., causing diseases in various crops, was investigated, by applying southern subtractive hybridization (SSH) technique. The homologues of *chit42* and *prb1*, two genes considered to be essential for mycoparasitism in other *Trichoderma* spp. were expressed at higher levels by *T. hamatum* in medium containing glycerol differed significantly from *T. atroviride*, suggesting that substantial differences might exist in mycoparasitism in these two *Trichoderma* spp. The sequence, Northern and Southern analyses of the subtraction products revealed 19 novel *T. hamatum* genes upregulated during mycoparasitism, representing a substantial increase in the number of *T. hamatum* genes. Four sequences had no significant similarity to any sequences in the GenBank and they may be perhaps restricted to mycoparasites to facilitate mycoparasitism. The SSH technique was found to be useful for identifying genes upregulated during mycoparasitism (Carpenter et al. 2005). *Trichoderma* spp. with significant biocontrol potential against *Sclerotinia sclerotiorum*, causal agent of carnation stem rot disease, were identified as *T. asperellum* (NVTA1, NVTA2), *T. harzianum* (NVTH1, NVTH2), T. *citrinoviride* (NVTC1,

NVTC2) and *T. erinaceum* (NVTE1). Presence of both cellobiohydrolase (*cbh1*) and endochitinase (*ech42*) genes was detected in NVTA2 strain of *T. asperellum*. The crude metabolite from NVTH2 inhibited the mycelial growth of *S. sclerotiorum* most effectively (Vinodkumar et al. 2017).

Mycoparasitism of *Sclerotinia sclerotiorum* by *T. harzianum* was investigated, using green fluorescent protein (GFP)-expressing *T. harzianum* ThzID1-M3 mutant. A specific PCR primer 1 probe set for detecting the GFP-expressing isolate was employed for monitoring its presence. Quantitative real-time PCR along with epifluorescence microscopy and image analysis was applied to study the dynamics of colonization of sclerotia in nonsterile soil. It was possible to quantify the amounts of ThzID1-M3 DNA and wild-type *S. sclerotiorum* DNA from individual sclerotia using real-time PCR assay. Epifluorescence from transformants was quantified using computer image analysis for estimating colonization on a per-sclerotium basis. Colonization of sclerotia by *T. harzianum* on agar plates was observed, using confocal laser scanning microscopy (CLSM) to detect the GFP-fluorescing hyphae of ThzID1-M3 mutant. This method, although highly labor intensive, provided high spatial resolution of colonization dynamics. Both techniques quantified colonization of sclerotia by *T. harzianum* over a period of time. The real-time PCR provided a more precise assessment of the extent of sclerotial colonization by the BCA and it could be more easily applied to sample entire sclerotia (Kim and Knudsen 2011). *T. harzianum* effectively suppressed the development of bean foot rot disease caused by *Fusarium solani* through mycoparasitism. A transcriptome analysis was performed, using expressed sequence tags (ESTs) and quantitative real-time PCR (RT-qPCR) in order to study the mechanism of disease suppression by *T. harzianum*. A cDNA library from *T. harzianum* mycelium (ALL42 isolate) grown on cell walls of *F. solani* (CEFs) was constructed and analyzed. High quality sequences 2,927 of 3,845 were selected and 37.7% were identified as unique genes. The ontology analysis indicated that majority of the annotated genes were involved in metabolic processes (80.9%), followed by cellular processes. Twenty genes that encoded proteins with potential role in biological control were investigated. RT-qPCR analysis showed that none of these genes were expressed, when *T. harzianum* was challenged with the pathogen. These genes showed different patterns of expression during in vitro interaction between *T. harzianum* and *F. solani* (Steindorff et al. 2012).

3.1.5.1.2 Antibiosis

The fungal biocontrol agents (BCAs) may produce different kinds of enzymes and antimicrobial compounds that may facilitate their successful establishment in the soil environment to interact with the microbial plant pathogens. Some BCAs may suppress the pathogen development via more than one mechanism. Mycoparasitism and antibiosis may overlap and one mechanism may operate predominantly, depending on the pathogen, substrate or environment prevailing in the soil. Different steps in the process of parasitism, such as recognition of the host, attachment and subsequent penetration

and killing of the host cells have been recognized. During parasitism, *Trichoderma* spp. secretes hydrolytic enzymes that hydrolyze cell wall of the host (pathogen) (Woo et al. 2006). The proteolytic activity of *T. harzianum* preceded lysis of the protein matrix of the pathogen cell and for inactivation of the hydrolytic enzymes secreted by the pathogen, resulting in decrease in its pathogenicity. It is possible to select the isolate(s) of the BCA with greater potential for secretion of hydrolytic enzymes from different natural sources such as compost or disease suppressive agricultural soil or through transformation of the BCA isolate with multiple copies of the genes involved in the biosynthesis of the hydrolytic enzymes (Narayanasamy 2013).

T. harzianum inhibited the mycelia growth of *F. oxysporum* f.sp. *lycopersici (Fol)*, causing Fusarium wilt disease in tomato. The culture filtrate (CF) with volatile and nonvolatile metabolites of *T. harzianum* inhibited all isolates of *Fol* tested (Mishra et al. 2010). Mechanisms of biocontrol activity of strains of *Trichoderma* sp. against *Sclerotium rolfsii* and *F. oxysporum* f.sp. *ciceris* were investigated. The BCA strains were plated on media amended with colloidal chitin and cell wall extracts of *S. rolfsii*. Chitinolytic activity was detected in all isolates of *Trichoderma* sp. tested. Two strains, with high activities of endochitinase and exochitinase, produced cellulase which might also contribute to the biocontrol potential of *Trichoderma* sp. (Anand and Reddy 2009). *Trichothecium roseum* MML003 with strong suppressive activity against rice sheath blight pathogen *Rhizoctonia solani* did not have either mycoparasitic activity or ability to produce siderophores and H_2O_2. The culture filtrate (CF) of *T. roseum* inhibited the mycelia growth and formation of sclerotia by *R. solani*. Sclerotial germination and viability were also substantially reduced by treatment with CF of *T. roseum*. Application of *T. roseum* resulted in significant suppression of sheath blight development under greenhouse conditions. Antifungal compounds produced by *T. roseum* might be involved in arresting the development of rice sheath blight disease (Jayaprakashvel et al. 2010).

The fungal BCAs may vary in the range of fungal pathogens against which they are effective. Isolates of *Trichoderm harzianum* have been shown to be effective against a wide spectrum of fungal pathogens, including *F. oxysporum* f.sp. *melonis*, causing melon Fusarium wilt disease. Isolates (31) of *Trichoderma* sp. were analyzed by random amplified polymorphic DNA (RAPD)-PCR technique. Five most effective isolates of *T. harzianum* (T-30, T-31, T-32, T-57 and T-78) were characterized by their ability to secrete hydrolytic enzymes viz., chitinases, glucanases and proteases. In the plate cultures, the greatest mycoparasitic activity was exhibited by the isolates T-30 and T-78, as reflected by the total and extracellular hydrolytic activities of N-acetyl glucosaminidase (NAGases), chitinase and ß-1,3-glucanase, which were greater than other isolates tested. The expression of genes encoding for NAGases (*exc1* and *exo2*) or glucanases (*bgn13.1*) activities and their respective enzyme activities in vitro were measured. Different profiles of gene expression between various *T. harzianum* isolates were related to the activity values and dual plate confrontation test. The high NAGase activity detected in T-30 and T-38 corresponded to the levels of expression of the gene *exc1* for T-30, but not for T-78. The high NAGase activity of the isolate T-28 might be due to a higher expression of *exc1* over previous hours before sampling. The high chitinase activity of T-78, both total and extracellular, could be linked to the levels of expression of genes *chit42* and *chit33*, as both were highest for this isolate. These two isolates exhibited the maximum activity of ß-1,3-glucanase. These corresponded with the expression level of gene *bgn13.1* for T-30. The isolates T-30 and T-78 exhibited the greatest mycoparasitic potential against *F. oxysporum* f.sp. *melonis* (López-Mondejar et al. 2011).

Molecular mechanism of biocontrol activity of *Trichoderma virens* against *Rhizoctonia solani* was investigated. The gene encoding for chitinase (*chit42*) in *T. virens* was disrupted or overexpressed. Decrease or increase in biocontrol activity of the transformants matched with the disruption or over-expression of *chit42* gene in the cotton-*Rhizoctonia solani* pathosystem. As the differences in the biocontrol activity of transformants and wild-type strain were less, it might be possible for the involvement of other factors in the biocontrol efficiency of *T. virens* against *R. solani* (Baek et al. 1999). Variable results on the role of chitinase in the biocontrol potential of *T. harzianum* were obtained depending on the BCA-pathogen combination. Disruption of *ech42* of *T. harzianum* did not alter the biocontrol activity against *Pythium ultimum*, but the biocontrol actvity was greater in the transformants than in the wild strain of *T. harzianum* against *R. solani* (Woo et al. 1999). Investigations on molecular genetics of the fungal BCAs, were focused on the role of genes encoding the enzymes involved in the biocontrol activity. Identification of enzymes required for biocontrol activity, was difficult due to redundancy of the cell wall-degrading enzymes (CWDEs) encoding genes in the genome of *Trichoderma*. *T. longibrachiatum* was effective in suppressing development of damping-off disease caused by *Pythium ultimum*. *T. longibrachiatum* was transformed with the gene *egl1*, encoding for ß-1,3-glucanase. The transformants overexpressing *egl1* were slightly more effective in reducing the disease incidence in cucumber than the wild-type strain. As the antagonistic efficiency was not enhanced in the transformants, it is likely that biosynthesis of several CWDEs might be necessary to increase the biocontrol efficiency to significant levels (Migheli et al. 1998). The effectiveness of *T. harzianum* in suppressing development of many fungal pathogens such as *Rhizoctonia solani* and *Sclerotinia sclerotiorum* might be due to the release of lytic enzymes primarily chitinases, glucanases and proteases in susceptible hosts (Chet and Chernin 2002). The involvement of proteases in the biocontrol activity of *Trichoderma* sp. was indicated. *T. harzianum* T334 produced low levels of protease constitutively. Mutants of T334 were generated by UV-irradiation. Some mutants of T334 were more effective against *P. debaryanum* and *R. solani* than the wild-type strain. The mutants were able to produce greater amounts of extracellular trypsin- and chymotrypsin-like proteases with manifold levels of activities of the wild-type strain T334 (Sezekeres et al. 2004). The aspartic proteases were reported to have a major role in the biocontrol potential of *Trichoderma*. The gene *SA76*, encoding an aspartic protease, was cloned for 3′ rapid amplification of cDNA ends from T88. The Northern blot analysis indicated that *SA76* was induced in response to different fungal cell walls. Analysis of *SA76* expression confirmed that aspartic protease activity was induced in a simulated parasitism by the presence of cell walls of *R. solani*, *Phytophthora sojae*, *F. oxysporum* and *Sclerotinia sclerotiorum*. The enhanced activity was due to induction at the transcription level, because the transcripts accumulated abundantly shortly after induction (Liu and Yang 2007).

Involvement of endochitinase of fungal BCAs, as a mechanism of biocontrol activity, has been demonstrated. The effectiveness of a 42-kDa endochitinase encoded by *ThEn42* gene from *T. harzianum* against *Rhizoctonia solani* AG-8 and/ or *R. oryzae*, causing barley root rot was assessed. Purified endochitinase strongly inhibited both *R. solani* and *R. oryzae*. By contrast, the endochitinase showed only moderate level of inhibition against *Gaeumannomyces graminis* var. *tritici (Ggt)*, causal agent of wheat take-all disease and it was ineffective against *Fusarium graminearum*, *F. pseudograminearum* and *F. culmorum*, causal agents of wheat head blight disease (Wu et al. 2006). Proteomic, genomic and transcriptomic methods were applied for the isolation and characterization of a novel plant cell wall (PCW)-*Trichoderma* gene coding for a plant cell wall-degrading enzyme (CWDE). A proteomic analysis, using a three-component (*Trichoderma* spp.-tomato-pathogen) system facilitated the identification of a differentially expressed *T. harzianum* endopolygalacturonase (endo-PG). A specific spot (0303) remarkably increased only in the presence of *R. solani* and *Pythium ultimum* and corresponded to an expressed sequence tag (EST) from a *T. harzianum* T34 cDNA library that was constructed in the presence of PCW polymers and used to isolate the *Thpg1* gene. The *Thpg1*-silenced transformants had lower PG activity, less growth on pectin medium and reduced capacity to colonize tomato roots. The results were confirmed by real-time PCR assay, which revealed the presence of a pathogen in the system triggering the expression of *Thpg1* (Morán-Diez et al. 2009).

A nonpathogenic strain of *F. oxysporum* S6 isolated from soil suppressive to *Sclerotinia sclerotiorum* exhibited antagonistic activity in dual culture assay. The toxic non-volatile metabolic compounds from S6 strain were isolated by chromatographic technique. After purification, they were identified as cyclosporine A by spectroscopic methods. The antibiotic inhibited mycelial growth and suppressed sclerotial formation. The antifungal activity against *S. sclerotiorum* was correlated with the presence of cyclosporine A by dilution plate method. When the sclerotia were placed at the center of BCA colony, the percentage of germination of sclerotia was significantly reduced due to infection of sclerotia by the BCA. In the greenhouse assessment, the number of surviving soybean plants significantly increased, when the BCA and the pathogen were coinoculated. The results indicated that the antifungal activity of *F. oxysporum* S6 against *Sclerotinia sclerotiorum* was primarily due to secretion of cyclosporine A by the BCA (Rodriguez et al. 2006). Similar association of antagonistic activity of *Trichoderma* spp. against *S. sclerotiorum* infecting potato with nonvolatile toxic compounds was reported by Ojaghian (2011). *T. harzianum* T23 produced viridofungin A (VFA) in culture. Bioautography assay showed that three fractions F223, F323 and F423 were produced by T23 strain, whereas the T16 strain formed two fractions F416 and F516 in culture. VFA appeared to have wider antifungal activity against *Verticillium dahliae, Phytophthora infestans* and *Sclerotinia sclerotiorum*. VFA was fungistatic, rather than fungicidal (El-Hasan et al. 2009).

Mycofumigation with antimicrobial volatiles produced by fungal BCAs has been applied for suppression of fungal pathogens. The fungi *Muscodor albus* and *M. roseus* were employed for mycofumigation to enhance sugar beet stand and to decrease severity of diseases caused by *Rhizoctonia solani, Pythium ultimum* and *Aphanomyces cochlioides*. Five classes of compounds viz., alcohols, esters, ketones, acids and lipids were the key components of the mycofumigant gas volatiles. These compounds were tested either singly or as mixtures in vitro against *Pythium ultimum, F. oxysporum* f.sp. *betae, Rhizoctonia solani, Phytophthora cinnamomi, Verticillium dahliae* and *Sclerotinia sclerotiorum*. No single class of the natural volatiles from *M. albus* was toxic individually to the test pathogens. The most effective single compound belonged to the esters group (Strobel et al. 2001). In a later investigation, the efficiency of five different formulations containing *M. roseus* was assessed for the control of sugar beet Pythium damping-off and eggplant Verticillium wilt diseases. The Stabileze formulation was effective in reducing consistently disease severity and population of *Verticillium dahliae* in vivo. The results indicated that mycofumigation efficacy could be maximized by selecting an appropriate formulation (Stinson et al. 2003). The efficacy of *Muscodor albus* in suppressing the development of *Rhizoctonia solani* in greenhouse soilless-growing mix was assessed. The treatment showed only local effect essentially, indicating the inability of volatiles to move through the growing mix. The temperature range of 4–22°C was suitable for fumigation activity of *M. albus*. The ability of M. albus to suppress damping-off disease developing in broccoli seedlings declined rapidly after its incorporation in the growing mix. In treated mix, disease incidence remained at low levels, regardless of planting time after treatment, suggesting that biofumigation could eliminate R. *solani* effectively. In the case of pepper, *M. albus* provided high level of protection against *Phytophthora capsici*. Improved growth of pepper was attributed to the suppression of other deleterious microorganisms that often contaminate commercial growing mix (Mercier and Manker 2005). In a later investigation, 12 isolates of *M. albus* were found to produce volatile compounds that were biologically active (Strobel et al. 2007).

Muscodor albus strain MFC2 was evaluated for its efficacy in protecting kale (*Brassica oleracea*) against *Pythium ultimum*, causing damping-off disease under greenhouse conditions. Kale seeds were sown in soils infested with *P. ultimum*,

followed by inoculation with *M. albus* culture. Seedling emergence in pots inoculated with the BCA and pathogen was equal to a level close to that in the control without the pathogen. The volatiles from *M. albus* appeared to have no harmful effect on plant development and the biocontrol activity of *M. albus* might be due to volatiles from the BCA (Worapong and Strobel 2009). Results of another investigation showed that volatiles from *M. albus* suppressed the development of damping-off of broccoli seedlings, when pots containing soil or soilless potting mix infested with *Rhizoctonia solani* were placed in the presence of active *M. albus* culture without physical contact in closed containers. Gas chromatographic analysis revealed that isobutyric acid and 2-methyl-1-butanol were released from the treated soil/substrates. Production of isobutyric acid showed positive correlation with the extent of disease control. Amounts of isobutyric acid released from the soil were several folds greater than that released from potting mix. In addition, higher populations of the BCA were required to achieve effective control of damping-off disease in soilless potting mix than in soil, suggesting that soil environment was better for the biological activity or viability of *M. albus* than the soilless potting mix (Mercier and Jiménez 2009). The endophyte *Oidium* sp. isolated from *Terminalia catappa* (tropical chestnut) effectively suppressed the mycelial growth of *Pythium ultimum*. The BCA produced primarily esters of propianic acid, 2-methyl-y-butanoic acid and 3-methyl-yl-butanoic acid. Addition of exogenous volatile naphthalene-1-oxybis caused substantial synergistic increase in the antibiotic activity of the volatile organic compounds (VOCs) of *Oidium* sp. against *P. ultimum*. The development of the pathogen was entirely arrested, resulting in ultimate death of the pathogen. The results suggested that the VOCs of different endophytic fungi might act both additively and synergistically to suppress the development of different pathogens colonizing the same plant species (Strobel et al. 2008).

Different mechanisms may operate in the interaction between *T. harzianum* strain SQR-T037 and *F. oxysporum* f.sp. *cucumerinum*, causing Fusarium wilt in cucumber continuous cropping (CCC) system. The allelochemicals exuded from cucumber cause stress and these chemicals have to be biodegraded for better growth of cucumber plants. The allelochemicals isolated from cucumber rhizosphere included 4-hydroxy-benzoic acid, vanillic acid, ferulic acid, benzoic acid, 3-phenylpropionic acid and cinnamic acid. The allelochemicals were entirely degraded by SQR-T037 after 170 h of incubation. Inoculation of SQR-T037 in the CCC soil also led to degradation of allelochemicals exuded from cucumber roots. The degradation of allelochemicals was accompanied by significant decrease in disease index and increase in dry weights of cucumber plants in pot experiments, following application of *T. harzianum*. The results indicated that alleviation of allelopathic stress could be attributed to the activity of the BCA strain (Chen et al. 2011). In the further investigation, the antifungal compounds produced by *T. harzianum* SQR-T037 were characterized. One antifungal compound was purified and identified as 6-pentyl-α-pyrone (6PAP), using mass spectrometry and nuclear magnetic resonance spectroscopy. The antifungal activity of 6PAP at different concentrations (50, 150, 350 and 450 mg/l) was assayed using growth inhibition tests in petriplates. Antifungal activity increased with increase in concentrations of 6PAP. At 350 mg/l, 6PAP inhibited mycelial growth and spore germination by 73.7% and 79.6%, respectively, compared with control. In addition, 6PAP at 150 mg/l decreased sporulation and fusaric acid production by *Foc* by 88% and 52.68% respectively. Application of 6PAP to cucumber continuously cropped soil reduced pathogen population by 41.2% and the incidence of cucumber Fusarium wilt disease by 78 to 89.6%, in addition to stimulation of cucumber plant growth due to elimination of allelochemicals from soil (Chen et al. 2012).

3.1.5.1.3 Competition for Nutrients and Space

A biocontrol agent, to be effective in suppressing soilborne pathogens, should have rhizosphere competence (the ability to grow in the rhizosphere) and compete with other microorganisms, including microbial pathogens for nutrients and space. Three-way interactions among plants, pathogens and BCAs are complex and variable, depending on the environment existing in the soil and microclimte around the plants. Competition and rhizosphere competence may result in replacement of endogenous fungi on root surface. *Trichoderma* sp. could suppress the growth of endogenous fungi on agar medium and mask the presence of these fungi. *T. virens*-treated root segments taken from soil heavily infested with propagules of *Macrophomina phaseolina*, when plated on agar medium yielded *T. virens* from the root tissues at room temperature. But at 40°C, the BCA did not grow, the pathogen readily developed, because of its tolerance to higher temperature. It was not possible to detect the BCA without suppressing the development of the pathogen (see Figure 3.7) (Howell 2003).

Several nonpathogenic strains of *F. oxysporum,* that have an important role in microbial ecology of soil, have been isolated from soil, particularly from soils naturally suppressive to diseases caused by soilborne pathogens. Soils suppressive to

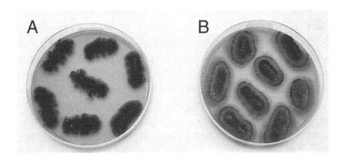

FIGURE 3.7 Efficiency of competition between and rhizosphere competence of *Trichoderma virens* (Tv) and *Macrophomina phaseolina* in soil incubated at 25°C and 4°C Cultures of cotton roots from plants grown in *M. phaseolina*-infested soil and incubated at 25°C; (A): roots of nontreated cotton seed yielding only *M. phaseolina* and B: roots of Tv-treated seed yielding only the biocontrol agent; incubation at 40°C favored development of the pathogen, but not the biocontrol agent.

[Courtesy of Howell 2003 and with kind permission of the American Phytopathological Society, MN]

Fusarium wilts supported large populations of nonpathogenic *Fusarium* spp. of which *F. oxysporum* strain Fo47 has been shown to be efficient biocontrol agents. In order to generate mutants, insertional mutagenesis was employed to tag genes involved in the biocontrol activity of *F. oxysporum* Fo47. The mutants of Fo47 were evaluated for their antagonistic activity against *F. oxysporum* f.sp. *lini*, causing linseed wilt disease. The biocontrol activity of Fo47 was primarily due to competition with the pathogen involving their saprophytic ability. The mutants were characterized by their saprophytic traits. Mutants 83 and 94, the most significantly effective in their biocontrol activity, had the same ability to grow and elongate on MMA-nitrate medium, as the wild-type strain Fo47. The mutants 83 and 94 exhibited significant differences with respect to their antagonistic ability. Mutant 83 inoculated in the ratio of 10:1 was as effective as the parent Fo47 strain inoculated in the ratio of 1,000:1, whereas mutant 94 inoculated in the ratio of 1,000:1 was no more effective than strain Fo47 inoculated in the ratio of 10:1. The results indicated that the mutants were not impaired in saprophytic phase. As the mutants were either less or more antagonistic than the wild-type strain, the biocontrol activity of the strains Fo47 was not dependent entirely on its saprophytic ability of the nonpathogenic Fo47 strain (Trouvelot et al. 2002).

Fusarium oxysporum F2, a nonpathogenic strain, effectively suppressed the development of symptoms of Verticillium wilt disease induced by *Verticillium dahliae* in eggplant under greenhouse and field conditions. Parasitism or antibiosis was not involved in the interaction between the pathogen and the biocontrol agent (BCA). In order to investigate the mechanism of biocontrol activity, the pathogen strain and BCA F2 strain were transformed respectively with the EGFP and DsRed2 reporter genes to facilitate visualization of their presence on the root surface of eggplant. In addition, the real-time PCR analysis was applied to monitor the ramification of both fungi into the vascular system. The strain F2 colonized the root surface along the intercellular junctions excluding *V. dahliae* from the same ecological niche. The QPCR analysis also showed that application of F2 reduced the levels of *V. dahliae* vascular colonization, as well as disease severity. The results of the split-root experiment showed that the strain F2 did not trigger the defense mechanisms of eggplant against *V. dahliae*. The results seemed to support the view that the mechanism of biocontrol activity of the strain F2 against *V. dahliae* was through competition for space or nutrients on the root surface of eggplants (Pantelides et al. 2009). In a further study, the F2 strain of *F. oxysporum* was applied by stem injection of a conidial suspension, as root drenching might adversely affect the native beneficial microbial community. Stem injection of the strain F2 at 7 days prior to transplanting the seedlings on soil infested with *V. dahliae* microsclerotia, resulted in reduced disease severity, compared to untreated control plants. Ramification of F2 into the plant vascular system of eggplant stems was visualized by inoculatting an EGFP-transformed F2 strain. The transformed F2 strain colonized the plant vascular tissues effectively over a long period of time as determined by the levels of DNA. The

QPCR analysis showed that application of F2 strain reduced significantly the DNA contents of *V. dahliae* in stem tissues, compared to the untreated control plants (Gizi et al. 2011).

3.1.5.1.4 Prevention of Colonization of Host Tissues by Pathogens

In some interactions between soilborne microbial pathogens and biocontrol agents (BCAs), the biocontrol activity may be due to prevention of the pathogen gaining access to the susceptible host (root) tissues. Treatment of cotton seeds with *Trichoderma (Gliocladium) virens* reduced colonization of cotton roots by *Rhizoctonia solani*, causing root rot disease, resulting in reduction in disease incidence and severity (Howell and Stipanovic 1995). Maintenance of adequate population levels of BCAs at target sites and timing of their application contribute significantly to the effectiveness of biocontrol of soilborne pathogens causing root diseases. The nonpathogenic strain Fo47 of *F. oxysporum* protected tomato roots against infection by *F. oxysporum* f.sp. *radicis-lycopersici* (FORL) causing tomato foot and root rot (TFRR) disease. When tomato seedlings were planted in sand amended with spores of Fo47, hyphae attached to the roots earlier than FORL, whereas root colonization by the pathogen was arrested at the stage of initial attachment to tomato roots. The percentage of spores of Fo47 germinating in the tomato root exudates in vitro was higher than that of FORL. By using different autofluorescent proteins as markers and observing under confocal laser scanning microscope, the pathogen and the BCA could be visualized simultaneously on tomato roots and colonization of root surface by them was quantified. The preferential germination of Fo47 spores in root exudates components was considered to reduce pathogen growth toward roots and consequently to reduce the number of FORL for attachment sites on roots (Bolwerk et al. 2005).

The rhizosphere competence and capacity to adapt and maintain adequate population of the biocontrol agent in the rhizosphere, in addition to timing of application of the BCA, are important requirement of the BCA for providing effective protection to host plants against soilborne pathogens like *Rhizoctonia solani*, causing stem and root disease of poinsettia. One application of the binucleate *Rhizoctonia* (BNR) isolate was not effective in suppressing the development of stem rot of poinsettia. In contrast, one application of BNR isolate, after transplanting rooted pointsettias, was more effective than the bacterial BCA *Burkholderia cepacia*. Root colonization by BNR isolate reached the maximum, when the *B. cepacia* was applied at propagation, followed by BNR application after transplantation. Both BNR isolates and *B. cepacia* were good colonizers of poinsettia roots and maintained initial high population levels up to 5 weeks, after application (Hwang and Benson 2002). In a later investigation, BNR isolates were found to effectively protect cotton against soilborne pathogens. The BNR isolates could be consistently recovered from hypocotyls and roots of soybean, indicating that colonization of root tissues was associated with the suppression of development of *R. solani*, infecting soybean plants (Khan et al. 2005). The yeast species *Candida valida, Rhodotorula*

glutinis and *Trichosporon asahii* showed significant biocontrol potential against *R. solani*, causing postemergence damping-off of sugar beet seedlings. Root colonization in plant assay indicated that *C. valida* and *T. asahii* colonized 95% of roots of sugar beet at 5 days after application, whereas *R. glutinis* colonized 90% of the roots after 8 days. All yeast species were present at all depths of rhizosphere soils adhering to the tap roots up to 10 cm. The yeast species had high levels of rhizosphere competence, as reflected by the extent of their colonization of roots. In addition, the yeast BCAs promoted plant growth, when applied individually or in combination (El-Tarbily 2004).

Naturally occurring root endophytic fungi, such as *Heteroconium chaetospira* and *Phialocephala fortinii* suppressed the development of Verticillium wilt disease of eggplant. Colonization patterns of *P. fortinii* and a dark septate endophytic (DSE) fungus were investigated for their antagonistic potential against *Verticillium longisporum*, causing Verticillium yellow disease in Chinese cabbage. Hyphae of *P. fortinii* and DSE taxon extensively colonized roots of Chinese cabbage seedlings without inducing visible symptoms. Hyphae of *P. fortinii* grew along the surface of roots and formed microsclerotia on or in the epidermal layer, whereas the hyphae of the DSE taxon heavily colonized some root cortical cells. *P. fortinii* suppressed the effects of postinoculated virulent strains of *Verticillium* in vitro. The DSE taxon was able to colonize Chinese cabbage roots and suppressed the development of Verticillium yellows. The protective values of DSE taxon against the disease were significantly higher, compared to other fungal endophytes as reflected by higher marketable value of the produce obtained from DSE taxon-treated plots (Narisawa et al. 2004). *Fusarium equisetti,* naturally occurring endophyte of barley roots, significantly reduced the severity of take-all disease caused by *Gaeumannomyces graminis* var. *tritici (Ggt)*. *F. equisetti* colonized barley roots endophytically and competed with other fungal root colonizers present in the rhizosphere. The BCA isolates reduced the mean root lesions length induced by *Ggt*. However, the suppressive effect by *F. equisetti* was not distinct (Maciá-Vicente et al. 2009).

Accumulation of phenolic acids, occurring in continuously cropped soils, is increased due to infection by soilborne fungal pathogens. The biocontrol potential of *Phomopsis liquidambari* against *Fusarium solani* in soil enriched with phenolic acids was assessed by inoculating the BCA in the soil. Inoculation of *P. liquidambari* significantly inhibited the reproduction of F. *solani*. The extent of degradation of soil phenolic acids by *P. liquidambari* was determined. No direct antagonistic relationship between *F. solani* and *P. liquidambari* was observed, implying the alleviated stimulation of phenolic acid was the major factor in suppressing the development of *F. solani*. Further, presence of glucose did not significantly impact the biocontrol activity of the BCA. Inoculation with *P. liquidambari* reduced the severity of disease caused by *F. solani* in peanut. The results showed that *P. liquidambari* could be employed to suppress *F. solani* in phenolic acids-rich continuous cropping soils (Xie et al. 2017).

3.1.5.1.5 Induction of Resistance in Plants to Diseases

Biotic biocontrol agents (BCAs) may have single or multiple modes of action, involving direct or indirect effect on the microbial plant pathogens, resulting in suppression of disease development. The BCAs have been employed as inducers of disease resistance in several agricultural and horticultural crops. Induction of resistance by BCAs has been considered as an important disease management strategy, since enhancement of genetic resistance of crops through conventional breeding or biotechnological methods, has been found to be difficult or time-consuming or not feasible. Fungal BCAs have been shown to induce resistance to crop diseases, in addition to other mechanisms of biocontrol activity against microbial pathogens. Inoculation of tomato with the nonpathogenic *Penicillium oxalicum* resulted in reduction in severity of Fusarium wilt disease caused by *F. oxysporum* f.sp. *lycopersici (Fol)* as reflected by area under disease progress curve (AUDPC) values and extent of dwarfing of plants. *P. oxalicum* was shown to induce systemic resistance in tomatoes to *Fol*. Histological examination showed that BCA-inoculated plants did not lose cambium, had lower number of bundles and less vascular colonization by *Fol*. Renewal or prolonged cambial activity in treated plants, leading to the formation of additional secondary xylem might result in reduced disease severity. As no adverse effect of treatment with the BCA in tomato cultivars susceptible or resistant to Fusarium wilt disease was observed, *P. oxalicum* could be employed for protecting tomatoes against infection by *F. oxysporum* f.sp. *lycopersici* (de Cal et al. 1997, 2000). Application of *Crinipellis perniciosa*-chitosan filtrate (MCp) significantly delayed the development of Verticillium wilt disease in tomato induced by *Verticillium dahliae*. Activation of synthesis of pathogenesis-related (PR) protein with tissue lignification of tomato leaves was observed (Cavalcanti et al. 2007).

Trichoderma viride with multiple mechanisms of biocontrol activity was evaluated for its effectiveness against *F. oxysporum* f.sp. *adzuki* and *Pythium arrhenomanes*, infecting soybean. In vitro experiments revealed its mycoparasitic nature, whereas the pot experiments showed the disease suppressive ability of the BCA against both diseases. Furthermore, enhancement of growth of shoot and root systems, as well as pod yield of soybean plants treated with *T. viride* was also observed. *T. viride* seemed to be an avirulent opportunistic symbiont in the soybean rhizosphere. Enhancement of resistance against secondary infection of soybean by the fungal pathogen was also discernible (John et al. 2010). The dynamics of expression of defense response genes in the root tissues of potato plantlets were studied, following treatment with *T. harzian* and challenge with *Rhizoctonia solani*. Gene expression analysis showed that genes for *PR1* at 168 h postinoculation (*hpi*) and phenylalanine ammonia lyase (*PAL*) at 96 hpi. In plants treated with *T. harzianum* strain Rifai MUCL 2907, induction of *PR1*, *PR2* and *PAL* at 48 hpi in plants inoculated with *R. solani* and induction of LOX at 24 hpi and *PR1*, *PR2*, *PAL* and *GST1* at 72 hpi in plants inoculated with both BCA and pathogen were observed. The results suggested that in the presence of BCA isolate, expression of LOX and GST1 genes might be primed in potato plantlets with *R. solani* at an early stage of infection (Gallou et al. 2009).

T. harzianum produces various metabolites with different functions that influence plants and pathogens to different extent. The role of oxidant-antioxidant metabolites of *T. harzianum* isolates in induction of resistance in sunflower against *Rhizoctonia solani* was investigated. Changes in the apoplast of sunflower challenged by *R. solani* in the presence or absence of *T. harzianum* NBRI-1055 were determined. Analysis of oxidative stress response revealed a reduction in hydroxyl radical concentration. The protection by the BCA strain against the pathogen was associated with accumulation of the reactive oxygen species (ROS) gene network, involving catalase (CAT), superoxide dismutase (SOD), glutathione peroxidase (GPx) and ascorbate peroxidase (APx). In NBRI-1055-treated plants challenged with *R. solani*, these enzymes registered maximum activity after different periods (7–8 days). The enhanced enzymatic activities were accompanied by inhibition of lipid and protein oxidation in *Trichoderma*-treated plants. In addition, synthesis of secondary metabolites of phenolic nature was stimulated by the BCA strain, reaching a fivefold increase in concentration. Strong antioxidant activity at 8 days postinoculation resulted in the systemic accumulation of phytoalexins. The results suggested that the mechanism of biocontrol activity of *T. harzianum* against *R. solani* might be related to neutralizing *R. solani*-induced oxidative stress (Singh et al. 2011). The multianalysis technique was applied to quantify endogenous levels of salicylic acid (SA), jasmonic acid (JA), abscisic acid (ABA) and 1-aminocyclopropane-1-carboxylic acid (ACC) and the ethylene (ET) precursor, in melon plants inoculated with *Glomus intraradices* and *T. harzianum* in the presence of Fusarium wilt pathogen *F. oxysporum* f.sp. *melonis (Fom)*. Infection by *Fom* activated defense response in plants, mediated by plant hormones SA, JA, ET and ABA, similar to the activation of *T. harzianum*. Both *T. harzianum* and *G. intraradices* attenuated plant response mediated by ABA and ET, elicited by the pathogen infection. Furthermore, *T. harzianum* attenuated the SA-mediated plant response. A synergistic effect of *T. harzianum* and *G. intraradices* in reducing the disease incidence was observed. But no such effect was noted in the hormonal disruption induced by the pathogen. The results suggested that the mechanisms of biocontrol activity of *T. harzianum* might be induction of the hormonal disruption induced by *F. oxysporum* f.sp. *melonis*, causing Fusarium wilt disease of melons, whereas the mechanisms involving *G. intraradices* appeared to be independently of SA and JA signaling (Martinez-Medina et al. 2010).

The interaction between *Trichoderma virens* and *Rhizoctonia solani*, incitant of cotton root rot disease, was investigated to evaluate the mechanism of disease suppression by the biocontrol agent (BCA). Defense-related compounds were induced in root tissues, following colonization by *T. virens*. The effect of seed treatment with BCA on elicitation of defense responses of cotton plants was studied. The role of terpenoid compounds in the control of cotton root rot disease was investigated by analyzing the extracts of cotton roots and hypocotyls grown from *T. virens*-treated seeds. Terpenoid synthesis and peroxidase activity were enhanced in the roots of treated plants, but not in the untreated controls. The terpenoid pathway intermediates deoxyhemigossypol (dHG) and hemigossypol (HG) strongly inhibited the development of *R. solani*, indicating that terpenoid production was the major contributor for the control of root rot disease. In addition, a strong correlation between the biocontrol activity and induction of terpenoid was revealed, when the strains of *T. virens* and *T. koningii* were compared. The results indicated that induction of resistance by *T. virens* occurred through activities of terpenoids acting as elicitors of defense responses in cotton (Howell et al. 2000). Treatment of roots with effective strains of T. virens resulted in elicitation of heat stable proteinaceous compounds. One of the compounds with MW of 3–5 kDa, was sensitive to proteinase K. SDS-PAGE analysis revealed the presence of several bands in the gel used for separating the proteins in the active material. One band showed cross-reaction with an antibody to ethylene-inducing xylanase from *T. viride*. Another band (18 K) induced production of terpenoids, in addition to increase in peroxidase activity, in cotton radicles and this protein showed maximum similarity to a serine proteinase from *Fusarium sporotrichoides* (Hanson and Howell 2004).

Trichoderma virens TRS106 protected tomato plants, against *Rhizoctonia solani* by inducing host resistance to the pathogen. Tomato plants treated with the strain TRS 106, showed limited lesion development on inoculation with *R. solani*. No direct inhibition of *R. solani* by TRS106 was evident in in vitro test. The strain TRS stimulated systemic defense responses in tomato plants by activating defense enzymes, including guaiacol peroxidase (GPX), syringaldazine peroxidase (SPX) and phenylalanine ammonia lyase (PAL). Simultaneously it enhanced accumulation of phenolics and H_2O_2, accompanied by decrease in lipid peroxidation in the leaves. Remarkable increases in the contents of 22 phenolics occurred in leaves of *Trichoerma*-treated tomato plants, both uninoculated and inoculated with *R. solani*. Some phenolics were present in a free form, while others were accumulated in bound forms glycosylated conjugates belonging to phenylpropanoids, hydroxybenzoic and cinnamic acid derivatives and flavonoids. Several of the detected phenolics, ferulic and salicylic acids, pyrocatechol and hesperidin were strongly toxic to *R. solani* in plate tests. The systemic mobilization of phenolic metabolism might be a possible factor of tomato defense response positively involved in biocontrol of *R. solani* by TRS106 strain. The results indicated the suitability of using *T. virens* TRS106 strain as a potential biofungicides in the integrated management of diseases caused by *R. solani* in various crops (Malolepsza et al. 2017).

The effectiveness of the biotic inducer of resistance, *T. harzianum* strain 382 and was assessed for the management of Phytophthora blight of pepper caused by *Phytophthora capsici*. The biotic inducer *T. harzianum* remained spatially separated from *P. capsisci* in split root and leaf blight bioassays. The results suggested that resistance induced by *T. harzianum* was systemic in nature (Khan et al. 2004). Treatment of pepper seeds with spores of *T. harzianum* significantly reduced stem necrosis caused by *P. capsici*. Drenching the roots of pepper plants with spore suspension of *T. harzianum* also provided similar level of protection against Phytophthora blight. The necrotic lesions yielded only the pathogen, but not the BCA, suggesting the absence of direct contact between the pathogen and the BCA. The percentage of *P. capsici* isolated at 9 days after inoculation was higher in untreated inoculated plants than in treated inoculated plants. *T. harzianum* applied in the subterranean part of plants, could induce defense response against *P. capsici* in the aerial parts of the plants. Concentrations of caspsidol in stems of treated and inoculated plants were > sevenfold greater than in nontreated inoculated plants at 6 days after inoculation. The capsidol concentration was reduced at later stages. Accumulation of capsidol in the earlier stages of BCA-pathogen interaction with pepper plants might contribute to enhancement of resistance to the pepper blight disease (Ahmed et al. 2000).

Trichoderma asperllum activated metabolic pathways in cucumber involved in plant signaling and biosynthesis, eventually leading to systemic accumulation of phytoalexins. Penetration of epidermis and subsequent ingress into the outer cortex of cucumber seedlings by *Trichoderma* required secretion of cell wall lytic enzymes. Two differentially secreted arabinofuranosidases were detected by SDS-PAGE procedure, when *T. asperellum* was grown in the presence of cucumber roots. Furthermore, an aspartyl protease was also detected. Differential mRNA display performed on *Trichoderma* mycelia, interacting and noninteracting with plant roots showed that another aspartyl protease was present along with differentially regulated clones. RT-PCR assays revealed that the proteases were induced in response to plant root attachment and were expressed in planta. The gene *papC* (similar to *papA* from *T. harzianum*) was induced in plate confrontation assay with *Rhizoctonia solani*. The expression studies indicated that *T. asperellum papA* was upregulated during the first 48 h of interaction by cell wall proximity. The gene *papB* did not appear to be regulated by the presence of the pathogen. The results suggested that the protease identified, might play a role in *Trichoderma* to function both as a mycoparasite and as a plant opportunistic symbiont (Viterbo et al. 2004). *T. harzianum* isolate T39 is versatile in its biocontrol activity against many plant pathogens. The mode of action of *T. harzianum* differed, based on pathogen involved. Activation of defense responses locally, as well as systemically in cucumber plants treated with T39 strain, was observed against different fungal pathogens (Elad 2000). Treatment of onion seeds with *T. harzianum* strains TR1C7 or TR1C8 induced acceleration of production of antifungal compounds, suppressing the development of black mold disease caused by *Aspergillus niger* (Özer 2011).

Plants have an immune system that is able to detect motifs or domains with conserved structural traits, typical of entire class of microorganisms, but not present in their hosts. These are named as pathogen- or microbe-associated molecular patterns (PAMPs or MAMPs). MAMP-triggered plant responses are elicited rapidly and involve iron fluxes across the plasma membrane, the generation of reactive oxygen species (ROS), nitric oxide, ethylene (ET) and also, but later, the deposition of callose and the synthesis of antimicrobial compounds by the host plant. Effective *Trichoderma* strains produce a wide range of MAMPs which may be either proteins or xylanase (Xyn2/Eix) by *T. viride*, cellulases by *T. longibrachiatum*, cerato-platanins (Sm1/Epl1) by *T. virens/T. atroviride*, swollenin TasSwo by *T. asperelloides*, endopolygalacturonase (ThPG1) by *T. harzianum* or secondary metabolites alamethicin, 18-mer peptaibols by *T. virens* and 6 pentyl-α-pyrone, harzianolide and harzianopyridone by different *Trichoderma* species (Hermosa et al. 2012). The potential of *Trichoderma* sp. to induce systemic resistance was not realized until the demonstration of induction of defense responses by *T. harzianum*, following colonization of bean roots (De Meyer et al. 1998) and penetration of roots of cucumber seedlings by *T. asperellum* resulting in triggering of induced systemic resistance (ISR) (Yedida et al. 1989). Development of ISR in plants colonized by *Trichoderma* has been demonstrated by inoculating leaves of plants with the pathogen, the roots of which were treated with the BCA. However, the effectiveness of ISR against soilborne pathogens is yet to be demonstrated. The growth-promoting activity of *T. atroviride* on tomato seedlings was suggested to be associated with reduced ethylene (ET) production, resulting from a decrease in its precursor 1-aminocyclopropane-1-carboxylic acid (ACC) through the microbial degradation of indole-3-acetic acid (IAA) in the rhizosphere and/ or through the ACC deaminase (ACCD) activity in the microorganisms. Putative sequences of ACCD were detected in *Trichoderma* genomes (Kubicek et al. 2011).

Nonpathogenic binucleate *Rhizoctonia* spp. (np-BNR) was reported to protect plants against damping-off and crown and root rot diseases caused by *Pythium* spp. and *Rhizoctonia solani*. The greenhouse and field assessments indicated that strain np-BNR strain 232-CG could induce systemic resistance (ISR) in the stem and cotyledons of bean to challenges with *R. solani* AG-4 or *Colletotrichum lindemuthianum*, causing root rot and anthracnose disease respectively (Xue et al. 1998). The mechanism of biocontrol activity of np-BNR was investigated in comparison with the chemical inducer benzothiadiazole (BTH) against *Rhizoctonia solani* and *Alternaria macrospora*, causing pre- and postemergence damping-off of cotton. Pretreatment of cotton seedlings with np-BNR isolates protected the plants effectively against a virulent strain of R. *solani* AG-4. Several isolates significantly reduced disease severity. The combination of BTH and np-BNR provided significant protection against seedling rot and leaf spot of cotton. However, the degree of disease reduction obtained with np-BNR treatment alone was comparable to effectiveness of combined treatment. The results indicated that np-BNR isolates could protect cotton from infections by both root and foliar pathogens and they were more efficient than the chemical inducer of resistance, BTH (Jabaji-Hare and Neate 2005). Nonpathogenic Fo47 strain of *F. oxysporum* reduced wilt disease incidence in tomato from 100 to 75%, when applied as seed treatment. As the presence of the BCA hyphae could be

observed only just below the crown region, the disease suppression might be due to induction of resistance in tomato to *F. oxysporum* f.sp. *radicis-lycopersici* (FORL). Induction of resistance in tomato by Fo47 functioned through a systemic acquired resistance (SAR)-like mechanism (Duijff et al. 1998; Bolwerk et al. 2005).

The biocontrol activity of nonpathogenic isolates of *F. oxysporum* (npFo) was investigated, through induction of systemic resistance (ISR) in asparagus against *F. oxysporum* f.sp. *asparagi (Foa)*. In the split-root system experiments, roots inoculated with npFo strain showed hypersensitive response and those subsequently inoculated with *Foa* exhibited resistance. Development of ISR in npFo-treated plants resulted in significant reduction in the number of necrotic lesions and reduced wilt disease severity, compared with untreated control plants. In hyphal-sandwich root inoculation experiments, activation of POX and PAL and lignin content were higher in npFo-treated plants and increased more rapidly than in npFo-nontreated plants, after *Foa* inoculation. Presence of antifungal compounds in the exudates of roots inoculated with *Foa* was observed for npFo-treated plants, but not for npFo-nontreated plants. The results indicated that the isolates of npFo might function as inducers of systemic acquired resistance (SAR) and defense responses against *Foa* invasion in asparagus (He et al. 2002). Mutagenesis via ultraviolet (UV) irradiation has been applied successfully to generate mutants for use as biocontrol agents against crop pathogens. Two nonpathogenic mutants (4/4 and 15/15) were obtained from the cucurbit wilt pathogen *F. oxysporum* f.sp. *melonis (Fom*, race 1,2) by a continuous dip-inoculation technique, following UV mutagenesis. The strain 15/15 induced mortality of susceptible cultivars to a lesser level, compared to the wild-type strain. The strain 4/4 could colonize 100% of the roots and 33 to 70% of lower stem tissues at 7 days after inoculation of seedlings. The nonpathogenic strain lacking only pathogenicity might more efficiently compete with the pathogenic strain, than other BCAs that might require different set of conditions for their survival and development in the soil environment (Freeman et al. 2002).

Fusarium solani Fs-K, an endophytic isolate was obtained from root tissues of tomato plants grown on the compost which suppressed soilborne pathogens. Strain Fs-K colonized root tissues and subsequently protected plants against *F. oxysporum* f.sp. *radicis-lycopersici* (FORL) and also elicited ISR against the foliar pathogen *Septoria lycopersici*. The attenuated expression of genes of PR-proteins like, PR-5 and PR-7 was detected in tomato roots inoculated with strain Fs-K, compared with noninoculated plants. The expression pattern of PR genes was either not affected or aberrant in leaves. A genetic approach, using mutant tomato plant lines, was used to determine the role of ethylene (ET) and jasmonic acid in the response of plant to infection by FORL, in the presence or absence of isolate Fs-K. Mutant lines *Never ripe (Nr)* and *epinastic (epi 1)*, both impaired on ethylene-mediated plant responses, inoculated with FORL, were not protected by Fs-K isolate, indicating that ethylene signaling pathway is required for the mode of action used by the endophytic BCA to induce resistance. In contrary, *def1* mutants defective in jasmonate biosynthesis, showed reduced susceptibility to FORL, in the presence of Fs-K, which suggested that jasmonic acid was not required for mediation of biocontrol activity of isolate Fs-K (Kavroulakis et al. 2007). In a later investigation, the nonpathogenic mutant generated from *F. oxysporum* f.sp. *melonis* (*Fom*) (rev157) could not protect muskmelon plants against infection by pathogenic strain *Fom* 24. The mutant rev157 was unable to protect nonhost flax plants against wilt pathogen *F. oxysporum* f.sp. *lini*. In contrast, the parental strain *Fom* 24 of the mutant rev157, was able to protect the flax plants against *F. oxysporum* f.sp. *lini*. The comparative molecular genetics of the pair of strains Fom24/rev157 might shed light on the identity of genes involved in the biocontrol activity of *F. oxysporum* (L'Haridon et al. 2007). Effective suppression of Verticillium wilt and Phytophthora blight caused by *Verticillium dahliae* and *Phytophthora capsici* respectively in pepper could be achieved by employing nonpathogenic isolate *F. oxysporum* Fo47. The isolate Fo47 inhibited the growth of *V. dahliae*, but not that of *P. capsici* in plate confrontation assay, indicating that at least part of the protective effect of the BCA strain against *V. dahliae* was due to anatagonism or competition for nutrition. In order to determine the role of induction of resistance as a mechanism of biocontrol activity of Fo47, three defense genes previously related to pepper resistance were monitored over time. The genes encoded a basic pathogenesis-related (PR)-1 protein (CABPR1) a class II chitinase (CACHI2) and a sesquiterpene cyclase (CASC1) involved in the synthesis of capsidol, a phytoalexin. These three genes were transiently upregulated in the roots of Fo47-treated plants in the absence of inoculation with the pathogen, but in the stem only CABPR1 was upregulated. In the plants inoculated with *V. dahliae* prior to the treatment with Fo47, three genes had a higher relative expression level than the control in both roots and stem of pepper plants, indicating the involvement of induction of resistance as another mechanism of the biocontrol activity of Fo47 strain in treated pepper plants (Veloso and Díaz 2012).

Pythium oligandrum, a mycoparasite, produces oligandrin, an elicitin-like protein, which was evaluated along with crude glucan obtained from the cell wall of *P. oligandrum* and crabshell chitosan for their ability to induce resistance in tomato root tissues to Fusarium root rot and wilt caused by *F. oxysporum* f.sp. *radicis-lycopersici* (FORL). These compounds were applied to decapitated tomato plants and induction of defense mechanisms on root tissues were monitored. Disease incidence was significantly reduced in plants treated with oligandrin and chitosan, whereas glucans did not have any effect on disease incidence. In oligandrin-treated tomato plants, restriction of fungal growth to outer root tissues, decrease in pathogen viability and formation of aggregated deposits accumulating at the surface of invading pathogen hyphae were the striking features of the defense responses. The results showed that oligandrin could induce systemic resistance in tomato and exogenous foliar application of fungal protein could sensitize susceptible tomato plants to react rapidly and efficiently to infection by *F. oxysporum* f.sp. *radicis-lycopersici*. Reduction

in disease incidence might be primarily through accumulation of fungitoxic compounds at sites of attempted pathogen penetration (Benhamou et al. 2001). In the later investigation, four elicitin-like proteins (POD-1, POD-2, POS-1 and oligandrin) were identified as elicitor proteins in P. *oligandrum*. Two groups of *P. oligandrum* isolates were differentiated, based on the nature of cell wall proteins (CWPs) as D-type containing POD-1 and POD-2 and the S-type isolate containing POS-1. The distribution of genes encoding elicitin-like proteins among ten *P. oligandrum* isolates was analyzed, using genomic fosmid library of the D-type isolate MMR2. Based on Southern blot analyses, the isolates were divided into the same two groups, as those based on the CWPs. The D-type isolates contained *pod-1* and *pod-2* and two oligandrin genes designated *oli-d1* and *oli-d2*, whereas S-type isolates had *pos-1* and one oligandrin gene *oli-s1*. These genes were single copies present only in *P. oligandrum*, but not in nine other *Pythium* spp. tested. All genes were expressed during colonization of tomato roots by *P. oligandrum*, as indicated by RT-PCR assays. It is possible that these genes encode potential elicitor proteins, resulting in the enhancement of resistance in plants against pathogens. Analyses of genetic relationships suggested that the D-type isolates might be derived from S-type isolates by genetic duplication and deletion events (Matsunaka et al. 2010). The ability of *Pythium oligandrum* to induce resistance in potato against black scurf caused by *Rhizoctonia solani* was assessed, using potato tuber disk assay. Treatment of tuber disks with cell wall protein fraction of *P. oligandrum* enhanced the expression of defense-related genes such as 3-deoxy-d-arabino-heptulosonate-7-phosphate synthase, lipoxygenase and basic PR-6 genes and reduced severity upon challenge with *R. solani*, compared with untreated controls. The results suggested that the biocontrol mechanisms employed by *P. oligandrum* against *R. solani* might involve induction of disease resistance, as well as mycoparasitism (Ikeda et al. 2012).

3.1.5.2 Bacterial Biological Control Agents

The bacterial biocontrol agents (BCAs) may act against soilborne microbial plant pathogens directly or indirectly by different mechanisms such as antagonism, antibiosis, competition for nutrients and space, colonization of specific sites required for establishing infection by the pathogens and inducing resistance to the pathogens by activating host plant defense systems. In addition, the plant growth-promoting rhizobacteria (PGPRs) may enhance plant growth to different extent. The PGPRs are included under the genera *Pseudomonas*, *Bacillus*, *Azospirillum*, *Rhizobium* and *Serratia*.

3.1.5.2.1 Antibiosis

Antagonistic activity of *Pseudomonas jassenii* against *Aphanomyces cochlioides* AC5 and *Pythium aphanidermatum* PA-5 was investigated. The antagonist produced two related secondary metabolites, 3-[(1R)-hydroxyoctyl]-5-methylene-2(5H) furanone (4,5-didehydroacterin) and 3-[(1R)-hydroxyhexyl-7-5-methylene-2 (H)-furanone. These compounds inhibited radial growth and also induced morphological abnormalities like hyperbranching and swellings in treated pathogen isolates, AC-5 and PA-5 respectively. Staining with rhodamine-phalloidin, which bound to plasma membrane-associated filamentous-actin (F-actin), revealed that tip-specific actin filaments were redmodelled into a plaque-like form at an early stage of interaction up to 24 h with any one of the secondary metabolites. At later stages (48 h), the plaques were eliminated, reflecting the disorganization of actin arrays in the morphologically abnormal hyphae of AC-5 and PA-5 isolates. A similar response of actin disorganization was seen in AC-5 and PA-5 hyphae following treatment with latrunculin B, an actin-assembly inhibitor produced by a sea sponge. The results showed that actin disorganization and inhibitory activities of the secondary metabolites of *P. jessenii* were similar to those induced by the actin-assembly inhibitor latrunculin (Deora et al. 2010). The native *Streptomyces* isolates C and S2 were evaluated for their biocontrol potential against the fungal pathogens, *Rhizoctonia solani* AG-2, *Fusarium solani* and *Phytophthora dreschleri* involved in sugar beet root rot disease. The in vitro antagonism assays showed the antagonistic nature of the isolate C with percentages of inhibition of mycelial growth being 45, 53 and 26% for *R. solani, F. solani* and *P. drechsleri*. Treatment with NaCl increased the biocontrol activity of soluble and volatile compounds of isolates C and S2. Both isolates showed protease, chitinase and α-amylase activity. In addition, both isolates were able to produce siderophores. Addition of salt enhanced the production of siderophores and activities of protease and α-amylase. In contrast, salt treatments reduced chitinase activity significantly. Production of salicylic acid (SA), ß-glucanase and lipase by isolate S2 and biosynthesis of cellulase by isolate C was significant both in the presence and absence of NaCl. Soil application of isolate C reduced root rot of sugar beet by the soilborne fungal pathogens. The results indicated that the isolates C and S2 had the potential for use as biocontrol agents for suppressing sugar beet root rot disease, particularly in the saline soils (Karimi et al. 2012).

The mechanism of biocontrol activity of *Streptomyces griseocarneus* strain Di944, isolated from the rhizosphere of field tomato plant, against damping-off and root rot pathogens affecting tomato plug transplants, was investigated. The biocontrol agent (BCA) inhibited *Rhizoctonia solani, Pythium* spp., *Phytophthora* spp., *F. oxysporum* f.sp. *lycopersici*, *F.oxysporum* f.sp. *radicis-lycopersici, F. solani, Thielaviopsis basicola* and *Verticilliium dahliae*, but not bacterial pathogens of tomato, as indicated by agar diffusion bioassay. The culture filtrates of *S. griseocarneus* contained an antifungal compound, a pentaene macrolide complex, designated rhizostreptin, which was detected also in the extracts from rhizospheres of tomato transplants grown from seeds treated with the BCA. The antifungal compound suppressed the development of damping-off caused by *R. solani*. Spore germination and mycelial growth of fungi pathogenic to tomato were inhibited by rhizostreptin at concentrations between 0.5 and 2.0 µg/ml. Rhizostreptin was more fungitoxic than other polyene macrolides, amphotericin B, nystatin, filipin and candicidin-type heptanes. Growth of *S. griseocarneus*

was substantial in mineral medium supplemented with cell wall components of *R. solani* and this indicated the ability of the BCA to utilize pathogen cell wall components as carbon and nitrogen sources. In addition, production of hydrolytic enzymes, chitinase, glucanase, phospholipase and proteinase by *S. griseocarneus* Di944 was observed in mineral medium supplemented with glucose and ammonium sulfate or cell wall components of *R. solani* as carbon sources. Secretion of extracellular antifungal pentaene macrolide and fungal cell wall-degrading enzymes (CWDEs) was considered to be the major mechanism by which the BCA was able to inhibit pathogenic fungi and oomycetes involved in the damping-off and root rot of tomato transplants (Sabaratnam and Traquair 2015).

Pseudomonas chlororaphis strain JP1015 and *P. fluorescens* strain JP2175, capable of inhibiting the growth of *Aspergillus flavus* in vitro, were evaluated for their antifungal activity in soil coculture. Growth of *A. flavus* was inhibited up to 100-fold by *P. chlororaphis* and *P. fluorescens* within three days following soil coinoculation. *A. flavus* propagule densities after 16 days remained 7- to 20-fold lower in soil treated with either bacterial strain. Under bench-scale wind chamber conditions, treatments of soil with *P. chlororaphis* and *P. fluorescens* reduced airborne spores dispersed across a 1 m distance by 75- to 1,000-fold and 10- to 50-fold, respectively, depending on the soil type and inoculum level. The results suggested the possibility of employing bacterial BCAs to reduce soil population of mycotoxigenic fungal pathogens like *A. flavus* that could infect crop plants via airborne propagules (Palumbo et al. 2010). The antagonistic activity of two *Rhizobium* strains PchAzm and PchS. Nir2 against *Rhizoctonia solani* was assessed. These strains reduced fungal growth observed in vitro. Further, these isolates reduced chickpea infection by *R. solani*, due to the direct effect of *Rhizobium* strains on the pathogen. Concomitantly reduction in infection was accompanied by enhanced level of defense-related enzymes, phenylalanine ammonia lyase (PAL) and peroxidase (POX). Phenol contents of the roots were increased following bacterization of plants. The results showed the direct effect of *Rhizobium* strains on pathogen and indicated the possible induction of resistance by the BCA strains (Hemissi et al. 2013).

Indigenous plant growth-promoting bacteria from solarized soil effective against *Monosporascus cannonballus*, incitant of root rot and vine decline of melon were evaluated for their antagonistic activity. Two bacterial species were identified as *Bacillus subtilis/amyloliquefaciens* (BsCR) and *Pseudomonas putida* (PpF4), based on phenotypic, physiological characteristics and analysis of 16S rDNA sequences. Antagonism by BsCR was characterized by a consistent inhibition of the pathogen growth in vitro. PpF4 strain strongly inhibited the development of perithecia of *M. cannonballus*. Under greenhouse conditions, BsCR alone and in combination with PpF4 consistently decreased the disease symptoms. BsCR and the combination of bacterial strains significantly increased root biomass in both inoculated and noninoculated control plants. Following seed treatment with BsCR, the accumulation and the isoenzyme induction of peroxidase in roots as biochemical marker for induction of resistance were detected, indicating that BsCR might reduce disease severity by the activation of the plant defense responses also. The results showed that the synergistic biocontrol activities of *B. subtilis* BsCR and *P. putida* PpF4 might be included as a component in the integrated management of root rot and vine decline of melon caused by *M. cannonballus* (Antonelli et al. 2013). Biological control potential of *B. subtilis* strain Z-14 against wheat take-all disease pathogen *Gaeumannomyces graminis* var. *tritici (Ggt)* was assessed by amending the medium with crude extract of the bacterial culture filtrate in petridish assay. The severity of take-all in wheat seedlings grown on petridishes was reduced by 91.3% and in potted plants by 69.8%, compared to *Ggt*-inoculated control plants. Treatment with crude extract significantly increased growth and fresh weight of roots. The culture filtrate of strain Z-14 was relatively thermostable with 88.2% of antifungal activity being retained at 100°C for 30 min. The pH range (3 to 8) did not significantly affect the antifungal activity of the culture filtrate, under basic conditions. The activity was not transferable to the organic solvent phase after treatment with organic solvent extractants. The culture filtrate showed a broad-spectrum of antifungal activities against fungal pathogens (Zhang et al. 2017a).

Although the plant growth-promoting rhizobacteria (PGPRs) have multiple mechanisms of biocontrol activities, production of different kinds of antibiotics appears to be the principal mechanism of action against microbial pathogens causing soilborne crop diseases. These BCAs secrete phloroglucinols, phenazines, pyoluteorin, pyrrolnitrin and rhamnolipids. Production of antibiotics by PGPRs has been demonstrated in different BCA-pathogen-crop plant species has been demonstrated. Among phloroglucinols, 2,4-diacetylphloroglucinol (2,4-DAPG) is commonly detected in the presence of *Pseudomonas* spp. Abundance of 2,4-DAPG-producing *Pseudomonas* spp. was correlated with natural suppression of wheat take-all disease caused by *Gaeumannomyces graminis* var. *tritici*. The fluorescent pseudomonads have been reported to suppress the development of root and seedling diseases caused by soilborne fungal pathogens. *P. fluorescens* CHA0 suppressed tobacco black root disease and tomato wilt and crown and root rot disease (Duffy and Defago 1997). Suppression of development of sugar beet damping-off disease by *Pseudomonas* spp. (Shanahan et al. 1992) and wheat take-all disease by *P. fluorescens* strains Q2-87 and Q8rl-96 (Raaijmakers and Weller 1998) has been reported. The effects of 2,4-DAPG on fungal pathogens, which can produce different spore forms and survival structures have been investigated. The responses of *Pythium* spp. infecting several crops to 2,4-DAPG were assessed. Variations in the sensitivity of 14 *Pythium* isolates obtained from different hosts to 2,4-DAPG were recorded. Different propagules of P. *ultimum* var. *sporangiferum* exhibited differences in their sensitivity to 2,4-DAPG. Zoospores were the most sensitive, followed by sporangia, the mycelium being the most resistant structure. Ultrastructural changes in the hyphal

tips of *P. ultimum* var. *sporangiferum* exposed to 2,4-DAPG were assessed, using the transmission electron microscope (TEM). Different extents of disorganization in hyphal tips of the pathogen were visualized. Localized alteration (proliferation or disruption) in plasma membrane organization, development of an extensive network of smooth membranous vesicles, degenerated cytoplasm bordered by a retracted plasma membrane and hyphal senescence accompanied by vacuolization and degeneration of its contents were the frequently observed alterations in the fungal pathogen structure. It seemed that 2,4-DAPG did not affect the cell wall structure and composition of hyphal tips of the pathogen, since B(1,3)-1, B(1,4)- and B(1,6)-glucans were present at the same concentrations in hyphal tips both in the presence or absence of 2,4-DAPG, as revealed by immunolocalization experiments, using the primary antibody (de Souza et al. 2003a, 2003b). The sensitivity of 76 plant pathogenic and/or saprophytic strains of *F. oxysporum* to 2,4-DAPG produced by *Pseudomonas fluorescens* was assessed. *F. oxysporum* strains (17%), including *F. oxysporum* f.sp. *melonis* (strain Fom38 and Fom1127) and *F. oxysporum* f.sp. *cubense* (strains Focub 1, 2 and 13) were relatively tolerant to high concentrations of 2,4-DAPG. Some tolerant strains (18) could metabolize 2,4-DAPG. In two tolerant strains, deacetylation of 2,4-DAPG to less fungitoxic derivatives, monoacetylphloroglucinol and phloroglucinols occurred. Fusaric acid produced by *F. oxysporum* strain might directly affect 2,4-DAPG biosynthesis by repressing the expression of the biosynthetic gene *phlA*. Fusaric acid-mediated repression of 2,4-DAPG synthesis in *Pseudomonas* spp. was strain-dependent, as fusaric acid blocked the 2,4-DAPG biosynthesis in strain CHA0, but not in the strain Q21-87 (Schouten et al. 2004.).

Pseudomonas fluorescens UP61 was effective in suppressing the development of *Sclerotium rolfsii*, infecting beans and *Rhizoctonia solani* infecting tomato. The strain UP61 produced three antibiotics, viz., 2,4-diacetylphloroglucinol (2,4-DAPG), pyrrolnitrin and pyoluteorin, contributing to its antagonistic activity. Molecular techniques like 16S rDNA sequencing, random fragment length polymorphism (RFLP), random amplified polymorphic DNA (RAPD) and rep-PCR assays and partial sequencing of the *phlD* gene, governing the biosynthesis of 2,4-DAPG showed similarity between the strain UP61 with other biocontrol agents isolated from other geographical locations that have been shown to produce these antibiotics (De La Fuente et al. 2004). A simple and rapid method was developed to detect the presence and assess the genetic diversity of *phlD*+ *Pseudomonas* strains directly in the rhizosphere samples, without the need for enrichment on nutrient media and prior isolation of the BCA strains. The denaturing gradient gel electrophoresis (DGGE) analysis of the 350-bp fragments of *phlD*+ allowed discrimination between genotypically different *phlD*+ reference strains and indigenous isolates. The DGGE analysis of the *phlD* gene allowed identification of new genotypic groups of specific antibiotic-producing *Pseudomonas* with different abilities to colonize the rhizosphere of sugar beet seedlings (Bergsma-Vlami et al. 2005).

Pseudomonas fluorescens strain KD did not produce 2,4-DAPG which is the main antibiotic involved in the biocontrol activity of several strains of *Pseudomonas* spp. However, strain KD effectively protected cucumber plants against damping-off disease caused by *Pythium ultimum*. The type III secretion system (TTSS) is employed by bacteria for pathogenic or symbiotic interactions with plant and animal hosts. The presence of TTSS genes in *P. fluorescens* KD strain was detected. In spite of the presence of pathogenic attribute, the strain KD was not pathogenic to cucumber. Inactivation of *hrcV* strongly reduced the biocontrol potential of the strain KD against *P. ultimum*. The reduced biocontrol efficacy was not due to a lower ecological fitness of *hrcV* mutant, since the mutant persisted in the potting mix and colonized the plant roots to the same level as that of the wild-type strain, regardless of whether the pathogen was present or not. The expression of the operon containing *hrcV* in the strain KD was strongly stimulated in vitro and in situ by *P. ultimum*, but not by cucumber (Rezzonico et al. 2005). The polar growth of *Aphanomyces cochlioides*, causing damping-off disease of sugar beet and spinach was inhibited by *P. fluorescens* strain ECO-001 isolated from *Plantago asiatica*. The antibiotic 2,4-DAPG secreted by ECO-001 induced excessive branching of hyphae of the pathogen. Confocal laser scanning microscopic (CLSM) observations revealed that both ECO-001 and synthetic DAPG severely disrupted the organization of F-actin filaments present in the apical cap adjacent to the plasma membrane, in a similar manner (Islam and Fukushi 2010).

Phenazines (Phzs) are the antibiotics produced by *Pseudomonas* spp., with biocontrol activity against fungal pathogens. *P. chlororaphis* PCL1391 was highly effective in suppressing production of microsclerotia by *Verticillium longisporum*, infecting cauliflower due to phenazine-1-carboxamide (PCN) formed by the BCA. A seven-gene operon *phzABCDEFG* was responsible for the biosynthesis of phenazine-1-carboxylic acid (PCA) and *phzH* encoded for an asparagine synthetase-like enzyme, capable of converting PCA to PCN. A *phzB* mutant (PCL1119) was phenazine-deficient and a *phzH* mutant (PCL1121) produced PCA, instead of PCN. PCL1119 was effective in inhibiting microsclerotial germination, as the wild-type strain, whereas PCL1121 was more effective, when compared to the wild-type strain. Further, the mutants of *P. chlororaphis* PCL1391 and *P. aeruginosa* 7NSK2 overproducing PCA were more effective in inhibiting microsclerotial germination and formation of secondary microsclerotia, when compared to wild-type strain. Some strains of *Pseudomonas* produced biosurfactants that might facilitate adhesion of the bacteria on the surface of fungal pathogens and act synergistically with the antibiotics to increase the effectiveness of biocontrol by the bacterial BCA (Debode et al. 2007). Phenazine-1-carboxamide (PCN) was shown to be crucial for the biocontrol activity of *Pseudomonas chlororaphis* strain PCL1391 against tomato wilt pathogen *F. oxysporum* f.sp. *radicis-lycopersici* (FORL). The strain PCL1391 also produced hydrogen cyanide (HCN), chitinase and protease. The expression of the biosynthetic operon for HCN was under the regulation of quorum sensing

(QS) (Chin-A-Woeng et al. 2001). Development of canola stem rot disease caused by *Sclerotinia sclerotiorum* was suppressed effectively by *Pseudomonas* sp. DF41. Two mutants *gacS* (DF41-469) and *lp* (DF469-1278) mutants were generated via transposon mutagenesis. The GacS/GacA system is known to control expression of genes required for the synthesis of secondary metabolites such as antibiotics in several *Pseudomonas* spp. The *gacS* mutant had an insertion in *gacS*, forming part of the GacS/GacA regulatory system. By contrast, mutation occurred in a gene involved in lipopeptides (LP) synthesis in the *lp* mutant. The wild-type strain DF41 produced a number of compounds including hydrogen cyanide (HCN), proteases, alginate and LP molecules that might contribute to biocontrol. All of these compounds were generated under Gac control. Furthermore, DF41 strain produced autoinducers, suggesting that this strain might employ quorum sensing as part of its lifestyle. The *gacS* and *lp* mutants could not protect canola against stem rot disease caused by *Sclerotinia sclerotiorum*. Further, both mutants were unable to sustain themselves in canola phyllosphere. The results indicated that suppression of stem rot disease development in canola by DF41 strain was dependent on LP production and presence of a functional Gac system (Berry et al. 2010). A *gacS* mutant of *Pseudomonas* sp. lost biocontrol activity against *Leptosphaeria maculans*, causal agent of blackleg of canola. The biocontrol activity could be restored in the mutant PA23-314 by incorporating the *gacS* gene. The phenazine mutant PA23-63 showed an antifungal and biocontrol activity to the same level as the wild-type strain (Selin et al. 2010).

Pseudomonas chlororaphis isolated from the rhizosphere of green pepper plants produced phenazines mainly phenazine-1-carboxylic acid (PCA) and 2-hydroxyphenazine (2.OH-PHZ) which had broad spectrum of antifungal activity against plant pathogens (Liu et al. 2007). The biocontrol activity of *P. aeruginosa* PNA1 against *Pythium splendens*, infecting bean and *P. myriotylum* infecting cocoyam was found to be due to production of phenazine-1-carboxylic acid (PCA) and phenazine-1-carboxamide (PCN). The tryptophan autotrophic mutants, FM13 deficient in phenazine production, were unable to protect the plants against oomycete pathogens. Exogenous supply of tryptophan restored the biocontrol activity of the mutant FM13, as reflected by reduction in disease severity in treated cocoyam plants (Tambong and Höfte 2001). *Pseudomonas* CMR56 and CMR12a strains isolated from the rhizosphere of cocoyam plants were highly efficient in protecting cocoyam plants against *Pythium myriotylum*. These two strains produced phenazines and surfactants. The strain CMR5c produced pyrrolnitrin and pyoluteorin also (Perneel et al. 2007).

Mechanisms of action of two bacterial biocontrol agents *Pseudomonas chlororaphis* PCL 1391 and *P. fluorescens* WCS365 against *F. oxysporum* f.sp. *radicis-lycopersici* (FORL), causal agents of tomato foot and root rot disease were investigated, using confocal laser scanning microscope (CLSM) and different autofluorescent proteins as markers. Tomato seedlings were bacterized with PCL1391 and WCS 365 strains and planted in sand system infested with

FORL. The BCA strains reached the root surface earlier and multiplied faster than the pathogen. The bacterial strains and pathogen hyphae colonized the same niches on tomato root, the intercellular junctions, probably due to chemotaxis toward and utilization of compounds in the exudates. By colonizing these sites and utilizing the nutrients in the exudates, the bacteria prevented colonization and penetration of root tissue by FORL. The strain PCL1391 produced phenazine-1-carboxamate (PCN), which altered the growth and morphology of pathogen hyphae both in vitro and in vivo (greenhouse conditions). The lack of PCN production in the strain PCL1391 resulted in a delay in the appearance of morphological abnormalities in pathogen hyphae. By contrast, the strain WCS 365 might suppress pathogen development via induced systemic resistance (ISR). However, there was no difference in biocontrol activity of the bacterial BCAs that could be associated with ISR. It is possible that the effects of ISR by WCS365 strain were nearly compensated by the antibiosis effect of the strain PCL1391. The extensive root colonization by both bacterial BCAs might represent a new mechanism in biocontrol by these *Pseudomonas* strains (Bolwerk et al. 2003). *Pseudomonas chlororaphis* strain PA23 effectively suppressed the development of disease symptoms induced by *Sclerotinia sclerotiorum* on canola and sunflower crops. The strain PA23 produced nonvolatile antibiotics phenazine and pyrrolnitrin, as well as volatile antibiotics nonanal, benzothiazole and 2-ethyl-1-hexanol. The role of nonvolatile antibiotics on root colonization and biocontrol potential of PA23 against *S. sclerotiorum* on sunflower was investigated. Application of strain PA23 alone or in combination with phenazine- and pyrrolnitrin-deficient *Tn* mutants resulted in significantly higher root bacterial number and suppression of Sclerotinia wilt disease (P = 0.05). The bacterial population decreased considerably and it seemed that the bacterial population was negatively correlated with the number of antibiotics produced by the strain PA23. The strains producing at least one antibiotic were able to maintain relatively higher population than nonproducers and an increase in bacterial population at 6 weeks after sowing was recorded for strains producing at least one antibiotic. The results did not indicate a clear role for phenazine or pyrrolnitrin in the antagonistic activity of *P. chlororaphis* PA23 (Athukorala et al. 2010).

Wheat-take-all disease caused by *Gaeumannomyces graminis* var. *tritici (Ggt)* is an economically important disease in China and other wheat-growing countries around the world. Bacteria were isolated from winter wheat from irrigated and rainfed fields. Samples from rhizosphere soil, roots, stems and leaves were used for isolation of bacteria on King's medium B. Of the 553 isolates, 105 isolates inhibited *Ggt* and they were identified as *Pseudomonas* spp. Based on the antagonistic potential, 13 strains were selected and they aggressively colonized the rhizosphere of wheat and suppressed the development of take-all disease. Three stem endophytic strains possessed genes for biosynthesis of phenazine-1-carboxylic acid (PCA), but none of them had genes for production of 2,4-DAPG, pyrrolnitrin or pyoluteorin. High-pressure liquid chromatography (HPLC) analysis of 2-day

old cultures confirmed that the strains HC9-07, HC13-07 and JC14-70 produced PCA, but not any other phenazines. All three strains produced PCA in the rhizosphere, as revealed by HPLC quantitative time-of-flight 2 mass-spectrometry analysis of the extracts from roots of spring wheat colonized by the endophytic strains of *Pseudomonas fluorescens* 2-79. Loss of PCA production by strain HC9-07 abolished its biocontrol activity. Analysis of DNA sequences within the key phenazine biosynthesis gene *phzF* and of 16S rDNA indicated that strains HC9-07, HC13-07 and JC14-07 were similar to *P. fluorescens* 2-79, which was a well-known producer of PCA (Yang et al. 2011).

Pseudomonas CMR12a, producing phenazines and cyclic lipopeptides (CLPs), suppressed the development of bean root rot disease caused by *Rhizoctonia solani*. The involvement of phenazines and CLPs in the antagonistic activity of the strain CMR12a was investigated, using two different anastomosis groups AG2-2 (intermediately aggressive) and AG4HG1 (highly aggressive) pathogen strains. The wild-type strain CMR12a drastically reduced the disease severity induced by both *R. solani* strains. A CLP-deficient and phenazine-deficient mutant of CMR12a could still protect bean plants with less efficiency, compared with the wild-type strain. The biocontrol activity was entirely abolished, when the mutant deficient in both phenazine and CLP was inoculated. Washing bacterial cells before application, significantly reduced the suppressive capacity of the BCA, indicating that metabolites produced during growth on plate were required for high level of biocontrol activity of CMR12a strain. The BCA induced pronounced branching of hyphal tips of both *R. solani* AGs. Phenazine-deficient mutant induced more branched and denser mycelium, whereas CLP-deficient mutant and mutants deficient in both CLPs and phenazine did not affect the vegetative growth of the pathogen. The results indicated that phenazine and CLPs might have a role in the suppression of pathogen development and consequent symptom expression in *R. solani*-infected bean plants (D'aes et al. 2011).

Pseudomonas fluorescens HC1-07 isolated from wheat phyllosphere suppressed the development of wheat take-all disease caused by *Gaeumannomyces graminis* var. *tritici (Ggt)*. The strain HC1-07 produced a cyclic lipopeptides (CLP) with a MW of 1,126 and the extracted CLP strongly inhibited the growth of *Ggt* and *Rhizoctonia solani* AG-8, causing Rhizoctonia root rot of wheat at concentration of 100 µg/ml. In order to determine the role of the CLP in biological control, plasposon mutagenesis was applied to generate two nonproducing mutants, HC1-07viscB and HC107prtR2. These two mutants were deficient in swimming and swarming mobility. Furthermore, HC1-07prtR2 mutant lost the ability to secrete exoprotease, whereas HC1-07viscB mutant retained wild-type levels of exoprotease production. Production of siderophores was not affected in both mutants. But the biofilm formation was reduced in both mutants, compared to the wild-type strain. The genomic complementation of the mutants restored the surface mobility and production of exoprotease to the wild-type strain levels and partially restored biofilm formation. The mutants were introduced individually into Quincy Virgin soil and the population sizes of each strain in the rhizosphere of wheat plants were determined at 2-week interval. Of the two mutants, HC1-07prtR2 was mildly, but significantly impaired in the ability to persist in the wheat rhizosphere. Complementation of the *prtR* mutation restored plant colonization to the wild-type strain levels (see Figure 3.8). The CLP-deficient mutants HC1-07viscB and HC1-07prtR2 and complemented mutants were tested for the biocontrol potential against *Ggt* and *R. solani* under controlled conditions. Strain HC1-07 applied as seed treatment significantly suppressed development of both wheat take-all and Rhizoctonia root rot diseases, compared with control (see Table 3.1). Both mutants were less efficient in suppressing the development of both diseases. Genetically complemented mutants regained the ability to suppress disease development. The results suggested that the viscosia-like CLP appeared to be a more important determinant of suppression of wheat take-all and Rhizoctonia root rot diseases (Yang et al. 2014).

The mechanism of biocontrol activity of *Pseudomonas fluorescens* Pf29Arp against the wheat take-all pathogen, *Gaeumannomyces graminis* var. *tritici* (*Ggt*) was investigated, using the confrontation assay in petriplates. In addition, disease development in roots, rates of bacterial and fungal root colonization, the transcript levels of candidate fungal pathogenicity genes and plant-induced genes were monitored during the 10-day infection process. The BCA inoculation of wheat roots reduced the development of *Ggt*-induced disease expressed as attack frequency and necrosis length. The growth rates of *Ggt* and Pf19Arp, monitored through qPCR assay of DNA contents with a part of the *Ggt* 18S rDNA gene and a specific Pf29Arp strain deletion probe, respectively,

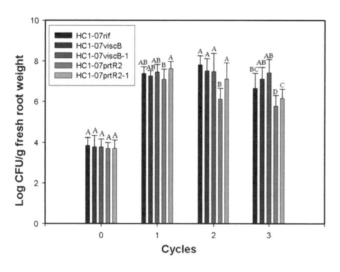

FIGURE 3.8 Colonization of wheat roots by *Pseudomonas fluorescens* strain HCl-07rif, cyclic lipopeptide-deficient mutants HCl-07prtR2-1 under growth chamber conditions Cycle 0: populations of each strain in soil at planting and populations determined at 2-week interval for cycle 1, 2, and 3; same letter above bars for the same cycle indicates that means are not significantly different (P = 0.05) as per Fischer's protected least significant difference test.

[Courtesy of Yang et al. 2014 and with kind permission of the American Phytopathological Society, MN]

increased throughout the interactions. Bacterial antagonism and colonization did not have any significant effect on root colonization by *Ggt*. Expression of fungal and plant genes was quantified in planta by qRT-PCR assay, during interaction. During the early stages of the tripartite interaction, several pathogen genes were downregulated by Pf29Arp, including two laccases, a ß-1,3-glucanase and a mitogen-associated protein kinase. The host plant glutathione-S-transferase gene was induced by *Ggt* alone and upregulated by pf29Arp in interaction with the pathogen. The results showed that the antagonism of BCA strain Pf29Arp might act through alteration of fungal pathogenesis and probably through the activation of host defenses (Daval et al. 2011). The interactions among bacterial BCA *Pseudomonas* spp., wheat-infecting root pathogens and the 10 representative Swiss agricultural soils with cereal cropping history were investigated. The Swiss agricultural soils that were highly suppressive to *Ggt* were often highly conducive to *P. ultimum* and vice-versa. No significant relationship could be established between soil suppressiveness and the abundance of *Pseudomonas* spp. carrying DAPG, PHZ and pyrrolnitrin biosynthetic genes and ability of the soil to support expression of the antimicrobial genes. Correlation analyses indicated that certain soil factors such as silt, clay and some nutrients might influence both abundance and expression of antimicrobial genes. The results suggested that *Pseudomonas* spp. producing DAPG, PHZ or pyrrolnitrin were abundantly present in Swiss soils and the expression of genes encoding antimicrobial compounds in the bacteria might have a role in the suppression of development of the soilborne diseases infecting wheat. However, the precise role of the BCAs in soil suppression could not be clearly understood (Imperiali et al. 2017).

The genes *phzM* and *phzS* genes in *Pseudomonas aeruginosa* PA01 coded for eight enzymes that modify phenazine into related derivatives. The gene *phzM* is located in upstream of *phzA1B1C1D1E1F1G1* operon which is involved in the production of pyocyanin antibiotic (Mavrodi et al. (2001). Purified pyocyanin from *P. aeruginosa* PA01 inhibited the mycelial growth of of *Macrophomina phaseolina* infecting peanut. Using the well-diffusion method, the effect of pyocyanin on disease suppression and biofilm formation by the rhizobial strain Ca 12 on radicles of peanut was assessed. Pyocyanin suppressed disease more effectively at high concentration. However, at lower concentration, pyocyanin increased CFUs of Ca12 strain on radicles of peanut seedlings. Application of pyocyanin-producing pseudomonads together with rhizobia contributed to the enhancement of nodulation ability and sustained the growth and productivity of peanut even in the presence of *M. phaseolina* (Khare and Arora 2011). *Pseudomonas chlororaphis* strains DF190 and PA23, *Bacillus cereus* strain DFE4 and *B. amyloliquefaciens* strain DFE16 were evaluated for the antagonistic activity against canola blackleg pathogen, *Leptosphaeria maculans* (anamorph, *Phoma lingam*). Application of the bacteria at 24 or 48 h prior to pathogen inoculation of the canola cotyledon was an important factor for suppression or prevention of formation of blackleg lesions. The strains PA23 and DF190 produced phenazines and pyrrolnitrin, whereas the strains

DFE4 and DFE16 produced lipopeptides, antibiotics iturin A, bacillomycin D and surfactin. The strains PA23 and DFE4 were tested as representative producers of each set of antibiotics for the split inoculation (SPI) of the extracts for induction of induced systemic resistance (ISR). The assays showed a small, but significant reduction in severity via a systemic response. The local SPI inoculation of the extract (for direct antagonism) showed significantly greater reduction of disease severity, which was also consistent with the SPI of the bacterial cells, establishing a more important role for the antifungal metabolites present in the culture extracts for the direct suppression of development of blackleg symptoms. The localized inhibition of pycnidiospores by the bacteria could be due to the successful colonization of the infection site which in turn most possibly function as a suitable delivery system for antifungal metabolites. The biocontrol activity of the mutant PA23-63, though deficient in phenazine production, was as effective as the wild strain in suppressing *L. maculans* development, indicating the nonrequirement of phenazines for its antagonistic activity. Strains of *P. chlororaphis* did not induce defense-related enzymes at the point of inoculation. Direct antifungal activity of *P. chlororaphis* strains appeared to be the dominant mechanism mediating control of blackleg disease of canola caused by *L. maculans* (Ramarathnam et al. 2011).

Hydrogen cyanide (HCN) is a volatile antibiotic produced as a secondary metabolite by *Pseudomonas* spp. HCN is highly toxic to most aerobic microorganisms, due to its ability to block the cytochrome oxidase pathway even at very low concentrations (pmol). Suppression of plant disease development is attributed to the action of HCN on some oomycetes. *P. fluorescens* CHA0 produces HCN, other antibiotics and siderophores. The mutants of CHA0 deficient in synthesis of HCN, antibiotics and exoenzymes were unable to protect tobacco plants against tobacco black rot pathogen *Thielaviopsis basicola*. HCN was considered to be primarily responsible for the antagonistic activity against the pathogen (Voisard et al. 1989). In *Pseudomonas fluorescens* strains Q2-87 and CHA0, HCN synthase required for HCN production was encoded by the *hcnABC* gene (Haas and Défago 2005). Primers targeting *hcnABC* genes with MultiAlin were employed for identification of isolates containing these sequences. A single amplicon of about 570-bp in length was amplified from the DNA of HCN-producing strains. No amplicon was generated by two negative HCN pseudomonads tested (Svercel et al. 2007). Bacterial species with antagonistic activity have been shown to produce cyclic lipopeptides (CLP) effective against fungal pathogens. A derivative of *Bacillius subtilis* BBG110, over-producing the CLP mycosubtilin exhibited enhanced suppressive activity, against *Pythium* spp. infecting tomato seedlings (Leclerc et al. 2005). Application of viscosinamide-producing *P. fluorescens* strain DR54 to sugar beet considerably increased seedling emergence and root length in soil infested with *Pythium ultimum* (Thrane et al. 2000). Sugar beet seeds were coated with CLP-producing *P. fluorescens* strains and subsequently germinated in nonsterile soil. The strain DR54 maintained a high and constant level of viscosinamide in the rhizosphere of plants growing from treated seeds for about 2

days, whereas strains 96.578 and DSS73 produced higher concentrations of CLPs tensin or amphisin. All three CLPs were detectable for several days in the rhizosphere. The results suggested that production of CLPs might occur only in specific in habitats like rhizosphere of germinating sugar beet seeds, rather than in the bulk soil (Nielsen and Søensen 2003). An antifungal polyketide 2,3-deepoxyl-2,3-didehydrorhizoxin (DDR) was produced by *Pseudomonas* spp. The DDR was effective in suppressing the development of wheat seedlings blight disease caused by *Fusarium culmorum* (Johansson and Wright 2003). *Bacillus* strains DE07, QST713 and F2B24 were evaluated for the antagonistic activity against *F. oxysporum*. The cellulases, cell-free supernatants and volatiles from BCA strains exhibited varying degrees of suppressive effects. Proteomic analysis of secreted proteins from EU07 and F2B24 revealed the presence of lytic enzymes, cellulases, protease, 1,4-ß-glucanase and hydrolases. All these enzymes contributed to degradation of seedling cell wall. Further, the proteins involved in the metabolism of protein folding, protein degradation, translocation, recognition and signal transduction cascade may play an important role in the suppression of *F. oxysporum* development (Baysal et al. 2013). The microbial population of a Korean soil suppressive to Fusarium wilt of strawberry was characterized. In this suppressive soil, infection of strawberry roots by *F. oxysporum* resulted in a response by microbial defenders, of which members of the Actinobacteria appeared to have a vital role. *Streptomyces* genes responsible for ribosomal synthesis of a novel heat-stable antifungal thiopeptide antibiotic inhibitory to *F. oxysporum* and the mode of action of the antibiotic against fungal cell wall biosynthesis were studied. The results revealed the role of natural antibiotics as weapons in the fight against the pathogen and antagonists in the rhizosphere that is required for plant health, vigor and development (Cha et al. 2016).

The biosurfactants produced by *Pseudomonas* spp. interact with cell surface of fungal plant pathogens and also can act on lipids creating pores on the membrane layer (Raaijmakers et al. 2006). *Pseudomonas fluorescens* SS101 strain effectively prevented infection of tomato leaves by *Phytophthora infestans*, causing late blight disease and also restricted the expansion of existing lesions and sporangial production. As the sporangia are required for primary and secondary infections by *P. infestans*, destructive effect of the strain SS101 on both lesions area and sporangia formation might result in reduction of rate of disease development and spread. The biosurfactants produced by SS101, massetolide A was involved in the antagonistic activity of the BCA. The massetolide A-deficient mutant 10.24 was significantly less effective in its biocontrol activity, compared with wild-type strain. Application of massetolide A on tomato leaves and roots effectively protected treated plants against *Phytophthora infestans*. The results indicated that the CLP massetolide A was an important component of biocontrol efficacy of strain SS101. The purified preparation of massetolide A provided significant protection to tomato against tomato late blight disease both locally and systemically via induced resistance, revealing the multifunctional capacity of the cyclic lipopeptide produced by *P. fluorescens* SS101 strain (Tran et al. 2007).

The role of biosurfactants in *Pseudomonas*-mediated suppression of viability of microsclerotia of *Verticillium longisporum*, infecting cauliflower was studied. The biosurfactant-deficient mutants of *Pseudomonas* CMR12a and *P. aeruginosa* PNA1 strains were less efficient in reducing viability of *Verticillium* microsclerotia, compared with the wildtype strain. The biosurfactants were effective at a BCA cell density of 2×10^9 CFU/ml. However, the biosurfactant production by the BCA strains did not fully account for *Verticilliium* microsclerotial germination. The biosurfactants had residual adverse effect on microsclerotial germination and/or formation of secondary microsclerotia (Debode et al. 2007). The potential of a biosurfactants produced by *Pseudomonas koreensis* was assessed by applying crude extract of the BCA cells. Assessment of the effect of the crude extract on *Pythium ultimum*, infecting tomato plants in hydroponic system showed that the incidence of the disease was significantly reduced. Application of the biosurfactants did not affect the indigenous microflora (Hultberg et al. 2010a). The effectiveness of the biosurfactants from *P. koreensis* 2.742 strain against *Phytophthora infestans* was assessed, using a detached leaf assay. The biosurfactants inhibited the motility of zoospores, but mycelial growth was not significantly affected. The sporangia production was also not reduced by the surfactant from *P. koreensis* (Hultberg et al. 2010b). Likewise, the biosurfactants of *P. koreensis* effectively suppressed the development of damping-off caused by *Pythium ultimum* in tomato (Hultberg et al. 2011).

3.1.5.2.2 Competition for Nutrients and Space

Competition for nutrients and space available in various substrates is a natural phenomenon for the survival of all microorganisms, including soilborne bacterial pathogens. Availability of essential micronutrients such as iron is a crucial factor for all microorganisms. Iron becomes a limiting factor in the rhizosphere, depending on the soil pH. The iron-binding ligands known as siderophores with high affinity for iron are produced by microorganisms for sequestration from the micro-environment. Ability to produce siderophores confers competitive advantages to bacteria for colonizing plant tissues and to exclude other microorganisms from the same ecological niche. The siderophores may be of two kinds: catechol type or hyroxamate type (Neilands 1981). The fluorescence of pseudomonads is attributed to the presence of an extracellular diffusible pigment known as pyoverdin (Pvd) or pseudobactin, which has affinity for Fe^{3+} ions. It is a siderophore (iron-carrier) of the producer strain. In iron-depleted media, Pvd-producing *Pseudomonas* spp. may inhibit the growth of fungi and bacteria with efficient siderophores. Under certain conditions, Pvd may function as a diffusible, bacteriostatic or fungistatic antibiotic (Kloepper et al. 1980).

Under greenhouse conditions, *Pseudomonas putida* strain B10 suppressed Fusarium wilt and take-all diseases of wheat. The suppressive effect was lost, when the soil was amended with iron which repressed siderophores production by the strain B10 (Kloepper et al. 1980). The strain B324 of *Pseudomonas* was inhibitory to all seven isolates of *Pythium ultimum* var.

sporangiferum, infecting wheat, in addition to induction of growth-promoting effect on wheat plants under iron-limiting conditions (Becker and Cook 1988). *Pseudomonas fluorescens* strain CV6 effectively suppressed the development of cucumber root rot disease caused by *Phytophthora drechsleri*. The BCA also produced substantial amount of siderophores, antifungal compounds and indole acetic acid (IAA). In addition, strain PN1 produced enzymes involved in induction of systemic resistance, indicating that its biocontrol activity against *P. drechsleri* was successfully achieved through multiple mechanisms operating during the interaction between the pathogen and biocontrol agent (Maleki et al. 2010). *Pseudomonas aeruginosa* (PN1–PN10) strains (10) isolated from the rhizosphere of chir-pine were evaluated in vitro and in vivo for their antagonistic activities against *Macrophomina phaseolina*, causing root rot. The strain PN1 produced siderophores, IAA, cyanogens and solubilized phosphorus, in addition to chitinase and ß-1,3-glucanase. In dual culture assay, the BCA inhibited the mycelial growth of *M. phaseolina* by 69%. PN1 showed strong chemotaxis toward root exudates, resulting in effective root colonization. The BCA PN1 strain showed strong antagonistic effect on *M. phaseolina* and also significantly promoted the growth of chir-pine seedlings (Singh et al. 2010).

Microorganisms may exert positive or negative effects on others for their survival in the rhizosphere and the nature of the interaction may, in turn, have influence on the development of diseases caused by root pathogens. *Pseudomonas putida* WCS 358 suppressed radish Fusarium wilt disease caused by *F. oxysporum* f.sp. *raphani* by competing for iron through the production of the siderophore pseudobactin. Addition of iron to nutrient solution fed to the radish plants, reduced the suppressive effect of the strain WCS358. The pseudobactin negative mutant WCS358 was as effective in suppressing development of Fusarium wilt disease as the wild-type strain, suggesting that the mutant might act on the pathogen through an alternative mechanism. Another strain RE8 of *P. putida* did not depend on the production of pseudobactin for its suppressive activity on wilt pathogen. The strain RE8 could induce systemic resistance in treated radish plants. When the mixture of WCS358 and RE8 strains was applied to the soil, the level of disease suppression was enhanced, compared to application of single strains. This additive effect on disease suppression could be due to pseudobactin-mediated competition for iron and induction of systemic resistance by these BCA strains. The results indicated the possibility of exploiting the complementary mechanisms of disease suppression by two compatible strains of *P. putida* on *F. oxysporum* f.sp. *raphani*, causing Fusarium wilt disease of radish (de Boer et al. 2003).

The competition between the pathogens and the biocontrol agents (BCAs) for available nutrients, as a mechanism of biocontrol activity has been investigated in other pathosystems also. The efficiency of strains of *Pseudomonas fluorescens* in suppressing the development of sugar beet root rot caused by *Pythium ultimum* was assessed. All strains of the BCA reduced the number of lesions and the root and soil populations of *Pythium*, whereas the strains of SBW25 and CHA0 increased lateral roots formed in treated plants. The strain SBW25 did not produce any antifungal compounds and its biocontrol activity was related to its greater colonization ability and rhizosphere competition (Naseby et al. 2001). *P. fluorescens* strain F113 protected the sugar beet roots effectively against *Pythium ultimum*. During colonization of alfalfa rhizosphere, the strain F113 produced variants that were characterized by the translucent and diffused colony morphology. The phenotypic variation in this strain appeared to be mediated by the activity of two site-specific recombinases Sss and XerD. The mutants with disruption in either of the genes involved in the biosynthesis of Sss or XerD, showed drastic reduction in rhizosphere colonization and the mutants had reduced chemotactic motility. Large number of variants (mutants), overexpressing the genes *sss* or *xerD* were generated. All isolated variants were more motile than the wild-type strain and appeared to contain mutations in the *gacA* and /or *gacS* genes. The highly motile mutants were more competitive than the wild-type strain, displacing it from the root tip within 2 weeks (Martínez-Granero et al. 2006).

3.1.5.2.3 Prevention of Pathogen Colonization of Host Roots

Colonization of host plant roots may be prevented by the bacterial biocontrol agents. Application of Pseudomonas *fluorescens* strain CH31 to cucumber significantly reduced root colonization by *Pythium aphanidermatum*, resulting in reduction in incidence of root rot disease in cucumber (Moulin et al. 1996). Likewise, treatment of sugar beet seeds with *P. putida* 40RNF strain as pellets significantly reduced the incidence of sugar beet damping-off disease caused by *Pythium ultimum* by 43% at 48 h after planting. Further, the number of sporangia was reduced by 68% in the soil around sugar beet plants growing from treated seeds. Seed treatment with *P. putida* 40RNF was effective as the fungicide hymexazole in reducing disease incidence in infested soil (Shah-Smith and Burns 1996). Based on the competitive tomato root tips colonization assay used to select efficient root colonizers, 24 isolates of *Pseudomonas rhodesia* isolates with equal or better colonizing ability as *P. fluorescens* WCS365 were selected. These isolates of *P. rhodesia* effectively reduced the incidence of tomato foot and root rot disease caused by *F. oxysporum* f.sp. *radicis-lycopersici* (Validov et al. 2007).

3.1.5.2.4 Induction of Resistance in Plants to Disease

Resistance to crop diseases caused by soilborne microbial plant pathogens may be induced by biotic and abiotic agents. Induced resistance has been broadly differentiated into two forms as systemic acquired resistance (SAR) and induced systemic resistance (ISR). In response to pathogen infection or chemical application 'SAR' process is initiated, whereas colonization of host plant roots by plant growth promoting rhizobacteria (PGPR) leads to ISR. The direct or indirect effects of treatment of wheat or apple with *Pseudomonas fluorescens* SS101 on the development of *Pythium* spp. were assessed by the split-root plant assay. The assays were performed in

orchard soils to explore the possible role of ISR. The strain SS101 or the massA mutant 10.24 significantly reduced infection by *Pythium* spp. on the apple or wheat root system raised in soil treated with the respective bacteria. Wheat root infection was reduced from 34% for plants grown in non-treated soil to 11% for the component of root systems grown in SS101- or mutant 10.24-treated soils. The frequencies of *Pythium* root infection in SS101 and 10.24 treatments did not show significant differences for the component of the root system physically separated from bacterially treated soil, compared to the nontreated control (Mazzola et al. 2007). The efficacy of *Pseudomonas fluorescens* strain FP7 in suppressing the development of turmeric rhizome rot disease caused by *Pythium aphanidermatum* was assessed. A combination of rhizome dip and soil drench of FP7 liquid formulation resulted in the reduction of disease incidence to the minimum (19.0%) under greenhouse conditions and to 10.18 and 13.29% under field conditions respectively in trial I and II. Twelve differentially expressed tripartite interaction between host-pathogen-bioagent through protein profiling could be identified. Further, mass spectrometry (MS) analysis revealed that proteins such as tryptophan synthase beta subunit-like phosphoglycerate kinase subunit beta, cysteine-rich peptide, ribosomal protein S3, clathrin assembly protein and disease resistance protein RPP13-like were differentially regulated. The differentially expressed proteins during tripartite interaction might be directly or indirectly involved in disease resistance in turmeric plants (Karthikeyan et al. 2017).

Induction of systemic resistance in lupine by *Pseudomonas fluorescens* and *P. putida* against Fusarium wilt disease caused by *F. oxysporum* f.sp. *lupine* (FOL) was investigated. Both BCAs significantly reduced the disease incidence under greenhouse and field conditions. A time-course of defense-related enzymes showed substantial increases in their activities in induced infected seedlings, compared to nontreated infected or healthy control plants. The extent of enhancement of enzyme activities varied among treatments. Increases in chitinase and ß-glucanase activities attained maximum levels at 12 and 18 days after inoculation with FOL, respectively. Furthermore, phenylalanine ammonia lyase (PAL) activity increased significantly at 8 days after inoculation. Accumulation of phenolic compounds and specific flavonoids occurred significantly, following infection by the pathogen in induced and/ or infected seedlings, compared with noninoculated control plants. The BCAs increased the growth parameters and seed yield, compared with untreated control plants (Abd El-Rhaman et al. 2012). *Didymella bryoniae*, causing gummy stem blight of watermelon is a serious threat to production in Mekong Delta of Vietnam. *Pseudomonas aeruginosa* strain 23 effectively suppressed directly the pathogen development by producing antibiotics locally and/or indirectly by stimulating the defense systems systemically. Foliar infection by *D. bryoniae* was significantly reduced by seed treatment with the BCA, indicating the induction of resistance by the strain 23 under field conditions. *P. aeruginosa* colonized watermelon plants endophytically, more actively in *D. bryoniae* infected plants than in healthy plants. Treatment with the BCA, inhibited penetration, which was associated with accumulation of H_2O_2, followed by enhanced peroxidase activity and production of new peroxidase isoforms. Suppression of watermelon gummy blight development was achieved by production of antibiotics and ISR under greenhouse and field conditions, following treatment with *P. aeruginosa* (Nga et al. 2010).

Strains of bacterial biological control agents with multiple mechanisms of action against the fungal pathogens are likely to be more effective in suppressing pathogen development than BCAs with single mode of action. *Pseudomonas chlororaphis* strain PA-23 could produce phenazine-1-carboxylic acid (PCA), which inhibited the mycelial growth of *Sclerotinia sclerotiorum*, incitant of canola stem rot disease. Ascospore germination in canola flower petals was inhibited by the strain PA-23. In addition, two applications of strain PA-23 induced resistance against infection by *S. sclerotiorum*. Enhanced accumulation of PR-proteins and oxidative enzymes, including chitinase and ß-1,3-glucanase by PA-23 in canola leaf tissues might account for reduction of pathogen infection. The combination of antibiotic production along with induction of systemic resistance might act in tandem, resulting in restriction of pathogen growth, colonization of canola tissues and consequent suppression of disease intensity significantly (Fernando et al. 2007). Two strains of *P. putida* WCS358r and *P. fluorescens* WCS374r could trigger ISR in *Eucalyptus urophylla* against bacterial wilt disease caused by *Ralstonia solanacearum*, when infiltrated into two lower leaves at 3–7 days before challenge inoculation with the pathogen, but not when the BCA strains were incorporated into the soil. The mutant strain WCS358r, deficient in the biosynthesis of pseudobactin (siderophores), did not induce ISR. The purified pseudobactin from WCS358r induced ISR, suggesting that pseudobactin 358 was the determinant of ISR in WCS358r strain. In contrast, the siderophores-deficient mutant of WCS374r was able to induce ISR to the same level as the wild-type strain. The purified siderophores from this strain induced ISR, indicating that both the siderophores and other uncharacterized ISR determinants of WCS374r could trigger ISR in *Eucalyptus* (Ran et al. 2005). Development of tobacco soft rot disease caused by *Pectobacterium carotovorum* subsp. *caratovorum* (*Pcc*) was reduced by *Pseudomonas chlororaphis* 06 and the mechanism of the biocontrol activity of the BCA was via induction of resistance in tobacco plants. The bacterial determinant involved in ISR was identified using high performance liquid chromatography (HPLC) and nuclear magnetic resonance (NMR) mass spectrometry. The active compound was 2R, 3R-butanediol, which induced systemic resistance against *Pcc*, but not against *P. syringae* pv. *tabaci* (Pst). The global sensor kinase GacS of *P. chlororaphis* 06 was a key regulator for ISR against *P. carotovorum* subsp. *carotovorum* through regulation of 2R, 3R-butanediol, a fermentation product, as well as to the sensor kinase GacS for its production (Han et al. 2006).

Pseudomonas fluorescens PICF7 strain, a native root endophyte of olive trees effectively suppressed the development of olive vascular disease caused by *Verticillium* dahliae.

The bacterial BCA could trigger a broad range of defense responses in root tissues of olive. The mechanism of induction of systemic resistance to olive wilt pathogen by PICF7 strain was studied. Root bacterization of olive plants with PICF7 strain was performed and aerial tissues were sampled at different time interval after bacterization. The suppression subtractive hybridization (SSH) cDNA library, enriched in upregulated genes was generated. Of the 376 expression sequence tags (ESTs), five involved in the defense responses, were selected to perform time course quantitative real-time PCR (qRT-PCR) experiments. Induction of olive genes potentially coding for lipoxygenase 2, catalase, 1-aminocyclopropane-1-carboxylic oxidase and PAL was confirmed at some stages of interaction. The results revealed that root colonization by the endophytic strain of *P. fluorescens* could trigger defense responses in this organ. In addition, root colonization by the BCA could induce a range of systemic defense responses in tissues away from the roots (Cabanás et al. 2014).

Bacillus amyloliquefaciens BS6 and *Pseudomonas chlororaphis* PA-23 were evaluated for their biocontrol potential against canola stem rot disease caused by *Sclerotinia scleortiorum* under greenhouse and field conditions. The BCAs were equally effective in suppressing the development of stem rot disease, as the fungicide Rovral Flo® (iprodione). Suppression of disease development by the BCAs was attributed to reduction of canola flower infections by *S. sclerotiorum* through direct antimicrobial activity and /or induction of plant defense enzymes (Fernando et al. 2007). The mechanism of biocontrol activity of *Bacillus pumilus* 7 Km isolated from wheat rhizosphere against wheat take-all disease pathogen *Gaeumannomyces graminis* var. *tritici* was investigated. The soil was drenched with bacterial suspension and changes in the defense-related enzymes and total phenolics contents were monitored. Disease severity was significantly reduced in the bacterized roots of wheat plants and plant growth was also promoted by treatment with the BCA. The activities of soluble peroxidase (SPOX), ionically cell wash bound peroxidase (CWPOX), ß-1,3-glucanase, ß-1,4-glucanase were increased in plants treated with *Bacillus pumilus*. Further, the phenolics contents were at higher levels in BCA-treated plants, compared to untreated control plants. The activities of enzymes reached the peak at 4–8 days after application of BCA to wheat roots. The disease suppressive activity of *B. pumilus* might be due to its ability to induce local/systemic defense system operating in the host root system (Sari et al. 2007). Seed bacterization and soil application of a mixture of *Bacillus subtilis* strains S2BC-1 and GIBC-Jamog reduced the incidence of tomato wilt disease caused by *F. oxysporum* f.sp. *lycopersici* significantly as indicated by localized and split-root experiments. Activities of chitinase and ß-1,3-glucanase were enhanced in root samples from treated tomato plants. High concentrations of peroxidase isoforms were detected in samples from localized and ISR experiments. The results indicated the possibility of operation of both antibiosis and induction of ISR mechanisms functioning in concert in tomato plants protected with the BCA strains (Shanmugam and Kanoujia 2011).

Elicitation of induced systemic resistance (ISR) by *Bacillus* sp. strain BS107 in tobacco against soft rot caused by *Pectobacterium carotovorum* subsp. *carotovorum (Pcc)* was observed. A determinant of ISR secreted by the strain BS107 was isolated from the cell-free culture supernatant and identified, using mass spectrometry and NMR analyses, as 2-amino-benzoic acid (2-AB) as the principal ISR determinant. Purified 2-AB displayed effective ISR activity against soft rot disease development in the tobacco leaves. Treatment of tobacco roots with 2-AB resulted in protection against *Pcc*. The pathogen was not inhibited at the concentration that induced resistance. Reverse transcription (RT)-PCR assays of tobacco leaves of plants treated with 2-AB on the roots, showed upregulation of the induced marker genes such as *PR1a, PR1c, PR2* and *PR-4*. The inducer determinant 2-AB was biosynthesized from chorismic acid which is a precursor of salicylic acid (SA) and SA is known to have a vital role in mediating plant defenses (Yang et al. 2010). *Bacillus vallismortis* strain EXTN-1 isolated from red pepper suppressed tomato bacterial wilt disease caused by *Ralstonia solanacearum. Bacillus subtilis* strain 816-6, *B. pumilus* strain 228-7, *Bacillus* sp. strain 113-3 and *Paenibacillus polymyxa* strain H321-5 were also evaluated for their ability to induce resistance to tomato bacterial wilt pathogen, along with the strain EXTN-1. The tomato seedlings roots were bacterized with the test strains and the plants were grown in perlite-hydroponic system, followed by challenge inoculation with *R. solancearum*. All bacterial strains suppressed bacterial wilt development to different extent. The strain EXTN-1 was the most effective in reducing infection to 65%, as against 95% in untreated control plants. The movement of the pathogen from the site of inoculation was hampered in plants treated with the strain EXTN-1 was probably due to a mechanism other than antibiosis, since no direct antagonistic activity of this strain against *R. solanacearum* could be detected. The strain EXTN-1 produced an elicitor which efficiently induced systemic resistance in many crops (Park et al. 2007). The bacterial mixture of *Bacillus atrophinus* S2BC-2 and *Burkholderia cepacia* TEPF-Sungal protected gladiolus plants against *F. oxysporum* f.sp. *gladioli,* causal agent of vascular wilt and corm rot disease to the maximum extent. Protection by these strains was attributed to induction of systemic resistance as deduced from stimulation of PR-proteins and activities of peroxidase (PO), polyphenol oxidase (PPO) and phenylalanine ammonia lyase (PAL), in addition to phytoalexins and /or chalcone synthase (Shanmugam et al. 2011).

The mechanisms underlying the interactions of *Streptomyces*- host plant-microbial plant pathogens have been investigated. The biocontrol potential of *Streptomyces coelicolor*, S. griseus, *S. albus, S. antibioticus* and *S. champavatti* against *Rhizoctonia solani* infecting tomato was assessed. *Streptomyces* spp. have been shown to induce systemic resistance in different pathosystems. Association of changes in H_2O_2 production, lipase peroxidation (LPO) and antioxidant enzymes with ISR induction was studied. Production of H_2O_2 at 2 days after inoculation (*dai*) of *R. solani* was 1.1-fold higher in BCA-treated tomato plants, compared with untreated inoculated control plants. Increases in catalase and ascorbate peroxidase activities were recorded at 4 *dai*, whereas enhancement of activities of guaiacol reductases and glutathione peroxidase were observed at 5 *dai*. Similarly, LPO reached the maximum at 6 *dai* in untreated inoculated plants, whereas there was a decrease of 1.3–1.5 fold in BCA-treated plants, compared with untreated inoculated plants. Priming by *Streptomyces* in tomato as a mechanism of biocontrol activity against *R. solani* was revealed, highlighting the capacity of *Streptomyces* spp. to activate plant defense responses in tomato plants through induction of antioxidant enzymes along with improved reactive oxygen species management (Singh et al. 2016). The endophytic *Streptomyces* spp., viz., *S. diastaticus, S. fradiae, S. olivochromogenes, S. collinus, S. ossamyceticus* and *S. griseus* were evaluated for their biocontrol potential against *Sclerotium rolfsii*, infecting chickpea. Stimulation of systemic resistance was monitored by treating the chickpea seeds, followed by inoculation with the pathogen. Substantial increase in defense-related enzymes such as PAL and PPO, along with the accumulation of total phenolics and flavonoids occurred in *S. griseus*-treated and pathogen-inoculated chickpea plants. Likewise, significant increase in superoxide dismutase (SOD), peroxidase (PO), ascorbate oxidase (APX) and guaiacol peroxidase (GPX) was recorded in the same treatment, which favored least lipid peroxidation and chickpea pathogenic stress. Further, in *S. griseus*-primed plants, collar tissues remained undamaged, as revealed by scanning electron microscopic observations. Real-time PCR analysis of genes encoding SOD, PO, PAL, APX, catalase, chitinase and ß-glucanase showed significant increases. The results indicated that the chickpea defense pathway might be triggered after perception of endophytes to synthesize various enzymes, resulting in enhancement of resistance in chickpea against *S. rolfsii* (Singh and Gaur 2017). The extracellular polysaccharide (EPS) produced by *Ralstonia solanacearum*, incitant of eggplant bacterial wilt, functions as an elicitor of resistance in eggplant against bacterial wilt disease. Eggplants treated with intact crude EPS showed a significant decrease in bacterial wilt incidence under both in vitro and greenhouse conditions. In the eggplant cultivars, EPS induced significant upregulation of guaiacol peroxidase and ascorbate peroxidase defense genes, providing significant enhancement of resistance to bacterial wilt disease (Prakasha et al. 2017).

3.1.6 Formulations of Biological Products

Several species/strains of fungi or bacteria have been shown to possess biocontrol potential against soilborne microbial plant pathogens. However, only some of them have been found to be suitable for development of formulation and commercialization as biological products or microbial biocides for use as alternatives to chemicals for the management of diseases affecting field or glasshouse crops.

3.1.6.1 Development of Formulations

Development of formulations of biotic biocontrol agents (BCAs) depends on several factors, such as the efficiency and adaptability of the BCAs that have been found to effectively suppress the development of the soilborne fungal/bacterial pathogens through single or multiple mechanisms of action on target pathogen(s). The knowledge of mechanisms of action of the selected BCA in combination with analysis of the sequence of corresponding genes can provide gene targets to develop high throughput screening procedures. The effectiveness of the BCA at low doses, production of antimetabolites or toxins against target pathogen, tolerance to fungicides or other plant protection chemicals and plant activators and adaptability to general crop management practices are considered as desirable attributes of the BCA to be advanced to formulation stage (Narayanasamy 2013).

Nonpathogenic *F. oxysporum* strain CS-20 with significant antagonistic activity against *F. oxysporum* f.sp. *lycopersici* was inhibited by the fungicide azoxystrobin and chlorothalonil, whereas mefenoxam (Ridomil Gold) and mefenoxam + copper did not affect the growth of the BCA. However, *F. oxysporum* CS-20 was found to be incompatible with all fungicides tested under greenhouse conditions, since the disease incidence was not reduced by the combination of the BCA and fungicides (Fravel et al. 2005). In another investigation, the possibility of combining soil application of *Coniothyrium minitans*, an effective biocontrol agent capable of suppressing the development of Sclerotinia stem rot of oilseed rape, caused by *Sclerotinia sclerotiorum*, with compound fertilizer (N, P and K), was examined. Simultaneous application of the BCA and fertilizer at various concentrations significantly reduced the number of apothecia formed by sclerotia of *S. sclerotiorum* in both pot and field-plot experiments. The compound fertilizer did not affect the ability of *C. minitans* to infect sclerotia of the pathogen in vitro or suppress carpogenic germination of sclerotia of *S. sclerotiorum*. The results indicated that *C. minitans* was compatible with fertilizer, when applied at planting of oilseed rape. Thus, application of the BCA along with fertilizer could save labor cost, resulting in greater monetary benefit to the grower and enhancement of production efficiency (Yang et al. 2011).

After assessment of the biocontrol potential and investigation of mechanisms of the biocontrol activity, the most effective strains/isolates are tested in pilot trials under conditions as close as possible to the natural field conditions under which the BCA is to be placed in practice. At this stage, it would be desirable to expose the BCA in an environment where several pathogens or pests will be available. This will enable verification of the spectrum of action of the bioproduct and to determine the effectiveness of the BCA in different environmental conditions which can be diverse enough to guarantee a wide range of applicability. It would be possible to select the effective isolates under these limiting conditions. Most of the less effective isolates of the BCA are eliminated at this stage, because of their narrow spectrum of action and inconsistency in their performance in various field trials conducted during different seasons. Endophytic plant symbionts include plant growth-promoting rhizobacteria (PGPRs), free living rhizosphere-competent fungi, like strains of *Trichoderma* spp. and mycorrhizal fungi like *Piriformaspora indica*. It is likely that the endophytic BCAs may have much longer periods of efficacy than nonendophytic organisms, because of their ability to grow within the plants and in the same environment in which host plants develop. In addition, some endophytic fungal strain may colonize shoots, roots or stems and they are able to increase resistance to water stress and salt or temperature tolerance. *T. harzianum* T22 enhanced expression of proteins involved in photosynthesis and starch accumulation (Shoresh and Harman 2008). *Trichoderma* strains may alleviate intrinsic stresses like loss of seed vigor and improve seed germination (Shoresh et al. 2010).

3.1.6.1.1 Preparation of Formulations

Strains of biocontrol agents with high level of antagonistic ability to suppress the development of the soilborne microbial plant pathogens by acting through one or more mechanisms are selected for preparing formulations. These strains should be genetically stable, effective at a wide range of temperatures, easy to mass-produce in culture on inexpensive media and be effective against several microbial pathogens. Furthermore, the putative BCA should be available in an easily distributable form and it should be nontoxic and nonpathogenic to human beings and other plants species grown in the same ecosystem. Different methods have been applied for preparing formulations containing fungal and bacterial biocontrol agents. Formulated microbial products may have biomass of the selected BCA and ingredients to improve its survival and effectiveness of the products as the major components. The microbial mass may be of two types: culture grown on semi-solid or liquid media. Liquid formulations may be flowable or aqueous suspensions consisting of biomass suspensions in water, oils or emulsions. Dry (powder) formulations may be in the form of wettable powder, dusts or granules. Wettable powders and dusts contain dry inactive or active ingredients and granules are free-flowing, aggregated product consisting of active or inactive ingredients (Schisler et al. 2004). The formulated product should be shelf-stable, retaining the biocontrol activity at a level similar to that of the freshly isolated microorganism. The shelf-life of a biocontrol product is the duration for which the microorganisms remain viable and antagonistic to the required level. The biocontrol product should be easy to prepare and apply and produce abundant viable propagules with long-shelf life, in addition to being efficacious and economical. The shelf-life of the product should be at least 12–18 months under unrefrigerated conditions to provide protection to crops at commercially acceptable level. Various aspects of product formulations are described in detail in earlier publications (Narayanasamy 2006, 2013).

Strains of *Pseudomonas syringae* pv. *syringae* produce syringomycin E (SRE), a cyclic lipodepsinonapeptide with potent antifungal activity. The potential of the compound as an organic-compatible agrofungicide and vegetable seed treatment against soilborne pathogen *Pythium ultimum* var. *ultimum* was assessed. A variant of *P. syringae* pv. *syringae* B301D with enhanced SRE-producing ability was isolated and grown in bioreactor with SRE yields averaging 50 mg/l in 40 h. SRE was extracted and purified through a large-scale chromatorgraphy system using organic-compatible processes and reagents. The minimum concentrations of the purified product required to inhibit 50% and 90% of *P. ultimum* oospore germination were determined as 31.3 and 250 µg/ml, respectively. Drench treatment of cucumber seeds in *P. ultimum*-infested potting medium (500 oospores/g) with 50 µg/ml SRE or water with no SRE resulted in 90.2 ± 4.5% and 65.7 ± 4.6% germination rates, respectively. Seed coating with 0.03% (w/w) SRE allowed 65.7 ± 4.6% seed germination on naturally infested soil, whereas 100.0 ± 0.0% of noncoated seeds were unable to germinate due to infection by *P. ultimum*. Organic-compatible SRE suitable for large-scale production, has the potential for use as organic fungicide seed protectant in organic systems of crop production (Kawasaki et al. 2016). Large-scale production of *Trichoderma* sp. formulations is confronted with many difficulties, although these BCAs have been employed for the management of several crop diseases. Development of *Trichoderma* sp. formulations in encapsulated granules (CG) for extended period of viability of conidia during storage was attempted. An ionic gelling method was followed for producing encapsulated granules containing sodium alginate matrix modified with different polymers. Granules characterization showed that there was interaction between the alginate matrix and polymers used in the formulation of the encapsulated granules, which ensured stability of formulations, maintaining the viability of *Trichoderma* during different stages of production and storage. After 14 months, the product stored at 28°C had viable concentration maintained above 10^6 CFU/g, at which the BCA might be effective against target pathogens (Locatelli et al. 2018).

3.1.6.1.2   Application Methods

The formulated bioproducts containing fungal or bacterial biocontrol agents (BCAs) may be applied on seeds, propagules, soils and/or foliage of crops to suppress development of soilborne microbial plant pathogens either as protective or curative treatments. The efficacy of formulated products in reducing the disease incidence or severity has to be compared with that of the unformulated fresh fungal or bacterial species to determine the extent of loss of biocontrol activity, if any, due to the process of formulation and storage for different duration. The commercial bioproducts are recommended to the growers, after their evaluation by state-owned research centers/universities.

3.1.6.1.2.1   Seed/Propagule Treatment   Seeds or vegetatively propagated plant materials (propagules) are treated with liquid, powder or granular formulations containing single or mixture of strains of fungal or bacterial BCAs. Seed treatment with biocontrol products is generally preferred, because of ease of application and lesser cost of treatment. Seed treatment not only suppresses the pathogen present on the spermosphere, but also protects the emerging young seedlings against infection by the soilborne pathogens. Various kinds of stickers, such as carboxymethyl cellulose (CMC, 1%), methocel (2%), polysulf (0.8%) and polyvinyl alcohol (20%) have been used for efficient adherence of the BCA to seeds or propagules.

The comparative efficacy of biocontrol agents and chemicals used for treating the seeds and soil was assessed for suppressing the development of damping-off and wilt diseases of spinach caused by *Pythium ultimum, Rhizoctonia solani* and *F. oxysporum* f.sp. *spinaciae* in organic production system. Two experimental seed treatments GTGI and GTGII (each comprised of a proprietary organic disinfectant and the latter also containing *T. harzianum* T22) provided protection as efficiently as the fungicide mefenoxam against *P. ultimum* in one trial and significant reduction of damping-off in the second trial. Natural II and Natural X (*Streptomyces* products) and Subtilex (*Bacillus subtilis*) applied as seed treatments suppressed damping-off significantly in one of two trials. For suppression of development of *R. solani*, GTGI and Natural II seed treatments reduced damping-off as effectively as a drench with the fungicide Terraclor (pentachloronitrobenzene). Soil drench with Prestop (*Gliocladium catenulatum*) suppressed postemergence wilt caused by *F. oxysporum* f.sp. *spinaciae* in both trials. Compost tea drench and seed treatment with Yield Shield (*Bacillus humilus*) each suppressed postemergence wilt in only one of two trials. None of the treatments was effective against all three pathogens and some treatments intensified symptoms of damping-off disease (Cummings et al. 2009).

Treatment of cotton seeds with *Trichoderma (Gliocladium) virens* or *Bacillus subtilis* reduced colonization of roots by *F. oxysporum* f.sp. *vasinfectum*, causing Fusarium wilt disease, resulted in lower levels of disease incidence and severity (Zhang et al. 1996). Corn damping-off disease caused by *Pythium ultimum, P. arrhenomanes* and *Fusarium graminearum* was more effectively suppressed by coating the seeds with *T. virens* isolate Gl3, leading to greater seedling stand and better plant growth due to significant reduction in root rot severity compared to the fungicide captan or other BCAs tested (Mao et al. 1997). Seed treatment with *Bacillus* sp. strain L324-94 suppressed the development of Pythium root rot caused by *Pythium irregulare* and *P. ultimum*, Rhizoctonia root caused by *Rhizocotonia solani* and wheat-take-all caused by *Gaeumannomyces graminis* var. *tritici*, when the treated seeds were directly drilled into soil in both winter and spring seasons (Kim et al. 1997). Antibiotic-producing isolate J-2 of *Streptomyces* sp. efficiently reduced damping-off of sugar beet caused by *Sclerotium rolfsii* by treating the seed a few weeks before planting. The BCA isolate was able to multiply and survive in the rhizosphere soil from naturally infested soil for more than 3 weeks (Errakhi et al. 2007). *Streptomyces rochei* strain PTL2 was highly antagonistic to *Rhizoctonia solani*, causing damping-off of tomato seedlings and also exhibited remarkable capacity for production of hydrogen cyanide, siderophores, 1-aminocyclopropane-1-caroboxylate deaminase and phytohormones, chitinolytic activity and inorganic phosphate solubilization. Talc-based formulation of the strain PTL2 showed highest protective activity by reducing the disease incidence to 14.1 from 89.3% (in control), whereas the fungicide Thiram®-treated plots had 16.7% infection by *R. solani*. Furthermore, the talc-based formulation increased the root length, shoot length and dry weight of tomato seedlings to the maximum extent. The ability to reduce disease incidence and enhance plant growth parameters, indicates the advantages of employing the strain PTL2 of *S. rochei* for large-scale application to protect tomato nurseries (Zamoum et al. 2017).

3.1.6.1.2.2 Treatment of Propagules and Transplants

Potato silver scurf disease caused by *Helminthosporium solani* primarily spread from infected to healthy tubers during storage. The minitubers of Red Norland potato were immersed in a suspension of formulated preparation of *Acremonium strictum* for 3 min and air-dried. The treated and untreated tubers were inoculated with *H. solani* by spraying conidial suspension. The BCA reduced sporulation and spore germination of *H. solani*. The BCA was effective, when applied as a protective treatment only, as *A. strictum* did not have any curative effect on already infected tubers (Rivera-Varas et al. 2007). Dipping of wounded roots of tomato seedlings in a suspension of cells of nonpathogenic *Agrobacterium radiobacter* strains K84 or K1026 or the commercial product 'Nogall' entirely inhibited crown gall formation by *A. tumefaciens* (Fakhouri and Khalaif 1996). *Pseudomonas aureofaciens* and *P. fluorescens* with wide spectrum of antagonistic activity, when applied as root dip to grapevine cuttings, showed differential effect, depending on grapevine cultivar. Disease incidence was reduced by 50 to 80% and disease severity index (DSI) by 75 to 86%. Both *P. aureofaciens* and *P. fluorescens* persisted on the root surfaces of treated grapevine cuttings and in nonsterile soil. Treatment of rooted raspberry seedlings with *P. aureofaciens* strain B4117 protected the seedlings effectively against infection by *A. tumefaciens*, infecting raspberry (Khmel et al. 1998). Soaking roots of grapevine seedlings in a cell suspension of nonpathogenic *Agrobacterium vitis* strain VAR03-1 for 24 h before a 1-h soak in a cell suspension of tumorigenic *A. vitis* and subsequent planting in infested soil resulted in reduction of crown gall incidence significantly (Kawaguchi et al. 2007). The effectiveness of treatment of roots of grapevine, rose and tomato with *A. vitis* strain VAR-03, for reducing infection and disease severity induced by respective pathogens *A. vitis, A. rhizogenes* and *A. tumefaciens* was demonstrated under field conditions. VAR-03 strain was efficient in suppressing development of grapevine crown gall and the BCA could multiply and establish a population averaging 10^6 CFU/g of roots in the rhizosphere of grapevine and persist on grapevine roots for 2 years (Kawaguchi et al. 2008b). Tomato bacterial wilt disease caused by *Ralstonia solanacearum* was reduced to 65% by root bacterization with *Bacillus vallismortis* strain EXTN-1, as against 95% infection in nonbacterized control plants (Park et al. 2007) The roots of banana plantlets were immersed in a suspension of cells of the endophytic actinomycete *Streptomyces griseorubrigenosus* at 1 hour prior to planting, followed by inoculation of Fusarium wilt disease. The pathogen inoculum was reduced by 50%. In addition, the plant growth was also improved by the BCA treatment (Cao et al. 2005). Tomato Fusarium wilt disease incidence could be reduced by dipping the roots in suspension of cells of *Achromobacter xylosoxydans* at a concentration

of 10^8 CFU/ml by about 50%. The symptoms induced by the phytotoxin produced by the pathogen *F. oxysporum* f.sp. *lycopersici* were not observed in BCA-treated tomato plants. In addition, BCA treatment enhanced plant growth significantly, compared to control plants (Moretti et al. 2008).

3.1.6.1.2.3 Soil Treatment

Biocontrol agents with required levels of biocontrol activity against target pathogen(s) have to remain viable as actively proliferating propagules under natural field conditions. The biocontrol agents either forming a natural component of the microflora or as introduced organisms have to compete with pathogens or other soil microflora for available nutrients and niches for establishment. Both fungal and bacterial BCAs applied to the soil have been found to be aggressive colonizers of the rhizospheres of plants to be protected. Some of the BCAs have multiple mechanisms of action on the fungal and bacterial pathogens, resulting in suppression of disease development and also enhanced plant growth and consequently higher yields. Some of the plant growth-promoting rhizobacteria (PGPRs) have the ability to induce systemic resistance in treated plants to soilborne diseases as well as to foliar diseases. The fungal biocontrol agents like *Trichoderma viride, T. harzianum* and *T. virens* have been more frequently employed for the management of a wide range of soilborne pathogens such as *Pythium* spp., *Phytophthora* spp., *Rhizoctonia solani, F. oxysporum, Sclerotinia* spp. and *Verticillium* spp. The nonpathogenic *F. oxysporum* Fo47, *Coniothyrium minitans* and *Pythium oligandrum* have also been found to be effective in suppressing the development of soilborne pathogens infecting several crops. Among the bacterial BCAs, strains of *Pseudomonas fluorescens, Bacillus subtilis* and *Agrobacterium radiobacter* have been effective in suppressing different soilborne pathogens and also in enhancing plant growth significantly. Soil application of BCAs is less preferred, because of the requirement of large quantities of the products. Further, uniform spread of the products to cover the infested areas of the soil may be difficult to achieve. The BCAs may be applied as a broadcast incorporation or as placement in the seed furrows at the time of sowing (Narayanasasmy 2013).

Nonpathogenic isolates of *F. oxysporum* isolated from wilt-suppressive soils were more effective in suppressing the development of Fusarium wilt diseases of tomatoes, watermelon and muskmelon, compared to *Burkholderia cepacia, Trichoderma virens, T. hamatum* and *Pseudomonas fluorescens* (Larkin and Fravel 1998). *Trichoderma koningii*, with multiple mechanisms of action, suppressed the saprophytic growth of *Gaeumannomyces graminis* var. *tritici* in natural soils and consequently reduced incidence of wheat take-all disease (Simon and Sivasithamparam 1989). *T. koningii* applied to the seed furrows in the field reduced crown rot infection in root by 40% and also increased yield of spring wheat by 65%. The combination of *T. koningii* and *Pseudomonas fluorescens* Q292-80 provided more effective protection, resulting in greater reduction in disease incidence and greater yield levels (Duffy et al. 1996). The ability of the hypovirulent binucleate Rhizoctonia (HBNR) to protect tomato plants against Fusarium crown and root rot (FCRR) throughout the growing period and their effects in reducing the population of *F. oxysporum* f.sp. *radicis-lycopersici* (FORL) in roots and stem under soilless systems were assessed under greenhouse conditions. In the greenhouse and soilless systems, HBNR isolates significantly reduced vascular discoloration and discoloration of total root system by 90 to 100% and by 73 to 89% respectively. Under field conditions, HBNR isolate WI significantly reduced vascular discoloration by 7%. Application of HBNR resulted in increases in marketable and total yields of tomatoes by 70% and 73% respectively, compared to untreated control plants (Muslim et al. 2003). A plant growth-promoting rhizobacterium (PGPR) *Bacillus* sp. strain HN09 isolated from neem tree rhizosphere inhibited the development of *F. oxysporum* f.sp. *radicis-lycopersici* (FORL) in vitro. Substantial level of disease suppression was achieved by soil supplementation with a preparation containing neem cake seeded with the strain HN09. Dry sterilization of neem cake before fermentation provided comparable protection against FCR disease, as that provided by unsterilized treatment, whereas moist sterilization treatment decreased the effectiveness significantly. Application of the bioformulation activated the tomato genes encoding PR-proteins, resulting in enhancement of resistance of tomato plants to Fusarium crown and root rot (FCRR) disease (Lin et al. 2017).

The bioplastic formulation Master-Bi® granules were inoculated with conidial suspension of a nonaflatoxigenic strain of *Aspergillus flavus* NRL30797 (ca 10^7 conidia/granule). Incubation of 20-g samples receiving a single Master-Bi® granule for 60 days, resulted in $10^{4.2}$ and $10^{5.2}$ propagules of *A. flavus*/g of soil in microbiologically active (nonsterile) and sterilized soil respectively. The bioplastic granules were highly stable, in addition to supporting the proliferation of the BCA. The results indicated that Master-Bi® had the potential for substituting the wheat grains, used earlier as carrier for *A. flavus* NRRL30797 (Accinelli et al. 2009). In a later investigation, the effect of field application of the BCA-inoculated bioplastic granules on soil population of aflatoxigenic strains of *A. flavus* was assessed. A rapid shift in composition of soil population of *A. flavus* with a significant decrease in relative abundance of indigenous aflatoxigenic isolates was observed. Application of bioplastic granules at 30-kg/ha was more efficient in replacing aflatoxigenic isolates was observed than a 15-kg/ha dosage. Soil application of 15- and 30-kg/ha bioplastic granules treated with the BCA, reduced aflatoxin contamination by 59% and 86% in 2009 and 80% and 92% in 2010, respectively (Accinelli et al. 2012).

Members of endophytic fungal genus *Trichoderma* are well established as plant-beneficial microorganisms functioning as the highly successful biological component in the form of biofertilizers, biocontrol agents and plant growth activators. Variable interactions among different lentil genotypes and *Trichoderma* strains were studied both in the presence and absence of the root pathogen *Aphanomyces euteiches*. Two commercial *Trichoderma* formulations, RootShield® (RS) and RootShield® Plus (RSP) based on *T. harzianum* T22 and *T. virens* G41 respectively were evaluated for their efficacy in

suppressing development of Aphanomyces root rot and plant growth promotion in 23 wild and cultivated lentil genotypes. No significant disease control was observed with either formulation in any lentil genotype. Significant genotype-specific plant growth promotion was recorded. The overall effect of *Trichoderma* treatment was markedly higher in the presence of the pathogen compared to pathogen-free conditions. In many cases, negative responses were evident, particularly in the absence of root infection by *A. euteiches*. The genotype PI572390 of *L. tomentosus* alone exhibited positive responses for most tested parameters. The results showed that lentil genotype played a major role in interactions among tested *Trichoderma* strains and the plant. The beneficial effect of *Trichoderma* could be realized under biotic stress (infection) and hence, blanket recommendation of the BCA is likely to be a negative strategy affecting the yield in fields free from the root pathogen(s) (Prashar and Vandenberg 2017).

Three endophytic bacterial strains, *Stenotophomonas maltophila* H8, *Pseudomonas aeruginosa* H40 and *Bacillus subtilis* H18, were evaluated for their efficacy in suppressing the development of root rot disease caused by *Rhizoctonia solani* in cotton. The BCAs were applied as soil drench or talc-based formulation on the pathogen-infested soil under greenhouse conditions. Soil drench was more effective than talc-based formulation. Increase in emergence and survival of seedlings with significant reduction in disease severity was observed. In addition, the BCAs enhanced growth parameters markedly. Gas chromatography-mass spectrometry analysis revealed the presence of bacterial bioactive compounds with a broad-spectrum antifungal activity and capacity to induce systemic resistance in cotton seedlings (Selim et al. 2017).

3.2 ASSESSMENT OF BIOCONTROL POTENTIAL OF ABIOTIC AGENTS

Abiotic agents of plant and animal origin and organic and inorganic compounds have been evaluated for their capacity to suppress development of soilborne microbial pathogens and the diseases induced by them in various crops.

3.2.1 NATURAL PRODUCTS OF PLANT AND ANIMAL ORIGIN

Natural plant products such as plant residues, green manures and composts are added to the soil after harvest or before planting next crop. The effects of plant residues and green manures on soil microflora and soilborne pathogens affecting various crops have been investigated.

3.2.1.1 Effects of Composts

Composts are the most frequently applied amendments to improve the soil fertility and to encourage the proliferation of microorganisms antagonistic to soilborne microbial plant pathogens. Composts have been reported to be effective, especially in controlled environment or container-based production systems. The inability to produce predictable and reproducible compost composition, has hampered a meaningful interpretation of the results obtained from the investigations directed toward standardization of compost treatments (Mazzola 2007). Application of composts, because of the significant beneficial effect by disease suppression, has been employed as a component of integrated management system for soilborne diseases. The composts carry antagonists that can suppress development of soilborne pathogens. In addition, they encourage the development of antagonists already present in the field soils. The microflora of three composts were investigated to isolate and test the biocontrol potential of microorganisms in the composts, against *Pythium ultimum*, incitant of damping-off of greenhouse-grown cucumber. A more diverse bacterial population was present in the compost from paper mill sludge (170 groups) than in composts from plant waste and manure (75 and 88 groups respectively). Among the bacteria and fungi tested, *Zygorhyncus moelleri* and *Bacillus marinus* were the most effective in reducing incidence of damping-off disease under greenhouse conditions, followed by *Penicillium thomii, Pseudomonas fluorescens, P. aeruginosa* and *Graphium putredinis. Z. moelleri* proved to be consistently effective in the second trial also, indicating its suitability for large-scale application (Carisse et al. 2003). The effectiveness of seed-colonizing microbial communities presents in municipal bio-solids compost in suppressing the development of Pythium damping-off in greenhouse-grown cucumbers was demonstrated by Chen and Nelson (2012).

Composts may serve as a food base for endogenous microorganisms, including plant pathogens or introduced biocontrol agents to sustain suppression based on the activities of microbial communities. Two composted swine wastes CSW1 and CSW2 were incorporated into peat moss-based potting mix at 4 to 20% (v/v) to determine the degree of suppression of pre-emergence damping-off of impatiens caused by *Rhizoctonia solani*. Disease incidence was reduced to a greater extent in potting mix amended with 20% CSW1, compared with CSW2. Mixes amended with CSW1 (20%), after 35 weeks or more of curing, were consistently suppressive to Rhizoctonia- and Pythium damping-off (Diab et al. 2003). Fish emulsion (2% and 4%) in naturally infested muck soil effectively and consistently suppressed damping-off in cucumber caused by *R. solani*, as well as in peat substrate (Abbasi et al. 2004). The composts consisting of dairy and greenhouse wastes significantly reduced the severity of cucumber Fusarium root rot and stem rot caused by *F. oxysporum* f.sp. *radicis-cucumerinum* (FORC). Strains of *Pseudomonas aeruginosa* isolated from the composts showed greatest degree of antagonism against FORC. Furthermore, internal stem colonization by FORC was also reduced by *P. aeruginosa*. The results indicated that suppressiveness of composts against FORC depended primarily on the production of antibiotics by *P. aeruginosa* (Bradley and Punja 2010). The biocontrol potential of two types of date palm composts and indigenous microorganisms present in them for their efficacy in suppressing the development of potato stem canker and black scurf caused by *R. solani* was assessed. The microorganisms inhibitory to *R. solani* in confrontation assay were further examined under light microscope. Mycelial lysis, mycoparasitism and/or formation of mycelial cords via anastomosis between hyphae

were observed. The compost extracts lost its antagonistic activity, after heating or filtration, indicating chemical components of the compost had no inhibitory effect on *R. solani*. Suppressive effect of the date palm compost was due to the antagonists. Under greenhouse conditions, incidence of stem canker and black scurf of potato was significantly reduced in peat-sand amended with compost, compared to the untreated control. Plant growth was not affected by the application of fungal antagonists from the compost, suggesting that antibiosis might be the principal mechanism of action of date palm compost in reducing the incidence of stem canker and black scurf caused by *R. solani* (El-Khaldi et al. 2016).

Household waste compost, composted cow manure and fresh *Brassica* tissues were evaluated for their biocontrol potential against tomato Verticillium wilt caused by *Verticillium albo-atrum*. The composts reduced the disease severity significantly. The biological activity increased with increase in organic matter input levels. The organic matter suppressive to soilborne pathogens either alone or in combination with chitin/chitosan soil amendments might be effective to achieve disease suppression to the desired levels (Giotis et al. 2009). The efficacy of onion waste compost (OWC), spent mushroom waste (SMC) amended with *Trichoderma viride* S17A for the control of Allium white rot (AWR) caused by *Sclerotium cepivorum* was assessed. Incorporation of OWC into the soil reduced viability of sclerotia and the incidence of AWR on onion plants in the glasshouse pot assays. On the other hand, SMC or *T. viride* reduced only the incidence of AWR. In the field trials, OWC reduced sclerotial viability and disease incidence as effectively as the fungicide. Addition of *T. viride* to SMC facilitated proliferation of the BCA in the soil and increased healthy onion bulb yield. The suppressive activity of OWC could be attributed to the presence of sulfur compounds in the compost (Coventry et al. 2006). Composted cotton-gin trash (CGT) showed highly suppressive effect on tomato southern blight disease caused by *Sclerotium rolfsii* and it could be used as an alternative to or in combination with conventional soil fumigation. Propagules density of *Trichoderma* spp. was higher in soils amended with CGT than in soils receiving synthetic fertilizers. The disease incidence was at low level in soil amended with CGT (23%), compared with synthetic fertilizer-applied (61%). Organic amendments enhanced soil fertility, in addition to suppression of soilborne diseases (Bulluck III and Ristaino 2002). Cow manure amendment significantly reduced the incidence of potato brown rot caused by *Ralstonia solanacearum* in Dutch sandy soils but did not affect pathogen population. However, cow manure reduced the bacterial pathogen densities in Egyptian sandy soils, most probably due to microbial competition, as a clear shift in population was detected by denaturing gradient gel electrophoresis (DGGE) technique (Messiha et al. 2007).

Compost teas are fermented watery extracts of composted materials that have beneficial effects on plants including antimicrobial activities. Compost teas may be employed after active aeration and with additives to enhance microbial population density required for effective suppression of plant pathogens. Aerated compost tea (AET) and nonaerated compost tea (NCT) with or without additives were applied as soil drenches for suppressing development of *Pythium ultimum*, causing damping-off in container system. ACT fermented with kelp and humic acid additives was suppressive to the maximum extent. Suppressiveness of the formulation was significantly reduced by heat treatment (Scheuerell and Mahaffee 2004). Nonaerated compost teas (NCTs) prepared from seaweed compost, shrimp powder compost and chicken, bovine and sheep manure composts decreased the percentage of necrotic tomato seedlings inoculated with *P. ultimum* from 100 to 42%, but did not reduce necrosis in *Rhizoctonia solani*-inoculated seedlings. Sterilization of NCT resulted in complete or partial inhibition of inhibitory effect on pathogen growth. The NCTs, when applied on tomato seedlings, promoted plant growth to different levels (Dionne et al. 2012). Microbe-fortified composts and compost tea preparations were evaluated for their efficacy. The composts and compost teas amended with *Anabaena oscillarioides* C12 and *Bacillus subtilis* B5, respectively enhanced tomato seed germination and growth parameters, in addition to significant reduction in severity of diseases caused by *Pythium debaryanum, P. aphanidermatum, R. solani* and *F. oxysporum* and also reduced the fungal load (Dukare et al. 2011).

Biochar is a heterologous material generated through a process (pyrolysis) carried out at temperatures ranging from 200°C to 900°C, under limited oxygen availability. Biochar has been generated from a wide range of organic materials such as crop residues, wood, municipal biowastes, sewage sludge, manure and animal bones. Biochar differs basically from charcoal by its final end use–agriculture and environmental management for biochar and fuel and energy for charcoal (Lehmann and Joseph 2009). Use of biochar as soil amendment has been shown to enhance crop yields significantly (Jeffery et al. 2011). The beneficial effects on plant growth and health may be related to its ability to stimulate development of beneficial microbes, in the bulk soil as well as in the rhizosphere (Thies et al. 2015). In addition, biochar application to soil may result in suppression of diseases caused by airborne and soilborne pathogens. The disease suppressive effects of biochar, the response of different crop plants and the mechanisms of action of biochar on development of soilborne microbial pathogens and diseases induced by them have been studied in some pathosystems (see Table 3.2) (Bonanomi et al. 2015).

Biochar may be an alternative to crop residues or composts, applied as soil amendments, since it selectively enhances the activity of beneficial microorganisms without stimulating pathogen populations and virulence. Biochar made from animal bones was used as a carrier for an effective delivery of biocontrol agents. Scanning electron microscopic observations revealed that *Pseudomonas chlororaphis, Bacillus pumilus* and *Streptomyces pseudovenezuelae* could extensively colonize the porous structure of biochar. They were highly effective in suppressing damping-off and Fusarium crown and root rot of tomato caused, respectively, by *Pythium aphanidermatum* and *F. oxysporum* f.sp. *radicis-lycopersici*, following soil application of biochar (Postma et al. 2013). Phytotoxic compounds may be released into agricultural soils from decomposed organic materials, including crop residues. By contrast, biochar can protect plant roots by absorbing phytotoxic compounds and thus indirectly protect the roots from pathogen infection. Asparagus decline was due to accumulation of allelopathic toxins and Fusarium crown and root rot (FCRR) caused by *F. oxysporum* f.sp. *radicis-lycopersici* (FORL) and *F. proliferatum* (Fp). A commercial formulation of biochar from hardwood dust was evaluated for its efficacy in reducing the adverse effects of allelopathy on arbuscular mycorrhizal (AM) root colonization and on FCRR disease. In the greenhouse experiments, addition of biochar at 1.5% and 3.0% (w/w) to asparagus field soil caused by proportional increases in root weight and linear reductions in the percentage of root lesions by FORL and Fp, compared with control. Concomitantly, there was a 100% enhancement in root colonization by AM fungi at incorporation rate of 3% of biochar. Under microplot conditions, biochar was added at 3.5% (w/w). Plots with biochar amendment, had plants with greater AM colonization in the first year of growth. This effect was due to absorption of phytotoxic and fungitoxic phenolic compounds (cinnamic-, coumaric- and ferulic-acids) released from decaying *Asparagus* crop residues. However, in the

TABLE 3.2
Effectiveness of Biochar Application in Suppressing Development of Soilborne Pathogens Infecting Different Plant Species

Plant pathogens	Plant host	Source material for biochar	References
Phytophthora cinnamomi	*Quercus rubra*	wood	Zwart and Kim (2012)
Pythium aphanidermatum	*Acer rubrum*	wood	Zwart and Kim (2012)
Plasmodiophora brassicae	*Brassica rapa*	Miscanthus	Knox et al. (2015)
F. oxysporum f.sp.	*Asparagus*	coconut, charcoal	Matsubra et al. (2002)
f.sp. *asparagi*	*officinalis*	carbonized chaff	Matsubra et al. (2002)
F. oxysporum f.sp. radicis-lycopersici	tomato	pig bone	Matsubra et al. (2002)
F. proliferatum	*Asparagus* spp.	*Asparagus* spp.	Quest Biochar
Rhizoctonia solani	cucumber	Eucalyptus wood	Jaiswal et al. (2014a)
	French bean	crop wastes	Jaiswal et al. (2014b)

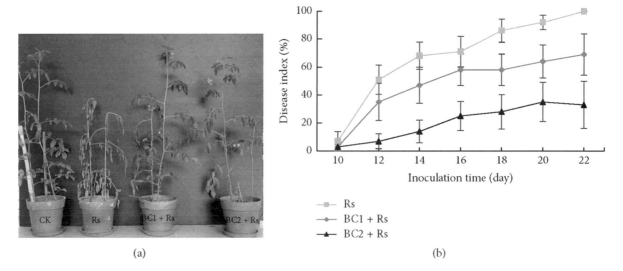

(a)   (b)

FIGURE 3.9 Comparative suppressive effect of biochars obtained from peanut shell (BC1) and wheat straw (BC2) on bacterial wilt disease incidence in tomato (a) and (b): Soil amendment with BC2 was more effective than BC1 in reducing bacterial wilt disease incidence.

[Courtesy of Lu et al. 2016 and with kind permission of Hindwai Publishing Corporation, International Journal of Agronomy, Open Access Journal]

subsequent years, biochar-treated plants were smaller in size, probably due to greater than average rainfall and the ability of biochar to retain moisture, creating conditions favoring root rot disease development. The results indicated variations in effectiveness of biochar in mitigating the deleterious effects of allelopathic residues in replant soils on *Asparagus* (Elmer and Pignatello 2011).

Induced systemic resistance (ISR) was suggested as mechanism of action of biochar against soilborne pathogens. Strawberry plants grown in substrates amended with biochar had higher expression of genes encoding three pathogenesis-related (PR)-proteins *FaPR1, Faolp2, Fraa3*), one gene encoding lipoxygenase (*Falox*) and one gene (*FaWRKY1*), encoding transacting factor that belongs to the WRKY family (Harel et al. 2012). The effects of two biochars obtained from peanut shell (BC1) and wheat straw (BC2) on severity of bacterial wilt disease of tomato caused by *Ralstonia solanacearum* and also on soil microbial properties were assessed. The biochar formulations were applied at 2% (w/w) to field soil infested with *R. solanacearum*. BC1 and BC2 treatments reduced the disease index of bacterial wilt by 28.6% and 65.7%, respectively, and increased level of resistance to bacterial wilt in tomato (see Figure 3.9 (a), (b)). BC2 was more effective in suppressing the disease severity. Biochar treatments significantly reduced the pathogen density by 51.63% in BC1-treatment and 68.22% in BC2 treatment, whereas *R. solanacearum* density increased by 80.43%, after pathogen inoculation (see Figure 3.10). Following inoculation of soil with the pathogen, contents of soil bacteria and actinomycetes were reduced in control (without biochar) plots. In contrast, the contents of biochar treated plots showed significant increases in soil bacteria (57.3% and 96.43% in BC1 and BC2 respectively). Soil actinomycetes populations also registered increases, whereas soil fungi populations were reduced. Biochar treatments increased soil neutral phosphatase and urease activity. Higher metabolic

capabilities were observed in biochar amendment treatments, indicating high substrate utilization by potential of microorganisms. Resistance of tomato grown in soils amended with biochars was closely related to the changes in soil microbial activity and community structure (Lu et al. 2016).

The mechanisms of suppression of soilborne diseases caused by microbial plant pathogens and enhancement of plant growth by biochar application are not clearly understood. The relationships between biochar-induced changes in the rhizosphere microbial community structure and composition and

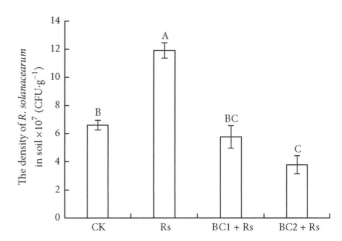

FIGURE 3.10 Efficacy of biochars BC1 and BC2 on density of *Ralstonia solanacearum* (*Rs*) in biochar-amended soil CK–unamended soil without *Rs* and biochars amended with Rs; BC1 + *Rs*–peanut biochars amended + pathogen infestation; BC2 + *Rs*–wheat biochar amendment + pathogen infestation; bars with different letters are significantly different in their efficiency (P <0.05) using Duncan's new multiple range tests.

[Courtesy of Lu et al. 2016 and with kind permission of Hindwai Publishing Corporation, International Journal of Agronomy, Open Access Journal]

activity associated with suppression of development of soilborne diseases were investigated. Biochar application suppressed Fusarium crown and root rot caused by *F. oxysporum* f.sp. *radicis-lycopersici* (FORL) and simultaneously improved the growth of treated plants. Furthermore, biochar reduced Fusarium colonization and survival in soil, and increased the culturable counts of several biocontrol and plant growth promoting microorganisms. Illuminia sequencing analyses of 16S rRNA gene revealed substantial differences in the rhizosphere bacterial taxonomical composition between biochar-amended and non-amended treatments. Furthermore, biochar amendment significantly increased the microbial and taxonomic and functional diversity, microbial activities and an overall shift in carbon-source utilization. High microbial taxonomic and functional diversity and activity in the rhizosphere was found to be associated with suppression of diseases caused by soilborne pathogens by biochar application (Jaiswal et al. 2017).

3.2.1.2 Effects of Plant Residues

The residues of cruciferous plants have been found to have the high disease suppressive effect, among various kinds of crop residues that have been evaluated. Cruciferous plants contain glucosinolates, sulfur-containing secondary metabolites. The glucosinolates are hydrolyzed by the enzyme myrosinase to produce isothiocyanates (ITCs). Many ITCs are volatile with inhibitory effects on a wide range of soilborne pathogens. Although incorporating *Brassica napus* crop residues was effective in suppressing soilborne fungal pathogens, the adoption of this approach was impracticable. The impact of *B. napus* seed meal on apple replant disease caused by *Rhizoctonia solani* and *Pythium* spp. was assessed under greenhouse conditions. Seed meal amendment of soil, irrespective of glucosinolate content, significantly suppressed apple root infection by *Rhizoctonia* spp. and nematode *Pratylenchus penetrans*. On the other hand, seed meal amendment of the cv. Dwarf Essex with high glucosinolate content did not consistently suppress soil populations of *Pythium* sp. Seed meal amendment enhanced *Pseudomonas* spp. and total bacterial populations. Dwarf Essex seed meal was toxic, when applied at the rate of 2% (v/v) (Mazzola et al. 2001). Among the individual isothiocyanates tested, propenyl- and ethyl-isothiocyanates were the most fungistatic and inhibited mycelia growth and completely suppressed conidial and chlamydospore germination of *F. oxysporum* pathogens infecting conifer seedlings (Smolinska et al. 2003).

Incorporation of lettuce residues into the pathogen-inoculated soil significantly reduced the incidence of root and stem rot disease of cucumber caused by *F. oxysporum* f.sp. *radicis-cucumerinum* (FORL) and also increased the total cucumber yield. Lettuce residue incorporation could form an effective component of integrated disease management (IDM) system (Pavlou and Vakalounakis 2005). In a later investigation, residues of various other plant species, *Diplotaxis tenuifolia* (wildrocket, WR), *Artemisia dracunculus* (tarragon), *Salvia officinalis* (sage) and *Brassica oleracea* var. *italica* (broccoli) were evaluated for their efficacy in suppressing the development of cucumber crown and root rot disease caused by

F. oxysporum f.sp. *radicis-lycopersici* (FORL). Disease incidence and severity of the disease in cucumber plants inoculated with FORL were reduced by 20 to 80%, when seedlings were planted in soils incorporated with residues of different plant species. Effective soil suppressiveness persisted, after repeated inoculation and planting in the same soil without additional treatment between inoculations. In addition, residues of WR induced soil suppressiveness in two additional tested soils, differing in their physical and chemical properties. Soil suppressiveness to Fusarium crown and root rot disease was induced, when cucumber seeds were sown in soils that were initially amended with WR residues and later infested with FORL chlamydospores. The results indicated the possibility of reducing incidence of soilborne diseases caused by *F. oxysporum* by incorporating suitable plant residues that contribute to development of soil suppressiveness (Klein et al. 2011). Incorporation of organic amendments into the soil may induce soil suppressiveness against specific soilborne pathogens. Soil suppressiveness could be induced by incubating sandy soil with debris of wildrocket (*Diplotaxis tenuifolia*, WR) under field conditions. The microbial dynamics in the roots of cucumber seedlings were investigated, following transplantation into WR-amended or nonamended soil, as influenced by inoculation with *F. oxysporum* f.sp. *radicis-cucumerinum* (FORC). Appearance of disease symptoms on plants grown on nonamended soil was discernible at 6 days after inoculation, whereas the symptoms appeared only at 14 days after inoculation of WR-amended soil. The pathogen propagules were quantified using real-time PCR assay. The pathogen population was significantly lower (66%) at 6 days after inoculation in WR-amended soil compared to unamended soil. The decrease in root colonization by FORL was correlated with a reduction in disease incidence at 21 days after inoculation and transplanting into suppressive soil. Quantitative analyses and mass-sequencing methods indicated a qualitative shift in the root's bacterial community composition in suppressive soil, rather than a change in bacterial community composition. The shift in bacterial community was related primarily to the increase in *Streptomyces humilus*, which was antagonistic to fungal pathogens (Klein et al. 2013).

The effects of green manure crops (buckwheat and canola) and crop sequences on potato scab caused by *Streptomyces scabies* and Verticillium wilt caused by *Verticillium dahliae* were assessed in the 2-year field trials. Tubers grown in buckwheat-treated soil had significantly lower Verticillium wilt ratings, whereas tuber yield was increased significantly. Potatoes grown in soil planted to corn or alfalfa in the previous year had significantly lower Verticillium wilt and potato scab ratings, as well as higher yields than potatoes grown in soil previously planted to potato. Green manure crops may selectively enrich the abundance or activity of antibiotic producers like *Streptomyces* spp. within the soil microbial community (Wiggins and Kinkel 2005). The comparative suppressive effects of Austrian winter pea (*Pisum sativum*) cv. Melrose, broccoli cv. Excelsior and Sudangrass amendment on Verticillium wilt severity and yield of Russet Burbank potato were assessed. The disease was consistently reduced by all three green manure types applied at the highest rate tested. A positive correlation between amendment rate and degree of disease reduction was also observed. Austrian winter pea was more effective than the other two green manure types even at a reduced rate of amendment applied (Ochiai et al. 2007). The efficacy of green manure *Brassica* crops in functioning as biofumigant was assessed for suppression of *F. oxysporum* f.sp. *conglutinans* (*Foc*) and *F. oxysporum* f.sp. *raphani* (*For*), causal agents of Fusarium yellows of cabbage and Fusarium wilt disease respectively in cabbage and radish. Green manure treatment carried out by growing nine cycles of biocidal plants, with a short crop cycle (30–35 days) did not reduce Fusarium wilt incidence on susceptible *Brassica* crops. The population of pathogen was partially increased as a result of incorporation of tissues of susceptible plants. In contrast, *Brassica* crops resistant to both *Foc* and *For* proved to be biocidal to both pathogens. The results indicated that biofumigation with *Brassica* spp. might not be effective for soil disinfestation on crops susceptible to the pathogen concerned. *Brassica* crops resistant to Fusarium yellows disease of cabbage might be grown for biofumigation to suppress the disease development (Lu et al. 2010). Sweet corn (cv. Jubilee Sweet corn and Jubilee Super Sweet corn) as green manure was evaluated for its efficacy in reducing incidence of potato Verticillium wilt disease. The sweet corn cultivars reduced disease incidence by 60 to 70% by reducing colonization of potato feeder roots and potato stem apices by the pathogens. Feeder root colonization was positively correlated with Verticillium wilt incidence (P < 0.05) and negatively correlated with yield. Further, corn green manures increased populations of several fungi like *Ulocladium* and *Fusarium equiseti*. When potato crop was grown consecutively for 2 years, the beneficial effect of sweet corn green manures was almost entirely lost. But, following two consecutive years of potato, a single sweet corn crop was enough to restore the level of disease suppression to the original level and increase potato yield by fourfold. Austrian pea, Sudangrass, oilseed rape, oats and rye grown as green manure crops, provided similar beneficial effects of disease suppression and enhancement of potato yields (Davis et al. 2010).

Disease suppressive effects of condensed distiller's soluble (CDS), a coproduct of ethanol production from corn (*Zea mays*) were investigated, using them as amendment in the pathogen-infested soils and peat-based substrate prior to planting. The CDS amendment (1% and 3% w/w) to sandy-loam soil, displayed a low level of toxicity to microsclerotia of *Verticillium dahliae* and reduced their germination by 46 to 63% after 1 week in the laboratory soil microcosm tests. The CDS contained moderate levels (~ 144 mmol/l) of volatile acetic acid and formic acid, as well as nonvolatile glycolic organic acids, some of which are known to be toxicants. In solution assays, the viability of *V. dahliae* microsclerotia treated for 24 h in 1, 2, 5 and 10% (v/v) CDS (pH 3.6–4.5) or mixture of organic acids at the same parent composition as in CDS, was reduced by 2, 7, 22 and 18% or 6, 32, 53 and 69% respectively. The mixture of organic acids with the same

volumetric ratios as 2% and 4% CDS entirely inhibited the growth of *Pythium ultimum*. Treatment of *P. ultimum*-infested muck soil with 2% CDS (v/w) reduced damping-off severity by 45 to 52% and increased the percentage of healthy seedlings from 164 to 180% over untreated control. Preplanting amendment of glycolic acid (0.075% and 0.15% w/w) to infested muck soil significantly increased the percentage of healthy cucumber seedlings by 107% and 122%, respectively, and decreased the damping-off severity by 33% and 40%, respectively, over control. The results suggested that organic acids in the CDS might have suppressive effects on soilborne fungal pathogen in sandy-loam and muck soils (Abbasi et al. 2009). Rice bran (RB), as amendment to correct soil pH, has been employed for reducing the incidence of potato common scab (PCS) disease caused by *Streptomyces scabiei* in Japan. The mechanism underlying the beneficial effect of rice bran was investigated. RB amendment-reduced PCS incidence by repressing the pathogenic *Streptomyces* population in young tubers. Amplicon sequencing analyses of 16S rRNA genes from rhizosphere microbiome revealed that RB amendment dramatically altered bacterial composition, resulting in enhancement of relative abundance of Gram-positive bacteria such as *Streptomyces* spp. and this was negatively correlated with PCS disease severity. Most actinomycete isolates were antagonistic to *S. scabiei* and *S. turgidiscabies* on R2 medium. Under field conditions, inoculation of potato plants with *Streptomyces* isolates reduced common scab incidence. The results suggested that the rice bran amendment differentially enhanced antagonistic bacterial populations in the potato rhizosphere (Tomihama et al. 2016).

Broccoli residue or crabmeal (chitin) amendments were added to the naturally infested Verticillium wilt-conducive soils from Salinas Valley of coastal California. Illumina sequencing of a 16S rRNA gene library generated from 160 bulk soil samples was employed to monitor changes in the soil prokaryote community. Under greenhouse conditions, extent of suppression of Verticillium wilt, plant height, soil microsclerotia density and soil chitinase activity were assessed using eggplant as assay host. In soil with high soil microsclerotia density, all amendments significantly reduced only in the broccoli-amended treatments. Error-corrected sequence variants (8,790) representing 1,917,893 different sequences were included in the analyses. The treatments had significant impact on soil microbiome community structure, but measures of α diversity did not vary between treatments. Community structure correlated with disease score, plant height, microsclerotia density and soil chitinase activity, suggesting that the prokaryote community might affect the disease-related response variables or vice-versa. Likewise, the abundance of 107 sequence variants correlated with disease-related response variables which included variants from genera with known antagonists of filamentous fungal pathogens. Generally, fungal genera with antagonistic activity were more abundant in amended soils than in unamended soils, and constituted up to 8.9% of all sequences in broccoli + crabmeal-amended soil. The results showed that substrate-mediated shifts in soil prokaryote communities might be associated with the transition of Verticillium wilt-conducive soils to Verticilliium wilt-supprressive soils (Inderbitzin et al. 2018).

Reductive soil disinfestation (RSD), performed under unaerobic conditions, is an ecofriendly alternative to chemical soil disinfestation (CSD) for the management of soilborne plant pathogens. Damping-off disease caused by *Rhizoctonia solani* is responsible for serious losses in cucumber. Accumulation of soilborne pathogens including *Rhizoctonia solani*, in addition to soil degradation occurred, due to intensive vegetable cultivation in the greenhouses in China. Effects of flooding soil along with incorporation of alfalfa (Al-RSD-F), alfalfa-RSD (Al-RSD) irrigated to the maximum field capacity and covered with plastic film, Al-RSD + T37 (*Tricoderma harzianum* T37) inoculated at the end of Al-RSD, ethanol (Et)-RSD incorporated into the soil covered with plastic film and ammonia (AW)-incorporated into the soil and covered with plastic film on cucumber damping-off development were assessed. Al-RSD treatment reduced the populations of *R. solani* and incidence of damping-off disease in cucumber seedlings. AW treatment was toxic to cucumber seedlings which could not survive under greenhouse conditions. The results showed that alfalfa-amended soil disinfestation with or without *Trichoderma* and ethanol soil disinfestion could be applied for effective suppression of damping-off disease and improving the quality of degraded greenhouse soils (Huang et al. 2016b). Reductive soil disinfestation (RSD) was performed using ethanol (Et)-RSD and alfalfa (Al-RSD) as organic carbons in soil infested by *R. solani*. At the conclusion of RSDs, *Chaetomium* (a fungal biocontrol agent) population increased, while *Rhizoctonia* and *Aspergillus* decreased significantly. Furthermore, some nitrification, denitrification and nitrogen-fixing genes were apparently increased in the RSD-treated soils. But the effect of Al-RSD was greaer than that of Et-RSD. Overall, Et-RSD could induce more antagonists belonging to Firmicutes under anaerobic condition. By contrast, Al-RSD could continuously stimulate some functional microorganisms (Lysobacter and Rhodanobacter), facilitating efficient nitrogen transformation actitivities in the soil in the following cropping season (Huang et al. 2016a).

Effectiveness of application of organic matter and reductive soil disinfestation (RSD) were evaluated for reducing the disease incidence/severity. Real-time PCR and MiSeq pyrosequencing were employed to monitor changes in microbial community during the progress of biocontrol activity of organic matter. Antagonists (Ant) significantly reduced pathogen population and also disease incidence. Soil microbial population and activity were also increased. By contrast, combination of RSD and antagonists (RSD + Ant) was more effective and facilitated to maintain the antagonist population and activity. The results showed that combination of organic matter and antagonists altered the soil microbial community to a greater extent and decreased incidence of damping-off disease in cucumber caused by *Rhizoctonia solani*, whereas combination of RSD and antagonists was useful in improving soil microbial community structure and stability, resulting in reduction in disease incidence (Huang et al. 2017). In a later investigation, the effects of RSD with ethanol (10 t/

ha), sugarcan bagasse (SB, 15 t/ha) and bean dregs (BD, 15 t/ha) were compared with chemical soil disinfestation (CSD) with dazomet (DZ, 0.5 t/ha) on suppression of soilborne pathogen development and soil fungal community structure. Quantitative PCR and high-throughput sequencing techniques were employed to determine the populations of microorganisms. BD treatment effectively alleviated soil acidification and salinization. Both RSD-related treatments and CSD significantly reduced the populations of *F. oxysporum* causing wilt diseases. Furthermore, RSD and CSD treatments harbored a distinct unique microbiome in the treated soils. The results indicated that BD treatment could considerably alleviate soil deterioration, improve soil microbial activity and support development of disease-suppressive microorganisms (Zhao et al. 2018).

3.2.1.3 Effects of Plant Products

Various secondary metabolites of plants such as essential oils and tissue extracts of plants have antifungal, antibacterial, antiviral and cytotoxic properties. A wide range of plant species has been screened for the presence of antimicrobial compounds, some of which have been shown to be effective under field conditions.

An immobile phytohormone, 24-epibrassinolide (EBL) was evaluated for its biocontrol potential in suppressing the development of Fusarium wilt disease of cucumber caused by *F. oxysporum* f.sp. *cucumerinum (Foc)*. Pretreatment with EBL of either the roots or shoots significantly reduced disease severity and as well as improved the plant growth, regardless of the treatment methods applied. EBL applications decreased the *Fusarium* populations on root surfaces and in nutrient solution, but increased the population of fungi and actinobacteria on root surfaces. The PCR-DGGE analysis showed that *Foc* inoculation had significant effect on the bacterial community on root surfaces, as expressed by diversity index and eveness index. But EBL applications alleviated these changes. In addition, several kinds of decomposing bacteria and growth-promoting bacteria were identified from root surfaces of *Foc*-inoculated plants and EBL-pretreated plants, respectively. The results, in general, indicated that the microbial community on root surface was affected by a complex interaction between phytohormone-induced resistance and plant pathogens (Ding et al. 2009). Formulations containing 1, 5 and 10% aqueous emulsions of clove oil, neem oil, mustard oil, synthetic cinnamon oil, pepper extract and cassia extract were evaluated for their efficacy in pathogen suppression. Treatment of soil with 5% and 10% aqueous emulsions reduced significantly populations of *Phytophthora nicotianae*. The population densities were reduced at 1 day after treatment to the level below the limit of detection (< 0.04 CFU/cm³). Soil was treated with 10% aqueous emulsions of two pepper extract-mustard oil emulsions and two cassia extract formulations. Populations of *P. nicotianae* in soil treated with one of the pepper-mustard oil formulations, were still not detectable at 21 days after application. The neem oil formulations were still not detectable at 21 days after application. The neem oil formulation and metalaxyl did not reduce pathogen populations at all rates tested. In the greenhouse assays, 10% aqueous emulsions of a pepper extract–mustard oil formulation, cassia extract and cinnamon oil formulation at 35 days after treatment suppressed the disease development in periwinkle by 87 to 93% of the disease incidence, compared with untreated, infested soil (control) (Bowers and Locke 2004). The efficacy of 14 essential oil products commercially available was assessed by in vitro and in vivo tests for suppressing the development of *Phytophthora capsici*, infecting zucchini (*Cucurbita pepo*) fruit. Oregano, palmarosa and red thyme essential oils (EOs) had the lowest EC_{50} values (< 0.15 µg/ml) for inhibiting the production and germination of sporangia and zoospores and mycelial growth of *P. capsici*. Populations of *P. capsici* in soil were significantly reduced by all three EOs. Zucchini fruits sprayed with red thyme (0.1 µg/ml) or oregano and palmarosa (0.2 µg/ml) were effectively protected against infection by *P. capsici*. Emergence of zucchini seedlings was affected by oregano but not by red thyme. Zucchini seedlings remained unaffected by *P. capsici* in soil treated with red thyme at 0.1 µg/ml, whereas all seedlings were killed in the untreated control soil. The results indicated the effectiveness of red thyme EO in protecting zucchini seedlings against infection by *P. capsici* (Bi et al. 2012).

The biocontrol potential of thymol and palmarosa (antibacterial agents produced respectively by *Thymus vulgaris* and *Cymbopogon maritinii),* against soilborne bacterial pathogen *Ralstonia solanacearum* was assessed. Thymol, palmarosa and lemongrass oil were applied at 400 mg and 700 mg/l of soil as a soil fumigant. At 7 days after application, pathogen population declined to undetectable levels in thymol, palmarosa oil and lemongrass oil treatments at both concentrations. Tomato seedlings planted in soil treated at 700 mg of essential oils were not infected by *R. solanacearum*. Further, all plants in thymol treatment were free of the pathogen. Thyme oil-producing plants such as thyme, creeping thyme and Greek Oregano were found to be symptomless carriers of the pathogen. Hence, these plant species should not be grown as rotation crops (Pradhanang et al. 2003). The ability of essential oil from *Pimenta racemosa* to suppress the development of tomato bacterial wilt caused by *Ralstonia solanacearum* (phylotype IIB/4NPB) was assessed. Lemongrass (chemotype 1)-, aniseed (chemotype 2)- and clove (chemotype 3)-scented chemotypes of *P. racemosa* var. *racemosa* essential oils were tested under in vitro conditions. The chemotype 3 displayed most effective suppressive activity on disease development, as no incidence of bacterial wilt was observed on tomato plants grown in soil treated with chemotype 3 at a concentration of 0.14%. In untreated control treatment, infection by bacterial wilt up to the extent of 62% was recorded. In addition, treatment of soil with chemotype 3 increased growth of tomato plants, compared with control plants. The results indicated the effectiveness of essential oil from *P. racemosa* var. *racemosa* for advancing to large scale application for managing an important bacterial disease affecting tomato and other crops (Deberdt et al. 2018). Field evaluation of the biocontrol activity of essential oils against bacterial wilt of tomato caused by *Ralstonia solanacearum* was carried out. Thymol

and palmarosa at 0.7% concentration were applied to field soil at 2 h after infesting with the pathogen, followed by sealing off the soil with plastic mulch for 3 or 6 days. Tomato seedlings were planted 7 days later. Both thymol and palmarosa oil treatments reduced bacterial wilt incidence significantly in susceptible cv. SolarSet (12%), compared with untreated control plots (65.5%). Thymol can be artificially synthesized, and it is available commercially, resulting in reduction in chemical cost, compared to cost of production of essential oils from plants. Further, thymol could be applied through drip irrigation system and this may reduce the cost of application further. As thymol has a wide spectrum of activity against bacteria, fungi and nematodes, it has the potential for large scale application for the management of bacterial wilt disease affecting tomato (Ji et al. 2005).

The biocontrol potential of leaf extracts of *Moringa oleifera* either alone or in combination with *Trichoderma* Kd63, *Trichoderma* IITA508 or *Bacillus subtilis* was assessed in vitro, greenhouse and field tests for the suppression of development of *Sclerotium rolfsii* causing damping-off and stem rot in cowpea. Under field conditions, seed treatment with *Moringa* combined with *Trichoderma* soil sprinkle reduced disease incidence and severity by more than 70% with significant yield increase in cowpea (Adandonon et al. 2006). Similar approach of combining leaf extract and fungal and bacterial biocontrol agents was applied for suppressing development of pre- and postemergence damping-off of pepper caused by *Pythium aphanidermatum*. Among the 66 medicinal plant species tested, zimmu leaf extract (*Allium sativum* x *A. cepa*) was found to be the most effective in suppressing pathogen growth in vitro and the presence of 22 antimicrobial compounds was detected in zimmu leaf extract. Assessment under pot culture conditions, showed that seed treatment with a combination of *T. viride, P. fluorescens* and zimmu leaf extract protected the pepper seedlings effectively and also promoted plant growth and increased yield as well (Muthukumar et al. 2010).

3.2.1.4 Effects of Animal Products

Various animal products such as cattle, chicken, duck and pig manures are applied to the soil as organic amendments and their usefulness as nutrients, enhancer of soil fertility and activator of antagonistic microorganisms have been discussed earlier. Chitosan, among the animal products, has been evaluated for its biocontrol potential more frequently and extensively, for suppressing the development of crop diseases caused by soilborne microbial plant pathogens. The shells of marine crustaceans such as crabs and shrimps are very affordable sources for commercial production of chitin. They are available as waste from seafood processing industry. Chitosan, a deacetylated form of chitin, is a natural biodegradable fiber (polymer), whose principal characteristics correspond to its polycationic nature. Low molecular weight chitosan possesses high antimicrobial activity, which increases with decreasing MW. Use of chitosan as a film coating as a delivery system for fertilizers, plant protection products and micronutrients for crop growth promotion forms an important application.

Chitosan is dissolved in an acid solution to activate its anti-microbial properties. Direct suppressive effect of chitosan on fungal pathogen is due its fungicidal properties. In addition, chitosan can also function as an inducer of resistance in plants against pathogens by activating host defense responses.

Chitosan was applied as a soil amendment either alone or in combination with other treatments. In soilless tomato, root rot caused by *F. oxysporum* f.sp. *radicis-lycopersici* was effectively suppressed by chitosan amendment (Lafontaine and Benhamou 1996). Infection in forest tree seedlings in the nurseries by *Fusarium acuminatum* and *Cylindrocladium* was drastically reduced by applying chitosan as soil amendment (Laflamme et al. 1999). Chitosan was applied as seed treatment for suppression of diseases caused by *F. oxysporum* in many crops (Rabea et al. 2003). The biocontrol potential of different organic amendments, along with crab shell (chitin) against cotton Verticillium wilt pathogen *Verticillium dahliae* was assessed. Crab shell was the most effective in reducing disease severity by 72% in pots, whereas soybean stalk and alfalfa reduced the disease severity by 60% and 56%, respectively. Crab shell stimulated the proliferation of antagonists effective against *V. dahliae* in the rhizosphere. The extracts of crab shell chitin were inhibitory to *V. dahliae* in vitro. Changes induced by crab shells in the composition and structure of microorganisms in the rhizosphere might contribute to the suppression of development of Verticillium wilt disease of cotton (Huang et al. 2006). Chitosan was inhibitory to *F. oxysporum* f.sp. *radicis-lycopersici* in vitro. Ultrastructural investigations, using transmission electron microscope (TEM), revealed marked changes in the sensitive fungal cells. Confocal laser microscopy showed that Rhodamine-labeled chitosan entered rapidly into the conidia via an energy-dependent process. The results suggested that chitin application might be combined with other compatible biocontrol agents to enhance the level of effectiveness of disease suppression (Palma-Guerrero et al. 2008). Pearl millet downy mildew caused by *Sclerospora graminicola* is both soilborne (oospores remaining in soil) and airborne (zoospore and sporangia spreading via wind). The efficacy of seed priming with chitosan in reducing the incidence of downy mildew disease was determined. Chitosan treatment of seeds at 2.5 g/kg increased seed germination and seedling vigor, compared with untreated seeds. Chitosan did not affect the sporulation and release of zoospores from sporangia. In seedlings growing from chitosan-treated seeds, activities of defense-related enzymes, chitinase and peroxidase were stimulated, following challenge inoculation with the pathogen. Assessment of effects of chitosan treatments under greenhouse and field conditions showed 79% and 76% protection, respectively, against downy mildew disease in pearl millet (Manjunatha et al. 2008).

Oligochitosan, obtained through hydrolysis or degradation of chitosan, is water soluble and more effective than chitosan, in suppressing disease development due to its ability to induce resistance in treated plants and to inhibit pathogen growth directly. Among the nine fungal pathogens tested for their sensitivity to oligochitosan, *Phytophthora capsici*, lacking chitin in the cell wall, was the most sensitive, with EC_{50} and mean inhibitory concentration (MIC) values of 100 and 580 µg/ml, respectively. Oligochitosan at low concentrations inhibited different stages in the life cycle of *P. capsici*, including production of zoospores, zoospore release, cystospore germination and it induced leakage of electrolytes from the pathogen mycelium. Oligochitosan might act on the cell membrane by altering osmotic pressure. Most drastic structural alteration in the hyphae treated with oligochitosan (10 µg/ml) was the disruption of endomembrane system, especially vacuole and secretory vesicles, such as plasmamembranes in the hyphal tips as revealed by observations under electron microscope. The polycationic nature of oligochitosan might contribute to its antifungal and multiple modes of action, including induction of resistance to diseases in plants treated with oligochitosan (Xu et al. 2007). The efficacy of a heterogenous chitosan suspension (MCp) and commercial plant activator acibenzolar-S-methyl (ASM) for inducing resistance in cocoa against *V. dahliae* was assessed. The MCp and ASM enhanced the level of protection to susceptible cocoa cv. SIAL70 against Verticillium wilt. MCp treatment reduced the disease severity to a level equivalent of 80% of ASM protection level. Local induced resistance was basically associated with activities of peroxidase (POX) and polyphenol oxidase (PPO) in leaves and with lignin deposition at 13 days after application. Induction of resistance locally was indicated by enhanced activities of chitinase and ß-1,3-glucanase in the leaves at 4–18 days after treatment with MCp and ASM. Increase in lignin deposition level following treatment with MCp and ASM and challenge with *V. dahliae,* was considered as the defense strategy operating in cocoa against *V. dahliae* (Cavalcanti et al. 2008).

Comparative biocontrol potential of chitosan and salicylic acid (SA) against soilborne fungal pathogens infecting tomato was assessed. Chitosan (0.5–4 mg/ml) and SA (1–25 mM) inhibited mycelial growth of *F. oxysporum* f.sp. *lycopersici (Fol), F. oxysporum* f.sp. *radicis-lycopersici* (FORL), *V. dahliae, Rhizoctonia solani, Colletotrichum coccodes, Pythium aphanidermatum* and *Sclerotinia sclerotiorum* in a concentration-dependent manner. *P. aphanidermatum* and *S. sclerotiorum* were the most sensitive to both resistance inducers (RIs). When applied as soil drench, single treatments with chitosan (4 mg/ml) or SA (10 mM) provided varied degree of protection to tomato plants against *V. dalhliae, Fol* and FORL. Disease severity to *V. dahliae, Fol* and FORL, respectively, was reduced to the extent of 42.1 to 73.68%, 60.86 to 78.26% and 45 to 50% following chitosan- and SA-based treatments. In addition, treatments with chitosan or SA enhanced growth of tomato plants, compared to untreated control plants (Jabnoun-Khiareddine et al. 2015). Chitosan can be used as a vehicle and for protecting other antimicrobial compounds such as essential oils, which are volatile and can ensure better persistence of the active ingredient of thyme (*Thymus vulgaris*) and tea tree (*Melaleuca alternifolia*) which are applied against plant pathogens. The effectiveness of chitosan in reducing foot and root rot of wheat caused by *Fusarium graminearum* was assessed by seed treatment in combination with essential oils. The pathogen growth on seeds was reduced without affecting seed germination. Disease severity was reduced in wheat seedlings. Chitosan seed treatment increased the contents of resistance markers such as phenols and activities of defense-related enzymes like phenylalanine ammonia lyase (PAL), polyphenol oxidase (PPO), peroxidase (PO) and chitinase in the seedlings. Greenhouse and field experiments also indicated that chitosan could enhance the level of resistance of plants growing from chitosan-treated seeds (Orzali et al. 2017).

Chitosan has been used for seed treatment and soil application. Treatment of seeds with chitosan (low MW 5–20 kDa) induced resistance in tomato plants against *Phytophthora infestans* (Kiprushkina et al. 2017). Seed soaking followed by foliar application of chitosan (0.25–2.0 g/l) were effective in suppressing development of wilt and root rot diseases of bean caused by *Fusarium solani* and *Rhizoctonia solani* (El-Mohamedy et al. 2017). In vitro and in vivo experiments revealed the efficacy of chitosan in suppressing bacterial wilt of potato. Tubers were soaked for 30 min (0.5–2.0 g/l) and spraying on leaves of potato effectively protected plants against infection by *Ralstonia solanacearum*. In addition, chitosan treatments promoted growth of potato plants, compared with untreated control (Farag et al. 2017). In another investigation, chitosan was mixed with a biofertilizer containing phosphate and potassium rocks, supplemented with addition of *Cunninghamella elegans* containing chitin and they were incorporated in the soil prior to planting green peppers or tomato. The treatment protected green peppers, but not tomato against *R. solanacearum* infection (Stamford et al. 2017).

3.2.2 Synthetic Organic Compounds

Different kinds of synthetic organic compounds have been evaluated for their biocontrol potential against soilborne microbial plant pathogens infecting various crops. Most of the organic compounds seem to act on the pathogens indirectly by activating host defense responses. The compounds activate the same spectrum of systemic acquired resistance (SAR) genes to levels comparable to those induced by the biotic inducers of disease resistance.

3.2.2.1 Salicylic Acid

The molecular mechanism underlying systemic acquired resistance (SAR) has been elucidated using the model plant *Arabidopsis thaliana*. The SAR regulatory protein nonexpressor of PR (*NPR1*) gene is activated by salicylic acid (SA) through redox changes. These changes, in turn, drive systemic expression of antimicrobial PR proteins and facilitate their secretion by upregulating protein secretory pathway genes (Mou et al. 2003; Wang et al. 2005). Further, the long distance signaling in *Arabidopsis* appeared to depend on a peptide signal system mediated by the Asp protease constitutive disease resistance (CDR1) (Xia et al. 2004). The reactive oxygen species (ROS)-mediated systemic signaling network also contributed to SAR. The SA-induced defense expression via nonexpressor of PR gene-1 (*NPR1*), a key mediator of SAR

functions in both dicotyledons and monocotyledons (Dong 2004). Interaction between wilt pathogen, *F. oxysporum (Fo)* and *Arabidopsis thaliana (At)* was investigated to understand the mechanism of activation of host defense responses directed against wilt pathogens. The expression of salicylate- and jasmonate-responsive defense genes in *Fo*-challenged roots of *At* plants, as well as in the roots of plants whose leaves treated with salicylate or jasmonate were analyzed. The genes (*PR1.PDF1.1* and *CHIB*) encoding proteins with defense functions or transcription factors (AtERF1, AtERF2, AtER4 and ATMYC2) known to positively or negatively regulate defenses against *Fo* were not activated in Fo-inoculated roots. In contrast, the jasmonate responsive gene *PDF1.2* was induced in the leaves of plants, whose roots were challenged with *Fo*, but salicylate-responsive *PR1* gene was not induced in the leaves of inoculated plants. Exogenous SA application prior to inoculation, however, activated *PR1* and *BGL2* defense gene expression in leaves and provided enhanced level of resistance to *Fo*, as indicated by foliar necrosis and subsequent plant death. Exogenous SA treatment of the foliar tissues did not activate defense gene expression in the roots of treated plants. The results suggested that salicylate-dependent defenses may function in foliar tissues to reduce the development of pathogen-induced wilting and necrosis. Although jasmonate application induced defense gene expression in leaves, it did not increase the resistance level to *Fo* (Edgar et al. 2006).

The effects of salicylic acid application were assessed to elucidate the mechanism of biocontrol activity of nonpathogenic *F. oxysporum (npFo)* against *F. oxysporum* f.sp. *asparagi (Foa)*, causing Fusarium wilt disease of asparagus (*Asparagus officinalis*). Split-root system was employed, where one-half of the root system of asparagus seedling was drenched with salicylic acid and the other half of the root system was examined for the activation of defense responses. Treatment of asparagus root with SA by soil drench (20 mg/l) at 2 days before challenge inoculation with *Foa*, was sufficient to protect the plants systemically. Diphenyleneiodonium chloride (DPI), an SA synthesis inhibitor, prevented induction of resistance to *Foa* by *npFo*. In vitro assays showed lack of inhibitor effect of SA on conidial germination and mycelial growth of *Foa*. SA-treated plants showed enhancement of systemic resistance with significant reduction in disease severity of the roots inoculated with *Foa*, compared with untreated control plants. SA-activated peroxidase and phenylalanine ammonia lyase (PAL), as well as lignification upon *Foa* infection, in a manner similar to that observed with *npFo* pretreatment. In addition, pretreatment of asparagus roots with SA or *npFo* primed the plants for a potential defense response to *Foa*. The results suggested the involvement of an SA-dependent SAR pathway in the *npFo*-induced potential defense responses and resistance in asparagus to *F. oxysporum* f.sp. *asparagi* (He and Wolyn 2005). The mechanism of activity of SA in suppressing tomato wilt disease caused by *F. oxysporum* f.sp. *lycopersici (Fol)* was studied. SA at a concentration of 200 mM was supplied through root feeding and foliar spray on tomato plants. Endogenous accumulation of free SA in tomato roots was detected by high performance liquid chromatography (HPLC) technique and its identity was confirmed by LC-MS/MS analysis. The endogenous level of SA in the roots increased at 168 h after application to about ten times higher than in untreated control plants. Similar increase in SA content in leaves also was observed, following foliar application of SA. The activities of PAL and PO were significantly stimulated, following SA application through either root or foliage. SA-treated tomato plants challenged with *Fol*, showed significant reduction in the intensity of vascular browning, leaf yellowing and wilting. However, the mycelial growth of *Fol* was significantly inhibited as shown by in vitro assays. The results indicated that SA-induced SAR might be responsible for suppression of wilt disease development in tomato, as the principal mechanism of biocontrol activity of SA (Mondal et al. 2005).

Perception of both general and specific pathogenesis-associated molecules by plants triggers defense responses via signal transduction cascades and transcriptional activation of numerous genes. In chickpea (*Cicer arietinium*), putative genes potentially involved in defense responses, including the rapid synthesis of PR-proteins, presence of an oxidative burst and synthesis of putative cell wall-strengthening proteins and antimicrobial proteins have been identified (Coram and Pang 2006). The responses of three chickpea genotypes treated with defense signaling compounds SA, methyl jasmonic acid (MeJ) and aminocyclopropane carboxylic acid (ACC) to *Ascochyta rabiei* were investigated, using microarray technique, followed by validation with QRT-PCR assay. Of the 715 experimental microarray features, 425 (59.4%) were differentially expressed (DE) at least in one condition. According to treatment applied, 69, 15.8 and 57.6% were differentially expressed respectively by ACC, MeJ and SA. The coregulation of transcripts between treatments for each genotype with varying levels of disease resistance showed large proportions of transcripts were independently regulated by ACC, MeJ or SA. Of the coregulated transcripts, the ACC-SA category contained the most for all genotypes, lending support to the view of cross talk and overlap occurring between signaling pathways (Salzman et al. 2005; Jalali et al. 2006).

3.2.2.2 Benzothiadiazole

Benzo (1, 2, 3)-thiadiazole-7-carbothioic acid-*S*-methyl ester (BTH) is a nontoxic functional analogue of salicylic acid and its mechanism of suppression of development of diseases caused by microbial plant pathogens has been investigated. The biocontrol potential of BTH and an avirulent strain of *Pseudomonas syringae* pv. *maulicola* (*Psm*) in inducing systemic acquired resistance (SAR) in canola (*Brassica napus*) against *Leptosphaeria maculans* was assessed. Application of BTH enhanced resistance against virulent strains of *Psm* and *L. maculans* to a great extent, then localized preinoculation of plants with avirulent strain. Pretreatments of plants with BTH and avirulent strain resulted in enhancement of defense-related responses to a greater level than in untreated plants. Development of SAR in *B. napus* plants expressing a bacterial salicylate hydroxylase transgene (*NahG*) that is

known to metabolize SA to catechol was significantly compromised. The plants accumulated only reduced levels of PR gene transcripts, compared with nontransformed plants (control). The results indicated that BTH induced SAR including long-lasting and broad host range resistance, associated with PR gene activation and requirement of SA (Potlakayala et al. 2007). The effectiveness of benzothiadiazole (BTH) in inducing resistance in cocoyam (*Xanthosoma sagittifolium*) against root rot pathogen *Pythium myriotylum* was assessed. Under controlled conditions, BTH (0.2 mg/ml) applied on leaves induced resistance to the pathogen effectively, resulting in significant reduction in disease incidence and severity. The activities of peroxidase (PO) and polyphenol oxidase (PPO) and total phenol contents registered increases. The enhancement of peroxidase activity was correlated with two new isoforms in a white (sensitive) cultivar inoculated, after stimulation. In a yellow (resistant) cultivar, stimulation was characterized by the appearance of one isoform. Quantitative analysis of phenolic compounds by HPLC showed an increase of hydroxycinnamomic and flavonoid derivatives after inoculation. Presence of a new caffeoylshikimic acid derivative was also detected after stimulation, following inoculation of both cultivars. The pattern of induction for resistance to *P. myriotylum* appeared to be cultivar-dependent (Mboubda et al. 2010).

Acibenzolar-S-methyl (ASM), a derivative of BTH was evaluated for its biocontrol potential against *Phytophthora cactorum*, incitant of crown rot and red stele disease of strawberry. The comparative efficacy of ASM and chitosan in suppressing the development of crown rot and red stele disease in strawberry was assessed. Both ASM and chitosan reduced crown rot symptoms and the suppressive effect was enhanced, when the interval between treatment and challenge inoculation with *P. cactorum* was increased from 2 to 20 days. Increase in concentration of ASM from 10 to 1,000 µg a.i./plant did not provide any additional advantage. Inoculation of alpine strawberry plants (*Fragaria vesca* var. *alpina*) cv. Alexandria with *P. fragariae* var. *fragariae*, after treatments with ASM, chitosan or fosetyl-Al (fungicide) showed that ASM provided effective protection to alpine strawberry, whereas chitosan was ineffective in protecting the plants. No significant difference was observed between the effectiveness of ASM and fosetyl-Al in protecting alpine strawberry plants against the pathogen (Eikeno et al. 2003). The effect of resistance inducers may differ, based on the method of application. In the greenhouse experiments, soaking seeds of cotton in ASM solutions (25 or 50 µg/ml) for 3–5 h, before planting the seeds in soils naturally infested with *Thileaviospsis basicola*, incitant of black root disease of cotton, resulted in consistent reduction in disease severity on tap roots by 20 to 30%, under field conditions. ASM was applied as sprays over seeds during sowing (in-furrow spray) as seed soaking or as foliar spray over seedlings. Seed-soaking reduced symptom severity by 33%, whereas in-furrow spray reduced disease severity by 24%, increased the number of relatively healthy roots by 35% and increased boll number by 29%. Foliar sprays were ineffective in reducing severity of disease. The results indicated the need for selecting suitable method of application of ASM for providing effective protection against cotton black root disease (Mondal et al. 2005). The comparative protective effects of ASM (BTH), methyl jasmonate (MeJ) and K_2HPO_4 against *Monosporascus cannonballus*, causing melon decline disease was assessed under pot and field conditions for 2 years. Seed treatment with MeJ significantly reduced root rot symptom severity and wine decline, whereas plants from seeds treated with BTH and K_2HPO_4 were slightly more resistant to the pathogen. Greenhouse assessments in 2006 using soil naturally infested with *M. cannonballus* showed that MeJ seed treatment followed by foliar application, decreased disease severity. Both MeJ and BTH treatments reduced root rot and vine decline in 2007, but K_2HPO_4 was ineffective. The resistance inducers differentially activated the synthesis of a number of PR protein isozymes, markers of induced systemic resistance (ISR) in the root system. Application of MeJ, as inducer of resistance in melon to root rot and vine decline in melon appeared to be a feasible strategy to manage the disease (Aleandri et al. 2010).

Acibenzolar-S-methyl (ASM) showed direct inhibitory activity (up to 40% inhibition) against *Rhizoctonia solani* AG-4, causing soybean root rot disease. At a concentration of 0.08 and 0.5 g/l, ASM induced systemic resistance responses in soybean hypocotyls, resulting in reduction of rotting severity. Reduction in disease severity by ASM application was correlated with stimulation of chitinase activity. The protective effect of ASM against *R. solani* AG-4 was, possibly due to a combination of induced resistance and its direct antifungal activity against the pathogen. Further, ASM adversely affected growth of soybean plants. A dose-dependent inhibition of root growth was observed, following seed treatment with ASM. However, the growth retardation of soybean attributed to ASM treatment was overcome and the treated plants recovered and attained normal growth condition, in the case of plants treated with lower dose of ASM (0.08 g/l). The need for determining optimum dose of ASM was revealed by the results of investigation (Faessel et al. 2008). The comparative effectiveness of ASM and chitosan in suppressing the development of Verticillium wilt disease of potato was assessed under in vitro and greenhouse conditions. ASM did not significantly reduce the mycelial growth of *V. dalhliae*, whereas chitosan inhibited the mycelial growth (5.4 to 16.9%), depending on the concentration. Potato tubers were dipped in different concentrations of the elicitors of resistance and planted in soil artificially infested with *V. dahliae*. ASM and chitosan were sprayed on the foliage of potato seedlings at 100 μg a.i./plant at 15, 25 and 40 days after planting. All treatments reduced Verticillium wilt severity and increased fresh tuber weight. ASM was more effective in reducing disease severity and enhancing yield level than chitosan. The results showed that ASM and chitosan had the potential to effectively reduce Verticillium wilt disease by inducing systemic resistance and increase tuber yield (Amini 2015).

Ceratocystis fimbriata causes mango wilt disease. Induction of systemic resistance to the pathogen in mango by applying acibenzolar-S-methyl (ASM) and potassium phosphate (Phi), a salt of phosphorous acid with systemic mobility in treated plants was investigated. The effects of treatment with the inducers were assessed by microscopic and biochemical analyses. Disease development in test plants was monitored using fluorescence and light microscopy. High performance liquid chromatography (HPLC) procedure was employed to quantify secondary metabolites in the stem sections of treated and control plants. Spraying mango plants with ASM and Phi resulted in reduction of internal necrosis and disease development. Both chemicals induced defense responses in the stem tissues against *C. fimbriata* infection. HPLC analysis showed that concentration of two alkaloids (theobromine and 7-methylxanthinine) and 10 phenolic compounds (catechin, epicatechin, epigallocatechin, gallic acid, myricetin, p-coumaric acid, p-hydroxybenzoic acid, phloridzin, sinapinic acid, salicylhydroxamic acid) were in present in higher concentrations in stem tissues of plants treated with ASM or Phi, compared with untreated control plants. By contrast, higher concentrations of secondary metabolites were detected in the stem tissues at early stages of infection by *C. fimbriata*, particularly in plants treated with resistance inducers. The results suggested that phenylpropanoid pathway in the stem tissues of treated mango plants might be induced, following infection by the pathogen (Araujo et al. 2015).

Acibenzolar-S-methyl (ASM) was evaluated for its ability to induce resistance in tomato against the bacterial wilt disease caused by *Ralstonia solanacearum (Rs)*. Tomato plants were treated with ASM (25 μg/ml) as foliar spray and soil drench (12.5 μg/ml) and inoculated with different populations of *Rs* and suitable controls were maintained. Growth reduction in untreated and inoculated plants was observed. Application of ASM significantly reduced bacterial wilt disease incidence, when inoculated with low concentration of *Rs*, suggesting that ASM might be effective only when the pathogen inoculum level was low (low disease pressure conditions) (Haciasalihoglu et al. 2007). The mechanism of action of ASM and *Pseudomonas fluorescens* Pf2 in suppressing development of bacterial wilt disease in tomato was studied. Treatment of tomato seedlings with either Pf2 strain or ASM significantly reduced bacterial wilt disease severity by 58% and 56%, respectively. An increase in the effectiveness of disease suppression was observed, when both Pf2 and ASM were combined providing 72% reduction in disease severity. The seedling biomass showed increase due to ASM treatment, relative to the control. Significant changes in the activities of polyphenol oxidase (PPO), ß-glucosidase (B-GL) and peroxidase (PO) were recorded in tomato plants treated with ASM or Pf2. Under field conditions, application of ASM as foliar spray or soil drench was effective in suppressing bacterial wilt development in tomato (Abo-Elyousr et al. 2012).

3.2.2.3 ß-Aminobutyric Acid

The biocontrol potential of the nonprotein amino acid ß-aminobutyric acid (BABA) against fungal pathogens, infecting different crops has been assessed. The protective effect of BABA against potato late blight disease caused by *Phytophthora infestans* was demonstrated, using two potato cultivars, Bintje and Pampena with different levels of horizontal resistance to late blight disease. Foliar treatment at 30 days after emergence provided protection (60%) in cv. Pampena against *P. infestans*. BABA treatment stimulated the expression of defense molecules, such as glucanases, chitinases and phenolic compounds (Altamiranda et al. 2008). Potato cultivars with different levels of resistance to *P. infestans* received four applications of BABA throughout the crop season and they produced tubers with greater resistance to *P. infestans* and also to *Fusarium solani* infection of tubers, compared with those from untreated potato plants. Tuber slices from treated plants, inoculated with *P. infestans* showed an increase in contents of phenolics and phytoalexins. Aspartyl protease StAP1 accumulation was also higher in tubers from BABA-treated plants and inoculated with *F. solani*. Infected tubers of BABA-treated plants showed minor fungal proteolytic activity than those from nontreated plants. Application of BABA improved the plant growth, in addition to protection against tuber infection by fungal pathogens (Olivieri et al. 2009).

The protective efficacy of ß-aminobutyric acid (BABA) against the white mold pathogen *Sclerotinia sclerotiorum* infecting artichoke was investigated, using cvs. C3 and Explorer. Soil treatment by drenching with BABA induced high level of resistance in artichoke plantlets of both cvs. C3 and Explorer with similar levels of protection. A consistent increase in peroxidase activity paralleled with the differential induction of an alkaline isoenzyme with a pI of 8.6 also was detected. The results indicated that BABA-induced resistance and an augmented ability to express basal defense responses were positively correlated in a more pronounced manner in the cv.C3 (Marcucci et al. 2010). In another investigation, the ability of BABA to protect *Brassica napus* plants against infection by *Leptosphaeria maculans* was assessed. BABA showed direct inhibitory effect against *L. maculans* in in vitro tests. The EC_{50} value of BABA was similar to that of the fungicide tebuconazole. Both spore germination and hyphal growth were inhibited. Suppression of disease progression in plants and antifungal activity in vitro were weaker for α-aminobutyric acid and negligible for γ-aminobutyric acid. In contrast to benzothiadiazole (BTH), another resistance inducer, the effect of BABA on disease development was nearly independent of the timing of treatment, indicating possible antifungal activity in planta. By contrast, quantification of multiple hormones and an expression analysis indicated that BABA treatment induced synthesis of salicylic acid (SA) and expression of SA marker gene PR-1. However, no evidence for priming SA responses to *L. maculans* could be obtained. The antifungal activity of BABA against *L. maculans* could be another possible mechanism of action of BABA to provide protection to canola against microbial pathogens (Šašek et al. 2012).

3.2.2.4 Glycerol

Glycerol is an environment-friendly, nontoxic, edible and biodegradable sugar alcohol. Glycerol and its derivative glycerol-3-phosphite (G3P) were considered to have the ability to participate in plant defense-against stresses. Exogenous application of glycerol as foliar spray was evaluated for its potential in suppressing infection by *Phytophthora capsici* in cocoa. Glycerol applied over a period of 4 days on cocoa leaves increased endogenous level of G3P and decreased the level of oleic acid (18:1). Reactive oxygen species (ROS) known as a defense activation marker, were produced and the expression of many pathogenesis-related genes was induced. The effect of glycerol application on G3P and 18:1 fatty acid content, and gene expression levels, in cocoa leaves were dosage dependent. Spray application of glycerol at 100 mM concentration was sufficient to stimulate the defense response without causing any observable damage and resulted in a significantly decreased lesion formation by *P. capsici*. However, at higher concentration (500 mM), chlorosis and cell death were observed on treated leaves (Zhang et al. 2015).

3.2.2.5 Ethanol

Biological soil disinfestation (BSD), an environmentally safe approach, is increasingly becoming acceptable for management of soilborne diseases, because of rising concerns related to environmental risks. The efficacy of soil disinfestation using ethanol was evaluated for the control of Fusarium wilt disease of tomato caused by *Fusarium oxysporum* f.sp. *lycopersici* (*Fol*). Survival of bud cells and chlamydospores of *Fol* declined significantly in soil saturated with diluted ethanol solution in vitro. In the field trials, artificially added nonpathogenic *F. oxysporum* and indigenous *F. oxysporum* were both strongly suppressed in soil saturated with 1% ethanol solution. Wheat bran treatment was not as effective as ethanol treatment. Artificially added *F. oxysporum* could not be detected in three of four sites amended with wheat bran. Ethanol application did not show any suppressive activity in preautoclaved soil, indicating the requirement of presence of native microorganisms for disease suppression. The ethanol-mediated biological soil disinfestation (Et-SBD) transiently increased the number of all aerobic bacteria, but the number of fungi and aerobic bacteria was stable. Bacterial community structure in the soil treated with Et-SBD showed slight, but apparent differences, compared with soils that received irrigation or other treatments, as indicated by polymerase chain reaction (PCR)-denaturing gradient gel electrophoresis (DGGE) analysis. The results indicated the potential of ethanol-based soil disinfestation for suppressing soilborne fungal diseases affecting various crops (Momma et al. 2010).

3.2.3 INORGANIC CHEMICALS

The beneficial effects of inorganic chemicals on plant growth and disease suppression by enhancing host resistance have been reported in some pathosystems. Phosphates are considered to generate an endogenous systemic acquired resistance (SAR) signal, because of calcium sequestration at points of phosphate application (Reuveni et al. 1994). Phosphites (Phi) are alkali metal salts of phosphorous acid with the ability to protect plants against microbial plant pathogens. The effects of treatment of potato seed tubers and foliage with Phi on development of *Phytophthora infestans*, (late blight and tuber rot), *Fusarium solani* (dry rot) and *Rhizoctonia solani* (black scurf) infecting potatoes were assessed, using cvs. Shepody and Kennebec. Protection provided by Phi was variable, being high against *P. infestans*, intermediate against *F. solani* and low against *R. solani*. In addition, seed tubers treated with calcium or potassium phosphites (CaPhi and KPhi) at 1% commercial product, emerged earlier than untreated controls. When Phi applied as foliar sprays four times at different doses, high levels of protection against *P. infestans* were observed on both cultivars. Higher protection was recorded in Kennebec, when CaPhi was applied, whereas the protective effect of KPhi was greater in cv. Shepody. Expression of ß-1,3-glucanase was induced at different times after treatment, but no correlation between ß-1,3-glucanases expression and level of foliar protection could be established. By contrast, Phi did not induce any negative effect on plant growth. Leaves of treated plants were darker green than those in untreated plants. Increase in Rubisco protein and a delay on setting of leaf senescence in the Phi-treated leaves were also seen (Lobato et al. 2008). The comparative effectiveness of phosphite (Phostrol) and metalaxyl-m (Ridomil Gold 480EC) applied as planting in-furrow treatment was assessed for reducing the incidence of potato pink rot caused by *Phytophthora erythroseptica* under field conditions in 2005 and 2006. Inoculum of a metalaxyl insensitive isolate of *P. erythroseptica* was applied either in-furrow as vermiculite slurry at planting or as a zoospore drench in soils adjacent to plants in late August. After harvest, disease incidence and severity and effect on tubers were determined. The mean percentages of diseased tubers were 1.7% (2005) and 1.3% (2006) for the fungicide-treated plots and 10.1% (2005) and 3.1% (2006) for phosphite-treated plots. The potato cv. Shepody was significantly more susceptible to pink rot (9.9% and 3.3% diseased tubers in 2005 and 2006 respectively) than Russet Burbank (3.4% and 1.2% in 2005 and 2006 respectively). The results indicated that metalaxyl was more effective than phosphite in reducing the incidence of pink rot affecting potato tubers (Al-Mughrabi et al. 2007).

The protective ability of potassium nitrate (KNO_3) was determined for suppressing the development of soybean stem rot caused by *Phytophthora sojae*. Application of KNO_3 (4–30mM) prior to challenge inoculation with *P. sojae*, reduced the incidence of disease in two soybean cultivars. The extent of disease suppression depended on concentration of potassium in the plants of both cultivars. Observations under scanning electron microscope revealed marked accumulation of potassium at penetration stopping sites of *P. sojae* in the cortex layer of soybean plants treated with 30 mM of KNO_3, compared to untreated control plants. Reduction in release of zoospores of *P. sojae* was observed in the presence of 0.4–30 mM KNO_3, indicating some direct effect of the chemical on the pathogen development. The results indicated the potential of KNO_3 as a strategy to reduce fungicide use for managing soybean stem rot disease caused by *P. sojae* (Sugimoto et al. 2009). The biocontrol activity of potassium sorbate (PS), potassium bicarbonate (PB) and dipotassium hydrogen phosphate (DPHP) against soilborne pathogens *F. oxysporum* f.sp. *lycopersici (Fol), F. oxysporum* f.sp. *radicis-lycopersici*

(FORL), *F. solani (Fs), Verticillium dahliae (Vd), Rhizoctonia solani, Pythium aphanidermatum* and *Sclerotinia sclerotiorum* was assessed. The effects on growth were variable, depending on the sensitivity of different fungal pathogens tested. Single treatment of soils with PS (0.25%), PB (50 mM) and DPHP (50 mM) reduced wilt diseases to different extent. PS treatment reduced wilt incidence by 50, 78, 26 and 65% respectively, compared to *Vd-, Fol-* and FORL-inoculated control plants. In addition, PS treatment improved plant growth significantly. PB-based treatment resulted in reduction to an extent of 60, 86 and 30% in *Fol,* FORL and *Fs* infections and severity, but it had no effect on *Vd* wilt disease. DPHP suppressed Fusarium wilt by 65.2%. PB treatment reduced Rhizoctonia root rot also to some extent. The results indicated the need for selecting suitable potassium salt for suppressing different diseases affecting tomato (Jabnoun-Khiareddine et al. 2016).

The potential of silicon (Si) applied as soil amendment in inducing resistance in banana (*Musa acuminata*) against *Cylindrocladium spathiphylli*, incitant of banana toppling disease was assessed. Banana plantlets, inoculated by dipping the root system in the suspension of pathogen conidia, were planted on desilicated ferrosol and amended with 2 mM of soluble Si under greenhouse conditions and control without Si amendment. Image analysis program WinRHIZO was applied at 7, 14 and 21 days after inoculation. Root necrosis (lesions) was reduced by about 50% at 14 days after inoculation in the plants on Si-amended soil, compared with control plants. Furthermore, Si amendment also improved plant growth because of the suppression of pathogen development by enhancing the level of resistance of banana plants supplied with Si. The results indicated the possibility of using Si amendment of soil as an ecofriendly disease management strategy and an alternative to chemical control (Vermeire et al. 2011). The extent of protection provided by silicon to tomato against crown and root rot disease caused by *F. oxysporum* f.sp. *radicis-lycopersici* was assessed using sand culture system. Hoagland's nutrient solution with (100 mg Si/l, Si^+) was used as a nutrient source for tomato plants without Si (Si^-) and control plants were also maintained. At 8 weeks after transplantation, the plants were inoculated with three inoculum levels (0, 10^6 and 10^7 conidia/plant). Disease severity was significantly reduced by Si treatment at 4 weeks after inoculation. Si contents of roots and stems of treated tomato plants were significantly greater, compared to control plants. The increase in the Si contents of the roots was positively correlated with reduction in disease severity in roots, crown and stem of treated tomato plants. The results suggested that the decrease in disease severity due to Si treatment, might be because of the delay in the onset of initial infection of roots and movement of the pathogen from roots to stem (Huang et al. 2011).

The effects of silicon on development of symptoms of Fusarium wilt disease caused by *F. oxysporum* f.sp. *cubense* (*Foc*) on banana plants were assessed, using seedlings of Grand Nain (resistant) and Maca (susceptible) cultivars grown in plastic trays amended with 0.39g Si (Si^+) and without amendment (Si⁻). The soil was inoculated with *Foc* at 60 days after transplanting banana seedlings. The Si concentrations in the roots and rhizome- pseudostem significantly increased by 30.26% and 58.82%, respectively, compared with Si⁻ treatment. The Si-treated plants showed a reduction in disease severity determined, based on area under reflex leaf symptoms progress curve, the area under root symptom progress curve and the area under asymptomatic fungal colonization of tissues progress curve, compared to Si⁻ plants. The area under darkening rhizome-pseudostem progress curve (AUDRPPC) of Maca significantly increased by 15.98% for the Si⁻ treatment, compared with Si⁺ treatment. The area under relative lesion length progress curve (AURLLPC) of Maca plants significantly reduced by 45.54% for the Si⁺ treatment. Grand Nain plants showed no difference in AUDRPPC and AURLLPC values due to treatment with Si. The plants in the Si⁻ treatment of cvs. Maca and Grand Nain plants did not show difference in AUDRPPC and AURLLPC values. The results showed that amendment with Si in soil for growing susceptible banana cultivars showed potential for reducing wilt severity in Si-deficient soils (Fortunato et al. 2012a). In further investigation, plants of banana cv. Grand Nain and Maca were grown in plastic pots with soil amended with Si and inoculated with *F. oxysporum* f.sp. *cubense (Foc)* race 1. Relative lesion lengths (RLLs) and asymptomatic fungal colonization were reduced to a great extent at 40 days after inoculation in the resistant Grand Nain, then in the susceptible Maca. Reduction in severity of symptoms was the greatest in resistant variety. The activities of PAL, PPO, POX, chitinases and ß-1,3-glucanase were enhanced in the roots of banana plants grown on soil with Si amendment (Fortunato et al. 2012b). *Phytophthora pistacia* causes pistachio gummosis, one of the major diseases impacting production adversely. The biocontrol potential of sodium and potassium silicate in suppressing the disease development was assessed. The silicon salts inhibited mycelial growth, sporangial production, cyst germination and fungal biomass. Further, zoospore release was significantly reduced by potassium silicate, but not by sodium silicate. Under pot experiments, broad bean (annual host crop) and seedlings of pistachio grown on sterilized soil, following application of silicon salts were inoculated with *P. pistacia*. Significant reduction was observed in disease incidence, percentage of roots colonized by the pathogen and mortality of plants, compared with nontreated control plants was observed. The silicon salts did not have significant effect on pH and electrical conductivity of soil. Coupled plasma mass spectrometry (ICP-MS) analyses showed that silicon concentration was increased in Si-treated broad bean and pistachio plants. The results showed the potential of silicon salts could be exploited for suppressing the development of an economically important disease of pistachio (Mostowfizadeh-Ghalamfarsa et al. 2017).

The biocontrol potential of silver nanoparticles (AgNPs) synthesized with aqueous extract of *Artemisia abisinthium* against *Phytophthora* spp. was assessed. The AgNPs (10 μg /ml) inhibited mycelia growth of *Phytophthora parasitica*, *P. infestans*, *P. palmivora*, *P. cinnamomi*, *P. tropicalis*, *P. capsici* and *P. katsurae*. The AgNPs were highly inhibitory to mycelial growth, zoospore germination, germ tube elongation and zoospore production of *P. parasitica* and *P. capsici*, as revealed by detailed dose-response analyses. Under greenhouse conditions, AgNP treatment prevented infection by the pathogen and improved plant survival. Further, no adverse effects of AgNPs in plant growth were evident. The results indicated that AgNPs could be applied for effective suppression of diseases caused by *Phytophthora* spp. as an alternative to fungicides (Ali et al. 2015). *Stenotrophomonas* spp. (earlier known as *Pseudomonas maltophila*) have wide distribution and they have potential for use as plant growth promoter and biocontrol agents. Biosynthesis of gold and silver nanoparticles (AgNPs) was achieved using *Stenotrophomonas* sp. BHU-S7 (MTCC5978) strain. Biosynthesis of AgNPs by this strain was monitored by UV-visible spectrum, showing surface plasmon resonance (SPA) peak at 440 nm. The antifungal activity of AgNPs (~12 nm size) was assessed. *Sclerotium rolfsii* exposed to AgNPs failed to germinate on potato dextrose agar (PDA) medium, as well as in soil system. Furthermore, treatment with AgNPs reduced collar rot disease incidence in chickpea caused by *S. rolfsii* under greenhouse conditions. Induction of phenolics, altered lignification and H_2O_2 production in plants treated with AgNPs indicated possible involvement of induced systemic resistance (ISR) as a possible mechanism of action of AgNPs against *S. rolfsii* (Mishra et al. 2017).

REFERENCES

Abbasi PA, Conn KL, Lazarovits G (2004) Suppression of Rhizoctonia and Pythium damping-off of radish and cucumber seedlings by addition of fish emulsion to peat mix or soil. *Canad Plant Pathol* 26: 177–187.

Abbasi PA, Lazarovits G (2005) Effects of AG3 phosphonate formulations on incidence and severity of Pythium damping-off of cucumber seedlings under growth room, microplot, and field conditions. *Canad J Plant Pathol* 27: 420–429.

Abbasi PA, Lazarovits G (2006) Seed treatment with phosphonate (AG3) suppresses Pythium damping-off of cucumber seedlings. *Plant Dis* 90: 459–464.

Abbasi PA, Lazarovits G, Weselowski B, Lalin I (2009) Organic acids in condensed distiller's soluble: toxicity to soilborne plant pathogens and role in disease suppression. *Canad J Plant Pathol* 31: 88–95.

Abd El-Rahman SA, Mazen MM, Mahmoud NM (2012) Induction of defence-related enzymes and phenolic compounds in lupine (*Lupinus albus* L.) and their effects on host resistance against Fusarium wilt. *Eur J Plant Pathol* 134: 105–116.

Abo-Elyousr KAM, Ibrahim YE, Balabel NM (2012) Induction of disease defensive enzymes in response to treatment with acibenzolar-S-methyl (ASM) and *Pseudomonas fluorescens* Pf2 and inoculation with *Ralstonia solanacearum* race 3, biovar (phylotype II). *J Phytopathol* 160: 382–389.

Accinelli C, Mencarelli M, Saccá M, Vican A, Abbas HK (2012) Managing and monitoring of *Aspergillus flavus* and corn using bioplastic-based formulations. *Crop Protect* 32: 30–35.

Accinelli C, Saccá ML, Abbas HK, Zablotowicz RM, Wilson JR (2009) Use of granular bioplastic formulation of *Aspergillus flavus*. *Bioresources Technol* 100: 3977–4004.

Adandonon A, Avenling TAS, Labuschagne N, Tami M (2006) Biocontrol agents in combination with *Moringa oleifera* extract for integrated control of Sclerotium-caused cowpea damping-off and stem rot. *Eur J Plant Pathol* 115: 409–418.

Addy HS, Askora A, Kawasaki T, Fujie M, Yamada T (2012) Utilization of filamentous phage ØRSMS to control bacterial wilt caused by *Ralstonia solanacearum*. *Plant Dis* 96: 1204–1209.

Ahmed AS, Izziyyani M, Sanchez CP, Candela ME (2003) Effect of chitin on biological control activity of *Bacillus* spp. and *Trichoderma harzianum* against root rot disease in pepper (*Capsicum annuum*) plants. *Eur J Plant Pathol* 109: 633–637.

Ahmed AS, Sánchez CP, Candela ME (2000) Evaluation of induction of systemic resistance in pepper plants (*Capsicum annuum*) to *Phytophthora capsici* using *Trichoderma harzianum* and its relation with capsidol accumulation. *Eur J Plant Pathol* 106: 817–824.

Akrami M, Ibrahimov AS, Zafari DM, Valizadeh E (2009) Control of Fusarium rot of bean by combination of *Trichoderma harzianum* and *Trichoderma asperellum* in greenhouse conditions. *Agric J* 4: 121–123.

Aleandri MP, Reda R, Tagliavento V, Magro P, Chilosi G (2010) Effect of chemical resistance inducers on the control of Monosporascus root rot and vine decline of melon. *Phytopathol Mediterr* 49: 18–26.

Alejo-Iturvide F, Márquez-Lucio M, Morales-Ramirez I, Vázquez-Garciadueñas S, Olalde-Portugal V (2008) Mycorrhizal protection of chilli plants challenged by *Phytophthora capsici*. *Eur J Plant Pathol* 120: 13–20.

Ali M, Kim B, Belfield D, Norman D, Brennan M, Ali GS (2015) Inhibition of *Phytophthora parasitica* and *P. capsici* by silver nanoparticles synthesized using aqueous extract of *Artemisia absinthium*. *Phytopathology* 105: 1183–1190.

Al-Mughrabi KI, Peters RD, (Bud) Platt HW, Moreau G, Vikram A, Poirier R, MacDonald I (2007) In-furrow application of metalaxyl and phosphite for control of pink rot (*Phytophthora erythroseptica*) of potato in New Burnswick, Canada. *Plant Dis* 91: 1305–1309.

Altamiranda EAG, Andreu AB, Daleo GR, Olivieri FP (2008) Effect of ß-aminobutyric acid (BABA) on protection against *Phytophthora infestans* throughout potato. *Austr Plant Pathol* 37: 421–427.

Amini J (2015) Induced resistance in potato plants against Verticillium wilt invoked by chitosan and acibenzolar-S-methyl. *Austr J Crop Sci* 9: 570–576.

Amira MB, Lopez D, Mohamed AT et al. (2017) Beneficial effect of *Trichoderma harzianum* strain Ths97 in biocontrolling *Fusarium solani*, causal agent of root rot disease of olive trees. *Biol Contr* 110: 70–78.

Amith KN, Momol MT, Kloepper JW, Marios JJ, Olson SM, Jones JBC (2004) Efficacy of plant growth promoting rhizobacteria, acibenzolar-S-methyl, and soil amendment for integrated management of bacterial wilt of tomato. *Plant Dis* 88: 669–673.

Anand S, Reddy J (2009) Biocontrol potential of *Trichoderma* sp. against plant pathogens. *Internatl J Agric Sci* 1: 30–39.

Antonelli M, Reda R, Aleandri MP, Varvaro L, Chilosi G (2013) Plant growth-promoting bacteria from solarized soil with the ability to protect melon against root rot and vine decline caused by *Monosporascus cannonballus*. *J Phytopathol* 161: 485–496.

Araujo L, Silvva Bispo WM, Rios VS (2015) Induction of the phenylpropanoid pathway by acibenzolar-S-methyl and potassium phosphate increases mango resistance to *Ceratocystis fimbriata* infection. *Plant Dis* 99: 447–459.

Arfaoui A, Sifi B, Boudabous A, El Hardani I, Cherif M (2006) Identification of *Rhizobium* isolates possessing antagonistic activity against *Fusarium oxysporum* f.sp. *ciceris*, the causal agent of Fusarium wilt of chickpea. *J Plant Pathol* 88: 67–75.

Arjona-Girona I, Lopez-Herrera CJ (2018) Study of a new biocontrol fungal agent for avocado white root rot. *Biol Contr* 117: 6–12.

Athukorala SN, Fernando WGD, Rashid KY, de Kievity T (2010) The role of volatile and nonvolatile antibiotics produced by *Pseudomonas chlororaphis* strain PA23 in its root colonization and control of *Sclerotinia sclerotiorum*. *Biocontr Sci Technol* 20: 875–890.

Baek JM, Howell CR, Kenerley C (1999) The role of extracellular chitinase from *Trichoderma virens* GV29-8 in the biocontrol of *Rhizoctonia solani*. *Curr Genet* 35: 41–50.

Bardin SD, Huang HC, Liu L, Yanke LJ (2003) Control by microbial seed treatment of damping-off caused by *Pythium* sp. on canola, safflower, dry pea, and sugar beet. *Canad J Plant Pathol* 25: 268–275.

Baysal F, Benitez MS, Kleinhenz MD, Miller SA, McSpadden Gardener BB (2008) Field management effects on damping-off and early season vigor of crops in a transitional organic cropping system. *Phytopathology* 98: 562–570.

Baysal O, Lai D, Xu H-H et al. (2013) A proteomic approach provides new insights into the control of soilborne plant pathogens by *Bacillus* species. *PLoS ONE* 8(1): e53182

Baz M, Lahbabi D, Samri S, Val F, Hamelin G, Madore I, Bouarab K, Beaulieu C, Ennaj MM, Barakate M (2012) Control of potato soft rot caused by *Pectobacterium carotovorum* and *Pectobacterium atrosepticum* by Moroccan actinobacteria isolates. *World J Microbiol Biotechnol* 28: 303–311.

Becker JO, Cook RJ (1988) Role of siderophores in suppression of *Pythium* species and production of increased growth response of wheat by fluorescent pseudomonads. *Phytopathology* 78: 778–782.

Beirner BP (1967) Biological control and its potential. *World Rev Pest Contr* 6: 7–20.

Belgrove A, Steinberg C, Viljoen A (2011) Evaluation of nonpathogenic *Fusarium oxysporum* and *Pseudomonas fluorescens* for Panama wilt disease control. *Plant Dis* 95: 951–959.

Benhamou K, Garand C, Goulet A (2002) Ability of nonpathogenic strain Fo47 to induce resistance against *Pythium ultimum* infection in cucumber. *Appl Environ Microbiol* 68: 4044–4060.

Benhamou N, Bélanger RA, Rey P, Trilly Y (2001) Oligandrin, the elicitin-like protein produced by the mycoparasite *Pythium oligandrum* induces systemic resistance to Fusarium crown and root rot in tomato plants. *Plant Physiol Biochem* 39: 681–696.

Benitez MS, Mc Spadden Gardener BB (2009) Linking sequence to function in soil bacteria: sequence-directed isolation of novel bacteria contributing to soilborne disease suppression. *Appl Environ Microbiol* 75: 915–924.

Bergsma-Vlami M, Prins ME, Staats M, Raajmakers JM (2005) Assessment of genotype diversity of antibiotic-producing Pseudomonas species in the rhizosphere by denaturing gradient gel electrophoresis. *Appl Environ Microbiol* 71: 993–1003.

Berry C, Fernando WGD, Loewen PC, de Kievit TR (2010) Lipopeptides are essential for *Pseudomonas* spp. DF41 biocontrol of *Sclerotinia sclerotiorum*. *Biol Contr* 55: 211–218.

Berta G, Sampo S, Gamalero E, Massa N, Lemanceau P (2005) Suppression of Rhizoctonia root rot of tomato by *Glomus mosseae* BEG12 and *Pseudomonas fluorescens* A6R1 is associated with their effect on pathogen growth and on root morphogenesis. *Eur J Plant Pathol* 111: 279–288.

Bertagnolli BL, Daly S, Sinclair JB (1998) Antimycoctic compounds from plant pathogen *Rhizoctonia solani* and its antagonist *Trichoderma harzianum*. *J Phytopathol* 146: 131–135.

Bi V, Jiang H, Hausbeck MK, Hao JS (2012) Inhibitory effects of essential oils for controlling *Phytophthora capsici*. *Plant Dis* 96: 797–803.

Bianco PA, Bruno I, Fortusini A, Belli G (1988) Cross-protection tests on herbaceous hosts with *Grapevine fan leaf virus* (GFLV). *Riv Patol Veg* 24: 81–88.

Bitsadze N, Siebold M, Koopmann B, von Tiedemann A (2015) Single and combined colonization of *Sclerotinia sclerotiorum* by the fungal mycoparasite *Coniothyrium minitans* and *Microsphaeropsis ochracea*. *Plant Pathol* 64: 690–700.

Bolwerk A, Lagopodi AL, Lugtenberg BJJ, Bloemberg GV (2005) Visualization of interactions between a pathogenic and a beneficial *Fusarium* strain during biocontrol of tomato foot and root rot. *Molec Plant-Microbe Interact* 18: 710–721.

Bolwerk A, Lagopodi AL, Wijffjes AHM, Lamers GEM, Chin-A-Wong TFC, Lugtenberg BJJ, Bloemberg GV (2003) Interactions in the tomato rhizosphere of two *Pseudomonas* biocontrol strains with the phytopathogenic fungus *Fusarium oxysporum* f.sp. *radicis-lycopersici*. *Molec Plant-Microbe Interact* 11: 983–993.

Bonanomi G, Ippolito F, Scala F (2015) A 'black' future for plant pathology? Biochar as a new soil amendment for controlling plant diseases. *J Plant Pathol* 97: 223–234.

Boukaew S, Chuenchit S, Petcharat V (2011) Evaluation of *Streptomyces* spp. for biological control of Sclerotium root and stem rot and Ralstonia wilt of chill pepper. *BioControl* 56: 365–374.

Bowers JH, Locke JC (2004) Effect of formulated plant extracts and oils on population density of *Phytophthora nicotianae* in soil and control of Phytophthora blight in the greenhouse. *Plant Dis* 88: 11–16.

Bradley GG, Punja ZK (2010) Composts containing fluorescent pseudomonads suppress Fusarium root and stem rot. *Austr J Agric Res* 25: 121–137.

Bulat SA, Lübeck M, Mironenko N, Jensen DF, Lübeck PS (1998) UP-PCR analysis and ITS-1 ribotyping of strains of *Trichoderma* and *Gliocladium*. *Mycol Res* 102: 933–943.

Bull CT, Shetty KG, Subbarao KV (2002) Interaction between myxobacteria, plant pathogenic fungi and biocontrol agents. *Plant Dis* 86: 889–896.

Bulluck LR III, Ristaino JB (2002) Effect of synthetic and organic soil fertility amendments on southern blight, soil microbial communities and yield of processing tomatoes. *Phytopathology* 92: 181–189.

Buttimer C, AcAuliffe O, Ross RP, Hill C, O'Mahony J, Coffey A (2017) Bacteriophages and bacterial plant diseases. *Front Microbiol* 8, Article 34

Cabanás CG, Schiliro E, Valverde-Corredor A, Mercado-Blanco J (2014) The biocontrol endophytic bacterium *Pseudomonas fluorescens* induces systemic defense responses in aerial tissues upon colonization of olive roots. Front Microbiol.

Camp AR, Dillard HR, Smart CD (2005) Efficacy of *Muscador albus* for the control of Phytophthora blight of sweet pepper and butternut squash. *Plant Dis* 92: 1488–1492.

Campbell R (1989) *Biological Control of Microbial Plant Pathogens*, Cambridge University Press, UK.

Cao L, Qui Z, You J, Tan H, Zhou S (2005) Isolation and characterization of endophytic streptomycete antagonists of Fusarium wilt pathogen from surface-sterilized banana roots. *FEMS Microbiol Lett* 247: 147–152.

Carisse O, El Bassam S, Benhamou N (2001) Effect of *Microsphaeropsis* sp. strain P130A on germination and production of sclerotia of *Rhizoctonia solani* and interaction between the antagonist and the pathogen. *Phytopathology* 91: 782–791.

Carisse O, Bernier J, Benhamou N (2003) Selection of biological agents from composts for control of damping-off of cucumber caused by *Pythium ultimum*. *Canad J Plant Pathol* 25: 258–267.

Carpenter MA, Stewart A, Ridgway HJ (2005) Identification of novel *Trichoderma hamatum* genes expressed during mycoparasitism using subtractive hybridization. *FEMS Microbiol Lett* 251: 105–112.

Cavalcanti FR, Resende MLV, Carvalho CPS, Silveira JAG, Oliveira JTA (2007) An aqueous suspension of *Crinipellis perniciosa* mycelium activates tomato defense responses against *Xanthomonas vesicatoria*. *Crop Protect* 26: 729–738.

Cavalcanti FR, Resende MLV, Ribeiro Junior PM, Pereira RB, Oliveira JTA (2008) Induction of resistance against *Verticilliium dahliae* in cacao by a *Crinipellis perniciosa* suspension. *J Plant Pathol* 90: 273–280.

Cha J-Y, Han S, Hong H-J et al. (2016) Microbial and biochemical basis of Fusarium wilt suppressive soil. *Internatl Soc Microbiol Ecol (ISME)* J 10: 119–129.

Chandel S, Allan EJ, Woodward S (2010) Biological control of *Fusarium oxysporum* f.sp. *lycopersici* in tomato by *Brevibacillus brevis*. *J Phytopathol* 158: 470–478.

Chen L-H, Cui Y-Q, Yang X-M, Zhao D-K, Shen Q-R (2012) An antifungal compound from *Trichoderma harzianum* SQR-T037 effectively controls Fusarium wilt of cucumber in continuously cropped soil. *Austr Plant Pathol* 41: 239–245.

Chen L-H, Yang X, Raza W, Li J, Liu Y, Qiu M, Zhang F, Shen Q (2011) *Trichoderma harzianum* SQR-T037 rapidly degrades allelochemicals in rhizospheres of continuously cropped cucumbers. *Appl Microbiol Biotechnol* 89: 1653–1663.

Chen M-H, Nelson EB (2012) Microbial-induced carried competition in the spermosphere leads to pathogen and disease suppression in municipal biosolids compost. *Phytopathology* 102: 588–596.

Chet I, Chernin L (2002) Biocontrol microbial agents in soil. In: Bitton G (ed.), *Encyclopedia of Environmental Microbiology*, Volume 1, Wiley, New York, pp. 450–465.

Chin-A-Woeng TFC, Thomas-Oates JE, Lugtenberg BJJ, Bloemberg GV (2001) Introduction of the *phzH* gene of *Pseudomonas chlororaphis* PCL1391 extends the range of biocontrol activity of phenazine-1-carboxylic acid-producing *Pseudomonas* spp. strains. *Molec Plant-Microbe Interact* 14: 1006–1015.

Chitrampalam P, Wu BM, Koike ST, Subbarao KV (2011) Interactions between *Coniothyrium minitans* and *Sclerotinia minor* affect biocontrol efficacy of *C. minitans*. *Phytopathology* 101: 358–366.

Chung W-C, Wu R-S, Hsu C-P, Huang H-C, Huang JW (2011) Application of antagonistic rhizobacteria for control of Fusarium seedling blight and basal rot of lily. *Austr Plant Pathol* 40: 269.

Clarkson JP, Payne T, Mead A, Whipps JM (2001) Selection of fungal biological control agents of *Sclerotium cepivorum* for control of white rot by sclerotial degradation in UK soil. *Plant Pathol* 51: 735–745.

Colburn GC, Graham JH (2007) Protection of citrus rootstocks against *Phytophthora* spp. with a hypovirulent isolate of *Phytophthora nicotianae*. *Phytopathology* 97: 958–963.

Cook RJ (1987) Research briefing panel on biological control in managed ecosystems. Committee of Science, Engineering and Public Policy. *Natl Acad Eng Inst Medicine*, Natl Acad Press, Washington, DC, 12 pp.

Coons GH, Kotila JE (1925) The transmissible lytic principle (bacteriophage) in relation to plant pathogens. *Phytopathology* 15: 357–370.

Coram TE, Pang ECK (2006) Expression profiling of chickpea genes differentially regulated during a resistance response to *Ascochyta rabiei*. *Plant Biotechnol J* 4: 647–666.

Cordier C, Gianinazzi S, Gianinazzi-Pearson V (1998) Colonization pattern of root tissues by *Phytophthora nicotianae* var. *parasitica* in tomato by an arbuscular mycorrhizal fungus. *Molec Plant-Microbe Interact* 11: 1017–1028.

Coşkuntuna A, Özer N (2008) Biological control of onion basal rot disease using *Trichoderma harzianum* and induction of antifungal compounds on onion set following seed treatment. *Crop Protect* 27: 330–336.

Coventry E, Noble R, Mead A, Marin FR, Perez JA, Whipps JM (2006) Allium white rot suppression with composts and *Trichoderma viride* in relation to sclerotia viability. *Phytopathology* 96: 1009–1020.

Cummings JA, Miles CA, du Toit LJ (2009) Greenhouse evaluation of seed and drench treatments for organic management of soilborne pathogens of spinach. *Plant Dis* 93: 1281–1292.

D'aes J, Hua GKH, De Meyer K et al. (2011) Biological control of Rhizoctonia root rot on bean by phenazine- and cyclic lipopeptides- producing *Pseudomonas* CMR12a. *Phytopathology* 101: 996–1004.

Daval S, Lebreton L, Gazengel K, Boutin M, Guillerm-Erckelboudt A-Y, Sarniguet A (2011) The biocontrol bacterium *Pseudomonas fluorescens* Pf29Arp strain affects the pathogenesis-related gene expression of the take-all *Gaeumannomyces graminis* var. *tritici* on wheat roots. *Molec Plant Pathol* 12: 839–854.

Davis JR, Huisman OC, Everson DO, Nolte P, Sorenson LH, Schneider AT (2010) Ecological relationships of Verticillium wilt of potato using corn as a green manure crop. *Amer J Potato Res* 87: 195–208.

Dawe AL, Nuss DL (2001) Hypoviruses and chestnut blight: exploiting viruses to understand and modulate fungal pathogens. *Annu Rev Genet* 35: 1–29.

Debode J, De Meyer K, Perneel M, Pannecoucque J, De Backer G, Höfte M (2007) Biosurfactants are involved in the biological control of *Verticillium* microsclerotia by *Pseudomonas* spp. *J Appl Microbiol* 103: 1184–1196.

de Boer M, Bom P, Kindt F et al. (2003) Control of Fusarium wilt of radish by combining *Pseudomonas putida* strains that have different disease-suppressive mechanisms. *Phytopathology* 93: 626–632.

de Cal A, Garcia-Lepe R, Melgarejo P: (2000) Induced resistance by *Penicillium oxalicum* against *Fusarium oxyporum* f.sp. *lycopersici*: histological studies of infected and induced tomato stems. *Phytopathology* 90: 260–268.

de Cal A, Pascual S, Melgarjejo P (1997) Involvement of resistance induction by *Penicillium oxalicum* in the biocontrol of tomato wilt. *Plant Pathol* 46: 72–79.

De La Fuente L, Thomashow L, Weller D et al. (2004) *Pseudomonas fluorescens* UP61 isolated from birdsfoot trefoil rhizosphere produces multiple antibiotics and exerts a broad-spectrum of biocontrol activity. *Eur J Plant Pathol* 110: 671–681.

De Meyer G, Bigirimana J, Elad Y, Höfte M (1998) Induced systemic resistance in *Trichoderma harzianum* T39 biocontrol of *Botrytis cinerea*. *Eur J Plant Pathol* 104: 279–286.

de Souza JT, Arrnould C, Deuvlot C, Lemanceau P, Gianinazzi-Pearson V, Raaijmakers JM (2003a) Effect of 2,4-diacetylphloroglucinol on *Pythium*: Cellular responses and variation in sensitivity among propagules and species. *Phytopathology* 93: 966–975.

de Souza JT, de Boer M, de Waard P, van Beek TA, Raaijmakers JM (2003b) Biochemical, genetic and zoosporicidal properties of cyclic lipopeptides surfactants produced by *Pseudomonas fluorescens*. *Appl Environ Microbiol* 69: 7161–7172.

de Souza JT, Weller DM, Raaijmakers JM (2003c) Frequency, diversity and activity of 2,4-diacetylphloroglucinol-producing *Pseudomonas* spp. in Dutch take-all decline soils. *Phytopathology* 93: 54–63.

Deberdt P, Davezies I, Coranson-Beaudu R, Jestin A (2018) Efficacy of leaf oil from *Pimentosa racemosa* var. *racemosa* in controlling bacterial wilt of tomato. *Plant Dis* 102: 124–131.

Deora A, Hatano E, Tahara S, Hashidoko Y (2010) Inhibitory effects of furanone metabolites of a rhizobacterium *Pseudomonas jessenii* on phytopathogenic *Aphanomyces cochlioides* and *Pythium aphanidermatum*. *Plant Pathol* 59: 84–99.

Diab HG, Hu S, Benson DM (2003) Suppression of *Rhizoctonia solani* on *Impatiens* by enhanced microbial activity in composted swine wash-amended potting mixes. *Phytopathology* 93: 1115–1123.

Diehl K, Resenburg P, Lentzsch P (2013) Field application of non-pathogenic *Verticillium dahliae* genotypes for regulation of wilt in strawberry plants. *Amer J Plant Sci* 4: 24–32.

Ding J, Shi K, Zhou YH, Yu JQ (2009) Microbial community responses associated with the development of *Fusarium oxysporum* f.sp. *cucumerinum* after 24-epibrassinolide application to shoots and roots in cucumber. *Eur J Plant Pathol* 124: 141–150.

Dionne A, Tweddell RJ, Antoun H, Avis TJ (2012) Effect of non-aerated compost teas on damping-off pathogens of tomato. *Canad J Plant Pathol* 34: 51–57.

Dodd JC, Boddington CL, Rodriguez A, Gonzalez-Chavez C, Mansur I (2000) Mycelium of arbuscular mycorrhizal fungi (AMF) from different genera form, function, and detection. *Plant Soil* 226: 131–151.

Dong XN (2004) NPR1, all things considered. *Curr Opin Plant Biol* 7: 547–552.

Doster MA, Cotty PJ, Michailides TJ (2014) Evaluation of the atoxigenic *Aspergillus flavus* strain AF36 in pistachio orchards. *Plant Dis* 98: 948–956.

Duffy BK, Défago G (1997) Zinc improves biocontrol of Fusarium crown and root rot of tomato by *Pseudomonas fluorescens* and represses production of pathogen metabolites inhibitory to bacterial antibiotic biosynthesis. *Phytopathology* 87: 1250–1257.

Duffy BK, Simon A, Weller DM (1996) Combination of *Trichoderma koningii* with fluorescent *Pseudomonas* for control of take-all on wheat. *Phytopathology* 86: 188–194.

Duijff BJ, Pouhair D, Olivain C, Alabouvette C, Lemanceau P (1998) Implication of systemic induced resistance in suppression of Fusarium wilt of tomato by *Pseudomonas fluorescens* WCS417r and by nonpathogenic *Fusarium oxysporum* Fo47. *Eur J Plant Pathol* 104: 903–910.

Dukare AS, Prasanna R, Dubey SC, Nain L, Chaudhary V (2011) Evaluating novel microbe-amended composts as biological agents in tomato. *Crop Protect* 30: 436–442.

Eastwell KC, Sholberg PL, Sayler RJ (2006) Characterizing potential bacterial biocontrol agents for suppression of *Agrobacterium vitis*, the causal agent of crown gall disease in grapevine. *Crop Protect* 25: 1191–1200.

Edgar CI, McGrath KC, Dombrecht B et al. (2006) Salicylic acid mediates resistance to the vascular wilt pathogen *Fusarium oxysporum* in the model host *Arabidopsis thaliana*. *Austr Plant Pathol* 35: 581–591.

Eikeno H, Stensvand A, Tronsmo AM (2003) Induced resistance as a possible means to control diseases of strawberry caused by *Phytophthora* spp. *Plant Dis* 87: 345–350.

Elad Y (2000) Biological control of foliar pathogens by means of *Trichoderma harzianum* and potential mode of action. *Crop Protect* 19: 709–714.

El-Hasan A, Walker F, Schöne J, Buchenauer H (2009) Detection of viridofungin A and other antifungal metabolites excreted by *Trichoderma harzianum* active against different plant pathogens. *Eur J Plant Pathol* 124: 457–470.

El-Khaldi R, Daami-Remadi M, Cherif M (2016) Biological control of stem canker and black scurf of potato by date palm compost and its associated fungi. *J Phytopathol* 164: 40–51.

Elmer WH, Pignatello JJ (2011) Effects of biochar amendments on mycorrhizal association and Fusarium crown rot and root rot of asparagus in replant soil. *Plant Dis* 95: 960–966.

El-Mohamedy RSR, Shafeek MR, Abd El-Samad EE, Salama DM, Rizak FA (2017) Field application of plant resistance inducers (PRIs) to control important root rot diseases and improvement in growth and yield of green bean (*Phaseolus vulgaris* L.). *Austr J Crop Sci* 11: 496–505.

El-Tarbily KA (2004) Suppression of *Rhizoctonia solani* diseases by sugar beet antagonistic and plant growth-promoting yeasts. *J Appl Microbiol* 96: 69–75.

Erdogan D, Benlioglu K (2010) Biological control of Verticillium wilt on cotton by the use of fluorescent *Pseudomonas* spp. under field conditions. *Biol Contr* 53: 39–45.

Erjavec J, Ravnikar M, Brzin J et al. (2016) Antibacterial activity of wild mushroom extracts on bacterial wilt pathogen *Ralstonia solanacearum*. *Plant Dis* 100: 453–464.

Errakhi R, Bouteau F, Lebrihi A, Barakate M (2007) Evidences for biological control capacities of *Streptomyces* spp. against *Sclerotium rolfsii* responsible for damping-off disease in sugar beet (*Beta vulgaris* L.). *World J Microbiol Biotechnol* 23: 1503–1509.

Errakhi R, Lebrihi A, Barakate M (2009) In vitro and in vivo antagonism of actinomycetes isolated from Moroccan rhizospherical soils against *Sclerotium rolfsii*: A causal agent of root rot on sugar beet (*Beta vulgaris* L.). *J Appl Microbiol* 107: 672–681.

Ezziyyani M, Requena ME, Egen-Gilabert C, Candela ME (2007) Biological control of Phytophthora root rot of pepper using *Trichoderma harzianum* and *Streptomyces rochei* in combination. *J Phytopathol* 155: 342–349.

Faessel L, Nasser N, Lebeau T, Walter BC (2008) Effects of a plant defense inducer, acibenzolar-S-methyl on hypocotyl rot of soybean caused by *Rhizoctonia solani* AG-4. *J Phytopathol* 156: 236–242.

Fakhouri WD, Khalaif H (1996) Biocontrol of crown gall disease in Jordan. *Disrat Ser B, Pure Appl Sci* 23: 17–22.

Farag SMA, Elhalag KMA, Mohamed H et al. (2017) Potato bacterial wilt suppression and plant health improvement after application of different antioxidants. *J Phytopathol* 165: 522–537.

Fernando WGD, Nakkeeran S, Zhang Y (2005) Biosynthesis of antibiotics by plant growth-promoting rhizobacteria and its relation in biocontrol of plant diseases. In: Siddiqui ZA (ed.), *PGPR Biocontrol and Biofertilizers*, Springer, Germany, pp. 67–109.

Fernando WGD, Nakkeeran S, Zhang Y, Savchuk S (2007) Biological control of *Sclerotinia sclerotiorum* (Lib.) de Bary by *Pseudomonas* and *Bacillus* species on canola petals. *Crop Protect* 26: 100–107.

Ferrigo D, Causin R, Raiola A (2017) Effect of potential biocontrol agents selected among grapevine endophytes and commercial products on crown gall disease. *BioControl* 62: 821–833.

Filion M, St-Arnaud M, Jabaji-Hare SH (2003) Quantification of *Fusarium solani* f.sp. *phaseoli* in mycorrhizal bean plants and surrounding mycorrhizosphere soil using real-time polymerase chain reaction and direct isolation on selective media. *Phytopathology* 93: 229–235.

Fortunato AA, Rodrigues FA, Baroni JCP, Soares GCB, Rodriguez MAD, Pereira OL (2012a) Silicon suppresses Fusarium wilt development in banana plants. J Phytopathol.

Fortunato AA, Rodriguez FA, do Nascimento KJT (2012b) Physiological and biochemical aspects of resistance of banana plants to Fusarium wilt potentiated by silicon. *Phytopathology* 102: 957–966.

Fravel DR, Deahl KL, Stommel JR (2005) Compatibility of biocontrol fungus *Fusarium oxysporum* strain CS−20 with selected fungicides. *Biol Contr* 34: 1654–169.

Freeman S, Zuveibil A, Vintal H, Mayon M (2002) Isolation of nonpathogenic mutants of *Fusarium oxysporum* f.sp. *melonis* for biological control of Fusarium wilt in cucurbits. *Phytopathology* 92: 164–168.

Fujiwara A, Fujisawa M, Hamasaki R, Kawasaki T, Fujie M, Yamada (2011) Biocontrol of *Ralstonia solanacearum* by treatment with lytic bacteriophages. *Appl Environ Microbiol* 77: 4155–4162.

Fujiwara A, Kawasaki T, Usami S, Fujie M, Yamada T (2008) Genomic characterization of *Ralstonia solanacearum* phage ORSA1 and its related prophage (ØRSX) in strain GM11000. *J Bacteriol* 190: 143–156.

Funhashi F, Parke JL (2016) Effects of soil solarization and *Trichoderma asperellum* on soilborne inoculum of *Phytophthora ramorum* and *Phytophthora pini* in container nurseries. *Plant Dis* 100: 438–443.

Gallo A, Mulé G, Favilla M, Altomare C (2004) Isolation and characterization of a trichodiene synthase homologous gene in *Trichoderma harzianum*. *Physiol Molec Plant Pathol* 65: 11–24.

Gallou A, Cranebrouck S, Declerck S (2009) *Trichoderma harzianum* elicits defense response genes in roots of defence genes. *Plant Pathol* 61: 281–288.

Gao K, Liu Z, Mendgen K (2005) Mycoparasitism of *Rhizoctonia solani* by endophyte *Chaetomium spirale* ND35: Ultrastructure and cytochemistry of interaction. *J Phytopathol* 153: 289–290.

Gerlagh M, Goosen-van de Geijn HM, Hoogland AE, Vereijken PFG (2003) Quantity aspects of infection of *Sclerotinia sclerotiorum* sclerotia by *Coniothyrium minitans*–Timing of application, concentration, and quality of conidial suspension of the mycoparasite. *Eur J Plant Pathol* 109: 489–502.

Ghabrial SA (1998) Origin, adaptation and evolutionary pathway of fungal viruses. *Virus Genes* 16: 119–131.

Giachero ML, Marquez N, Gallou A, Luna CM, Declerck S, Ducasse DA (2017) An in vitro method for studying the three-way interaction between soybean, *Rhizophagus irregularis*, and the soilborne pathogen *Fusarium virguliforme*. Front Plant Sci.

Giotis C, Markelou E, Theodoropoulous A et al. (2009) Effect of soil amendments and biological control agents (BCAs) on soilborne root diseases caused by *Pyrenochaeta lycopersici* and *Verticillium albo-atrum* in organic greenhouse tomato production system. *Eur J Plant Pathol* 123: 387–400.

Gizi D, Stringlis JA, Tjamos SE, Paplomatas EJ (2011) Seedling vaccination by stem-injecting a conidial suspension of F2, a nonpathogenic *Fusarium oxysporum* strain suppresses Verticillium wilt of eggplant. *Biol Contr* 58: 382–392.

Goates BJ, Mercier J (2011) Control of common bunt of wheat under field conditions with a biofumigant fungus *Muscodor albus*. *Eur J Plant Pathol* 131: 403–407.

Grimme E, Zidack NK, Sikora RA, Strobel GA, Jacobsen BJ (2007) Comparison of *Muscodor albus* volatiles with a biorational mixture for control of seedling diseases of sugar beet and root knot nematode in tomato. *Plant Dis* 91: 220–225.

Grondona I, Hermosa R, Tejada M et al. (1997) Physiological and biochemical characterization of *Trichoderma harzianum*, biological control agent against soilborne fungal plant pathogens. *Appl Environ Microbiol* 68: 3189–3198.

Haas D, Défago G (2005) Biological control of seedborne pathogens by fluorescent pseudomonads. Nat Rev Microbiol.

Haciasalihoglu G, Ji P, Longo LM, Olson S, Momol TM (2007) Bacterial wilt-induced changes in nutrient distribution and biomass and the effect of acibenzolar-S-methyl on bacterial wilt in tomato. *Crop Protect* 26: 978–982.

Hage-Ahmed K, Moyses A, Voglgruber A, Hadacek F, Steinkellner S (2013) Alterations in root exudation of intercropped tomato mediated by the arbuscular mycorrhizal fungus *Glomus mosseae* and the soilborne pathogen *Fusarium oxysporum* f.sp. *lycopersici*. *J Phytopathol* 161: 763–773.

Han SH, Lee SJ, Moon JH et al. (2006) GacS-dependent production of 2R-3R butanediol by *Pseudomonas chlororaphis* 06 is a major determinant for eliciting systemic resistance against *Erwinia carotovora*, but not against *Pseudomonas syringae* pv. *tabaci*. *Molec Plant-Microbe Interact* 19: 924–930.

Hanson LE, Howell CR (2004) Elicitors of plant defense responses from biocontrol strains of *Trichoderma virens*. *Phytopathology* 94: 171–176.

Harel YM, Elad Y, Rav-David D et al. (2012) Biochar mediates systemic response of strawberry to foliar fungal pathogens. *Plant Soil* 357: 245–257.

Harman GE, Petzoldt R, Comis A, Chen J (2004) Interactions between *Trichoderma harzianum* strain T22 and maize inbred line Mo17 and effects of these interactions on disease caused by *Pythium ultimum* and *Colletotrichum graminicola*. *Phytopathology* 94: 147–153.

He CY, Hsiang T, Wolyn DJ, (2002) Induction of systemic disease resistance and pathogen defence responses in *Asparagus officinalis* inoculated with nonpathogenic strains of *Fusarium oxysporum*. *Plant Pathol* 51: 225–230.

He CY, Wolyn DJ (2005) Potential role of salicylic acid in induced resistance of asparagus roots to *Fusarium oxysporum* f.sp. *asparagi*. *Plant Pathol* 54: 227–232.

Hemissi I, Mabrouk Y, Mejri S, Saidi M, Sifi B (2013) Enhanced defence responses of chickpea plants against *Rhizoctonia solani* by preinoculation with rhizobia. *J Phytopathol* 161: 412–418.

Hermosa R, Viterbo A, Chet I, Monte E (2012) Plant-beneficial effects of *Trichoderma* and of its genes. *Microbiology* 158: 17–25.

Horinouchi H, Muslim A, Suzuki T, Hyakumachi M (2007) *Fusarium equiseti* GF191 as an effective biocontrol agent against Fusarium crown and root rot of tomato in rockwool systems. *Crop Protect* 26: 1514–1523.

Howell CR (1987) Relevance of mycoparasitism as biological control of *Rhizoctonia solani* by *Gliocladium virens*. *Phytopathology* 77: 992–994.

Howell CR (2003) Mechanisms employed by *Trichoderma* species in the biological control of plant disease: The history and evolution of current concepts. *Plant Dis* 87: 4–10.

Howell CR, Hanson LE, Stipanovic RD, Puchkhaber LS (2000) Induction of terpenoid synthesis in cotton roots and control *Rhizocotonia solani* by seed treatment with *Trichoderma virens*. *Phytopathology* 90: 248–252.

Howell CR, Stipanovic RD (1995) Mechanisms of biocontrol of *Rhizoctonia solani*-induced cotton seedling disease by *Gliocladium virens*: Antibiosis. *Phytopathology* 85: 469–472.

Hu X, Roberts DP, Xie L et al. (2015) Components of a rice-oilseed rape production system augmented with *Trichoderma* sp. Tri-1 control *Sclerotinia sclerotiorum* on oilseed rape. *Phytopathology* 105: 1325–1333.

Huang C-H, Roberts PD, Datnoff LE (2011) Silicon suppresses Fusarium crown and root rot of tomato. *J Phytopathol* 159: 546–554.

Huang HC (1977) Importance of *Coniothyrium minitans* in survival of sclerotia of *Sclerotinia sclerotiorum* in wilted sunflower. *Canad J Bot* 55: 289–295.

Huang HC, Erickson RS (2007) Effect of seed treatment with *Rhizobium leguminosarum* on Pythium damping-off seedling height, root nodulation, root biomass, shoot biomass and seed yield of pea and lentil. *J Phytopathol* 155: 31–37.

Huang J, Li H, Yuan H (2006) Effect of organic amendments on Verticillium wilt of cotton. *Crop Protect* 25: 1167–1173.

Huang X, Cui H, Yang L, Lan T, Zhang J, Cai Z (2017) The microbial changes during the biological control of cucumber damping-off disease using biocontrol agents and reductive soil disinfestation. *BioControl* 62: 97–100.

Huang X, Liu L, Wen T, Zhang J, Cai Z (2016a) Changes in the soil microbial community after reductive soil disinfestation and cucumber cultivation. *Appl Microbiol Biotechnol* 100 : 5581–5593.

Huang X, Liu L, Wen T, Zhang J, Shen Q, Cai Z (2016b) Reductive soil disinfestations combined or not with *Trichoderma* for the treatment of a degraded and *Rhizoctonia solani*-infested greenhouse soil. *Sci Hortic* 206: 51–61.

Huang Y, Xie X, Yang L, Zhang J, Li G, Jiang D (2011) Susceptibility of *Sclerotinia sclerotiorum* strains different in oxalate production to infection by the mycoparasite *Coniothyrium minitans*. *World J Microbiol Biotechnol* 27: 2799–2805.

Hultberg M, Alsberg T, Khalil S, Alsanius B (2010a) Suppression of disease in tomato infected by *Pythium ultimum* with a biosurfactant produced by *Pseudomonas koreensis*. *BioControl* 55: 435–440.

Hultberg M, Bengtsson T, Liljeroth E (2010b) Late blight on potato is suppressed by the surfactant-producing *Pseudomonas koreensis* 2.74 and its biosurfactant. *BioControl* 55: 543–550.

Hultberg M, Holmkvist A, Alsanius B (2011) Strategies for administration of biosurfactant-producing pseudomonads for biocontrol in closed hydroponic systems. *Crop Protect* 30: 995–999.

Huss B, Walter B, Fuchs M (1989) Cross-protection between *Arabis mosaic virus* and *Grapevine fan leaf virus* isolates in *Chenopodium quinoa*. *Ann Appl Biol* 114: 45–60.

Hwang J, Benson DM (2002) Biocontrol of *Rhizoctonia* stem and root rot of poinsettia with *Burkholderia cepacia* and binucleate *Rhizoctonia*. *Plant Dis* 86: 47–53.

Ikeda S, Shimizu A, Shimizu M, Takahashi H, Takenaka S (2012) Biocontrol of black scurf on potato by seed tuber treatment with *Pythium oligandrum*. *Biol Contr* 60: 297–304.

Imperiali N, Dennert F, Schneider J et al. (2017) Relationship between root pathogen resistance, abundance and expression of *Pseudomonas* antimicrobial genes, and soil properties in representative Swiss agricultural soils. Front Plant Sci.

Inderbitzin P, Ward J, Barbella A et al. (2018) Soil microbiomes associated with Verticillium wilt-suppressive broccoli and chitin amendments are enriched with potential biocontrol agents. *Phytopathology* 108: 31–43.

Iriarte FB, Obradovic A, Wernsing MH et al. (2012) Soil-based systemic delivery and phyllosphere in vivo propagation of bacteriophage: two possible strategies for improving bacteriophage efficacy for plant disease control. *Bacteriophage* 2: 214–224.

Islam MT, Fukushi Y (2010) Growth inhibition and excessive branching in *Aphanomyces cochlioides* induced by 2,4-diacetylphloroglucinol is linked to disruption of filamentous actin cytoskeleton in the hyphae. *World J Microbiol Biotechnol* 26: 1163–1170.

Isnaini M, Keane PJ (2007) Biocontrol and epidemiology of lettuce drop caused by *Sclerotinia minor* at Baccus Marsh, Victoria. Austr *Plant Pathol* 36: 95–304.

Jabaji-Hare S, Neate SM (2005) Nonpathogenic binucleate *Rhizoctonia* spp. and benzothiadiazole protect cotton seedlings against Rhizoctonia damping-off and Alternaria leaf spot in cotton. *Phytopathology* 95: 1030–1036.

Jabnoun-Khiareddine H, Abdallah R, El-Mohamedy R et al. (2016) Comparative efficacy of potassium salts against soilborne and airborne fungi and their ability to suppress tomato wilt and fruit rots. J Microbiol Biochem Technol.

Jabnoun-Khiareddine H, El-Mohamedy RSR, Abdel-Kareem F et al. (2015) Variation in chitosan and salicylic acid efficacy towards soilborne and airborne fungi and their suppressive effect on tomato wilt severity. *J Plant Pathol Microbiol* 6org 110.4172/2157-7471.1000325

Jacob S, Sajjalaguddam RR, Kumar KVK, Varshney R, Sudini HR (2016) Assessing the prospects of *Streptomyces* sp. RP1A-12 in managing groundnut stem rot disease caused by *Sclerotium rolfsii* Sacc. *J Gen Plant Pathol* 82: 96–104.

Jaime MDLA, Hsiang T, McDonald MR (2005) Effects of *Glomus intraradices* and onion cultivar on *Allium* white rot development in organic soils in Ontario. *Canad J Plant Pathol* 30: 543–553.

Jaiswal AK, Elad Y, Graber ER, Frenkel O (2014) *Rhizoctonia solani* suppression on and plant growth promotion in cucumber as affected by biochar pyrolysis temperature, feedstock and concentrations. *Soil Biol Biochem* 69: 110–118.

Jaiswal AK, Elad Y, Paudel I, Graber ER, Cytryn E, Frenkel O (2017) Linking below ground microbial composition, diversity and activity to soilborne disease suppression and growth promotion of tomato amended with biochar. *Sci Reports* 7: 44382

Jalali B, Bhargava S, Kamble A (2006) Signal transduction and transpirational regulations of plant defense responses. *J Phytopathol* 154: 65–74.

Jayaprakashvel M, Selvakumar M, Srinivasan K, Ramesh S, Mathivanan N (2010) Control of sheath blight in rice by thermostable secondary metabolites of *Trichoderma roseum* MML003. *Eur J Plant Pathol* 126: 229–239.

Jeffery S, Verheijn FGA, van der Velde M, Bastos AC (2011) A quantitative review of effects of biochar application to soils on crop productivity using meta-analysis. *Agric Ecosyst Environ* 144: 175–187.

Ji P, Momol MT, Olson SM, Pradhanang PM, Jones JB (2005) Evaluation of thymol as biofumigant for control of bacterial wilt of tomato under field conditions. *Plant Dis* 89: 497–500.

Johansson PM, Wright SAI (2003) Low-temperature isolation of disease suppressive bacteria and characterization of a distinctive group of pseudomonads. *Appl Environ Microbiol* 69: 6464–6474.

John RP, Tyagi RD, Prévost D, Brar SK, Poleur S, Surampalli RY (2010) Mycoparasite *Trichoderma viride* as a biocontrol agent against *Fusarium oxysporum* f.sp. *adzuki* and *Pythium arrhenomanes* and as growth promoter of soybean. *Crop Protect* 29: 1452–1459.

Jones EE, Clarkson JP, Mead A, Whipps JM (2004) Effect of inoculum type and timing of application of *Coniothyrium minitans* on *Sclerotinia sclerotiorum* influence on apothecial production. *Plant Pathol* 53: 621–628.

Jones EE, Whipps JM (2002) Effects of inoculum rates and sources of *Coniothyrium minitans* on control of *Sclerotinia sclerotiorum* disease in glasshouse lettuce. *Eur J Plant Pathol* 108: 527–538.

Jousset A, Rochat L, Lanoue A, Bonkowski M, Keel C, Scheu S (2011) Plants respond to pathogen infection by enhancing the antifungal gene expression of root-associated bacteria. *Molec Plant-Microbe Interact* 24: 352–358.

Kamal MM, Linbeck KD, Savocchia S, Ash GJ (2015) Biological control of Sclerotinia stem rot of canola using antagonistic bacteria. *Plant Pathol* 64: 1375–1384.

Kanematsu S, Arakawa M, Oikawa Y et al. (2004) A *Reovirus* causes hypovirulence of *Rosellinia necatrix*. *Phytopathology* 94: 561–568.

Kanematsu S, Sasaki A, Onoue M, Oikawa Y, Ito T (2010) Extending the fungal host range of a *Partitivirus* and a *Mycoreovirus* from *Rosellinia necatrix* by inoculation of protoplasts with virus particles. *Phytopathology* 100: 922–930.

Kang D-S, Min K-J, Kwak A-M, Lee S-Y, Kang H-W (2017) Defense response and suppression of Phytophthora blight disease of pepper by water extract from spent mushroom substrate of *Lentinula edodes*. *Plant Pathol J* 33: 264–275.

Kanini GS, Katsifas EA, Savvides AL, Hatzinikolaou DG, Karagouni AD (2013) Greek indigenous streptomycetes as biological control agents against soilborne fungal plant pathogen *Rhizoctonia solani*. *Appl Microbiol* 114: 1468–1479.

Karimi E, Sadeghi A, Dehaji PA, Dalvand Y, Omidvari M, Nezhad MK (2012) Biocontrol activity of salt tolerant *Streptomyces* isolates against phytopathogens causing root rot of sugar beet. *Biocontr Sci Technol* 22: 333–349.

Karthikeyan SRP, Manikandan R, Durgadevi D et al. (2017) Biosuppression of turmeric rhizome rot disease and understanding the molecular basis of tripartite interaction among *Cucurcuma longa*, *Pythium aphanidermatum* and *Pseudomonas fluorescens*. *Biol Contr* 111: 23–31.

Kavroulakis N, Ntougias S, Zervakis GI, Ehaliotis C, Haralampidis K, Papadopoulou KK (2007) Role of ethylene in the protection of tomato plants against soilborne fungal pathogens conferred by an endophytic *Fusarium solani*. *J Expt Bot* 58: 3853–3864.

Kawaguchi A, Inoue K (2012) New antagonistic strains of nonpathogenic *Agrobacterium vitis* to control grapevine crown gall. *J Phytopathol* 160: 509–518.

Kawaguchi A, Inoue K, Ichinose Y (2008a) Biological control of crown gall of grapevine, rose and tomato by nonpathogenic *Agrobacterium vitis* to control grapevine crown gall. *J Phytopathol* 160: 509–518.

Kawaguchi A, Inoue K, Ichinose Y (2008b) Biological control of crown gall of grapevine, rose and tomato by nonpathogenic *Agrobacterium vitis* strain VAR03-1. *Phytopathology* 98: 1218–1225.

Kawaguchi A, Inoue K, Nasu H (2007) Biological control of grapevine crown gall by nonpathogenic *Agrobacterium vitis* strain VAR03-1. *J Gen Plant Pathol* 73: 133–138.

Kawaguchi A, Inoue K, Tanina K (2013) Evaluation of the nonpathogenic *Agrobacterium vitis* strain ARK-1for crown gall control in diverse plant species. *Plant Dis* 99: 409–414.

Kawasaki T, Nagata S, Fujisawara A et al. (2007) Genomic characterization of filamentous integrative bacteriophage ØRSS1 and ØRSM1 which infect *Ralstonia solanacearum*. *J Bacteriol* 189: 5792–5802.

Kawasaki Y, Nischwitz C, Grilley MG, Jones J, Brown JD, Takemotoa JY (2016) Production and application of syringomycin E as an organic fungicide seed protectant against Pythium damping-off. *J Phytopathol* 164: 801–810.

Kawasaki T, Shimizu M, Satsuma H et al. (2009) Genomic characterization of *Ralstonia solancearum* phage ØRSB1, a T-7-like wide host range phage. *J Bacteriol* 191: 422–427.

Khan FU, Nelson BD, Helms T (2005) Greenhouse evaluation of binucleate *Rhizoctonia* for control of *Rhizoctonia solani* in soybean. *Plant Dis* 89: 373–379.

Khan J, Ooka JS, Miller SA, Madden LV, Hoitnik HA (2004) Systemic resistance induced by *Trichoderma harzianum* 382 in cucumber against Phytophthora crown rot and leaf blight. *Plant Dis* 88: 280–288.

Khare E, Arora NK (2011) Dual activity of pyocyanin from *Pseudomonas aeruginosa* antibiotic against phytopathogen and signal molecule for biofilm development by rhizobia. *Canad J Microbiol* 57: 708–713.

Khastini RO, Ogawara T, Sato Y, Narisawa K (2014) Control of Fusarium wilt in melon by the fungal endophyte *Cadophora* sp. *Eur J Plant Pathol* 139: 339–348.

Khmel IA, Sorokino TA, Lamanova NB et al. (1998) Biological control of crown gall in grapevine and raspberry by two *Pseudomonas* species with wide spectrum of antagonistic activity. *Biocontr Sci Technol* 8: 45–57.

Kim DS, Cook RJ, Weller DM (1997) *Bacillus* sp. 324-92 for biological control of three root diseases of wheat grown with reduced tillage. *Phytopathology* 87: 551–558.

Kim H-S, Sang MK, Jeun Y-C, Hwang BK, Kim KD (2008) Sequential selection and efficacy of antagonistic rhizobacteria for controlling Phytophthora blight of pepper. *Crop Protect* 27: 436–443.

Kim SG, Khan Z, Jeon YH, Kim YH (2009) Inhibitory effect of *Paenibacillus polymyxa* GBR-462 on *Phytophthora capsici*, causing Phytophthora blight in chilli pepper. *J Phytopathol* 157: 329–337.

Kim TG, Knudsen GR (2009) Colonization of *Sclerotinia sclerotiorum* by a biocontrol isolate of *Trichoderma harzianum* and effects on myceliogenic germination. *Biocontr Sci Technol* 19: 1081–1085.

Kim TG, Knudsen GR (2011) Comparison of real-time PCR and microscopy to evaluate sclerotial colonization by a biocontrol fungus. *Fungus Biol* 115: 317–325.

Kiprushkina EJ, Shestopalova LA, Pekhotina AM, Nikitina OV (2017) Protective stimulator properties of chitosan in the vegetation and strong tomatoes. *Prog Chem Appl Chitin Deriv* 23: 73–81.

Klein E, Katan J, Gamliel A (2011) Soil suppressiveness to Fusarium disease following organic amendments and solarization. *Plant Dis* 95: 1116–1123.

Klein E, Ofek M, Katan J, Minz D, Gamliel A (2013) Soil suppressiveness to Fusarium disease: Shifts in root microbiome associated with reduction of pathogen root colonization. *Phytopathology* 103: 23–33.

Klironomos JN (2003) Variation in plant response to native and exotic arbuscular mycorrhizal fungi. *Ecology* 84: 2292–2301.

Kloepper JW, Leong J, Teintze M, Schroth MN (1980) *Pseudomonas* siderophores: a mechanism explaining disease suppressive soils. *Curr Microbiol* 4: 317–320.

Knox OGG, Oghoro CO, Burnett FJ, Fountaine JM (2015) Biochar increases soil pH, but is as ineffective as lime at controlling clubroot. *J Plant Pathol* 97: 149–152.

Kobayashi Y, Kobayashi A, Maeda M, Takenaka S (2012) Isolation of antagonistic *Streptomyces* sp. against a potato scab pathogen for field-cultivated wild oat. *J Gen Plant Pathol* 78: 62–72.

Komar V, Vigne E, Demaget G, Lemaire O, Fuchs M (2008) Cross-protection as control strategy against *Grapevine fan leaf virus* in naturally infected vineyards. *Plant Dis* 92: 1689–1694.

Kotasthane A, Agrawal T, Kushwah R, Rahatkar OV (2015) In vitro antagonism of *Trichoderma* sp. against *Sclerotium rolfsii* and *Rhizoctonia solani* and their response towards growth of cucumber, bottlegourd and bittergourd. *Eur J Plant Pathol* 141: 523–543.

Kotila JE, Coons GH (1925) Investigations on the black leg disease of potato. *Michigan Agric Exp Sta Tech Bill* 67: 3–29.

Kubicek CP, Herrera-Estrella A, Seidi-Seiboth V et al. (2011) Comparative genome sequences analysis underscored mycoparasitism as the ancestral life style of *Trichoderma*. *Genome* 12: R40.

Kurze S, Bahl H, Dahl R, Berg G (2001) Biological control of fungal strawberry diseases by *Serratia plymuthica* HRO-C48. *Plant Dis* 85: 529–534.

L'Haridon F, Aimé S, Alabouvette C, Olivain C (2007) Lack of biocontrol capacity in a nonpathogenic mutant of *Fusarium oxysporum* f.sp. *melonis*. *Eur J Plant Pathol* 118: 239–246.

Laflamme P, Benhamou N, Bussieres G, Desseureault M (1999) Differential effect of chitosan on root rot fungal pathogens in forest nurseries. *Canad J Bot* 77: 1460–1468.

Lafontaine JP, Benhamou N (1996) Chitosan treatment: an emerging strategy for enhancing resistance of greenhouse tomato plants to infection by *Fusarium oxysporum* f.sp. *radicis-lycopersici*. *Biocontr Sci Technol* 6: 111–124.

Lahlali R, Hijiri M (2010) Screening, identification, and evaluation of biocontrol fungal endophytes against *Rhizoctonia solani* AG3 on potato plants. *FEMS Microbiol Lett* 311: 152–159.

Lahlali R, Peng G, Gossen BD et al. (2013) Evidence that the biofungicide Serenade (*Bacillus subtilis*) suppresses clubroot on canola via antibiosis and induced systemic resistance. *Phytopathology* 103: 245–254.

Larkin RP, Fravel DR (1998) Efficacy of various fungal and bacterial biocontrol organisms for control of Fusarium wilt of tomato. *Plant Dis* 82: 1022–1028.

Larkin RP, Fravel DR (2002) Effects of varying environmental conditions on biological control of Fuarium wilt of tomato by nonpathogenic *Fusarium* spp. *Phytopathology* 92: 1160–1166.

Larsen J, Graham JH, Cubero J, Ravinskov S (2012) Biocontrol traits of plant growth suppressive arbuscular mycorrhizal fungi against root rot in tomato caused by *Pythium aphanidermatum*. *Eur J Plant Pathol* 133: 361–369.

Leclerc V, Brechet M, Adam A, Guez JS, Wathelet B, Ongena M, Thonart P, Gancel F, Chollect-Imbert M, Jacques P (2005) Mycosubtilin overproduction by *Bacillus subtilis* BBG100 enhances the organisms antagonistic and biocontrol activities. *Appl Environ Microbiol* 71: 4577–4584.

Lee J, Kim S, Park TH (2017) Diversity of bacteriophage infecting *Pectobacterium* from potato fields. *J Plant Pathol* 99: 453–460.

Legin R, Bass P, Etianne L, Fuchs M (1993) Selection of mild virus strains of fan leaf degeneration by comparative field performance of infected grapevines. *Vitis* 32: 103–110.

Lehmann J, Joseph S (2009) *Biochar for Environment Management: Science and Technology*, Routledge, London, UK.

Li GD, Huang HC, Arysa SN, Erickson RS (2005) Efficiency of *Coniothyrium minitans* and *Trichoderma atroviride* in suppression of Sclerotinia blossom blight of alfalfa. *Plant Pathol* 54: 204–211.

Li I, Ma J, Li Y, Wang Z, Gao T, Wang Q (2012) Screening and partial characterization of *Bacillus* with potential applications in biocontrol of cucumber Fusarium wilt. *Crop Protect* 35: 29–35.

Lin F, Liu N, Lai D et al. (2017) A formulation of neem cake seeded with *Bacillus* sp. provides control over tomato Fusarium crown and root rot. *Biocontr Sci Technol* 27: 393–407.

Liu HM, He YJ, Jiang HX et al. (2007) Characterization of a phenazine-producing strain *Pseudomonas chlororaphis* GP72 with broad spectrum antifungal activity from green pepper rhizosphere. *Curr Microbiol* 54: 302–306.

Liu Y, Yang Q (2007) Cloning and heterologous expression of aspartic protease SA76 related to biocontrol in *Trichoderma harzianum*. *FEMS Lett* 277: 173–181.

Lobato MC, Olivieri FP, Altamiranda EAG et al. (2008) Phosphite compounds reduce disease severity in potato seed tubers and foliage. *Eur J Plant Pathol* 122: 349–358.

Locatelli GD, dos Santos GF, Botelho PS, Finkler CLL, Bueno LA (2018) Development of *Trichoderma* sp. formulations in encapsulated granules (CG) and evaluation of conidia shelf-life. *Biol Contr* 117: 21–29.

López-Mondéjar R, Ros M, Pascual JA (2011) Mycoparasitism-related gene expression of *Trichoderma harzianum* isolates to evaluate their efficacy as biocontrol agent. *Biol Contr* 56: 59–66.

Lu P, Gilardi G, Gullino ML, Garibaldi A (2010) Biofumigation with *Brassica* plants and its effect on the inoculum potential for Fusarium yellows of *Brassica* crops. *Eur J Plant Pathol* 126: 387–402.

Lu Y, Rao S, Huang F, Cai Y, Wang G, Cai K (2016) Effects of biochar amendment on tomato bacterial wilt resistance and soil microbial amount and activity. *Internatl Agron*, Article ID 2938282 (10 pages)

Lübeck M, Jensen DF (2002) Monitoring of biocontrol agents based on *Trichoderma* strains following three applications to glasshouse crops by combining dilution plating with UP-PCR fingerprinting. *Biocontr Sci Technol* 12: 371–380.

Lutz MP, Michel V, Martinez C, Camps C (2012) Lactic acid bacteria as biocontrol agents of soilborne pathogens. *IOBC-WPRS Bull* 78: 285–288.

Maciá-Vicente JG, Rosso LC, Ciancio A, Jansson H-B, Lopez-Llorca LV (2009) Colonization of barley roots by endophytic *Fusarium equiseti* and *Pochonia chlamydosporia*: Effects on plant growth and disease. *Ann Appl Biol* 155: 391–401.

Madi L, Katan T, Katan J, Henis Y (1997) Biological control of *Sclerotium rolfsii* and *Verticillium dahliae* by *Talaromyces flavus* is mediated by different mechanisms. *Phytopathology* 87: 1054–1060.

Maleki M, Mostafaee S, Mokhtarnejad L, Farzaneh M (2010) Characterization of *Pseudomonas fluorescens* strain CV6 isolated from cucumber rhizosphere in Varamin as a potential biocontrol agent. *Austr J Crop Sci* 4: 676–683.

Mallmann WL, Hemstreet CJ (1924) Isolation of an inhibitory substance from plants. *Agric Res* 28: 599–602.

Malolepsza U, Nawrocka J, Szczech M (2017) *Trichoderma virens* 106 inoculation stimulates defence enzyme activities and enhances phenolic levels in tomato plants, leading to lowered *Rhizoctonia solani* infection. *Biocontrol Sci Technol* 27: 180–189.

Mandal B, Mandal S, Csinos AS, Martinez N, Culbreath AK, Pappu HT (2008) Salicylic acid-induced resistance to *Fusarium oxysporum* f.sp. *lycopersici* in tomato. *Plant Physiol Biochem* 47: 642–649.

Manjunatha G, Roopa KS, Geetha N, Prashanth N, Shetty HS (2008) Chitosan enhances disease resistance in pearl millet against downy mildew caused by *Sclerospora graminicola* and defense-related enzyme activation. *Pest Manag Sci* 64: 1250–1257.

Mao W, Lewis JA, Hebbar PK, Lamsden RD (1997) Seed treatment with a fungal or bacterial antagonist for reducing corn damping-off caused by species of *Pythium* and *Fusarium*. *Plant Dis* 81: 450–454.

Marcucci E, Aleandri MP, Chilosi G, Magro P (2010) Induced resistance by ß-aminobutyric acid in artichoke against white mould caused by *Sclerotinia sclerotiorum*. *J Phytopathol* 158: 659–667.

Markovic O, Markovic N (1998) Cell cross-contamination in cell cultures: the silent and neglected danger. *In Vitro Cell Dev Biol* 34: 1–8.

Martinez-Granero F, Rivilla R, Martin M (2006) Rhizosphere selection of highly motile phenotypic variants of *Pseudomonas fluorescens* with enhanced competitive colonization ability. *Appl Environ Microbiol* 72: 3429–3434.

Martínez-Medina A, Pascual JA, Pérez-Alfocea F, Albacete A, Roldan A (2010) *Trichoderma harzianum* and *Glomus intraradices* modify the hormone disruption induced by *Fusarium oxysporum* infection in melon plants. *Phytopathology* 100: 682–688.

Masunaka A, Sekiguchi H, Takahashi H, Takenaka S (2010) Distribution and expression of elicitin-like protein genes of the biocontrol agent *Pythium oligandrum*. *J Phytopathol* 158: 417–426.

Matsubra Y, Hasegawa N, Fukui H (2002) Incidence of Fusarium root rot in asparagus seedlings infected with arbuscular mycorrhizal fungus as affected by several soil amendments. *J Jpn Soc Hortic Sci* 71: 370–374.

Matsubra Y, Ohba N, Fukui H (2001) Effect of arbuscular mycorrhizal fungal infection on the incidence of Fusarium root rot in asparagus seedlings. *J Jpn Soc Hortic Sci* 70: 202–206.

Maung CEH, Choi TG, Nan HH, Kim KY (2017) Role of *Bacillus amyloliquefaciens* Y1 in the control of Fusarium wilt disease and growth promotion of tomato. *Biol Contr Sci Technol* 27: 1400–1415.

Mavrodi OV, McSpadden Gardener BB, Mavrodi DV, Bonsall RF, Weller DM, Thomashow LS (2001) Genetic diversity of *phlD* from 2,4-diacetylphloroglucinol-producing fluorescent *Pseudomonas* spp. *Phytopathology* 91: 35–43.

Mavrodi OV, Walter N, Elateek S, Taylor CG, Okubara PA (2012) Suppression of Rhizobacteria and Pythium root rot of wheat by new strains of *Pseudomonas*. *Biol Contr* 62: 93–102.

Mazzola M (2007) Manipulations of rhizosphere-based communities to induce suppressive soils. *J Nematol* 39: 213–220.

Mazzola M, Granatstein DM, Elfting DC, Mullinox K (2001) Suppression of specific apple root pathogens by *Brassica napus* seed meal amendment, regardless of glucosinolate content. *Phytopathology* 91: 673–679.

Mazzola M, Zhao X, Cohen MF, Raaijmakers JM (2007) Cyclic lipopeptide surfactant production by *Pseudomonas fluorescens* SS101 is not required for suppression of complex *Pythium* spp. populations. *Phytopathology* 97: 1348–1355.

Mboubda HDF, Djocgoue PF, Omokolo ND, El Hadrami I, Boudjek T (2010) Benzo-(1,2,3)-thiadiazole-7-carbothioic-S-methyl ester (BTH) stimulates defense reactions in *Xanthosoma sagittifolium*. *Phytoparasitica* 38: 71–79.

Mc Laren DL, Huang HC, Rimmer SR (1996) Control of apothecial production of *Sclerotinia sclerotiorum* by *Coniothyrium minitans* and *Talaromyces flavus*. *Plant Dis* 80: 1373–1378.

McSpadden Gardener B, Gutierrez L, Joshi R, Edema R, Lutton E (2005) Distribution and biocontrol potential of *phlD*+ pseudomonads in corn and soybean fields. *Phytopathology* 95: 715–724.

Melzer MS, Ikeda SS, Boland GJ (2002) Interspecific transmission of double-stranded RNA and hypovirulence from *Sclerotinia sclerotiorum* to *S. minor*. *Phytopathology* 92: 780–784.

Meng Q, Hao JJ (2017) Optimizing the application of *Bacillus velezensis* BAC03 in controlling the disease caused by *Streptomyces scabies*. *BioControl* 62: 535–544.

Mercier J, Jiménez JI (2009) Demonstration of the biofumigation activity of *Muscodor albus* against *Rhizoctonia solani* in soil and potting mix. *BioControl* 54: Article 797.

Mercier J, Manker DC (2005) Biocontrol of soilborne diseases and plant growth enhancement in greenhouse soilless mix by the volatile-producing fungus *Muscodor albus*. *Crop Protect* 24: 355–362.

Messiha NAS, van Bruggen AHC, van Diepeningen AD et al. (2007) Potato brown rot incidence and severity under different management regimes in different soil types. *Eur J Plant Pathol* 119: 367–381.

Metcalf DA, Dennis JJC, Wilson CR (2004) Effect of inoculum density of *Sclerotium cepivorum* on the ability of *Trichoderma koningii* to suppress white rot of onion. *Plant Dis* 88: 287–291.

Migheli Q, Gonzalez-Candelas L, Dealsessi L, Camponogara A, Ramon-Vidal D (1998) Transfromants of *Trichoderma longibrachiatum* overexpressing the ß-1,3-endoglucanase gene *egl1* show enhanced biocontrol of *Pythium ultimum* in cucumber. *Phytopathology* 88: 673–677.

Mishra KK, Kumar A, Pandey KK (2010) RAPD-based genetic diversity among differen isolates of *Fusarium oxysporum* f.sp. *lycopersici* and their comparative biocontrol. *World J Microbiol Biotechnol* 26: 1079–1085.

Mishra S, Singh BR, Naqvi AH, Singh HB (2017) Potential of synthesized silver nanoparticles using *Stenotrophomonas* sp. BHU-S7 (MTCC5978) for management of soilborne and foliar phytopathogens. *Sci Rep* 7: 45154

Mohandas S, Manjula R, Rawal RD, Lakshmikantha HC, Chakraborty S, Ramachandra YL (2010) Evaluation of arbuscular mycorrhiza and other biocontrol agents in managing *Fusarium oxysporum* f.sp. *cubense* infection in banana cv. Neypoovan. *Biocontr Sci Technol* 20: 165–181.

Momma N, Momma M, Kobara Y (2010) Biological soil disinfestation using ethanol on *Fusarium oxysporum* f.sp. *lycopersici* and soil microorganisms. *J Gen Plant Pathol* 76: 336–344.

Mondal AH, Nehl DB, Allen SJ (2005) Acibenzolar-S-methyl induces systemic resistance in cotton against black root rot caused by *Thielaviopsis basicola*. *Austr Plant Pathol* 34: 499–507.

Morán-Diez E, Hermosa R, Ambrosino P et al. (2009) The ThPG1 endopolygalacturonase is required for the *Trichoderma harzianum*-plant beneficial interaction. *Molec Plant-Microbe Interact* 22: 1021–1031.

Moretti M, Gilandi G, Gullino ML, Garibaldi A (2008) Biological control of potential of *Achromobacter xylosoxydans* for suppressing Fusarium wilt of tomato. *Internatl J Bot* 4: 369–375.

Moruzzi S, Firrao G, Polano C, Borselli S, Loschi A, Eramcora P, Loi N, Martini M (2017) Genomic-assisted characterization of *Pseudomonas* sp. strain Pf4, a potential biocontrol agent in hydroponics. *Biocontr Sci Technol* 27: 969–991.

Mostowfizadeh-Ghalamfarsa R, Hussaini K, Ghasemi-Fiasei R (2017) Activity of two silicon salts in controlling the pistachio gummosis-inducing pathogen *Phytophthora pistaciae*. *Austr J Plant Pathol* 46: 323–332.

Mou Z, Fan W, Dong X (2003) Inducers of plant systemic acquired resistance regulate NPR1 function through redox changes. *Cell* 113: 935–944.

Moulin F, Lemanceau P, Alabouvette C (1996) Suppression of Pythium root rot of cucumber by a fluorescent pseudomonad is related to reduced root colonization by *Pythium aphanidermatum*. *J Phytopathol* 144: 125–129.

Murillo-Williams A, Pedersen P (2008) Arbuscular mycorrhizal colonization response to three seed-applied fungicides. *Agron J* 103: 795–800.

Murugaiyan S, Bae JY, Wu J et al. (2010) Characterization of filamentous bacteriophage PE226 infecting *Ralstonia solanacearum* strains. *J Appl Microbiol* 110: 296–303.

Muslim A, Horniouchi H, Hayakumachi M (2003) Control of Fusarium crown root rot of tomato with hypovirulent binucleate *Rhizoctonia* in soil and rockwool systems. *Plant Dis* 87: 739–747.

Muthukumar A, Eswaran A, Nakkeeran S, Sangeetha G (2010) Efficacy of plant extracts and biocontrol agents against *Pythium aphanidermatum* incitant of chilli damping-off. *Crop Protect* 29: 1483–1488.

Nakayama T (2017) Biocontrol of powdery scab of potato by seed tuber application of an antagonistic fungus, *Aspergillus versicolor* isolates from potato roots. *J Gen Plant Pathol* 83: 253–263.

Naraghi L, Heyadri A, Rezee S, Razavi M, Jahanifar H (2010) Study on antagonistic effects of *Talaromyces flavus* on *Verticillium albo-atrum*, the causal agent of potato wilt disease. *Crop Protect* 29: 658–662.

Naraghi L, Heydari A, Askari H, Pourrahim R, Marzban R (2014) Biological control of *Polymyxa betae*, fungal vector of rhizomania disease of sugar beets in greenhouse conditions. *J Plant Protect Res* 54: 109–114.

Narayanasamy P (2002) *Microbial Plant Pathogens and Crop Disease Management*, Science Publishers, Enfield, NH.

Narayanasamy P (2006) *Postharvest Pathogens and Disease Management*, John Wiley, Hoboken, NJ.

Narayanasamy P (2013) *Biological Management of Diseases of Crops*, 1 and 2, Springer Science + Business Media B.V., Heidelberg, Germany.

Narayanasamy P (2017) *Microbial Plant Pathogens – Detection and Management in Seeds and Propagules*, Wiley-Blackwell, Chichester, UK.

Narisawa K, Usuki F, Hashiba T (2004) Control of Verticllium yellows in Chinese cabbage by the dark septate endophytic fungus LtVB3. *Phytopathology* 94: 412–418.

Naseby DC, Way JA, Bainton NJ, Lynch JM (2001) Biocontrol of *Pythium* in the pea rhizosphere by antifungal metabolite-producing and non-producing *Pseudomonas* strains. *J Appl Microbiol* 90: 421–429.

Neilands JB (1981) Microbial iron compounds. *Annu Rev Biochem* 50: 715–731.

Nel B, Steinberg, Labuschagne N, Viljoen A (2006a) Isolation and characterization of nonpathogenic *Fusarium oxysporum* isolates from the rhizosphere of healthy banana plants. *Plant Pathol* 55: 207–216.

Nel B, Steinberg C, Labuschagne N, Viljoen A (2006b) The potential of nonpathogenic *Fusarium oxysporum* and other biological control organisms for suppressing Fusarium wilt of banana. *Plant Pathol* 55: 217–223.

Nga NTT, Giau NT, Long NT et al. (2010) Rhizobacterially induced protection of watermelon against *Didymella bryoniae*. *J Appl Microbiol* 109: 567–582.

Nguyen XH, Naing KW, Lee YS, Kim KY (2015) Isolation of butyl 2,3-dihydroxybenzoate from *Paenibacillus elgii* HOA73 against *Fusarium oxysporum* f.sp. *lycopersici*. *J Phytopathol* 163: 342–352.

Nielsen TH, Sørensen J (2003) Production of cyclic lipopolypeptides by *Pseudomonas fluorescens* strains in bulk soil and in the sugar beet rhizosphere. *Appl Environ Microbiol* 69: 861–868.

Norman JR, Atkinson D, Hooker JE (1996) Arbuscular mycorrhizal fungal-induced alteration to root architecture in strawberry and induced resistance to the root pathogen *Phytophthora fragariae*. *Plant Soil* 185: 191–198.

Ochiai N, Powelson ML, Dick RP, Crowe FJ (2007) Effects of green manure type and amendment rate in Verticillium wilt severity and yield of Russet Burbank potato. *Plant Dis* 91: 400–406.

Ojaghian MR (2011) Potential of *Trichoderma* spp. and *Talaromyces flavus* for biological control of potato stem rot caused by *Sclerotinia sclerotiorum*. *Phytoparasitica* 39: 185–193.

Olivieri FP, Lobato MC, Altamiranda EG et al. (2009) BABA effects on the behavior of potato cultivars infected by *Phytophthora infestans* and *Fusarium solani*. *Eur J Plant Pathol* 123: 47–56.

Orio AGA, Brücher E, Ducasse DA (2016) A strain of *Bacillus subtilis* subsp. *subtilis* shows a specific antagonistic activity against the soilborne pathogen *Setophoma terrestris*. *Eur J Plant Pathol* 144: 217–223.

Orzali L, Corsi B, Forni C, Riccioni L (2017) *Chitosan in Agriculture: A New Challenge for Managing Plant Disease*

Oskiera M, Szcech M, Stepowska A, Smolinska U, Bartoszewski G (2017) Monitoring of *Trichoderma* species in agricultural soil in response to application of biopreparations. *Biol Contr* 113: 65–72.

Ospina-Giraldo MD, Royse DJ, Chen X, Romaine CP (1999) Molecular phylogenetic analysis of biological control strains of *Trichoderma harzianum* and other biotypes of *Trichoderma* spp. associated with mushroom green mold. *Phytopathology* 89: 308–313.

Özer N (2011) Screening for fungal antagonists to control black mold disease to induce the accumulation of antifungal compounds in onion after seed treatment. *BioControl* 56: 237–247.

Ozgonen H, Erkilic A (2007) Growth enhancement and Phytophthora blight (*Phytophthora capsici* Leonian) control by arbuscular mycorrhizal fungal inoculation in pepper. *Crop Protect* 26: 1682–1688.

Pal KK, McSpadden Gardener B (2006) Biological control of plant pathogens. *The Plant Health Instructor*. 10.1094/PHI-A-2006-1117-02.

Palma-Guerrero J, Jansson H-B, Salinas J, Lopez-Llorca LV (2008) Effect of chitosan on hyphal growth and spore germination of plant pathogenic and biocontrol fungi. *J Appl Microbiol* 104: 541–553.

Palumbo JD, O'Keefe TL, Kattan A, Abbas HK, Johnson BJ (2010) Inhibition of *Aspergillus flavus* in soil by antagonistic *Pseudomonas* strains reduces the potential for airborne spore dispersal. *Phytopathology* 100: 532–538.

Pantelides IS, Tjamos SE, Stringlis IA, Chatzipavlidis I, Paplomatas EJ (2009) Mode of action of a nonpathogenic *Fusarium oxysporum* strain against *Verticillium dahliae* using real-time QPCR analysis and biomarker transformation. *Biol Contr* 50: 30–36.

Park K, Paul D, Kim YK et al. (2007) Induced systemic resistance by *Bacillus vallismortis* EXTN-1 suppressed bacterial wilt in tomato caused by *Ralstonia solanacearum*. *Plant Pathol J* 23: 22–25.

Pastrana AM, Basallote-Ureba MJ, Aguado A, Akdi K, Capote N (2016) Biological control of strawberry soilborne pathogens *Macrophomina phaseolina* and *Fusarium solani*, using *Trichoderma asperellum* and *Bacillus* spp. *Phytopathol Mediterr* 55: 109–120.

Pavlou GC, Vakalounakis DJ (2005) Biological control of root and stem rot of greenhouse cucumber caused by *Fusarium oxysporum* f.sp. *radicis-cucumerinum* by lettuce soil amendment. *Crop Protect* 24: 135–140.

Perneel M, Heyrman J, De Maeyer R, Raaijmakers JM, Devos P, Höfte M (2007) Characterization of CMR5C and CMR12a, novel fluorescent *Pseudomonas* strains from the cocoyam rhizosphere with biocontrol activity. *J Appl Microbiol* 103: 1007–1020.

Postma J, Clematis F, Nijhuis EH, Someus E (2013) Efficacy of four phosphate-mobilizing bacteria applied with an animal bone charcoal formulation in controlling *Pythium aphanidermatum* and *Fusarium oxysporum* f.sp. *radicis-lycopersici* in tomato. *Biol Contr* 67: 284–291.

Potlakayala SD, Reed DW, Covello PS, Fobert PR (2007) Systemic acquired resistance in canola is linked with pathogenesis-related genes expression and requires salicylic acid. *Phytopathology* 97: 794–802.

Pozo MJ, Azcón-Aguilar C, Dumoas-Gaudot E, Barea JM (1999) ß-1,3-glucanase activities in tomato roots inoculated with arbuscular mycorrhizal fungi and /or *Phytophthora parasitica* and their possible involvement in bioprotection. *Plant Sci* 141: 149.

Pradhanang PM, Mamol MT, Olson SM, Jones JB (2003) Effects of essential oils on *Ralstonia solanacearum* population density and bacterial wilt incidence in tomato. *Plant Dis* 87: 423–433.

Prakasha A, Grice ID, Vinay Kumar KS, Sadashiv MP, Shankar HN, Umesha S (2017) Extracellular polysaccharide from *Ralstonia solanacearum*: A strong inducer of eggplant defense against bacterial wilt. *Biol Contr* 110: 107–116.

Prashar P, Vandenberg A (2017) Genotype-specific responses to the effects of commercial Trichoderma forumulations in lentil (Lens culinaris ssp. culinaris) in the presence and absence of the oomycete pathogen Aphanomyces euteiches. *Biocontr Sci and Technol* 27: 1123–1144.

Puopolo R, Raio A, Zoina A (2010) Identification and characterization of *Lysobacter capsici* strain PG4: A new plant health-promoting rhizobacterium. *J Plant Pathol* 92: 157–164.

Raaijmakers JM, de Bruijn I, de Kock MD (2006) Cyclic lipopeptide production by plant associated *Pseudomonas* species diversity, activity, biosynthesis and regulation. *Molec Plant-Microbe Interact* 19: 699–710.

Raaijmakers JM, Weller DM (1998) Natural plant protection by 2,4-diacetylphloroglucinol-producing *Pseudomonas* spp. in take-all decline soils. *Molec Plant-Microbe Interact* 11: 144–152.

Rabea EL, El-Badawy MT, Stevens CV, Smagghe G, Steurbant W (2003) Chitosan as antimicrobial agent: Applications and modes of action. *Biomacromolecules* 4: 1457–1465.

Rahman MME, Hossain DM, Suzuki K et al. (2016) Suppressive effects of *Bacillus* spp. on mycelia, apothecia and sclerotia formation of *Sclerotinia sclerotiorum* and potential as biocontrol of white mold of mustard. *Austr Plant Pathol* 45: 103–117.

Ramarathnam R, Fernando WGD, de Kievit T (2011) The role of antibiosis and induced systemic resistance mediated by strains of *Pseudomonas chlororaphis*, *Bacillus cereus* and B. *amyloliquefaciens* in controlling blackleg disease of canola. *BioControl* 56: 225–235.

Ran LX, Li ZN, Wu GJ, van Loon LC, Bakker PAHM (2005) Induction of systemic resistance against bacterial wilt in *Eucalyptus urophylla* by fluorescent *Pseudomonas* spp. *Eur J Plant Pathol* 113: 59–70.

Rashad YM, Al-Askar AA, Ghoneem KM, Saber WIA, Hafez EE (2017) Chitinolytic *Streptomyces griseorubens* E44G enhances the biocontrol efficiency against Fusarium wilt disease of tomato. *Phytoparasitica* 45: 227–237.

Ren L, Li G, Han YC, Jiang DH, Huang HC (2007) Degradation of oxalic acid by *Coniothyrium minitans* and its effects on production and activity of ß-1,3-glucanase of this mycoparasite. *Biol Contr* 43: 1–11.

Reuveni R, Agapov V, Reuveni MC (1994) Foliar spray of phosphates induces growth increases and systemic resistance to *Puccinia sorghi* in maize. *Plant Pathol* 43: 245–250.

Rezzonico F, Binder C, Défago G, Möenne-Loccoz Y (2005) The type III secretion system of biocontrol *Pseudomonas fluorescens* KD targets the phytopathogenic chromista *Pythium ultimum* and promotes cucumber protection. *Molec Plant-Microbe Interact* 18: 991–1001.

Rivera-Varas V, Freeman TA, Gudmestad NC, Secor GA (2007) Mycoparasitism of *Helminthosporium solani* by *Acremonium strictum*. *Phytopathology* 97: 1331–1337.

Roberts DP, Lakshman DK, McKenna LF, Emche SE, Maul JE, Bauchan G (2016) Seed treatment with ethanol extract of *Serratia marcescens* is compatible with *Trichoderma* isolates for control of damping-off of cucumber caused by *Pythium ultimum*. *Plant Dis* 100: 1278–1287.

Rodriguez MA, Cabrera G, Godeas A (2006) Cyclosporine A from nonpathogenic *Fusarium oxysporum* suppressing *Sclerotinia sclerotiorum*. *J Appl Microbiol* 100: 575–586.

Rodriguez-Diaz M, Rodeles-Gonzales B, Pozo-Clemente C, Martinez-Toledo MU, Gonzalez-Lopez J (2008) A review on the taxonomy and possible screening of traits of plant growth-promoting rhizobacteria. In: Ahmad I, Pichtel J, Hayat S (eds.), *Plant-Bacteria Interactions, Strategies, and Techniques to Promote Plant Growth*, Wiley-VCH, Weinhem, pp. 106–146.

Sabaratnam S, Traquair JA (2015) Mechanism of antagonism by *Streptomyces griseocarneus* (strain Di944) against fungal pathogens of greenhouse-grown tomato transplants. *Canad J Plant Pathol* 37: 197–211.

Sabate DC, Brandan CP, Petroselli G, Erra-Balsells R, Audisio MC (2017) Decrease in the incidence of charcoal root rot in common bean (*Phaseolus vulgaris* L.) by *Bacillus amyloliquefaciens* B14, a strain with PGPR properties. *Biol Contr* 113: 1–8.

Sadeghi A, Koobaz P, Azimi H, Karimi E, Akbari AR (2017) Plant growth promotion and suppression of *Phytophthora drechsleri* damping-off in cucumber by cellulase-producing *Streptomyces*. *BioControl* 62: 805–819.

Salzman R, Brady J, Finlayson S, Buchnan C et al. (2005) Transcriptional profiling of sorghum induced by methyl jasmonate, salicylic acid, aminocyclopropane carboxylic acid reveals cooperative regulation and novel gene response. *Plant Physiol* 138: 352–368.

Sari E, Etebarian HR, Aminian H (2007) The effects of *Bacillus pumilus*, isolated from wheat rhizosphere on resistance on wheat seedling roots against the take-all fungus *Gaeumannomyces graminis* var. *tritici*. *J Phytopathol* 155: 720–727.

Sarrocco S, Mikkelsen L, Vergara M, Jensen DF, Lübeck M, Vannacci G (2006) Histopathological studies of sclerotia of phytopathogenic fungi parasitized by a GFP-transformed *Trichoderma virens* antagonistic strain. *Mycol Res* 110: 179–187.

Sasaki A, Kanematsu S, Onoue M, Oikawa Y, Nakamura H, Yoshida K (2007) Artificial infection of *Rosellinia necatrix* with purified virus particles of a member of the genus *Mycoreovirus* reveals its uneven distribution in single colonies. *Phytopathology* 97: 278–286.

Šašek V, Nováková M, Dobrev RI, Valentová O, Burketová L (2012) ß-aminobutyric acid protects *Brassica napus* plants from infection by *Leptosphaeria maculans*–resistance induction or a direct antifungal effect? *Eur J Plant Pathol* 133: 279–289.

Scheuerell SJ, Mahiffee WF (2004) Compost tea as a container medium drench for suppressing seedling damping-off caused by *Pythium ultimum*. *Phytopathology* 94: 1156–1163.

Schisler DA, Slininger PJ, Behle RW, Jackson MA (2004) Formulations of *Bacillus* spp. for biological control of plant diseases. *Phytopathology* 94: 1267–1271.

Schouten AO, Berg GVD, Edel-Herman V et al. (2004) Defence responses of *Fusarium oxysporum* to 2,4-diacetylphloroglucinol, a broad-spectrum antibiotic produced by *Pseudomonas fluorescens*. *Molec Plant-Microbe Interact* 17: 1201–1211.

Selim HMM, Gomaa NM, Essa AMM (2017) Applications of endophytic bacteria for the biocontrol of *Rhizoctonia solani* (Cantharellales: Ceratobasidiaceae) damping-off in cotton seedlings. *Biocontr Sci Technol* 27: 81–95.

Selin S, Negres J, Wendehenne D, Ochatt S, Gianinazzi S, van Tuinen D (2010) Stimulation of defense reactions in *Medicago truncata* by antagonistic lipopeptides from *Paenibacillus* sp. strain B2. *Appl Environ Microbiol* 76: 7420–7428.

Sezekeres A, Kredics L, Antal Z, Kevei F, Manczinger L (2004) Isolation and characterization of protease overproducing mutants of *Trichoderma harzianum*. *FEMS Microbiol Lett* 233: 215–222.

Shah-Smith DA, Burns RD (1996) Biological control of damping-off of sugar beet by *Pseudomonas putida* applied as seed pellets. *Plant Pathol* 45: 572–582.

Shanahan P, O'Sullivan DJ, Simpson P, Glennon JD, O'Gara F (1992) Isolation of 2,4-diacetylphloroglucinol from a fluorescent pseudomonad and investigation of physiological parameters influencing its production. *Appl Environ Microbiol* 58: 353–358.

Shanmugam V, Kanouja N (2011) Biological management of vascular wilt of tomato caused by *Fusarium oxysporum* f.sp. *lycopersici* by plant growth-promoting rhizobacterial mixture. *Biol Contr* 57: 85–93.

Shanmugam V, Kanoujia N, Singh M, Singh S, Prasad R (2011) Biocontrol of vascular wilt and crown rot of gladiolus caused by *Fusarium oxyporum* f.sp. *gladioli* using plant growth-promoting rhizobacterial mixture. *Crop Protect.* 2011.02.033

Shishido M, Miwa C, Usami T, Amemiya, Johson HB (2005) Biological control efficiency of Fusrium wilt of tomato by nonpathogenic *Fusarium oxysporum* B-B2 in different amendments. *Phytopathology* 95: 1072–1080.

Shoresh M, Harman GE (2008) Molecular basis of maize responses to *Trichoderma harzianum* T22 inoculation: A proteomic approach. *Plant Physiol* 147: 2147–2163.

Shoresh M, Harman GE, Mastouti F (2010) Induced systemic resistance and plant responses to fungal biocontrol agents. *Annu Rev Phytopathol* 48: 21–43.

Simon A, Sivasithamparam K (1989) Pathogen suppression: a case study in biological suppression of *Gaeumannomyces graminis* var. *tritici* in soil. *Soil Biol Chem* 21: 331–337.

Singh BN, Singh A, Singh SP, Singh HB (2011) *Trichoderma harzianum*-mediated reprogramming of oxidative stress response in root apoplast of sunflower enhances defense against *Rhizoctonia solani*. *Eur J Plant Pathol* 131: 121–134.

Singh N, Kumar S, Bajpai VK, Dubey RL , Maheshwari DK, Kang SC (2010) Biological control of *Macrophomina phaseolina* by chemotactic fluorescent *Pseudomonas aeruginosa* PN1 and its plant growth-promoting activity in chir-pine. *Crop Protect* 29: 1142–1147.

Singh SP, Gupta R, Gaur R, Srivastava AK (2016) *Streptomyces* spp. alleviate *Rhizoctonia solani*-mediated oxidative stress in *Solanum lycopersicum*. *Ann Appl Biol* 168: 232–242.

Singh SP, Gaur R (2017) Endophytic *Streptomyces* spp. underscore induction of defense regulatory genes and confers resistance against *Sclerotium rolfsii* in chickpea. *Biol Contr* 104: 44–56.

Sivasithamparam K, Ghisalberti EL (1996) Secondary metabolism in *Trichoderma* and *Gliocladium*. In: Kubicek CP, Harman GE (eds.), *Trichoderma* and *Gliocladium*, Volume 1, Taylor and Francis, London, pp. 139–161.

Smolinska U, Morra MJ, Knudsen GR, James RL (2003) Isothiocyanates produced by Brassicaceae species as inhibitor of *Fusarium oxysporum*. *Plant Dis* 87: 407–412.

Sodeghi A, Koobaz P, Azimi H, Karimi E, Akbari AR (2017) Plant growth promotion and suppression of *Phytophthora dreschleri* damping-off in cucumber by cellulase-producing *Streptomyces*. *BioControl* 62: 805–819.

Soltanzadeh M, Nejad MS, Bonjar GHS (2016) Application of soil-borne actinomycetes for biological control against Fusarium wilt of chickpea (*Cicer arietinum*) caused by *Fusarium solani* f.sp. *pisi*. *J Phytopathol* 164: 967–978.

Sotoyama K, Akutsu K, Nakajima M (2017) Suppression of bacterial wilt of tomato by soil amendment with mushroom compost containing *Bacillus amyloliquefaciens* IOMC7. *J Gen Plant Pathol* 83: 51–55.

Srinivasan K, Gilardi G, Garibaldi A, Gullino ML (2009) Bacterial antagonists from used rockwood soilless substrates suppress Fusarium wilt of tomato. *J Plant Pathol* 91: 147–154.

Srivastava R, Khalid A, Sing US, Sharma AK (2010) Evaluation of arbuscular mycorrhizal fungus, fluorescent *Pseudomonas* and *Trichoderma harzianum* formulations against *Fusarium oxysporum* f.sp. *lycopersici* for the management of tomato wilt. *Biol Contr* 53: 24–31.

Stamford NP, Santos LRC, dos Santos AP, de Souza KR, da Silva Oliveira W, da Silva EVN (2017) Response of horticultural crops to application of bioprotector and biological control of Ralstonia wilt in Brazilian Ultisol. *Austr J Crop Sci* 11: 284–289.

Steddom KC, Menge JA (1999) Continuous application of the bio-control bacterium *Pseudomonas putida* 06909, improves bio-control of *Phytophthora parasitica* on citrus. *Phytopathology* 89: 575 (Abst).

Steindorff AS, Silva RN, Coelho ASG, Nagata T, Noronha ET, Ulhoa CJ (2012) *Trichoderma harzianum* expressed sequence tags for identification of genes with putative roles in mycoparasitism against *Fusarium solani*. *Biol Contr* 61: 134–140.

Stergiopoulos I, de Wit PJ (2009) Fungi effector proteins. *Annu Rev Phytopathol* 47: 233–263.

Stinson AM, Zidack NK, Strobel GA, Jacobsen BJ (2003) Mycofumigation with *Muscodor albus* and *Muscodor roseus* for control of seedling diseases of sugar beet and Verticillium wilt of eggplant. *Plant Dis* 87: 1349–1354.

Strobel GA, Driske E, Sears J, Markworth C (2001) Volatile anti-microbials from *Muscodor albus*, a novel endophytic fungus. *Microbiology* 147: 2943–2950.

Strobel GA, Kluck K, Hess WM, Sears J, Ezra D, Vargas PN (2007) *Muscodor albus* E-6, an endophyte of *Ganzma ulmifolia* making volatile antibiotics: isolation, characterization and experimental establishment in the host plant. *Microbiology* 153: 2613–2620.

Strobel GA, Spang S, Kluck K, Hess WM, Sears J, Livinghouse T (2008) Synergism among volatile organic compounds resulting in increased antibiosis in *Oidium* sp. *FEMS Microbiol Lett* 283: 140–145.

Stutz E, Défago G, Kern H (1986) Naturally occurring fluorescent pseudomonad involved in suppression of black root of tobacco. *Phytopathology* 76: 181–185.

Sugimoto T, Watanabe K, Furiki M et al. (2009) Stem rot disease of soybeans, the growth rate and zoospore release of *Phytophthora sojae*. *J Phytopathol* 157: 379–389.

Sundaramoorthy S, Raguchander T, Ragupathi, N, Samiyappan R (2012) Combinatorial effect of endophytic and plant growth-promoting rhizobacteria against wilt disease of *Capsicum annuum* L. caused by *Fusarium solani*. *Biol Contr* 60: 59–67.

Svercel M, Duffy B, Défago G (2007) PCR amplification of hydrogen cyanide biosynthetic locus *hcnAB* in *Pseudomonas* spp. *J Microbiol Meth* 70: 209–213.

Tambong JT, Höfte M (2001) Phenazines are involved in biocontrol of *Pythium myriotylum* on cocoyam by *Pseudomonas aeruginosa* PNA1. *Eur J Plant Pathol* 107: 511–521.

Thies J, Rilling M, Graber ER (2015) Biochar effects on the abundance, activity and diversity of soil biota. In: Lehmann J, Joseph S (eds.), *Biochar for Environmental Management: Science and technology*, Earthscan, London, 327–389.

Thongkamngam T, Jaenaksorn T (2017) *Fusarium oxysporum* (F221-B) as biocontrol agent against plant pathogenic fungi in vitro and in hydroponics. *Plant Protect Sci* 53: 85–95.

Thrane C, Nielsen TH, Nielsen MN, Sørensen J, Olsson S (2000) Vicosinamide-producing *Pseudomonas fluorescens* DR54 exerts a biocontrol effect on *Pythium ultimum* in sugar beet rhizosphere. *FEMS Microbiol Ecol* 33: 139–146.

Thygesen K, Larsen J, Bødker L (2004) Arbuscular mycorrhizal fungi reduce development of pea root rot caused by *Aphanomyces euteiches* using oospores as pathogen inoculum. *Eur J Plant Pathol* 110: 411–419.

Tian T, Li S-D, Sun M-H (2014) Synergistic effect of dazomet soil fumigation and *Clonostachys rosea* against cucumber wilt. *Phytopathology* 104: 1314–1321.

Timmusk S, van West P, Gow NAR, Huffstutler RP (2009) *Paenibacillus polymyxa* antagonizes oomycete plant pathogens *Phytophthora palmivora* and *Pythium aphanidermatum*. *J Appl Microbiol* 106: 1473–1481.

Tjamos EC, Tsitsigiannis DI, Tjamos SE, Antoniou PP, Katinakis P (2004) Selection and screening of endorhizosphere bacteria from solarized soils as biocontrol agents against *Verticillium dahliae* of solanaceous hosts. *Eur J Plant Pathol* 110: 35–44.

Tomihama T, Nishi Y, Mori K et al. (2016) Rice bran amendment suppresses potato common scab increasing antagonistic bacterial community levels in the rhizosphere. *Phytopathology* 106: 719–728.

Tran H, Ficke A, Ashimwe T, Höfte M, Raaijmakers JM (2007) Role of the cyclic lipopeptides massetolide A in biological control of *Phytophthora infestans* and in colonization of tomato plants by *Pseudomonas fluorescens*. *New Phytol* 175: 731–742.

Trotta A, Varese GC, Gnavi E, Fusconi A, Sampó S, Berta G (1996) Interaction between the soilborne root pathogen *Phytophthora nicotianae* var. *parasitica* and the arbuscular mycorrhizal fungus *Glomus mosseae* in tomato plants. *Plant Soil* 185: 199–207.

Trouvelot S, Olivain C, Recorbet G, Migheli Q, Alabouvette (2002) Recovery of *Fusarium oxysporum* Fo47 mutants affected in their biocontrol activity after transposition of the *Fot1* element. *Phytopathology* 92: 936–945.

Uppal AK, El-Hadrami A, Adam LR, Tenuta M, Daayf F (2008) Biological control of potato Verticillium wilt under controlled and field conditions using selected bacterial antagonists and plant extracts. *Biol Contr* 44: 90–100.

Utkhade R (2006) Increased growth and yield of hydroponically grown greenhouse tomato plants inoculated with arbuscular mycorrhizal fungi and *Fusarium oxysporum* f.sp. *lycopersici*. *BioControl* 51: 393–400.

Validov S, Kamilova F, Qi S et al. (2007) Selection of bacteria able to control *Fusarium oxysporum* f.sp. *radicis-lycopersici* in stonewood substrate. *J Appl Microbiol* 102: 461–471.

Veloso J, Díaz J (2012) *Fusarium oxysporum* Fo47 confers protection to pepper plants against *Verticillium dahliae* and *Phytophthora capsici* and induces its expression of defence genes. *Plant Pathol* 61: 281–288.

Vermeire ML, Kablan L, Dorel M, Delvaux B, Risede J-M, Legreve A (2011) Protective role of silicon on the banana-*Cylindrocladium spathiphylli* pathosystem. *Eur J Plant Pathol* 131: 621–630.

Vigo C, Norman JR, Hooker JE (2000) Biocontrol of the pathogen *Phytophthora parasitica* by arbuscular mycorrhizal fungi is a consequence of effects of infection foci. *Plant Pathol* 49: 509–514.

Vinodkumar S, Indumathi T, Nakkeeran S (2017) *Trichoderma asperellum* (NVTA2) as a potential antagonist for the management of stem rot in carnation under protected cultivation. *Biol Contr* 113: 58–64.

Viterbo A, Chet I (2006) *TasHydI*, a new hydrophobin gene from the biocontrol agent *Trichoderma asperellum*, is involved in plant root colonization. *Molec Plant Pathol* 7: 249–258.

Viterbo A, Harel M, Chet I (2004) Isolation of two aspartyl proteases from *Trichoderma asperellum* expressed during colonization of cucumber roots. *FEMS Microbiol Lett* 238: 151–158.

Voisard C, Keel C, Haas D, Défago G (1989) Cyanide production by *Pseudomonas fluorescens* helps suppress black root rot of tobacco under gnotobotic conditions. *EMBO J* 8: 351–358.

Wang D, Weaver ND, Kesarwani M, Dong X (2005) Induction of protein secretory pathway is required for systemic acquired resistance. *Science* 308: 1036–1040.

Wang H, Li W, Chen Q, Huang Y et al. (2012) A rapid microbioassay for discovery of antagonistic bacteria for *Phytophthora parasitica* var. *nicotianae*. *Phytopathology* 102: 267–271.

Widmer TL, Shishkoff N (2017) Reducing infection and secondary inoculum of *Phytophthora ramorum* on *Viburnum tinus* roots grown in potting medium amended with *Trichoderma asperellum* isolate 04–22. *Biol Contr* 107: 60–69.

Wiggins BE, Kinkel LL (2005) Green manures and crop sequences influence potato diseases and pathogen inhibitory activity of indigenous streptomyce. *Phytopathology* 95: 178–185.

Wilson PS, Ketola EO, Ahvenniemi PM, Lehtonen MJ, Valkonen JPT (2008) Dynamics of soilborne *Rhizoctonia solani* in the presence of *Trichoderma harzianum*: Effects on stem canker, black scurf, and progeny tubers of potato. *Plant Pathol* 57: 152–161.

Woo SL, Donzelli B, Scala F et al. (1999) Disruption of the *ech-42* (endochitinase-encoding) gene affects biocontrol activity in *Trichoderma harzianum* P1. *Molec Plant-Microbe Interact* 12: 419–429.

Woo SL, Scala F, Ruocco M, Lorito M (2006) The molecular biology of the interactions between *Trichoderma* spp., phytopathogenic fungi and plants. *Phytopathology* 96: 181–185.

Worapong J, Strobel GA (2009) Biocontrol of a root rot of kale by *Muscodor albus* strain MFC2. *BioControl* 54: 301–306.

Wu Y, Von Wettstein D, Kannangara CG, Nirmala J, Cook RJ (2006) Growth inhibition of the cereal root pathogens *Rhizoctonia solani* var. *tritici* by recombinant 42 kDa endochitinase from *Trichoderma harzianum*. *Biol Contr Sci Technol* 16: 631–646.

Xia Y, Suzuki H, Borevitz J et al. (2004) An extracellular aspartic protease function in *Arabiopsis* disease resistance signaling. *EMBO J* 23: 980–988.

Xie J, Wei D, Jiang D, Fu Y, Li G, Ghabrial S, Peng Y (2006) Characterization of debilitation-associated mycovirus infecting the plant pathogenic fungus *Sclerotinia sclerotiorum*. *J Gen Virol* 87: 241–249.

Xie X-G, Dai C-C, Li X-G, Wu J-R, Wu Q-Q, Wang X-X (2017) Reduction of soilborne pathogen *Fusarium solani* reproduction in soil enriched with phenolic acids by inoculation of endophytic fungus *Phomopsis liquidambari*. *BioControl* 62: 111–123.

Xu J, Zhao X, Han X, Du Y (2007) Antifungal activity of oligochitosan against *Phytophthora capsici* and other plant pathogenic fungi in vitro. *Pestic Biochem Physiol* 87: 220–228.

Xue L, Charest PM, Jabaji-Hare SH (1998) Systemic induction of peroxidases, 1,3-ß-glucanases, chitinases and resistance in bean plants by binucleate *Rizoctonia* species. *Phytopathology* 88: 359–365.

Yamada T, Kawasaki T, Nagata A et al. (2007) New bacteriophages that infect the phytopathogen *Ralstonia solanacearum*. *Microbiology* 153: 2630–2539.

Yang L, Li G, Zhang J, Jiang D, Chen W (2011) Compatibility of *Coniothyrium minitans* with compound fertilizer in suppression of *Sclerotinia sclerotiorum*. *Biol Contr* 59: 221–227.

Yang L, Xie J, Jiang D, Fu J, Li G, Lin F (2008) Antifungal substances produced by *Penicillium oxalicum* strain P 4–1, potential antibiotics against plant pathogenic fungi. *World J Micorbiol Biotechnol* 24: 909–915.

Yang M-M, Mavrodi DV, Mavrodi OV, Thomashow LS, Weller DM (2017) Construction of a recombinant strain of *Pseudomonas fluorescens* producing both phenazine-1-carboxylic acid. *Eur J Plant Pathol* 149: 683–694.

Yang M-M, Mavrodi DV, Mavrodi OV et al. (2011) Biological control of take-all by fluorescent *Pseudomonas* spp. from Chinese wheat fields. *Phytopathology* 101: 1481–1491.

Yang M-M, Wen S-S, Mavrodi DV et al. (2014) Biological control of wheat root diseases by the CLP-producing strain *Pseudomonas fluorescens* HC1-07. *Phytopathology* 104: 248–256.

Yang M-M, Xu L-P, Xue Q-Y et al. (2012) Screening potential bacterial biocontrol agents towards *Phytophthora capsici* in pepper. *Eur J Plant Pathol* 134: 811–820.

Yang P (2017) The gene *task1* is involved in morphological development, mycoparasitism and antibiosis of *Trichoderma asperellum*. *Biol Contr Sci Technol* 27: 620–635.

Yang P, Sun Z-X, Liu S-Y, Lu H-X, Zhou Y, Sun M (2013) Combining antagonistic endophytic bacteria in different growth stages of cotton for control of Verticillium wilt. *Crop Protect* 47: 17–23.

Yang S-Y, Park MR, Kim IS, Kim YC, Yang JW, Ryu C-M (2010) 2-aminobenzoic acid of *Bacillus* sp. BS10 as an IST determinant against *Pectobacterium carotovorum* subsp. *carotovorum* Scc1 in tobacco. *Eur J Plant Pathol*.

Yang W, Xu Q, Liu H-X et al. (2012) Evaluation of biological control agents of Ralstonia wilt on ginger. *Biol Contr* 62: 144–151.

Yedida L, Benhamou N, Chet I (1989) Induction of defense responses in cucumber plants (*Cucumis sativus* L.) by the biocontrol agent *Trichoderma harzianum*. *Appl Environ Microbiol* 65: 1061–1070.

Yu X, Li B, Fu Y et al. (2010) A geminivirus-related DNA mycovirus that confers hypovirulence to a plant pathogenic fungus. *Proc Natl Acad Sci USA* 107: 8387–8392.

Zamoum M, Goudjal Y, Sabaou N, Mathieu F, Zitouni A (2017) Development of formulations based on biocontrol of *Rhizoctonia rochei* strain PTL2 spores for biocontrol of *Rhizoctonia solani* damping-off of tomato seedlings. *Biocontr Sci Technol* 27: 723–738.

Zeng W, Wang D, Kirk W, Hao J (2012) Use of *Coniothyrium minitans* and other microorganisms for reducing *Sclerotinia sclerotiorum*. *Biol Contr* 60: 225–232.

Zhang D, Gao T, Li H, Lei B, Zhu B (2017a) Identification of antifungal substances secreted by *Bacillus subtilis* Z-14 that suppress *Gaeumannomyces graminis* var. *tritici*. *Biocontr Sci Technol* 27: 237–251.

Zhang JX, Howell CR, Starr JL (1996) Suppression of *Fusarium* colonization of cotton roots and Fusarium wilt by seed treatment with *Gliocladium virens* and *Bacillus subtilis*. *Biocontr Sci Technol* 6: 175–187.

Zhang JX, Xue AG (2010) Biocontrol of Sclerotinia stem rot (*Sclerotinia sclerotiorum*) of soybean using novel *Bacillus subtilis* strain SB24 under controlled conditions. *Plant Pathol* 59: 382–391.

Zhang N, Pan R, Shen Y et al. (2017b) Development of a novel bio-organic fertilizer for plant growth promotion and suppression of rhizome rot of ginger. *Biol Contr* 114: 97–105.

Zhang Y, Smith P, Maximova SN, Guiltinan MJ (2015) Application of glycerol as a foliar spray activates the defence response and enhances disease resistance of *Theobroma cacao*. *Molec Plant Pathol* 16: 27–37

Zhao J, Zhou X, Jiang A et al. (2018) Distinct impacts of reductive soil disinfestation and chemical soil disinfestation on soil fungal communities and memberships. Appl Microbiol Biotechnol.

Zwart DC, Kim S-H (2012) Biochar amendment increases resistance to stem lesions caused by *Phytophthora* spp. in tree seedlings. *HortScience* 47: 1736–1740.

4 Management of Soilborne Microbial Plant Pathogens
Chemical Application

The impact of crop diseases on quantitative and qualitative yields may vary, depending on the pathogen characteristics and responses of the host plant species to the pathogen(s) prevailing environment. Various aspects of host-pathogen interactions are investigated primarily to gather relevant information that will be useful for reducing infection by soilborne microbial pathogens and spread of the disease(s) induced by them. Strategies applicable against diseases caused by soilborne microbial pathogens may be of two types, acting either indirectly or directly on the pathogens. The strategies such as use of disease-free seeds and propagules, cultural practices and host resistance aim at preventing introduction of pathogens and/or build-up pathogen populations and their effects are not distinctly perceptible, but the beneficial effects may be realized over several years. By contrast, use of biocontrol agents (BCAs) and chemicals has direct adverse impact on the development of pathogens and consequently on the progress of disease symptoms. The effect of chemical application is highly recognizable and more effective against some pathogens, which may not be contained by other disease management strategies. However, there are certain limitations for universal application of chemicals, as the preferred method of disease control, necessitating the search for alternative and more economically feasible method(s). Application of chemicals against crop diseases will be acceptable, only if, the net profit for the growers can be increased. Generally, application of chemicals against diseases affecting cereals is limited. In contrast, high value crops such as fruits and vegetable crops are protected by frequent application of chemicals. Blemish-free market demands and economic value per unit of crops such as apples, citrus, grapes and potatoes may justify intensive use of chemicals against microbial plant pathogens. Such repeated applications at high doses of chemicals are, however, considered to be the primary reason for development of resistance in microbial pathogens to the chemicals. Host resistance and cultural practices are the tactics generally applicable for agronomic crops, keeping the use of chemicals to the minimum level, because of the lower net profit from these crops (Narayanasamy 2002, 2017).

The chemicals used against soilborne microbial plant pathogens include fungicides, antibiotics, bactericides, nematicides and herbicides. There are certain situations wherein application of chemicals cannot be avoided. If a high yielding, but susceptible cultivar is to be grown in environments conducive for disease incidence and spread, protection of the cultivar with chemicals becomes indispensable, despite the expected effects of chemicals on the products, pathogen and the environments. During occurrence of a disease in epidemic proportions, large scale application of chemicals is recommended to minimize further spread to new locations. Furthermore, glasshouse crops are raised in confined areas, providing conditions favorable for rapid spread of the diseases (s) and these crops are necessarily protected with intensive chemical use. Seed crops and seed stocks have to be protected by frequent chemical applications to satisfy the tolerance limits prescribed by the certification and quarantine programs. Several other factors also may influence the need for chemical application. Cultivars with horizontal resistance (polygenic resistance) may need less chemicals than cultivars with vertical resistance, as in potato-late blight pathosystem. Crops infected by more aggressive (virulent) strains of a pathogen have to be protected by increasing the frequency of chemical application. When the microenvironment is conducive (with high levels of nitrogen application, no-till practice and irrigation practices) for disease development, the need for chemical protection increases. Furthermore, under favorable environmental conditions, the chemical at a recommended dose may be less effective for reducing disease intensity to required level. Calendar-based application of chemicals may be less effective than need-based application as suggested by disease forecasting systems, resulting in saving of costs chemicals and labor to some extent. Among the chemicals applied for the management of soilborne pathogens, fungicides are used more commonly against fungal pathogens than antibiotics, which are more commonly applied against bacterial diseases. The chemicals may be classified based on their mode of action, target site of activity and nature of active ingredients (a.i.). The chemicals may be grouped as fumigants, protectants or eradicants (therapeutatns), based on their ability to act on the pathogen, when applied prior to or after infections, respectively. They may be classified as nonsystemic (contact) and systemic based on their ability to move to other plant parts from the site of application to act on the pathogens present in other plant tissues.

4.1 ASSESSMENT OF ACTIVITY OF CHEMICALS AGAINST FUNGAL PATHOGENS

The fungicides are applied as seed treatment, soil application and foliage treatment against fungal pathogens. The fungicidal activity against the test fungal pathogen(s) is determined under in vitro, greenhouse/growth room and field conditions.

4.1.1 Laboratory Tests

4.1.1.1 Agar-Amended Assay

The effects of fungicides on mycelial growth, sporulation, spore germination and germ tube elongation formed at different stages of life cycle of target pathogens, are assessed. The growth media are amended with the test fungicides at different concentrations. The concentration of the fungicide that reduces the mycelial growth/spore germination by 50% (EC_{50}) is determined. The fungicide that has low EC_{50} value is considered to be more effective than other fungicides that have higher EC_{50} values. Baseline sensitivities of isolates of soilborne fungal pathogens have been determined by growing them on suitable medium amended with different concentrations of the test fungicides.

The effective concentrations (EC_{50}) of azoxystrobin, dimethomorph and fluazinam against *Phytophthora capsici*, *P. citrophthora* and *P. parasitica* were determined. Dimethomorph had the lowest EC_{50} values for inhibition of mycelial growth (<0.1 to 0.38 µg/ml). Azoxystrobin and fluazinam had higher EC_{50} values (>3,000 µg/ml). By contrast, azoxystrobin and fluazinam were more effective in inhibiting sporangium formation. Germination of encysted zoospores was the most sensitive to dimethomorph (EC_{50} and EC_{90} values ranging from 3.3 to 7.2 and 5.6 to 21 µg/ml, respectively), intermediate in sensitivity to fluazinam and metalaxyl and lowest in sensitivity to azoxystrobin and fosetyl-Al. The results suggested that azoxystrobin, dimethomorph and fluazinam might provide control of *Phytophthora* spp. to a level comparable to metalaxyl and fosetyl-Al used earlier (Matheron and Porchas 2000a, 2000b). The responses of mefenoxam-treated cucurbit plants to infection by isolates of *P. capsici* that were sensitive (S), intermediately sensitive (IS) and highly insensitive (I) were recovered from a cross between IS parents or from naturally infected cucumber, squash and pepper plants. Pumpkin or yellow squash seedlings were treated with water or mefenoxam at either 19.17 µg/ml or 57.51 µg/ml at 24 h prior to inoculation. All isolates of *P. capsici* were highly virulent on water-treated seedlings at 4 days after inoculation. The results suggested that in vitro screening of mefenoxam sensitivity using a single high rate of mefenoxam (100 ppm) might provide information useful for predicting the response of natural populations of *P. capsisci* to mefenoxam (Lamour and Hausbeck 2003).

Variations in the sensitivity of field isolates of fungal pathogens have been revealed in the case of *P. capsici*. Of the 120 isolates tested for sensitivity to mefenoxam at 100 mg/l, 8 isolates were resistant [based on relative colony diameter (RCD)]; 60 isolates were sensitive and 52 isolates were intermediately sensitive. The sensitive isolates were present in two fields, where mefenoxam-containing fungicides had not been applied at all earlier. Intermediately sensitive or resistant isolates were recovered in the four fields, where mefenoxam had been applied earlier. The in vitro tests were performed to determine comparative baseline sensitivity levels of field isolates of *P. capsici*, causing Phytophthora blight and crown rot in summer squash and pepper in South Carolina, USA

(Keinath 2007). In a later study, the sensitivity levels of 40 isolates of *P. capsici* to mandipropamid, dimethomorph and cyazofamid were assessed, using agar medium amended with different concentrations of the test fungicides. The EC_{50} values for inhibiting mycelial growth, zoospore germination and sporangial production were determined. The results showed that *P. capsici* populations in Georgia did not develop resistance to mandipropamid and dimethomorph, whereas majority of the isolates in certain asexual stages showed resistance to cyazofamid, indicating the need for testing the effectiveness of the fungicides against all stages in the asexual phase of the life cycle to determine the efficiency of the fungicide(s) (Jackson et al. 2012). Tetramycin with high level and broad-spectrum fungicidal activity was evaluated for its efficacy in suppressing the development of *P. capsici* on leaves of pepper. Formation of sporangia and the discharge of zoospores were inhibited by lower concentrations of tetramycin, (approximately 5 µg/ml on V8 medium), whereas the mycelial growth could be inhibited only at higher concentration of the fungicide. The frequency distribution curves for tetramycin sensitivity were unimodal with mean values for the fungicide concentration that reduced mycelial growth, sporangia formation and zoospore discharge by 50%, compared with control of 1.18 ± 0.91, 0.64 ± 0.42 and 0.63 ± 0.30 µg/ml respectively. In addition, no correlation was recorded between tetramycin and other fungicides including mandipropamid, azoxystrobin, mefenoxam, fluazinam, fluopicolide and famoxadone. Tetramycin possessed both protective and curative activities against *P. capsici* in vitro, and its protective activity was more effective than its curative activity. Under the greenhouse conditions, tetramycin at 60 and 90 µg/ml provided protective control efficacy of 47.1 to 56.4% and curative efficacy of 43.3 to 52.7%, indicating the possibility of applying this fungicide to reduce incidence of Phytophthora blight of pepper caused by *P. capsici* (Ma et al. 2018). The relationship between pathotypes and sensitivity of 395 single-spore isolates of *Phytophthora sojae* to metalaxyl was investigated. The pathotypes were identified by inoculation of soybean cultivars with 14 different *Rps* genes. Ninety-six percent of the isolates were virulent against more than four *Rps* genes, indicating the presence of multivirulent isolates in China. All 223 test isolates of *P. sojae* were sensitive to metalaxyl, but the EC_{50} values for the fungicide increased by 18-fold in the last 20 years. The results suggested that incorporation of *Rps* genes and application of metalaxyl might be an effective integrated disease management approach for Phytophthora root and stem rot of soybean caused by *P. sojae* (Tian et al. 2016).

The sensitivity of *Thielaviopsis basicola* and *Phytophthora parasitica* var. *nicotianae* to aluminum (Al) was determined over a range of pH. Toxicity of monomeric Al species to production of sporangia of *P. parasitica* var. *nicotianae* and chlamydospores of *T. basicola* was quantified. Pathogen colonies were grown in 5% carrot broth and washed with water after incubation period. After 2 days, the colonies were stained with Al-specific fluorescent stain lumogallion. *P. parasitica* var. *nicotianae* was sensitive to multiple monomeric Al species, whereas sensitivity of *T. basicola* to Al was pH-dependent, suggesting that only Al^{3+} might be responsible for suppression of *T. basicola*. Chlamydospore production by *T. basicola* was inhibited at pH <5.0 and Al levels >20 μM. By contrast, sporangia production by *P. parasitica* was inhibited at Al levels as low as 2 μM, across all pH levels tested. Aluminum accumulation was observed in sporangia and zoospores of *P. parasitica* var. *nicotianae* and in nonmelanized chlamydospores of *T. basicola*, but not in cell walls of either pathogen. The differential sensitivity of these pathogens might indicate its differential response to aluminum (Fichtner et al. 2006).

The demethylation inhibitor (DMI) fungicide epoxiconazole was evaluated for its efficacy in suppressing the development of stem rot caused by *Sclerotinia sclerotiorum* in oilseed rape and other crop hosts. Epoxiconazole when applied as a preventive treatment (5 and 15 μg/ml) reduced disease severity by 98.5 to 100% and it was more effective than dimethachlon. However, its curative effect was less than that of dimethachlon. No positive cross-resistance was observed between epoxiconazole and carbendazim or dimethachlon. The results indicated the need for ascertaining the nature of the fungicide activity as preventive or curative to obtain effective suppression of disease development (Li et al. 2015). Likewise, fluazinam was found to be highly effective as preventive, rather than curative treatment against *S. sclerotiorum* and did not induce any cross-resistance with carbendazim or dimethachlon (Liang et al. 2015a). Similar protective effect of fluazinam against *S. sclerotiorum* was observed under field conditions (Wang et al. 2016). Fungicides with site-specific mode of action, such as quinine outside-inhibiting (QoI), azoxystrobin and pyraclostrobin and the carboximide fungicide boscalid were evaluated for their fungicide activity against *Ascochyta rabiei*, causing Acochyta blight of chickpea under in vitro conditions. The effect of salicylhydroxamic acid (SHAM), alternative oxidase inhibitor of alternative respiration of *A. rabiei* was also assessed in the presence and absence of azoxystrobin. Five of nine isolates of *A. rabiei* had significantly higher EC_{50} values in the absence of SHAM, in the medium amended with azoxystrobin, indicating the potential of *A. rabiei* to use alternative respiration to overcome the fungicide toxicity in vitro (Wise et al. 2008). The QoI fungicide pyraclostrobin and SHAM were investigated for their activity against *S. sclerotiorum* with wide host range including several crops. SHAM at 20 μg/ml significantly increased the inhibitory effect of pyraclostrobin on mycelial growth of the pathogen in potato dextrose agar (PDA) medium and also increased the control efficacy of the fungicide in planta. The results indicated that SHAM should not be incorporated into the media for in vitro assay of *S. sclerotiorum* sensitivity to pyraclostrobin and other QoI fungicides (Liang et al. 2015b). The efficacy of three fungicides, fluopicolide, mandipropamid and oxathiapiprolin in inhibiting the development of *Phytophthora nicotianae*, causal agent of black shank disease of tobacco was assessed. All three fungicides inhibited the mycelial growth of the isolates tested with a mean of EC_{50} values of 0.09, 0.04 and 0.001 μg/ml, respectively. The EC_{50} values for inhibition of sporangial formation for these fungicides were 0.15, 0.03 and 0.0002 μg/ml, whereas the zoospore germination was inhibited at EC_{50} values of 0.16, 0.04 and 0.002 μg/ml. Oxathiapiprolin was the most effective in inhibiting the pathogen at three different stages in its life cycle. Determination of base line sensitivity levels at different stages in the life cycle of the pathogen might facilitate monitoring of resistance development in pathogen populations in a geographical location (Qu et al. 2016).

The fungicidal activity of disinfectant and natural products has been assessed for restricting development of soilborne fungal diseases. The effect of chlorine dioxide (ClO_2), used as a disinfectant against waterborne pathogens was evaluated by mixing different concentrations in solutions containing nitrogen and hard water with equal concentrations of ammonia, nitrate and synthetic hard water and a divalent metal ion solution with equal concentrations of Cu, Fe, Mn and Zn at pH 5 and 8. Spore germination of *Fusarium oxysporum* f.sp. *narcissi* and *Thielaviospsis basicola* was inhibited by ClO_2 to different extent. High concentration of ClO_2 was required at pH 8 than at 5.0 to achieve lethal dose, resulting in 50% mortality of spores (LD_{50}). The concentration of ClO_2 to give 50% mortality ranged from 0.5 to7.0 mg/l for conidia of *F. oxysporum* f.sp. *narcissi*, 0.5 to 11.9 mg/l for *T. basicola* and 15.0 to 45.5 mg/l for aleuriospores of *T. basicola* (Copes et al. 2004). The condensed distiller's soluble (CDS), a coproduct of ethanol production from corn contain both volatile (acetic and formic) and nonvolatile (glycolic) organic acids. The viability of *Verticillium dahliae* microsclerotia treated for 24 h in 1, 2, 5 and 10% (v/v) CDS (pH 3.6–4.5) or a mixture of organic acids with the same percent composition as the CDS was reduced by 2, 7, 22 and 48% or 6, 32, 53 and 63%, respectively. A mixture of organic acids with the same volumetric ratios as 2% and 4% CDS completely inhibited the growth of *Pythium ultimum* after culture plugs treatment for 24 h (Abbasi et al. 2009).

The stimulatory effects of low doses of fungicides on aggressiveness of fungal pathogens have been reported. Hormesis is a toxicological concept of dose-response relationships characterized by low dose-stimulation and high-dose inhibition. Hormesis refers to a reversal response between low and high doses of the chemical. Under field conditions, a portion of the population of the pathogen strain/isolate is likely to be exposed to low dosages of a fungicide due to evaporation, drift, or uneven distribution of the active ingredients. Furthermore, it is possible that the concentration of fungicides may be reduced/degraded, reaching sublethal concentration due to rain or other environmental conditions. The effects of sublethal dose of mefenoxam on isolates of *Pythium aphanidermatum*, incitant of Pythium damping-off of geranium were assessed. The mefenoxam resistant isolate of *P. aphanidermatum* showed an increase of mean radial growth (10%) with mefenoxam at 1×10^{-10} μg/ml, compared to growth on nonamended medium. Geranium seedlings were treated with 1/8 concentrations of mefenoxam and inoculated with pathogen colonized agar plugs. Significant increases in AUDPC values were recorded on geranium seedlings treated with mefenoxam at 1×10^6 and 1×10^{-10} μg/ml concentrations. The results indicated the stimulatory effect of sublethal doses of mefenoxam on isolates resistant to the fungicide, resulting in higher level of disease severity in geranium seedlings (Garzón et al. 2011).

The benzimidazole fungicide carbendazim was extensively applied for the control of *S. sclerotiorum*, infecting oilseed rape and other crops, resulting in emergence of isolates with resistance to the fungicide and yet its use continued. Stimulatory effect of subtoxic doses of carbendazim on *S. sclerotiorum* was assessed, using seven field isolates of the pathogen, showing resistance to the fungicide. All field-resistant isolates with EC_{50} values greater than 1,000 μg/ml exhibited stimulation of pathogenicity toward detached leaves of rapeseed at subtoxic concentration of carbendazim. Assays, using detached leaves and potted plants, showed that carbendazim at 0.2 to 5 μg/ml, more consistently stimulated development of lesions/symptoms than in the control infected by two resistant lines AH-17 and LJ-86 of *S. sclerotiorum*. On potted plants, percent stimulation of pathogenicity ranged from 18.8 to 22.0% for AH-17 and from 15.1 to 23.2 % for LJ-86. Detached leaf assessments also showed similar stimulations of pathogenicity of the two isolates. Secretions of oxalic acid and tolerance to oxidative stresses of H_2O_2 and paraquat after exposure to subtoxic doses of carbendazim did not change significantly, indicating the possible mechanism of stimulatory effect of subtoxic concentration of carbendazim on *S. sclerotiorum* (Di et al. 2015). In the further investigation, the time course of the stimulatory effect of carbendazim on *S. sclerotiorum* was studied. At 12 hours postinoculation (*hpi*), the initial necrotic lesions were visible in rapeseed leaves treated with carbendazim at 0.2 and 1 μg/ml, whereas no disease symptom appeared in nontreated control. At 18 hpi, carbendazim stimulation on pathogenicity was more clearly observed. Examination under scanning electron microscope (SEM) revealed no recognizable variation in the development of disease symptoms at 8 hpi. Greenhouse experiments showed that spraying carbendazim at 400 μg/ml on potted rapeseed plants, resulted in significant stimulation of pathogenicity for inoculation at 1, 3, 5 and 7 days after application (DAA). The stimulatory activity was lost for inoculation at 14 DAA. Mycelia in potato dextrose agar (PDA) amended with carbendazim at 400 μg/ml were more virulent than nontreated control. However, after additional growth of the mycelium on fungicide-free PDA for two days resulted in loss of stimulatory effect entirely, indicating that carbendazim was indispensable for stimulation of virulence. Biochemical analyses showed that cell wall-degrading enzymes such as cellulase, pectinase and polygalacturonase were not involved in pathogenicity stimulations. The results indicated the need for understanding positive and negative effects of fungicides to harness the benefits of chemical application (Di et al. 2016a, 2016b).

The stimulatory effects of flusilazole, a demethylation inhibitor (DMI) fungicide, on the virulence of *S. sclerotiorum* were assessed. Flusilazole sprayed at concentrations 0.02–0.5 μg/ml, induced significant stimulation of virulence of *S. sclerotiorum* on rapeseed plants grown in pots. The stimulation magnitudes reached maximum levels of 11.0% and 10.7% respectively for two isolates GD-7 and HN-24 of

S. sclerotiorum. Stimulatory effect on virulence could be detected as early as 18 h postinoculation, indicating a direct stimulation mechanism, rather than an overcompensation of initial inhibition by the fungicide. In order to differentiate the effects of the fungicide and rapeseed plants, mycelia grown on flusilazole-amended potato dextrose agar (PDA) medium, were inoculated on leaves of rapeseed plants without spraying the fungicide. Mycelial growth of *S. sclerotiorum* on PDA amended with flusilazole at 0.005 to 0.16 μg/ml was inhibited by 10.11 to 48.7% for GS-7 isolates and by 4.1 to 24.9% for isolate HN-24. Flusilazole in PDA at 0.04 and 0.08 μ/ml caused slightly deformed mycelia and twisted hyphal tips, as revealed by observations under scanning electron microscope (SEM). However, following inoculation of leaves of potted rapeseed plants, virulence of the isolates, after initial inhibition, showed significant enhancement of virulence to a level greater than that of nontreated control, reaching the maximum stimulation magnitudes of 16.2% and 19.8%, respectively, in isolates GS-7 and HN-24. Analysis of physiological mechanism of virulence stimulation showed that tolerance to hydrogen peroxide did not increase significantly for mycelia grown on flusilazole-amended PDA, excluding the possibility of tolerance to reactive oxygen species as potential mechanism of virulence stimulation in *S. sclerotiorum* (Lu et al. 2018).

Sensitivity of *Rhizoctonia solani*, causing soybean root rot to succinate dehydrogenase inhibitor (SDHI)–penflufen and sedaxane and demethylation inhibitor (DMI) fungicides–ipconazole and prothioconazole, was assessed. All isolates of *R. solani*, regardless of collection date, were extremely sensitive (EC_{50} <1 μg/ml) to SDHI fungicides but were either extremely or moderately sensitive (EC_{50} 1 < to <10 μg/ml) to the DMI fungicides. Variations in the sensitivity to all four active ingredients were observed within and among the different anastomosis groups composing both isolate groups. The in vitro sensitivity was not always correlated with in vivo assessments in the greenhouse. All isolates were effectively suppressed by seed treatment with the fungicides in the greenhouse trials. The results showed that no shift in sensitivity of the two groups of isolates occurred, as indicated by in vitro and greenhouse assessments (Ajayi-Oyetunde et al. 2017). *F. oxysporum* f.sp. *niveum* (*Fon*) causes Fusarium wilt disease of watermelon. Application of prothioconazole and thiophanate-methyl™ significantly reduced the disease incidence under field conditions. The effects of the fungicides in different stages of the life cycle of the pathogen and the possible development of resistance in the pathogen isolates to the fungicides were investigated. In vitro assessments showed that based on the mycelial growth, all isolates (100) were sensitive to prothioconazole with EC_{50} values ranging from 0.75 to 5.69 μg/ml with average values of 1.62 μg/ml. By contrast, 33% and 4% of the isolates were resistant to thiophanate-methyl at 10 and 100 μg/ml respectively. Microconidial germination assays showed that 36% and 64% of the isolates of *Fon* were sensitive or intermediately sensitive to prothioconazole at 100 μg/ml, but the fungicide did not inhibit spore germination at 10 μg/ml. The results showed that prothioconazole might have less chance of inducing resistance in the isolates of *F. oxysporum* f.sp. *niveum*, whereas isolates with resistance to thiophanate-methyl were already present in the fields in Georgia, USA (Petkar et al. 2017).

4.1.1.2 Microbioassay

In order to select chemicals highly inhibitory to *Phytophthora* spp., the microbioassay involving the use of a 96-well format was developed for high-throughput capability and a standardized method for quantification of initial zoospore concentrations for maximum reproducibility. Zoospore suspensions could be quantified using a suspension containing 0.7–1.5 x 10^5/ml at 620 nm. Subsequent mycelial growth was monitored by measuring optical density (OD) at 620 nm at 24-h intervals for 96 h. The assay was useful to determine effective concentration values for 50% growth reduction by test fungicides. *Phytophtora nicotianae* was grown in Roswell Park Memorial Institute mycological broth (RPMI). Suspension at 1,000 zoospores/ml was used as optimal initial concentration. The EC_{50} values of azoxystrobin, fosetyl-aluminum, etridiazole, metalaxyl, pentachloronitrobenzene (PCNB), pimarcin and propamocarb were compared with those obtained by measuring linear growth of mycelia on fungicide-amended medium. The microbioassay was rapid and the results could be reproduced (Kuhajek et al. 2003). The sensitivity levels of 89 isolates of *Phytophthora cactorum*, incitant of crown rot and leathery rot of strawberry to quinone-outside-inhibiting (QoI) fungicide azoxystrobin were assessed, using lima bean agar amended with the fungicide and salicylhydroxamic acid (SHAM) in microtiter plates. The effective dose to reduce mycelial growth by 50% (ED_{50}) ranged from 0.16 to 12.52 µg/ml for leathery rot isolates and 0.10 to 15 µg/ml for crown rot isolates. Differences between sensitivity distributions for zoospores of *P. cactorum* were determined and the zoospores were more sensitive to another QoI fungicide, pyraclostrobin than to azoxystrobin. Sensitivities to azoxystrobin and pyraclostrobin were moderately, but significantly correlated (Rebollar-Alviter et al. 2007). Of the 320 isolates of *P. aphanidermatum*, *P. irregulare* and *P. myriotylum* tested, isolates sensitive and resistant to mefenoxam were detected in the same greenhouse and 12 of 26 greenhouses had more than one *Pythium* spp. The isolates and species of *Pythium* were identified by sequencing the internal transcribed spacer (ITS) rDNA gene region. Levels of sensitivity of *Pythium* spp. were determined using microtiter plate wells containing clarified V8 agar medium amended with 100 µg a.i. /ml of mefenoxam. Majority of isolates (52%) were resistant to mefenoxam. All isolates of *P. myriotylum* were sensitive to mefenoxam, whereas sensitivity of isolates of *P. aphanidermatum* and *P. irregulare* varied. Sensitivity to mefenoxam was not related to the virulence of *Pythium* species (Lookabaugh et al. 2015).

4.1.1.3 Colorimetric Bioassay

The colorimetric bioassay involving the use of 96-well microtiter plates, Almar Blue (AB) dye, a stable water-soluble, redox indicator was employed for evaluating sensitivity of isolates of *V. dahliae* to fungicides with different mode of action. The AB dye could be used for measuring cell proliferation, viability and cytoxicity levels. The redox reaction of AB occurs as a result of cellular metabolism of viable cells as continued growth maintains a reduced environment. Reduction causes the dye to change the color from blue (oxidized form) to fluorescent pink color (reduced form) which can be visually recognized or measured precisely using spectrometer or fluorometer using a range of serial dilutions. Metalaxyl, mancozeb, benomyl, thiophanate-methyl, fosetyl-Al, vacomil (mancozeb + 8% metalaxyl) etridiazole, mancozeb + copper hydroxide (Mankocide), Phyton 27 (copper sulfate), 9% dimethomorph + 60% mancozeb (Acrobat) were tested against 10 isolates of *V. dahliae*. The incubation time, spore density and media type were the important parameters that had to be optimized. V8 juice broth, a spore density of 10^7 spores/ml at incubation temperature of 37°C for 24 h. were employed for the test. Greater percentage of inhibition of growth was recorded for AB assay, compared to conventional agar-amended (AA) assay. Three-dimensional growth and germ tube production from germinating spores could be recorded in AA assay. Both mycelial growth and spore germination were inhibited by fosetyl-Al, dithiocarbamate and dimethocarb which are multisite inhibitors (Rampersad 2011).

4.1.1.4 Nanoparticles Assay

Silver nanoparticles (AgNPs) possess strong antifungal and antibacterial activity. AgNPs are primarily composed of zerovalent silver (Ag^0) clusters with a size range of 5–100 nm in diameter. AgNP preparations may consist of nanospheres, nanotubes, triangular crystals or a combination of these particles. Silver nanoparticles synthesized using plant extracts have inhibitory effect on plant pathogenic fungi and bacteria. The potential of silver nanoparticles (AgNPs) synthesized with aqueous extracts of *Artemisia absinthium* was assessed against *Phytophthora* spp. that cause several economically important crop diseases. In vitro dose-response tests conducted in microplates showed that AgNPs at 10 µg/ml could inhibit mycelial growth of *P. parasitica*, *P. infestans*, *P. palmivora*, *P. cinnamomi*, *P. capsici*, *P. tropicalis* and *P. kasturae*. AgNPs strongly inhibited mycelial growth, zoospore germination, germ tube elongation and zoospore production by *P. parasitica* and *P. capsici*, as revealed by detailed dose-response analysis. Furthermore, AgNP treatment accelerated encystment of zoospores. Greenhouse experiments showed that AgNP treatment prevented infection by *Phytophthora* sp. and improved survival of plants. No adverse effects on plant growth, following AgNP treatment were discernible. The results showed that use of AgNPs, a simple and economical method could be applied for the management of diseases of economically important crops caused by *Phytopthhora* spp. (Ali et al. 2015).

The efficacy of nonfungicide chemicals have been evaluated for their antifungal activity against plant pathogens. Nursery production depends on recycled irrigation water, which is the primary source of inoculum of *Phytophthora* spp. and the pathogen propagules may spread through the irrigation water. The effectiveness of chlorine to be used as a disinfectant was investigated. Zoospores of seven species and eight isolates of *Phytophthora* spp. were exposed for 2 min to free, available chlorine at 0.25, 0.5, 1.0, 2.0 and 4.0 mg/l. Zoospores, mycelial fragments and culture plugs of *P. nicotianae* also were exposed to chlorine concentration ranging from 0.25 to 8.0 mg/l for periods ranging from 0.25 to 8.0 mg/l for periods ranging from 15 s to 8 min. Further, chlorinated water was assayed at monthly interval during 2000–2001 at two commercial nurseries and at monthly interval in the nurseries in Virginia, for chlorine and survival of pythiaceous species, using a selective medium. Zoospores of all pathogen species tested did not survive end point–free chlorine at 2 mg/l, while mycelial fragments of *P. nicotianae* survived at 8 mg/l to a limited extent. Mycelial plugs treated with chlorine at 8 mg/l produced a few sporangia. *Phytophthora* spp. could be recovered only from nursery irrigation water with levels of free chlorine at 0.77 mg/l or lower. The results of the investigation may be used as a guideline for improving the chlorination protocols (Hong et al. 2003).

Suppressive potential of algaecides on the development of *P. capsici*, incitant of several economically important crop diseases, was assessed under controlled conditions. Zoospore mobility and mortality in response to commercial algaecides were determined. Cucumbers were infected at all temperatures (2–32°C), except at 2°C tested and the highest infection percentage was observed in cucumbers incubated in suspensions of zoospores held at ≥ 19°C. Fewer fruits were infected (<40% at >19°C, 0% at ≤ 12°C), when water contained 1 x 10^2 zoospores/ml. Almost 100% of fruits were infected, when water contained ≥ 5 x 10^3 zoospores/ml at ≥ 2°C. Infection percentage declined with increase in age of zoospore suspension, although occurred, when 5-day-old suspension was infective. Commercial algaecides increased zoospore mortality significantly, showing promise for use as disinfectant of irrigation water. Use of infested irrigation water should be avoided throughout the growing season (Granke and Hausbeck 2010).

4.1.2   METHODS OF APPLICATION OF CHEMICALS

Development of diseases caused by soilborne microbial pathogens may be restricted by applying chemicals as seed treatment, root-dip of transplants/propagules treatment, soil application and foliage treatment. The frequency and timing of application have to be determined, based on the concentration of chemicals and prevailing environmental conditions.

4.1.2.1   Treatment of Seeds/Propagules

Soilborne fungal pathogens may colonize spermosphere and remain as dormant mycelium or as resting spores/overwintering structures such as sclerotia or chlamydospores. Systemic fungicides may be more effective, because of their systemic action and ability to move from the site of application to different organs and act on the pathogen(s) that have already invaded other plant tissues including root tissues. Seed treatments with formulations containing fluquinconazole or fluquinconazole plus prochloraz, decreased wheat take-all disease caused by *Gaeumannomyces graminis* var. *tritici* and

also increased grain yield, especially in the second wheat crop in which take-all was building-up, and in the third wheat crop, when take-all disease was most severe. Seed treatment was less effective in the fourth wheat crop, when take-all appeared to be in a decline mode in plots that had no seed treatment throughout the experiment. The results suggested that seed treatment with fungicide might delay take-all build-up and progress of take-all decline (Dawson and Bateman 2001). The fungicide silthiofam was applied as seed treatment to contain the incidence and spread of wheat take-all disease caused by *Gaeumannomyces graminis* var. *tritici* (Ggt). Seed treatment reduced the proportion of diseased roots throughout both phases of primary and secondary infection that may otherwise result in epidemic. The potential of silthiofam to affect secondary infection from either source of infected roots were not affected. Seed treatment controlled primary infection of seminal roots from particular inoculum but not secondary infection from either seminal or adventitious roots. The reduction in disease incidence in silthiofam-treated plants was observed, following secondary infection phase of the epidemic was not due to long-term activity of the chemical but to the manifestation of disease control early in the epidemic (Bailey et al. 2005).

The efficacy of 13 fungicides in inhibiting the mycelial growth of *A. rabiei*, causing Ascochyta blight of chickpea was assessed. Of these fungicides, treatment of seeds with benomyl + thiram, carbendazim and carbendazim + chlorothalonil reduced infection by *A. rabiei* by more than 85%, on vacuum infiltration of naturally infected seeds with the fungicides. Coating the seed with polymers did not increase the protective effect of the fungicides. Azoxystrobin, chlorothalonil and mancozeb were the most effective under field conditions. With increased disease pressure, the efficacy of the fungicides declined (Demirci et al. 2003). The effectiveness of mefenoxam and metalaxyl in suppressing the development of damping-off disease of pumpkin caused by *P. capsici* was assessed under greenhouse condition. Seed treatment with mefenoxam (0.42 ml of Apron XL LS/kg of seed) and metalaxyl (0.98 ml of Allegiance FL/kg seed) significantly reduced pre- and postemergence damping-off of pumpkin seedlings of cv. Dickinson Hybrid-401 and Hyrid-698) tested. Under field conditions, the average seedling stands at 35 days after seeding were 76.7, 74.7 and 44.9% for mefenoxam, metalaxyl and untreated control, respectively (Babadoost and Islam 2003). The comparative efficacy of triazole and resistance compounds, in reducing incidence of black root rot of cotton disease caused by *T. basicola* by seed treatment was assessed. In naturally infested soil, seed treatment with myclobutanil was effective in reducing root and hypocotyl discoloration over a wide range of pathogen densities. Higher dose of myclobutanil (42 g a.i./100 kg seed) provided greater protection than low rates (21 g a.i/100 kg seed) in some experiments. Acibenzolar-S-methyl (ASM), a resistance inducer, applied as seed treatment, reduced black root rot or colonization by *T. basicola* on seedlings in artificially infested soils. When combined, the fungicide and resistance inducer were more effective than when applied separately under controlled conditions. Similar results were obtained, when both chemicals were used as seed treatment, in soils with low (24 CFU/g soil) and high (154 CFU/g soil) population of the pathogen. The cotton plants showed the lowest root discoloration, due to combined treatment of seeds (Toksoz et al. 2009).

The effectiveness of treatment of soybean seeds with fungicides, followed by planting in two soil types was assessed under growth chamber conditions. Seed treatment with broad-spectrum fungicides trifloxystrobin + metalaxyl and mefenoxam + fludioxonil + azoxystrobin provided effective protection against *Pythium* spp. and *R. solani*, resulting in highest plant stands at 21°C, 25°C and 28°C, prevailing at different months of planting. Metalaxyl and pentachloronitrobenzene + carboxin were also effective in reducing disease incidence (Urrea et al. 2013). Seed treatment of pearl millet seeds with cyazofamid was evaluated for its potential for reducing incidence of downy mildew disease caused by *Sclerospora graminicola*. Seed treatment with cyazofamid at 2.0 mg/ml, followed by a single foliar spray (1–10 mg/ml) provided effective protection against infection by *S. graminicola* of emerging seedlings and prevented spread of the sporangia formed on the aerial plant parts. Cyazofamid did not have any systemic action but had only moderate translaminar activity with marginal reduction of disease incidence due to its curative effect. Loss of fungicidal activity over time was very low, indicating its stable, residual and rainfastness activity. As no phytotoxic symptoms could be seen on treated plants, the potential of cyazofamid for suppressing the downy mildew disease of pearl millet could be exploited (Jogaiah et al. 2007). Fungicides were evaluated for their effectiveness in reducing incidence of damping-off disease of pea caused by *R. solani*. The fungicides carbendazim, carbendazim + thiram, captan, iprodione, iprodione + thiram, metalaxyl-M + fludioxonil, pencycuron, procyamidone and tolyfluamid were applied as seed dressers either alone or in combination. Seeds were planted in artificially infested or uninfested soil (control). Seedling emergence was reduced in soil infested with *R. solani*. Development of damping-off disease was most effectively suppressed by carbendazim, pencycuron, iprodione and carbendazim + thiram (da Silva et al. 2013).

Asexually propagated planting materials (propagules), are likely to be infected asymptomatically, if they are taken from mother plants that are infected late in the cropping season. Furthermore, the propagules obtained from disease-free mother plants also have to be protected by chemicals against soilborne pathogens. Potato seed tubers may remain symptomless, but carry fungal, bacterial and viral pathogens. Preplanting treatments of seed tubers with sodium hypochlorite solution (as NaOCl, 50 ppm for 8 min), thiophanate-methyl (TPM, as Esout™, 50 g a.i 100 kg seed tubers) or a combination treatment with NaOCl, followed by TPM were evaluated for their efficacy to reduce incidence of black scurf disease caused by *R. solani* and common scab caused by *Streptomyces scabies* on tubers of potato cv. Russet Burbank. Treatment with NaOCl + TPM reduced *R. solani* infection on progeny tubers at harvest and also after storage. Combined chemical application might have killed majority of *R. solani*

sclerotia present on tuber surface and also suppressed mycelial growth of *R. solani*. No phytotoxic symptoms on tubers treated with single or combinations of fungicides were recorded (Errampalli and Johnston 2001). The effectiveness of potato seed tuber treatment against tuber rot caused by *Phytophthora infestans* was assessed. Treatment of seed tubers with thiophanate-methyl + mancozeb increased plant emergence and uniform stand of potato crop. Foliar fungicides dimethomorph + mancozeb, cymoxanil + mancozeb, or propamocarb + chlorothalonil provided protection to seed tubers. However, additional application of these fungicides on foliage was required for effective suppression of development of late blight on potato plants at later stages of crop growth. Treatment of sprouts with oxadixyl or Ridomil protected them more effectively than treatment of whole tubers with deep incision or cut tubers (Singh and Pundhir 2004). Potato seed tubers were treated with fungicides singly or in combination for effective protection of daughter tubers against silver scurf disease caused by *Helminthosporium solani*. Using low-volume spray, fludioxonil and prochloraz-Zn were the most effective showing control efficiencies of 88% and 82% respectively. Propineb and mancidan, applied as dust treatment had control efficiencies of 78% and 79% respectively. Treatment with azoxystrobin and imazil were less effective with control efficiencies of 68% and 43%, respectively. The fungicides did not improve the tuber yields in all three locations tested (Tsror and Pertz-Alon 2004). Thiabendazole (TBZ), fludioxonil, mancozeb, thiophanate-methyl (TPM), azoxystrobin, sorbic acid and sodium carbonate were evaluated for their efficiency in reducing incidence of silver scurf disease in potato seed tubers at four locations, following tuber treatment. Incidence of the disease in progeny tubers was significantly reduced to different levels. Fludioxonil, fludioxonil + quintozene, azoxystrobin, or TPM + mancozeb treatments were significantly more effective in reducing disease incidence in all locations at harvest in the spring in two years of trials. TPM, sorbic acid, mancozeb and sodium carbonate were not effective in reducing silver scurf incidence at any of the locations tested. Furthermore, isolates of *H. solani* resistant to TBZ and TPM were recovered also from the fields (Geary et al. 2007).

Treatment of alfalfa seeds with mefenoxam (Apron XL) for suppressing the development of Phytophthora root rot (PRR) caused by *P. capsici* and Aphanomyces root rot (ARR) caused by *Aphanomyces euteiches* was not effective and also it was not acceptable for organic production system. Hence, a seed coating method involving use of aluminosilicate (natural zeolite) at 0.33 g of zeolite/g seed was evaluated under growth chamber conditions. The mineral seed coating proved to be more effective in providing protection against PRR, compared to Apron XL, with a mean of 89% and 38% healthy seedlings respectively in these treatments, whereas the control treatment had only 15% healthy seedlings. Likewise, mineral seed coating was more effective in protecting the alfalfa seedlings against ARR disease. In addition, mineral coating was more effective against seed rot and damping-off caused by *P. ultimum* and *P. paroecandrum*, as indicated by greater percentage of healthy seedlings. In growth chamber assays using naturally infested field soils with a range of disease pressure, seed treatment with mineral coating was more effective than Apron XL treatment. Furthermore, mineral seed coating did not affect growth of and nodules produced by *Sinorhizobium meliloti*, providing an additional advantage of using mineral coating for treating the alfalfa seeds against important seedling diseases, which could affect the production system seriously (Samac et al. 2015). The effectiveness of phosphite applied as liquid and soluble capsule implants via stem injections for controlling *Phytophthora cinnamomi*, infecting *Banksia grandis* and *Eucalyptus marginata*, was assessed. At 4 weeks after application of chemical, excised branches were underbark inoculated with *P. cinnamomi*. In *B. grandis*, phosphite implants and liquid injections reduced lesion length significantly, compared to untreated control and MEDICAP MD® implants. In *E. marginata*, phosphite implants and liquid injections significantly reduced lesion length, compared to control, PHOSCSA® and MEDICAP MD® implants. The results showed that both liquid and phosphite implants were able to provide more effective protection to *B. grandis* and *E. marginata* against *P. cinnamomi* (Scott et al. 2005).

4.1.2.2 Treatment of Aerial Plant Organs

The effectiveness of fosetyl-Al and metalaxyl in suppressing the development of citrus gummosis caused by *Phytophthora citrophthora* and *P. nicotianae* was compared with other fungicides azoxystrobin, dimethomorph, fluazinam and zoxamide. The number and average size of cankers were reduced by treatment with dimethomorph, fosetyl-Al or metalaxyl, compared with nontreated control and those treated with azoxystrobin or fluazinam, when the bark removed from treated trees were inoculated with *P. citrophthora* on the cambium surface at 5, 30, or 60 days after treatment (DAT), lesion development was inhibited significantly on the bark strips treated with dimethomorph, fosetyl-Al or metalaxyl to a greater extent than that detected on bark treated with azoxystrobin, fluazinam or zoxamide. By contrast, with *P. nicotianae* inoculated at 5 or 30 DAT, reduction in lesion size on bark strips treated with dimethomorph, fosetyl-Al or metalaxyl was significantly greater than that detected in bark treated with azoxystrobin or fluazinam. Reduction of lesion development on the cambium surface compared with outer bark surface, when inoculated with *P. citrophthora*, did not differ significantly from 5 to 30 DAT for bark tissue treated with azoxystrobin, dimethomorph, fosetyl-Al, or metalaxyl. The results indicated that among the fungicides, dimethomorph protected citrus plants more effectively against infection by *Phytophthora* spp. causing gummosis disease (Matheron and Porthas 2002).

The effectiveness of cyazofamid with far less minimum inhibitory concentration (MIC) than mancozeb and metalaxyl used earlier, was evaluated for its effectiveness against *P. infestans*, causing potato late blight disease. The fungicide reduced sporangia formation at 6.3 mg/l on susceptible plants by 100%. Cyazofamid showed translaminar and curative activity with no cross-resistance with other commonly used fungicides under field conditions (Mitani et al. 2002).

Phosphorus acid was applied on the foliage of potato cultivars at different times and rates to determine its effectiveness in reducing tuber rots caused by *P. infestans*, *P. erythroseptica* and *P. ultimum* by inoculating harvested tubers. Mean incidence and severity of late blight/tuber rot in tubers inoculated with US-8 and US-11 isolates of *P. infestans* were significantly less in plants treated with phosphorus acid. With two sprays of phosphorus acid, late blight, tuber rot in tuber-resistant cv. Umatilla Russet was significantly less than in cv. Ranger Russet. Incidence and severity of late blight/ tuber rot did not differ significantly between the rates 7.49 and 9.37 kg a.i/ha at both locations (Othello and Mount Vernon). Incidence of pink rot was significantly less in inoculated tubers from plots treated with three sprays of phosphorus acid than tubers from untreated control plots in 2002 and 2003. Pink rot incidence, but not severity, was reduced by three sprays to a great extent. Application timing had similar effect on pink rot incidence, but not severity of disease. The effect of phosphorus acid on Pythium tuber rot was not clearly observed (Johnson et al. 2004). The efficacy of azoxystrobin on potato black dot and inoculum of the pathogen *Colletotrichum coccodes* in the seed tuber was assessed under field conditions during two potato seed generations (1 and 3) of susceptible cvs. Norkotah Russet and Russet Burbank during 2002–2004). Severity of black dot was reduced by azoxystrobin by 19 to 81% and 22 to 81% on above- and below-ground stem sections respectively. Both cultivars treated with azoxystrobin, had 9 to 26% less infected progeny tubers than the non-treated plants, indicating the potential of azoxystrobin to reduce black dot severity on both stems and progeny tubers. The incidences of *C. coccodes* in generation 1 mother tubers of Norkotah Russet and Russet Burbank were 2% and 16%, respectively, in 2003, and 0% and 30%, respectively, in 2004. Disease incidence showed increases in the next generation tubers. Yield reduction in the potato cultivars was not significant in the two generations due to black dot disease. The results suggested that the effect of inoculum source of *C. coccodes* on disease severity might be cultivar-specific (Nitzan et al. 2005).

Fungicide application frequency and timing, in addition to the concentration, are important factors influencing the effectiveness of disease suppression. In order to suppress the development of stem rot of potato caused by *S. sclerotiorum*, fungicides were sprayed on potato foliage at row closure (between rows) and at full bloom of primary inflorescences during 2003–2005. Thiophanate-methyl (TPM), fluazinam and boscalid applied at full label rates at full bloom of primary inflorescences were more effective, than when applied at row closure. Mean percentage of control for the fungicides combined was 43, 48 and 30%, relative to the control, respectively, following fungicide application at row closure stage, whereas the mean percentage of control was 77, 83 and 80% for fungicides applied at full bloom of primary inflorescences. The results indicated the need for determining the timing of fungicide application to derive maximum benefit of the fungicides (Johnson and Atallah 2006). Rates and timings of application of chlorothalonil and azoxystrobin separately and in combination were determined for effective suppression of *A.*

rabiei, which was a major limiting factor for production under low and high disease pressure conditions. Fungicides did not have any effect on disease severity under low disease pressure and dry weather conditions. But a single spray of fungicides under high pressure and wet conditions, disease severity was reduced, but had no effect on yield. Two applications (early + midflowering stages) of chlorothalonil at 1 kg a.i /ha or two applications of azoxystrobin at 125 g a.i/ha or chlorothalonil + azoxystrobin reduced the disease and also increased the yield. A strong relationship between disease severity and yield of seed was observed. Application of azoxystrobin at early + midflowering stages reduced seed infection by *A. rabiei* to 7 to 9% from 30 to 48% in control. Other seedborne fungal pathogens *Botrytis cinerea, S. sclerotiorum* and *Fusarium* were not affected by these fungicides (Chongo et al. 2003). Ascochyta blight disease of chickpea caused by *A. rabiei* reduced the yield significantly in Canada. The fungicides chlorothalonil, azoxystrobin, pyraclostrobin, mancozeb and boscalid were evaluated for their efficacy, application frequency, timing and rotation of different fungicides to control the disease effectively. Disease severity was reduced by the fungicides to levels below 25% of the untreated control, which had 21 to 99% disease severity. Increasing the frequency of fungicide application was beneficial, when the disease pressure was high. Preflowering application had a positive effect by suppressing Ascochyta blight severity, increasing the seed yield or 1,000-seed weight in almost all experiments. Higher numbers of applications were not always correlated with higher seed yield and better seed quality. Different fungicide rotations had less impact on disease management than the timing and the number of applications, but under moderate to high disease pressure, including strobilurin fungicides in the rotation was beneficial (Banniza et al. 2011).

Azoxystrobin, pyraclostrobin, mefenoxam and phosphite were evaluated for their effectiveness against strawberry leathery rot caused by *P. cactorum*, using pot-grown plants and attached fruit sprayed to run-off level at pre- and post-inoculation. Zoospore suspensions (10^5/ml) were applied. Inoculated plants with fruits were placed in a moist chamber for 12 h to ensure infection. Fungicides were applied at either 2, 4, or 7 days before inoculation or 13, 24, 36 or 48 h after inoculation. Azoxystrobin and pyraclostrobin showed protectant activity for up to 7 days before inoculation, but only slight curative activity was recorded, when applied 13 h after inoculation. Phosphite and mefenoxam also provided protection up to 7 days, as well as curative activity for at least 36 h. No significant differences in the protective activity of azoxystrobin and pyraclostrobin (QoI fungicides) and mefenoxam could be observed (Rebollar-Alviter et al. 2007). The effects of azoxystrobin application timings, based on crop growth stage and soil temperature thresholds on incidence of Rhizoctonia damping-off and Rhizoctonia crown and root rot of sugar beet caused by *R. solani* AG4 and AG2-2 were investigated. Soil temperature thresholds of 10, 15 and 20°C were selected for fungicide application timings and used to determine the possibility of selecting the optimum time of applying azoxystrobin. Applying the fungicide at attainment

of specific soil temperature thresholds did not improve efficacy of azoxystrobin in controlling damping-off and crown rot, compared with application timings based on either planting date, seedling development, or leaf stage in a susceptible (E-17) and resistant (RH-5) cultivar. Application rate and split application timings of azoxystrobin had no significant effect on severity of crown and root rot. The results suggested that other environmental factors such as soil moisture might interact with soil temperature to influence disease development, making it difficult to clearly determine the effects of soil temperature on disease development in sugar beet (Kirk et al. 2008).

The impact of application timing for 12 fungicides, including organic and conventional compounds, on the development of tomato late blight disease caused by *P. infestans* clonal lineages (US-22, US-23 and US-24) was assessed using detached tomato leaf assay. Fungicide applications made 2 days after inoculation did not reduce late blight on detached leaves in all treatments, except Bravo Ultrex (US-23 only) and Phostrol (US-22 only). Preventive application of Bravo Ultrex, Ridomil Gold SL, Revus, Zoni X and low and high rates of EF400 significantly controlled late blight caused by US-22, US-23 and US-24 isolates. Phostrol, low rate of Mycostat and high rate of Champ significantly controlled late blight caused by the US-23 isolate. Late blight caused by US-24 isolate was significantly reduced, compared with US-22 and US-23 isolates for all fungicide treatments applied after inoculation, as well as for all preventive fungicide treatment, with the exception of Bravo, Ridomil and Revus. In the whole-potted plant assays with US-23 isolate, late blight was significantly controlled by preventive application of Bravo Ultrex, Ridomil Gold SL and high rate of EF400. Late blight disease was not suppressed by Zonix, low rate of EF400, Phostrol or low and high rates of Champ. The results showed that late blight disease of tomato caused by different lineages of *P. infestans* could be suppressed by preventive application of fungicides. Late blight disease caused by US-24 clonal lineage may require less fungicide application than US-22 or US-23 to reduce disease severity (Johnson et al. 2015).

The effects of calcium chloride ($CaCl_2$) and calcium nitrate [Ca $(NO)_3$] on the mycelial growth of *P. sojae* and on stem blight disease development in two soybean cultivars were assessed under in vitro and greenhouse conditions. Calcium chloride (20–30 mM) or calcium nitrate (30 mM) inhibited the mycelial growth slightly, but lower concentrations (0.4–4.0 mM) increased the pathogen growth. Application of 4 mM calcium chloride or >4 mM calcium nitrate prior to inoculation, inhibited infection of two soybean cultivars by *P. sojae*. Disease suppression was considered to be due to the response of plant tissues rather than the inhibitory effect of calcium chloride and calcium nitrate. Calcium contents in treated plants increased at the time of inoculation. The extent of disease reduction was related to an increased calcium uptake by plants. Results indicated that the effective element in reducing Phytophthora stem rot was calcium. Differences in the extent of disease severity between the soybean cultivars appeared to be due to operation of different mechanisms of calcium uptake. The release of zoospores by isolates of *P. capsici* was reduced by a concentration of 4–30 mM of calcium chloride and calcium nitrate on lima bean agar, but lower concentration (0.4 mM) of calcium chloride and calcium nitrate induced higher rate of zoospore release. The results suggested that applying solutions of calcium at >4 mM might be a possible approach to reduce the incidence of Phytophthora stem blight disease in soybean crops (Sugimoto et al. 2005).

Phosphite (Phi)-based fungicides have been shown to be effective in suppressing the development of diseases caused by *Phytophthora* spp. and other oomycetes. Foliar and postharvest applications of Phi-based fungicides were effective against *P. infestans* causing late blight and tuber rot disease of potatoes. Optimization of usage of Phi-based fungicides during growing season and storage might result in improvement of disease-free potato production. The efficiency of translocation of Phi to tubers following foliar and tuber treatments of potato crops was assessed, using high-performance ion chromatography (HPIC) procedure. The quantity of Phi present in tubers increased with total amount of Phi-based fungicides applied during growing season. Foliar applications of Phi resulted in an uneven distribution of Phi in the three tuber regions analyzed, high concentration being identified in the tuber cortex (32.5–166.4 µg/g fresh tissue) and medulla regions followed by the skin area. Postharvest treatment of tubers resulted in different type of distribution of Phi, with the highest concentrations of Phi found in the skin (411.0–876.6 µg/g fresh tissue), followed by the cortex and medulla regions. As the pathogen could infect both aerial plant parts and tubers, the protective measures should be directed toward both potato plants and seed tubers to have Phi concentration in plant tissues, in excess of 100 µg/g of fresh tissues (Borza et al. 2017).

The fungicides cyprodonil and fludioxonil were evaluated for their potential in suppressing the development of gummy stem blight disease of watermelon caused by *Didymella bryoniae* under field conditions. The baseline sensitivity of isolate of *D. bryoniae* and the efficacy of the fungicides in suppressing development of isolates resistant to QoI fungicides and boscalid were determined. In the autumn of 2008, 2009 and 2011, field-grown watermelon plants inoculated with isolates of *D. bryoniae* resistant to QoI fungicides and boscalid were treated with chlorothalonil, cyprodonil-fludioxonil alternated with chlorothalonil, cyprodonil-difenconazole alternated with chlorothalonil, tebuconazole alternated with chlorothalonil, or water. In 2008, both cyprodonil treatments reduced the disease severity, compared with control (water) treatment and chlorothalonil alone. In 2008 and 2009, cyprodonil-fludioxonil reduced disease severity, compared with boscalid-pyraclostrobin and in 2008, cyprodonil-difenconazole and tebuconazole also reduced disease severity. Application of cyprodonil-fludioxonil effectively reduced the disease and also delayed development of resistance in isolates of *D. bryoniae* to cyprodonil and fludioxonil. However, possible emergence of isolates of *B. cinerea* resistant to cyprodonil and fludioxonil, infecting watermelon, had to be monitored carefully (Keinath 2015). A formulated mixture of benzothiostrobin

and fluazinam at 1:1 had synergistic activity against *S. sclerotiorum*, causing Sclerotinia stem rot of oilseed rape in vitro. Under field conditions, benzothiostrobin alone or formulated with fluazinam at 1:1 (150 g a.i/ha) was more effective in controlling Sclerotinia stem rot disease in rapeseed. The results indicated the potential of benzothiostrobin for suppressing the development of the disease seriously affecting oilseed rape crops (Xu et al. 2015). Chemical application for the management of rice sheath blight caused by *R. solani* was employed as an important disease management strategy to obtain economically sustainable yields, as the availability of reliable sources of resistance was limited. The isolates of *R. solani* became less sensitive to boscalid than fluxapyroxad introduced into the market. This fungicide showed excellent protective and curative activity against *R. solani* by providing 82.6 to 94.2% protective or curative control. Fluxapyroxad was applied as spray twice (at 100 g a.i /ha) at 15 and 30 days after planting. After second application, the control efficacy of the fungicide was 83.4 to 88%, whereas boscalid could provide only lower level of control efficacy (51.7–57.0%) at a dosage of 600 g a.i/ ha. Fluxapyroxad showed the potential for more effective disease suppression than boscalid or jinggangmycin (Chen et al. 2014).

The suppressive activity of metalaxyl, fosetyl-Al, dimethomorph and cymoxanil against Phytophthora crown rot of peach trees caused by *P. cactorum* was assessed. Application of fosetyl-Al, or metalaxyl by painting the trunk (150 g/l) of 3-year-old PR204 trees inhibited the pathogen development, but dimethomorph or cymoxanil was not effective. Application of metalaxyl as soil drench suppressed canker development, when treated trees were subsequently inoculated with *P. cactorum*. Fosetyl-Al significantly reduced pathogen development, compared with cymoxanil, dimethomorph and control, but it was not as effective as metalaxyl. Furthermore, cymoxanil and dimethomorph did not influence canker development. Metalaxyl and fosetyl-Al were effective for at least 21 days after application. Strips or trunk bark removed from trees drenched with the fungicides 20 days earlier, were inoculated with the pathogen on the cambium side. Metalaxyl was the most effective fungicide and fosetyl-Al significantly reduced the development of the pathogen, compared with dimethomorph, cymoxanil and untreated control strips. Pathogen colonization of the strips treated with dimethomorph was significantly less than in untreated control strips. By contrast, cymoxanil did not inhibit the pathogen growth. The results indicated that metalaxyl or fosetyl-Al might be used as trunk paint or soil drench to suppress the development of crown rot of peach trees (Thomidis and Elena 2001). Phytophthora branch canker in Clementine trees (*Citrus clementina*) is responsible for considerable losses in Spain. Fosetyl-Al, metalaxyl and mefenoxam were evaluated for their efficacy in reducing lesion expansion. The fungicides were applied as foliar sprays, drip chemigation or paint treatment. None of the fungicides had curative effect on the lesion development. The residual effect was better on young than on the mature trees. Painting treatments were more effective in reducing lesion expansion in general than drip chemigation or foliar sprays. However, painting treatment was found to be laborious and uneconomical. As a long-term strategy foliar spray or drip chemigation was suggested for consideration (Alvarez et al. 2008).

Foliar spray of validamycin A (VMA) or validoxylamine A (VAA) at >10 mg/ml effectively suppressed the development of tomato Fusarium wilt caused by *F. oxysporum* f.sp. *lycopersici*, although they did not show any antifungal activity in vitro. In pot experiments, the effect of foliar application of VMA or VAA at 100 μg/ml lasted for 64 days. Plants sprayed with VMA or VAA accumulated salicylic acid (SA) and showed elevated expression of SAR marker genes *P4* (PR-1), *Tag* (PR-2) and *NP24* (PR-25). Foliar spray of VMA also controlled late blight and powdery mildew of tomato. Protection provided by VMA and VAA lasting up to 64 days after treatment, was broad-spectrum and induced expression of PR genes, all essential indicators of SAR, suggesting that VMA and VMA might function as plant activators, rather than as antibiotics (Ishikawa et al. 2005).

4.1.2.3 Pre-Plant Dip Treatment of Crowns/Roots

Root and heart rot caused by *P. cinnamomi* is one of the limitations for pineapple production. The effectiveness of pre-plant dips of crowns in potassium phosphate and phosphorous acid was assessed for reducing the incidence/severity of the disease using pineapple hybrids MD2 and 73-50 and cv. Smooth Cayenne. High volume spray at planting was much less effective, compared with the preplant dip treatment. Smooth Cayenne was more resistant to heart rot than MD2 and 73-50. Further, Smooth Cayenne was more responsive to treatment with potassium phosphate. Based on the cumulative scores of heart rot incidence over time, MD was considered to be more susceptible to heart rot than 73-50 and it was more responsive to phosphorous acid. The roots contained highest levels of phosphonate at one month after planting and the levels declined during the next two months. Pre-plant dipping of crowns was highly effective during the first few months, but protection declined to levels that were not sufficient to maintain the health of the mother plant root system until harvest, when favorable weather conditions for disease development prevailed (Anderson et al. 2012). Oxathiapiprolin (Zorvec) was applied (0.35 l/ha) onto tobacco seedlings one week prior to transplanting in conjunction with directed applications of the fungicide (0.7 l/ha) at first cultivation and lay-by (last cultivation) reduced black shank caused by *P. nicotianae* significantly, compared to the control. Further, application of oxathiapiprolin at 14 l/ha through transplant water, followed by directed sprays at first cultivation and lay-by at 0.7 l/ha, reduced black shank significantly compared with the control. These treatments did not differ in the extent of disease reduction from mefenoxam treatment. All treatments with oxathiapiprolin increased tobacco yield, compared to untreated control (Ji et al. 2014).

4.1.2.4 Treatment of Soil

Chemicals and biofumigants may be applied to the soil prior to sowing seeds or transplanting seedlings to disinfest the field soil as the in-furrow application or with irrigation water or drenching the soil with fungicides and other chemicals. Soilborne pathogens can infect crops at different growth stages from preemergence to heading stage. Nonsystemic and systemic chemicals may be applied in the soil to interact with the pathogens present in the soil or infected plant residues left after harvest. The systemic fungicides may penetrate into the tissues of germinating seed, emerging seedlings and interact with pathogen structures, if infection has already been initiated. Soil drenches with fungicides and chemicals have been carried out to reduce soilborne pathogen populations and consequent reduction in disease incidence/severity. Crown rot and root rot of *Capsicum* caused by *P. capsici* causes significant losses. Metalaxyl was drenched to restrict the outbreak of the disease in glasshouse crops in southern Italy. More than 80% of the isolates of *P. capsici* showed moderate levels of resistance indicating the loss of effectiveness of metalaxyl application, as soil drenches (Agosteo et al. 2000). Metalaxyl applied in-furrow provided effective protection for longer periods to soybean against root and stem rot caused by *P. sojae*, than when applied as seed treatment. However, seed treatment with metalaxyl was able to protect soybean cultivars with horizontal resistance to *P. sojae* effectively (Dorrance and McClure 2001). As soil disinfestation prior to planting with methyl bromide (MB) has been banned in many countries, the efficacy of application of fungicides in reducing sudden wilt (vine decline) of melon caused by *Monosporascus cannonballus* was assessed. Azoxystrobin, prochloraz and pyroclostrobin + boscalid suppressed the development of sudden wilt disease under field conditions. Fludioxonil applied at high rates was effective, but it induced phytotoxic symptoms. Fluazinam was less effective. The results indicated that during short fall seasons, two applications of the fungicide might be sufficient. But for long spring seasons, three applications might be required to protect the melon crops effectively and boscalid had to be limited to first application before fruit set (Pivonia et al. 2010).

The efficacy of two phosphonate-containing fungicides (FNX-100 and FNX-2500) against Phytophthora crown rot of pumpkin caused by *P. capsici* was assessed. Pumpkin cultivar, treatment (type of fungicide product), phosphonate concentration and application method significantly influenced the effectiveness of Phytophthora crown rot control. FNX-100 only suppressed the disease development in pumpkin. The highest level of disease control obtained for cv. Phantom was with drench application of 3.0% FNX-100, whereas for cv. Spooktacular, all three concentrations of FNX-100 (1.0, 2.0 and 3.0% v/v) applied as soil drench suppressed or significantly reduced the severity of crown rot. On zucchini, FNX-100 was more effective than FNX-2500 in controlling crown rot. Zucchini cultivar or FNX-100 concentration did not have significant influence on disease control with phosphonates. FNX-100 used as drench provided satisfactory protection to pumpkin and zucchini against Phytophthora crown rot. By contrast, FNX-2500, a foliar fungicide containing phosphate/phosphonate and copper, manganese and zinc was not effective in reducing disease severity. The results showed that selection of cultivar and fungicide type might be critical factors in achieving desired level of disease suppression (Yandoc-Ables et al. 2007). The baseline sensitivity of *P. capsici*, causing Phytophthora blight disease of pepper to oxathiapiprolin, the first member of a new class of isoxazoline fungicides was determined to assess the effects on pathogen mycelial growth, sporangia formation and zoospore germination, representing different growth stages of the pathogen life cycle. The asexual life stages of *P. capsici* appeared to be more sensitive to oxathiapiprolin. In field assessment of the fungicidal activity, oxathiapiprolin applied at different rates through drip irrigation in tubes, or by soil drench plus foliar sprays, reduced the disease and increased pepper yield significantly. The results indicated the potential of oxathiapiprolin in suppressing development of pepper Phytophthora blight disease (Ji and Csinos 2015).

The effect of foliar fungicides applied for soil treatment against late blight disease of potato caused by *P. infestans* was assessed. Mancozeb, metiram, cyazofamid and other fungicides were applied at 24 h prior to soil infestation with a suspension of zoospores and sporangia of *P. infestans*. Spore viability in soil treated with various fungicides was determined using buried healthy whole tubers and by assaying infested soil applied to freshly cut tuber disks. Mancozeb, metiram and cyazofamid were more effective in protecting the tubers and tuber disks against infection by *P. infestans* than other fungicides. Whole tuber infections were significantly less in soils treated with mancozeb, metiram, fluazinam and fenamidone than when treated with distilled water. Infection of buried tubers and tuber disks was prevented for 3 to 5 days, following a single application of mancozeb, or metiram under field conditions. Tuber disk assay was more effective than buried whole tuber method in assessing the effectiveness of the test fungicides. However, both methods might be useful in understanding the different modes of action of the fungicides, as whole tubers were not infected, when protected by some fungicides, but tuber disks were vulnerable to infection, even after treatment with fungicides (Porter et al. 2006).

Potato tubers are infected by *Phytophthora erythroseptica* and *P. ultimum* causing pink rot and leak tuber disease respectively. Mefenoxam was effective in suppressing both pathogens. Mefenoxam was applied as a single in-furrow application at planting, followed by an additional sidedress application at 3 weeks after planting. It was also tested as a single foliar application, when tubers were 7–8 mm diameter and repeated 14 days later. The recommended label rate plus two additional lower application rates were used for each method. Mefenoxam was more effective in controlling pink rot relative to leak tuber over all application methods for tubers challenge-inoculated after harvest. Mefenoxam applied as in-furrow at planting and sidedress application was the most effective treatment for reducing pink rot infection (89%). However, this method provided only modest level of leak tuber control (35%) and this disease was not suppressed by foliar application of mefenoxam at any of the dosages tested. By contrast, pink rot infection was reduced by 10 to 50% by foliar application of mefenoxam. The differences in

the effectiveness of control of pink rot and leak tuber disease might be due to different modes of entry into the tubers–direct entry by *P. erythroseptica* and through wounds by *P. ultimum* (Taylor et al. 2004).

The effectiveness of metalaxyl-M (Ridomil Gold 480 EC) and phosphite (Phostrol) applied at planting as in-furrow treatment was assessed for reducing incidence of pink rot disease of tubers caused by *P. erythroseptica* in potato cvs. Shepody and Russet Burbank during 2005–2006. Inoculum of metalaxyl-sensitive isolate was applied as a vermiculite slurry at planting or as a zoospore suspension drenched adjacent to the potato plants. After harvest the percentage of tubers infected by *P. erythroseptica* was determined for each treatment. Metalaxyl-M applied in-furrow (1.5% and 1.2%) was significantly more effective than phosphite (9.6% and 2.8%) based on tuber weight reduction. The mean percentage of diseased tubers was 1.7% (2005) and 1.3% (2006) for metalaxyl-M-treated plots as against 10.1% (2005) and 3.1% (2006) for phosphite-treated plots. The potato cv. Shepody was significantly more susceptible to pink rot than cv. Russet Burbank. The results showed that metalaxyl-M applied in-furrow at planting could be an adoptable practice for the management of potato pink rot disease (Al-Mughrabi et al. 2007). The fungicides were evaluated for their efficacy in suppressing the development of infection of potato stems by using plant inoculation assay. Azoxystrobin, pyraclostrobin, fluoxastrobin and mandipropamid + difenoconazole were significantly more effective in reducing stem infection, when applied prior to inoculation. Postinoculation treatment with these fungicides was not effective. Azoxystrobin applied by chemigation at 50 and 67 days after planting (DAP) reduced the infected stem surface area covered with sclerotia at 79 DAP than nontreated potato plants in both treatments in two years of experimentation. However, the reduction in sclerotia-covered stem surface was not significantly different in treated and untreated ones at 102 and 140 DAP. The fungicide treatment did not eliminate latent infection of potato plants in the field (Ingram et al. 2011). The efficacy of mefenoxam and phosphorous acid in reducing pink rot disease of potato tubers caused by *P. erythroseptica* was investigated in replicated small-plot and replicated split commercial field trials. Fungicides were applied in-furrow at planting, or as 1, 2, or 3 foliar applications via ground sprayer, irrigation system (chemigation), or fixed-wing aircraft. The efficacy of the fungicides was assessed either by determining natural infections in the field or by inoculating eyes of harvested tubers using a mefenoxam-sensitive and -resistant isolates of *P. erythroseptica* via postharvest challenge inoculation. In the field trials, both in-furrow and two foliar applications of mefenoxam suppressed tuber rot in the field and significantly controlled tuber rot in storage. Phosphorous acid also reduced tuber rot in the field following two or three foliar applications. Phosphorous acid controlled mefenoxam-sensitive and -resistant isolates of *P. erythroseptica* during storage for 187 days, whereas mefenoxam could not control the resistant isolate. In replicated split commercial field trials, two aerial sprays of phosphorous acid were as effective as three applications in reducing pink rot incidence in tubers inoculated after harvest. Three aerial applications were equally effective as three chemigation applications in replicated commercial field trials in 2008 but provided significantly greater protection than chemigation in 2009 (Taylor et al. 2011).

The effect of preplant fumigation to reduce resident populations of *M. cannonballus*, causing root rot and vine decline of melon was investigated. Preplant fumigation with methyl iodide injected as a hot gas at 448.4 kg/ha through drip irrigation tape in preformed, tarped beds consistently reduced the percentage of roots infected, compared with nonfumigated controls. The reduction in infection was equal to or better than those achieved with an equivalent rate of methyl bromide. Chloropicrin applied in water at 249.0 kg/ha through buried drip irrigation tape to either tarped or nontarped beds significantly reduced the percentages of both roots infected and roots on which perithecia were produced, compared with nonfumigated controls (Stanghellini et al. 2003). The effectiveness of fludioxonil + metalaxyl-M (F1) and hyemexazol (F2) along with four irrigation levels to fill the soil depths of 0–90 cm to capacity (I_1), 25% (I_2), 50% (I_3) and water-free (I_4) conditions was determined for the control of Fusarium wilt of melon caused by *F. oxysporum* f.sp. *melonis* (Fom) in a split-plot design. The F1 (250 ml/ha) and F2 (500 ml/ha) were drip-chemigated at planting and again 15 days after planting. Both fungicides applied through chemigation were equally effective in reducing wilt disease and increasing yields, compared with control in both years 2011 and 2012. However, the highest yield was obtained in plots with no water stress (I_1 treatment) (Ozbahce 2014). Vapam (dithiocarbamate) applied as fumigant reduced primary and secondary infection of canola by *Plasmodiophora brassicae* by 12–16 folds at application rates varying from 0.4–1.6 ml/l soil. Vapam also reduced club root severity and improved seed yield of canola and other growth parameters, under field conditions. At moisture levels (10–30% v/v) vapam application had significant effect on both disease severity and infection rates and plant growth parameters. Application of vapam could effectively reduce disease severity (Hwang et al. 2014).

Brassica spp. plant tissues contain glucosinolates (GSLs), which when hydrolyze release isothiocyanates (ITCs) with biocidal properties. Biofumigation may be achieved by incorporating fresh plant material (green manure), seed meals or dried plant material treated to preserve ITC activity. The effects of the fungicides mefenoxam and fresh Brassica tissues and Brassica pellets on the development of *P. nicotianae* isolates from pepper and tomato plants were assessed. Sensitivity of *P. nicotianae* isolates to mefenoxam at 50% and 90% effective concentrations (EC_{50} and EC_{90}) varied considerably. Tomato isolates were more resistant to mefenoxam than pepper isolates. Similar variations in the pathogen isolates concerned were observed in the *Brassica* tissue assay. *Brassica nigra* was the most effective in inhibiting the mycelial growth of *P. nicotianae*. The effectiveness of *B. carinata*, *Sinapis alba* and *B. oleracea* varied, depending on the dose. Isolates differed in their sensitivity to the compounds released by the Brassica pellets. Greenhouse experiments showed that treatments with

mefenoxam and Brassica pellets effectively suppressed the development of disease in pepper plants. Mefenoxam application might be a potential risk for the emergence of resistant isolates, the presence of which should be monitored carefully (Morales-Rodríguez et al. 2014). The effect of metam sodium (sodium *N*-methyldithiocarbamate, trade name Vapam) and application methods, including watering, soil surface covering and soil incorporation on canola clubroot disease caused by *P. brassicae* was assessed. Metam sodium (0.4–1.6 ml/l) soil increased canola seedling emergence and plant health, and reduced root hair infection, gall weight and clubroot severity under greenhouse conditions. Fungicide application improved plant growth and reduced clubroot severity. Incorporation of metam sodium into the soil followed by covering with plastic sheet after application, improved fungicide efficiency. The results showed that soil fumigation with metam sodium could reduce canola clubroot severity and also improve plant growth in the subsequent canola crop also (Hwang et al. 2018).

Chemical fumigants have been evaluated for the effect on soilborne fungal pathogens. Four soil treatments, viz., methyl bromide (MB) + chloropicrin (50:50) as positive control, chloropicrin + 1,3-dichloropropene (DCP) (60:40) solarization for 6 weeks and metam-sodium were applied in soil naturally infested by *F. oxysporum* f.sp. *vasinfectum*, causal agent of Fusarium wilt of cotton (Pima, *Gossypium barbadense*; upland *G. hirsutum*). Two cultivars each of Pima and Upland cotton with varying resistance to race 4 were grown in the experimental plots. Plant mortality was lowest in MB + chloropicrin, solarization and chloropicrin + DCP treatments and highest in the nontreated and metam-sodium treatments. Vascular discoloration was reduced in MB + chloropicrin and solarization treatments, compared with the nontreated control, metam-sodium and chloropicrin + DCP treatments. Populations of *F. oxysporum* in soil were reduced only in MB + chloropicrin in soil only in MB + chloropicrin, chloropicrin + DCP and solarization treatments. Solarization for 6 weeks and chloropicrin + DCP (60:40, 295 1 a.i/ha) were effective in reducing Fusarium wilt disease in cotton in heavy clay soil (Bennett et al. 2011). The efficacy of methyl iodide (MeI) and chloropicrin (Pic) in suppressing the development of soilborne pathogens infecting ginger was assessed for use as an alternative to methyl bromide (MB), which has been banned in several countries. In vitro assessments of the inhibitory activity showed that MeI at 24 mg/kg soil was the most effective in reducing populations of *Pythium* spp., *F. oxysporum*, *Ralstonia solanacearum* and *Meloidogyne incognita* by >90%. Treatments with MeI + Pic at 12 mg/kg soil (1:3 and 1:5) also reduced the populations of the pathogens by >82%. In the field trails, MeI at 30 or 40 g/m² and MeI + Pic (1:3) at 40 g/m² provided highly effective long-term protection against all target pathogens. MeI at 20 g/m² or Pic at 40 g/m² was as effective as methyl bromide in reducing the disease incidence. The results showed that injecting MeI at 30 and 40 g/m² or MeI + Pic (1:3) at 40 g/m², followed by covering with virtually impermeable film could be an effective alternative for soil fumigation for the control of major soilborne pathogens of ginger (Li et al. 2014).

The efficacy of new formulations of phosphonates (AG3) was assessed for suppressing the development of damping-off disease of cucumber caused by *Pythium* spp., using cucumber seedlings grown in peat-based mix (P-mix), muck soil (Mu-soil) and sandy loam (SL-soil). Aqueous solutions of phosphonate formulations (0.035%–0.280% a.i) were added to pathogen-infested P-mix or SL or Mu-soil as a preplanting amendment or postplanting drench treatments. Preplanting amendment with powder or liquid phosphonate formulation provided control of damping-off and disease suppression increased with concentration of phosphonate. Higher concentrations of phosphonates (0.140% or 0.280% a.i) provided complete control of the disease (100%), whereas lower rates (0.035% or 0.070% a.i) reduced disease with an average of 34 to 75% healthy plants, compared with untreated control with less than 5% healthy plants. The AG3 liquid formulation, as a preplanting treatment also provided effective control in naturally infested Mu-soil and artificially infested SL-soil. Postplanting drench with AG3 was equally effective in all three substrates. Treatment of AG3 at 2 weeks prior to planting in Mu-soil was as effective as that added just prior to planting. AG3 was effective in suppressing the Pythium damping-off of cucumber seedlings under microplot and field conditions also. The results indicated that the amount and frequency of irrigation and rainfall might affect the extent of disease suppression by AG3 phosphonate in the field (Abbasi and Lazarovits 2005). Phosphonate formulation (AG3) as seed treatment was evaluated for the control of Pythium damping-off disease of cucumber plants under controlled and field conditions. Seeds were soaked in phosphonate solution for 10 min and planted in peat-based mix or sandy-loam soil inoculated with *P. aphanidermatum* or *P. ultimum* or into muck soil naturally infested by *P. irregulare, P. ultimum* and other *Pythium* spp. Under growth room conditions, seed treatment with phosphonate provided protection to an extent of more than 80% in all infested substrates. Seeds treated with phosphonate could be stored up to 18 months prior to planting without loss of protection by the fungicide. In microplots containing naturally infested muck soil, phosphonate seed treatment decreased the percentage of diseased cucumber plants and increased total fresh weights, compared with untreated control and phosphonate post-planting drenching. At 6 weeks after planting in field plots, cucumber stands were 63% in phosphonate-treated seed in *Pythium*-infested muck soil, compared with 18% healthy plants in untreated control plots and 53% in the post-planting drenching treatment. Tests for assessing phosphonate toxicity showed that germination of radish and bok choy was reduced, but not that of corn, cucumber, soybean, sugar beet, tomato and wheat. The results showed that phosphonate seed treatment was a cost-effective strategy for reducing the incidence of damping-off disease in cucumber (Abbasi and Lazarovits 2006). The effectiveness of pre- and post-inoculation application of cyazofamid and metalaxyl along with phosphonate in reducing damping-off was assessed, after applying the fungicide at 3 days before soil infestation with the pathogen suspension. Phosphonate and cyazofamid were also applied at 24 h after inoculation. All three chemicals showed similar levels of protectant activity. Phosphonate and cyazofamid showed curative activity also under field conditions (Miyake et al. 2015).

The effectiveness of rhamnolipid and saponin biosurfactants was assessed in suppressing the development of *P. capsici*, incitant of pepper Phytophthora blight disease in recirculating production systems. The efficacy of amending the recirculating nutrient solutions with either a rhamnolipid or a saponin biosurfactant (150 and 200 µg a.i/ml, respectively), which differentially killed the zoospores, arresting the spread of the pathogen entirely was assessed. Disease development was effectively suppressed in both ebb and flow and top-irrigated cultural systems, with either an organic potting mix or rockwool as the planting substrate. In the control, all plants were killed within 6–7 weeks, following hypocotyl inoculation of a single plant in the system, which functioned as the source of secondary inoculum. Injecting rhamnolipid biosurfactant into the irrigation line during every irrigation also provided complete protection to all plants (100% disease-free). The results indicated that biosurfactants had the potential for use as alternatives for synthetic surfactants in the recirculating systems (Nielsen et al. 2006). Effects of aluminum-containing salts on the development of carrot cavity spot disease caused by *Pythium sulcatum* and potato dry rot caused by *Fusarium sambucinum* were investigated. Various aluminum-containing salts inhibited the mycelial growth of the pathogens to different extent. Aluminum chloride and aluminum sulfate were the most effective not only in in vitro assays, but also in planta tests. At 5 mM concentration, aluminum sulfate reduced dry rot and cavity spot by 28% and 100% respectively. Aluminum chloride (5 mM) reduced dry rot by 25%, whereas aluminum lactate (5 mM) reduced cavity spot lesions by 86%. The results indicated the potential of aluminum salts as alternatives to the synthetic fungicides for the control of carrot cavity spot and potato dry rot diseases (Kolaei et al. 2013).

Verticillium wilt caused by *Verticillium* spp. and crown rot caused by *P. cactorum* are important soilborne diseases affecting strawberry production seriously. Soil fumigant chloropicrin, 1,3-dichloropropene, dazomet, metam potassium and dimethyl disulfide were evaluated for their efficacy in suppressing the soilborne pathogens in combination with different plastic films, as alternatives for methyl bromide (MB) in strawberry nurseries. Chloropicrin, 1,3-dichloropropene and dazomet suppressed disease development to the level that was obtained with methyl bromide. In addition, 1,3-dichloropropene and methyl bromide applied at 50% rate under virtually impermeable film provided effective protection to the strawberry nurseries (De Cal et al. 2004). The efficacy of soil fungicide applications, including subsurface drip chemigation, was assessed for the control of summer squash Phytophthora crown and root rot (PCRR) caused by *P. capsici*, since foliar fungicide application provided limited protection to the crop. Soil drenches and foliar applications of various fungicides (11) were compared for their efficacy in controlling PCRR disease under field conditions by applying at 7-day interval. Death of cv. Coughar plants following inoculation with

P. capsici isolate 12889 occurred at all growth stages from first true leaf to full maturity in field trials. Plant death at 42 days postinoculation (dpi) differed significantly among fungicides and application methods (<0.0001). The fungicide application method had also significant effect on infection. In general, soil drenches were more effective than foliar applications at limiting plant death, but none of the treatments entirely arrested symptom development. Mean plant death at 42 dpi was 41% for soil drenches and 92% for foliar applications. Drenches of fluopicolide, mandipropamid, or dimethomorph reduced plant death to <10% and prevented yield loss associated with root rot and crown rot. Foliar application of fungicides did not reduce plant mortality, compared with untreated inoculated control and yield losses were also not reduced by foliar application under field conditions. Crown rot was less severe under greenhouse conditions and disease progress was slower, following soil drenches, compared with foliar applications. Some fungicide treatments were more effective on cv. Leopard, which was less susceptible to *P. capsici* than the cv. Coughar. The results showed that soil drenches of fungicides could be adopted for effective suppression of Phytophthora crown and root rot disease in summer squash (Meyer and Hausbeck 2013).

The effectiveness of dimethyl sulfide (DMDS) applied either alone or in combination with other fumigants was assessed for suppressing the development of *R. solani* and *F. oxysporum* infecting tomatoes in commercial greenhouses. The comparative effects of DMDS and metam sodium (MS) on the populations of dominant pathogens were investigated, using quantitative PCR assay. Prior to soil fumigation, the fungal diversity was estimated in the soil via libraries, which identified *F. oxysporum* and *R. solani*, as the most dominant soilborne plant pathogens, whereas *Cladosporium* sp., a known opportunistic airborne tomato pathogen was also detected. DMDS at two dose rates reduced the population of *F. oxysporum* and *R. solani* drastically and the effect lasted for the whole crop season. Metam sodium showed inhibitory effect on *F. oxysporum* that was alleviated at 120 days post-fumigation. The effect of fumigants on *Cladosporium* sp. was only transient and population level recovered by 60 days postfumigation. The results indicated the high level of fumigant activity of DMDS even at a low dose (56.4 g/m²) on the soilborne pathogens, *F. oxysporum* and *R. solani* infecting tomatoes in the greenhouse soils. The qPCR assay for estimation of population levels of these pathogens was found to be a valuable tool to determine the level of risk for disease severity and subsequent yield losses likely to occur (Papazlatani et al. 2016). Chloropicrin, as a soil fumigant, was evaluated for its efficacy in reducing soil populations of *Spongospora subterranea*, incitant of powdery scab disease of potato tubers under field conditions in Minnesota and North Dakota. Potato cultivars (16) with different levels of susceptibility to disease were planted in plots treated with chloropicrin at rates ranging from 0 to 201.8 kg a.i/ha. The qPCR assay was employed to quantify DNA contents of *S. subterranea* in soil. The effect of chloropicrin on root colonization by the pathogen in two potato cultivars with contrasting levels of susceptibility was

assessed. Chloropicrin at 70.1 to 201.8 kg a.i/ha significantly reduced the initial inoculum of *S. subterranea* in the soil, but increased the intensity of disease on roots and tubers of susceptible cultivars. The DNA contents of the pathogen in the roots of bioassay plants decreased with increasing chloropicrin dosage. But the trend was similar in all potato cultivars. Chloropicrin fumigation increased the tuber yield significantly and this increase in cvs. Shepody and Umatilla was associated with the amount of root galls induced by the pathogen. Factors such as environment affecting inoculum efficiency and host susceptibility might contribute to development of powdery scab disease in potato (Bittara et al. 2017).

Herbicides are applied primarily to eradicate different weed species and some of them are likely to be alternative hosts for soilborne microbial plant pathogens, facilitating their survival and perpetuation. The effects of herbicides lactofen, glyphosate and imazethapyr, commonly used in soybean, on the development of sudden death syndrome (SDS) and the pathogen, *F. oxysporum* f.sp. *glycines (Fog)* were assessed on four soybean cultivars with varying resistance to the disease and tolerance to glyphosate. Inhibition of conidial germination, mycelial growth and sporulation of Fog in vitro were the vital functions of the pathogen affected by glyphosate and lactofen. Under growth chamber and greenhouse conditions, significant increase was observed in disease severity and frequency of isolation of *Fog* from roots of all cultivars treated with imazethapyr or glyphosate, compared with untreated control. In contrast, lactofen application reduced disease severity and isolation frequency of *Fog* in treated plants. Sudden death syndrome disease severity and isolation frequency were less in resistant than in susceptible soybean cultivars. Results suggested that soybean cultivars resistant or tolerant to glyphosate showed similar responses to infection by *F. oxysporum* f.sp. *glycines* after herbicide application (Sanogo et al. 2000). In the later investigation, effects of interaction between glyphosate-tolerant soybean and herbicides (glyphosate, imazethapyr, lactofen and pendimethalin) on damping-off and root rot caused by *R. solani* were assessed under greenhouse and field conditions. The herbicides were applied at recommended rate in soils-infested with *R. solani* and glyphosate-tolerant (Pioneer 93B01 and Pioneer 9344) and glyphosate-sensitive (BSR1) soybean cultivars were planted. The soybean cultivars responded differentially to various herbicides, especially to pendimethalin. Plant stands were reduced by *R. solani* alone or in combination with different herbicides, compared with noninoculated control. In BSR101, *R. solani* + pendimethalin + imazethapyr treatment significantly reduced plant stand, compared with *R. solani* + pendimethalin treatment. Root rot severity was generally low in both years (1998 and 1999) of experimentation. Generally, glyphosate-tolerant and –sensitive soybean cultivars reacted in a similar manner to most herbicide treatments in respect of root rot and damping-off disease incidence/severity (Harikrishnan and Yang 2002). The role of herbicide application on the incidence/severity of soilborne diseases was investigated, as a strategy for root rot disease of barley caused by *R. solani* AG-8 and *R. oryzae*. The effect of glyphosate application timings on the severity of the disease was assessed under naturally infested field conditions during 2007–2009. Crop volunteer plants and weeds were sprayed with glyphosate at 42, 28, 14, 7 and 2 days prior to planting. As the herbicide application interval increased, there were significant increases in shoot length, length of the first true leaf and number of healthy seminal roots and the disease severity also was reduced. The activity of *R. solani* as determined by tooth-pick assay and real-time PCR assay, declined over time in all treatments after planting barley. The herbicide application interval of 13 to 37 days was found to be required for reducing disease severity to the maximum extent (Babiker et al. 2011).

Glyphosate, applied as an herbicide to kill weeds and previous crops, acts on treated plants and inherent defense responses of the treated plant. *Rhizoctonia* sp. with wide host range could colonize dying roots rapidly. As a result, severity of root rot caused by *R. solani* might increase in subsequent crops planted in close proximity to the dying or dead roots (Babiker et al. 2011). The cover crops and cereals planted in the previous fall infected by *Rhizoctonia* spp. might serve as a green bridge for the pathogen on cereal roots to colonize roots of onion planted later, resulting in severe stunting of onion seedlings. The importance of timing of application of glyphosate on the green bridge effect was assessed. The wheat cover crops were killed with glyphosate application at 27, 17 and 3 days before onion seeding in 2012 and 19, 10 and 3 days before seeding in 2014. As the interval between herbicide application and onion planting increased from 3 days to 19 or 27 days, the number of patches with stunted onion plants decreased by >55%, total area of stunted plants decreased by 59 to 63% and patch severity index decreased by 54 to 65%. Likewise, *R. solani* AG8 DNA contents in the soil sampled from the dead cover crop rows declined as the interval between glyphosate application and onion seeding increased in the 2012 trial, but not in the 2014 trial. By contrast, *R. solani* AG3 and AG-8 DNA contents in soils sampled from the cover crop rows were significantly positively correlated with the number of patches of stunted onion plants, total area of stunted patches and patch severity index in the 2012 trial. The results indicated that increasing the interval between herbicide application to the cover crop and onion planting might be a feasible approach for the management of root rot and stunting in onion crops (Sharma-Poudyal et al. 2016).

4.1.2.5 Fungicide Application Based on Weather-Based Advisories

Peanut crops are affected seriously by stem rot disease caused by *Sclerotium rolfsii*. Fungicides were applied at 14-day interval to protect peanut against diseases. In order to extend spray intervals, the efficacy of fungicide application on a 14-day schedule and 21-day schedule was assessed. Both schedules provided similar levels of suppression of stem rot caused by *S. rolfsii*. Disease suppression decreased in plots treated on a 28-day schedule, but disease severity was less than in untreated plots. Similar trend was recorded in yield levels for different spray schedules (Brenneman and Culbreath 1994). Later several spray advisories were formulated based on

environmental conditions that favor Sclerotinia blight of peanut in Virginia and North Carolina (Langston et al. 2002). These advisories are based on air and soil temperatures, precipitation, relative humidity, vine growth and canopy closure. Fungicide application had significant effect on the suppression of stem rot disease and yields in Georgia. Furthermore, application of fungicides according to advisories based on soil temperature, rainfall and host growth provided similar or better suppression of stem rot disease than by traditional calendar-based fungicide treatment (Woodward et al. 2013).

4.1.2.6 Dissipation Dynamics of Fungicides

Indiscriminate use of agricultural chemicals with high residual activity and potential mobility has created unwelcome environmental problems. Soil, an important component of the environment is known to be a sink for herbicides, fungicides and other agricultural chemicals, including fertilizers that have significant adverse effect on the soil microbial community structure and components, in addition to polluting subsoil water. The harmful residues in the soil may reach waterbodies, making the water unsuitable for irrigation and drinking purposes. Fungicides belonging to different classes/ groups are applied on seeds, plants, soil and harvest produce during storage. Glasshouse crops like tomatoes receive different fungicides either singly or in combination. It is essential to investigate the dissipation dynamics of the fungicides to assess the quantum of residues and persistence of the fungicides in the substrate (soil) and produce, to determine the risk of using fungicide(s).

Propamidine, a systemic fungicide was used to protect tomatoes against diseases caused by microbial plant pathogens. The residue level, rate of dissipation and half-life ($t_{1/2}$) of propamidine in tomato were determined, using reverse phase high performance liquid chromatography (RP-HPLC). Residue of propamidine was extracted from tomato using methanol and determined by RP-HPLC, with UV detection at 262 nm. The average recoveries of the samples fortified with propamidine at concentration range of 25 to 300 mg/kg, ranged from 87.072 to 106.34% with a relative standard deviation ranged between 0.169 to 3.503%. Initial deposit ranged from 2.45 to 5.75 mg/kg. The dissipation of propamidine in tomato followed the first order kinetic equation. The dissipation rate constants in tomato treated with recommended and double-recommended dose applied at 4 times and 2 times, ranged from 0.110 to 0.151 days and the corresponding half-life values were from 4.589 to 6.300 days. At the day 14 after last application, the residue concentrations of propamidine in tomato ranged from 0.42 to 0.54 mg/kg from the two blocks for all treatments. The propamidine residues dissipated below the limit of detection of 0.07 mg/kg at 28 days after treatment. The results indicated that because of the low toxicity, propamidine may not pose any residual toxicity problem at 14 days after application, and tomato fruits could be safely used for consumption (Kansaye et al. 2013).

The dissipation kinetics and behavior of fungicides after washing, peeling, simmering and canning of tomatoes were investigated and the values were expressed as processing factor (PF). Two cultivars of tomatoes Marissa and Harzfeuer were treated with azoxystrobin, boscalid, chlorothalonil, cyprodinil, fludioxonil, or pyroclostrobin at single and double dose of the fungicides. The risk assessment defined as hazard quotient was performed. The Quick, Easy, Cheap, Effective, Rugged and Safe (QuEchERS) method was used for sample preparation by liquid chromatography coupled with tandem mass spectrometry (LC-MS/MS). The average initial residues of the fungicides for variety Marissa and Harzfeuer were in the range of 0.158–1.076 and 0.217–1.143 mg/kg, respectively. The concentrations of fungicides in tomatoes decreased to 0.090–0.451 and 0.121–0.568 mg/kg, indicating that up to 99% of the initial deposits dissipated during 21 days after treatment. The dissipation rates were initially faster but slowed down over time, showing a nonlinear trend that fitted with first-order kinetic model (see Figure 4.1). The half-life values $(t_{1/12})$, theoretical dissipation time $(t_{0.01})$ to reach concentration of 0.01 mg/kg and dissipation rate constants (k) of the fungicides were 2.49 to 5.00 days for cv. Marissa and 2.67 to 5.32 days for cv. Harzfeuer. The shortest half-life time for cyprodinil and the longest for chlorothalonil were recorded in both tomato varieties. The residues dissipated below detectable level at 21 days after application except for chlorothalonil and azoxystrobin. Most fungicides dissipated faster in cv. Marissa than in cv. Harzfeuer. The results showed that chlorothalonil was fairly persistent with long residual activity. Hence, longer

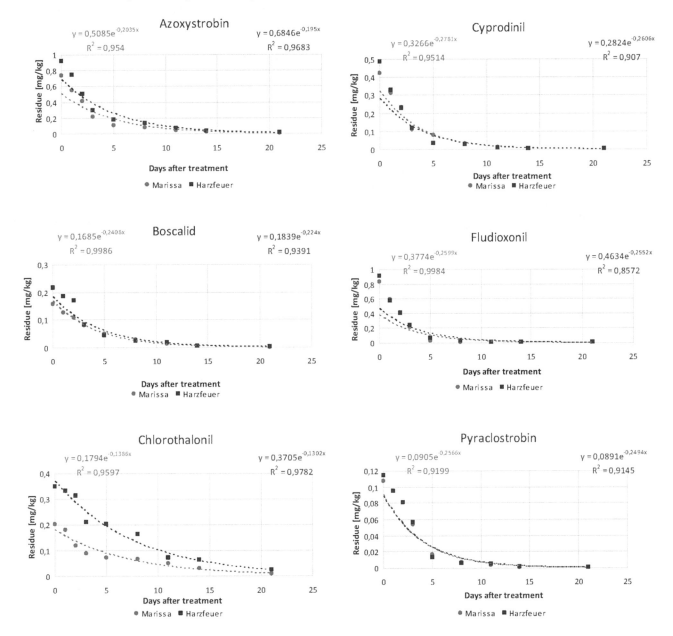

FIGURE 4.1   Dissipation kinetics of active ingredients of six fungicides from fruits of two tomato cultivars Marissa and Harzfeuer the concentrations of azoxystrobin, boscalid, chlorothalonil, cyprodinil, fludioxonil and pyroclostrobin decreased from the range of 0.158–1.076 mg/kg to 0.090–0.541 mg/kg in cv. Marissa and from the range of 0.217–1.143 mg/kg to 0.121–0.568 mg/kg in cv. Harzfeuer over a period of 21 days.

[Courtesy of Jankowska et al. 2016 and with kind permission of Springer Science, Germany]

safety waiting periods have to be adopted for safe consumption of fungicide-treated tomatoes (Jankowska et al. 2016).

Fluazinam residues in the environment may cause dermatitis and occupational asthma. Hence, the dissipation dynamics of fluazinam in edible raw food and in environment were studied. Fluazinam residues were analyzed by the modified Quick, Easy, Cheap, Effective, Rugged and Safe (QUEChERS) method and gas chromatography coupled with electron capture detector (GC-ECD). Mean recoveries and relative standard deviations (RSD) in potato plants were determined. Potatoes and soil at three spiking levels were 85.1 to 99.5% and 0.7 to 2.8% respectively. The limits of quantification were 0.01 mg/kg. The dissipation dynamics of fluazinam were determined in field trials in two provinces of China. In potato plants, fluazinam had a half-life of 2.5 days in Hebei and 3.6 days in Anhui. The half-life of fluazinam in soil was 4.7 days in Hebei and 13 days in Anhui. Terminal residue in soil samples ranged from 0.0925 to 0.949 mg/kg. Fluazinam was not detected in potato at preharvest intervals of 5, 7 and 10 days. The results indicated that fluazinam could be applied as per recommended dosage and timing (Feng et al. 2015). The pattern of dissipation of mandipropamid applied on cabbage at different harvest times was investigated under greenhouse conditions. The pre-harvest residue limit (PHRL) was determined, as it is the criterion to ensure the safety of the terminal fungicide residue during cabbage production. Tissues of Korean cabbage were collected at 0, 1, 3, 5, 7 and 10 days after mandipropamid application and analyzed. Recoveries ranged from 88.2 to 92.2% after fungicide application at concentrations of 0.4 and 2.0 mg/kg. The limit of quantitation (LOQ) was 0.04 mg/kg. Mandipropamid residues in Korean cabbage gradually decreased over time. The dissipation rate of the residue would be affected by intrinsic degradation of the compound along with dilution, resulting from the fast growth of Korean cabbage. The decay pattern was well fitted by simple first-order kinetics. Biological half-life values for mandipropamid ranged from 3.9 to 4.0 days in two field conditions. The PHRLs of mandipropamid in Korean cabbage recommended were 11.07–12.19 and 5.76–6.05 mg/kg for 10 and 5 days prior to harvest respectively (Gun et al. 2016).

Persistence of fungicides, pyraclostrobin, iprodione, tebuconazole and cyprodinil, following two sprays on strawberries in the greenhouse was determined for making risk assessments for these fungicides. The QuEChERS procedure was employed for sample preparation and high-performance liquid chromatography (HPLC)-tandem mass spectrometry and gas chromatography-tandem mass spectrometry were used for sample analyses. The range for the average recoveries of the four fungicides was 86.2 to 105.4% and the relative standard deviation range was 2.7 to 6.1%. The half-life values of pyroclostrobin, iprodione, tebuconazole and cyprodinil, after single application were 3.7, 3.6, 3.3 and 2.8 days respectively. Compared with a single application, a second application of the fungicides increased the average concentrations of residues left on the strawberries. Risk assessments of the four fungicides were performed, by comparing national estimated daily intakes against acceptable daily intakes under good agricultural practice conditions. Strawberries sprayed with pyraclostrobin, iprodione and tebuconazole were safe for consumption after two applications at recommended dosage. In contrast, the first application of cyprodinil was safe, whereas the second application might result in high health risk even at recommended dosage of 720 g a.i/ha and hence, the second application should not be taken up (Wang et al. 2015). Cyazofamid, member of phenylimidazole group, has broadspectrum activity against oomycetes and plasmodiophoromycetes with different site of action by inhibition of respiration. The residual toxicity of and groundwater contamination by cyazofamid were investigated, using two texturally different soils of Tarai region in India, at two fortification levels (100 and 200 g a.i/ha). Recovery of the fungicide was accomplished by RP-HPLC and detection at 279 nm. Degradation pattern fitted to monophasic first-order kinetics in soils with half-life values ranging from 4.3 to 4.95 days. Persistence of the fungicide at higher level in sandy loam, compared to silty clay loam soil. Linear correlation was observed in the range of 0.1–10.0 μg/ml of cyazofamid. The fungicide could not be recovered in the soil beyond 15 cm of soil depth. Maximum concentration of the fungicide remained at 5–10 and 10–15 cm soil depth. Average recoveries from soils fortified at 0.5–5.0 μg/g ranged from 78 to 86%. Limit of detection (LOD) and quantitation (LOQ) values were 0.01μg/ml and 0.05 μg/g respectively. The results showed that cyazofamid had short half-life in soils and low potential to be leached down by heavy rainfall conditions and the possibility of groundwater contamination was also less (Singh and Tandon 2015).

4.1.3 Resistance of Fungal Pathogens to Fungicides

Application of fungicides is an important and effective disease management strategy adopted for protecting various crops. The frequency of fungicide treatment varies depending on the economic importance of the crop concerned. Generally, the high-value crops–fruits and vegetables–receive more frequent applications of fungicides, compared to cereal crops which are marginally protected because of the low net return to the grower. Conditions, demanding repeated and continued application of the same fungicide, lead to development of fungicide resistance in fungal pathogens. Indiscriminate and excessive use of chemicals also results in hazards to human health and environment and adverse effects on nontarget microorganisms that may be beneficial for crop growth. Development of tolerance/resistance in fungal pathogens to chemicals has been a great concern for the growers and agencies involved in the development of agrochemicals, leading to the withdrawal or replacement of fungicides to which tolerant/resistant isolates/strains have been identified. Resistance/tolerance to fungicide may be defined as reduction in response of a microbial plant pathogen to a fungicide or chemical as a result of application for suppression of development of the pathogen concerned. Development of resistance in pathogen occurs at a faster rate against certain fungicides, whereas no single pathogen is known to develop resistance to different groups of fungicides. Resistance to nonsystemic (contact) fungicides in fungal pathogens used prior to 1970s was rarely detected. As the use of site-specific, systemic fungicides with narrow spectrum of activity increased, recovery of strains/isolates with reduced sensitivity or resistance to these fungicides, increased in number. Fungicide resistance may be of two kinds: (i) field resistance characterized by the development of resistant strains identified by their reduced sensitivity to the fungicide, when used as per label recommendation and (ii) resistance of isolates identified based on determination of baseline sensitivity and mean inhibitory concentration (MIC) under in vitro conditions (Staub 1991).

Fungicides can be of two kinds: (i) protectant (nonsystemic) fungicides and (ii) systemic fungicides. Protectant fungicides are frequently nonsystemic and unable to suppress established infections. By contrast, systemic fungicides are translocated to different plant parts from the site of application and protect or eradicate established infections. Systemic fungicides have close association with the biochemistry and physiology of the treated plants. Hence, their modes of action are specific and seek out a biochemical target site lethal in the pathogen, but not in treated plants. The systemic fungicides may inhibit steps in sterol biosynthesis, respiration, methionine biosynthesis, tubulin function or signal transduction. Resistance to these major modes of action of the fungicides has been identified in several fungal pathogen populations and lack of alternative modes of action of the fungicide(s) seriously limits their application. Resistance of fungal pathogens is a phenomenon of natural selection and the potential of pathogens to generate resistant mutants is a key factor in risk assessment. This risk may be determined by generating resistant mutants of target pathogen either by artificial mutagenesis following treatment of spores with chemical mutagen or UV-irradiation or exposing successive generation to increasing fungicide concentrations. Stable mutants are examined for determining whether resistance is under single or multiple genes control which may significantly influence risk analysis (Hollomon 2015).

4.1.3.1 In Vitro Assessment of Fungicide Resistance

Sensitivity of fungal pathogens to fungicides may be assessed by collecting the isolates occurring in the field under natural conditions and exposing them to different concentrations of the test fungicide(s). Isolates of *P. capsici* (150) obtained from 17 fields at 8 grower locations, were tested in agar-amended with mefenoxam. Among the isolates from all locations, 30% were sensitive; 10% were intermediate and 59% were resistant to mefenoxam. Isolates resistant to mefenoxam were present in 82% of the fields sampled and the proportion of resistant isolates in individual fields ranged from 28 to 100%. The mean EC_{50} value for mefenoxam-resistant isolates was 366.5 µg/ml, as against 0.12 to 1.1 µg/ml for sensitive isolates. Fields receiving mefenoxam alone had the highest proportion of resistant isolates, whereas fields applied with combination of other fungicides had much lower proportion of resistant isolates of *P. capsici* (Parra and Ristaino 2001). Isolates of *P. capsici* (75) obtained from five pepper and one squash fields were evaluated for their sensitivity to mefenoxam. The relative fitness of resistant and sensitive isolates was contrasted in vitro by their respective rates of colony growth and their ability to produce sporangia in unamended V8 juice agar medium. The frequency of resistant isolates in North Carolina populations was 63% much higher than resistance levels in areas where mefenoxam was not widely applied. All isolates from three fields, including two pepper and one squash field were resistant to mefenoxam. Response to mefenoxam remained stable during in vitro and in planta assessments. Resistance isolates grew on amended media at rates >80 to 90% and >100% of the nonamended control at 100 µg/ml and 5 µg/ml, respectively. Sensitive isolates could not develop at 5 or 100 µg/ml of mefenoxam incorporated in the medium. When measured by rate of colony growth, sporulation in vitro, or aggressiveness in plant, fitness of resistant isolates were not reduced. The results suggested that mefenoxam-resistant isolates of *P. capsici* were as virulent and fit as sensitive isolates of the pathogen (Café-Filho and Ristaino 2008).

Isolates of *P. capsici* (150), causing devastating disease of several economically important crops were examined for their sensitivity to iprovalicarb to determine the risk of development of resistance in the pathogen. The EC_{50} values of the isolates for mycelial growth inhibition by the fungicide averaged 0.3923 µg/ml, with a unimodal distribution. Highly resistant isolates of *P. capsici* were identified with a resistance factor >100. Resistance of the isolates to iprovalicarb remained stable through 10 transfers on iprovalicarb-free medium and most resistant isolates had the same fitness level (mycelial growth, zoospore production and virulence), as their corresponding parents, indicating iprovalicarb resistance was dependent on general growth characteristics. Cross-resistance was observed among all tested carboxylic acid amide (CAA) fungicides, including iprovalicarb, flumorph, dimethomorph and mandipropamid, but not with non-CAA fungicides, including azoxystrobin, cymoxanil, etridiazole, metalaxyl and zoxamide. The results indicated the possible moderate level of resistance risk when CAA fungicides are applied against *P. capsici* (Lu et al. 2010). In another investigation, 572 isolates of *P. capsici* from seven provinces in China were evaluated for their sensitivity to three carboxylic acid amide (CAA) fungicides, dimethomorph, flumorph and pyrimorph. Baseline sensitivity was determined, using 90 isolates which were not exposed to CAA fungicides. All other isolates (480) were completely inhibited at a single discriminatory dose of 0.5 µg/ml of dimethomorph. Four CAA-resistant mutants were generated by repeated exposure to dimethomorph in vitro. The four resistant mutants had similar fitness as the parents, based on the parameters of hyphal growth, sporulation in vitro and pathogenicity. Resistant mutants induced visible lesions on pepper stems or roots treated with recommended dose of dimethomorph. CAA-fungicides were considered to inhibit phospholipid biosynthesis and the primary target of the fungicide might be the choline phosphotransferase (CPT), which is referred to as amino alcohol phosphotransfereases (AAPTs). Two CPT (AAPT1 and AAPT2) genes in *P. capsici* were sequenced and analyzed. Analysis of cDNA sequences showed that AAPT1 and AAPT2 genes spanned 1538- and 1459-bp and were interrupted by five and three introns respectively. No differences in the amino acid sequences of parental wild-type isolate and CAA-resistant mutants were detected, suggesting that amino acid sequences may not contribute to development of resistance in *P. capsici* isolates to CAA fungicides (Sun et al. 2010). The baseline sensitivity and risk for resistance development to zoxamide were determined for 158 isolates of *P. capsici*, occurring under natural conditions. Effective concentration for 50% inhibition of mycelia growth showed a skewed unimodal distribution. Zoxamide resistant mutants were obtained either by treating mycelial culture and zoospores with UV irradiation or adapting the culture on zoxamide-amended medium in plates. The frequency of resistance selection averaged 1.8×10^{-7}. Resistant isolates were also derived by selfing or crossing two sexually compatible isolates, resulting in a mean selection frequency of 0.47. The resistance factor (RF) for zoxamide was 25 to 100 in *P. capsici* mutants. The mutants, through 10 transfers, maintained high levels of resistance factor (RF) (between 14 and 134) and had almost equal fitness as their wild-type parents in mycelial growth, sporulation and virulence. No cross-resistance was detected between zoxamide and either flumorph, metalaxyl, azoxystrobin, or etridiazole. The results showed that isolates of *P. capsici* might develop resistance to zoxamide with a predicted moderate level of risk for resistance development (Bi et al. 2011).

Flumorph, a carboxylic acid amide (CAA) fungicide was investigated for possible development of resistance in isolates of *P. capsici*. Baseline sensitivity of 83 isolates of *P. capsici* showed a unimodal distribution of effective concentration for 50% inhibition of mycelial growth ranging from 0.716 to 1.363 with a mean of 1.033 ± 0.129 µg/ml. Flumorph-resistant mutants (13) of *P. capsici* were generated through ultraviolet irradiation. Most of these mutants and their progenies showed high levels of fitness with similar rate of mycelial growth, sporulation and virulence as the wild-type isolates. Slight alterations in resistance to flumorph occurred through 10 transfers on agar media. Cross-resistance between flumorph and other fungicides, dimethomorph and iprovalicarb was observed, but not with non-CAA fungicides, cymoxanil, metalaxyl, azoxystrobin and cyazofamid. Inheritance of resistance to flumorph was investigated, through self-crossing and sexual hybridization, 619 progenies were generated. Segregation of the progeny from self-crossed isolate PCAS1 (flumorph-resistant) was 1:15 in the first generation and 0:1 or 1:15 in the second generation. In sexual hybridization, segregation of progeny was 0:1 and 1:7 for R x S hybridization. The results indicated that resistance of *P. capsici* isolates to flumorph was controlled by two dominant genes (Meng et al. 2011). The conventional fungicide-amended agar method employed to assess the proportion of fungicide-resistant isolates in a pathogen population is time-consuming and labor-intensive. A mutation (Q1077K) in cellulase synthase 3 was shown to confer resistance to pyrimorph (a CAA fungicide). The competitive ability of pyrimorph-sensitive and- resistant isolates of *P. capsici* was investigated. In addition, a real-time PCR-based assay was developed to quantify the frequency of the pyrimorph resistant allele Q1077K in *P. capsici* populations

and to evaluate the competitive fitness of pyrimorph-resistant isolates. Mixed zoospore suspensions of resistant (R) and sensitive (S) isolates at five ratios (1R:9S, 3R:7S, 5R:5S, 7R:3S and 9R:1S) were applied to carrot agar in vitro test and to the soil surface around pepper plants in in planta test. The proportion of resistant isolates was measured by a conventional assay in which single zoospore isolates recovered after five successive transfers or 10 successive disease cycles were grown on agar media with discriminatory concentration of pyrimorph. The results of the conventional assay and real-time PCR assay were similar. Both assays showed that the competitive ability of resistant isolates was similar to or less than that of the sensitive isolates. The real-time PCR assay has the potential for high-throughput analysis and monitoring the development of pyrimorph resistance in field populations of *P. capsici* (Pang et al. 2014).

Pathogen isolates with resistance to fungicides may arise following exposure to high-risk fungicides or they may be introduced from external sources through seeds/propagules. The possibility of watermelon seeds being source of fungicide-resistant inoculum of *D. bryoniae*, causal agent of gummy stem blight (GSB) of watermelon and other cucurbits was explored by direct plating, sweat box and blotter assays. Three isolates of *D. bryoniae* were isolated from two seed lots and they were evaluated for their levels of sensitivity/insensitivity to boscalid, tebuconazole, azoxystrobin and thiophanate-methyl (TM), which were commonly applied to protect watermelon plants against GSB disease. The isolates were resistant to thiophanate-methyl, indicating the presence of fungicide-resistant isolates of *D. bryoniae* in naturally infected watermelon seeds distributed commercially. The results indicated the need for adopting suitable disease management strategies to prevent introduction and spread of pathogen isolates resistant to fungicides through seeds to new locations (Thomas et al. 2014). In order to study the genetic modifications in response to development of resistance to quinine-outside-inhibiting (QoI) fungicides in *A. rabiei*, a fragment of the cytochrome *b* gene (*cytb*) from resistant isolates was characterized. The sequenced genomic DNA fragment contained a group I intron immediately after codon 131. A mutation in the codon for amino acid 143 (GLT → GTC) was detected, introducing an amino acid substitution from glycine to alanine (G143A) which was associated with QoI resistance. Based on the mutation, the diagnostic PCR assay was developed, using mismatch amplification mutation assay. Of the 70 *A. rabiei* isolates showing resistance to QoI fungicide, 38 isolates were found to have the amino acid substitution. The PCR assay had the potential for application to screen the *A. rabiei* isolates for their sensitivity/resistance to QoI fungicide (Delgado et al. 2013). The QoI fungicide azoxystrobin (AZO) initially suppressed the development of gummy stem blight (GSB) pathogen *D. bryoniae* effectively in the southeastern United States. Azoxystrobin inhibited mitochondrial respiration of the pathogen by binding to the outer quinine-oxidizing pocket of the cytochrome bc1 (*cytb*) enzyme complex. Resistance to azoxystrobin was detected later in *D. bryoniae* isolates from several watermelon-growing regions.

Comparison of the DNA sequences of partial *cytb* genes of four azoxystrobin-resistant (AZO-R) and four sensitive (AZO-S) isolates of *D. bryoniae* revealed the amino acid substitution of glycine by alanine at 143 codon (G143A) in the AZO-R isolates tested. Allele-specific primers were designed to detect resistant or sensitive allele at codon 143 of the *cytb* gene, which amplified a 165-bp PCR product from genomic DNA of nine AZO-R and not from nine AZO-S isolates. The results indicated that the PCR assay could be employed to detect the presence of azoxystrobin-resistant isolates of *D. bryoniae* in the field (Finger et al. 2014).

Development of resistance to mefenoxam in *P. nicotianae* and fitness of resistance isolates to induce disease in the nurseries were studied. Of the 95 isolates, 25 isolates were highly resistant to mefenoxam with EC_{50} values ranging from 235.2 to 466. 3 µg/ml; four were intermediately resistant with EC_{50} values ranging from 1.6 to 2.9 µg/ml and 60 isolates were sensitive with EC_{50} values less than 0.04 µg/ml. Mefenoxam provided effective protection to *Pelargonium* x *hortorum* cv. White Orbit against sensitive isolates, but not to any resistant isolate. Resistant isolates outcompeted sensitive ones, when inoculated on *Lupinus* Russell Hybrids in the absence of mefenoxam. Resistant isolates had greater sporulation capacity and greater infection rates than the sensitive ones. The results clearly showed that resistance to mefenoxam might pose serious challenge to the continued application of the fungicide in herbaceous and perennial plant species in nurseries (Hu et al. 2008). The baseline sensitivity of *R. solani* to thifluzamide, a succinate dehydrogenase inhibitor (SDHI) fungicide and the risk of the pathogen developing resistance to the fungicide were investigated, using 227 isolates of *R. solani* from 12 provinces of China, collected during 2007–2011. One insensitive isolate GD-1 was collected from the field and the EC_{50} values of other sensitive isolates (226) had a unimodal frequency distribution with a mean of 0.035 µg/ml. Using the fungicide-amended medium or UV irradiation nine resistant mutants were generated in the laboratory. The mutants were stable through 10 transfers on potato dextrose agar (PDA) medium. Fitness of the most resistant mutants was lower than that of sensitive isolates, indicating a lower competitive ability of the mutants than the sensitive isolates in the field. Cross-resistance was observed between thifluzamide and other SDHI fungicides fenfuram, carboxin, penflufen and boscalid, but not between thifluzamide and carbendazim, difenconazole, propiconazole, SYP-2815 (QoI fungicide developed in China), fluazinam, jinggangmycin, pyrimorph or mepronil. The results suggested that the risk of *R. solani* developing resistance to thifluzamide might be low to moderate (Mu et al. 2017). Resistant isolates of *P. capsici* appeared following extensive application in China. Metalaxyl-resistant (R) and – sensitive (S) isolates were differentiated based on the sporangium germination. Mandipropamid, a carboxylic acid amide (CAA) fungicide was highly inhibitory to *P. capsici* with EC_{50} values of 0.075 and 0.004 µg/ml, respectively, against mycelial growth and sporangium germination. The QoI fungicides azoxystrobin and trifloxystrobin were less effective in inhibiting mycelial growth. However, they provided over 80% protection against pepper Phytophthora blight (PPB) disease. Propamocarb and cymoxanil, although had no inhibitory effect on mycelial growth or sporangium germination, had 70% efficacy on controlling PPB. The fungicide mixture fluopicolide (62.5 g/l) + propamocarb (62.5 g/l) offered >88% efficacy in controlling PPB disease. The results showed that no cross-resistance existed between metalaxyl and other tested fungicides, which could be used as alternative to metalaxyl for management of diseases caused by *P. capsici* (Qi et al. 2012).

Phenylamide fungicides are extensively applied for the control of diseases caused by oomycetes. Mefenoxam was effective in suppressing development of potato late blight disease caused by *P. infestans*. Isolates of *P. infestans* collected during 2009–2012 in Wisconsin were evaluated for their sensitivity to mefenoxam. In addition, distribution of clonal lineages and mating types were also studied. The usefulness of analysis of glucose-6-phosphate isomerase (Gpi) allozyme was assessed for predicting mefenoxam sensitivity of isolates of *P. infstans* to enable growers to apply suitable measures to reduce disease incidence and spread. Among the 143 isolates, three clonal lineages, US-22, US-23 and US-24 were identified. The US-22 was of A2 mating type and sensitive to mefenoxam with Gpi 100/122. By contrast, US-23 and US-24 belonged to A1 mating type and primarily intermediately sensitive to mefenoxam with Gpi 100/100 and 100/111, respectively. Thus, the mefenoxam sensitivity could be predicted based on the allozyme assay, because of the close correlation and the unique Gpi patterns for each lineage. As the new clonal lineages of *P. infestans* might appear, it was considered essential to continue close monitoring of late blight pathogen isolates for their sensitivity to fungicides (Seidl and Gevens 2013).

Phytophthora spp., infecting ornamentals could be recovered from irrigation water in six states of the southeastern United States. Isolates (1,483) from forest and suburban streams in Georgia and Virginia were screened for sensitivity to mefenoxam, using 48-well tissue culture plates. *P. nicotianae*, *P. hydropathica* and *P. gonapodyides* were the most abundant among the isolates obtained from plants, irrigation water and streams. Only 6% of isolates associated with plants and 9% from irrigation water were insensitive to mefenoxam at 100 µg/ml. About 78% of insensitive isolates associated with plants were *P. nicotianae* and most of these (67%) were from herbaceous annual plants. Most of the insensitive isolates recovered from irrigation water were *P. gonapodyides*, *P. hydropathica*, *P. megasperma* and *P. pini* and 83% of insensitive isolates from streams were *P. gonapodyides*. The results suggested that mefenoxam could be applied for the management of Phytophthora diseases affecting ornamental plants (Olson et al. 2013). The genetic basis of development of insensitivity/resistance in isolates of *P. infestans* to phenylamide fungicide mefenoxam was studied. Crosses between insensitive isolates of *P. infestans* were made. The F1 progeny showed expected semi-dominant phenotypes for mefenoxam insensitivity and suggested involvement of multiple loci, complicating insensitivity to mefenoxam. The primary effect of phenylamide compounds was found to inhibit ribosomal RNA synthesis by the pathogen. The subunits of RNA polymerase 1 (RNApol I) were sequenced from sensitive and insensitive isolates and F1 progeny. Single nucleotide polymorphisms (SNPs) specific to insensitive field isolates were identified in the gene encoding the large subunit of RNApol I. Examination of field isolates of *P. infestans* showed that SNPT1145A was associated with mefenoxam insensitivity to a great extent (86%). Isolates lacking this association belonged predominantly to one genotype of *P. infestans*. Incorporation of the allele governing sensitivity RPA190 in the sensitive line, resulted in the generation of transgenic lines insensitive to mefenoxam. The results established that sequence variation in RPA190 contributed to insensitivity in *P. infestans* to mefenoxam (Randall et al. 2014).

Carrot cavity spot disease is induced by *Pythium violae, P. irregulare, P. dissotocum, P. ultimum* and *P. sulcatum* and their dominance may differ, depending of the location. On mefenoxam-amended media, 93% of 197 isolates were sensitive to the fungicide. But two of five isolates of *P. irregulare* were highly resistant and about half of the isolates of *P. intermedium* and *P. sylvaticum* and a single isolate of *P. violae* were highly or intermediately resistant to mefenoxam. *P. dissotocum, P. irregulare, P. sulcatum* and three of seven isolates of *P. intermedium* were insensitive to fluopicolide, whereas *P. sylvaticum, P. ultimum, P. violae* and some isolates of *P. intermedium* were insensitive. All isolates of *Pythium* spp. were sensitive to zoxamide (EC$_{50}$ <1 µg a.i/ml). The results indicated the need for assessing the sensitivity of the isolates of *Pythium* spp. causing carrot cavity spot disease and determining their distribution in different locations for selecting the fungicide that may be applied in different locations to contain the disease incidence and spread (Lu et al. 2012). Forest nurseries were treated with mefenoxam and fosetyl-Al commonly to protect the seedlings against damping-off and root rot. In vitro and greenhouse tests were performed to assess the extent of presence of isolates of *Pythium irregulare, P. sylvaticum* and *P. ultimium* insensitive to these fungicides. *P. irregulare* was about three times less sensitive to mefenoxam (0.20 µg/ml) than *P. sylvaticum* (0.06 µg/ml) and *P. ultimum* (0.06 µg/ml). In addition, two resistant isolates of *P. ulitmum* were identified (≥ 311 µg/ml). All *Pythium* spp. were sensitive to fosetyl-Al to the same level (1,256 to 1,508 µg/ml) and none of the isolates was found to be resistant to fosetyl-Al. Under greenhouse conditions, fosetyl-Al and phosphorous acid were consistently effective in suppressing damping-off caused by *P. dissotocum, P. irregulare* and *P. 'vipa'*. Depending on *Pythium* spp., causing the disease, mefenoxam also provided acceptable level of protection. The results indicated that rotating mefenoxam with other fungicides might be an effective strategy to contain the Pythium damping-off disease, affecting forest nurseries. *P. dissotocum, P. irregulare* and *P. 'vipa'* were equally pathogenic and caused similar levels of disease incidence on Douglas-fir seed planted in nontreated, inoculated control plots. *P. dissotocum* reduced seedling survival by 90%, whereas *P. irregulare* and *P. 'vipa'* were responsible for reduction of 94% and 96% respectively. Treatment with

phosphorous acid and fosetyl-Al consistently provided effective protection against damping-off, regardless of *Pythium* spp. involved and the levels of protection provided by both fungicides were similar. By contrast, mefenoxam provided similar level of protection against infection by *P. dissotocum* and *P. irregulare*, but not against *P. 'vipa'*. All other fungicide treatments were ineffective in suppressing development of symptoms induced by *Pythium* spp. Biological treatment with *Trichoderma harzianum* + *T. virens* was also effective against damping-off disease caused by *Pythium* spp. None of the fungicides induced any observable phytotoxic symptoms in seedlings (Weiland et al. 2014).

Sensitivity of *A. rabiei* (causing Ascochyta blight of chickpea) populations to QoI fungicides was assessed. Isolates of *A. rabiei* resistant to these fungicides were detected in the northern Great Plain of the United States. Hence, 78 isolates of *A. rabiei* collected during 1983–2007 were screened for their sensitivity to determine the baseline sensitivity to the demethylation-inhibiting foliar fungicide prothioconazole. Baseline EC$_{50}$ values for mycelial growth for prothioconazole ranged from 0.0526 to 0.2958 µg/ml, with a mean of 1.783 µg/ml. Isolates collected from 2007 to 2009 from North Dakota chickpea fields exposed to prothioconazole sensitivity were screened, using the same assay. Mean EC$_{50}$ values for these isolates were 0.3544 µg/ml, 0.3746 µg/ml and 0.7829 µg/ml, respectively. These values indicated a decrease in the sensitivity of the isolates by about twofold (in 2007–2008) to 4.4-fold (in 2009) to prothioconazole over the years. By contrast, sensitivity of the pathogen isolates to thiabendazole was not altered to any significant level (Wise et al. 2011). Development of resistance in *Ascochyta pinodes* (*Mycosphaerella pinodes*) causing Ascochyta blight in pea, lentil and chickpea to DMI fungicide, prothioconazole and succinate dehydrogenase-inhibiting (SDHI) fungicides, boscalid and fluxapyroxad was investigated. The EC$_{50}$ values and the single discriminatory fungicide concentration were determined. These parameters could be used to monitor shifts in sensitivity of pathogen isolates of *A. pinodes, A. lentis* and *A. rabiei* to make disease management decisions (Lonergan et al. 2015).

Isolates of *S. sclerotiorum* infecting canola were evaluated for their genotypic and phenotypic diversity. Genetic diversity was assessed with eight microsatellite markers and mycelial compatibility groups (MCGs). Phenotypic diversity was determined, based on the levels of sensitivity to three fungicides, production of oxalate and microsclerotia, growth rate and pathogenicity on two canola cultivars. The populations of *S. sclerotiorum* from China and United States differed significantly for all phenotypic traits, except for sensitivity to fungicides fluazinam and virulence. Chinese isolates had several unique traits. Despite the phenotypic differences, heritabilities of the phenotypic traits were similar for both populations. Significant correlations were found among five phenotypic traits. Cross-resistance to benomyl and iprodione was detected. Virulence was not significantly correlated with any other phenotypic trait and had the least heritability. However, both populations were virulent to the same level on both susceptible or moderately resistant canola cultivars

(Attanayake et al. 2013). Isolates of *S. sclerotiorum* (3,701) from *Sclerotinia*-infected oilseed rape stems were tested for their sensitivity to carbendazim (MBC) and dimethachlon. Sensitivity tests showed that 375 isolates were resistant to carbendazim with a resistance frequency of 10.1%. The proportion of carbendazim-resistant isolates ranged from 0 to 44%, depending on the location in Jiangsu Province, China. The field isolates with carbendazim resistance showed mycelial growth, sclerotial production, pathogenicity and osmotic sensitivity comparable to sensitive isolates. The results suggested that carbendazim resistant isolates might have enough parasitic fitness as their parental isolates. In addition, positive cross-resistance between carbendazim and dimethachlon and other dicarboximide fungicides such as iprodione and procymidone was observed (Wang et al. 2014).

Isolates of *S. sclerotiorum* (2,424) were evaluated for the resistance to the dicarboximide fungicide dimethachlon, which was widely used for suppressing stem rot disease development in oilseed rape and soybean in China. The discriminatory dose of 5 μg/ml was used to determine dimethachlon-resistance based on mycelial growth inhibition. In one of the five provinces, Hunan, 3 of 268 isolates (1.2%) from oilseed rape plants were resistant to dimethachlon and the resistance ratios for the three resistant isolates were 4.56, 32.70 and 105.53 respectively. In Heilongjiang province, 8 of 243 isolates (3.29%) from soybean plants showed resistance to dimethachlon, the resistance ratios being 5.57 to 94.80 (2011); 11 of 409 isolates (2.69%) had resistance ratios of 3.21 to 9.69 (2012). Cross-resistance between dimethachlon and other fungicides iprodione, procyamidone and N-phenylcarbamate fungicide, diethofencarb was recorded, whereas there was no cross-resistance with carbendazim, tebuconazole, kersoxin-methyl, thiram and boscalid. Dimethachlon-resistant isolates from fields were more sensitive to osmotic pressure and had lower growth rate on potato dextrose agar (PDA) medium. They were less virulent on leaves of oilseed rape, indicating the adverse effect of resistance on pathogen fitness (Zhou et al. 2014). Resistance of *S. sclerotiorum* isolates to thiophanate-methyl (TM), fluazinam and procymidone, that were extensively applied in Brazil for the control of white mold disease of common bean, was investigated. None of the 282 isolates was resistant to fluazinam or procymidone with EC_{50} values of 0.003–0.007μg/ml for fluazinam and 0.11–0.72 μg/ml for procymidone. One isolate showing resistance to thiophanate-methyl had EC_{50} value of >100 μg/ml, whereas the TM sensitive isolates had EC_{50} values varying from 0.38 to 2.23μg/ml. The TM-resistant isolate had a L240F mutation in the ß-tubulin gene. Mutation at codon 240 inducing resistance in *S. sclerotiorum* to a benzimidazole fungicide was identified for the first time. The high-resolution melting analysis allowed differentiation of TM-sensitive and -resistant isolates by specific melting peaks and curves. The TM-resistant isolate had mycelial growth, sclerotia production and virulence in equivalence to the sensitive isolates, indicating that with similar fitness, the resistant genotypes of *S. sclerotiorum* might be able to compete well with sensitive isolates under field conditions. The results showed that resistance of *S. sclerotiorum* to

TM, fluazinam and procymidone might be rare. But constant monitoring the possible emergence of fungicide resistant isolates should be in place (Lehner et al. 2015).

Boscalid-resistant (BR) *S. sclerotiorum* mutants, obtained by fungicide induction, were characterized by in vitro tests. No cross-resistance was detected between boscalid and dimethachlon, fluazinam or carbendazim. But positive cross-resistance existed between boscalid and carboxin and negative cross-resistance was detected between boscalid and kresoxim-methyl. BR mutants had slower growth rate, loss of ability to produce sclerotia, reduced virulence and lower oxalic acid content. But the BR mutants had higher mycelial respiration and succinate dehydrogenase (SDH) activity. In addition, BR mutants had reduced sensitivity to salicylhydroxamic acid (SHAM), but not to oxidative stress. The results indicated that the risk of resistance to boscalid in *S. sclerotiorum* was low to moderate. All BR mutants had the same point-mutation A11V(GCA to GTA) in the iron sulphur protein subunit (SDHB), as indicated by DNA sequence analysis. Expression of the cytochrome *b* (*cytb*) gene was reduced to different degrees in BR mutants and this might be correlated with negative cross-resistance between boscalid and kresoxim-methyl (Wang et al. 2015). Extensive application of a single fungicide selects fungicide resistance in fungal pathogens like *S. sclerotiorum*. Hence, resistance detection is important for risk assessment and integrated management of the disease(s) caused by the pathogen. Because of the importance of Sclerotinia stem rot and development of carbendazim resistance, a rapid technique was needed for detecting the F220Y mutant genotype of carbendazim-resistant isolates of *S. sclerotiorum*. The point mutation at codon 200 (TTC → TAC, F200Y) conferred moderate resistance in *S. sclerotiorum* to carbendazim. The mutant genotype F200Y was detected mainly by determining the minimum inhibitory concentration (MIC) requiring 3 to 5 days. The loop-mediated isothermal amplification (LAMP) assay was employed for rapid detection of F200Y mutant genotype of carbendazim resistant isolates of *S. slcerotiorum*. Specific LAMP primers were designed and concentrations of LAMP components were optimized as 62 to 63°C for 45 min. The assay generated specific products only from isolates of F200Y mutant genotype, but not from other carbendazim-resistant mutants of the pathogens, *B. cinerea* and *Fusarium graminearum*. Inclusion of the loop backward (LB) primer reduced the reaction time to 15 min. The results of LAMP and MIC assays were similar. The LB-accelerated LAMP assay detected F200Y mutant genotype using the sclerotia produced on rape stems of plants artificially inoculated in the field. The results demonstrated the usefulness of LAMP assay for detection and differentiation of carbendazim-resistant isolates of *S. sclerotiorum* infecting oilseed rape (Duan et al. 2016).

Two heptapeptide derivatives 77-3 and 77-12 showed minimum inhibitory concentration (MIC) values between 3.8 and 7.5 μg/ml against *F. sambucinum*, causing dry rot of potato tubers. Incubation of conidia or mycelia of *F. sambucinum* showed that the treated conidia/hyphal cells were stained by the membrane impermeant dye SYTOX Green, indicating disruption of pathogen membranes by the synthetic peptide. Conidial viability treated with peptide 77-3 at 10 μg/ml showed a reduction (91 ± 3.6%) in viability in 15 min. The small peptides could act synergistically with thiabendazole (TBZ) against TBZ-resistant *F. sambucinum* (Gonzalez et al. 2002). In the later investigation, *Fusarium* spp. associated with potato tuber dry rot disease were frequently identified as *F. oxysporum* (30.3%), *F. equiseti* (19.3%), *F. sambucinam*, *F. solani* and *F. acuminatum*. Other species isolated were *F. sporotrichioides, F. torulosum, F. tricinctum* and *F. graminearum*. *F. sambucinum* isolates were the most virulent. All isolates (228) of *Fusarium* spp. were sensitive to difenconazole, with an EC_{50} values of <5 mg/l. By contrast, isolates of *F. sambucinum* and *F. oxysporum* were sensitive to fludioxonil (EC_{50} >100 mg/l). All isolates of *F. oxysporum* were sensitive to thiabendazole (EC_{50} <5 mg/l), except those of *F. sambucinum* (EC_{50} >100 mg/l). The results indicated the importance of establishing accurate identity of *Fusarium* spp. associated with tuber dry rot disease to determine the appropriate fungicide for suppressing the development of the disease (Gachango et al. 2012).

4.1.3.2 Greenhouse Assessment of Fungicide Resistance in Pathogens

Development of insensitivity/resistance in fungal pathogens to fungicides has been assessed under greenhouse conditions. Prolonged use of phosphite may lead to development of resistance in *P. cinnamomi*, infecting avocado. *P. cinnamomi* isolates (30) were obtained from a range of sites with different phosphite-use histories, including phosphite-treated and -untreated avocado orchards and phosphite-treated and -untreated native vegetation sites. The colonizing potential of three isolates was assessed by different inoculation methods against a range of host tissues treated and untreated with phosphite, including mycelia stem inoculation on clonally propagated *Leucadendron* sp., mycelial root inoculation of lupin seedling and zoospore inoculation of *Eucalyptus sieberi* cotyledons. Isolates from avocado orchards with long history of phosphite use, were on an average more extensive colonizers of the phosphite-treated *Leucadendron* sp., lupin seedling roots and *E. sieberi* cotyledons. But these isolates did not colonize untreated plant tissue (*Leucadendron* sp.) more extensively than the isolates from sites with no history of phosphite use and no isolate was resistant to phosphite. Selection of reduced sensitivity to phosphite in planta might have occurred within sexual clonal lineages of *P. cinnamomi* in sites with prolonged application of phosphite (Dobrowolski et al. 2008). *Pythium* spp. commonly occurring in the greenhouse in Oman were identified as *P. aphanidermatum, P. spinosum, P. myriotylum* and *P. catenulatum*, based on the sequences of the internal transcribed spacer region (ITS) of ribosomal RNA. Phylogenetic analysis of *P. aphanidermatum* isolates showed that most of the isolates obtained from cocopeat clustered separately from isolates obtained from soil and roots of cucumber plants. Resistance of isolates of *P. aphanidermatum* (27) to hymexazol was assessed. Most of the isolates were sensitive (0.9 to 31.2 mg/l) and one isolate was resistant (142.9

mg/l) to hymexazol. The results suggested that development of resistance to hymexazol in greenhouse populations of *P. aphanidermatum* should be constantly monitored to prevent increase in the number of isolates resistant to the fungicide (Al-Balushi et al. 2018).

4.1.3.3 Field Assessment of Fungicide Resistance in Pathogens

Existence of isolates of fungal pathogens resistant to fungicides may be detected under field conditions. Tubers from potato plants treated with in-furrow and foliar application of mefenoxam were inoculated with eight isolates of *P. erythroseptica* exhibiting different levels of sensitivity to the fungicide. Isolates of *P. erythroseptica* with EC_{50} values of 0.02–0.04 μg/ml and >1.0 μg/ml were, respectively, considered as sensitive and insensitive to mefenoxam. The isolates with EC_{50} values of 1.1–5.3 μg/ml were grouped as low intermediate, EC_{50} values of 26–74 μg/ml as high intermediate and EC_{50} values >100 μg/ml as resistant. The biological activity of these isolates was assessed by quantifying the extent of disease control. *P. erythroseptica* isolates in the resistant group infected a significantly greater percentage of untreated tubers than isolates in any other group. Mefenoxam reduced infection frequency of sensitive isolates by about 37%. Mefenoxam did not reduce incidence of pink rot disease caused by resistant isolates. The isolates of resistant and high intermediate groups were more aggressive than sensitive isolates in the absence of mefenoxam pressure and significantly even in the presence of mefenoxam. The results suggested that pink rot disease might become more severe in the fields with *P. erythroseptica* populations with EC_{50} values >1.0 mg/ml of mefenoxam (Taylor et al. 2006). In the further study, the competitive ability of the mefenoxam-resistant and -sensitive isolates of *P. erythroseptica* was assessed under no selection pressure (untreated) and also under the influence of mefenoxam and non-mefenoxam (phosphorous acid) fungicide. Isolates of *P. erythroseptica* were combined in four ratios of mefenoxam-resistant (R) to mefenoxam-susceptible (S) (0R:0S, 1R:1S, 3R:1S and 1R:3S) and subsequently infested the soil at the time of planting. In-furrow application of mefenoxam to the soil was carried out, soon after infestation with the pathogen. Phosphorous acid was applied twice at tuber initiation and 14 days after tuber initiation. *P. erythroseptica* isolates were recovered from tuber infected by pink rot in the field at harvest. At 3–4 weeks after harvest, isolates were evaluated for their sensitivity to mefenoxam in vitro. Tests were performed in vivo, by challenge-inoculating the tubers with zoospore suspension in the four-ratio mentioned above. The field plots were not infested with *P. erythroseptica* at planting. The isolates of the pathogen were as fit as sensitive isolates in the absence of selection pressure or in the presence of phosphorous acid treatment. Under mefenoxam selection pressure, mefenoxam-resistant *P. erythroseptica* isolates were more pathogenically fit than sensitive isolates. The results suggested that mefenoxam-based fungicides could be used for the management of pink rot disease of potato, since the resistant isolates could remain stable in most agroecosystems (Chapara et al. 2011).

4.1.3.4 Distribution of Pathogen Isolates Resistant to Fungicides

Fungal pathogens are capable of producing isolates/strain that can readily overcome the challenges posed due to the release of disease resistant cultivars of various crops, adverse environmental conditions, or application of different kinds of chemicals that have different modes of action on the pathogens. Fungicide resistant isolates of a pathogen species emerge following repeated and extensive use of a single fungicide over the years. Selection or mutation may be the principal mode by which fungicide insensitive/resistant isolates are formed. It is essential to determine the distribution of fungicide resistant isolates of pathogen in the location(s) to plan for measures for fungicide resistance management. The distribution of isolates of *A. rabiei* sensitive and resistant to the QoI fungicides, azoxystrobin and pyraclostrobin was monitored using spore germination assay in vitro. Of the 403 isolates of *A. rabiei*, the EC50 values of 98 isolates were determined for each isolate- fungicide combination. A discriminatory dose of 1 μg/ml of azoxystrobin was established, using 305 isolates. The isolates (65%) collected from North Dakota during 2005–2007 and from Montana in 2007 were found to exhibit a mean of 100-fold decrease in sensitivity to both azoxystrobin and pyraclostrobin, compared to sensitive isolates. They were considered to be resistant to azoxystrobin and pyraclostrobin. Under greenhouse conditions, QoI resistant isolates of *A. rabiei* caused significantly higher disease severity than sensitive isolates on azoxystrobin- or pyraclostrobin-treated chickpea plants. The results suggested that the level of protection provided by azoxystrobin or pyraclostrobin against infection by resistant isolates of *A. rabiei* might be inadequate (Wise et al. 2009).

Variations in mefenoxam resistance of 257 isolates of *P. capsici* collected from vegetables at 22 sites in four regions of New York and pathogen population genetic structure were studied. Isolates were assayed for mefenoxam resistance and genotyped for mating type and five microsatellite loci. Mefenoxam-resistant isolates were present at high frequency in the Capital District and Long Island, but none was found in Western New York or Central New York. Presence of both A1 and A2 mating types was recorded at all sites (12) and 126 distinct multilocus genotypes were also recorded. Only one of these multilocus genotypes could be observed at more than one site. Approximately 24% and 16% of the variations in the population could be attributed to differences among regions and sites respectively. The results indicated high diverse populations of *P. capsici* in New York, but the gene flow among regions and fields appeared to be restricted. Hence, population in each field had to be considered as an independent population and effective measures have to be applied to prevent movement of inoculum from field to field and region to region (Dunn et al. 2010). Shifts in population structure are a major concern for the management of diseases caused by fungal pathogens. Isolates (127) of *P. nicotianae*, incitant of

tobacco black shank disease were evaluated for mefenoxam resistance in Virginia State, USA. None of them was insensitive; 98% of the isolates were highly sensitive or sensitive and the remaining 2% of the isolates were intermediately sensitive, indicating the effectiveness of mefenoxam in suppressing the black shank disease development and spread (Parkunan et al. 2010). Similar results were reported on the effects of mefenoxam on *P. parasitica* var. *nicotianae* from Guizhou province of China. Among the 179 isolates screened for sensitivity to mefenoxam, none of the isolates was resistant, whereas 41.9% of the isolates were intermediately resistant and 58.1% were sensitive to mefenoxam. The results obtained showed that application of mefenoxam could be a component of the integrated management system applicable to tobacco black shank disease (Wang et al. 2013)

During official plant health survey on *Rhododendron, Viburnum* and *Camellia* in ornamental nurseries, 94 isolates of *Phytophthora ramorum* were obtained from different geographic regions in northern Spain from 2003 to 2008. The pathogen isolates were evaluated for sensitivity to phenylamide fungicides, mating type and genetic diversity. All Spanish isolates belonged to the EU1 lineage. The genotypes EU1MG37, EU1MG39 and EU1MG40 were isolated from *Rhododendron* from one region; EU1MG38 and EU1MG41 were isolated from Camellia from two different regions. Isolates of the genotype EU1MG38 were resistant to metalaxyl and mefenoxam. The level of genetic diversity within the Spanish population of *P. ramorum* was found to be limited and indicated a relatively recent clonal expansion. Among the five genotypes of *P. ramorum* infecting ornamental plants only one genotype isolates showed resistance to the fungicides, indicating the need for accurate identification of the isolates of *P. ramorum* to decide whether or not to apply phenylamide fungicides (Pérez-Sierra et al. 2011). Application of fungicides continued to be the principal management strategy adopted for suppression of development of stem rot disease caused by *S. sclerotiorum* on oilseed rape and other crops in China. The benzimidazole fungicide carbendazim was extensively applied to control *S. sclerotiorum* during 1980s and 1990s. As a result, high levels of resistance to carbendazim in the isolates were recorded in the late 1990s and became widespread in eastern China in the following years. In order to overcome this situation, dicarboximide fungicides dimethachlon, iprodione and procymidone were recommended from 2000. After a decade, low levels of resistance to dimethachlon were reported in northwestern and central China, necessitating the use of alternative fungicides like pyraclostrobin (Liang et al. 2015a).

4.1.3.5 Fungicide Resistance Induced by Mutation in Fungal Pathogens

Fungicide resistant isolates of pathogens may be obtained by selecting the isolates by growing them on medium amended with increasing concentrations of the test fungicide(s). Fungicide resistant isolates may also be generated through mutagenesis, using chemicals or UV irradiation. Populations of *H. solani* infecting potato tubers, were exposed frequently to thiophanate-methyl (TM), but not to benomyl. Of the 238 isolates, 182 were resistant to both TM and benomyl and 56 isolates were sensitive to both fungicides. None of the isolates showed differential reaction to these fungicides. Two-point mutations were detected in ß-tubulin gene of *H. solani* genome of isolates from California. However, only the mutation at the 455-nucleotide position (codon 198) was strongly associated with TM resistance. This mutation resulted in a change of glutamic acid to alanine at codon 198. Independent of fungicide resistance, a mutation at the 836 sites were detected in all isolates from California. The PCR-based restriction fragment length polymorphism (RFLP) assay was applied to identify thiophanate-methyl resistance in *H. solani*. The resistant isolates showed two DNA fragments of 390- and 480-bp, whereas sensitive isolates had three fragments of 62-, 390- and 420-bp. The isolates of *H. solani* identified as resistant or sensitive to thiophanate-methyl by PCR-RFLP technique were classified exactly similarly by the results of fungicide-amended medium assay (Cunha and Rizzo 2003).

Sensitivity of isolates of *P. cinnamomi* to mefenoxam was assessed, based on EC_{50} values (50% inhibition of mycelial growth) and lesion development on the cotyledons of lupine seedlings. Mefenoxam provided effective protection to lupin seedlings against sensitive isolates of *P. cinnamomi*. Isolates with reduced sensitivity were generated by both UV mutagenesis and mycelium adaptation to mefenoxam. The mutants did not grow as rapidly as the wild-type sensitive isolates (Hu et al. 2010). Different fungicides were applied for suppressing the development of Sclerotinia stem rot caused by *S. sclerotiorum* in oilseed rape crops in China. Following widespread occurrence of benzimidazole resistance, dimethachlon, one of the dicarboximide fungicides was more frequently applied for the control of the disease. Hence, the need for assessing the levels of sensitivity of isolates of *S. sclerotiorum* in Jiangsu Province was recognized. Single-sclerotium isolates (1,066) were evaluated for their sensitivity to the pathogen based on the resistance factor (expressed as ratios between the EC_{50} values of resistant isolates and the average EC_{50} values of sensitive isolates). Dimethachlon-resistant mutants were derived from 10 wild-type sensitive isolates of *S. sclerotiorum* via in vitro selection for resistance to dimethachlon. The resistance factors for the mutants ranged from 198 to 484 and they were considered as highly resistant to dimethachlon. Two different mechanisms seemed to be involved in development of fungicide resistance. Moderate sensitivity (insensitivity) and high resistance under in vitro were differentiated. Positive cross-resistance was observed between dimethachlon and other dicarboximide fungicides, iprodione and procymidone in the resistant isolates. The field-collected dimethachlon-insensitive and laboratory-induced dimethachlon-resistant isolates appeared to have mycelial growth, sclerotial production and pathogenicity comparable to wild-type parental isolates. The dimethachlon-resistant isolates showed fitness as the sensitive isolates; appropriate precautionary measures have to be taken against resistance development in natural pathogen populations (Ma et al. 2009). In a later investigation, mutation in the genome of *S. sclerotiorum* in the highly conserved amino acids of succinate dehydrogenase (SDH) complex was

examined. Boscalid resistance of isolates of *S. sclerotiorum* was attributed to the mutation H132R in the highly conserved SdhD subunit protein of the SDH complex. Only one-point mutation, A11V in SdhB (GCA to GTA change in SdhB) was detected earlier in *S. sclerotiorum* boscalid-resistant (BR) mutants. Replacement of the *SdhB* gene in boscalid-sensitive (BS) *S. sclerotiorum* strain with the mutant *SdhB* gene conferred resistance. The BR and GSM (SdhB gene in the wild-type strain replaced by the mutant *SdhB* gene) mutants were more sensitive to osmotic stress and lacked the ability to produce sclerotia. It is important to note that the point mutation was not located in the highly conserved sequence of the iron-sulfur subunit of SDH. The results suggested that different mechanisms of boscalid resistance of the pathogen isolates might function in non-conserved and conserved domains of *S. sclerotiorum* (Wang et al. 2015a).

P. capsici causes crown, root and fruit rot of vegetable crops including pepper and tomato. Fluopicolide was effective in suppressing development of the disease. In vitro assessment of sensitivity of *P. capsici* isolates (126) to fluopicolide was taken up. Values of effective concentrations for 50% inhibition of mycelial growth ranged from 0.08 to 0.24 µg/ml and were distributed as a unimodal curve, indicating that all isolates were sensitive to fluopicolide. Mutants resistant to fluopicolide were selected from five isolates by screening zoospores on fluopicolide-amended media (at 5 µg/ml) at a mutation frequency above 1.0×10^{-7}. The mutants were clustered with either intermediate (resistance factor ranging from 3.53 to 77.91) or high (RF = 2481.40 to 7034.79) resistance. Fluopicolide resistance in the isolates was stable through 10 cycles of mycelia transfers on fungicide-free medium. Similar fitness levels, were recorded among the mutants, based on the parameter's zoospore production, cyst germination and virulence as in their wild-type parent isolates with few exceptions. Risk of development of resistance in field populations of *P. capsici* under field conditions was considered to be moderately high level. Hence, emergence of resistant isolates has to be closely monitored (Lu et al. 2010).

Quinine-outside inhibitor (QoI) fungicides were applied for the control of *A. rabiei*, causal agent of Ascochyta blight disease of chickpea in North Dakota, USA. In order to identify the mutation in *A. rabiei* genome, a fragment of cytochrome *b* gene was characterized. The sequenced genomic DNA fragment contained a group I intron immediately after codon 131. The size of the cytochrome *b* gene was estimated to be over 4.6-kb. Multiple alignment analysis of cDNA and protein sequences revealed a mutation that changed the codon for amino acid 143 from CGT to GCT, introducing an amino acid substitution from glycine to alanine (G143A), which was frequently associated with QoI resistance. The diagnostic PCR assay was developed, using mismatch amplification mutation assay. *A. rabiei* isolates (70) were assayed to validate the procedure. Based on this assay 38 of 70 isolates of *A. rabiei* were found to be resistant to QoI fungicide. The PCR protocol provided a rapid, simple and efficient screening method for differentiating QoI fungicide resistant and sensitive isolates of *A. rabiei* (Delgado et al. 2013). The gene conferring resistance in *F. oxysporum* f.sp. *niveum* isolates to the thiophanate-methyl, was identified. Sequence analysis of a portion of the ß-tubulin gene of eight isolates resistant or sensitive to the thiophanate-methyl showed that fungicide resistance was associated with a point mutation at nucleotide position 200, resulting in a substitution of phenylalanine by tyrosine. The results indicated that thiophanate-methyl should be used judiciously or alternatives like prothioconazole may be recommended (Petkar et al. 2017).

4.1.3.6 Molecular Characterization and Mechanisms of Fungicide Resistance

Molecular methods are useful to detect, differentiate and characterize the fungicide-sensitive and -resistant isolates of fungal pathogens. Recombination, following sexual reproduction, may play a role in the survival and evolution of sensitivity to the fungicide mefenoxam in the populations of *Phtophthora capsici*, infecting cucurbitaceous crops. Mefenoxam insensitive isolates (63) of *P. capsici* were recovered in 1998 from a squash field, where mefenoxam had been applied. Isolates (234) were collected in 1998 and 1999 from untreated fields. These isolates were characterized, using fluorescent amplified fragment length polymorphisms (AFLP) markers and all isolates were screened for compatibility and mefenoxam sensitivity. Of the isolates collected in 1998 and 1999 (92% and 72%), respectively, had unique multilocus AFLP genotypes with no identical isolates recovered between years. The AFLP markers (72) were clearly resolved in both 1998 and 1999 sample sets and fixation indices for the 37 polymorphic AFLP loci indicated little differentiation between years. No difference in the frequency of occurrence of resistant isolates was recorded during the 2 years without mefenoxam selection. The results showed that oospores might play a key role in overwintering and the frequency of mefenoxam sensitivity might not decrease in an agriculturally significant period (2 years), if mefenoxam selection pressure was absent (Lamour and Hausbeck 2001).

Tridemorph was applied to suppress development of corn smut disease caused by *Ustilago maydis*. Mutants of *U. maydis* with low resistance to tridemorph were isolated in a mutation frequency of 7×10^6, after UV irradiation and selection on media amended with tridemorph at 25 µg/ml. Genetic analysis of nine mutant isolates resulted in the identification of two unlinked chromosomal loci U/tdm-1 and U/tdm-2. The mutations in U/tdm-1 were responsible for low levels of resistance to tridemorph, based on effective concentration causing 50% inhibition of mycelial growth (EC_{50}) or minimal inhibitory concentration (MIC). Haploid strains carrying both U/tdm mutations had higher levels of fungicide resistance, indicating interallelic interaction between nonallelic genes. Cross-resistance studies with inhibitors of C-14 demethylase showed that the U/tdm mutations were responsible for increased sensitivity to the triazoles, triadimefon, triadimenol, propiconazole and flusilazole and the pyridine pyrifenox. Study of the gene effect on the fitness of *U. maydis* showed that U/tdm mutations appeared to be pleiotropic, having more or less adverse effects on pathogen growth rate in liquid culture and pathogenicity in

young corn plants (Markoglou and Ziogas 2000). In another investigation, the role of mutations in the sterol 14α demethylase (14DM) target site was studied by cloning this gene and establishing its identity by complementation in an appropriate mutant. The azole-resistant mutant, Erg40 of *U. maydis*, which was totally blocked at the 14 α demethylation step in sterol biosynthesis appeared to be suitable for such expression investigations. Transformation of Erg40 with a plasmid containing the yeast 14α demethylase (*CYP51A1*) gene removed from the block in sterol biosynthesis and generated azole-sensitive transformants. Presence of the yeast gene could not be detected in the transformants. The results suggested that changes in sterol biosynthesis occurred simply from the transformation of extracellular DNA. Subsequent sequence analysis revealed a mutation in the 14α demethylase gene of Erg40. Azole resistance in Erg40 seemed not simply controlled by the mutation, but might involve additional regulatory function (Butters et al. 2000). Development of mutants of *U. maydis* resistant to azoxystrobin, a QoI fungicide, was observed. Mutants with high resistance to azoxystrobin (AZO), an inhibitor of mitochondrial transport of cytochrome bc1 complex were isolated in a mutation frequency of 2.3×10^7, after nitrosoguanidine mutagenesis and selection on media amended with 1 µg/ml of azoxystrobin, in addition to 0.5 mM salicylhydroxamine (SHAM), a specific inhibitor of cyanide-resistant (alternative) respiration. Oxygen uptake in the whole cells was strongly inhibited in the wild-type strain by azoxystrobin (1.5 µg/ml), in addition to SHAM (1 mM), but not in the mutant isolates. Genetic analysis with nine such mutants resulted in the production of progeny phenotypes, which did not follow Mendelian segregation, but satisfied the criteria of non-Mendelian cytoplasmic inheritance (Ziogas et al. 2002).

Resistance to single-site fungicides occurs more commonly in fungal pathogens. Boscalid is a broad-spectrum fungicide of the class succinate dehydrogenase inhibitors (SDHIs), which includes carboxin, flutolanil and fluopyram. SDHI group interferes with the activity of succinate quinone reductases (complex II), which is a functional part of the tricarboxylic acid (TCA) cycle in mitochondrial electron transport chain (Keon et al. 1991). Boscalid lost its effectiveness against *D. bryoniae* (anamorph, *Stagonosporasis cucurbitacearum*), causing gummy stem blight disease of cucurbits. Isolates of *D. bryoniae* (103) were tested for their sensitivity to boscalid. Among these isolates, 82 and 7 were found to be very highly resistant (B^{VHR}) and highly resistant (B^{HR}) to boscalid, respectively. Cross-resistance studies showed that the B^{VHR} isolates were only highly resistant to penthiopyrad ($B^{VHR}P^{HR}$), while B^{HR} isolates appeared to be sensitive to a new SDHI fungicide penthiopyrad (B^{HR}-P^S). The molecular mechanism of fungicide resistance in these two phenotypes (B^{VHR}-P^{HR} and B^{HR}-P^S were investigated. In addition, their sensitivity to SDHI fungicide fluopyram was also determined. A 456-bp cDNA amplified fragment of the succinate dehydrogenase iron sulfur gene (DbSDHB) was initially cloned and sequenced from two sensitive (B^S-P^S), two B^{VHR}-P^{HR} and one B^{HR}-P^S isolate of *D. bryoniae*. Comparative analysis of the DbSDHB protein revealed that a highly conserved histidine residue involved in the binding of SDHIs present in the wild-type isolates, was replaced by tyrosine (H277Y) or arginine (H277R) in the B^{VHR}-P^{HR} and B^{HR}-P^S variants respectively. Both SDHB mutants were as sensitive to the fungicide as the wild-type isolates. A selection of fungicides for effective management of *D. bryoniae* has to consider the genotype-specific cross-resistance relationships between SDHI boscalid and penthiopyrad, and the lack of cross-resistance between these fungicides and fluopyram (Avenot et al. 2012).

The genetic basis of insensitivity of isolates of *P. infestans* to the phenylamide fungicide mefenoxam was investigated. Crosses were made between insensitive isolates and intermediate insensitive isolates of *P. infestans*. Expected level of semi-dominant phenotypes in the F1 for mefenoxam insensitivity was observed, suggesting the involvement of multiple loci and this complicated the positional cloning of the gene(s) conditioning insensitivity to mefenoxam. The primary effect of phenylamide compounds was inhibition of ribosomal RNA synthesis by the pathogen. The subunits of RNA polymerase I (RNA pol I) were sequenced from sensitive and insensitive isolates and F1 progeny. Single nucleotide polymorphisms (SNPs) specific to insensitive field isolates were identified in the gene encoding large subunit of RNA pol I. In a survey of field isolates of *P. infestans*, SNPT114A (Y382F) showed an association of 86% with mefenoxam insensitivity. Isolates lacking this association belonged predominantly to one genotype of *P. infestans*. Transfer of the allele governing sensitivity in RPA190 to a sensitive line resulted in the generation of transgene lines insensitive to mefenoxam. The results indicated that sequence variation in RPA190 contributed to insensitivity to mefenoxam in *P. infestans* (Randall et al. 2014). In a later investigation, the genetic basis of inheritance of field resistance of *P. infestans* to mefenoxam was studied. Mefenoxam has a negative effect on the synthesis of ribosomal RNA by the pathogen. A sensitive isolate became tolerant after a single passage on mefenoxam-amended medium. All sensitive isolates tested (from three diverse genotypes) acquired resistance to mefenoxam equally rapidly. On the contrary, isolates that were resistant to mefenoxam in the initial assessment (stably resistant) did not show any increase following subsequent exposure to the fungicide. The isolates that acquired resistance grew more slowly on mefenoxam-free medium than the isolates with no exposure to mefenoxam. The acquired resistance of the sensitive isolates declined slightly with subsequent passage through nonamended medium. Many differentially expressed genes were genotype-differentially expressed genes and they were genotype-specific, but one set of genes was differentially expressed in all genotypes. The differentially expressed genes included a phospholipase 'Pi-PLD-like 3', two ATP-binding cassette superfamily (ABC) transporters, and a mannitol dehydrogenase, that were upregulated. The function of these genes might contribute to a resistance phenotype (Childers et al. 2015).

Involvement of fungal histidine kinases (HKs) in osmotic and oxidative stress responses, hyphal development, fungicidal sensitivity and virulence has been reported. Members of HK class III signal through the high-osmolar glycerol mitogen-activated protein kinase (HOGMAPK). The *Shk1* gene (SSIG_12694.3), which encodes a putative class III HK, from *S. sclerotiorum*, was characterized. Disruption of *Shk1* resulted in resistance to phenylpyrrole and dicarboximide fungicides and increased sensitivity to hyperosmotic stress and H_2O_2-induced oxidative stress. The *Shk1* mutant showed significant reduction in vegetative hyphal growth and was unable to produce sclerotia. Quantitative real-time PCR assay showed that expression of SsHOG1 (the last kinase of the Hog pathway) and glycerol accumulation were regulated by the *Shk1* gene, but PHK (p21-activated kinase) was not. In addition, the *Shk1* mutant showed no change in virulence. All the deficiencies were restored by genetic complementation of the *Shk1* deletion mutant with the wild-type *Shk1* gene (Duan et al. 2013).

4.1.3.7 Inheritance of Resistance to Fungicides

Resistance to fungicides in the fungal pathogens may be inherited by the progeny, as in the case of other gene-dependent characteristics. Field isolates of *Gibberella pulcaris* (anamorph, *F. sambucinum*), incitant of potato dry rot disease, produced trichothecene mycotoxins, capable of causing ailments in humans and animals. They were found to be resistant to thiabendazole (TBZ), a benzimidazole fungicide. Resistance to TBZ was characterized by deriving homokaryons from single ascospores that were produced in vitro from crosses of thiabendazole-resistant field isolates of *G. pulicaris* with sensitive isolates of opposite mating type. The ß-tubulin gene was amplified from these homokaryons, using degenerate oligonucleotides synthesized to conserved sequences of the ß-tubulin gene derived from related fungi. An open-reading frame (ORF) containing three putative introns and a deduced amino acid sequence of 446 amino acids exhibited a high level of homology to the ß-tubulin gene of other fungi. Differences in nucleotide and amino acid sequences were seen. However, none of the differences could be linked to thiabendazole resistance. Linkage analysis confirmed that the isolated ß-tubulin gene was a single-copy gene and it was not linked to thiabendazole resistance of *G. pulicaris* (Kawachuk et al. 2001).

P. capsici isolates (1,511) collected from farms with no history of exposure to carboxylic acid amide (CAA) fungicides in 32 provinces in China during 2006 to 2013 were assayed for mating type. Of these, 403 isolates were evaluated for their sensitivity to dimethomorph (DMM) and metalaxyl. The EC_{50} values ranged from 0.126 to 0.339 μg/ml. Both A1 and A2 mating types were detected in the same farm, indicating the possibility of sexual reproduction, facilitating recombination of genetic characteristics. The segregation of DMM resistance and sensitivity among 337 progenies obtained from hybridization and self-crossing in vitro, indicated that the resistance of *P. capsici* to DMM was controlled by two dominant genes. Progenies (18) derived from hybridization, differed in DMM sensitivity and in fitness. Some progenies were as fit as the parental isolates. As the mating types A1 and A2 were distributed in 1:1 ratio, sexual reproduction in *P. capsici* was expected to be high. The results indicated that the control of resistance by two dominant genes and the fitness of hybrid progeny, the risk of *P. capsici* populations developing DMM resistance in China was substantial (Bi et al. 2014). Inheritance of resistance in *P. nicotianae* to mefenoxam was studied. Three sexual crosses involving isolates of *P. nicotianae* with differing sensitivity to mefenoxam were performed. Mefenoxam sensitivity of the pathogen isolates, based on the effects on mycelial growth in both mefenoxam-amended clarified V8 agar and non-amended agar was determined. With both parents of same phenotype (resistant or sensitive), all F1 progenies had the parental phenotype and no segregation for a major effect gene was observed in the level of sensitivity to mefenoxam. However, variations in mycelial growth of the progeny could be seen, indicating the segregation of minor-effect genes. Cross between mefenoxam-resistant isolate 3A4 and a sensitive parent resulted in segregation for mefenoxam resistance in a ratio of 1:1 (resistant: sensitive), indicating that mefenoxam resistance was controlled by a single dominant gene. Mating type was not linked to the mefenoxam-resistance gene locus. One RAPD marker linked in trans to the MEX locus was obtained by bulked segregant analysis (BSA) and it was converted to a sequence characterized region marker (SCAR). The SCAR marker was found to be a useful tool for differentiation of homozygous resistant isolates from sensitive isolates of *P. nicotianae* (Hu and Li 2014).

4.1.4 Mechanisms of Action of Fungicides

Fungicides or nonfungicide chemicals are classified, based on the nature of active ingredients (a.i.). Commercially available fungicides include active ingredients, inert diluents, wetting agents, stickers and emulsifiers. They are formulated as wettable powder, emulsifiable concentrates (ECs) or dusts. Special formulations with stickers are prepared for seed treatment. Generally, the protectant fungicides, applied prior to infection, do not have systemic action, as they are not absorbed by the plant system and translocated to different organs. The fungicidal activity of the systemic fungicides is site-specific on the target pathogen and it is lost after short periods, exposure to high temperature or heavy rainfall. However, many organic chemicals with systemic action developed later, retain the fungicidal activity for longer periods and have rainfastness. The major problem with systemic fungicides is development of resistance, sooner or later, in the pathogen isolates, making them ineffective after their use for a few crop seasons. The fungicides may act on fungal pathogens in several ways, resulting in reduction in mycelial growth, sporulation, spore germination and germ tube formation. One or more physiological functions of the pathogen may be inhibited by the fungicide. The dithiocarbamates (thiram and mancozeb) have a broad-spectrum of fungitoxicity and inhibit pathogen cellular functions. The inhibition of 6-phosphoglucanate dehydrogenase and glucose-6-phosphate dehydrogenase by different dithiocarbamic acid derivatives has been suggested to be a possible mode of action (Martin and Woodcock 1983). Quinones may act on the fungal pathogens by adversely affecting the oxidation-reduction reactions and inhibit extracellular

pectolytic and cellulolytic fungal enzymes. Aromatic compounds (pentachloronitrobenzene, PCNB) may generally inhibit the growth rate and sporulation, but do not inhibit spore germination (Heitefuss 1989).

Systemic fungicides, when applied on the plant/soil, may be absorbed and translocated upward (acropetal) and/or downward (basipetal) by the plant tissues. They may act directly on the fungal pathogens or through its metabolites, resulting in reduction of disease severity away from the site of application. The fungicide or its metabolites can be detected in the tissues of treated plants. Systemic fungicides have selective and specific activity on the fungal pathogens. The mobility of systemic fungicides within the plants, essentially differentiates them from nonsystemic protectants, which are not selective in their action against fungal pathogens. Systemic fungicides may have two types of movements within plant tissues. The movement of chemicals in the direction of the transpiration stream is termed apoplastic movement, which is also named as acropetal movement. By contrast, the fungicides may move along with photosynthates in the phloem toward roots and such movement is designated symplastic movement or basipetal movement. Nonliving structures such as cell walls, cuticle, xylem vessels and tracheids are interconnected forming a continuous system through which fungicides may be translocated from the roots to aerial plant parts. The movement of fungicides is passive and the amount of systemic fungicide present in the plant organs is related to the capacity of the organ to transpire. Thus, mature leaves may have greater quantities of systemic fungicide than young leaves and fruits that transpire negligibly. Translaminar movement, translocation of systemic fungicide from one leaf to another is frequently observed in most systemic fungicides. This movement results in the action against the pathogen in lower leaf surface, even though the fungicide is applied generally on the upper leaf surface. Plant age is another factor affecting movement. The symplast consists of protoplasts, plasmodesmata and sieve cells of phloem forming a coherent network through which sugars and other products of photosynthesis move downward. Systemic fungicides are also transported downward in a manner similar to photosynthates (Narayanasamy 2002).

Systemic fungicides may adversely affect one or more of the cellular functions of the fungal pathogens such as cell wall synthesis, cell membrane synthesis and function, energy production and intermediary metabolism, lipid synthesis and nucleic acid synthesis and nuclear functions (see Table 4.1). Fungicidal application is taken up against major diseases affecting a particular crop, which may differ depending on the age of the crop and environmental conditions that significantly influence disease severity. Effectiveness of treatment of maize (corn) seeds with Cruiser Extreme 250R (fludioxonil + azoxystrobin + mefenoxam + thiamethoxam) was assessed for suppressing development of crown rot disease. Root, mesocotyl and crown rot severity, colonization by *Fusarium* spp. and chlorophyll fluorescence (CF) were determined at different growth stages. Measurements of CF decreased significantly with increased disease severity and incidence of *Fusarium* spp. at V2 and V4 growth stages, indicating seedling disease

TABLE 4.1

Modes of Action of Different Groups of Fungicides on Target Fungal Pathogens

Group name	Common name	Trade name(s)	Mode of action	Mobility
Phenylamide	mefenoxam or metalaxyl	Ridomil Gold EC	nucleic acid	locally systemic
		Ridomil Gold GR	synthesis	
Benzimidazole	thiophanate-methyl	Topsin M	mitosis and cell division	locally systemic
Carboxamide	penthiopyrad	Fontelis	respiration	locally systemic
	boscalid	Endura	respiration	systemic
	flutanil	Artisan	respiration	systemic
Strobilurin-QoI	azoxystrobin	Abound	respiration	locally systemic
	fluoxastrobin	Evito	respiration	locally systemic
	pyroclostrobin	Headline	respiration	locally systemic
	trifloxystrobin	Absolute	respiration	locally systemic
Dinitroanaline	fluazinam	Omega	respiration	protectant
Dicarboximide	iprodione	Rovral	lipids and membranes	locally systemic
Aromatic	Dichloran	Botran	lipids and membranes	protectant
Hydrocarbon	PCNB	PCNB	lipids and membranes	protectant
Demethylation Inhibitor (DMI)	cyproconazole	Alto	sterol synthesis	systemic
	metconazole	Quash	sterol synthesis	locally systemic
	propiconazole	Tilt	sterol synthesis	locally systemic
	tebuconazole	Folicur	sterol synthesis	locally systemic
Inorganic	copper salts	Kocide	multisite activity	protectant
Dithiocarbamate	mancozeb	mancozeb	multisite activity	protectant
Chloronitrile	chlorothalonil	mancozeb	multisite activity	protectant
Phosphonate	phosphorous acid	Phostrol	—	systemic
	Potassium phosphite	Fosphite	—	systemic

Source: Adapted from Woodward et al. (2013).

adversely affected photosynthesis (Rodriguez-Brljevich et al. 2010). Incidence of Phytophthora blight disease caused by *P. capsici* in vegetable crops grown in the greenhouse and field was observed for the first time in Canada. A commercial formulation Vikron® disinfectant at 0.25, 0.5, 1 and 2% was found to be highly effective in preventing infection entirely (100%) by inhibiting zoospore germination of *P. capsici*. The fungicides fluazinam, mandipropamid, cyazofamid and fluopicolide were also effective in suppressing disease development in the greenhouse tomato, cucumber and pepper plants for a 14-day period (Cerkauskas et al. 2015).

The fungicides azoxystrobin, dimethomorph, fluazinam, fosetyl-Al and mefenoxam were used to treat the soil for suppressing the development of root, crown and fruit rot of pepper caused by *P. capsici*. Dimethomorph and mefenoxam reduced soil populations of *P. capsici* and inhibited lesion development (Matheron and Porchas 2000a, 2000b). The mechanism of action of flumorph was investigated by analyzing changes in hyphal morphology, cell wall deposition patterns, F-actin organelle and other organelles of *Phytophthora melonis* infecting melon. Flumorph did not inhibit synthesis of cell wall materials, but disrupted the polar deposition of newly synthesized cell wall materials during cystospore germination and hyphal growth, as suggested by calcofluor white staining. After exposure to flumorph, zoospores were able to

switch into cystospores accompanied by the formation of a cell wall, whereas cystospores failed to induce the isotropic-polar switch and did not produce germ tubes, but continued the isotropic growth phase. The most characteristic change induced by flumorph, was the development of periodic swelling (beaded morphology) and the disruption of tip growth. Newly synthesized cell wall materials were deposited uniformly throughout the diffuse expanded region of the hyphae, in contrast to their normal polarized patterns of deposition. These changes were due to F-actin disruption, identified with the fluorescein isothiocyanate (FITC)-phalloidin staining. The disruption of F-actin was accompanied by disorganized organelles: swelling of subapical hyphae associated with a nucleus vesicle did not undergo polarized secretion of the apical hyphae, but diffused around nuclei for the subapical growth. Thus, the cell wall was thickened with periodic expansion along the hyphae. After removal of flumorph, normal tip growth and organized F-actin were reported. The results suggested that flumorph may be involved in the impairment of cell polar growth through directly or indirectly disrupting the organization of F-actin (Zhu et al. 2007).

The mode of action of flumorph in inhibiting the development of *P. infestans*, incitant of potato late blight disease was studied. Transgenic *P. infestans* strains, expressing lifeact-enhanced green fluorescent protein (lGFP), which facilitated monitoring the actin cytoskeleton during hyphal growth, were employed. Flumorph was an effective oomicide with growth inhibiting activity. Observations under the microscope showed that low flumorph concentrations induced hyphal tip swellings accompanied by accumulation of actin plaques in the apex as observed in the tips of nongrowing hyphae. At higher concentrations, more pronounced swelling accompanied by an increase in hyphal bursting events occurred. However, actin filaments in hyphae that remained intact, were indistinguishable from the untreated nongrowing hyphae. In contrast, in the hyphae treated with the actin depolymerizing drug latrunculin B, no hyphal bursting was observed, but the actin filaments were entirely disrupted. The results indicated that actin was not the primary target of flumorph in *P. infestans* (Hua et al 2015). The potential of fungicides for postinfection control was assessed for suppressing late blight disease of tomato caused by *P. infestans*. Metalaxyl-M and cymoxanil showed the most effective curative action against *P. infestans*. Among the CAA fungicides, dimethomorph was more efficient in suppressing further development of late blight disease. The mixtures of fungicides with different modes of action such as dimethomorph + ametocardin, dimethomorph + pyraclostrobin and fosetyl-Al + propamocarb showed synergistic effect by suppressing disease development more efficiently (Pirondi et al. 2017).

Systemicity of fungicides was monitored by treating three terminal, fully expanded leaves of primary lateral branches of peanut cv. Tifrunner with prothioconazole + tebuconazole (Provost at 0.29 kg a.i / ha), azoxystrobin (Abound, 0.31 kg a.i /ha) or flutolanil (Moncut, 0.79 kg a.i /ha) under field conditions. All fungicides protected newly formed acropetal leaves against *S. rolfsii*, whereas prothioconazole + tebuconazole also inhibited symptom development to some extent in nontreated basipetal leaves. None of the fungicides prevented infection of pods. Under greenhouse conditions, prothioconazole + tebuconazole or prothioconazole (Proline, 0.18 kg a.i/ ha) applied to main stems of cv. Georgia Green, protected the leaves from nontreated cotyledonary branches sampled at 14 days after last treatment. However, no inhibition of pathogen development in nontreated roots, stems or pods could be observed. The results showed that protection of peanut plants against *S. rolfsii* by systemic fungicides whose systemic antifungal activity differed, depending on the cultivar and they lacked basipetal movement as indicated by absence of protection of roots and pods in plants treated with foliar application of fungicides tested (Augusto and Brenneman 2012).

The investigations on biochemical mode of action of quinone-outside inhibitor (QoI) fungicides showed that QoI fungicides display a single site mode of action. The fungicidal activity of QoI fungicides relies on their ability to inhibit mitochondrial respiration by binding at the QoI site (the outer, quinoloxidation site) of the cytochrome bc1 enzyme complex (complex III). The inhibition blocks the transfer of electrons between cytochrome b and cytochrome c1, leading to an energy deficiency in the fungal pathogen cells by arresting the production of ATP and ultimately leading to fungal death. The QoI target, cytochrome bc1, is an integral membrane protein complex essential for fungal respiration. In eukaryotes, it comprises 10 to 11 different peptides with a combined molecular mass of about 240 kDa and acts as a structural and functional dimer. The location of the quinol/quinone binding sites of *bc1*, both of which are located within cytochrome *b* subunit was revealed by X-ray crystallography, using bound inhibitors. Despite differences in binding between the different QoI inhibitors, their fit to the enzyme pocket was similar. The toxophore is similar in all compounds and always contains a carbonyl oxygen moiety that is considered to be responsible for binding to the enzyme (Fernandez-Ortuño et al. 2010). The QoI fungicides effectively inhibit sporulation, spore germination and mycelial growth of the target fungal pathogens and thus exhibit wide spectrum of suppression of many soilborne pathogens. Pyraclostrobin, a QoI fungicide, showed significant preventive, curative and eradicative activities against soilborne pathogens like *S. sclerotiorum* (Liang et al. 2015).

Pyrimorph, (Z)-3[4-tert-butylphenyl)-acryloyl] morpholine, a systemic antifungal agent that belongs to carboxylic acid amide (CAA) fungicides, which include mandipropamid, dimethomorph and pyrimorph showed high antifungal activity against *P. infestans, P. capsici* and *R. solani*. The mode of action of pyrimorph on *P. capsici* was investigated. The respiratory chain cytochrome bc1 complex (cyt*bc1*) is a major target of several antibiotics and fungicides. All cyt*bc1* inhibitors target the ubiquinol oxidation (Qp) or ubiquinone reduction (QN) site. The primary cause of resistance to bc1 inhibitors is target site mutations, creating a requirement of novel agents that act on alternative sites within cyt*bc1* to overcome resistance development in the target fungal pathogens. The mechanism of action of pyrimorph was studied, using mitochondria isolated from the pathogen *P. capsici*. Pyrimorph blocked mitochondrial electron transport by affecting the function of cyt*bc1*. Pyrimorph inhibited the activities of both purified 11-subunit mitochondrial and 4-subunit bacterial bci1 with IC_{50} values of 85.0 µM and 69.2 µM, respectively, indicating that the fungicide targeted the essential subunits of cytbc1 complexes. An array of biochemical and spectral methods indicated that pyrimorph might act on an area of the Qp site and belong to the category of a mixed type noncompetitive inhibitor in respect of the substrate ubiquinol (Xiao et al. 2014).

The dicarboximide fungicide, iprodione [3-(3,5-dichlorophenyl)-N-isopropyl-2,4-dioxoimidazol-idine-1-carboximide], a contact fungicide has been applied for protection of many crops against diseases. Iprodione inhibited sterol synthesis and hyphal development by interfering with signal transduction process as fludioxonil (Ochiai et al. 2002). The cell membrane is a selectively permeable wall that separates cell content from the outside environment. Membranes perform in all living cells many essential functions, including prevention of passage of large molecules and maintenance of cell shape and water potentials. They are involved in signal transduction (Alberts et al. 2002). Fungicides negatively impact the membrane of both pathogens and beneficial microorganisms, resulting in alterations in the substance and composition of soil microbial communities. The structure of lipids, the basic components of cell membranes was modified by aromatic hydrocarbon (AH) group of fungicides, adversely affecting the functionality of microbial membrane systems. Dichloran (2,6-dichloro-4-nitro aniline) is phototoxic and acted on fungi included in Basidiomycetes and Deuteromycetes. The cell membranes of treated fungi became sensitive to solar radiation, which destroyed the structure of linoleic acid, a common membrane lipid. Another AH fungicide ingredient, etridiazole [5-ethoxy-3(trichloromethyl)-1, 2,4-thiadiazole) caused the hydrolysis of cell membrane phospholipids into free fatty acids and lysophosphatides resulting in the lysis of membranes in the fungi (Radzuhn and Lyr 1984). Demethylation-inhibiting (DMI) fungicides inhibit biosynthesis of sterol, another important component of cell membrane in fungi. Triadimefon demethylated at C-14, introduced a double bond at C-22 and reduced a double bond at C-24 in the carbon skeleton of sterols in the fungal membrane, causing dysfunction and cell lysis (Pring 1984). Some fungicides may affect signal transduction, which occurs at the level of membranes, involving function of certain proteins. Phenylpyrrole fungicidal ingredient fludioxonil [4-(2-2-difluoro-1,3-benzodioxol-4-yl)-1H-pyrrole-3-carbonitrile] is a nonsystemic fungicide, known to interfere with signal transduction on pathways of target fungi. Inhibition of spore germination, germ tube elongation and mycelium growth was attributed to fludioxonil-related interference in the osmoregulatory signal transmission pathway of fungal pathogen (Rosslenbroich and Stuebler 2000; Kim et al. 2007).

Mitosis and cell division in target fungi are impacted by the methyl benzimidazole carbonates (MBC) fungicides. These fungicides exhibited inhibitory effects on polymerization of tubulin into microbtubules. They bind on ß-tubulin in microtubules, inhibiting their proliferation and suppressing their dynamic instability. Microtubules are the cytoskeletal polymers in eukaryotic cells and play a vital role in many cellular functions. MBC fungicides suppress assembly of spindle microtubules and disturb chromosomal alignment (Koo et al. 2009). Some fungicides affect fungal respiration at the level of the enzyme complex system and other fungicides may impact respiration through other targets. Fluazinam [Chloro-N-(3-chloro-5-trifuloro-methyl-2-pyridyl)-α-α-α-trifluoro-2-6-dinitro-p-toluidine] triggers very unusual uncoupling activity in target cells. The metabolic state of their mitochondria was inhibited after exposure to fluazinam, which may be caused by the conjugation of chemical with glutathione in mitochondria (Guo et al. 1991). Consequently, ATP production was inhibited and downstream cellular mechanisms was interrupted. Uncoupling activity of eight fluazinam derivatives was observed, suggesting that fluazinam had complicated ramification on fungal metabolic pathways (van Wingaarden et al. 2010).

Fungicides like chlorothalonil and mancozeb have multiple activity against a wide spectrum of microbial plant pathogens, due to their impacts on multiple biochemical sites. Chlorothalonil (tetrachlroiso-phthalonitrile) could block the transformation of alternative special structure of glutathione

and reduce enzyme activities, which used special conformation of glutathione as their reaction centers. Mancozeb [manganese-ethylene bis-(dithiocarbamate) complex with zinc salt] affects metabolism of target cells (Yang et al. 2011). Hormesis is a toxicological concept of dose-response relationships characterized by low- dose stimulation and high-dose inhibition. Hormesis is a reversal response between low and high doses demonstrated in the case of metalaxyl in *P. infestans* (Zhang et al. 1997) and mefenoxam on *P. aphanidermatum* (Garzón et al. 2011). Growth and virulence of metalaxyl-resistant *P. infestans* was stimulated by low concentrations of metalaxyl (Zhang et al. 1997). Significant increases in Pythium damping-off of geranium with mefenoxam at 1×10^{-6} and 1×10^{-1} µg/ml were recorded. Mycelial growth of propamocarb and mefenoxam-resistant strains of *P. aphanidermatum, P. irregulare* and *P. ultimum* could be stimulated by propamocarb at 1 µg/ml and in some resistant strains of *P. aphanidermatum* with propamocarb even at 1,000 µg/ml (Moorman and Kim 2004). The stimulation of sublethal doses of mefenoxam on mycelial growth of mefenoxam-resistant isolate of *P. aphanidermatum* was about 10%, whereas the penetration stimulation on virulence toward geranium seedlings was as high as 65% (Garzón et al. 2011). These investigations showed that application of inappropriate concentrations of fungicides might increase rather than decrease severity of the target disease(s). Hormetic effects of fungicides on soilborne pathogens, especially stimulation of mycelial growth and virulence have consequential effects on disease management using fungicides. Under field conditions, some of the pathogen populations are likely to be exposed to low doses of a fungicide, due to evaporation, drift or uneven distribution of the active ingredients. Subtoxic doses of a fungicide may induce moderate stimulatory effects on aggressiveness of pathogens, leading to enhancement of disease severity. The hormetic activity of trifloxystrobin, a quinone outside inhibitor (QoI) fungicide, on *S. sclerotiorum* was investigated. Trifloxystrobin at 0.0001, 0.0005 and 0.001 µg/ml exerted significant stimulatory effects on the virulence on potted rapeseed plants, the isolates HB15 and SX11 showing highest stimulation. The results suggested that direct stimulation was likely to be the underlying mechanisms of hormetic actions of trifloxystrobin. The biochemical mechanisms indicated that the cell wall-degrading enzymes, such as cellulase, pectinase and polygalacturonase were not involved in pathogenicity stimulation by trifloxystrobin (Di et al. 2016b).

Foliar application of the antibiotic validamycin A (VMA) or validoxylamine A (VAA) at ≥10 µg/ml effectively suppressed the development of tomato wilt disease caused by *F. oxysporum* f.sp. *lycopersici*. Plants sprayed with VMA or VAA accumulated salicylic acid (SA) and showed elevated expression of systemic acquired resistance (SAR) marker genes *P4* (PR-1), *Tag* (PR-2) and *NP24* (PR-5). The disease suppressive activity of VMA and VAA lasted up to 64 days after treatment and had broad-spectrum of activity. They induced expression of PR genes, all known markers of SAR. The results suggested that VMA and VAA functioned more as resistance inducers rather than as antibiotic directly acting on the pathogen (Ishikawa et al. 2005).

4.1.5 Effects of Chemical Application on Nontarget Organisms

Although chemicals have provided more effective protection to the crops against several microbial plant pathogens, the adverse effects on the environment and nontarget microorganisms cannot be ignored. Furthermore, the residues of chemicals remaining in the harvested produce add to the level of nonacceptance of wider use of chemicals as a component of disease management systems for various crops. Assessment of effects of fungicides on the beneficial microorganism is essential for determining the hazards associated with chemical use in agriculture. Microorganisms are either functionally or nutritionally connected with each other and alterations in a component of a microbial community may influence the structure and composition of entire community of plant-associated microorganisms. Investigations have been carried out to assess the effects of fungicides on soil microbial community (Yang et al. 2011).

Etridiazole, belonging to aromatic hydrocarbons (AH) group, causes hydrolysis of cell membrane phospholipids into free fatty acids and lysophosphatides, resulting in the lysis of membranes in fungi. The fungicides reduced the nitrification rate of ammonium-oxidizing bacteria in soil, which might change the components of microbial community in the treated soil (Rodgers 1986). Dimethomorph might influence the activity of bacteria involved in nitrogen cycling, with impact on nitrification and ammonification via its effect on different bacterial ecotypes and changes in bacterial community structure (Cycoń et al. 2010). Phenylamides (PA) fungicide, metalaxyl affect the nucleic acid synthesis by inhibiting the activity of RNA polymerase I system of the fungal pathogens. In addition to development of resistance in the fungal pathogens, the fungicides adversely affect the bacteria associated with N cycling in the soil (Monkiedje and Spiteller 2005). Chlorothalonil with multisite activity might influence bacterial growth in the soil, which might result in ecological consequences on nitrogen cycling (Chen et al. 2001). Mancozeb, another fungicide with multisite activity might influence bacterial growth in soil, affecting both carbon and nitrogen cycling in soil (Cernohlavkova et al. 2009). It is essential to consider the modes of action of fungicides and their potential effects on nontarget soil microorganisms involved in the maintenance of soil fertility and plant growth, while selecting fungicides to be applied in the soil against soilborne microbial plant pathogens.

4.2 ASSESSMENT OF ACTIVITIES OF CHEMICALS AGAINST BACTERIAL PATHOGENS

Crop diseases caused by soilborne bacterial pathogens are less numerous, compared with those caused by fungal pathogens. As the bacterial pathogens do not have well-defined life stages like fungal pathogens, the options for selecting the vulnerable stage for chemical application are limited. Further, the number of chemicals available specifically for the control of bacterial pathogens is also fewer. Hence, nonchemical methods, which are preventive in nature and act by reducing the sources of inoculum, such as cultural practices, and supply of required nutrition and enhancement of host resistance have been investigated. Furthermore, effectiveness of application of biotic and abiotic biocontrol agents has to be examined. Some chemicals with antibiotic activity have been used for suppressing development of diseases caused by soilborne bacterial pathogens.

4.2.1 Use of Disinfectants

The effectiveness of chemical disinfectants, ethanol (70%), sodium hypochlorite (1%), copper sulfate (2%), peracetic acid (5%), hydrogen peroxide (10%), benzoic acid (1%, MennoClean), trisodium phosphate (1%) and caffeine (0.2%) was assessed by treating tubers for suppressing the development of potato blackleg disease caused by *Dickeya solani*. All disinfectants effectively killed the bacterial pathogen in axenic cultures within 5 min. All disinfectants, except hydrogen peroxide, killed *D. solani* and trisodium phosphate, killed *D. solani* in spiked potato extracts. All disinfectants except trisodium phosphate reduced pathogen population on the periderm of potato tubers inoculated by dipping in a suspension of pathogen cells. In the greenhouse experiments, phytotoxicity due to all disinfectants, except hydrogen peroxide and caffeine, was observed, as reflected by reaction in sprouting tubers (10–100%). Sodium hypochlorite and benzoic acid selected for their effectiveness prevented development of soft rot symptoms under conditions favoring disease development. Inoculated tubers treated with sodium hypochlorite and benzoic acid reduced blackleg disease incidence from 50 to 0% and 5%, respectively, and symptomless infection of potato stems from 93 to 0% and 5%, respectively. Combination of both compounds was highly phytotoxic. The results showed that sodium hypochlorite and benzoic acid (MennoClean) could form a component of integrated management system for potato blackleg disease (Czajkowski et al. 2013).

The effectiveness of common agricultural disinfectants sodium hypochlorite, quaternary ammonium and hydrogen peroxide in eliminating potato bacterial ring rot pathogen *Clavibacter michiganensis* subsp. *sepedonicus* was assessed, based on their ability to act on the bacteria in biofilms. Generally, the bacteria in the biofilm state are more resistant to disinfection by chemical treatment. Artificial biofilms were grown on different surface materials commonly used in commercial potato storage facilities (concrete, mild steel, rubber, polycarbonate and wood) to test the effect of surface type on biofilm susceptibility to disinfection. Sodium hypochlorite was the most effective disinfectant on wood surfaces and hydrogen peroxide was the most effective on mild steel surface. The other disinfectants were not significantly effective on any of the substrates tested. When artificially grown biofilms and those grown naturally on potato surface were transferred to and dried onto coupons of different surface materials, they were significantly more difficult to inactivate than in situ grown biofilms. The resistance of plant pathogenic bacteria in the biofilm state, particularly when spread and dried onto surfaces of agricultural machines and other equipments, to commonly applied disinfectants, is a factor of epidemiological importance, demanding consideration for effective management of potato ring rot disease (Howard et al. 2015).

4.2.2 Use of Bactericides

The efficacy of many chemicals has been assessed for their application as bactericides. The bactericides used against *Ralstonia solanacearum*, incitant of bacterial wilt disease, for protecting geranium (*Pelargonium hortorum*) plants were able to slow down disease progress, but they were not able to protect plants from infection and subsequent mortality. Potassium salts of phosphorus acid were effective in protecting plants from infection, when applied as drench. The active ingredient of the potassium salt was phosphorous acid (H_3PO_3), which inhibited the growth of *R. solanacearum* in vitro. Phosphorous acid was considered to be a bacteriostatic compound in soil. The plants were not protected from aerial infection on wounded surfaces. Phosphorous acid protected plants from infection by either race 1 or 3 of *R. solanacearum* (Norman et al. 2006). The bactericidal activity of calcium carbonate ($CaCO_3$) on *R. solanacearum*, causing tobacco bacterial wilt disease was assessed under field conditions, in addition to in vitro assays. Soils treated with $CaCO_3$ particles (<1 mm) significantly reduced the survival of *R. solanacearum* and increased the pH approximately by 1.5 units, compared with untreated control. The pH range for the pathogen survival was 6.0–7.0, the optimum being 6.5. The growth of *R. solanacearum* was significantly inhibited by Ca^{2+}, but not by Cl of $CaCl_2$ and $NaCl_2$. With increase in Ca^{2+} concentration, the pectinase activity decreased markedly, whereas no detectable adverse effect could be observed in production of extracellular polysaccharide (EPS). Under field conditions, $CaCO_3$ reduced disease incidence. Combination of organic manure and $CaCO_3$ increased soil pH and reduced pathogen population by about 100 folds. Ca^{2+} contents of tobacco grown in treated soil correlated with $CaCO_3$ application to soil. The results suggested that $CaCO_3$ application had the potential for use as soil amendment for suppressing the development of tobacco bacterial wilt disease (He et al. 2014). The inhibitory activity of methyl gallate (MG) against *R. solanacearum* was investigated. Methyl gallate showed strong inhibitory activity against *R. solancearum*. Scanning electron microscope (SEM) observations showed that cell wall structure of the pathogen was damaged and morphological changes and plasmolysis were also revealed. Biochemical analysis indicated that methyl gallate inhibited protein synthesis and succinate dehydrogenase (SDH) activity of the pathogen markedly. Treatment of the pathogen with methyl gallate at 20 µg/ml showed that *R. solanacearum* could counteract the inhibitory effect of the chemical and regain its growth via stress responses, such as elevated Na^+K^+-ATPase activity and production of a large quantity of exopolysaccharides. However, the pathogenicity of *R. solanacearum* was abolished by treatment with methyl gallate and this was attributed to suppression of extracellular

enzymes such as pectinase and cellulase. Further, methyl gallate, at higher concentrations inhibited pathogen respiration, which adversely affected the energy metabolism of *R. solanacearum* (Fan et al. 2014).

The antibacterial activity of synthetic peptides CAMEL, Iseganan and Pexiganan against soft rot bacterial pathogen, *Pectobacterium carotovorum* (Pc) and *P. chrysanthemi* (Pch) was assessed, based on minimal inhibitory concentration (MIC) and minimal bactericidal concentration (MBC). CAMEL was the most effective peptide in inhibiting the growth of *Pectobacterium* spp. at concentrations ranging from 2 to 8 µg/ml. CAMEL inhibited the bacterial multiplication and exhibited greater tissue maceration capacity assessed on potato tuber slice, indicating the pathogenicity of *Pectobacterium* spp. (Kamysz et al. 2005).

Potato common scab disease caused by *Streptomyces scabiei* is one of the major yield-limiting constraints, because of reduction in both quantity and quality of tubers. The synthetic auxin 2,4-dichlorophenoxyacetic acid (2,4-D) is an herbicide commonly applied to eradicate weed populations that compete with crops for nutrients and space available in the field. 2,4-D was reported to be effective in reducing disease incidence, when applied at or shortly after tuber initiation during the period of known susceptibility, while treatment at later stage of crop growth provided little or no control of the disease. In a later investigation, application of 2,4-D as early as 5 days after average plant emergence provided greater disease suppression for potato cultivars, Russet Burbank and Desiree. Early application of 2,4-D provided sufficient material to induce resistance that lasted throughout the tuber susceptibility period with no need for a second application. Further, application of 2,4-D at concentration required for disease suppression did not result in accumulation of the chemical and levels at harvest were within the Australian maximum residue limit. Phytotoxic symptoms in potato were seen, when the concentration of 2,4-D was 200 mg/l or more. Tuber yield was also reduced with minimal significant threshold rates of 8.3–23 0.6 mg/l of 2,4-D, whereas the incidence of common scab disease was reduced in pot trials. At 10.8–41.0 mg/l, 2,4-D minimized disease severity in both pot and field trials, except in one pot trial. Disease control due to 2,4-D was associated with decreased sensitivity of tubers to thaxtomin A, the toxin produced by *S. scabiei* required for disease development. The amount of residual 2,4-D left in tubers at harvest differed with cultivars, Russet Burbank accumulating greater concentration of 2,4-D than cv. Desiree. The results showed that application of 2,4-D at low rates (<100 mg/l) could be adopted as a component of integrated disease management system applicable for potato common scab disease (Thompson et al. 2013).

REFERENCES

Abbasi PA, Lazarovits G (2005) Effects of AG3 phosphonate formulation on incidence and severity of Pythium damping-off of cucumber seedlings under growth room microplot, and field conditions. *Canad J Plant Pathol* 27: 420–429.

Abbasi PA, Lazarovits G (2006) Seed treatment with phosphonate (AG3) suppresses Pythium damping-off of cucumber seedlings. *Plant Dis* 90: 459–464.

Abbasi PA, Lazarovits G, Weselowski B, Lalin I (2009) Organic acids in condensed distiller's solubles: toxicity to soilborne plant pathogens and role in disease suppression. *Canad J Plant Pathol* 31: 88–95.

Agosteo GE, Ralldino F, Cacciola SO (2000) Resistance of *Phytophthora capsici* to metalaxyl in plastic-house capsicum crops in southern Italy. *EPPO Bull* 30: 257–261.

Ajayi-Oyetunde O, Butts-Wilmsmeyer CJ, Bradley CA (2017) Sensitivity of *Rhizoctonia solani* to succinate dehydrogenase inhibitor and demethylation inhibitor fungicides. *Plant Dis* 101: 487–495.

Al-Balushi ZM, Agrama H, Al-Mahmoodi IH, Maharachchikimbura SSN, Al-Sadi A (2018) Development of resistance to hymexazol among *Pythium* species in cucumber in greenhouse in Oman. *Plant Dis* 102: 202–208.

Alberts B, Johnson A, Lewis J (2002) *Molecular Biology of the Cell*, 4th Edition, Garland Science, New York, USA.

Ali M, Kim B, Belfield KD, Norman D, Brennan M, Ali GS (2015) Inhibition of *Phytophthora parasitica* and *P. capsici* by silver nanoparticles synthesized using aqueous extract of *Artemisia absinthium*. *Phytopathology* 105: 1183–1190.

Al-Mughrabi KI, Peters RD, (Bud) Platt HW et al. (2007) In-furrow applications of metalaxyl and phosphite for control of pink rot (*Phytophthora erythroseptica*) of potato in New Burnswick, Canada. *Plant Dis* 91: 1305–1309.

Alvarez LA, Vicent A, Soler M, De la Roca E, García-Jiménez J (2008) Comparison of application methods of systemic fungicides to suppress branch cankers in Clementine trees causes by *Phytophthora citrophthora*. *Plant Dis* 92: 1357–1363.

Anderson JM, Pegg KG, Scott C, Drenth A (2012) Phosphonate applied as a pre-plant dip controls *Phytophthora cinnamomi* root and heart rot in susceptible pineapple hybrids. *Austr Plant Pathol* 41: 59–68.

Attanayake RN, Carter PA, Jiang D, del Rio-Mendoza L, Chen W (2013) *Sclerotinia sclertotiorum* populations infecting canola from China and the United States are genetically and phenotypically distinct. *Phytopathology* 103: 750–761.

Augusto J, Brenneman TB (2012) Assessing systemicity of peanut fungicides through bioassay of plant tissues with *Sclerotium rolfsii*. *Plant Dis* 96: 330–337.

Avenot HF, Thomas A, Gitaitis R, Langston DB Jr., Stevenson KL (2012) Molecular characterization of boscalid and penthiopyrad-resistant isolates of *Didymella bryoniae* and assessment of their sensitivity to fluopyram. *Pest Manag Sci* 68: 645–651.

Babadoost M, Islam SZ (2003) Fungicide seed treatment effects on seedling damping-off of pumpkin caused by *Phytophthora capsici*. *Plant Dis* 87: 63–68.

Babiker EM, Hulbert SH, Schroeder KL, Paulitz TC (2011) Optimum timing of preplant application of glyphosate to manage Rhizoctonia root rot in barley. *Plant Dis* 95: 304–310.

Bailey DJ, Paveley N, Pillinger C, Spink J, Gilligan CA (2005) Epidemiology and chemical control of take-all on seminal and adventitious roots of wheat. *Phytopathology* 95: 62–68.

Banniza S, Armstrong-Cho CL, Gan Y, Chongo G (2011) Evaluation of fungicide efficacy and application frequency for the control of Ascochyta blight in chickpea. *Canad J Plant Pathol* 33: 135–149.

Bennett RS, Spurgeon DW, De Tar WR, Gerik JS, Hutmacher RB, Hanson BD (2011) Efficacy of four soil treatments against *Fusarium oxysporum*. *Plant Dis* 95: 967–976.

Bi Y, Cui X, Lu X, Cai M, Liu X, Hao JJ (2011) Baseline sensitivity of natural population and resistance of mutants in *Phytophthora capsici* to zoxamide. *Phytopathology* 101: 1104–1111.

Bi Y, Hu J, Cui X et al. (2014) Sexual reproduction increases the possibility that *Phytophthora capsici* will develop resistance to dimethomorph in China. *Plant Pathol* 63: 1365–1373.

Bittara FG, Secor GA, Gudmestad NC (2017) Chloropicrin soil fumigation reduced *Spongospora subterranea* soil inoculum levels but does not control powdery scab disease on roots and tubers of potato. *Amer J Potato Res* 94: 129–147.

Borza T, Peters RD, Wu Y et al. (2017) Phosphite uptake and distribution in potato tubers following foliar and postharvest applications of phosphite-based fungicides for late blight control. *Ann Appl Biol* 170: 127–139.

Brenneman TB, Culbreath AK (1994) Utilizing a sterol demethylation inhibiting fungicide in an advisory to manage foliar and soilborne pathogens of peanut. *Plant Dis* 78: 866–872.

Butters JA, Zhou MC, Hollomon DW (2000) The mechanism of resistance to sterol 14 α demethylation inhibitors in a mutant (*Erg40*) of *Ustilago maydis*. *Pest Manag Sci* 56: 257–263.

Café-Filho AC, Ristaino JB (2008) Fitness of isolates of *Phytophthora capsici* resistant to mefenoxam from squash and pepper fields in North Carolina. *Plant Dis* 92: 1439–1443.

Cerkauskas RF, Ferguson G, Mac Nair C (2015) Management of Phytophthora blight (*Phytophthora capsici*) on vegetables in Ontario: some greenhouse and field aspects. *Canad J Plant Pathol* 37: 285–304.

Cernohlavkova J, Jarkovsky J, Hoffman J (2009) Effects of fungicides mancozeb and dinocap on carbon and nitrogen mineralization in soils. *Ecotoxicol Environ Safety* 72: 80–85.

Chapara V, Taylor RJ, Pasche JS, Gudmestad NC (2011) Competitive parasitic fitness of mefenoxam-sensitive and -resistant isolates of *Phytophthora erythroseptica* under fungicide selection pressure. *Plant Dis* 95: 691–696.

Chen SK, Edwards CA, Subler S (2001) Effects of fungicides benomyl, captan, and chlorothalonil on soil microbial activity and nitrogen dynamics in laboratory incubations. *Soil Ecol Biochem* 17: 313–316.

Chen Y, Yao, Yang X, Zhang A-F, Gao T-C (2014) Sensitivity of *Rhizoctonia solani* causing sheath blight to fluxapyroxad in China. *Eur J Plant Pathol* 140: 419–428.

Childers R, Danies G, Myers K, Fei Z, Small IM, Fry WE (2015) Acquired resistance to mefenoxam in sensitive isolates of *Phytophthora infestans*. *Phytopathology* 105: 342–349.

Chongo G, Buchwaldt L, Gossen BD et al. (2003) Foliar fungicides to management of Ascochyta blight (*Ascochyta rabiei*) of chickpea in Canada. *Canad J Plant Pathol* 25: 135–142.

Copes WE, Chastaganer GA, Hummel RL (2004) Activity of chlorine dioxide in a solution of ions and pH against *Thielaviopsis basicola* and *Fusarium oxysporum*. *Plant Dis* 88: 188–194.

Cunha MG, Rizzo DM (2003) Development of fungicide cross-resistance in *Helminthosporium solani* populations from California. *Plant Dis* 87: 798–803.

Cycoń M, Piotrowska-Seget Z, Kozdroj J (2010) Responses of indigenous microorganisms to a fungicidal mixture of mancozeb and dimethomorph added to sandy soils. *Internatl Biodeter Biodegrad* 64: 316–323.

Czajkowski R, de Boer WJ, van der Wolf JM (2013) Chemical disinfectants can reduce potato blackleg caused by 'Dickeya solani'. *Eur J Plant Pathol* 136: 419–432.

da Silva PP, de Freitas RA, Nascimento WM (2013) Pea seed treatment for *Rhizoctonia solani* control. *J Seed Sci* 35: 17–20.

Dawson WAJM, Bateman GL (2001) Fungal communities on roots of wheat and barley and effects of seed treatments containing fluquinconazole applied to control take-all. *Plant Pathol* 59: 75–82.

De Cal A, Martinez-Treceño A, Lopez-Aranda JM, Melgarej P (2004) Chemical alternatives to methyl bromide in Spanish strawberry nurseries. *Plant Dis* 88: 210–214.

Delgado JA, Lynnes TC, Meinhardt SW et al. (2013) Identification of the mutation responsible for resistance to QoI fungicides and its detection in *Ascochyta rabiei* (teleomorph–*Didymella rabiei*). *Plant Pathol* 62: 688–697.

Demirci F, Bayraktar H, Babliogullu I, Dolar PS, Maden S (2003) In vitro and in vivo effects of some fungicides against chickpea blight pathogen *Ascochyta rabiei*. *J Phytopathol* 151: 519–524.

Di Y, Lu X-M, Zhu Z-Q, Zhu F-X (2016a) Time course of carbendazim stimulation on pathogenicity of *Sclerotinia sclerotiorum* indicates a direct stimulation mechanism. *Plant Dis* 100: 1454–1459.

Di Y-L, Cong M-L, Zhang R, Zhu F-X (2016b) Hormetic effects of trifluoxystrobin on aggressiveness of *Sclerotinia sclerotiorum*. *Plant Dis* 100: 2113–2118.

Di Y-L, Zhu Z-Q, Lu X-M, Zhu F-X (2015) Pathogenicity stimulation of *Sclerotinia sclerotiorum* by subtoxic doses of carbendazim. *Plant Dis* 99: 1342–1346.

Dobrowolski MP, Shearer BL, Colquhoun IJ, O'Brien PA, StJ, Handy GE (2008) Selection of decreased sensitivity to phosphite in *Phytophthora cinnamomi* with prolonged use of fungicide. *Plant Pathol* 57: 928–936.

Dorrance AE, McClure SA (2001) Beneficial effects of fungicide seed treatments for soybean cultivars with partial resistance to *Phytophthora sojae*. *Plant Dis* 85: 1063–1068.

Duan Y, Ge O, Liu S, Wang J, Zhou M (2013) A two-component histidine kinase *Shk1* controls stress response, sclerotial formation and fungicide resistance in *Sclerotinia sclerotiorum*. *Molec Plant Pathol* 14: 708–718.

Duan Y, Yang Y, Wang Y et al. (2016) Loop-mediated isothermal amplification for rapid detection of the F200Y mutant genotype of carbendazim-resistant isolates of *Sclerotinia sclerotiorum*. *Plant Dis* 100: 976–983.

Dunn AR, Milgroom MG, Meitz JC et al. (2010) Population structure and resistance to mefenoxam of *Phytophthora capsici* in New York State. *Plant Dis* 94: 1461–1468.

Errampalli D, Johnston HW (2001) Control of tuber-borne black scurf (*Rhizoctonia solani*) and common scab (*Streptomyces scabies*) of potatoes with a combination of sodium hypochlorite and thiophanate methyl preplanting seed tuber treatment. *Canad J Plant Pathol* 23: 68–77.

Fan W-W, Yuan G-Q, Li Q-Q, Lin W (2014) Antibacterial mechanisms of methyl gallate against *Ralstonia solanacearum*. *Austr Plant Pathol* 43: 1–7.

Feng X, Wang K, Mu Z, Zhao Y, Zhang H (2015) Fluazinam residue and dissipation in potato tubers and vines and in field soil. *Amer J Potato Res* 92: 567–572.

Fernandez-Ortuño D, Torés JA, de Vicente A, Pérez-Garcia (2010) The QoI fungicides–the rise and fall of a successful class of agricultural fungicides. In: Carrise O (ed.), *Fungicides*, www.intechopen.com, ISBN: 978-953-307-266-1, pp. 203–220.

Fichtner EJ, Hesterberg DL, Smyth TJ, Shew HD (2006) Differential sensitivity of *Phytophthora parasitica* var. *nicotianae* and *Thielaviopsis basicola* to monomeric aluminum species. *Phytopathology* 96: 212–220.

Finger MJ, Parkunan V, Ji P, Stevenson KL (2014) Allele-specific PCR for the detection of azoxystrobin resistance in *Didymella bryoniae*. *Plant Dis* 98: 1681–1684.

Gachango E, Hanson LE, Rojas A, Hao JJ, Kirk WW (2012) *Fusarium* spp. causing dry rot of seed potato tubers in Michigan and their sensitivity to fungicides. *Plant Dis* 96: 1769–1774.

Garzón GD, Molineros JE, Yánez JM, Flores FJ, Jimenez-Gasco MDM, Moorman GW (2011) Sublethal doses of mefenoxam enhance Pythium damping-off of geranium. *Plant Dis* 95: 1233–1238.

Geary B, Johnson DA, Hamm PB, James S, Rykbost KA (2007) Potato silver scurf affected by tuber seed treatments and locations and occurrence of fungicide-resistant isolates of *Helminthosporium solani*. *Plant Dis* 91: 315–320.

Gonzalez CF, Provin EM, Zhu L, Ebbole DJ (2002) Independent and synergistic activity of synthetic peptides against thiabendazole-resistant *Fusarium sambucinum*. *Phytopathology* 92: 917–924.

Granke LL, Hausbeck MK (2010) Effects of temperature, concentration, age, and algaecides on *Phytophthora capsici* zoospore infectivity. *Plant Dis* 94: 54–60.

Gun CM, Geun AK, Ppeum KG et al. (2016) Dissipation pattern of a fungicide mandipropamid in Korean cabbage at different harvest times under greenhouse conditions. *Kor J Hortic Sci Technol* 34: 644–654.

Guo Z, Miyoshi H, Komyoji T, Haga T, Fujita T (1991) Uncoupling activity of a newly developed fungicide fluazinam [3-chloro-N-(3-chloro-2,6-dinitro-4-fluoromethylphenyl)-5-trifluoromethyl-2-4-fluoromethyl) 5-trifluoro methyl-2-pyridinamine]. *Biochimica et Biophysica Acta* 1056: 89–92.

Harikrishnan R, Yang XB (2002) Effects of herbicides on root rot and damping-off caused by *Rhizoctonia solani* in glyphosate-tolerant soybean. *Plant Dis* 86: 1369–1373.

He K, Yang S-Y, Li H, Wang H, Li Z-L (2014) Effects of calcium carbonate on the survival of *Ralstonia solanacearum* in soil and control of tobacco bacterial wilt. *Eur J Plant Pathol* 140: 665–675.

Heitefuss R (1989) *Crop and Plant Protection: The Practical Foundations*, Ellis Hardwood Ltd., Chichester, UK.

Hollomon DW (2015) Fungicide resistance: Facing the challenge. *Plant Protect Sci* 51: 170–176.

Hong CX, Richardson PA, Kong P, Bush EA (2003) Effects of chlorine on multiple species of *Phytophthora* in recycled nursery irrigation water. *Plant Dis* 87: 1183–1189.

Howard RJ, Harding MW, Daniels GC, Mobbs SL, Lisowski SLI, De Boer SH (2015) Efficacy of agricultural disinfectants on biofilms of the bacterial ring rot pathogen *Clavibacter michiganensis* subsp. *sepedonicus*. *Canad J Plant Pathol* 37: 273–284.

Hu J, Li Y (2014) Inheritance of mefenoxam resistance in *Phytophthora nicotianae* populations from a plant nursery. *Eur J Plant Pathol* 139: 545–555.

Hu JH, Hong CX, Stromberg EL, Moorman GW (2008) Mefenoxam sensitivity and fitness analysis of *Phytophthora nicotianae* isolates from nurseries in Virginia, USA. *Plan Pathol* 57: 728–736.

Hu JH, Hong CX, Stromberg EL, Moorman GW (2010) Mefenoxam sensitivity in *Phytophthora cinnamomi* isolates. *Plant Dis* 94: 39–44.

Hua C, Kots K, Ketelaar T, Govers F, Meijer HJG (2015) Effect of flumorph on F-actin dynamics in the potato late blight pathogen *Phytophthora infestans*. *Phytopathology* 105: 419–423.

Hwang SF, Ahmed HU, Strelkov SE et al. (2018) Effects of rate and application method on the efficacy of metam sodium to reduce clubroot (*Plamodiophora brassicae*) of canola. *Eur J Plant Pathol* 150: 341–349.

Hwang SF, Ahmed HU, Zhou Q et al. (2014) Efficacy of vapam fumigant against clubroot (*Plasmodiophora brassicae*) of canola. *Plant Pathol* 63: 1374–1383.

Ingram J, Cummings TF, Johnson DA (2011) Response of *Colletotrichum coccodes* to selected fungicides using a plant inoculation assay and efficacy of azoxystrobin applied by chemigation. *Amer J Potato Res* 88: 309–317.

Ishikawa R, Shirouzu K, Nakashita H et al. (2005) Foliar spray of validamycin A or validoxylamine A controls tomato Fusarium wilt. *Phytopathology* 95: 1209–1216.

Jackson KL, Yin J, Ji P (2012) Sensitivity of *Phytophthora capsici* on vegetable crops in Georgia to mandipropamid, dimethomorph and cyazofamid. *Plant Dis* 96: 1337–1342.

Jankowska M, Kaczynski P, Hrynko I, Lozowicka B (2016) Dissipation of six fungicides in greenhouse-grown tomatoes with processing and health risk. *Environ Sci Pollution Res* 23: 11885–11900.

Ji P, Csinos AS (2015) Effect of oxathiapiprolin on asexual life stages of *Phytophthora capsici* disease development on vegetables. *Ann Appl Biol* 166: 229–235.

Ji P, Csinos AS, Hickman LL, Hargett U (2014) Efficacy and application methods of oxathiapiprolin for management of black shank on tobacco. *Plant Dis* 98: 1551–1554.

Jogaiah S, Mitani S, Nagaraj AK, Shetty HS (2007) Activity of cyazofamid against *Sclerospora graminicola*, a downy mildew disease of pearl millet. *Pest Manag Sci* 63: 722–727.

Johnson ACS, Jordan SA, Gevens AJ (2015) Efficacy of organic and conventional fungicides and impact of application timing on control of tomato late blight caused by US-22, US-23, and US-24 isolates of *Phytophthora infestans*. *Plant Dis* 99: 641–647.

Johnson DA, Attallah ZK (2006) Timing of fungicide applications for managing Sclerotinia stem rot of potato. *Plant Dis* 90: 755–758.

Johnson DA, Inglis DA, Miller JS (2004) Control of potato tuber rots caused by oomycetes with foliar applications of phosphorous acid. *Plant Dis* 88: 1153–1159.

Kamysz W, Krolicka A, Bogucka K, Ossowski T, Lukasiak J, Lojkowska E (2005) Antibacterial activity of synthetic peptides against plant pathogenic *Pectobacterium* species. *J Phytopathol* 153: 313–317.

Kansaye L, Zhang J, Wu H, Gao B-W, Zhang X (2013) Dissipation of propamidine fungicide residues in greenhouse tomato. *J Agric Sci* 5 (5)

Kawachuk LM, Hutchison LJ, Verhaege CA. Lynch DR, Bains PS, Holley D (2001) Isolation of the ß-tubulin gene and characterization of thiabendazole resistance in *Gibberella pulicaris*. *Canad J Plant Pathol* 24: 233–238.

Keinath AP (2007) Sensitivity of populations of *Phytophthora capsici* from South Carolina to mefenoxam, dimethomorph, zoxamide, and cymoxanil. *Plant Dis* 91: 743–748.

Keinath AP (2015) Baseline sensitivity of *Didymella bryoniae* to cyprodinil and fludioxonil and field efficacy of these fungicides against isolates resistant to pyroclostrobin and boscalid. *Plant Dis* 99: 815–822.

Keon JPR, White GA, Hargreaves JA (1991) Isolation, characterization and sequence of a gene conferring resistance to the systemic fungicide carboxin from the maize smut pathogen *Ustilago maydis*. *Curr Genet* 19: 475–481.

Kim JH, Campbell BC, Mahoney N, Chan KI, Molyneux RJ, May GS (2007) Enhancement of fludioxonil fungicidal activity by disrupting cellular glutathione homeostasis with 2,5-dihydroxybenzoic acid. *FEMS Microbiol Lett* 270: 284–290.

Kirk WW, Wharton PS, Schafer KL et al. (2008) Optimizing fungicide timing for the control of Rhizoctonia crown and root rot of sugar beet using soil temperature and plant growth stages. *Plant Dis* 92: 1091–1098.

Kolaei EA, Cenatus C, Tweddell RJ, Avis TJ (2013) Antifungal activity of aluminium-containing salts against the development of carrot cavity spot and potato dry rot. *Ann Appl Biol* 163: 311–317.

Koo BS, Park HH, Kalme S et al. (2009) α- and ß-tubulin from *Phytophthora capsici* KACC40483: molecular cloning, biochemical characterization, and antimicrotubule screening. *Appl Microbiol Biotechnol* 82: 513–524.

Kuhajek JM, Jeffers SN, Slattery M, Wedge DE (2003) A rapid microbioassay for discovery of novel fungicides for *Phytophthora* spp. *Phytopathology* 93: 46–53.

Lamour KH, Hausbeck MK (2001) The dynamics of mefenoxam insensitivity in a recombining population of *Phytophthora capsici* characterized with amplified fragment length polymorphism markers. *Phytopathology* 91: 553–557.

Lamour KH, Hausbeck MK (2003) Susceptibility of mefenoxam-treated cucurbits to isolates of *Phytophthora capsici* sensitive and insensitive to mefenoxam. *Plant Dis* 87: 920–922.

Langston DB Jr., Phipps PM, Stipes RJ (2002) An algorithm for predicting outbreaks of Sclerotinia blight of peanut and improving the timing of fungicide sprays. *Plant Dis* 86: 118–126.

Lehner MS, Paula Junior TJ et al. (2015) Fungicide sensitivity of *Sclerotinia sclerotiorum*: A thorough assessment of using discriminatory dosage, EC50, high-resolution melting analysis, and description of new point mutation associated with thiophanate-methyl resistance. *Plant Dis* 99: 1537–1543.

Li J-L, Li X-Y, Di Y-L, Liang H-J, Zhu F-X (2015) Baseline sensitivity and control efficacy of DMI fungicide epoxiconazole against *Sclerotinia sclerotiorum*. *Eur J Plant Pathol* 141: 237–246.

Li Y, Chi L, Mao L et al. (2014) Control of soilborne pathogens of *Zingiber officinale* by methyl iodide and chloropicrin in China. *Plant Dis* 98: 384–388.

Liang H-J, Di Y-L, Li J-L, Yu H, Zhu F-X (2015a) Baseline sensitivity of pyraclostrobin and toxicity of SHAM to *Sclerotinia sclerotiorum*. *Plant Dis* 99: 267–273.

Liang H-J, Di Y-L, Li J-L, Zhu F-X (2015b) Baseline sensitivity and control efficacy of fluazinam against *Sclerotinia sclerotiorum*. *Eur J Plant Pathol* 142: 691–699.

Lonergan E, Pasche J, Skoglund L, Burrows M (2015) Sensitivity of *Ascochyta* species infecting pea, lentil, and chickpea to boscalid, fluxapyroxad, and prothioconazole. *Plant Dis* 99: 1254–1260.

Lookabaugh EC, Ivors KL, Shew BB (2015) Mefenoxam sensitivity, aggressiveness and identification of *Pythium* species causing root rot on floriculture crops in North Carolina. *Plant Dis* 99: 1550–1558.

Lu X, Zhang R, Cong M, Li J, Zhu F (2018) Stimulatory effects of flusilazole on virulence of *Sclerotinia sclerotiorum*. *Plant Dis* 102: 197–201.

Lu XH, Davis RM, Livingston S, Nunez J, Hao JJ (2012) Fungicide sensitivity of *Pythium* spp. associated with cavity spot of carrot in California and Michigan. *Plant Dis* 96: 384–388.

Lu XH, Zhu SS, Bi Y, Liu XL, Hao JJ (2010) Baseline sensitivity and resistance-risk assessment of *Phytophthora capsici* to iprovalicarb. *Phytopathology* 100: 1162–1168.

Ma D, Zhu J, He L, Cui K, Mu W, Liu F (2018) Baseline sensitivity and control efficacy of tetramycin against *Phytophthora capsici* isolates in China. *Plant Dis* 102: 863–868.

Ma H-X, Feng X-J, Chen Y, Chen C-J, Zhou M-G (2009) Occurrence and characterization of dimethachlon insensitivity in *Sclerotinia sclerotiorum* in Jiangsu province of China. *Plant Dis* 93: 36–42.

Markoglou AN, Ziogas BN (2000) Genetic control of resistance to tridemorph in *Ustilago maydis*. *Phytoparasitica* 28: 349–360.

Martin H, Woodock D (1983) *The Scientific Principles of Plant Protection*, 7th edition, Edward Arnold, London.

Matheron ME, Porchas M (2000a) Impact of azoxystrobin, dimethomorph, fluazinam, fosetyl-Al and metalaxyl on growth, sporulation, and zoospore cyst germination of three *Phytophthora* spp. *Plant Dis* 84: 454–458.

Matheron ME, Porchas M (2000b) Comparison of five fungicides on development of root, crown and fruit rot of Chile pepper and recovery of *Phytophthora capsici* from soil. *Plant Dis* 84: 1038–1043.

Matheron ME, Porchas M (2002) Comparative ability of six fungicides to inhibit development of Phytophthora gummosis on citrus. *Plant Dis* 86: 687–690.

Meng QX, Cui XL, Bi Y, Wang Q, Hao JJ, Liu XL (2011) Biological and genetic characterization of *Phytophthora capsici* mutants resistant to flumorph. *Plant Pathol* 60: 957–966.

Meyer MD, Hausbeck MK (2013) Using soil-applied fungicides to manage Phytophthora crown and root rot on summer squash. *Plant Dis* 97: 107–112.

Mitani S, Araki S, Yamaguchi T, Takii Y, Ohshima T, Matsuo N (2002) Biological properties of the novel fungicide cyazofamid against *Phytophthora infestans* on tomato and *Pseudoperonospora cubensis* on cucumber. *Pest Manag Sci* 58: 139–145.

Miyake N, Nagi H, Kato S, Matsusaki M, Ishikawa H, Kageyama K (2015) Detection of damping-off of cape gooseberry caused by *Pythium aphanidermatum* and its suppression with phosphonate. *J Gen Plant Pathol* 81: 192–200.

Monkiedje A, Spiteller M (2005) Degradation of metalaxyl and mefenoxam and effects on the microbiological properties of tropical and temperate soils. *Internatl J Environ Res Publ Health* 2: 272–285.

Moorman GW, Kim SH (2004) Species of *Pythium* from greenhouses in Pennsylvania exhibit resistance to propamocarb and mefenoxam. *Plant Dis* 88: 630–632.

Morales-Rodríguez C, Palo C, Palo E, Rodríguez-Molina MC (2014) Control of *Phytophthora nicotianae* with mefenoxam, fresh *Brassica* tissues and *Brassica* pellets. *Plant Dis* 98: 77–83.

Mu W, Wang Y, Bi X et al. (2017) Sensitivity determination and resistance risk assessment of *Rhizoctonia solani* to SDHI fungicide thifluzamide. *Ann Appl Biol* 170: 240–250.

Narayanasamy P (2002) *Microbial Plant Pathogens and Crop Disease Management*, Science Publishers, Enfield, USA.

Narayanasamy P (2017) *Microbial Plant Pathogens – Detection and Management in Seeds and Propagules*. Wiley-Blackwell, Chichester, UK.

Nielsen CJ, Ferrin DM, Stanghellini ME (2006) Efficiency of biosurfactants in the management of *Phytophthora capsici* in pepper in recirculating hydroponic systems. *Canad J Plant Pathol* 28: 450–460.

Nitzan N, Cummings TF, Johnson DA (2005) Effect of seed-tuber generation, soilborne inoculum, and azoxystrobin application on development of potato black dot caused by *Colletotrichum coccodes*. *Plant Dis* 89: 1181–1185.

Norman DJ, Chen J, Yuen YMF, Mangravita-Novo A, Byrne D, Walsh L (2006) Control of bacterial wilt of geranium with phosphorous acid. *Plant Dis* 90: 798–802.

Ochiai N, Fujimura M, Oshima M et al. (2002) Effects of iprodione and fludioxonil on glycerol synthesis and hyphal development in *Candida albicans*. *Biosci Biotechnol Biochem* 66: 2209–2215.

Olson HA, Jeffers SN, Ivors KL et al. (2013) Diversity and mefenoxam sensitivity of *Phytophthora* spp. associated with the ornamental horticulture industry in the southeastern United States. *Plant Dis* 97: 86–92.

Ozbahce A (2014) Chemigation for soilborne pathogen management on melon growth under drought stress. *Austr Plant Pathol* 43: 299–306.

Pang Z, Shao J, Hu J et al. (2014) Competition between pyrimorph-sensitive and pyrimorph-resistant isolates of *Phytophthora capsici*. *Phytopathology* 104: 269–274.

Papazlatani C, Rousidou C, Katsoula A et al. (2016) Assessment of the impact of the fumigant dimethyl disulfide on the dynamics of major fungal plant pathogens in greenhouse soils. *Eur J Plant Pathol* 146: 391–400.

Parkunan V, Johnson CS, Bowman BC, Hong CX (2010) Population structure, mating type and mefenoxam sensitivity of *Phytophthora nicotianae* in Virginia tobacco fields. *Plant Dis* 94: 1361–1365.

Parra G, Ristaino JB (2001) Resistance to mefenoxam and metalaxyl among field isolates of *Phytophthora capsici* causing Phytophthora blight of bell pepper. *Plant Dis* 85: 1069–1075.

Pérez-Sierra A, Álvarez LA, Vercauteren A, Heungens K, Abad-Campos P (2011) Genetic diversity, sensitivity to phenylamides and aggressiveness of *Phytophthora ramorum* on Camellia, Rhododendron, and Viburnum plants in Spain. *Plant Pathol* 60: 1069–1076.

Petkar A, Langston DB, Buck JW, Stevenson KL, Ji P (2017) Sensitivity of *Fusarium oxysporum* f.sp. *niveum* to prothioconazole and thiophanate-methyl and gene mutation conferring resistance to thiophanate-methyl. *Plant Dis* 101: 366–371.

Pirondi A, Brunelli A, Muzzi E, Collina M (2017) Post-infection activity of fungicides against *Phytophthora infestans* on tomato (*Solanum lycopersicum* L.) *J Gen Plant Pathol* 83: 244–252.

Pivonia S, Gerstl Z, Maduel A, Levita R, Cohen R (2010) Management of Monosporascus sudden wilt of melon by soil application of fungicides. *Eur J Plant Pathol* 128: 201–209.

Porter LD, Cummings TF, Johnson DA (2006) Effects of soil-applied late blight foliar fungicides on infection of potato tubers by *Phytophthora infestans*. *Plant Dis* 90: 964–968.

Pring RJ (1984) Effects of triadimefon on the ultrastructure of rust fungi infecting leaves of wheat and broad bean (*Vicia faba*). *Pesticide Biochem Physiol* 21: 127–137.

Qi R, Wang T, Zhao W, Li P, Ding J, Gao Z (2012) Activity of ten fungicides against *Phyophthora capsici* isolates resistant to metalaxyl. *J Phytopathology* 160: 717–722.

Qu T, Shao Y, Csinos AS, Ji P (2016) Sensitivity of *Phytophthora nicotianae* from tobacco to fluopicolide, mandipropamid, oxathiapiprolin. *Plant Dis* 100: 2119–2125.

Radzuhn B, Lyr H (1984) On the mode of action of the fungicide etridiazole. *Pesticide Biochem Physiol* 22: 14–23.

Rampersad SN (2011) A rapid, colorimetric bioassay to evaluate fungicide sensitivity among *Verticillium dahliae* isolates. *Plant Dis* 95: 248–255.

Randall E, Young V, Sierotzki H et al. (2014) Sequence diversity in the large subunit of RNA polymerase I contributes to mefenoxam insensitivity in *Phytophthora infestans*. *Molec Plant Pathol* 15: 664–676.

Rebollar-Alviter A, Madden LV, Ellis MA (2007) Pre- and post-infection activity of azoxystrobin, pyraclostrobin, mefenoxam, and phosphite against leathery rot of strawberry caused by *Phytophthora cactorum*. *Plant Dis* 91: 559–564.

Rebollar-Alviter A, Madden LV, Jeffers SN, Ellis MA (2007) Baseline and differential sensitivity to two QoI fungicides among isolates of *Phytophthora cactorum* that cause leathery rot and crown rot on strawberry. *Plant Dis* 91: 1625–1637.

Rodgers GA (1986) Potency of nitrification inhibition following their repeated application to soil. *Biol Fertil Soils* 2: 105–108.

Rodriguez-Brljevich C, Kanobe C, Shanahan JF, Robertson AE (2010) Seed treatments enhance photosynthesis in maize seedlings by reducing infection with *Fusarium* spp. and consequent disease development in maize. *Eur J Plant Pathol* 126: 343–347.

Rosslenbroich HJ, Stuebler D (2000) *Botrytis cinerea*–history of chemical control and novel fungicides for its management. *Crop Protect* 19: 557–561.

Samac DA, Schraber S, Barclay S (2015) A mineral seed coating for control of seedling diseases of alfalfa suitable for organic seed production systems. *Plant Dis* 99: 614–620.

Sanogo S, Yang XB, Schern H (2000) Effects of herbicides on *Fusarium solani* f.sp. *glycines* and development of sudden death syndrome in glyphosate tolerant soybean. *Phytopathology* 90: 57–66.

Scott PM, Barber PA, Hardy GE St J (2005) Novel phosphite and nutrient application to control *Phytophthora cinnamomi* disease. *Austr Plant Pathol* 44: 431–436.

Seidl AC, Gevens AJ (2013) Characterization and distribution of three new clonal lineages of *Phytophthora infestans* causing late blight in Wisconsin from 2009 to 2012. *Amer J Potato Res* 90: 551–560.

Sharma-Poudyal D, Paulitz TC, du Toit LJ (2016) Timing of glyphosate application to wheat cover crops to reduce onion stunting caused by *Rhizoctonia solani*. *Plant Dis* 100: 1474–1481.

Singh N, Tandon S (2015) Dissipation kinetics and leaching of cyazofamid fungicide in texturally different soils. *Internatl J Environ Sci Technol* 12: 2475–2484.

Singh RP, Pundhir VS (2004) Possible method of eradicating tuber-borne inoculum of *Phytophthora infestans* (Mont.) de Bary. *J Mycol Plant Pathol* 34: 91.

Stanghellini ME, Ferrin DM, Kim DH et al. (2003) Application of preplant fumigants via drip irrigation systems for the management of root rot of melons caused by *Monosporascus cannonballus*. *Plant Dis* 87: 1176–1178.

Staub T (1991) Fungicide resistance–Practical experience with anti-resistance strategies and the role of integrated use. *Annu Rev Phytopathol* 29: 421–442.

Sugimoto T, Aino M, Sugimoto M, Watanabe K (2005) Reduction of Phytophthora stem rot disease on beans by the application of CaCl2 and Ca(NO3)2. *J Phytophthol* 153: 536–543.

Sun H, Wang H, Stammler G, Ma J, Zhou M (2010) Baseline sensitivity of populations of *Phytophthora capsici* from China to three carboxylic acid amide (CAA) fungicides and sequence analysis of choline phosphotransferases from a CAA-sensitive isolate and CAA-resistant laboratory mutants. *J Phytopathol* 58: 244–252.

Taylor RJ, Pasche JS, Gudmestad NC (2006) Biological significance of mefenoxam resistance in *Phytophthora erythroseptica* and its implications for the management of pink rot of potato. *Plant Dis* 90: 927–934.

Taylor RJ, Pasche JS, Gudmestad NC (2011) Effect of application method and rate on residual efficacy of mefenoxam and phosphorous acid fungicides on the control of pink rot of potato. *Plant Dis* 95: 997–1006.

Taylor RJ, Salas B, Gudmestad NC (2004) Differences in etiology affect mefenoxam efficacy and the control of pink rot and leak tuber diseases of potato. *Plant Dis* 88: 301–307.

Thomas A, Langston DB Jr., Walcott RR, Gaitaitis RD, Stevenson KL (2014) Evidence for fungicide resistant seedborne inoculum for gummy stem blight of watermelon. *Seed Sci Technol* 42: 92–96.

Thomidis T, Elena K (2001) Effects of metalaxyl, fosetyl-Al, dimethomorph and cymoxanil on *Phytophthora cactorum* of peach tree. *J Phytopathol* 149: 97–101.

Thompson HK, Tegg RS, Davies NW, Ross JJ, Wilson CR (2013) Determination of optimal timing of 2,4-dichlorophenoxyacetic acid foliar applications for common scab control in potato. *Ann Appl Biol* 163: 242–256.

Tian M, Zhao L, Li S et al. (2016) Pathotypes and metalaxyl sensitivity of Phytophthora sojae and their distribution in Heilongjiang, China 2011–2015. *J Gen Plant Pathol* 82: 132–141.

Toksoz H, Rothrock CS, Kirkpatrick TL (2009) Efficacy of seed treatment chemicals for black root rot caused by *Thielaviopsis basicola* on cotton. *Plant Dis* 93: 354–362.

Tsror L, Peretz-Alon I (2004) Control of silver scurf on potato by dusting or spraying seed tubers with fungicides before planting. *Amer J Potato Res* 81: 291–294.

Urrea K, Rupe JC, Rothrock CS (2013) Effect of fungicide seed treatments, cultivars and soils on soybean stand establishment. *Plant Dis* 97: 807–812.

van Wijngaarden RPA, Arts GHP, Belgers JDM et al. (2010) The species sensitivity distribution approach compared to a microcosm study: A case study with the fungicide fluazinam. *Ecotoxicol Environ Safety* 73: 109–122.

Wang H-C, Chen X-J, Cai L-T et al. (2013) Race distribution and distribution of sensitivities to mefenoxam among isolates of *Phytophthora parasitica* var. *nicotianae* in Guizhou province of China. *Crop Protect* 52: 136–140.

Wang J, Bradley CA, Stenzel O, Pedersen DK, Reuter-Carlson U, Chilvers MI (2017) Baseline sensitivity of *Fusarium virguliforme* to fluopyram fungicide. *Plant Dis* 101: 576–582.

Wang Y, Duan Y, Wang J, Zhou M (2015a) A new point mutation in the iron-sulfur of succinate dehydrogenase confers resistance to boscalid in *Sclerotinia sclerotiorum*. *Molec Plant Pathol* 16: 653–661.

Wang Y, Duan Y-B, Zhou M-G (2015b) Molecular and biochemical characterization of boscalid resistance in laboratory mutants of *Sclerotinia sclerotiorun*. *Plant Pathol* 64: 101–108.

Wang Y, Duan Y-B, Zhou M-G (2016) Baseline sensitivity and efficacy of fluazinam in controlling Sclerotinia rot of rapeseed. *Eur J Plant Pathol* 144: 337–343.

Wang Y, Hou Y-P, Chen C-J, Zhou M-G (2014) Detection of resistance in *Sclerotinia sclerotiorum* to carbendazim and dimethachlon in Jiangsu province of China. *Austr Plant Pathol* 43: 307–312.

Wang ZW, Cang T, Qi P et al. (2015c) Dissipation of fair fungicides on greenhouse strawberries and an assessment of their risks. *Food Control* 55: 215–220.

Weiland JE, Santamaria L, Grünwald NJ (2014) Sensitivity of *Pythium irregulare*, *P. sylvaticum*, and *P. ultimum* from forest nurseries to mefenoxam and fosetyl-Al, and control of Pythium damping-off. *Plant Dis* 98: 937–942.

Wise KA, Bradley CA, Markell S et al. (2011) Sensitivity of *Ascochyta rabiei* populations to prothioconazole and thiabendazole. *Crop Protect* 30: 1000–1005.

Wise KA, Bradley CA, Pasche JS, Gudmestad NC (2009) Resistance to QoI fungicides in *Ascochyta rabiei* from chickpea in the Northern Great Plains. *Plant Dis* 93: 528–536.

Wise KA, Bradley CA, Pasche JS, Gudmestad NC, Dugan FM, Chen W (2008) Baseline sensitivity of *Ascochyta rabiei* to azoxystrobin, pyraclostrobin and boscalid. *Plant Dis* 92: 295–300.

Woodward JE, Brenneman TB, Kemerai RC Jr (2013) Chemical control of peanut diseases: Targeting leaves, stems, roots and pods with foliar-applied fungicides

Xiao Y-M, Esser L, Zhou F et al. (2014) Studies on inhibition of respiratory cytochrome bc1 complex by the fungicide pyrimorph suggest a novel inhibitory mechanism. *PLoS ONE* 9 (4): e93765

Xu C, Liang X, Hou Y, Zhou M (2015) Effects of the novel fungicide benzothiostrobin on *Sclerotinia sclerotiorum* in the laboratory and on Sclerotinia stem rot in rape field. *Plant Dis* 99: 969–975.

Yandoc-Ables C, Rosskopf EN, Lamb EM (2007) Management of Phytophthora crown rot in pumpkin and zucchini seedlings with phosphonates. *Plant Dis* 91: 1651–1656.

Yang C, Hamel C, Vujanovic V, Gan Y (2011) Fungicides: Modes of action and possible impact on nontarget organisms. *ISRN Ecology* 2011, Article ID 130289 (8 pages)

Zhang S, Panaccione SG, Gallegly ME (1997) Metalaxyl stimulation of growth of isolates of *Phytophthora infestans*. *Mycologia* 89: 289–292.

Zhou F, Zhang X-L, Li J-L, Zhu F-X (2014) Dimethachlon resistance in *Sclerotinia sclerotiorum* in China. *Plant Dis* 98: 1221–1226.

Zhu SS, Liu XL, Liu PF et al. (2007) Flumorph is a novel fungicide that disrupts microfilament organization in *Phytophthora melonis*. *Phytopathology* 97: 643–649.

Ziogas BN, Makoglou AN, Tzima A (2002) A non-Mendelian inheritance of resistance to strobilurin fungicides in *Ustilago maydis*. *Pest Manag Sci* 58: 908–916.

5 Management of Soilborne Microbial Plant Pathogens

Development of Integrated Disease Management Systems

Investigations on various aspects of soilborne microbial plant pathogens and the diseases induced by them in different crops aim at providing information for the development of methods to eradicate or restrict the incidence and/or severity and also spread of the diseases within and outside the geographical location/field. As the first step, highly sensitive and reliable methods are employed to establish the identity of the pathogen(s) involved in the disease(s) and quantify their populations precisely along with their pathogenic potential (virulence). Studies on the biology of the pathogens provide information on the nature of survival structures formed by the pathogens during the life cycle, genetic diversity and differential sensitivity of strains/isolates to environmental conditions and chemicals, and availability of various sources of infection, in addition to the primary (crop) host to determine the extent of protective cover needed for crop cultivars that may vary in the levels of susceptibility/resistance to the target pathogen(s). Short-term and long-term strategies have been developed, the effectiveness of which differ based on the cultural practices adopted in different locations primarily for improving yield levels of cultivars. It has been necessary to continue the search for new disease management strategies, as the predominance of one pathogen(s) changed as new cultivars with improvement in yield and quality of produce were released for large-scale cultivation, without much concern for the level of resistance to disease(s). Prior to the 19th century, cultural methods were essentially applied for the management of diseases. From the latter half of the 20th century, chemicals with pathogen-suppressive activity were used increasingly either in conjunction with or instead of cultural practices. Development of cultivars with built-in resistance to major disease(s) was given more importance because of growing awareness of harmful effects of excessive use of chemicals on the environment and chemical residue accumulation in harvested produce. However, limitations of the approach of disease management through breeding for disease-resistant cultivars was realized when pathogen races/strains capable of overcoming the effects of resistance gene(s) introgressed into the susceptible cultivars emerged over wide areas. Effectiveness of biological control agents – biotic and abiotic – was demonstrated in due course and this approach has become acceptable because of its ecofriendly nature, receiving the attention of researchers and growers involved in the production and supply of agricultural and horticultural produce free of residues (Narayanasamy 2002, 2017).

Plant pathologists entrusted with research and extension activities provided the basic concept of integration of two or more compatible strategies to enhance the level of protection to plants that could individually provide against microbial pathogens. The basic concept of 'integrated pest management (IPM)' proposed by the entomologists Smith and van den Bosch (1967) was based on economic injury (EI) level and economic threshold (ET) as suggested earlier by Stern et al. (1959). This concept was found to be unsuitable for the management of crop diseases, as insect populations could be visually quantified but not populations of microbial plant pathogens. In contrast, plant pathologists require precise laboratory methods, spore traps, immunoassays, or nucleic acid-based assays for estimating pathogen populations. From the very inception of plant pathology discipline, crop disease management has relied upon integrated disease management (IDM), using basic information on the loss potential, pathogen biology, ecology and epidemiology. Furthermore, the ET levels prescribed for insects as pests cannot be used for the insects functioning as vectors of viruses. Populations remaining below ET levels are sufficient to spread plant viruses to significant levels. It has, therefore, not been possible for plant pathologists to adopt the concept of EI and ET levels of IPM meant for insect management. In addition, crop disease management generally did not depend at any time on levels of chemical uses that were commonly associated with insect control, whereas use of fungicides or other chemicals was kept at minimal levels, crop disease management relying more on integration of strategies such as sanitation, use of disease-free seeds/propagules, crop rotation, adopting suitable sowing/planting date and growing disease resistant/tolerant cultivars, if available. IDM has been used increasingly in plant pathology investigations, whereas the term 'pest' is largely confined to insect control. Different strategies compatible with each other are identified and sequenced to provide a comprehensive protective umbrella to reduce disease incidence/severity and further spread.

5.1 COMPONENTS OF INTEGRATED DISEASE MANAGEMENT SYSTEMS

Integrated disease management (IDM) systems for containing soilborne microbial pathogens aim to combine/sequence the strategies that have shown potential consistently in (i) reducing the pathogen population/inoculum sources (other than crops) in the soil, (ii) restricting the movement of the pathogens, (iii) suppressing the progress of development of disease symptoms in infected plants, (iv) enhancing the levels of resistance in crop cultivars and (v) eradicating the pathogen in the soil using chemicals that exert minimal effects on nontarget, beneficial microorganisms. Above all, estimation of pathogen populations in the soil prior to sowing/planting and monitoring the composition of isolates and sensitivity to chemicals and other environmental conditions have to be taken up using sensitive and reliable techniques such as PCR-based methods that can rapidly provide results. The sensitivity of detection and accuracy of quantification of microbial pathogens has been enhanced by applying loop-mediated isothermal amplification (LAMP) and biosensor-based gold nanoparticles. Information on all sources of inoculum of microbial pathogens and different kinds of spore forms produced at different stages in the life cycle of fungal pathogens should be gathered. IDM, a holistic approach for managing soilborne pathogens, depends on the understanding of the complex environment existing in the soil and the interactions among plants, pathogens, microbial community structure and composition.

Crop management practices are applied essentially with the aim of improving yield and quality of the produce to derive high returns. Disease management strategies, therefore, have to be dovetailed in order to reduce the negative impact of the diseases on crop yield and quality. Cereals, pulses, oilseeds, vegetables and fruit and other crops share some common practices. The integrated management systems for diseases caused by soilborne microbial pathogens affecting these crops are discussed in this chapter.

5.2　MANAGEMENT OF DISEASES OF AGRICULTURAL CROPS

5.2.1　Wheat Diseases

Soilborne pathogens infect roots, crowns and stems of cereal crops. Various investigations have been carried out to reduce the incidence and severity of diseases to realize high yields by combining two or more compatible strategies that have been found to be individually effective against disease(s) caused by them.

5.2.1.1　Wheat-Take-All Disease

Gaeumannomyces graminis var. *tritici (Ggt)* causes the wheat-take-all disease occurring in most wheat-growing countries around the world. Wheat take-all has been intensively studied by researchers investigating cereal diseases caused by soilborne pathogens. Planting wheat in the same field year after year resulted in a progressive decline in incidences of take-all disease. Significant reductions in take-all severity after the sixth wheat crop was attributed to the enhancement of populations of microorganisms antagonistic to *Ggt*. The take-all decline persisted as long as monocropping of wheat was continued. Wheat cultivars differed in their ability to support appreciably development of native populations of pseudomonads capable of producing antibiotic

2,4-diacetylphloroglucinol (2,4-DAPG) that suppressed the development of *Ggt* and led to a consequent reduction in disease severity (Mazzola et al. 2004). Crop rotation (sequence) and soil solarization may differentially encourage the proliferation of biological control agents (BCAs), resulting in increased effectiveness of soil suppressiveness (Chellemi et al. 2016). Colonization of rhizosphere by *Pseudomonas* spp. antagonistic to *G. graminis* var. *tritici* may enhance the soil suppressiveness. Various studies have revealed the importance of nutrient management for reducing incidence and severity of diseases. Nutrient deficiencies at any time during crop growth increased the severity of wheat-take-all disease. Hence, ensuring the availability of adequate N, P, K and sulfur (S) at sowing was found to be a critical factor to reduce the risks of take-all disease. Yields were generally higher with ammonium chloride than with ammonium nitrate, urea or ammonium sulfate. Phosphorus deficiency in soil favors development of wheat take-all disease. Phosphorus application was required regularly to reduce the susceptibility of wheat plants to take-all disease (Christensen and Hart 2008).

The incorporation of organic amendments (OA) into the soil generally encourages microbial antagonists, leading to mitigation of disease severity. Take-all severity on roots of wheat and barley caused by *Gaeumannomyces graminis* var. *tritici* was significantly reduced in organically-managed rather than conventionally-managed soils. Fluorescent *Pseudomonas* spp., especially phlD[+] (2,4-diacetylphloroglucinol)-producing strains, key factors in take-all decline were present in lower population densities in organically-managed soils compared to conventionally-managed soils. In addition, organic amendments adversely affected the initial establishment of introduced phlD[+] *Pseudomonas fluorescens* strain Pf32-gfp but not its survival. The efficacy of biocontrol of take-all disease by introduced strain Pf32-*gfp* was significantly stronger in conventionally-managed soil than in organically-managed soil. The results suggested that phlD[+] *Pseudomonas* spp. might not have a vital role in the suppressiveness of soils included in this investigation. Other bacterial species might inhibit the activity of the take-all pathogen occurring in some soils (Hiddink et al. 2005b). Compatibility of the BCAs with fungicides may permit the application of both components of the integrated disease management system. The fungal BCA *Penicillium radicum* promoted growth of wheat plants and also inhibited development of soilborne pathogens, including *G. graminis* var. *tritici*. A combination of *P. radicum* and the fungicide fludiconazole significantly reduced the incidence of take-all disease (Wakelin et al. 2006). Crop rotation was shown to be a highly effective strategy for suppression of wheat-take-all disease development, as survival of the pathogen in the absence of the primary wheat host plants was poor. A 1-year break from wheat or barley appeared to be sufficient to reduce the risk of take-all incidence to an insignificant level. Oats, corn, bean, vegetable, oilseed crops and annual legumes for seed, when grown in wheat fields infested with *Ggt*, were effective in breaking the continuous availability of susceptible wheat plants, leading to significant reduction in take-all disease incidence (Christensen and Hart 2008). The continuous winter wheat treatment with burning and plowing was compared with a 3-year no-till rotation of winter wheat-spring barley-winter canola. Take-all disease and inoculum increased from years 1 to 4 in the continuous winter wheat treatment, with burning and plowing applied as residue management treatments. The decline in take-all was observed after 4 years of monocropping (Paulitz et al. 2010).

5.2.1.2 Rhizoctonia Bare Patch Disease

Soil suppressiveness against soilborne pathogens appears to be nonspecific. Rhizoctonia bare patch disease of cereals by *Rhizoctonia solani* AG-8 became a major root disease in no-till cropping systems. Rhizoctonia bare patch disease increased when tillage was eliminated, and the disease became a major limiting factor for adoption of no-till technology for wheat cultivation. As several pulse and oilseed crops are susceptible to *R. solani*, the effect of mixed cropping was investigated. Mixed cropping of triticale wheat with clover and of barley with Brussels sprouts did not enhance soil suppressiveness to *R. solani* (Hiddink et al. 2005a). An 8-year crop rotation trial showed that crop rotation had no effect on the development of bare patch disease during the first 5 years. However, from years 6 to 8, both soft white and hard white classes of spring wheat (*Tricticum aestivum*) grown in a 2-year rotation with spring barley had an average of only 7% of total land area with bare patches, compared with 15% in continuous annual soft white wheat and hard white wheat (monoculture system). In addition, average yield on both soft white and hard white wheat during these 2 years were greater when grown in rotation with barley than in monoculture (Schillinger and Paulitz 2006). The influence of crop management practices, tillage, crop rotation and residue management on bare pitch disease development was investigated in field trials conducted for 6 years in east-central Washington state. The continuous winter wheat treatment with burning and plowing was compared with a 3-year no-till rotation with winter wheat-spring barley-winter canola and three straw management treatments, burning, straw removal and leaving straw stubble standing after harvest. Inoculum of *R. solani* AG-8 was significantly lower in tilled treatment compared to no-till treatments. Residue management treatments had no effect on *Rhizoctonia* inoculum. The percentage of wheat crown, root and seminal roots exhibiting symptoms of Rhizoctonia root rot (bare patch) was variable across years and year interaction with treatments. In 2003 and 2006, continuous winter wheat had low level of crown and seminal root infection symptoms (see Figure 5.1) and the lowest root rot rating, compared to other treatments (see Figure 5.2) (Paulitz et al. 2010).

5.2.2 Rice Sheath Blight Disease

Rice sheath blight disease, caused by *R. solani*, is one of the major rice diseases limiting production seriously in most of the rice-growing countries inducing characteristic greenish-gray lesions on leaf sheaths and flagleaf sheaths covering the panicles (see Figure 5.3) (Raguchander). Development of rice cultivars with resistance to sheath blight has not yet yielded

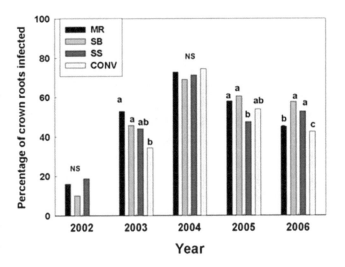

FIGURE 5.1 Effects of crop rotation/residue management treatments on percentage of crown roots with symptoms of infection by *R. solani* in April-grown winter wheat crops Treatments: winter wheat-spring barley-canola no-till rotation with mechanical stubble (MR), stubble burning (SB), or stubble left standing (SS) and moldboard plowed (CONV); treatments within each year with the same letters are not significantly different (P = 0.05) as per Tukey's mean separation test.

[Courtesy of (Paulitz et al. (2010)) and with kind permission of the American Phytopathological Society, MN]

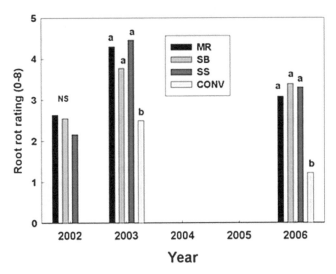

FIGURE 5.2 Effects of crop rotation/residue management treatments on Rhizoctonia root rot ratings induced by *R. solani* on April-grown winter wheat seedlings Treatments: winter wheat-spring barley-canola no-till rotation with mechanical stubble (MR), stubble burning (SB), or stubble left standing (SS) and moldboard plowed (CONV); treatments within each year with the same letters are not significantly different (P = 0.05) as per Kruskal-Wallis mean separation test; data were not recorded for 2004 and 2005.

[Courtesy of (Paulitz et al. (2010)) and with kind permission of the American Phytopathological Society, MN]

fruitful results. Application of fungicides is found to be an economically acceptable strategy and mainstay. The effectiveness of fungal and bacterial biocontrol agents has been assessed in several investigations. Talc-based formulations

a. Greenish grey lesion on panicle   b. Greenish grey lesion on sheath   c. Sclerotia on sheath

FIGURE 5.3 *R. solani* causes rice sheath blight disease inducing symptoms in leaf sheath and panicles a: formation of greenish grey lesions on flagleaf sheath; b: formation of greenish grey lesions on leaf sheath; and c: production of sclerotia on the leaf sheath later.

[Courtesy of Dr. T. Raguchander, Tamil Nadu Agricultural University, Coimbatore, India]

containing *P. fluorescens* PF1 and FP7, either alone or in combination, reduced the sheath blight incidence to different extents depending on the strain or strain mixture of the bacterial BCA. Promotion of plant growth by the BCAs was an additional advantage (Nandakumar et al. 2001). Level of protection by the bacterial BCA could be significantly enhanced by combination of *P. fluorescens* strains PF1 and FP7 with chitin amendment (Radjacommare et al. 2002). Talc-based formulations of *P. fluorescens* and *Trichoderma viride* reduced sheath blight incidence significantly under field conditions and the level of protection against sheath blight disease was comparable with the fungicide carbendazim (Mathivanan et al. 2005). In order to increase the effectiveness of the suppression of rice sheath blight disease, a combination of *Bacillus subtilis* NJ-18 and 50% kresoxim-methyl (225 g a.i/ha) was found to be more effective in suppressing the disease development than kresoxim-methyl alone or Jinggangmycin at 120 g a.i/ha (Yang et al. 2009). The efficacy of integrating *Brassica juncea* cover crops, tolerant rice cultivars and application of the fungicide azoxystrobin in reducing the incidence and severity of rice sheath blight disease caused by *R. solani* was assessed during 2011–2013 under field conditions. *B. juncea* with biofumigation activity was able to reduce sheath blight severity consistently in all 3 years of experimentation and also increased the yield in 2013 over that of fallow control. Rice cv. Presido had less sheath blight severity and produced greater yields than cv. Cocodrie in 2012 and 2013. Azoxystrobin application at half label rate (0.08 kg a.i/ha) reduced sheath blight severity significantly, and yield was also increased in 2012 and 2013. Increase in azoxystrobin dose to 0.16 kg a.i/ha did not contribute to any additional reduction in disease severity or yield level. The results demonstrated that cover cropping with *B. juncea* integrated with planting a tolerant rice variety and applying half label rate of azoxystrobin could be adopted by growers for effective management of rice sheath blight caused by *R. solani* (Handiseni et al. 2015).

5.2.3 Cotton Diseases

5.2.3.1 Verticillium Wilt Disease

Verticillium dahliae, incitant of Verticillium wilt disease of cotton, has worldwide distribution and wide host range, including several economically important crops, which can serve as sources of inoculum for cotton crops. *V. dahliae* exists in the form of two distinct strains, causing defoliation (D) and nondefoliation (ND) symptoms in infected cotton plants. The effectiveness of various organic amendments and bacterial BCAs in inhibiting germination of microsclerotia (MS) of the pathogen was assessed. The efficacy of disease suppression depended on plant species, incubation time, pathogen strain and soil disinfestation. All organic amendments reduced disease incidence and severity in soils infested with both D and ND isolates to different degrees. Debris of *Diplotaxis virgata* was the most effective in reducing viability of MS of *V. dahliae* consistently (Lopez-Escudero et al. 2007). In another investigation, crab shell chitin amendment was observed to stimulate the activities of antagonists capable of suppressing the development of *V. dahliae* (Huang et al. 2006). Seed bacterization with four strains of *Pseudomonas* spp. from the rhizosphere of cotton protected cotton plants against infection by *V. dahliae* (Erdogan and Benlioğlu 2010).

5.2.3.2 Seedling Diseases

Cotton seedlings are affected by *Pythium aphanidermatum*, *P. ultimum* and *Rhizopus oryzae*, causing damping-off at both pre- and postemergence stages, resulting in gappiness in the field. *Rhizoctonia solani* is also involved in seedling damping-off and it is more important as a pathogen-causing root rot disease in the later stages of crop growth. Options available for disease suppression are application of organic amendments, seed treatment with chemicals and maintenance of optimum soil moisture, in addition to application of biocontrol agents. *Trichoderma (Gliocladium) virens* Q strains were effective in suppressing development of *R. solani*, when combined with fungicide seed treatment (Howell et al. 1997). *T. virens* with multiple modes of action was not only parasitic on *R. solani* but also an inducer of resistance in cotton plants against the pathogen (Howell et al. 2000). Under field conditions, seed treatment with metalaxyl was partially effective against *P. aphanidermatum* and *P. ultimum* but not against *R. oryzae*. However, seed treatment with formulations of *Trichoderma virens* parent strains G6 and G6-5 and the hybrid strains of *T. virens* x *T. longibrachiatum* Tvl-30 and Tvl-35 were able to control the seedling diseases caused by all four pathogens (Howell 2002) effectively. Preemergence damping-off caused by *Pythium* spp. and *R. oryzae* could be controlled by seed treatment with *Trichoderma* spp. formulations. However, they were much less effective in controlling postemergence damping-off incited by *R. solani*, which could be controlled by fungicides effectively. Hence, combination of fungicides and *Trichoderma* spp. was evaluated for their efficacy in suppressing postemergence damping-off in pathogen-infested soil. The combination of the fungicide chloroneb plus *Trichoderma* spp. was the most effective treatment for controlling both phases of the disease. The treatment chloroneb + metalaxyl (Delta Coat AD) + *T. virens* strain G-6 was the next best in suppressing the pre- and postemergence damping-off of cotton seedlings (Howell 2007)

5.2.3.3 Root Rot Disease

Root rot disease caused by *Fusarium oxysporum* (FO) and *Pythium debaryanum* (PD) affects production level appreciably due to the mortality of the plants at different stages of crop growth. Mixtures of biocontrol agents (BCAs) and resistance inducers were evaluated for their efficacy in suppressing development of root rot disease in cotton plants. The combined treatment with *Trichoderma hamatum* + *Paecilomyces lilacinus* + salicylic acid (SA) + benzol (1,2,3) thiadiazole-7-carbothioic acid-*S*-methyl ester (BTH) protected the cotton plants most effectively, reducing disease index (DI) to the maximum extent. The results indicated the compatibility of the BCAs with inducers of resistance to disease to provide higher levels of protection to cotton plants against root rot disease (Abo-Elyousr et al. 2009). In a later investigation, integration of disease management strategies for the suppression of cotton root rot caused by *Rhizoctonia bataticola* was studied. The effects of intercropping system in cotton along with inorganic fertilizer and two bioagents *Azospirillum* and *Pseudomonas* sp. on root rot disease development were assessed. Cotton intercropping with *Sesbania aculeata* (1:1 ratio) reduced the incidence of root rot, whereas integration of *Azospirillum* and *Pseudomonas* with 50% recommended NPK rates reduced root rot incidence significantly. The combination of *S. aculeata* with 50% NPK and bioagents resulted in the lowest root rot incidence, in addition to increase in cotton yield (Marimuthu et al. 2013).

5.2.4 Soybean Diseases

5.2.4.1 Damping-Off/Root Rot Disease

Rhizoctonia solani (teleomorph-*Thanatephorus cucumeris*) causes damping-off, root rot, stem rot and foliar blight at different stages of soybean crop growth. Biofumigation using organic amendment of Brassicaceae green manure or seed meal incorporation into the soil is an ecologically suitable alternative to chemical fumigation. In order to combine the application of *Brassica carinata* seed meal (BCSM) and *Trichoderma* spp. for the suppression of *R. solani*, the level of tolerance of *Trichoderma* to glucosinolate-derived compounds released from BCSM was assessed. *Trichoderma* isolates were generally able to tolerate sinigin (10 μ mol) and they could grow over BCSM and the pathogen. The greatest inhibitory effect was recorded when BCSM was applied in combination with *Trichoderma*, regardless of the ability of BCSM to release isothiocyanates. The results indicated the effectiveness of combined application of BCSM and the BCA to derive higher degree of suppression of *R. solani* infecting soybean (Galletti et al. 2008). Rhizoctonia foliar blight (RFB) caused by *R. solani* AG1-IA, AG1-IB and AG2-3 in soybean is responsible for yield reduction of between 60 to 70% in Brazil and Japan. Preventive application of pyraclostrobin +

boscalid was the most effective in suppressing disease development among the fungicides tested. Plant resistance activators, SA at 2.5 mM a.i/l sprayed at 20 days before inoculation or acibenzolar-S-methyl (ASM) at 12.5 mg a.i/l at 10 days before inoculation, were effective in reducing disease incidence (Meyer et al. 2006).

5.2.4.2 Sclerotinia Stem Rot Disease

The effectiveness of integration of steam sterilization (pasteurization) of soil and application of the mycoparasite *Coniothyrium minitans* was assessed for the control of Sclerotinia stem rot disease of soybean. Soil was pasteurized in the autoclave (80°C for 3 min) and/or *C. minitans* subsequently was applied to the pasteurized soil. Either method applied separately was effective in reducing sclerotial viability. Similar results were obtained with two soil types tested. However, no increase in reduction of sclerotial viability was achieved by combining both pasteurization and BCA application. The effect of timing of application of *C. minitans* to the pasteurized soil was determined. Application of *C. minitans* to soil immediately after soil pasteurization favored sclerotial infection by *C. minitans*, but application of BCA at 7 days or more after soil pasteurization resulted in low recovery of BCA from sclerotia due to failure of sclerotial infection by the presence of other fungi capable of colonizing the sclerotia of *S. sclerotiorum* (Bennett et al. 2005). The biocontrol potential of cell suspensions or cell-free culture filtrates of two strains of *B. subtilis* SB01 and SB24 was assessed for suppressing Sclerotinia stem rot disease under field conditions. The BCA suspension was more effective in suppressing the development of *S. sclerotiorum*. The cell suspension applied on soybean leaves for up to 10 days significantly reduced the disease severity by about 20 to 90% at 5 days after inoculation with the pathogen. The cell suspension, capable of reducing disease severity by 40 to 70%, exhibited potential to be used as a component of integrated disease management system for Sclerotinia stem rot disease of soybean (Zhang et al. 2011). In a later investigation, *C. minitans* (applied as ContansR WG) was the most effective in reducing disease severity index (DSI) by 68.5% and the number of pathogen sclerotia in soil was reduced by 95.3%. Application of *Streptomyces lydicus* (ActinovateR AG) or *Trichoderma harzianum* T-22 also resulted in significant reductions in the disease severity index in treated soybean plants and the number of sclerotia of *S. sclerotiorum* in the soil was also reduced. Populations of the BCAs in the soil did not vary in the soil samples taken from 3 to 169 days, indicating the persistence of the BCA preparations (Zeng et al. 2012). The fungal and bacterial biocontrol agents (BCAs) have potential for suppressing the development of Sclerotinia stem rot disease and for replacing the chemical fumigants.

5.2.5 Chickpea Diseases

5.2.5.1 Fusarium Wilt Disease

Fusarium wilt disease of chickpea caused by *Fusarium oxysporum* f.sp. *ciceris* (Foc) race 5 drastically reduced grain yield because of high plant mortality in susceptible cultivars.

A combination of host resistance and soil solarization was evaluated for the extent of reduction in disease incidence. Soil temperature under polyethylene sheeting exceeded 60°C at 5 cm soil depth and reached 42°C at 20 cm soil depth. Substantial reductions in pathogen propagules due to solarization (> 1,000 propagules/g soil) was observed, whereas pathogen population increased in nonsolarized plots during the period of experimentation. In nonsolarized plots, virtually all plants of susceptible cultivar were killed. In contrast, none of the plants of resistant cv. JG74 was killed both in solarized and nonsolarized plots. In addition, soil solarization promoted plant growth and increased yield levels in both resistant and susceptible plants (ICRISAT 1980). The effectiveness of integration of appropriate sowing date, use of partially resistant chickpea cultivar and application of bacterial biocontrol agents in suppressing Fusarium wilt disease development was assessed. Advancing sowing date from early spring to winter significantly delayed disease onset, reduced final disease intensity and increased seed yield. *B. subtilis* GB03, nonpathogenic *F. oxysporum* Fo90105 and *P. fluorescens* RG26 were applied as seed and soil treatments. Under conditions conducive for development of Fusarium wilt disease, the extent of disease suppression depended primarily on choice of sowing date and, to a lesser extent, on host resistance and BCA treatment. Adoption of these strategies might effectively reduce disease incidence/intensity and increase seedling emergence, ultimately enhancing the yield level (Landa et al. 2004).

5.2.5.2 Ascochyta Blight Disease

Ascochyta blight disease of chickpea caused by *Ascochyta rabiei* (*Didymella rabiei*) is one of the major factors limiting production. The efficacy of applying integrated disease management approach by combining host resistance and chemical control to reduce incidence of Ascochyta blight disease was assessed. The contribution of host resistance was determined by including four commercially available chickpea cultivars with range of resistance to *A. rabiei*. The efficacy of chemical control in a highly susceptible cultivar was related to the conduciveness of the environment for pathogen development. Application of chemicals was effective, when disease pressure was low, but not under high disease pressure conditions. The contribution of genotype resistance to disease reduction in a moderately susceptible cultivar varied from < 10%, when weather favored development of severe epidemics to ≅ 60%. Spraying moderately resistant cultivar resulted in 95% control under low disease pressure conditions. The level of genotype resistance available in a highly resistant cultivar was sufficient to protect the plants under all weather conditions favoring epidemics, even without fungicide application. Application of tebuconazole or difenconazole at 3 days postinfection treatment (after rain or overhead irrigation) suppressed disease development as effectively as preventive application under field conditions and needed only few sprays (Shtienberg et al. 2000). Effectiveness of fungicide application differed as per the cultivar resistance level and host response differed depending on the virulence level of the strains of *A. rabiei* and

prevailing environmental conditions (Davidson and Kimber 2007). The possibility of combining host resistance and application of biocontrol agents (BCAs) to eliminate source of inoculum for reducing incidence of Ascochyta blight disease was examined. Chickpea lines CA-252 and CA-255 with partial resistance to Ascochyta blight disease could be grown in winter, when environmental conditions were highly conducive for disease development. Infected plant debris formed an important inoculum source for the subsequent chickpea crop. Application of nonamended suspension of the yeast species *Aureobasidium pullulans* to the postharvest debris resulted in reduction in Ascochyta lesions by 37.9% and 38.4% on chickpea plants relative to the untreated controls in 2004–2005 and 2006–2007 seasons, respectively. Ascospores released from debris were identified as *Didymella rabiei* accounting for high proportion of airspora released from chickpea debris. The results indicated that combination of strategies of host resistance and biological disinfestation of plant debris for restricting the incidence of Ascochyta blight disease of chickpea could be adopted (Dugan et al. 2009).

5.2.5.3 Damping-Off Disease

Damping-off disease caused by *Pythium ultimum* is responsible for seedling mortality in large numbers under favorable environmental conditions. The effectiveness of seed treatment with fungicides and BCAs was assessed, using Kabuli and desi chickpea types. BCAs *Bacillus pumilus* GB34 (Yield Shield), *B. subtilis* GB03 (Kodiak), *B. subtilis* MBI600 (Subtilex), *S. lydicus* WYEC108 (Actinovate), *S. griseoviridis* K61 (Mycostop), *Trichoderma harzianum* Rifai strain KRL-AG2 (T-22 Planter Box) and the fungicides fludioxonil (Maxim) and mefenoxam (Apron XL LS) were applied alone and in combination for assessing their efficacy in reducing the damping-off disease under greenhouse and field experiments. The desi cultivar had lower level of incidence of damping-off in greenhouse and field trials than Kabuli cultivar. Many biological seed treatments inhibited germination and growth of Kabuli cultivars in the absence of *P. ultimum* in the greenhouse but not in the field trials. Mefenoxam was the most effective seed treatment in the field trials at three locations in Montana, USA. Biological seed treatments were ineffective in reducing disease incidence even when combined with fungicides. The results indicated the usefulness of mefenoxam seed treatment for the control of chickpea damping-off disease (Leisso et al. 2009).

5.2.6 PEANUT DISEASES

5.2.6.1 Root Rot Disease

Peanut (groundnut) is severely affected by root rot disease caused by *Macrophomina phaseolina* or *Fusarium solani* in India and Argentina. Root rot disease affects peanut plants at any stage of crop growth. Infection is initiated in the roots which turn dark brown and the vascular tissues are invaded soon, resulting in restriction of supply of water and nutrients to the aerial plant parts. The affected plants are killed in due course (see Figure 5.4) (Narayanasamy). *Rhizobium*

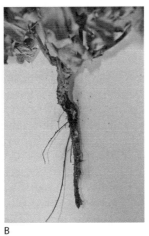

A   B

FIGURE 5.4 *Macrophomina phaseolina* causes root rot disease in peanut A: infected roots turn dark brown and infected plants are killed later B: close up view of infected root showing presence of sclerotia.

[Courtesy of Dr. P. Narayanasamy, Author]

spp. are known to fix atmospheric nitrogen following formation of root nodules in the leguminous crops and seed treatment of rhizobia is a regular practice to improve plant growth and consequently yield of peanut. The compatibility of the bacterial biocontrol agent *P. fluorescens* Pf1 effective against *M. phaseolina* with *Rhizobium* strain TNAU-14 was assessed under greenhouse and field conditions. Seed treatment and soil application of the BCA and *Rhizobium* effectively reduced root rot incidence and enhanced plant growth also (Shanmugam et al. 2002). The influence of crop rotation (maize, soybean and peanut), tillage systems (no-till, reduced and conventional tillage) on potential BCAs effective against *Fusarium solani* was assessed in two long-term field experiments in Argentina. Populations of BCAs were higher in both trials, when maize was the preceding crop and crops were under conservation tillage. However, the relationship between cultural management and root rot incidence was weak. When maize was the preceding crop, and peanut was under no-till, disease incidence was higher, under the same management conditions in the second location. Analysis of soil samples taken at sowing and harvest and root rot incidence in peanut showed that root rot incidence in peanut and populations of potential BCAs was negatively correlated under no-till. The BCAs identified were actinomycetes, *Trichoderma* spp. *Gliocladium* spp. Incidence of root rot was low under no-till and disc harrowing was associated with high populations of plant-associated bacteria. However, no such correlations could be seen, when soybean preceded peanut. The incidence of root rot was low, in spite of the presence of two populations of BCAs in the soil compared with that in maize as previous crop (Gil et al. 2008, 2010). In a later investigation, the effectiveness of *B. subtilis* formulation applied as seed treatment (10 g/kg) and soil incorporation (2.5 kg/ha) followed by foliar application of leaf extract of *Adhatoda vasica* (10%) at 30, 40 and 60 days after sowing reduced peanut root rot caused

by *Macrophomina phaseolina* and also another important disease, late leaf spot caused by *Phaeoisariopsis personata*. In addition, the combined treatment increased the pod yield significantly under field conditions (Shifa et al. 2018).

5.2.6.2 Sclerotinia Stem Blight Disease

Sclerotinia stem blight disease of peanut is caused by *Sclerotinia minor* which overwinters in soil as sclerotia that are viable for several years. Long-term crop rotations with cotton and maize (corn) were effective only to a limited extent, as the sclerotia were viable for 4 years (Melzer et al. 1997). An advanced breeding line N92056 C and cultivars Tamrun 98 and Perry showed moderate to high levels of tolerance to *S. minor* and produced high yields, compared with the susceptible cv. NC7. Acibenzolar-S-methyl (ASM) was applied as foliar spray along with the fungicide fluazinam. The fungicide suppressed the disease development effectively and the yield was also increased. Application of ASM was not effective in enhancing resistance of peanut plants to stem blight disease (Lemay et al. 2002). The effectiveness of soil application of *C. minitans* (as Contans WG) for suppressing the development of Sclerotinia blight by eliminating sclerotia in the soil was assessed. Field experiments were conducted for 5 years, including eight short-term experiments. The number of sclerotia of *S. minor* in the soil and incidence of the disease were reduced, compared with untreated control. Integration of application of *C. minitans* at 2 kg/ha in consecutive years with planting moderately resistant peanut cultivars and fungicide application might effectively reduce incidence of Sclerotinia stem blight disease (Partridge et al. 2006).

5.2.6.3 Stem Rot Disease

The effect of integration of cultural practices, fungicide application and use of biocontrol agents was investigated for reducing incidence of stem rot disease caused by *Sclerotium rolfsii* in South Africa. The fungicides difenconazole, carbendazim, flusilazole and chlorothalonil alone or in combination with *Trichoderma harzianam* were evaluated for their efficacy. Difenconazole significantly reduced the growth rate of *S. rolfsii*, but not of *T. harzianum*. Use of inversion plough significantly reduced infection of peanut plants by *S. rolfsii*. Lower plant density increased the disease incidence in the field. The results indicated that combined application of difenconazole and *T. harzianum* could be an effective strategy to reduce stem rot disease incidence (Cilliers et al. 2003). The effect of planting patterns (single or twin rows) and seeding rate on the efficacy of the fungicide azoxystrobin was investigated. Stem rot incidence was significantly greater in single rows planted at high seeding rates tested. Row pattern did not affect azoxystrobin efficacy and disease incidence was reduced in twin rows treated with fungicide, to one half (50%) of the incidence in single rows treated with fungicide. Symptoms were more severe in single than in twin rows and at higher seeding rates. The results suggested that plant spacing/density could be a critical factor affecting the efficacy of fungicides and suitable spacing should be adopted to realize the benefit of fungicide application (Sconyers et al. 2007).

5.2.6.4   Collar Rot Disease

Aspergillus niger or *Sclerotium rolfsii* may be involved in the collar rot disease of peanut in different locations. *S. rolfsii* is also seedborne, in addition to being soilborne and the disease is also called as southern blight disease in the United States. *Pseudomonas aeruginosa* GSE18 and GSE19 strains isolated from six habitats of peanut, were more effective than other isolates (391) tested, in suppressing the development of *A. niger*. When applied as seed treatment, strain GSE18 was more efficient than the strain GSE19 and reduced preemergence rotting and postemergence wilting by 60%. The strain GSE18 effectively colonized peanut plant rhizosphere both in the native and infested soils. *P. aeruginosa* strain GSE18 was tolerant to thiram and in combination with the fungicide provided more effective protection against collar rot disease. The results indicated that the BCA and thiram could be integrated to derive greater level of protection to peanut plants against infection by *A. niger* (Kishore et al. 2005).

5.2.7   CANOLA DISEASES

5.2.7.1   Sclerotinia Stem Rot Disease

Sclerotia of *Sclerotinia sclerotiorum* in the soil and plant debris have a long period of viability. Hence, crop rotation has limited applicability for disease suppression to a significant extent. Compatibility between application of biocontrol agents and fungicides was investigated. *Coniothyrium minitans* alone, *C. minitans* + benomyl (50 µg/µl), *C. minitans* + vinclozolin (100 µg/ml) and vincolzolin (500 µg/ml) alone were evaluated during 1997–2004. All treatments, applied as aerial sprays on leaves, reduced stem rot incidence significantly under field conditions. No difference in the efficacy of the treatments was discernible (Li et al. 2006). In a later investigation, the compatibility between *C. minitans* and commercial compound NPK fertilizer at different concentrations significantly reduced the number of apothecia produced by sclerotia of *S. sclerotiorum* in both pot and field experiments. No adverse effects of compound fertilizer were observed on the ability of *C. minitans* to infect sclerotia of *S. sclerotiorum* in vitro or to suppress carpogenic germination of sclerotia of *S. sclerotiorum*. The results showed that the BCA and the fertilizer could be combined and the integration of the strategies would save labor, cost and increase production efficiency of oilseed rape (Yang et al. 2011).

5.2.7.2   Rhizoctonia Root Rot Disease

Root rot caused by *R. solani* may affect canola and lupin plants at different stages of crop growth. The effectiveness of host resistance, fungicide application and treatment with BCAs was assessed under glasshouse conditions. Canola (8) and lupin (8) cultivars/selections were evaluated for their levels of resistance to damping-off and root rot caused by *R. solani* AG2-1, 2-2, 4 and 11. All canola cultivars were highly susceptible to AG2-1, but Rocket, Spectrum and 44C11 were more tolerant than other cultivars. Spectrum and 44C73 were more resistant to AG-4 than other canola cultivars.

Among cultivars of lupin, cv. Cedra 6150, E 16, Mandelup and Quilinock were more resistant to AG-4 than other cultivars/selections. Seed treatment with fungicide Cruiser OSR (difenconazole, fludioxonil, metalaxyl-M, thiamethoxam) and SA-combination (iprodione, metalaxyl, thiram) significantly protected canola and lupin seedlings, as reflected by higher seedling survival percentages and decrease in hypocotyl/root rot and increase in healthy seedlings. SA-combination was more effective than Cruiser OSR. The binucleate nonpathogenic *Rhizoctonia* AGs (CA, Bo, K and I) increased survival of lupin seedlings inoculated with *R. solani* AG2-2 and 4. The strains AG I and K were effective in protecting canola plants inoculated with AG-4. The results indicated that combination of cultivars tolerant to *R. solani*, seed treatment with fungicides and application of binucleate Rhizoctonia AGs might suppress more effectively the damping-off and root rot disease caused by *R. solani* in canola and lupin (Lamprecht et al. 2011).

5.2.7.3   Blackleg Disease

Leptosphaeria maculans and *L. biglobosa* are associated with canola blackleg disease in many countries, accounting for significant production losses. *L. biglobosa* is weakly virulent and placed in pathogenicity group 1 (PG-1), while *L. maculans* is aggressive, highly virulent and classified into pathogenicity groups PG-2, PG-3 and PG-4. The results of field experiments suggested that isolates of PG-1 could be employed to reduce the severity of blackleg induced by isolates of *L. maculans* (Chen and Fernando 2006). In a later investigation, the contribution of components of integrated disease management system was investigated. Canola once in the 4-year rotation comprising pea, wheat and flax was established in a split-plot design with fungicide application as subplot treatments. Blackleg incidence and severity was increased in rotations comprising more than one canola crop every 4 years, regardless of cultivar resistance, although blackleg resistant cultivar was affected to a much less extent than the susceptible cultivar. Fungicide application had limited value to maintain yield levels. The results suggested that the resistant canola cultivar could be grown once in 4 years with limited yield reduction. However, with increased amount of infected residue added to the soil in canola rotations, severity of infection was intensified, due to increases in the risk of inoculum carryover. Hence, one crop of canola in 4 years might be a practical proposition for the management of blackleg disease (Kutcher et al. 2013).

5.2.7.4   Clubroot disease

Clubroot disease incited by *Plasmodiophora brassicae* emerged as a serious constraint threatening canola production in western Canada. Various strategies for disease containment such as limiting the soil pH, crop rotation with nonhosts and bait crops, adjustment of sowing/planting dates, sanitation of farm equipments and deployment of canola cultivars tolerant/resistant to clubroot disease have been examined for their applicability and integration to enhance the effectiveness of protection against the disease. However, most of the strategies proved to be economically not feasible, although some of them were able to reduce disease incidence in areas meant for intensive production for short periods. Genetic resistance combined with appropriate cultural practices that prevent accumulation of resting spores may be adopted. Further, infested soil carried on both machinery and seed should be curtailed by applying sanitation methods to avoid movement of the pathogen within and/ or outside the field (Hwang et al. 2014).

5.3 MANAGEMENT OF DISEASES OF HORTICULTURAL CROPS

Cultivation of horticultural crops requires many practices, that are different from those applied for agricultural crops, for production of seeds/propagules and nutrient requirements and crop protection measures vary considerably. Many of them are multiplied via asexual propagation methods using grafting and budding in the nurseries or raised in the nurseries and later transplanted in the main fields. As they are high-value crops, fungicides and other chemicals are more intensively applied than in the case of agricultural crops, resulting in possible development of resistance in the microbial pathogens to chemicals.

5.3.1 TOMATO DISEASES

5.3.1.1 Damping-Off Diseases

Tomato seedlings in the nursery are affected by preemergence and postemergence damping-off diseases caused by *Pythium* spp., *R. solani* and *Sclerotium rolfsii*. Integration of host resistance, deployment of biocontrol agents and application of fungicides has been suggested for effectively reducing the incidence and severity of damping-off disease. In the rockwool system, seed treatment with bacterial biocontrol agents *P. fluorescens, P. putida, P. marginalis, P. corrugata and P. viridifolora* reduced the incidence of damping-off caused by both *P. aphanidermatum* and *P. ultimum* (Gravel et al. 2005). Soil application of *B. subtilis* RB14-C provided protection to the tomato seedlings against damping-off caused by *R. solani* (Szezech and Shoda 2006). For the suppression of development of damping-off caused by *P. aphanidermatum*, application of *P. fluorescens* isolate CW2 in combination with fungicides azoxystrobin, metalaxyl-M or pyraclostrobin was evaluated. The fungicides were fungitoxic to *P. ulimum*, but did not inhibit the multiplication of the BCA in in vitro assays. Under greenhouse conditions, the fungicides were applied at 5 and 10 μg/ml concentrations. The degree of effectiveness varied significantly depending on the fungicide. Combined application of *P. fluorescens* and fungicides resulted in significant enhancement of disease suppression and plant growth as well, as indicated by shoot and root dry weights. Metalaxyl-M alone, or in combination with *P. fluorescens* was more effective in protecting tomato seedlings than other fungicides (Salman and Abuamsha 2012). *R. solani* and *Sclerotium rolfsii* affect seedlings in the nursery and also mature plants after transplantation in the main field. Two bacterial isolates *Brukholderia cepacia* T1A-2B and *Pseudomonas* sp. T4B-2a

were applied along with two commercial biofungicides containing *B. subtilis* (BSF4) and *Trichoderma asperellum* (Tv1). The BCAs were applied in the soil, proximal to the plant crowns and main roots via drip irrigation system under field conditions for 2 years. The fungicides tolchlofos-methyl, azoxystrobin, fosetyl-Al and fosetyl-Al + propamocarb were also applied like the BCAs. The BCAs were equally effective as the fungicides, except tolchlofos-methyl, which was the most effective among all treatments tested. Application of chemicals and BCAs via drip irrigation system was found to be an effective delivery system for applying chemicals and BCA preparations (De Curtis 2010).

5.3.1.2 Fusarium Wilt Disease

Fusarium oxysporum f.sp. *lycopersici (Fol)* and *F. oxysporum* f.sp. *radicis-lycopersici* (FORL) cause wilt and root and crown rot diseases, respectively, in tomato. Different strategies, such as solarization, use of BCAs and fungicide application have been evaluated for their efficacy in reducing disease incidence/ severity. Soil solarization using photo-selective low-density polyethylene film for a period of 32 – 92 days prior to planting tomatoes significantly reduced population density of *Fol* and FORL in the upper 5 cm of soil. The increases in temperatures in the solarized soils at depths of 5, 15 and 25 cm were 5.7, 7.1 and 5.0°C, respectively, over nonsolarized soil (Chellemi et al. 1994). In a later investigation, plastic films were found to be more effective and consistent in increasing soil temperature, compared to biodegradable plastics. Plastic films used for solarization significantly suppressed weed development, in addition to the negative impact on *Fol* population. Although less effective, biodegradable solarizing materials may have to be considered seriously, because of the difficulties associated with disposal of plastic materials (Bonanomi et al. 2008). Application of *Streptomyces griseoviridis* strain K161 formulated as Mycostop® alone or combined with soil solarization was evaluated for suppressing the development of *F. oxysporum* under field conditions in 2001 and 2002. The biofungicides were very effective in suppressing *F. oxysporum* f.sp. *lycopersici (Fol)* in artificially infested soil. Soil spraying was more effective than soil irrigation for the control of *Fol*, but not against *F. oxysporum* f.sp. *radicis-lycopersici* (FORL), when applied alone. Soil solarization effectively controlled *Fol*, but it was slightly less effective, when combined with *S. griseoviridis* application. However, possible additive effect of solarization and BCA application might be derived, as indicated by significant increase in tomato fruit mass and higher yield (Minuto et al. 2006).

As the use of fungicides for the control of tomato Fusarium wilt disease was unavoidable, the effect of combined application of biocontrol agents tolerant to fungicide and fungicide was determined. The nonpathogenic strain *F. oxysporum* strain CS-20 reduced the incidence of Fusarium wilt disease of tomato significantly. The compatibility of *F. oxysporum* CS-20 with fungicides recommended for tomato was assessed. Mefenoxam and mefenoxam + copper did not affect the development of the BCA. Other fungicides azoxystrobin, chlorothalonil, mancozeb, mancozeb + copper inhibited the growth of the strain CS-20 in vitro assays, but they were less toxic in greenhouse tests. The results suggested that the biocontrol agent could be combined with the fungicide(s) that have inhibitory effect only on the pathogen to enhance control efficiency that could not attained individually (Fravel et al. 2005). The biocontrol potential of two BCAs, *Bacillius megaterium* (C96) and *Bacillus cepacia* (C91) against *F. oxysporum* f.sp. *radicis-lycopersici* (FORL) was assessed. In the initial assessments, these isolates reduced disease incidence by 75% and 88%, respectively. Both strains of BCA were highly tolerant to the fungicide carbendazim. Application of carbendazim alone at > 75 µg/ml reduced disease symptoms by over 50% under inoculated conditions. Combined application of carbendazim at a low concentration (1 µg/ml) in combination with *B. cepacia* C91 reduced disease symptoms by 46%, compared with the BCA alone. Combination of *B. megaterium* C96 and carbendazim at 10 µg/ml significantly reduced symptoms by 84%, compared with inoculated controls and by 77%, compared with carbendazim treatment alone. The combined treatment slightly outperformed application of carbendazim at 100 µg/ml and BCA applied without fungicide also provided acceptable level of protection (Omar et al. 2006).

5.3.1.3   Late Blight Disease

Late blight disease caused by *Phytophthora infestans* has negative impact on the yield and quality of tomatoes especially in the greenhouse crops. Soil solarization using polyethylene mulch provided consistent, effective and highly significant suppression of late blight disease development, the control efficacy being 83.6 ± 5.6%. The combined effect of polyethylene mulch and fungicide was additive. The type of polyethylene (bicolor aluminized, clear or black) did not affect the efficacy of suppression of late blight disease. The disease suppressive effect of polyethylene mulch appeared to be due to a reduction in leaf wetness duration. Further polyethylene mulching might result in reduced sporulation by *P. infestans*. Additional benefits of prevention of weed infestation and saving up to 30% of irrigation water also were realized by tomato growers in Israel (Shtienberg et al. 2010).

5.3.1.4   Bacterial Wilt Disease

Bacterial wilt disease of tomato is caused by *Ralstonia solanacearum*, which has a wide host range including many crops and weeds that can serve as sources of inoculum, resulting in epidemics. Different combinations of disease management strategies have been evaluated for their compatibility and enhancement of level of control of the bacterial wilt disease. The effect of combining the bacterial BCAs, plant resistance inducer and soil amendment with S-H mixture containing agricultural and industrial wastes such as bagasse, rice husk, oyster shell powder, urea, potassium nitrate, calcium superphosphate and mineral ash, was assessed. *Pseudomonas putida* 89B61 strain significantly reduced disease incidence, when applied to the transplants at transplantation and one week prior to inoculation with *R. solanacearum* race 1, biovar 1. Combination of acibenzolar-S-methyl (Actigard)

and *P. putida* reduced bacterial wilt incidence, compared with untreated control. Combination of ASM and S-H mixture also reduced disease incidence significantly in two experiments (Anith et al. 2004). The efficacy of combining chitosan, a known resistance inducer and bacterial BCA, *Paenibacillus polymyxa* MB02-1007 in suppressing bacterial wilt disease development was assessed. Chitosan was applied as seed treatment at 10 mg/ml for 2 h, followed by overnight drying or soil drench at 10 mg/l significantly reduced disease incidence by 48% and 72%, respectively. *P. polymyxa* as seed treatment and soil drench reduced disease incidence by 88% and 82%, respectively. Both chitosan and the BCA promoted plant growth (Algam et al. 2010).

The effectiveness of ASM applied on susceptible and moderately resistant tomato cultivars in reducing bacterial wilt disease, was assessed. ASM significantly enhanced the level of resistance in moderately resistant cvs. Neptune and BHN466, but it was ineffective in susceptible cvs. Equinox and FL47. The results suggested that ASM-mediated resistance might be partially due to prevention of internal spread of *R. solanacearum* to upper stem tissues of inoculated tomato plants. Similar results were obtained under field conditions also, where the ASM treatment of moderately resistant cultivar resulted in higher yield of tomato fruits, indicating the effectiveness of combining host resistance and resistance inducer ASM (Pradhanang et al. 2005). The efficacy of integrating ASM and the essential oil, thymol as preplant fumigant was assessed for suppressing the development of bacterial wilt disease. Thymol was applied, after infestation with *R. solanacearum* through drip irrigation lines under polyethylene mulch at the rate of 73 kg /ha during 2004–2005. ASM was applied as foliar spray at 25 mg/l. Thymol application significantly reduced disease incidence to 26% and 22.6%, respectively, in 2004 and 2005, as against > 95% infection in untreated control plots. Combined treatment with thymol and ASM reduced bacterial wilt disease and also the root knot nematode, *Meloidogyne arenaria* population. The disease suppression in tolerant tomato cultivar 7514 was significant, resulting in increased yield of tomato (Ji et al. 2007; Hong et al. 2011).

The potential of combining soil solarization and the biocontrol agent, *P. fluorescens* for suppressing bacterial wilt disease development was investigated. Soil solarization using transparent plastic sheets as mulch during summer months effectively reduced disease incidence. *P. fluorescens* preparation was incorporated uniformly in soils infested with *R. solanacearum*, followed by solarization for 8–10 weeks at two locations in Himachal Pradesh, India. Soil temperature was increased by 8.9°C and 10°C in those two locations. The BCA population increased in the solarized soil in the first year and declined subsequently. The incidence of bacterial wilt was reduced by 43 to 63% in solarized soil. A progressive decline in disease incidence was observed with increase in the increase of duration of solarization. A period of 10 weeks was the optimum duration of solarization to effectively reduce the terminal wilt phase of bacterial wilt disease (Ambadar and Sood 2010). The combination of host resistance and soil amendment (SA) was evaluated for its efficacy in reducing tomato bacterial wilt incidence. Amendment with urea or slaked lime alone or combined was evaluated at 30°C. The mixture of urea and slaked lime showed the best suppressive effect in three soils and also under field conditions. Resistant eggplant (EG203) and tomato (Hawaii 7996) rootstocks, selected based on stable resistance against representative strains of *R. solanacearum* at seedling stage, significantly reduced disease incidence under field conditions. EG203 grafted plants showed disease incidence up to 2.8%, as against 24.4 to 92.9% in nongrafted plants. Integration of use of Hawaii 7996 as the rootstock and soil amendment provided greater level of protection against bacterial wilt disease of tomato (Lin et al. 2008). In a later investigation, integration of host resistance and use of biocontrol agent was examined in order to improve the effectiveness of control of bacterial wilt disease in tomato. Two endophytic bacterial isolates of *Bacillus* sp. and *Serratia marcescens* were introduced into tomato seedlings via cutting of hypocotyl, substrate drenching and seed microbiolization procedures. The bacterial isolates introduced into tomato cultivars Drica, Caraibe, Yoshimatsu and Santa Clara protected the plants against bacterial wilt disease, as reflected by reduction in AUDPC values of the disease by 16 to 65%. The results indicated that the endophytic bacteria in combination with resistance of tomato cultivars could be employed for effective suppression of bacterial wilt disease in tomatoes (Barretti and Mage 2012).

5.3.2 POTATO DISEASES

5.3.2.1 Late Blight and Tuber Blight Disease

Phytophthora infestans, a pathogen of historical importance, can infect both aerial plant organs and also tubers formed in the soil, causing epidemics of late blight and tuber blight disease in potato. The pathogen is tuberborne, soilborne, waterborne and airborne, proving its versatility as pathogen with high potential. Different disease management strategies have been applied alone or in different combinations to restrict disease incidence and spread of the disease via forecasting models, genetic resistance, cultural practices and chemical application with varying degrees of success. Use of disease-free certified seed tubers is the basic step to effectively eliminate initial sources of inoculum of *P. infestans* (Frost et al. 2013). Crop sanitation involving elimination of all infected plant debris, volunteer plants, weed hosts and proper disposal of the plant materials has to be strictly applied. Choosing appropriate planting date resulted in significant reduction in the incidence of late blight under natural conditions in West Bengal, India (Basu 2009). Strip-cropping with cereal or grass-clover mix significantly reduced disease incidence by 4 to 20%, depending on the plot size (Bouws and Finckh 2008). Different kinds of polyethylene mulches were evaluated for their efficacy in reducing tuber infection by *P. infestans*. Black polyethylene film and copper hydroxide-treated agricultural textile reduced tuber blight incidence, compared with the control (Glass et al. 2001). Fungal and bacterial BCAs have been evaluated for their efficiency in suppressing

late blight and tuber blight disease. Combined application of *Pseudomonas putida* and *Serratia plymuthica* effectively suppressed development of late blight disease through different mechanisms. *Phytophthora infestans* US-8 clonal lineage was highly aggressive. *P. putida* induced resistance in potato plants against the pathogen, whereas *S. plymuthica* inhibited the pathogen development via antibiosis (Daayf et al. 2003). *Trichoderma harzianum* was the most effective among six antagonists tested, in reducing the infection of tubers by *P. infestans*, causing tuber blight under field conditions. Tuber treatment with *T. harzianum* and *Pseudomonas* provided effective protection to potato plants and tubers against infection by *P. infestans* (Basu 2009). The effectiveness of resistance inducer, ß-aminobutyric acid (BABA) applied alone or in combination with fungicides for reducing late blight incidence was reported by Liljeroth et al. (2010). Phosphites (Phi), alkali salts of phosphorous acid, when applied as foliar spray reduced late blight significantly (Lobato et al. 2008).

5.3.2.2 Verticillium Wilt Disease

Verticillium wilt disease is induced by *Verticillium dahliae*, which is capable of infecting several crops and weed species with potential to serve as sources of inoculum for potato crops. Use of fumigants and potato cultivars resistant to the disease has been the principal strategies for containing the Verticillium wilt disease. As the fumigation with chemicals is constrained, application of organic amendments to limit the disease incidence/severity gained importance. The sweet corn cultivars Jubilee Sweet corn and Jubilee Supersweet corn, applied as green manure, reduced disease incidence by 60 to 70%. These corn cultivars did not have direct influence on the pathogen, but colonization of on potato feeder roots and in potato tissues of stem pieces by *V. dahliae* was significantly reduced. Feeder root colonization was positively correlated with Verticillium wilt disease incidence and negatively correlated with yield. Further, populations of *Ulocladium* and *Fusarium equiseti* with antagonistic activity were stimulated (Davis et al. 2010a). Austrian peas, Sudangrass, oilseed rape, oats and rye as green manures, exerted similar beneficial effects by reducing disease incidence and increasing potato yields (Davis et al. 2010b). *P. fluorescens* biotype F isolate DF37 and the extract of Canada milkvetch effectively reduced disease incidence under field conditions was reported by Uppal et al. (2008).

5.3.2.3 Stem Canker and Black Scurf Disease

Rhizoctonia solani, incitant of potato black scurf disease, has wide host range, including many crops, which may provide inoculum for infecting potato crops. The 2-year rotation including spring barley and 3-year rotation including barley and red clover in addition to potato were evaluated. Disease incidence/severity was reduced to a greater extent in the 3-year rotational soils, compared with 2-year rotational soils (Peters et al. 2003). In a later investigation, a cropping system was formulated specifically for suppressing soilborne diseases affecting potato. The disease suppressive (DS) system included diverse crops such as *Brassica* spp. and Sudangrass, green manure crops and fall cover crops. High crop diversity resulted in greatest reduction in stem and stolon canker and black scurf caused by *R. solani* and also common scab disease induced by *Streptomyces scabiei* under both irrigated and nonirrigated conditions, compared to other systems of soil management (Larkin et al. 2011). Compatibility of biocontrol agents and chemicals was examined for suppressing development of *R. solani*. The pathogen *R. solani*-specific chemicals, pencycuron and flutalonil, when combined with the BCA *Verticillium biguttatum* showed additive effects on black scurf disease control. Combination of *V. biguttatum* and cymoxanil or propamocarb resulted in reduction of incidence of black scurf, as well as tuber rot caused by *Pythium* sp. or Phytophthora sp. The results showed *V. biguttatum* was compatible with pathogen-specific fungicides or might extend its control spectrum in combination with oomycete-specific fungicides (van den Boogert and Luttikholt 2004).

5.3.2.4 Scab Diseases

Potato scab and common scab diseases are due to infection by *Streptomyces scabies* and *S. turgidiscabies*, respectively. Effects of green manures (buckwheat and canola) and crop sequences (rotations) on potato scab disease incidence were assessed. Potatoes grown in soil planted to corn or alfalfa in the previous year had lower severity scab disease and increase in yield was an additional benefit of the green manure incorporation in soil. The streptomycete community that had antagonistic activity reached higher population levels in green manure-applied plots than ones left as fallow. Possibly, green manures might selectively enrich the abundance or activity of antibiotic producers within soil microbial community (Wiggins and Kinkel 2005). The effectiveness of green manures in suppressing development of common scab caused by *S. turgidiscabies* was assessed under field conditions. Lopsided oat followed by lopsided oat or woolly pod vetch was significantly more effective in suppressing the disease severity than oat and continuous potato cultivation. The results suggested that lopsided oat could be raised as fallow green manure for reducing common scab severity and increasing tuber yield (Sakuma et al. 2011).

5.3.2.5 Blackleg and Soft Rot Disease

Bacterial pathogens *Pectobacterium atrosepticum (Pa)*, *P. carotovorum* subsp. *carotovorum (Pcc)* and *Dickeya* spp. cause blackleg of potato plants and soft rot of potato tubers. *Pa* and *Pcc* cause blackleg symptoms, whereas all three pathogens induce soft rot symptoms in tubers. D. *dianthicola* is associated with blackleg in tropical and subtropical regions, whereas *P. carotovorum* subsp. *brasiliensis* was reported to be responsible for blackleg in Brazil and *P. carotovorum* subsp. *wasabiae* as the incitant of blackleg in New Zealand. As different pathogens are involved in the blackleg disease of potato in different geographical locations, the effects of different disease management strategies are variable depending on the location. However, methods based on avoiding contamination during harvest, storage and handling and use of certified disease-free seed tubers are emphasized to avoid introduction and buildup of inoculum in the soil during crop growth. To ensure freedom from infection, testing tubers should include peel tissues to detect lenticels and wound infections and the stolon end including vascular tissues using sensitive detection techniques. Crops should be dedicated, as seed and suitable tolerance levels have to be applied for certification. Traditional breeding methods have not been successful in developing potato cultivars with resistance to *Pectobacterium* spp. and *Dickeya* spp. Although several bacterial species were tested for their efficacy in suppressing soft rot of tubers, none of them was found to be effective to be advanced to commercial production. Until cultivars with acceptable levels of resistance are produced, disease suppression has to depend primarily on avoiding contamination of seed tubers and use of disease-free seed tubers (Czajkowski et al. 2011).

5.3.3 Pepper Diseases

5.3.3.1 Verticillium Wilt Disease

The efficacy of the bioproduct Polyversum® containing *Pythium oligandrum* and the fungicides benomyl and propamocarb hydrochloride was assessed for suppressing the development of *Verticillium dahliae*. Benomyl was the most effective in reducing disease development by 94.6% and 88.2%, respectively, when applied as pre- and postinoculation treatments. Polyversum was more effective (66.6%), when applied prior to inoculation and its effectiveness was equal to that of propamocarb hydrochloride, but less effective than benomyl (Rekanovic et al. 2007). In a later investigation, combined application of dazomet (DZ) and *Trichoderma asperllum* provided consistent protection against Verticillium wilt disease and also increased the yield by 40.1%, resulting in higher return to the grower (Slusarski and Pietr 2009).

5.3.3.2 Crown and Root Rot Disease

The efficacy of combination of host resistance and fungicides in suppressing the development of crown and root rot caused by *Phytophthora capsici* was assessed. Pepper cultivars Paladin and Red Knight resistant and susceptible, respectively, were inoculated with *P. capsici*, and the fungicides were applied as foliar sprays at 7- or 14-day intervals in the greenhouse. Paladin plants had less mortality than Red Knight. Under field conditions, fluopicolide- or mandipropamid-treated plants showed less mortality than the untreated control plants of both varieties. Interactions among fungicide-cultivars, application method-cultivar, application method-fungicide, fungicide-cultivar-application method were significant. Treatments applied as drenches were more effective than foliar spray. Combining resistant pepper cultivar and fungicide applied as drenches might be an effective option for suppression of crown and root disease caused by *P. capsici* (Foster and Hausbeck 2010).

5.3.4 Cucurbit Fusarium Wilt Disease

Fusarium wilt and root rot disease of cucurbitaceous crops are caused by formae speciales of *F. oysproum* f.sp. *radicis-cucumerinum* (FORC) infects cucumber, while *F. oxysporum* f.sp. *melonis* (FOM) infects melon. The disease suppressive effects of soil amendment were assessed. Greenhouse compost (GC) significantly reduced cucumber plant mortality due to FORC to 13%, as against 63% in untreated control treatment. The compost was as effective as the fungicide treatment (Punja et al. 2002). Strains of *Pseudomonas aeruginosa* isolated from the composts showed high antagonistic activity against FORC. The internal stem colonization by *F. oxysporum* f.sp. *radicis-cucumerinum* of cucumber plants, treated with the BCA, was reduced (Bradley and Punja 2010). Citrus wastes (60%) amended with *Trichoderma harzianum* T-78 reduced watermelon Fusarium wilt significantly and also promoted growth of melon plants. The results indicated that the combination of citrus wastes and *T. harzianum* could be a viable alternative to peat, and minimize fungicide use for the management of melon wilt disease (Lopez-Mondejar et al. 2010). Soil solarization for 5 years consecutively alone or in combination with calcium cyanamide as soil amendment was evaluated for suppressing Fusarium wilt of melons. Solarization reduced population of FOM in sterile soil and disease incidence by 82 to 90% in 3 of 5 trials. Reduction in disease incidence (%) was proportional to the duration of solarization, when the soil temperatures remained at above 42°C at 25 cm depth. Calcium cyanamide amendment at 80 g/m² did not have any influence on the efficiency of soil solarization (Tamietti and Valentino 2006).

5.3.5 Allium White Rot Disease

Sclerotium cepivorum causes the white rot disease in Allium crops with worldwide distribution. Soil solarization was highly effective in suppressing the white rot disease by reducing soil populations of *S. cepivorum* to negligible levels and the beneficial effect was extended to the second crop also with significant improvement in yield. Spraying tebuconazole to the stem bases was equally effective as soil solarization. Combination of soil solarization and tebuconazole application resulted in almost doubling of yield level and increase in bulb quality, even under high disease pressure conditions (Melero-Vara et al. 2000). In a later investigation, an integrated disease management system was formulated by combining host resistance, fungal BCA and fungicide application for the control of Allium white rot (AWR) disease. Two isolates of *Trichoderma viride* L4 and S17A were evaluated for their efficacy using 23 bulb onion accessions or tebuconazole-based seed treatment or composted onion waste. All onion accessions were equally susceptible to AWR disease. However, when combined with S17A isolate of *T. viride*, disease incidence was reduced up to two-thirds in all accessions. Isolates L4 and S17A, tebuconazole or composted onion waste reduced the percentage of infection by 50%. Combination of *T. viride* with either tebuconazole or composted onion waste enhanced disease suppression and there was practically no incidence of AWR disease in some treatments. The results showed that use of *T. viride*, tebuconazole seed treatment and application of composted onion waste as soil amendment had potential for effectively suppressing the development of AWR disease caused by *S. cepivorum* (Clarkson et al. 2006). Compatibility of *Trichoderma viride* Karsten strain C52 with organic amendments, nitrogen fertilizer, fungicides and diallylsulphide (DADS, a stimulant of germination of sclerotia of *S. cepivorum*) was investigated. Addition of two blended pellet products containing poultry manure and other organic matter to sandy soil stimulated proliferation of *Trichoderma atroviride* in sandy soil. Urea at twice field application rate reduced mycelial growth as well as spore germination and elongation of germ tube of the BCA. However, *T. atroviride* was less sensitive to field rate of urea in the field soil. Volatiles at half field rate did not affect mycelial growth of *T. atroviride* C52 in vitro. When DADS was applied to the soil at 4, 6 and 8 weeks before application of *T. atroviride* C52, the BCA populations were not affected. The results indicated the possibility of integrating the compatible strategies to suppress the development of Allium white rot disease caused by *S. cepivorum* (McLean et al. 2012).

5.3.6 Sugar Beet Root and Crown Rot Disease

Rhizoctonia solani AG2-2 causes crown and root rot disease of sugar beet, affecting production seriously. Post emergence damping-off is also due to the pathogen under high soil moisture and water-logged conditions. The combined effect of application of the antagonist *Bacillus* sp. strain MSU-127 and fungicide on suppression of development of the disease was investigated under field conditions. The fungicides azoxystrobin and tebuconazole were applied in furrows as sprays at planting or as band sprays directed at the crown at the 4-leaf stage and 8-leaf stage, while the BCA was applied at 4-leaf stage only. The strain MSU-127 provided long-term protection to sugar beet plants equal to the low rate of azoxystrobin (7.6 kg a.i/ha) or tebuconazole in both years (1996 and 1997). The efficacy was increased by combined application of MSU-127 and azoxystrobin at low rate. Maximum reduction in disease development and higher increase in root and sugar yield (15.9%) could be achieved by combining the BCA and fungicide. The sugar yield was further increased to 23%, when the BCA-fungicide combination was applied twice at 4- and 8-leaf stages during growing season (Kiewnick et al. 2001). The effect of combination of host resistance and crop rotations on the incidence of root and crown rot of sugar beet caused by *R. solani* AG2-2 IIIB was investigated in four field trials. Crop rotation with various proportions of maize and continuous sugar beet crop were included in the trials. Within crop rotations, cultivation methods varied in the form of soil tillage, intercrops, or both sugar beet cultivars. Crop rotation had the main impact on disease severity and sugar yield. With increasing proportion of maize, sugar beet yield decreased, whereas cultivation method had only a minor impact. The results indicated that crop rotation of sugar beet with nonhost crop and growing resistant cultivar could decrease disease severity/incidence effectively (Buhre et al. 2009).

5.3.7 Cabbage Diseases

5.3.7.1 Yellows Disease

Cabbage yellows disease caused by *F. oxysporum* f.sp. *conglutinans* affects production level appreciably. The inhibitory activities of the bacterial biocontrol agent (BCA), *P.*

fluorescens strain LRB3W1 and the fungicide benomyl were directed toward the pathogen. Benomyl did not reduce the severity of the disease at low concentration (1 or 10 µg/ml). However, the disease severity was significantly suppressed, when benomyl at low concentration was combined with *P. fluorescens*. The combined treatment was more effective than when the BCA was applied alone. The survival of *P. fluorescens* was not affected by benomyl. The results showed that the lower levels of efficiency of the BCA and the fungicide could be improved by combining them, because of the compatibility with each other (Someya et al. 2007).

5.3.7.2 Damping-Off Disease

Damping-off of Chinese cabbage is caused by *R. solani*, which can infect many other crops also. Combining organic amendments and soil solarization resulted in reduced incidence of the diseases induced by them. The granular formulation, PBGG containing *Pseudomonas boreopolis*, Brassica seed pomace, glycerin and sodium alginate (1%) applied to the soil infested with *R. solani*, significantly reduced the percentage of colonization of cabbage seeds by the pathogen. In addition, PBGG also stimulated the proliferation of *Streptomyces padanus* strain SS-07 and *S. xantholiticus* strain SS09 with antagonistic activity against *R. solani*. Soil application of *S. padanus* or *S. xantholiticus* alone or in combination with 1% PBGG significantly reduced the percentage of colonization of cabbage seed by *R. solani*, compared to untreated control. *S. padanus* was the most effective in suppressing the development of damping-off disease. Under field conditions, soil amendment with PBGG (1%) alone or in combination with *S. padanus* or *S. xantholiticus* effectively reduced the incidence of damping-off disease. Treatment with PBGG or actinomycetes did not induce any adverse effect on cabbage plants. Application of PBGG with actinomycetes has the potential for large scale adoption for the management of damping-off disease caused by *R. solani* in Chinese cabbage (Chung et al. 2005).

5.3.7.3 Clubroot Disease

The green cabbage (*Brassica oleraceae* var. *capitata*) cultivars Kilaton, Tekila, Kilaxy and Kilaherb were resistant to the pathotype 6 of *Plasmodiophora brassicae* with disease incidence varying from 0 to 3.8%, while cv. Bronco was susceptible (64–100% infection). Application of the fungicide fluazinam reduced clubroot severity on cv. Bronco by 6% at one of the three sites tested. Cultivar resistance was more effective in reducing clubroot than application of fluazinam. Napa cabbage (*Brassica rapa* subsp. *pekinensis*) cultivars, Yuki, Deneko, Bilko, Emiko and China Gold were resistant to clubroot with infection percentage varying from 0 to 13%, whereas cv. Mirako was highly susceptible (87–92%). The results showed that cultivars with resistance to clubroot and desirable agronomic characteristics could be combined for reducing disease incidence (Saude et al. 2012).

5.3.8 Lettuce Diseases

5.3.8.1 Sclerotinia Rot Disease

Sclerotinia sclerotiorum, causing lettuce Sclerotinia rot disease, accounts for appreciable loss in production. The effect of iprodione on *C. minitans*, used as a biological control agent against *S. sclerotiorum*, was assessed. The soil was infested with an iprodione-tolerant strain of *S. sclerotiorum*, before planting lettuce crops in the glasshouse tests. Disease incidence was reduced significantly by *C. minitans* and it was further reduced by a single application of iprodione, regardless of whether the isolate of *C. minitans* was tolerant to iprodione or not. In another experiment, disease incidence could be effectively suppressed by a combination of *C. minitans* and a single application of iprodione. The combination provided protection to the level that was equivalent to application of prophylactic sprays of iprodione at 2-week intervals. The fungicide did not affect the ability of *C. minitans* to infect the sclerotia and to spread into the plots, where only the fungicide was applied. The results indicated the effectiveness of integration of soil application of *C. minitans* and foliar application of iprodione for effective suppression of lettuce Sclerotinia rot caused by *S. sclerotiorum* (Budge and Whipps 2001).

5.2.8.2 Verticillium Wilt Disease

Verticillium wilt disease caused by *Verticillium dahliae* may be spread through seeds of spinach grown as a rotation crop and introduced into the soil, where lettuce is cultivated. After introduction of *V. dahliae*, it becomes a problem for lettuce grower. Due to the reluctance of seed companies to take into account the effect on lettuce producers and clean spinach seeds, the externalities have important implications for the management of Verticillium wilt disease in lettuce crops. Planting broccoli or not planting spinach is the option available for lettuce grower. Management of such migratory pathogens and the diseases caused by them requires special attention. Application of cultural practices, such as use of clean seeds, crop rotation with tolerant/resistant crops and biofumigants may be effective in reducing disease incidence/severity (Carroll et al. 2018).

5.3.9 Banana Panama (Fusarium) Wilt Disease

Banana Panama (Fusarium) wilt disease caused by *F. oxysporum* f.sp. *cubense* (*Foc*) is one of the most destructive diseases affecting banana all over the world. Infection is initiated in the roots by soilborne inoculum and *Foc* invades the vascular tissues in internal tissues of corm. However, disease symptoms are expressed in the leaves which turn yellow and later pseudostem splitting is observed as the external symptoms. Vascular discoloration can be seen only when the corm is cut open (see Figure 5.5) (Selvarajan). As the pathogen spreads to long distances through infected corms/suckers (planting materials), it is very difficult to eliminate the asymptomatic planting materials. Preventive methods such as use of pathogen-free plantlets produced through tissue culture technique, restriction of movement of handlers and contaminated equipments

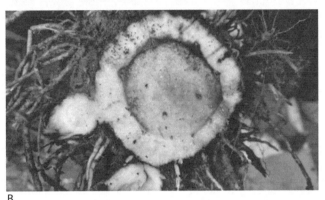

FIGURE 5.5 *Fusarium oxysporum* f.sp. *cubense* (*Foc*) causes Fusarium wilt disease of banana A: yellowing of leaves as external visible symptom of infection of banana plants by *Foc* B: presence of distinct vascular discoloration in corm as confirmatory internal symptom induced by *Foc*.

[Courtesy of Dr. R. Selvarajan, ICAR-National Research Centre for Banana, Tiruchirappali, Tamil Nadu, India]

and proper disposal of infected plants may reduce inoculum buildup in the soil. The effectiveness of integration of compatible biological and cultural methods was investigated. Application of nonpathogenic, endophytic *F. oxysporum (Fo)* strains, *Trichoderma harzianum* EcoT®, silicon application and mulching using macadamia alone or in combination were evaluated under greenhouse and field conditions. *Fo* strains were effective in reducing disease incidence. Amendment with potassium silicate to plants exposed to *Fo* strains, improved plant health of cold-stressed banana plants, reduced disease severity by more than 50% and shoot yellowing and wilting by 80%, compared to those treated with *Fo* strains alone. Under field conditions, plants treated with combinations of nonpathogenic *Fo, T. harzianum*, silicon and mulch had significantly enhanced plant growth, compared with plants receiving single treatment. Nonpathogenic strain N16 was more effective than strain N7 in suppressing disease development and improving shoot height. The results showed that integration

of nonpathogenic Fo strains, silicon and mulching treatments could provide more effective protection to banana plants against Fusarium wilt disease (Kidane and Laing 2010).

5.3.10 STRAWBERRY DISEASES

Strawberry crown rot and wilt disease is induced by *Phytophthora cactorum* and leathery rot and root rot disease are due to infection by *P. fragariae* var. *fragariae*, affecting production seriously under favorable environmental conditions. Soil solarization reduced root necrosis significantly and increased root weight of plants. Infection of strawberry by *Pythium* spp., *Rhizoctonia* sp. and *Cylindrocarpon* spp. was also reduced, in addition to infection by *P. fragariae* var. *fragariae* due to solarization. Combined application of solarization and fungicide mefenoxam did not enhance the effectiveness of disease suppression to a greater level, than solarization applied alone. The beneficial effect of solarization in suppressing root rot diseases was retained for 2 or more years after solarization (Pinkerton et al. 2002). The effectiveness of raised bed solarization (RBS) alone or in combination with application of chicken manure (CM) amendment, methyl bromide (MB), TeloDrip (1,3-dichloropropene + chloropicrin), short RBS combined with reduced doses of metham sodium (MS) and TeloDrip was assessed for the suppression of development of *Phytophthora cactorum* and *R. solani* infecting strawberry. In field trials conducted in two cropping seasons of 2002 and 2004, raised bed solarization (RBS) for 7 weeks alone or with CM amendment (10 t/ha), MS (50 ml/m^2) at 2 weeks after RBS and MB (50 g/m^2) significantly reduced soilborne diseases of strawberry due to both pathogens. TeloDrip was less effective in reducing disease incidence. Further, all treatments effectively controlled four weed species commonly growing in strawberry fields. RBS treatment alone or combined with CM amendment enhanced total marketable yields to levels equivalent to that was obtained with MB treatment. In addition, only RBS and CM amendment were as effective as methyl bromide in the second year. The results indicated that raised bed solarization (RBS) practice had the potential to replace methyl bromide which has been banned in many countries (Benlioğlu er al. 2005).

The effectiveness of soil solarization or the fungal BCA *Trichoderma* spp. alone and in combination for suppressing the development of crown rot and wilt disease caused by *Phytophthora cactorum* was assessed under field conditions for three consecutive years (2000–2003). Solarization using clear 50 μm low density polyethylene mulch was performed during summer. *Trichoderma* sp. was applied via drip and dip methods to the soil at 7 days before planting. Strawberry roots were dipped in the suspensions of *Trichoderma* spp. prior to planting. Soil solarization reduced the soil populations of *P. cactorum* by 100% in the first year, by 47% in the second year and by 55% in the third year, compared to untreated controls. The BCA treatment reduced soil pathogen population and also reduced leathery rot incidence on strawberry fruit by 76% in the first year and by 38% in the second year, relative to the untreated control plots. The combination of soil solarization and *Trichoderma* spp. reduced the pathogen population to a greater extent of 88.9, 97.6 and 99.0% in the first, second and third year, respectively. The results revealed that soil solarization in combination with *Trichoderma* sp. application effectively reduced disease incidence and also enhanced the yield of strawberry fruits, proving to be a promising alternative to chemical application (Porras et al. 2007).

5.3.11 TOBACCO DISEASES

5.3.11.1 Black Rot Disease

Tobacco black rot caused by *Thielaviopsis basicola* is one of the economically important diseases warranting application of effective plant protection measures. Application of fumigants containing chloropicrin has been the principal strategy of controlling the disease, but they are expensive and cause negative impact on the environment. Hence, experiments were conducted for 2 years to evaluate performance of resistant cultivar AC Gayed and moderately susceptible cv. Delgold and fumigation with chemicals. Infection by *T. basicola* reduced the yield significantly in susceptible cv. Delgold. The interaction between host genotype and fumigation was significant for most traits studied, indicating that the two genotypes responded in different ways. Yield from nonfumigated AC Gayed was higher than that of nonfumigated Delgold. Yield of nonfumigated AC Gayed was not significantly different from the yield of AC Gayed treated with Vortex Plus (1,3-dichloropropene + methyl isothiocyanate + chloropicrin). By contrast, the yield of nonfumigated Delgold was lower than fumigated Delgold. The results indicated that host resistance contributed more for reducing disease incidence than fumigation with chemicals (Haji and Brandle 2001).

5.3.11.2 Black Shank Disease

The effect of combination of host resistance and chemical application on suppression of tobacco black shank disease caused by *Phytophthora parasitica* var. *nicotianae* was investigated. Tobacco cultivars (25) were evaluated for their level of resistance to race 1 and 0 of *P. parasitica* var. *nicotianae*. The genotypes RJR75 and SP227 showed highest levels of resistance to both races 1 and 0. Mefenoxam was effective in controlling the disease induced by race 1. Fungicide application was essential at early stage of crop growth (a few days before or after transplanting). Cultivars carrying the *Php* gene produced newer and shorter adventitious roots than the cultivars possessing partial resistance to all races of *P. parasitica* var. *nicotianae*. The results indicated that use of mefenoxam especially at early stage of crop growth and planting a cultivar with high partial resistance or possessing the *Php* genes, when race 1 or race 0, respectively, could be effective for reducing loss due to *P. parasitica* var. *nicotianae* (Antonopoulos et al. 2010). The possibility of using riboflavin (0.2 mg/ml) to enhance the biocontrol activity of *B. subtilis* Tp55 for suppressing development of *P. parasitica* var. *nicotianae* was indicated. In addition, riboflavin could activate the defense-related enzymes in treated plants and enhance

effectiveness of disease suppression by the BCA (Zhang et al. 2017).

Crop disease management systems may be developed for various crops grown in different agroecosystems, based on the knowledge on the pathogen biology, crop management practices adopted in the location/region, prevalence of environmental conditions that favor the pathogen proliferation, leading to epidemics and the availability of cultivars with resistance to major diseases occurring in the location(s) concerned. Use of clean seeds/propagules, preference to cultivars with moderate level of resistance to high yielding susceptible varieties, deploying biotic and/or abiotic biocontrol agents that have the ability to adapt to field conditions, in addition to appropriate cultural practices to keep the field sanitation to the required level by eliminating all sources of infection, may be considered for integration of as many strategies that are compatible with each other to enhance the returns to the grower to the maximum extent by effectively suppressing the incidence and spread of diseases caused by soilborne microbial pathogens

REFERENCES

Abo-Elyousr KAM, Hasehm M, Ali EH (2009) Integrated control of cotton root rot disease by mixing fungal biocontrol agents and resistance inducers. *Crop Protect* 28: 295–301.

Algam SAE, Xie G, Li B, Yu S, Su T, Larsen J (2010) Effects of *Paenibacillus* strains and chitosan on plant growth promotion and control of Ralstonia wilt on tomato. *J Plant Pathol* 92: 593–600.

Ambadar VK, Sood AK (2010) Effect of solarization on tomato wilt incidence and population dynamics of *Pseudomonas fluoresens* and *Ralstonia solanacearum*. *J Mycol Plant Pathol* 40: 120–123.

Anith KN, Momol MT, Kloepper JW, Marois JJ, Olson SM, Jones JB (2004) Efficacy of plant growth-promoting rhizobacteria, acibenzolar-S-methyl and soil amendment for integrated management of bacterial wilt in tomato. *Plant Dis* 88: 669–673.

Antonopoulos DF, Melton T, Mila AL (2010) Effects of chemical control, cultivar resistance and structure of cultivar root system on black shank incidence of tobacco. *Plant Dis* 94: 613–620.

Barretti PB, and Mage R (2012) Combination of endophytic bacteria and resistant cultivars improves control of Ralstonia wilt of tomato. *Austr Plant Pathol* 41: 189–195.

Basu A (2009) Employing ecofriendly potato disease management allows organic tropical India production systems to prosper. *Asian J Food Agro-Indsutr* 2 (Suppl Issue): S 80–87.

Benlioğlu S, Boz O, Yildiz A, Kasakvalic G, Benglioğlu K (2005) Alternative soil solarization treatments for the control of soilborne diseases and weeds of strawberry in the Western Anatolia of Turkey. *J Phytopathol* 153: 423–430.

Bennett AJ, Leifert C, Whipps JM (2005) Effect of combined treatment of pasteurization and *Coniothyrium minitans* on sclerotia of *Sclerotinia sclerotiorum* in soil. *Eur J Plant Pathol* 113: 197–209.

Bonanomi G, Chiurazzi M, Caporaso S, Del Sorbo G, Moschetti G (2008) Soil solarization with biodegradable materials and its impact on soil microbial communities. *Soil Biol Biochem* 40: 1989–1998.

Bouws H, Finckh MR (2008) Effects of strip intercropping of potatoes with nonhosts on late blight severity and tuber yield in organic production. *Plant Pathol* 57: 916–927.

Bradley GG, Punja ZK (2010) Composts containing fluorescent pseudomonas suppress Fusarium root and stem rot development in greenhouse cucumber. *Canad J Microbiol* 56: 896–905.

Budge SP, Whipps JM (2001) Potential for integrated control of *Sclerotinia sclerotiorum* in glasshouse lettuce using *Coniothyrium minitans* and reduced fungicide application. *Phytopathology* 91: 221–227.

Buhre C, Kluth C, Burcky K, Marlander B, Varrelmann M (2009) Integrated control of root and crown rot in sugar beet: Combined effects of cultivar, crop rotation and soil tillage. *Plant Dis* 93: 155–161.

Carroll CL, Carter CA, Goodhue RE, Lawell CYCL, Subbarao KV (2018) A review of control options and externalities for Verticillium wilt. *Phytopathology* 108: 160–171.

Chellemi DO, Gamliel A, Katan J, Subbarao KV (2016) Development and deployment of system-based approaches for the management of soilborne plant pathogens. *Phytopathology* 106: 216–225.

Chellemi DO, Olson SM, Mitchell DJ (1994) Effects of soil solarization and fumigation on survival of soilborne pathogens of tomato in northern Florida. *Plant Dis* 78: 1167–1172.

Chen Y, Fernando WGD (2006) Induced resistance to blackleg (*Leptosphaeria maculans*) disease of canola (*Brassica napus*) caused by a weakly virulent isolate of *Leptosphaeria biglobosa*. *Plant Dis* 90: 1059–1064.

Christensen NW, Hart JM (2008) *Combating Take-all of Winter Wheat in Western Oregon*, Oregon State Univ Extn Ser, pp. 1–8.

Chung WC, Huang JW, Huang HC (2005) Formulation of a soil fungicide for control of damping-off of Chinese cabbage (*Brassica chinensis*) caused by *Rhizoctonia solani*. *Biol Contr* 32: 287–294.

Cilliers AL, Pretorius ZA, Van Wyk PS (2003) Integrated control of *Sclerotium rolfsii* on groundnut in South Africa. *J Phytopathol* 151: 249–258.

Clarkson JP, Scruby A, Mead A, Wright C, Smith B, Whipps JM (2006) Integrated control of Allium white rot with *Trichoderma viride*, tebuconazole, and composted onion. *Plant Pathol* 55: 375–386.

Czajkowski R, Perombelon MCM, van Veen JA, van der Wolf JM (2011) Control of blackleg and tuber rot caused by *Pectobacterium* and *Dickeya* species: A review. *Plant Pathol* 60: 999–1013.

Daayf F, Adam L, Fernando WGD (2003) Comparative screening of bacteria for biological control of potato late blight (strain US-8) using in vitro, detached-leaves and whole-plant testing systems. *Canad J Plant Pathol* 25: 276–284.

Davidson JA, Kimber RBE (2007) Integrated disease management of Ascochyta blight in pulse crops. *Eur J Plant Pathol* 119: 99–110.

Davis JR, Huisman OC, Everson DO, Nolte P, Sorensen LH, Schneider AT (2010a) The suppression of Verticillium wilt of potato using corn as a green manure crop. *Amer J Potato Res* 87: 195–208.

Davis JR, Huisman OC, Everson DO, Nolte P, Sorensen LH, Schneider AT (2010b) Ecological relationships of Verticillium wilt suppression of potato by green manures. *Amer J Potato Res* 87: 315–326.

De Curtis F, Lima G, Vitullo D, De Cicco V (2010) Biocontrol of *Rhizoctonia solani* and *Sclerotium rolfsii* on tomato by delivering antagonistic bacteria through a drip irrigation system. *Crop Protect* 29: 663–670.

Dugan FM, Akamatsu H, Lupien SL, Chen W, Chilvers ML, Peever TL (2009) Ascochyta blight of chickpea reduced by application of *Aureobasidium pullulans* (anamorphic Dothioraceae, Dothideales) to postharvest debris. *Biocontr Sci Technol* 19: 537–545.

Erdogan O, Benlioglu K (2010) Biological control of Verticillium wilt of cotton by use of fluorescent *Pseudomonas* spp. under field conditions. *Biol Contr* 53: 39–45.

Foster JM, Hausbeck MK (2010) Managing Phytophthora crown and root rot in bell pepper using fungicides and host resistance. *Plant Dis* 94: 697–702.

Fravel DA, Deahl KL, Strommel JR (2005) Compatibility of the biocontrol fungus *Fusarium oxysporum* strain CS-2- with selected fungicides. *Biol Contr* 34: 165–169.

Frost KE, Groves RL, Charkowski AO (2013) Integrated control of potato pathogens through seed certification and provision of clean seed potatoes. *Plant Dis* 97: 1268–1280.

Galletti S, Sala E, Leoni O, Burzi PL, Cerato C (2008) *Trichoderma* spp. tolerance to *Brassica carinata* seed meal for a combined use in fumigation. *Biol Contr* 45: 319–327.

Gil SV, Meriles JM, Haro R, Casini C, March GJ (2010) Crop rotation and tillage systems as a proactive strategy in the control of peanut fungal soilborne diseases. *BioControl* 53: 658–668.

Gil SV, Pedelini R, Oddino C, Zuga M, Marinelli A, March GJ (2008) The role of potential biocontrol agents in the management of peanut root rot in Argentina. *J Plant Pathol* 90: 35–41.

Glass JR, Johnson KB, Powelson ML (2001) Assessment of barriers to prevent the development of potato tuber blight caused by *Phytophthora infestans*. *Plant Dis* 85: 521–528.

Gravel V, Martinez C, Antoun H, Twedell RJ (2005) Antagonistic microorganisms with the ability to control Pythium damping-off of tomato seeds in rockwool. *BioControl* 50: 370–377.

Haji HM, Brandle JE (2001) Evaluation of host resistance and soil fumigation for the management of black root rot of tobacco in Ontario. *Plant Dis* 85: 1145–1148.

Handiseni M, Jo Y-K, Zhou X-G (2015) Integration of *Brassica* cover crops with host resistance and azoxystrobin for management of rice sheath blight. *Plant Dis* 99: 883–885.

Hiddink GA, Termorshuizen AJ, Raajmakers JM, van Bruggen AHC (2005a) Effect of mixed and single crops on disease suppressiveness of soils. *Phytopathology* 95: 1325–1332.

Hiddink GA, van Bruggen AHC, Termorshuizen AJ, Raajmakers JM, Semenor AV (2005b) Effect of organic management of soils on suppressiveness of *Gaeumannomyces graminis* var. *tritici* and its antagonist *Pseudomonas fluorescens*. *Eur J Plant Pathol* 113: 417–435.

Hong JC, Momol MT, Ji P, Olson SM, Colee J, Jones JB (2011) Management of bacterial wilt in tomatoes with thymol and acibenzolar-S-methyl. *Crop Protect* 30: 1340–1345.

Howell CR (2002) Cotton seedling preemergence damping-off incited by *Rhizopus oryzae* and *Pythium* spp. and its biological control with *Trichoderma* spp. *Phytopathology* 92: 177–180.

Howell CR (2007) Effect of seed quality and combination of fungicide-*Trichoderma* spp. seed treatments on pre- and post-emergence damping-off cotton. *Phytopathology* 97: 66–71.

Howell CR, De Vay JE, Garber RH, Baston WE (1997) Field control of cotton seedling diseases with *Trichoderma virens* in combination with fungicide seed treatments. *J Cotton Sci* 1: 15–20.

Howell CR, Hanson LE, Stipanovic RD, Puckhaber LS (2000) Induction of terpenoid synthesis in cotton roots and control of *Rhizoctonia solani* by seed treatment with *Trichoderma virens*. *Phytopathology* 90: 248–252.

Huang J, Li H, Yuan H (2006) Effects of organic amendments on Verticillium wilt of cotton. *Crop Protect* 25: 1167–1173.

Hwang S-F, Howard RJ, Strelkov SE, Gossen BD, Peng G (2014) Management of clubroot (*Plasmodiophora brassicae*) on canola (*Brassica napus*) in western Canada. *Canad J Plant Pathol* 36 (Suppl 1).

International Crops Research Institute for Semi-Arid Tropics (ICRIAST) (1980) Soil solarization. www.icrisat.org/what-we-do/learning.opportunities/tsu.pdfs/Soil Solarization/pdf

Ji P, Momol MT, Rich JR, Olson SM, Jones JB (2007) Development of an integrated approach for managing bacterial wilt and root knot on tomato under field conditions. *Plant Dis* 91: 1321–1326.

Kidane EG, Laing MD (2010) Integrated control of Fusarium wilt of banana (*Musa* spp.). *Acta Hortic* 879: 315–321.

Kiewnick S, Jabcobsen BJ, Eckhoff JLBK-A, Bergman JW (2001) Integrated control of Rhizoctonia crown and root rot of sugar beet with fungicides and antagonistic bacteria. *Plant Dis* 85: 718–722.

Kishore GK, Pande S, Podile AR (2005) Biological control of collar rot disease with broad-spectrum antifungal bacteria associated with groundnut. *Canad J Microbiol* 51: 123–132.

Kutcher HR, Brandt SA, Smith EG, Ulrich D, Malhi SS, Johnston AM (2013) Blackleg disease of canola mitigated by resistant cultivars and four year-rotations in Western Canada. *Canad J Plant Pathol* 35: 209–221.

Lamprecht SC, Tewoldemedhin YT, Calitz FJ, Mazzola M (2011) Evaluation of strategies for the control of canola and lupin seedling diseases caused by *Rhizoctonia* anastomosis groups. *Eur J Plant Pathol* 130: 427–439.

Landa BB, Navas-Cortes JA, Jimenz-Diaz RM (2004) Integrated management of Fusarium wilt of chickpea with sowing date, host resistance and biological control. *Phytopathology* 94: 946–960.

Larkin RP, Honeycutt CW, Griffin TS, Olanya OM, Halloran JM, He Z (2011) Effects of different potato cropping system approaches and water management on soilborne diseases and soil microbial communities. *Phytopathology* 101: 58–67.

Leisso RS, Miller PR, Burrows ME (2009) The influence of biological and fungicidal seed treatments on chickpea (*Cicer arietinum*) damping-off. *Canad J Plant Pathol* 32: 38–46.

Lemay AV, Bailey JE, Shew BB (2002) Resistance of peanut to Sclerotinia blight and the effect of acibenzolar-S-methyl and fluazinam on disease incidence. *Plant Dis* 86: 1315–1317.

Li GQ, Huang HC, Miao HJ, Erickson RS, Jiang DH, Xiao YN (2006) Biological control of Sclerotinia disease of rapeseed by aerial application of the mycoparasite *Coniothyrium minitans*. *Eur J Plant Pathol* 114: 345–355.

Liljeroth E, Bengtsson T, Wuk L, Andersson E (2010) Induced resistance in potato to *Phytophthora infestans*: Effects of BABA in greenhouse and field tests with different potato varieties. *Eur J Plant Pathol* 127: 171–183.

Lin C-H, Hsu S-T, Tzeng K-C, Wang J-F (2008) Application of a preliminary screen to select locally adapted resistant rootstock and soil amendment for integrated management of tomato bacterial wilt in Tainan. *Plant Dis* 92: 909–916.

Lobato MC, Olivieri FP, Altamiranda EAG et al. (2008) Phosphite compounds reduce disease severity in potato seed tubers and foliage. *Eur J Plant Pathol* 122: 349–358.

Lopez-Escudero FJ, Mwanza C, Blanco-Lopez MA (2007) Reduction of *Verticillium dahliae* microsclerotia viability in soil by dried plant residues. *Crop Protect* 26: 127–133.

Lopez-Mondejar R, Bernal-Vicente A, Ross M et al. (2010) Utilisation of citrus compost-based growing media amended with *Trichoderma harzianum* T-78 in *Cucumis melo* L. seedling production. *Bioresources Technol* 101: 3718–3723.

Marimuthu S, Ramamoorthy V, Samiyappan R, Subbian P (2013) Intercropping system with combined application of *Azospirillum* and *Pseudomonas fluorescens* reduces root rot incidence caused by *Rhizoctonia bataticola* and increases seed cotton yield. *J Phytopathol* 161: 405–411.

Mathivanan N, Prabavathy VR, Vijayanandaraj VR (2005) Application of talc formulations of *Pseudomonas fluorescens* Migula and *Trichoderma* viride Pers. Ex S.F. Gray decrease the sheath blight disease and enhance the plant growth and yield in rice. *J Phytopathol* 153: 697–701.

Mazzola M, Funnell DL, Raajmakers JM (2004) Wheat cultivar-specific selection of 2,4-diacetylphloroglucinol-producing *Pseudomonas* species from resident soil populations. *Microbiol Ecol* 48: 338–348.

McLean KL, Hunt JS, Stewart A, Wite D, Porter IJ, Villata O (2012) Compatibility of a *Trichoderma atroseptica* biocontrol agent with management practices of *Allium* crops. *Crop Protect* 33: 94–100.

Melero-Vara, Pardos-Ligero AM, Basallote-Ureba MJ (2000) Comparison of physical, chemical and biological methods of controlling garlic white rot. *Eur J Plant Pathol* 106: 581–588.

Melzer MS, Smith EA, Boland GJ (1997) Index of plant hosts of *Sclerotinia minor*. *Canad J Plant Pathol* 19: 272–280.

Meyer MC, Bueo CJ, de Souza NL, Yorinori JT (2006) Effect of doses of fungicides and plant resistance activators on the control of *Rhizoctonia solani* AG1-IA in vitro development. *Crop Protect* 25: 848–854.

Minuto A, Spadro D, Garibaldi A, Gullino ML (2006) Control of soilborne pathogens of tomato using a commercial formulation of *Streptomyces griseovirides* and solarization. *Crop Protect* 25: 468–475.

Nandakumar R, Babu S, Viswanathan R, Sheela J, Raguchander T, Samiyappan R (2001) A new bioformulation containing plant growth-promoting rhizobacteria mixture for the management of sheath blight and enhanced grain yield. *BioControl* 46: 493–510.

Narayanasamy P (2002) *Microbial Plant Pathogens and Crop Disease Management*, Science Publishers, Enfield, USA.

Narayanasamy P (2017) *Microbial Plant Pathogens–Detection and Management in Seeds and Propagules*, Volume 2, Wiley-Blackwell, Chichester, UK.

Omar I, O'Neill TM, Rossall S (2006) Biological control of Fusarium crown and root rot of tomato with antagonistic bacteria and integrated control when combined with the fungicide carbendazim. *Plant Pathol* 55: 92–99.

Partridge DE, Sutton TB, Jordan DL, Curtis VL, Bailey JE (2006) Management of Sclerotinia blight of peanut with the biological control agent *Coniothyrium minitans*. *Plant Dis* 90: 957–963.

Paulitz TC, Schroedder KL, Schillinger WF (2010) Soilborne pathogens of cereals in an irrigated cropping system: Effect of tillage, residue management, and crop rotation. *Plant Dis* 94: 61–68.

Peters RD, Sturz AV, Carter MR, Sanderson B (2003) Developing disease suppressive soils through crop rotation and tillage management practices. *Soil Tillage Res* 72: 181–192.

Pinkerton JN, Ivors KL, Reeser PW, Bristow PR, Windom GE (2002) The use of soil solarization for the management of soilborne plant pathogens in strawberry and raspberry production. *Plant Dis* 86: 645–651.

Porras M, Barrau C, Arroyo FT, Santos B, Blanco C, Romero F (2007) Reduction of *Phytophthora cactorum* in strawberry fields by *Trichoderma* spp. and soil solarization. *Plant Dis* 91: 142–146.

Pradhanang PM, Ji P, Momol MT, Olson SM, Mayfield JL, Jones JB (2005) Application of acibenzolar-S-methyl enhances host resistance in tomato against *Ralstonia solanacearum*. *Plant Dis* 89: 989–993.

Punja ZK, Rose S, Yip R (2002) *Biological Control of Root Diseases*, Proc Canadian Greenhouse Conference, pp. 1–5.

Radjacommare R, Nandakumar R, Kandan A et al. (2002) *Pseudomonas fluorescens*-based formulation for the management of sheath blight disease and leaf folder insect in rice. *Crop Protect* 21: 671–677.

Rekanovic E, Milijasevic S, Todorovic B, Potocnik I (2007) Possibilities of biological and chemical control of Verticillium wilt in pepper. *Phytoparasitica* 35: 436–441.

Sakuma F, Maeda M, Takahashi M, Kondo N (2011) Suppression of common scab of potato caused by *Streptomyces turgidiscabies* using lopsided oat green manure. *Plant Dis* 95: 1124–1130.

Salman M, Abuamsha R (2012) Potential for integrated biological and chemical control of damping-off disease caused by *Pythium ultimum* in tomato. *BioControl* 57: 711–718.

Saude C, Mekeown A, Gossen BD, McDonald MR (2012) Effect of host resistance and fungicide application on clubroot pathotype 6 in green cabbage and napa cabbage. *Hortic Technol* 22: 311–319.

Schillinger WF, Paulitz TC (2006) Reduction of Rhizoctonia bare patch in wheat and barley rotations. *Plant Dis* 90: 302–306.

Sconyers IE, Brenneman TB, Stevenson KL, Mullinix BG (2007) Effects of row pattern, seeding rate, and inoculation date on fungicide efficacy and development of peanut stem rot. *Plant Dis* 91: 273–278.

Shanmugam V, Senthil N, Raguchander T, Ramanathan A, Samiyappan R (2002) Interactions of *Pseudomonas fluorescens* with *Rhizobioum* for their effect on the management of peanut root rot. *Phytoparasitica* 30: 169–175.

Shifa H, Gopalakrishnan C, Velazhahan R (2018) Management of late leaf spot (*Phaeoisariopsis pernsonata*) and root rot (*Macrophomina phaseolina*) diseases of groundnut (*Arachis hypogaea* L.) with plant growth-promoting rhizobacteria, systemic acquired resistance inducers and plant extracts. *Phytoparasitica* 46: 19–30.

Shtienberg D, Elad Y, Bornstein Z, Ziv G, Grava A, Cohen S (2010) Polyethylene mulch modifies greenhouse microclimate and reduces infection of *Phytophthora infestans* in tomato and *Pseudoperonospora cubensis* in cucumber. *Phytopathology* 100: 97–104.

Shtienberg D, Vintal H, Brener S, Retig B (2000) Rational management of *Didymella rabiei* in chickpea by integration of genotype resistance and postinfection application of fungicides. *Phytopathology* 90: 834–842.

Slusarski C, Pietr SJ (2009) Combined application of dazomet and *Trichoderma asperellum* as an efficient alternative to methyl bromide in controlling the soilborne disease complex of bell pepper. *Crop Protect* 28: 668–674.

Someya N, Tsuchiya K, Yoshida T, Tsujimoto-Noguchi M, Sawada H (2007) Combined application of *Pseudomonas fluorescens* strain LRB3W1 with a low dosage of benomyl for control of cabbage yellows caused by *Fusarium oxysporum* f.sp. *conglutinans. Biocontr Sci Technol* 17: 21–37.

Stern VM, Smith RF, van den Bosch R, Hagen KS (1959) The integral concept. *Hilgardia* 29: 81–101.

Szezech M, Shoda M (2006) The effect of mode of application of *Bacillus subtilis* RB14-C on its efficacy as a biocontrol agent against *Rhizoctonia solani. J Phytopathol* 154: 370–377.

Tamietti G, Valentino D (2006) Soil solarization is an ecological method for the control of Fusarium wilt of melon in Italy. *Crop Protect* 25: 389–397.

Uppal AK, El-Hadrami A, Adam LR, Tenuta M, Daayf F (2008) Biological control of potato Verticillium wilt under controlled and field conditions using selected bacterial antagonists and plant extracts. *Biol Contr* 44: 90–100.

van den Boogert PHJF, Luttikholt AJG (2004) Compatible biological and chemical control systems for *Rhizoctonia solani* in potato. *Eur J Plant Pathol* 110: 111–118.

Wakelin S, Anstis ST, Warren RA, Ryder MH (2006) Role of pathogen suppression on the growth promotion of wheat by *Penicillium radicum. Austr Plant Pathol* 35: 253–258.

Wiggins BE, Kinkel LI (2005) Green manures and crop sequences influence potato diseases and pathogen inhibiting activity of indigenous streptomycete. *Phytopathology* 95: 178–185.

Yang D, Wang B, Wang J, Chen Y, Zhou M (2009) Activity and efficacy of *Bacillus subtilis* strain NJ-18 against rice sheath blight and Sclerotinia stem rot of rape. *Biol Contr* 51: 61–65.

Yang L, Li G, Zhang J, Jiang D, Chen (2011) Compatibility of *Coniothyrium minitans* with compound fertilizer in suppression of *Sclerotinia sclerotiorum. Biol Contr* 59: 221–227.

Zeng JX, Kirk W, Hao J (2012) Field management of Sclerotinia stem rot of soybean using biological control agents. *Biol Contr* 60: 141–147.

Zhang C, Gao J, Han T, Tian X, Wang F (2017) Integrated control of tobacco black shank by combined use of riboflavin and *Bacillus subtilis* strain Tpb55. *BioControl* 62: 835–845.

Zhang JX, Xue AG, Morrison MJ, Meng Y (2011) Impact of time between field application of *Bacillus subtilis* strains SB01 and SB24 and inoculation with *Sclerotinia sclerotiorum* on the suppression of Sclerotinia stem rot in soybean. *Eur J Plant Pathol* 131: 95–102.

Index